MARINE PARASITOLOGY

MARINE PARASITOLOGY

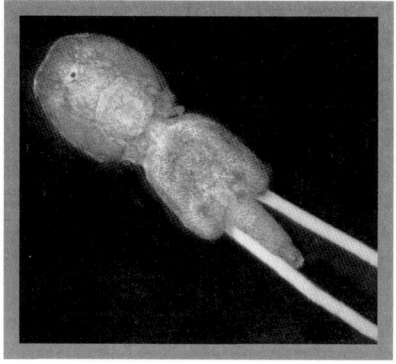

EDITOR

KLAUS ROHDE

© CSIRO 2005

All rights reserved. Except under the conditions described in the Australian Copyright Act 1968 and subsequent amendments, no part of this publication may be reproduced, stored in a retrieval system or transmitted in any form or by any means, electronic, mechanical, photocopying, recording, duplicating or otherwise, without the prior permission of the copyright owner. Contact CSIRO PUBLISHING for all permission requests.

National Library of Australia Cataloguing-in-Publication entry

>   Marine parasitology.
>   Bibliography.
>   Includes index.
>   ISBN 0 643 09025 8.
>
>   1. Parasitism. 2. Marine animals – Parasites.
>   I. Rohde, Klaus, 1932– . II. CSIRO.

578.65

Published exclusively in Australia, New Zealand and the Americas, and non-exclusively in other territories of the world (excluding Europe, the Middle East, Asia and Africa), by:
CSIRO PUBLISHING
PO Box 1139 (150 Oxford St)
Collingwood VIC 3066
Australia

Tel: (03) 9662 7666   Int: +(613) 9662 7666
Fax: (03) 9662 7555   Int: +(613) 9662 7555
Email: publishing.sales@csiro.au
Website: www.publish.csiro.au

Published exclusively in Europe, the Middle East, Asia (including India, Japan, China, and South-East Asia) and Africa, and non-exclusively in other territories of the world (excluding Australia, New Zealand and the Americas) by CABI Publishing, a Division of CAB International, with the ISBN 1 84593 053 3.

CABI Publishing
Wallingford
Oxon OX10 8DE
United Kingdom

Tel: 01491 832 111   Int: +44 1491 832 111
Fax: 01491 829 292   Int: +44 1491 829 292
Email: publishing@cabi.org
Website: www.cabi-publishing.org

Front cover:
Background: A heavy infection by *Cyamus ovalis* on skin of the Northern right whale, *Eubalaena glacialis*, from East Iceland. Photo by Jørgen Lutzen.

Inset: A mature female salmon louse, *Lepeophtheirus salmonis*, with newly extruded, whitish egg strings. Photo by Øivind Øines.

Set in 10 pt Adobe Minion and Stone Sans
Cover and text design by James Kelly
Typeset by Desktop Concepts Pty Ltd, Melbourne
Printed in Australia by Ligare

Contents

Contributors	xi
Preface	xxi
Acknowledgements	xxiii

Chapter 1 The nature of parasitism — 1

Definitions, and adaptations to a parasitic way of life — 1
Klaus Rohde

Marine parasites and the tree of life — 6
DTJ Littlewood

Chapter 2 Protistan parasites and Myxozoa — 11

Protistan biodiversity — 11
Peter O'Donoghue

'Sarcomastigophora' (amoebae and flagellates) — 17
Barbara F Nowak

Labyrinthomorpha (labyrinthomorphs) — 20
Susan M Bower

Haplosporidia (haplosporidians) — 23
Eugene Burreson

Apicomplexa (sporozoans) — 26
Kálmán Molnár

Microsporidia (microsporans) — 30
Elizabeth Moodie

Mikrocytos mackini (microcell) — 34
Susan M Bower

Ciliophora (ciliates) — 37
Jiří Lom

Myxozoa (myxozoans) — 41
Jiří Lom

Chapter 3 Helminth parasites — 47

'Turbellaria' (turbellarians) — 47
Lester Cannon

Monogenea Polyopisthocotylea (ectoparasitic flukes) — 55
Craig Hayward

Monogenea Monopisthocotylea (ectoparasitic flukes) — 63
Ian D Whittington

Aspidogastrea (aspidogastreans) — 72
Klaus Rohde

Digenea (endoparasitic flukes) — 76
Thomas H Cribb

Amphilinidea (unsegmented tapeworms) — 87
Klaus Rohde

Gyrocotylidea (unsegmented tapeworms) — 89
Willi Xylander

Eucestoda (true tapeworms) — 92
Janine N Caira and Florian B Reyda

Nematoda (roundworms) — 104
Gary McClelland

Acanthocephala (thorny or spiny-headed worms) — 116
Horst Taraschewski

Chapter 4 Crustacean parasites — 123

Copepoda (copepods) — 123
Geoff Boxshall

Isopoda (isopods) — 138
RJG Lester

Branchiura (fish lice) — 145
Geoff Boxshall

Tantulocarida (tantulocarids) — 147
Geoff Boxshall

Ascothoracida (ascothoracids) — 149
Mark J Grygier and Jens T Høeg

Cirripedia Thoracica and Rhizocephala (barnacles) — 154
Jens T Høeg, Henrik Glenner and Jeffrey D Shields

Amphipoda (amphipods) — 165
Jørgen Lützen

Chapter 5 Minor groups and fossils — 171

Fossil parasites — 172
Greg W Rouse

Porifera (sponges) John NA Hooper	174
Cnidaria and Ctenophora (cnidarians and comb jellies) Ferdinando Boero and Jean Bouillon	177
Mesozoa Orthonectida (orthonectids) Hidetaka Furuya	182
Mesozoa Dicyemida (dicyemids) Hidetaka Furuya	185
Myzostomida (myzostomids) Greg W Rouse	189
Polychaeta (bristle worms) Greg W Rouse	193
Hirudinea (leeches) Fredric R Govedich, Bonnie A Bain and Ronald W Davies	196
Cycliophora (wheel wearers) Iben Heiner and Reinhardt Møbjerg Kristensen	202
Nemertea (ribbon worms) Kirsten Jensen and Patricia S Sadeghian	205
Rotifera and *Seison* (rotifers) Wilko H Ahlrichs	211
Nematomorpha (horse-hair worms) Andreas Schmidt-Rhaesa	213
Acari (mites and ticks) Jacek Dabert	216
Pycnogonida (pycnogonids) David Staples	222
Insecta (insects) Kirsten Jensen and Ricardo L Palma	226
Tardigrada (water bears) Reinhardt Møbjerg Kristensen and Jesper Guldberg Hansen	230
Pentastomida (tongue worms) Wolfgang Böckeler	235
Mollusca (molluscs) Felix Lorenz	240
Echiura (spoon worms) Kirsten Jensen	246
Echinodermata (echinoderms) Greg W Rouse	248
Parasitic marine fishes David Woodland	250

Chapter 6 Behavioural aspects of parasitism — 259

Parasite induced changes in host behaviour and morphology — 259
Pierre Sasal and Frédéric Thomas

Cleaning mutualism in the sea — 264
Alexandra S Grutter

Chapter 7 Ecology — 279

Transmission of marine parasites — 280
David J Marcogliese

The ecological niches of parasites — 286
Klaus Rohde and Peter P Rohde

Marine hyperparasites — 293
Mark Freeman

Parasites in brackish waters — 298
C Dieter Zander

Metapopulation biology of marine parasites — 302
Serge Morand and Andrea Šimková

Structure of parasite communities — 309
Robert Poulin

Parasite populations and communities as non-equilibrium systems — 315
Klaus Rohde

Population and community ecology of larval trematodes in molluscan first intermediate hosts — 321
Armand M Kuris and Kevin D Lafferty

Chapter 8 Coevolution and speciation — 327

Coevolution in marine systems — 327
Eric P Hoberg

Speciation and species delimitation — 339
Steven A Nadler

Chapter 9 Zoogeography — 347

Latitudinal, longitudinal and depth gradients — 348
Klaus Rohde

Parasites as biological tags — 351
Ken MacKenzie

Parasites as indicators of historical dispersal — 355
Craig Hayward

Introduced marine parasites — 358
Mark E Torchin and Armand M Kuris

Deep-sea parasites 366
Rodney A Bray

Chapter 10 Economic and environmental importance 371

Mass mortalities in the oceans 371
Brian Jones

Effects of salmon lice on Atlantic salmon 374
Peter Andreas Heuch

Effects in finfish culture 378
Kazuo Ogawa

Effects in mollusc culture 391
Ryan B Carnegie

Effects in shrimp culture 398
Robin M Overstreet

Ecological aspects of parasites in the American lobster 404
Richard J Cawthorn

Parasites of marine mammals 408
Murray D Dailey

Marine birds and their helminth parasites 414
Eric P Hoberg

Effects of pollution on parasites, and use of parasites in pollution monitoring 421
Bernd Sures

Chapter 11 Medical importance 427

Cestode and trematode infections 427
David Blair

Anisakiasis 430
Kazuya Nagasawa

Zoonotic potential of Protozoa 434
Mark Freeman

Zoonotic aspects of trichinellosis 436
Lorry B Forbes

Marine schistosome dermatitis 439
John Walker

Infections by the rat lungworm, *Angiostrongylus cantonensis* 442
Robin M Overstreet

References 447

Index 559

Contributors

PD Dr Wilko H Ahlrichs
Zoosystematics and Morphology
University Oldenburg, Germany
Email: wilko.ahlrichs@uni-oldenburg.de

Dr Bonnie A Bain
Department of Biological Sciences
Northern Arizona University
Box 5640
Flagstaff, AZ 86011, USA
Email: bonnie.bain@invertebrate.ws or bonnie.bain@nau.edu

Associate Prof. David Blair
School of Tropical Biology, James Cook University
Townsville, Qld 4811, Australia
Email: david.blair@jcu.edu.au

PD Dr Wolfgang Böckeler
Zoological Institute University of Kiel
Am Botanischen Garten 9, D 24118 Kiel
Email: wboeckeler@zoologie.uni-kiel.de

Prof. F Boero
DiSTeBA, Università di Lecce, 73100
Lecce, Italy
Email: boero@unile.it

Prof. J Bouillon
Laboratoire de Biologie Marine, Université Libre de Bruxelles,
50 Ave. F. D. Roosevelt, 1050 Bruxelles, Belgium
Email: fa493994@skynet.be

Dr Susan Bower
Research Scientist
Fisheries and Oceans Canada, Pacific Biological Station
Nanaimo, British Columbia, V9T 6N7, Canada
Email: BowerS@dfo-mpo.gc.ca

Dr Geoff Boxshall, FRS
Department of Zoology, The Natural History Museum
Cromwell Road, London SW7 5BD, UK
Email: g.boxshall@nhm.ac.uk

Dr Rodney A Bray
Department of Zoology, The Natural History Museum,
Cromwell Road, London SW7 5BD, UK
Email: rab@nhm.ac.uk

Prof. E Burreson
Department of Environmental and Aquatic Animal Health
Virginia Institute of Marine Science, College of William and Mary
Gloucester Point, VA 23062, USA
Email: gene@vims.edu

Prof. Janine N Caira
Department of Ecology and Evolutionary Biology,
University of Connecticut
75 North Eagleville Road, Storrs, CT 06269-3043, USA
Email: Janine.Caira@uconn.edu

Dr Lester Cannon
Formerly Queensland Museum, PO Box 3300,
South Brisbane, Qld 4101, Australia
Email: jcannon@bigpond.net.au

Dr Ryan Carnegie
Department of Environmental and Aquatic Animal Health
Virginia Institute of Marine Science, College of William and Mary
PO Box 1346 (regular mail)/Route 1208 Greate Road (courier)
Gloucester Point, VA 23062, USA
Email: carnegie@vims.edu

Dr Richard J Cawthorn
Professor of Parasitology, Director & Senior Scientist,
AVC Lobster Science Centre, Atlantic Veterinary College
University of Prince Edward Island, 550 University Avenue
Charlottetown, Prince Edward Island,
C1A 4P3, Canada
Email: cawthorn@upei.ca

Prof. TH Cribb
Department of Microbiology and Parasitology
The University of Queensland, Qld 4072, Australia
Email: T.Cribb@mailbox.uq.edu.au

Dr J Dabert
Department of Animal Morphology
Adam Mickiewicz University

Umultowska 89, 61-614 Poznan, Poland
Email: dabert@amu.edu.pl

Dr Murray D Dailey
Research Parasitologist
The Marine Mammal Center, 1065 Fort Cronkhite
Sausalito, CA 94965, USA
Email: daileym@tmmc.org

Prof. Ronald W Davies
Department of Biological Sciences
University of Calgary
Calgary, Alberta T2N1N4, Canada
Email: ron.davies@shaw.ca

Dr Lorry B Forbes
Centre for Animal Parasitology, Canadian Food Inspection Agency
116 Veterinary Road, Saskatoon, SK S7N 2R3, Canada
Email: lforbes@inspection.gc.ca

Dr Mark Freeman
Department of Aquatic Biosciences
Graduate School of Agricultural and Life Sciences
The University of Tokyo, Yayoi 1-1-1, Bunkyo,
Tokyo 113-8657, Japan
and The Institute of Aquaculture, Stirling University, Stirling, Scotland, UK
Email: amaf100@mail.ecc.u-tokyo.ac.jp

Dr H Furuya
Department of Biology, Graduate School of Science, Osaka University
1-1 Machikaneyama, Toyonaka, Osaka 560-0043, Japan
Email: hfuruya@bio.sci.osaka-u.ac.jp

Dr H Glenner
Department of Cell Biology and Comparative Zoology
Institute of Biology, University of Copenhagen
Universitetsparken 15, DK-2100, Copenhagen, Denmark
Email: hglenner@bi.ku.dk

Dr Fredric R Govedich
Department of Biological Sciences
Northern Arizona University
Box 5640
Flagstaff, AZ 86011, USA
Email: fredric.govedich@invertebrate.ws or fredric.govedich@nau.edu

Dr Alexandra S Grutter
School of Integrative Biology
University of Queensland, Qld 4072, Australia
Email: alexandra.grutter@uq.edu.au

Dr Mark J Grygier
Research Scientist (*Sôkatsu gakugeiin*), Lake Biwa Museum
Oroshimo 1091, Kusatsu, Shiga 525-0001, Japan
Email: grygier@lbm.go.jp

Jesper Guldberg Hansen
Department of Invertebrate Zoology, Zoological Museum
University of Copenhagen
Universitetsparken 15, DK-2100 Copenhagen Ø, Denmark
Email: guldhans@mail.tele.dk

Dr Craig Hayward
School of Aquaculture and Aquafin CRC
University of Tasmania
PO Box 847,
Port Lincoln, SA 5606, Australia
Email: craig.hayward@utas.edu.au

Iben Heiner
Department of Invertebrate Zoology, Zoological Museum
University of Copenhagen
Universitetsparken 15, DK-2100 Copenhagen Ø, Denmark
Email: iheiner@zmuc.ku.dk

Dr Peter Andreas Heuch
Researcher, Section for Fish Health, National Veterinary Institute
PO Box 8156 Dep., 0033 Oslo, Norway
Email: peter-andreas.heuch@vetinst.no

Dr Eric P Hoberg
Curator, US National Parasite Collection, Animal Parasitic Diseases Laboratory, USDA,
Agricultural Research Service, BARC East 1180
Beltsville, MD 20705-2350, USA
Email: ehoberg@anri.barc.usda.gov

Prof. Jens T Høeg
Department of Cell Biology and Comparative Zoology
Institute of Biology, University of Copenhagen
Universitetsparken 15, DK-2100, Copenhagen, Denmark
Email: jthoeg@bi.ku.dk

Dr John NA Hooper
Head of Biodiversity Program & Director Queensland Centre for Biodiversity
Adjunct Professor, Natural Products Discovery, Griffith University, Brisbane;
Adjunct Assoc. Prof., School of Integrative Biology
University of Queensland, Brisbane.
Queensland Museum, PO Box 3300
South Brisbane, Qld 4101, Australia
Email: JohnH@qm.qld.gov.au

Dr Kirsten Jensen
Department of Ecology and Evolutionary Biology and Natural History Museum and
Biodiversity Research Center, University of Kansas
1200 Sunnyside Ave., Haworth Hall, Lawrence, KS 66045, USA
Email: kjensen@tapeworms.org

Dr JB Jones
Principal fish pathologist, Department of Fisheries, Government of Western Australia,
Animal Health Laboratories 3 Baron-Hay Court, South Perth, WA 6151 Australia
Adjunct Professor of Fish Health, Muresk Institute, Curtin University,
Adjunct Associate Professor, Murdoch University, School of Veterinary and
Biomedical Sciences.
Email: bjones@agric.wa.gov.au

Prof. Reinhardt Møbjerg Kristensen
Department of Invertebrate Zoology, Zoological Museum,
University of Copenhagen
Universitetsparken 15, DK-2100 Copenhagen Ø, Denmark
Email: rmkristensen@zmuc.ku.dk

Prof. Armand M Kuris
Department of Ecology, Evolution, and Marine Biology
University of California, Santa Barbara, CA 93106, USA
Associate Provost, College of Creative Studies, University of California
Santa Barbara, CA 93106, USA
Email: kuris@lifesci.ucsb.edu

Dr Kevin D Lafferty
US Geological Survey,
And Marine Science Institute, University of California,
Santa Barbara CA 93106, USA
Email: lafferty@lifesci.ucsb.edu

Prof. RJG (Bob) Lester
Microbiology and Parasitology,
The University of Queensland, Brisbane, Qld 4072, Australia.
Email: R.Lester@uq.edu.au

Dr DTJ Littlewood
Department of Zoology
The Natural History Museum
Cromwell Road, London SW7 5BD, UK
Email: T.Littlewood@nhm.ac.uk

Prof. Jiří Lom
Academy of Sciences of the Czech Republic,
Branisovská 31,
370 05 Ceské Budejovice, Czech Republic
Email: jlom@paru.cas.cz

Dr Felix Lorenz
Worldwide Seashells
Fr-Ebert-Strasse, 12-35418 Buseck, Germany
Email: Felix.Lorenz@t-online.de

Prof. Jørgen Lützen
Department of Zoomorphology
Institute of Zoology, University of Copenhagen
Universitetsparken 15, DK-2100 Copenhagen Ø, Denmark
Email: JLutzen@zi.ku.dk

Dr Ken MacKenzie
Research Fellow, Department of Zoology, School of Biological Sciences
The University of Aberdeen
Tillydrone Avenue, Aberdeen AB24 2TZ, UK
Email: k.mackenzie@abdn.ac.uk

Dr David J Marcogliese
Research Scientist
St Lawrence Centre, Environment Canada
105 McGill, 7th Floor, Montreal, Quebec H2Y 2E7, Canada
Email: david.marcogliese@ec.gc.ca

Dr Gary McClelland
Department of Fisheries and Oceans, Gulf Fisheries Centre
PO Box 5030, Moncton, New Brunswick EIC 9B6, Canada
Email: McClellandG@dfo-mpo.gc.ca

Dr K Molnár
Veterinary Medical Research Institute, Hungarian Academy of Sciences,
1143 Budapest, Hungária krt 21, Hungary
Email: kalman@vmri.hu

Dr Elizabeth Moodie
Zoology, University of New England
Armidale, NSW 2351, Australia
Email: emoodie@pobox.une.edu.au

Prof. Serge Morand
Centre de Biologie et de Gestion des Populations (CBGP)
Campus International de Baillarguet, CS 30016, 34988
Montferrier sur Lez, Cedex, France
Email: morand@ensam.inra.fr

Prof. Steven A Nadler
Department of Nematology
University of California, Davis, CA 95616, USA
Email: sanadler@ucdavis.edu

Prof. Kazuya Nagasawa
Department of Bioresource Science and Technology, Graduate School of Biosphere Science
Hiroshima University, Higashi Hiroshima 739-8528, Japan
Email: ornatus@hiroshima-u.oc.jp

Dr Barbara Nowak
School of Aquaculture
University of Tasmania, Locked Bag 1370
Launceston, Tas. 7250, Australia
Email: B.Nowak@utas.edu.au

Dr Peter O'Donoghue
Assoc. Professor, Reader in Protozoology
Department of Microbiology and Parasitology
School of Molecular and Microbial Sciences
University of Queensland,
Qld 4072, Australia
Email: p.odonoghue@uq.edu.au

Prof. Kazuo Ogawa
Department of Aquatic Bioscience, Graduate School of Life and Agricultural Sciences
University of Tokyo, Bunkyo, Tokyo 113-8657, Japan
Email: aogawak@mail.ecc.u-tokyo.ac.jp

Prof. Robin M Overstreet
Gulf Coast Research Laboratory, The University of Southern Mississippi
PO Box 7000, Ocean Springs, MS 39566, USA
Email: robin.overstreet@usm.edu

Dr Ricardo L Palma
Curator of insects, Museum of New Zealand Te Papa Tongarewa
PO Box 467, 169 Tory Street, Wellington, New Zealand
Email: ricardop@tepapa.govt.nz

Prof. Robert Poulin
Department of Zoology, University of Otago
PO Box 56, Dunedin, New Zealand
Email: robert.poulin@stonebow.otago.ac.nz

Florian B Reyda
Department of Ecology & Evolutionary Biology
University of Connecticut
75 North Eagleville Road, Storrs, CT 06269-3043, USA
Email: Florian.Reyda@uconn.edu

Prof. Klaus Rohde
School of Environmental Sciences and Resources Management
University of New England
Armidale, NSW 2351, Australia
Email: krohde@metz.une.edu.au

Peter Rohde
Centre for Quantum Computer Technology
Department of Physics
University of Queensland, Qld 4072, Australia
Email: rohde@physics.uq.edu.au

Dr Greg Rouse
South Australian Museum, Nth Terrace Adelaide SA, 5000 and
Earth and Environmental Science
University of Adelaide, SA 5005, Australia
Email:rouse.greg@saugov.sa.gov.au

Patricia S Sadeghian
Associate Curator of Invertebrate Zoology
Santa Barbara Museum of Natural History
2559 Puesta del Sol Road, Santa Barbara, CA 93105, USA
Email: psadeghian@sbnature2.org

Dr Pierre Sasal
Laboratoire de Parasitologie Fonctionnelle et Evolutive, UMR 5555 CNRS-UP, CBETM,
Université de Perpignan
Avenue Paul Alduy, 66860 Perpignan Cedex, France
Email: sasal@univ-perp.fr

Dr Andreas Schmidt-Rhaesa
Zoomorphology and Systematics
University Bielefeld, Germany
Email: a.schmidt-rhaesa@uni-bielefeld.de

Dr Jeffrey Shields
Virginia Institute of Marine Science, College of William and Mary
Gloucester Point, VA 23062, USA
Email: jeff@vims.edu

Dr Andrea Šimková
Department of Zoology and Ecology, Faculty of Science
Masaryk University, Kotlářská 2, 61137 Brno, Czech Republic
Email: simkova@sci.muni.cz

David A Staples
Marine Invertebrates, Sciences Department, Melbourne Museum
PO Box 6666E, Melbourne Vic. 3001, Australia
Email: dstaples@museum.vic.gov.au

HD Dr Bernd Sures
Zoologisches Institut–Ökologie/Parasitologie
University of Karlsruhe, 76128 Karlsruhe, Germany
Email: dc11@rz.uni-karlsruhe.de

Prof. Horst Taraschewski
Zoologisches Institut–Ökologie/Parasitologie
University of Karlsruhe, 76128 Karlsruhe, Germany
Email: dc20@rz-uni-karlsruhe.de

Dr Frédéric Thomas
Laboratoire de Génétique et Evolution des Maladies Infectieuses
UMR 2724 CNRS-IRD, 911 avenue Agropolis, BP 64501, 34394 Montpellier Cedex 5, France
Email: Frederic.THOMAS@mpl.ird.fr

Dr Mark E Torchin
Smithsonian Tropical Research Institute
Apartado 0843-03092
Balboa, Ancon, Republic of Panama
Email: torchinm@si.edu

Dr John Walker
Department of Medicine
University of Sydney, NSW 2006, Australia
Email: jwalker@mail.usyd.edu.au

Dr Ian D Whittington
Senior Research Scientist, Monogenean Research Laboratory, Parasitology Section
The South Australian Museum, North Terrace, Adelaide, SA 5000. Australia and Senior
Research Fellow, Marine Parasitology Laboratory, Earth and Environmental Sciences
University of Adelaide, North Terrace, Adelaide, SA 5005, Australia
Email: whittington.ian@saugov.sa.gov.au

Dr David Woodland
Honorary Fellow
Zoology, University of New England, Armidale, NSW 2351, Australia
Email: dwoodlan@metz.une.edu.au

Prof. Dr Willi Xylander
Director, Staatliches Museum für Naturkunde Görlitz
Postfach 300 154, 02806 Görlitz, Germany
Email: willi.xylander@smng.smwk.sachsen.de

Prof. C Dieter Zander
Biozentrum Grindel und Zoologisches Museum
University Hamburg
Martin-Luther-King-Platz 3, D-20146 Hamburg, Germany
Email: cedezet@zoologie.uni-hamburg.de

Preface

This book provides a concise but thorough account of our current knowledge about marine parasites. It is a text aimed at researchers and students, and can be used in introductory and advanced courses on marine biology, aquaculture, marine parasitology, general parasitology, invertebrate zoology, zoogeography and ecology. It is a text that will be of great use to postgraduate students. Seventy-five authors from around the world, all of them eminent in their field, have contributed in their area of expertise. They were asked to emphasise the many gaps still to be filled, and so provide a stimulating guide for future research. More than half of all animal species are parasitic, according to some estimates. Many parasites, including marine species, are of great ecological, medical or economic importance. In the marine environment, for example, the effects of parasites in aquaculture can be devastating. Total global aquaculture in 2004 was estimated to be worth more than US$55 billion, the greatest losses to it caused by parasites. Parasite species led to the collapse of European flat oyster aquaculture after 1979, and to devastating effects on oyster culture on the North American east coast over many years. Introduction of a monogenean ectoparasite into the Aral Sea, in Uzbekistan and Kazakstan, led to the total collapse of sturgeon and caviar fisheries in that region in the 1930s. Some marine parasites have considerable medical importance. A human roundworm, *Trichinella*, can be acquired from marine animals such as walrus, and others cause anisakiasis with sometimes serious effects. The ecological role of many marine parasites may be considerable, although it is little understood. Parasites can also be used as cheap biological tags, permitting distinction of host populations, and they are used for pollution monitoring.

Chapter 1 sets out definitions and describes adaptations, and the distribution of marine parasites in the protistan and animal kingdoms. Chapters 2 to Chapter 5 describe the various parasite groups. The sections differ a great deal in their approach, because the state of our knowledge of each group differs, often substantially. Some taxa, such as the phylum Cycliophora, which contains only one described species, have been described only recently. Others, for example the digenean trematodes, include thousands of species that have been studied in great detail over many years. In relation to ecological and behavioural studies, some taxa, for example the Rhizocephala, are known to have fascinating adaptations that result in modification of their hosts' behaviour. The ecology of parasitism in other taxa, such as the Rotifera, has hardly been studied.

Chapter 6 to Chapter 9 deal with behavioural, ecological, evolutionary and zoogeographical aspects. A detailed and well-illustrated section on cleaning symbiosis will be of particular interest to marine biologists, and the sections on parasite-induced changes in host behaviour, speciation and coevolution will be of particular interest to evolutionary biologists. The discussion of coevolution addresses a 'hot' topic, as do the various sections dealing with ecological aspects, such as metapopulation biology, community structure, transmission of parasites, ecological niches and parasite communities as non-equilibrium systems. Larval trematodes in snails have been extensively studied and the section by two of the foremost researchers in the field will be very much welcomed by ecologists.

One of the great mysteries in evolutionary biology is the explanation of latitudinal gradients in species diversity: why are there so many more species in the tropics than in colder environments? The chapter on zoogeography addresses this problem for marine parasites, and discussions of marine parasites as biological tags, as indicators of historical dispersal, of deep-sea parasites and of introduced parasites and their use as control agents for biological pests likewise deal with problems that have great interest.

The economic, environmental and medical importance of marine parasites is discussed in Chapter 10 and Chapter 11. Much emphasis is put on detailed and cutting-edge discussions of parasites important in aquaculture of finfish, molluscs and shrimp, and on the effects of parasite infections on marine mammals and birds. The section on lobster parasites emphasises that disease cannot be explained only by the parasites involved, but must consider various environmental factors as well. Biologists prominently involved in studying parasites threatening aquaculture have written the various sections, giving up-to-date accounts that cannot be ignored by anybody interested in the field. Few people realise how much damage parasites cause to marine birds and mammals, as described in the two detailed sections on these aspects. Also discussed are mass mortalities in the sea, and – based on long-term intensive studies – the effects of salmon lice on salmon populations. An account of the effects of pollution on parasites and the use of parasites in pollution monitoring is followed by discussions of medical aspects, including intestinal flatworms, roundworms (trichinellosis, anisakiasis, *Angiostrongylus*) and cercarial dermatitis.

There is no similar text that covers as broad a spectrum of marine parasitology in such depth as this. It is hoped that it will become the standard text and stimulate future work in the field by scientists and students.

Acknowledgements

The authors and editor wish to thank CSIRO Publishing's Ann Crabb for advice and encouragement, and Briana Elwood for her thorough work in producing the book, and the following colleagues for reviewing various sections (or, in some cases, providing substantial advice):

Rob Adlard, Queensland Museum Brisbane; Pål Arne Bjørn, Institute for Fisheries and Aquaculture, Tromsø; David Blair, James Cook University, Townsville; Geoff Boxshall, Natural History Museum, London; Dan Brooks, University of Toronto; Janine Caira, University of Connecticut; Doug Causey, Museum of Comparative Zoology, Harvard; Leslie Chisholm, South Australian Museum; Jane Cook, St Lawrence Centre Environment; Isabelle Cote, University of East Anglia; Murray Dailey, Marine Mammal Center, Sausalito, California; Bengt Finstad, Norwegian Institute for Nature Research, Trondheim; Mark Freeman, University of Tokyo; Terry Galloway, University of Manitoba; Boyko Georgiev, Bulgarian Academy of Sciences, Sofia; Ray Gibson, Liverpool John Moores University; Craig Hayward, University of Tasmania; Peter Heuch, Biologisk Institut, Oslo; Eric Hoberg, USDA, Beltsville, MD; Jens Chr. Holst, Institute for Marine Research, Bergen; Armand Kuris, University of California Santa Barbara; Jiří Lom, Czech Academy of Science; Greg Maguire, Curtin University Perth; Gary McClelland, Gulf Fisheries Centre, New Brunswick; Tom Mattis, Shaw Environmental and Infrastructure, Knoxville, Tennessee; Jason Melendy, Department of Fisheries and Oceans, Gulf Fisheries Centre; Libby Moodie, University of New England, Armidale; Lena Measures, Fisheries and Oceans Canada, Maurice Lamontagne Institute; Christian Melaun, University of Giessen; Jason Melendy, Department of Fisheries and Oceans, Gulf Fisheries Centre; EO Murdy, Office of International Science and Engineering, National Science Foundation, Arlington, VA;. Steve Nadler, University of California, Davis; Christian Neumann, Berlin Natural History Museum; Barbara Nowak, University of Tasmania; Peter O'Donoghue, University of Queensland; Kazuo Ogawa, University of Tokyo; Kim Overstreet, Gulf Coast Research Laboratory; Ricardo Palma, Museum of New Zealand Te Papa Tongarewa, Wellington; Sylvie Pichelin, University of Queensland; TW Pietsch, School of Fisheries, University of Washington, Seattle; IC Potter, Murdoch University, Perth; Robert Poulin, University of Otago; Heather Proctor, University of Alberta; Jack Randall, Bishop Muscum Hawai'i; Greg Rouse, South Australian Museum and University of Adelaide; Klaus Ruetzler, Smithsonian Institution, Washington; Jan Sauer, University of Giessen; Christine Schoenberg, University of Queensland; Jeffrey Shields, Virginia Institute of Marine Science; Frank Stephens, Curtin University, Perth; Horst Taraschewski, University Karslruhe; Anke Treuner, University of Tokyo; Jean-Paul Trilles, Université Montpellier-2; Wolfgang Vogelbein, Virginia Institute of Marine Science; John Walker, University of Sydney; Stephen Wesche, Queensland Museum, Brisbane; Ian Whittington, South Australian Museum. An anonymous reviewer checked the entire manuscript.

The School of Environmental Sciences and Resources Management, University of New England, kindly made resources available to the Editor.

Chapter 1

The nature of parasitism

This chapter consists of two sections. The first gives definitions of terms and a brief discussion of some adaptations to a parasitic way of life, such as body size, reproductive capacity, reduction and increase in complexity, and dispersal. The second discusses the distribution of parasites and in particular of marine parasites in the 'tree of life'. It is emphasised that life originated in an aqueous and probably marine environment, and that every marine organism is a potential host to parasites.

Definitions, and adaptations to a parasitic way of life
Klaus Rohde

Introduction
Biologists approach the study of parasites from different angles: some are interested in their physiology, others in their ecology, or medical and economic aspects, to mention only a few. Consequently, definitions of parasitism vary greatly, reflecting the research interests and biases of particular research workers. Baer (1952), Esch and Fernández (1993) and Rohde (1993, 2001) have discussed definitions and adaptations of parasites.

In most sections of this book, *parasitism* is understood to be a close association of two organisms, in which one – the parasite – depends on the other – the host – deriving some benefit from it. This benefit often is food. In some sections, however, the term *parasite* is used in a wider sense. For example, peritrich ciliates leading to severe problems in shrimp aquaculture are really fouling organisms that become pathogenic and fatal in heavy infections.

Many bacteria, viruses and fungi are parasitic. These organisms have traditionally been studied by microbiologists, and only protistan and metazoan parasites are considered to be objects of study by parasitologists *sensu strictu*, although the border between protistans and fungi is somewhat blurred. In this book, only protistan and metazoan parasites are discussed.

Related associations
Several types of associations are related to parasitism and cannot always be clearly distinguished from it. They include commensalism, phoresis (phoresy), mutualism, and symbiosis.

- *Commensalism* occurs where organisms use food supplied in the internal or external environment of a host without affecting the host in any way. Examples are certain barnacles on the skin of marine mammals.
- In *phoresis*, organisms use a host for transport or shelter. Examples again are certain barnacles on whales, or sea anemones on gastropod shells.

- A mutualistic association (*mutualism*) is one in which both organisms derive a benefit but the association is not obligatory. For example, cleaner fish and shrimp feed on parasites of hosts: the host is cleaned – an obvious benefit to it – and the cleaner derives food from the host; however, many cleaners can also feed on free-living animals and the association is therefore not obligatory.
- In a symbiotic relationship (*symbiosis*) the association is compulsory. The relationship of male and female echiurans or certain deep-sea fish, in which the dwarf male 'parasitises' the much larger female, can be considered symbiotic, because the male cannot live without the female and the female depends on the male for fertilising the eggs. The term symbiosis is also used to describe any association between organisms, such as parasitism, commensalism, mutualism, phoresis, and symbiosis *sensu strictu*.
- In *predation*, a predator kills and eats prey, in contrast to a parasite, which feeds on the host but often and usually does not kill it (although parasitic effects may sometimes be lethal in the long term).

A parasite, under certain conditions, may become a commensal, mutualist or predator. In many cases we know too little about a species to clearly state which kind of relationship it has with a host. Such species are included in many sections of this book.

Types of parasites, parasitic life cycles and hosts

Parasites can be divided into several different types:

- *Ectoparasites* are parasites that live on the surface of a host, while *endoparasites* live in the host's interior. For example, most trematodes are endoparasitic in the internal tissues and digestive tract, whereas almost all monogeneans live on the gills or skin of fish (see pp. 55–72).
- Most species of parasites are *obligate parasites*, which need a host for survival at least during certain stages of their life cycle. A few (e.g. some ciliates) are *facultative parasites* and are able to survive in the free environment during their whole life but can parasitise a host as well (see pp. 37–41).
- *Temporary parasites*, such as leeches, infect their hosts only for short periods, whereas *permanent parasites*, such as roundworms or trematodes in the digestive tract of many marine animals, infect hosts for a long time (see pp. 72–87 and 104–115).
- *Larval parasites* are parasitic only during their larval stage (e.g. praniza larvae of isopods) (see p. 144 and Fig. 6.5).
- *Periodic parasites*, such as leeches, visit their hosts in intervals (see pp. 196–202).
- *Hyperparasites* are parasites of parasites. In the marine environment, not many hyperparasites are known, but new ones are being described frequently. An example is the monogenean *Udonella* that infects copepod ectoparasites of fishes (see pp. 293–298).
- *Microparasites* (which include the protistans and some helminths) are small and have short generation times, reproduce in or on a host at high rates, the duration of infection is often shorter than the life span of hosts, and they induce immune responses in their vertebrate hosts. In contrast, *macroparasites* (arthropods and most helminths) are larger, do not multiply in or on the host, and have longer generation times than microparasites. They induce no or only weak immune responses depending on infection intensity, and infections are usually long lasting, leading to morbidity rather than mortality.

Parasite life cycles also can be divided into different types. Parasites with *direct life cycles* have only a single host. For example, most adult monogeneans are found on fish, whereas their larvae are free-living (see pp. 52–72). Parasites with *indirect life cycles* have several hosts. Adult digeneans, for instance, are found in various vertebrates and their larvae infect at least one and often several hosts (see pp. 76–87).

Similarly, hosts of parasites can be divided into different types. *Definitive* or *final hosts* harbour the sexually mature stage of a parasite; *intermediate hosts* harbour immature, developing stages; and *paratentic hosts* (also called *transport hosts*) harbour larval forms that do not develop within the host.

Adaptations

All animal species, whether free-living or parasitic, must be adapted to their habitats in order to survive. The conditions necessary for survival may be very restricted or they may be wide. Thus, the degree of specificity for certain environmental conditions varies greatly between species. The question arises whether all or perhaps a great range of parasites share similar adaptations to a parasitic way of life. And indeed, all parasites have certain characteristics in common.

Parasites are always smaller than their host, in spite of the existence of some very large parasites. Didymozoid trematodes, common parasites in many marine fish, may be very large. A species infecting the sunfish *Mola mola* reaches a length of 12 m; however, its diameter is very small and the host may be as heavy as one ton. Thus, based on a comparison of volume, the parasite still is minute relative to its host. Interestingly, free-living species are often considerably smaller than related parasitic species. Free-living turbellarians, for instance, are usually one to a few millimetres long, whereas parasitic flatworms are much larger, trematodes up to several centimetres, and cestodes up to several metres in length. Reasons for this may be twofold. First, parasites have an almost unlimited and secure food supply provided by the host. Second, selection may favour larger worms with larger reproductive organs because production of many offspring is even more important for parasitic than for free-living species since only a minute proportion of offspring ever manages to infect a host. However, selection may favour the smaller body size of parasites, because too large a size might damage the host on whose survival the parasite depends. In an evolutionary context there will therefore be a trade-off between selection favouring larger and selection favouring smaller body size.

As pointed out above, parasites depend for survival on the production of many offspring, and data show that reproductive capacity is often greater in parasitic species than in related free-living forms. This is well shown in the Platyhelminthes. In spite of the great variability in the number of offspring (as measured by the number of eggs/larvae produced plus multiplication at the larval stage), there are clear trends: free-living turbellarians produce about 10 or so eggs throughout their life on average, and there is no multiplication of larval stages; ectoparasitic Monogenea produce about 1000 eggs, without multiplication of larvae; endoparasitic trematodes may produce millions of eggs and larvae do multiply. Endoparasitic cestodes also may produce millions of eggs and may have multiplication of larvae. Larval multiplication is particularly well developed in the trematodes. A single egg produces one miracidium that infects a mollusc. In the mollusc intermediate host, the miracidium develops into a sporocyst which gives rise to numerous rediae. Rediae, in turn, produce numerous cercariae which leave the snail and infect the final host. Thus, a single egg can produce tens of thousands of offspring.

It is commonly believed that parasites have lost much of the complexity of free-living animals because they depend on the host for food and shelter and so supposedly do not need the same range of sensory receptors, a complex nervous system, sophisticated feeding organs and so on. This is indeed sometimes the case and an example of this is the rhizocephalan *Sacculina*, which parasitises marine crabs. The juvenile, free-living larva has all the characteristics of larval barnacles (to which the rhizocephalans are related) but the adult consists of a sac-like structure (the so-called externa) attached to the ventral surface of the crab's abdomen, and an extensive system of cytoplasmic processes that reach into the various host tissues, without any

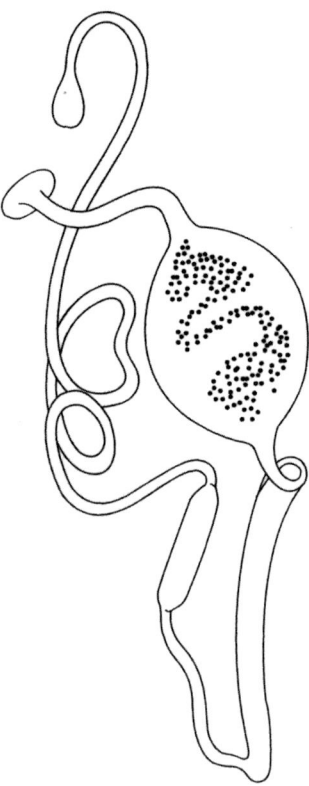

Figure 1.1 A parasitic snail from the body cavity of a sea cucumber. All external similarities with a snail (i.e. head with tentacles, eyes, foot and shell) have completely disappeared due to the endoparasitic way of life, leading to a worm-like appearance of the snail. Redrawn after Koehler and Vaney from Baer (1952).

morphological crustacean characteristics. Simplification of parasites is sometimes called 'sacculinisation', based on this example. Another example is illustrated in Figure 1.1, which shows a snail parasitic in a sea cucumber that has lost all characteristics of snails. Similarly, Figure 1.2 shows the gradual transformation of the copepod *Lernaeocera branchialis* from a free-living nauplius larva, with all the characteristics of other crustacean larvae, to the adult parasitic stage which has mouthparts inserted into a blood vessel of its fish host and a body completely transformed to a sac-like structure. Very often, however, sacculinisation is not evident and parasitic species may even be more complex than their free-living relatives. For example, detailed electron-microscopic studies of the aspidogastrean *Lobatostoma manteri* have shown that the juvenile/adult stage (3–5 mm long) has about 20 000–40 000 sensory receptors of about 12 different types, many more than are found in most free-living flatworms, the turbellarians. Its nervous system also has a much greater complexity than that of turbellarians (for details see Rohde 1999). Importantly, *Lobatostoma manteri* does not have a single free-living stage; therefore, the function of receptors and nerves must be related to the endoparasitic way of life of the parasite. No experimental evidence is available but it is possible that the receptors are important in finding mating partners and microhabitats within the final host, and in preventing damage to the delicate host tissue in the intermediate snail host. Comparison of monogeneans parasitic on the gills and body surface of fish to free-living turbellarians shows that the former have a variety of complex attachment and copulatory structures while the latter do not.

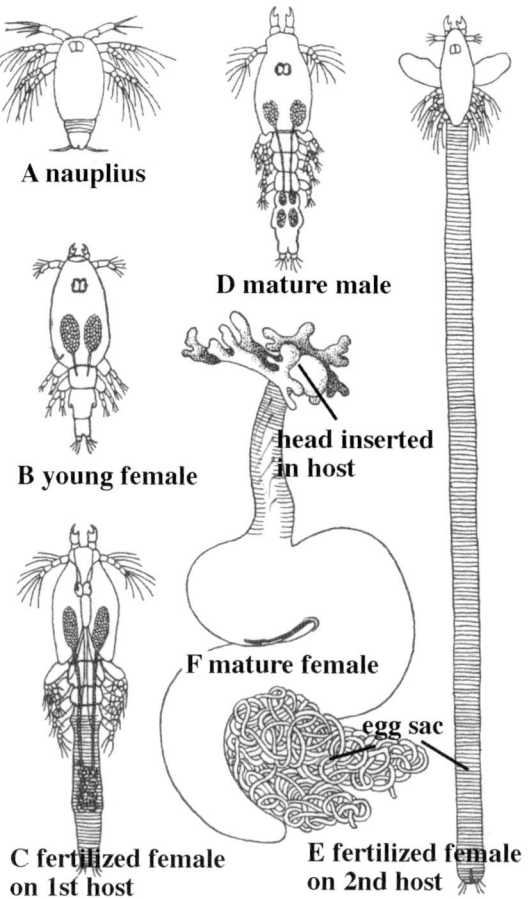

Figure 1.2 Transformation of *Lernaeocera branchialis*. **A**. Free-living larva. **C–F**. Adult stage. After Scott from Baer (1952).

All animal species disperse at some stage of their life cycle. This is essential for the survival of a species because populations restricted to a small area will become extinct if conditions deteriorate, and also because inbreeding in a small, restricted population may lead to loss of evolutionary versatility and to extinction. For parasites, dispersal must be particularly effective because hosts are difficult to find. Dispersal of parasites may result from host dispersal, as occurs when eggs of various helminths are spread by migrations of fish and mammal hosts. Alternatively, the parasites themselves may have dispersal mechanisms. Thus, eggs of various helminths are dispersed by migrations of fish or mammal hosts, whereas trematode larvae (cercariae), for instance, have various morphological adaptations such as finfolds or hair-like extensions of the tail that serve as flotation mechanisms, leading to dispersal by water currents.

Unique to parasites are mechanisms of infections, of which there is a great variety. A few examples are given here: trypanosomes are injected into the blood of fish by blood-sucking leeches; many trematodes are acquired by ingesting cysts (metacercariae); *Anisakis* is acquired by eating transport hosts, and *Diphyllobothrium* by eating infected intermediate hosts; and schistosomes penetrate the skin of marine birds. (For a fuller discussion see sections on the various parasite groups and pp. 280–286). Not infrequently, parasites induce behaviour changes in their host which increase the likelihood of infecting the next host in the life cycle (see pp. 259–264).

Many parasites either have asexual/parthenogenetic reproduction or are hermaphroditic, that is, even a single parasite individual can build up a population. This has definite advantages for them given that the likelihood of reaching a host is very small.

All species, whether parasitic or not, have restricted habitat requirements. In parasites this is expressed as host and site specificity. There is no universal parasite that infects all hosts and sites within a host equally, although the degree of specificity varies. Some trematodes, for example, infect a wide range of marine fish species whereas some monogeneans are restricted to a single host species and are always found on a small segment of the gills (see p. 292).

Besides the characteristics shared by many taxa and resulting from adaptations to a parasitic way of life, all parasites have special adaptations, including physiological ones, that are very specific, determined by the phylogenetic position of the parasite and by its host. Such specific adaptations are discussed in the various sections on parasite groups.

Important references
The following authors give detailed discussions of important aspects of parasitism, including definitions and adaptations: Baer (1952), Esch and Fernández (1993) and Rohde (1993, 2001).

Marine parasites and the tree of life
DTJ Littlewood

Introduction
It seems unlikely any species escapes being parasitised, except perhaps the very smallest organism that, simply because of its size, cannot house or provide sustenance to another. As such, much of biodiversity (species richness) may be viewed as a diversity of hosts for parasites. Nearly all organisms expend at least some energy in avoiding parasites or ameliorating their effects. Individuals may spend their life parasite-free, but few species have evolved without parasitism having at least some part in the natural selective processes that have shaped their present form and biology.

Parasitism is a highly successful, but often underrated and certainly underscored, life history strategy. A cryptic life stage or a fleeting association with a host may make it a difficult phenomenon to detect. Additionally, even when parasites are large, persistent and low prevalence levels may compromise their detection. Parasitism is a biological trait that is defined predominantly in ecological terms (see e.g. Combes 2001). It unites a plethora of species that may or may not be closely related phylogenetically. However, when viewed in the light of phylogeny, parasitism has evolved many times over. Even among predominantly free-living groups usually at least one lineage engaging in this life style can be detected. Frequently, whole clades of species (phyla, classes, orders and families) have evolved obligate parasitism. These groups are dealt with separately in this volume. Curiously, it seems that once parasitic, a lineage never gives rise to a free-living form; in evolutionary terms, parasitism is irreversible. Lineages that have successfully adopted parasitism are often species rich and are at least as diverse as the hosts they interact with. Parasitism itself is a driving force in genetic diversity (e.g. Frank 1993).

The evolutionary history of either the parasite or host cannot be viewed in isolation if we are to understand the nature and evolutionary ecology of parasitism (Poulin 1998). Complex life cycles, where more than one host species is involved, renders the subject of parasite evolutionary ecology even more problematical. In any case, the tree of life, or at least a phylogeny of parasites and their hosts, is an eminently suitable starting point; coevolutionary studies tracking both host and parasite phylogenies are proving to be even more powerful (Page 2003) (see also pp. 327–339).

The tree of life

An understanding of the interrelationships between organisms, through the construction of a phylogeny, provides us with a means to track the origins and radiation of novelties and adaptations through evolutionary time. With morphology alone, phylogeneticists find it difficult to determine relationships between, and sometimes within, plant and animal phyla. The advent of molecular and biochemical systematic tools, and the growth of phylogenetics as a discipline from the latter half of the 20th century onwards, has allowed many of these barriers to be overcome (Balter 1997), although in some cases these developments have fanned the flames of old controversies and generated some new ones. Modern molecular systematic methods are the foundation for understanding microbial and other relatively simple organisms, but have contributed also to our understanding of morphologically complex taxa. Studies of parasitic taxa that have converged on similar morphologies or that have poorly distinguishable life stages have benefited also from these new tools in resolving species boundaries and inter-relatedness. Indeed, whole clades owe their names to molecular evidence [e.g. the Ecdysozoa and the Syndermata (rotifers and acanthocephalans)]. Armed with molecular sequencing, bioinformatics, and increasingly faster computers that allow large quantities of data to be handled (including nucleotides, amino acids, proteins, genomes and, of course, morphology), researchers have begun to estimate the entire tree of life (e.g. Cracraft and Donoghue 2004). There is only one tree of life, recapitulating each and every branching point in the evolutionary history of life, but achieving its resolution, with every tip in its rightful place, is no simple task. One preliminary approximation, based on a multitude of published empirical studies and a consensus opinion of a subsample of researchers is shown in Figure 1.3 (after Pennisi 2003; with an online and interactive version also available).

Most major groups appear to be well resolved in the overall scheme but insufficient data, or even conflict between independent sets of data, have yielded unresolved branching points. Such *polytomies* are not uncommon, regardless of the taxonomic level under investigation, even among closely related species based on multiple data sets, and yet in spite of incongruence or poor nodal support, the tree of life is taking shape. The paucity of systematists, and the large collective effort required, limits the progress of the Tree of Life project, but for many purposes there is already sufficient useful resolution. Further websites from which interrelationships between organisms can be viewed include the Tree of Life Web Project and TreeBase.

Marine life
It is generally agreed that life itself probably arose in the marine, or at least a watery, environment (e.g. Whitfield 2004). Thus, it is not surprising to see how many organisms have representatives, or are indeed restricted to, living in seas and oceans. If we consider extinct as well as extant organisms, it is also clear that the marine environment has been abandoned and recolonised many times through evolutionary history. Mammals, turtles, snakes and crocodiles include obvious examples of terrestrial-born lineages that have colonised saltwater habitats since their appearance, sometimes bringing their parasites with them, although more usually being parasitised by many new species as they adapted to their different environment. Even groups and lineages we generally consider to be strictly terrestrial, or to have very little to do with the marine environment, include examples with some connection to the sea; for example, parasitic lice of marine birds renders the Paraneoptera as having both marine and marine parasitic representatives (Fig. 1.3). Even spiders, collembolans and beetles have marine representatives when we consider those intertidal examples that hide, trapped in air pockets in rock pools, at high tide. Organisms that we might not consider to be marine *per se* may at least interact with the marine environment and certainly, from a parasitological perspective, may be hosts in marine parasite life cycles (e.g. birds visiting and feeding in the intertidal zone).

8 Marine Parasitology

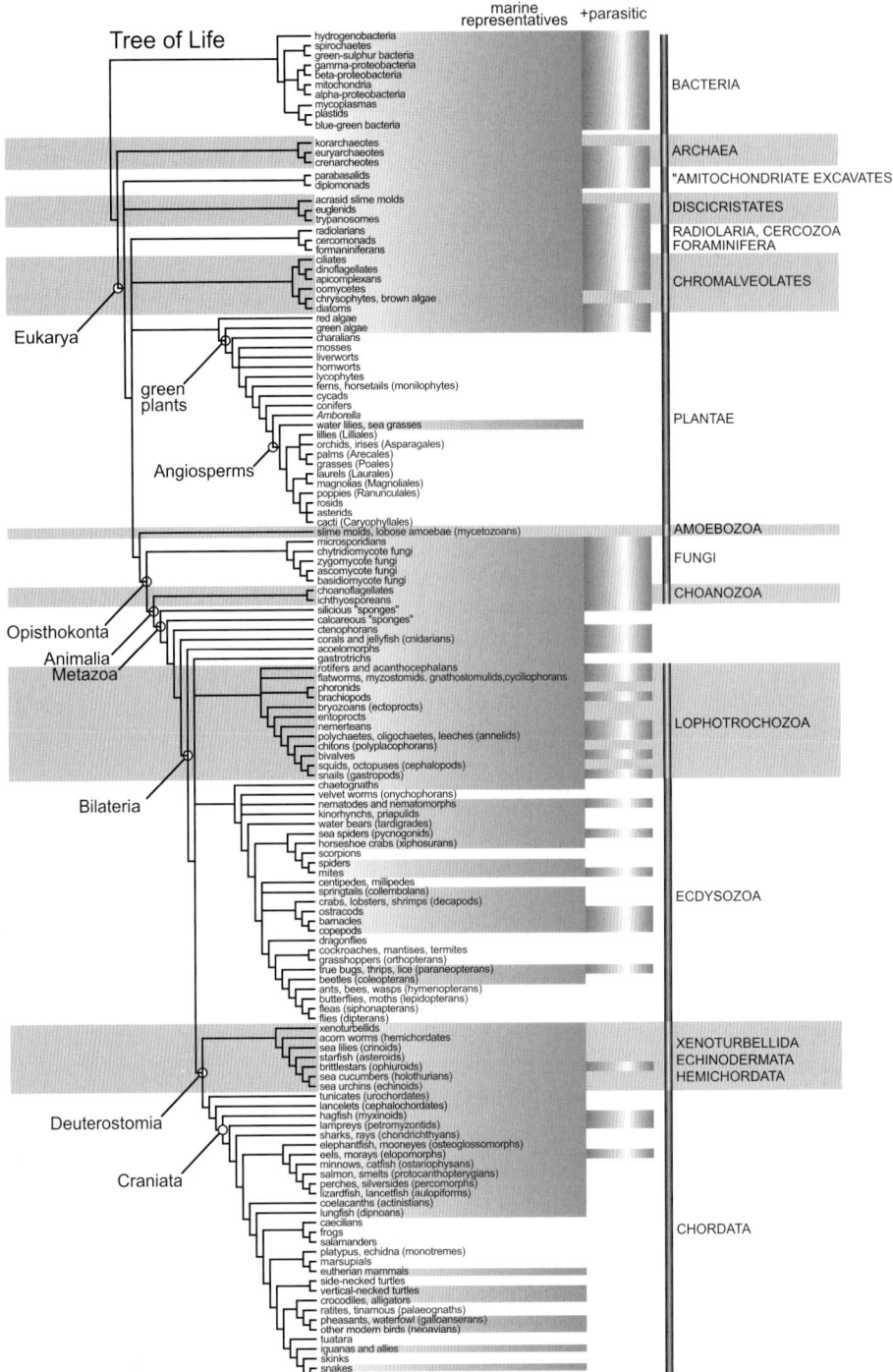

Figure 1.3 The tree of life based on a consensus of studies and opinions, originally published in Pennisi (2003) but modified and further annotated here, with an indication of which clades have marine representatives and which of these include parasitic forms. A web-based copy of the tree, with links to sites covering individual groups, is available at: *The American Association for the Advancement of Science* (2004) reproduced with permission.

Marine organisms as hosts

The enormous variety of marine environments and prevailing marine conditions that exist globally is equally matched by the diversity of organisms that make these habitats their home. Every marine organism is potentially a host to a parasite, but depending on locality, season, local ecology and parasite species in question, any candidate host may or may not be parasitised. Clearly, not every organism is a suitable host for every parasite, but to understand fully the taxonomic range of host-parasite interactions, we need to consider obligate and facultative interactions, and even the paratenic hosts that help to bridge trophic gaps. Beyond individual studies of hosts and their parasites, there exist few resources for quantifying the extent to which parasitic taxa occur within and between different host species. One such resource, covering five phyla (flatworms, nematodes, acanthocephalans, nematomorphs and mesozoans) is the Host-Parasite Database (Natural History Museum 2004).

Size of the host limits the upper, but usually not the lower, size of the parasites that manage to infect it. This contributes to the many factors determining which parasites we may expect to find in any given species at any time during its development. Larval stages are readily parasitised but not necessarily by the same taxa that infect adult stages. Host size, host ecology and parasite life history strategy constrain the taxonomic range of possible host-parasite combinations, not to mention host-specificity or host defence mechanisms. Small multicellular (metazoan) animals tend to be parasitised by unicellular organisms. Correspondingly, bigger animals become suitable hosts for bigger (unicellular and/or metazoan) parasites, and may even increase levels of prevalence within a species. The largest living animals, the cetaceans, are host to the longest nematodes and tapeworms and the sheer abundance of these parasites, measured as numbers or biomass, shows how host size may have a profound effect on the biology of the parasites that infect them. Nevertheless, in isolation, the tree of life of hosts indicates few patterns or trends that shed light on the evolution of marine parasitism.

Marine parasites and the origins of marine parasitism

Many of the lineages in the tree of life are exclusively marine, but considerably fewer are exclusively parasitic. The tree of life showing marine parasites (Fig. 1.3) indicates, perhaps obviously, that marine parasitism is generally restricted to marine hosts. However, the origins of marine parasites are not necessarily marine. Among the parasite lineages that have remained exclusively marine, or at least aquatic, since their divergence (e.g. copepods) we might safely deduce that parasitism originated in the marine environment. This appears to be true in most cases. Exceptions include mites of marine mammals that presumably had terrestrial origins and are themselves not truly marine, although they infect air pockets within marine hosts (e.g. nasal mites of seals and sea otters). To determine the origins of parasitism requires knowledge of sister group relationships within and between parasite lineages, the determination of free-living sister groups and the inference of ancestral life history strategies. Such an approach has been used to investigate the origins and radiation of parasitism in some groups (e.g. the Digenea, Cribb *et al.* 2003; the Nematoda, Blaxter 2003), but to a great extent most tips of the tree of life still lack sufficient resolution.

Although many lineages on the tree of life include marine parasitic representatives, there are a few groups of significant interest, including the various protistan, helminth and crustacean parasites. Among each of these groups are obligate parasite lineages that have radiated very successfully, in terms of biomass, species number, host diversity and geographic distribution. Estimating parasite numbers is as difficult as estimating the number of species on Earth. However, to illustrate their success, few species of aquatic vertebrate escape parasitism by at least one (and usually many more) species of digenean, cestode, monogenean, nematode and acanthocephalan (the helminths) and one or more crustacean species. The protistan parasites appear to be even more widespread and we know less about the invertebrates as hosts. Further study will reveal the true diversity of marine parasites, but given the small fraction of marine biodiversity studied so far, at least from a parasitological viewpoint, we can expect the number of parasites to represent a substantial and significant proportion.

Important references
Molecular data are increasingly used to clarify phylogenetic relationships, but morphology has to be considered as well. Important references in this field are by Balter (1997), Blaxter (2003) for nematodes, chapters in Page (2003), Pennisi (2003), Cracraft and Donoghue (2004) and Cribb *et al.* (2003) for the Digenea.

Chapter 2

Protistan parasites and Myxozoa

Coverage of 'lower' organisms in this book is restricted to protistan and metazoan parasites and excludes fungi and other organisms such as bacteria and viruses, many of which are *parasitic* as defined in the Introduction. The boundary between fungi and protistans is ill-defined, however. Protistan taxonomy, based on molecular phylogeny, is continually changing. For example, the 'Sarcomastigophora' (still treated as such in this chapter), has disintegrated into many separate, high ranking taxa. Among them, the Opalinata belongs to a class of Heterokonta, these themselves are a subphylum of the phylum Chromista. This Chapter includes groups that have commonly been considered to be protistan. The Chapter also includes the Myxozoa, which have the appearance of protistans and were for a long time considered to belong to that group. Recent DNA and ultrastructural studies have shown that they are metazoan, although their exact position within the metazoans has not been resolved. Like the Cnidaria, they possess nematocysts (or nematocyst-like structures), but molecular evidence points to a position among the bilaterian metazoan. All the groups discussed in this Chapter have considerable importance as agents of disease, particularly in aquacultured fish and molluscs. Some of them, in particular the Microsporidia and Apicomplexa, are important as infective agents in immunocompromised people. Species richness is practically unknown for Sarcomastigophora, Microsporidia, Ciliophora and Myxozoa: new species are being described continually. In Australia, for example, less than 5% of the thousands of fish species have been examined for these parasites, and sample sizes of those which have been examined, were small. Fish of African, South American and many Asian countries have been examined even less. It may well be that these parasites belong to the most speciose groups of parasites in the marine environment and, as such, have considerable ecological importance. Also, at least some of the vast number of marine invertebrates that are yet to be examined are likely to have protistan parasites, probably including large numbers of sarcomastigophorans and ciliophorans.

In view of the great morphological and taxonomic diversity of the protistans, this chapter begins with a brief overview of this kingdom before proceeding to the various sections dealing with the groups in greater detail.

Protistan biodiversity
Peter O'Donoghue

Introduction
The kingdom Protista (syn. Protoctista) comprises unicellular eukaryotic organisms which exist as structurally and functionally independent individual cells (including those species which are gregarious or form colonies). None have adopted multicellular somatic organisation characteristic of

metazoan organisms. Instead, Protista have developed relatively complex subcellular features (membranes and organelles) which enable them to survive the rigours of their environments. The kingdom is not considered a natural assemblage of organisms but rather a classification of convenience, containing motile protozoal protists as well as non-motile algal and fungal protists. They exhibit enormous diversity in form and function, and some 40 phyla have been recognised on the basis of their unique morphology and biology. It has been conservatively estimated that there are about 100 000 extant species of Protista and they are considered ubiquitous as free-living organisms in terrestrial and aquatic habitats and as commensals, mutualists or parasites of most animals and many plants. A brief guide to the major protistan parasites of marine hosts is given below.

Mastigophora (flagellates)

The Mastigophora exhibit locomotion by flagella for most of their life-cycle. They possess one or more flagella (sometimes called kinetids, mastigonts or undulipodia); each arises from a small basal body (centriole or kinetosome) and contains two single central microtubules and nine peripheral doublets (2 + 9 configuration). Multiflagellated forms may be isokont (equal flagella) or heterokont (unequal flagella) and some species have a trailing (recurrent) flagellum often associated with an undulating membrane. Flagellates reproduce usually by longitudinal binary (symmetrogenic) fission between, rather than across, rows of flagella. Most flagellates are free-living aquatic organisms but representatives of several families (esp. dinoflagellates and kinetoplastids) parasitise marine hosts. Dinoflagellates have a distinctive haploid (dinokaryon) nucleus where the chromosomes remain condensed during interphase and appear as beaded threads as a result of low levels of histones and associated proteins. Over 4000 species have been identified, most being free-living pelagic or planktonic autotrophs, while some 140 species are

Figure 2.1 (see opposite) Stylised drawings of representative protista parasitic in marine organisms (figures vary in scale, approximate length or diameter indicated). **A.** *Amyloodinium* trophont from fish skin (100 μm). **B.** Free-swimming gymnodinid-type dinospore (20 μm). **C.** *Trypanosoma* trypomastigote from fish blood (40 μm). **D.** *Cryptobia* trophozoite from fish gills (10 μm). **E.** *Neoparamoeba* trophozoite from fish gills (30 μm). **F.** *Labyrinthula* spindle cells from marine plant and free swimming zoospore (15 μm). **G.** *Haplosporidium* spore from hepatopancreas from oyster (5 μm). **H.** *Nematopsis* gamont from gut of crab (80 μm). **I.** *Haemogregarina* gamont in fish erythrocyte (12 μm). **J.** *Eimeria* oocyst from gut of fish (15 μm). **K.** *Calyptospora* oocyst from hepatocytes of fish (25 μm). **L.** *Haemohormidium* trophozoites in fish erythrocyte (2 μm). **M.** *Glugea* cyst from subcutaneous tissues of fish (50 μm). **N.** *Loma* spores from gills of fish (4 μm). **O.** Ultrastructure of *Pleistophora* spore from muscles of fish (4 μm). **P.** *Mikrocytos* microcells from digestive gland of oyster (1 μm). **Q.** *Cryptokaryon* trophont from skin of fish (100 μm). **R.** *Uronema* trophont from brain of fish (30 μm). **S.** *Epistylis* zooids from exoskeleton of decapod crustacean (50 μm). **T.** *Trichodina* trophont from skin of fish (60 μm). **U.** *Myxobolus* spore from gills of fish (15 μm). **V.** *Myxidium* spore from kidney of fish (10 μm). **W.** *Kudoa* spore from muscles of fish (10 μm). **X.** Triactinomyxid actinospore released from oligochaete (100 μm).

Key to abbreviations used in Figure 2.1: AD, adhesive disc; AF, anterior flagellum; AP, anchor-like protrusions; CC, caudal cilium; CI, cingulum; CS, central stem; CY, cytostome; CZ, contracted zooid; DE, denticles; DM, deuteromerite; EM, epimerite; ER, erythrocyte; EX, extracellular matrix; FF, free flagellum; FI, filiform projections; FO, filamentous ornaments; FV, food vacuole; GA, gamont; HA, haplosporosomes; HN, host cell nucleus; KI, kinetoplast; LF, longitudinal flagellum; LK, longitudinal kineties; MA, macronucleus; MC, microcells; MG, mastigonemes; MS, microspores; NK, Nebenkörper (parasome, a kinetoplastid endosymbiont); OC, oral ciliature; OO, oocyst; OR, orifice; PA, paroral membrane; PC, polar capsule; PE, peduncle; PG, postoral groove; PM, protomerite; PN, parasite nucleus; PP, polaroplast; PS, pseudopodia; PT, polar tube; PV, posterior vacuole; RF, recurrent flagellum; RP, radial pins; SB, Stieda body; SC, spindle cell; SD, sporopodia; SE, septum; SI, scuticum; SK, stalk; SM, sporoplasm; SO, somatic cilia; SP, sporocyst; ST, stomopode; SU, sulcus; SV, sporophorous vesicle; SW, spore wall; SZ, sporozoite; TF, transverse flagellum; TH, theca; TR, trophozoite; UM, undulating membrane; VC, valve cell; ZS, zoospore.

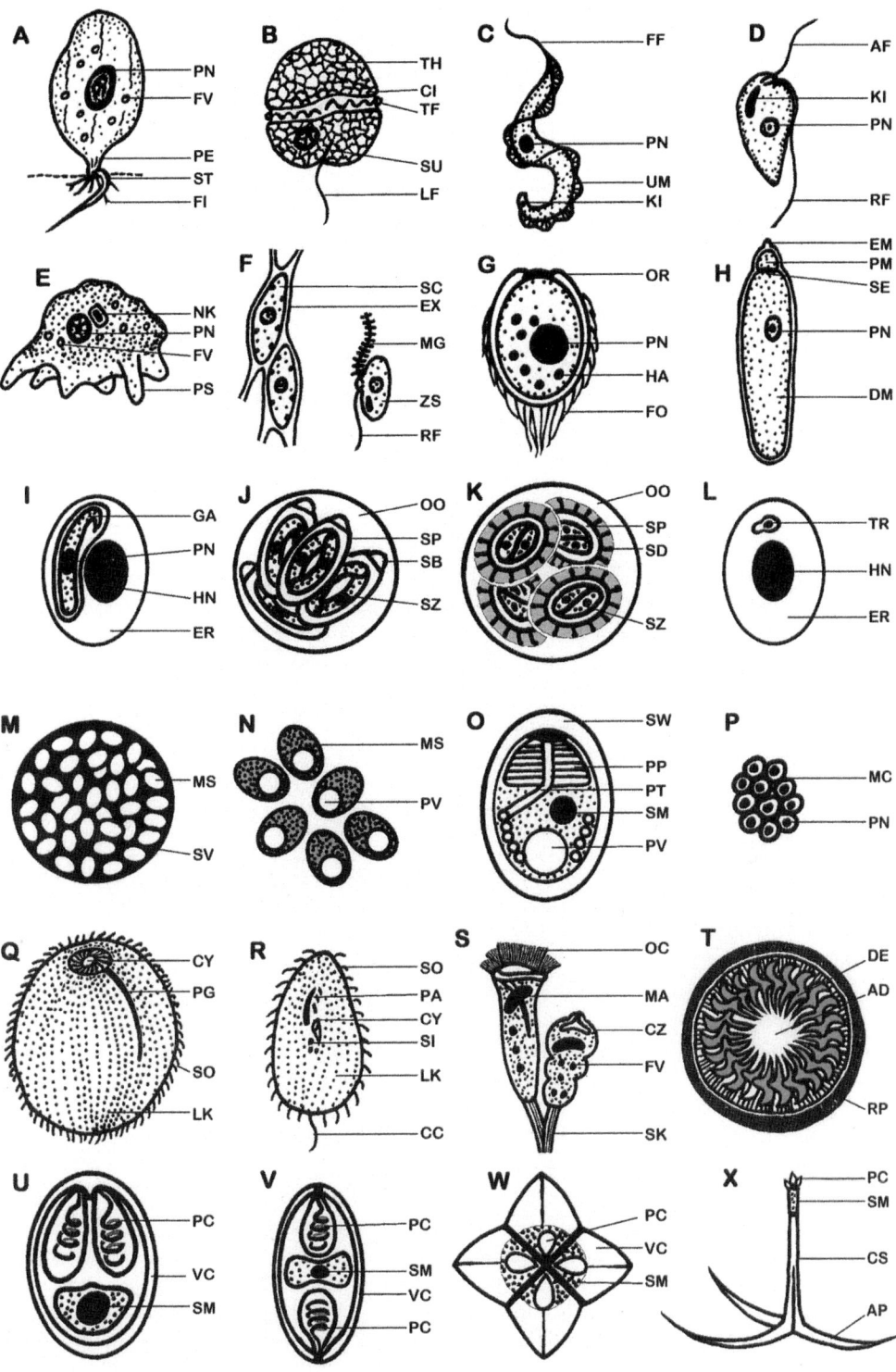

parasitic in zooplankton, filamentous algae or on the external surfaces (and gills) of crustaceans and fishes (Fig. 2.1A). Parasitic trophonts often develop elaborate holdfast attachments and may become macroscopic as they feed to repletion. They then drop from the host and form motile biflagellated dinospores, with a transverse flagellum lying in an equatorial cingulum and a posterior flagellum often lying in a longitudinal ventral sulcus (Fig. 2.1B). Many dinospores are encased in armour (theca) composed of cellulosic plates. Most species contain chloroplasts while others have coloured pigments; some pigments are neurotoxic to mammals when concentrated in the tissues of fish or filter-feeding shellfish.

Kinetoplastid flagellates are characterised by the possession of extranuclear DNA (kinetoplast) within the single large mitochondrion usually associated with the flagellar basal body. Over 500 species have been described and many species are parasitic in vertebrate and invertebrate hosts with simple monoxenous (one-host) or more complicated heteroxenous (two-host) life-cycles. Trypanosomes found in the bloodstream of fishes are transmitted by leech vectors. They form characteristic trypomastigotes with a single recurrent flagellum that adheres to the cell body and becomes an undulating membrane (Fig. 2.1C). Infections are usually chronic but some have been associated with tissue pathology and mortality. Bodonids have two unequal flagella arising from a deep flagellar pocket (Fig. 2.1D) and several species are endozoic or ectozoic parasites in fishes where they cause local irritation, degenerative changes and erratic behaviour.

Sarcodina (amoebae)

Amoebae exhibit locomotion by the formation of pseudopodia (false feet) or by distinct protoplasmic flow. Amoeboid movement is also used by many species to engulf and ingest food items (phagocytosis). They reproduce by binary fission where trophozoites undergo nuclear division (karyokinesis) then cytoplasmic division (cytokinesis). Rhizopod amoebae form broad lobopodia, filamentous filopodia or reticular anastomosing reticulopodia and they may be testate (producing a shell or test) or naked (without a test). Most species are free-living aquatic or terrestrial organisms although a small number of naked gymnamoebae are parasitic in animals (often as opportunistic histophages). Several *Neoparamoeba* and *Paramoeba* species, with a unique parasome (or Nebenkörper, recently revealed to be a kinetoplastid endosymbiont) adjacent to the nucleus (Fig. 2.1E), have been linked to disease and death in marine fish and invertebrates.

Labyrinthomorpha (slime nets)

Labyrinthomorpha do not produce orthodox pseudopodia but form elaborate networks where trophozoites are associated with, sometimes appearing to glide along, ectoplasmic slime channels (Fig. 2.1F) secreted by special organelles (called sagenogenetosomes or sagenogens). Many have recently been shown to undergo reproductive cycles involving the formation of heterokont biflagellated zoospores (Fig. 2.1F). About 30 species have been described as saprobic or parasitic on marine molluscs, algae and vascular plants, sometimes in association with wasting diseases.

Haplosporidia (haplosporidians)

Haplosporidia are characterised by the formation of unicellular spores (without polar filaments) that contain a single sporoplasm and several dense organelles (known as haplosporosomes). The spore wall has an orifice covered by an operculum or occluded by a lingula plug. The spores are covered with filamentous ornaments that sometimes appear as tails (Fig. 2.1G). No complete life cycle has been elucidated for any haplosporidian and the fate of spores is unknown. In the final host uninucleate cells undergo modified schizogony that gives rise to multinucleate plasmodia which develop into sporonts and eventually differentiate into spores. Some 40 species are found as histozoic or coelozoic parasites of aquatic molluscs, annelids, crustaceans and helminths, and several species cause significant oyster diseases throughout the world.

Apicomplexa (sporozoans)

The spore-forming parasites Apicomplexa possess a distinctive apical complex of organelles, comprising a conoid, polar ring, rhoptries, micronemes and subpellicular microtubules, which facilitate entry into host cells as they are obligate intracellular parasites for most of their life cycles. They undergo cyclic development involving three divisional processes: merogony (schizogony), gamogony and sporogony. Cell division may occur by fission (splitting of the maternal cell) or endogeny (internal formation of daughter cells). Over 8000 species have been described as monoxenous or heteroxenous parasites of vertebrate and invertebrate hosts. Representatives of most apicomplexan groups (gregarines, haemogregarines, coccidia and haematozoa) are found in marine hosts.

Gregarines form large extracellular gamonts which may be septate (cephaline) or aseptate (acephaline); the former being divided by a septum into an anterior protomerite and a posterior deutomerite (Fig. 2.1H). The conoid is modified, forming an anterior holdfast organelle (epimerite in septate species or mucron in aseptate species). Equal numbers of gametes are produced by male and female gamonts. Most species have monoxenous life cycles in the digestive tracts or body cavities of invertebrates and lower chordates although some have heteroxenous life-cycles cycling between molluscan and crustacean hosts involved in predator–prey relationships.

Haemogregarines are adeleorin coccidia which form small intracellular gamonts; microgamonts producing from 1 to 4 non-flagellated microgametes which associate pairwise with macrogametes (syzygy). Over 400 species have been recorded as heteroxenous parasites in vertebrate leucocytes and erythrocytes (Fig. 2.1I) with haematophagous invertebrates acting as vectors. Species in fish are transmitted by leeches and infections are usually mild and chronic although some have been associated with severe disease.

Coccidian parasites form non-motile resistant oocysts that contain infective sporozoites usually confined within secondary spores (sporocysts). The gamonts of eimerian coccidia develop separately and many flagellated microgametes are produced. Over 200 species have been described in fish predominantly on the basis of oocyst morphology and host occurrence. Most fish coccidia sporulate endogenously and the oocyst envelope is thin and fragile and never contains a micropyle. The sporocysts vary markedly in appearance, some with Stieda bodies (Fig. 2.1J), finger-like sporopodia (Fig. 2.1K), gelatinous coverings or geometric shapes. Infections may be confined to the gut or undergo extra-intestinal development leading to marked histopathological changes.

Haematozoa are small blood-borne parasites which undergo merogony and gamogony in vertebrate blood cells (Fig. 2.1L). They are transmitted by blood-sucking invertebrates where fertilisation occurs forming a motile zygote (ookinete). Gamonts do not exhibit syzygy and sporozoites are not enclosed within sporocysts. Two main groups are recognised in terrestrial vertebrates: pigment-forming haemosporidia with insect vectors and non-pigment forming piroplasms with arachnid vectors. Only around 10 species have been found in fish and they are transmitted by leech vectors.

Microsporidia (microsporans)

Microsporidia (also known as Microspora) are obligate intracellular parasites which lack mitochondria and form small unicellular spores. Over 1300 species have been described in invertebrates (especially insects) and lower (rarely higher) vertebrates. The parasites undergo cyclic merogony within host tissues followed by sporogony (often involving plasmotomy prior to sporoblastogenesis). Developmental stages may have single or paired nuclei (diplokaryotic) and they may be surrounded by a membranous sporophorous vesicle (pansporoblast) (Fig. 2.1M) or lie free in the host cell cytoplasm (Fig. 2.1N). All spores contain a unique coiled polar tube which can be extruded to inject the infective sporoplasm into host cells (Fig. 2.1O). Nearly 100 species occur in fish and infections may be disseminated throughout the tissues or they may cause focal lesions, inflammation and granulomas. Some species cause extensive hypertrophy of the host cell producing large xenomas.

Mikrocytos (microcells)
These enigmatic organisms are unicellular parasites characterised by the formation of small (1–2 μm) ovoid microcells with central nuclei (Fig. 2.1P). The classification of *Mikrocytos* is uncertain but they demonstrate many similarities to the haplosporidia. Several species have been described from the palps and mantle of molluscs (sometimes systemic) and infections have been associated with focal necrotic lesions and winter mortality in several oyster species.

Ciliophora (ciliates)
Ciliates are unique in that they possess two different types of nuclei (vegetative macronuclei and reproductive micronuclei), cilia (undulipodia) at some stage in their life cycle (kinetosomes and associated fibrils are organised into an infraciliature, even when cilia are absent) and the cell membrane is supported internally by membrane-bound alveoli. Asexual reproduction occurs by transverse (homothetogenic) binary fission across rows of cilia and some species exhibit sexual reproduction by conjugation. Most species are free-living aquatic or terrestrial organisms but many are commensals in vertebrate or invertebrate hosts and some are parasitic. About 150 species occur in fish, most as ectoparasites causing fouling, irritation and local lesions (occasionally penetrating wounds) and some as endoparasites causing variable damage at the tissue/organ level. Classification systems are based on multiple characters, including cilia organisation, kinetid ultrastructure, developmental cycles, life styles and habitats. Patterns of buccal (oral) and somatic (body) ciliation have been retained by many workers as user-friendly characters, although they may not reflect true phylogenetic relationships.Lower holotrichs exhibit little distinction between body and oral cilia. Several groups (gymnostomes, trichostomes and hypostomes) have been associated with skin, gill and internal lesions in freshwater fishes but relatively few species appear to be parasitic in marine fishes. Higher holotrichs have specialised oral cilia, usually comprising a paroral membrane adjacent to three membranelles. Several hymenostome species are notorious parasites of freshwater and marine fishes, causing white spot diseases. Large histophagous trophonts (Fig. 2.1Q) feed on epithelial tissues and form encysted stages (tomonts) in the external environment which produce hundreds of infective swarmers (tomites or theronts). Increasing numbers of scuticociliate species, with a non-ciliated scuticum or scutico-vestige (Fig. 2.1R), are being found to cause invasive systemic diseases in marine fishes and decapod crustaceans. Peritrichous ciliates have a conspicuous left-hand spiral of oral cilia and an antapical holdfast organelle. Many species are sessile for most of their life-cycles and they attach to substrates by means of a scopula or stalk (Fig. 2.1S). Many sessile species foul the external surfaces of fish and several have been implicated in hypoxic gill diseases. Other peritrichs are mobile and only attach temporarily to substrates using a concave adhesive disc reinforced with denticles (Fig. 2.1T). Most of these trichodinid species are ectoparasitic on fish and many cause skin lesions.

Myxozoa
Myxozoa form complex valved spores with polar capsules containing extrudible filaments (which are used for attachment to host cells and not to inject the infective sporoplasm). Their development involves multicellular differentiation of valvogenic, capsulogenic and sporoplasmic cells, which does not conform with the unicellular definition of Protista. Recent molecular studies suggest they are bilaterian metazoans but they continue to be documented with Protista

for historical reasons. Over 2770 species have been described, most as coelozoic or histozoic parasites in the organ cavities and tissues of fish although some are found in amphibia, reptiles and various invertebrates. Many infections are asymptomatic provoking little inflammation but some may cause tissue hyperplasia, unsightly cysts, erosive and necrotic lesions, myoliquefaction and deformities. Myxozoa are differentiated on the basis of spore morphology into bivalved (Figs 2.1U and V) and multivalved (Fig. 2.1W) species. The life cycles of several species (mainly from freshwater fishes) have been found recently to involve cyclic development between myxosporean stages in fishes and actinosporean stages in invertebrates, notably oligochaetes. Actinospores appear different and many have elongate protrusions to aid in flotation and anchorage (Fig. 2.1X).

Conclusion
The diversity of protistan organisms is well appreciated both in terms of their structural heterogeneity as well as their species richness, distribution and abundance. They are widespread in most environments and many species affect animal, water and soil health. Studies on marine protistan parasites must concentrate not only on the parasites themselves but also on host interactions culminating in disease. Detailed information is required on parasite morphology, development and virulence as well as host range, susceptibility and pathology in order to develop appropriate treatment, prevention or control strategies.

Important references
Comprehensive and well-illustrated texts on all the groups discussed here can be found in Margulis *et al.* (1990), Harrison and Corliss (1991), Lee *et al.* (2000), and Mehlhorn (2001). Lom and Dyková (1992) provided a detailed account of protistan parasites of fishes.

'Sarcomastigophora' (amoebae and flagellates)
Barbara F Nowak

Introduction
The 'phylum' Sarcomastigophora consists of a diverse group of protozoans. Historically, three 'subphyla' were included in it, the Mastigophora, Opalinata (now included as a class in the Heterokonta, phylum Chromista) and Sarcodina. Mastigophora are flagellates, their trophozoites use one or more flagella for locomotion. Members of the Subphylum Opalinata have numerous short flagella, which make them superficially similar to slowly swimming ciliates. The Subphylum Sarcodina includes the amoebae, characterised mainly by the use of pseudopodia for movement; flagella are uncommon and if they are present it is only in some developmental or sexual stages.

A single genus of the Opalinata contains marine species, whereas many species of flagellates and amoebas infect marine hosts. However, flagellates and amoebae have been little studied and estimates of species numbers are therefore impossible.

Morphology and diversity
The Subphylum Mastigophora includes the blood parasitic trypanosomatids, ectoparasitic bodonids and dinoflagellates, and the diplomonads, which are most commonly endocommensals of the intestine but can also colonise the gall bladder and other internal organs affecting fish health. Few flagellates are intracellular parasites. In the Subphylum Opalinata only the genus *Protoopalina* parasitises marine fish. Some species from this genus are symbiotic or commensals, living in the intestine or rectum of their hosts. Subphylum Sarcodina includes the amoebae,

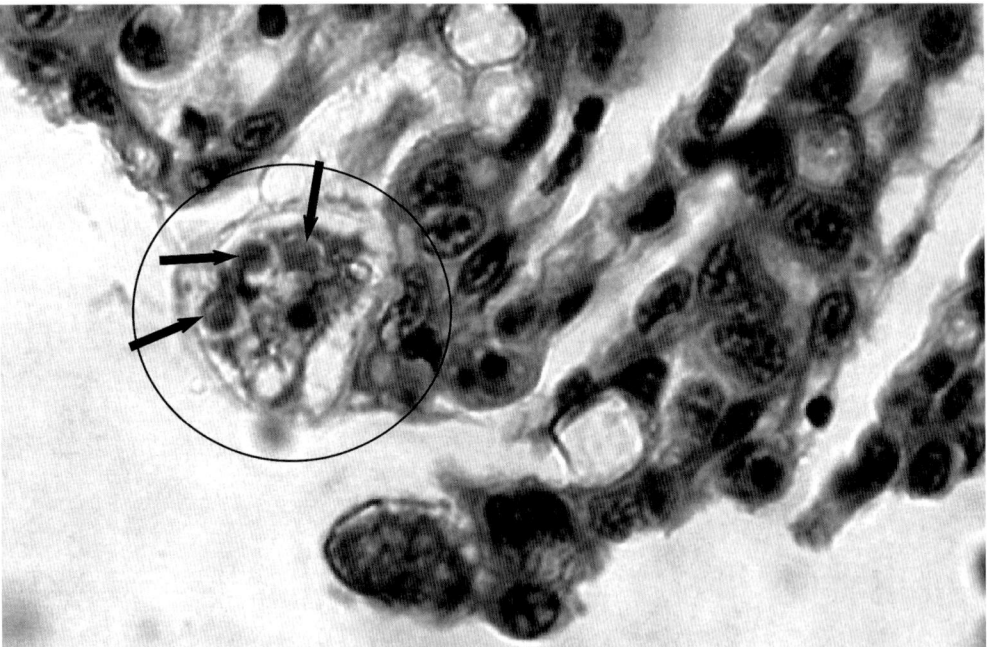

Figure 2.2 Histological section showing paramoeba (circled) on the gills of Atlantic salmon, *Salmo salar*. Note the presence of three parasomes (arrows).

most of which are amphizoic organisms, free living but able to colonise fish and cause significant pathology. A few species (e.g. *Entamoeba* or *Schizoamoeba*) are endocommensals, but can become parasitic.

Amoebae are usually difficult to identify and only now through the combination of molecular taxonomy, observation of living specimens and transmission electron microscopy can they be well characterised. Members of the Family Paramoebidae contain eukaryotic endosymbionts, known as parasomes (Fig. 2.2). Ultrastructural morphology of parasomes was first described from parasitic amoebae of chaetognaths, *Janickina pigmentifera* and *J. chaetognathi* (Hollande 1980). The endosymbiont was considered to be a kinetoplast, related to kinetoplastid flagellates and named *Perkinsiella amoebae* (Hollande 1980). A single, giant kinetoplast-mitochondrion is present in the usually binuclear symbiont (Dyková et al. 2003). The parasomes are usually associated with the amoeba nucleus. The number of parasomes may vary and in culture conditions it is usually reduced to one in each amoeba. The relationship between the amoeba and endosymbiont is stable and hereditary and neither of them can exist without the other. Recent phylogenetic analysis of the Small Subunit (SSU) rRNA gene sequence from different *Neoparamoeba* strains indicated a close relationship of the endosymbiont with the flagellate *Ichthyobodo necator* (see Dyková et al. 2003). The exact relationship between the endosymbiont and amoeba and the evolutionary origin of the endosymbiont are not known; however, Dyková et al. (2003) suggested that the symbiotic association was established in the early phase of kinetoplastid evolution.

Life cycles
Most parasitic flagellates have a simple, one-host life cycle. They reproduce by longitudinal binary fission. Some species can form resistant cysts. The life cycle of parasitic dinoflagellates includes a feeding stage (trophont), living on the host, a stage off the host which undergoes a

series of divisions (tomont) and a free-swimming infectious stage, called the dinospore or gymnospore. Kinetoplastids include the genus *Trypanosoma*, members of which live in the blood of hagfish, elasmobranchs and teleosts. These parasites have a complex life cycle, with several developmental stages within an intermediate host (the leech) including amastigote (stage without flagella), sphaeromastigote, epimastigote and trypomastigote. Infective parasites can be present in a leech for more than two years. In fish, trypanosomes undergo morphological changes, involving small, intermediate and large forms. Life cycles of amoebae are direct. Usually, they do not have distinct life stages; however, some species have flagellated stages or form protective cysts. Only intrusive species (Acanthoamoebidae) of marine Gymnoamoebae produce cysts (Page 1983). Amoebae reproduce by binary or multiple fission.

Effects on hosts and ecological importance

Most Sarcomastigophora are facultative ectoparasites. Some are commensals, for example intestinal amoebae and flagellates; however, these species include potential pathogens. Finally, trypanosomes are obligatory parasites. Most sarcomastigophorans do not seem to be host specific but susceptibility to infection can be species specific. Most of the parasitic species result in a significant host response, in some cases severe enough to affect the host even when the parasite is removed. Severity and rate of infection is temperature dependent. In the case of ectoparasites it can be also related to salinity, and changes in salinity have been used to control disease outbreaks. Whereas salinity usually affects the parasite, temperature has an effect not only on the parasite, but mostly on the host, with immunosuppression common in temperatures lower than the host optimum and stress in temperatures above the host optimum.

Species from the Family Paramoebidae have been shown to be parasitic to crustaceans, echinoderms and fish. *Neoparamoeba pemaquidensis* has been implicated as a cause of outbreaks of Amoebic Gill Disease (AGD) in salmonids cultured in the marine environment and other cultured marine fish species. Interestingly, AGD could not be found in wild fish species, even in wild fish cohabiting with infected Atlantic salmon in their cages (Douglas-Helders *et al.* 2002). This disease has a significant economic effect on salmonid aquaculture in Tasmania, Australia. AGD is defined as the presence of amoebae with parasomes that are in association with characteristic histological changes in gill tissue, including severe hyperplasia of lamellar epithelium and inflammatory response. *Neoparamoeba pemaquidensis* is widely distributed in temperate marine environments and has been isolated from water, sediments and biofouling invertebrates. Its presence is not confined to mariculture areas. No ultrastructural differences have been found so far between different isolates from fish and environmental samples (Dyková *et al.* 2000). Recently, another species of *Neoparamoeba*, isolated from fish gills and associated with AGD infections, has been described (Dyková *et al.* 2005). Further research is required to determine the role of *Neoparamoeba* species in AGD outbreaks.

Flagellates are mostly free living and can be autotrophic or heterotrophic. Autotrophic flagellates rarely become parasitic. Exceptions include *Amyloodinium pilularis* and *A. ocellatum*, both ectoparasitic dinoflagellates affecting marine fish in tropical aquaria. *Amyloodinium ocellatum* is unique, parasitising both elasmobranchs and teleosts (Lawler 1980). It is an obligate parasite causing one of the most significant diseases of temperate and warm-water marine fish over a wide range of temperatures and salinities. Dinoflagellates are considered to be algae by botanists; however, zoologists classify them as protozoans. Molecular studies suggest that *Pfiesteria*-like dinoflagellates and *Pfiesteria piscicida* are closely related to the parasitic *A. ocellatum* (see Litaker *et al.* 1999). The nature of interaction between *Pfiesteria* spp. and fish has been disputed in recent years. Initially, *Pfiesteria* spp. was reported to produce ichthyotoxin (Burkholder *et al.* 1992); however, attachment of the dinoflagellate to fish epithelium and its damage is the only proven cause of fish mortality (Vogelbein *et al.* 2002). *Ichthyobodo necator*, typically an ectoparasite of freshwater fish, has been reported from marine fish and salmonids after transfer to marine farms (Lom and Dyková 1992). *Ichthyobodo* isolates from the gills of Atlantic salmon reared in fresh, brackish and sea water were nearly 100% identical; however *Ichthyobodo* isolates from the gills of Atlantic cod (*Gadus morhua*) were not closely related to the species affecting salmonids (Todal *et al.* 2004), suggesting that this genus may be more diverse than suggested by their morphology.

From an ecological point of view, sarcomastigophorans form an interesting group, with examples of mixotrophic dinoflagellates, which derive some of their nutrition from autotrophy and some from their hosts. Many species are free living and become parasitic only under particular, as yet not fully understood conditions. The prevalence of some species of trypanosomes is stock specific and has been used as a parasite tag to differentiate between fish stocks for management. For example, *Trypanosoma murmanensis* in Atlantic cod from Newfoundland varied from 4% to 94% in different stocks (Khan *et al.* 1980). Overall, the ecological significance of sarcomastigophorans increases in artificial systems where they can cause substantial losses.

Important references
Lom and Dyková (1992) provide the most detailed description of morphology, taxonomy and life cycles of sarcomastigophorans. A detailed discussion on species descriptions of diplomonad flagellates from fish, using ultrastructural features and culture is covered by Poynton and Sterud (2002).

Labyrinthomorpha (labyrinthomorphs)
Susan M Bower

Introduction
The Labyrinthomorpha (a phylum included in the subkingdom Protozoa by Levine *et al.* 1980) that are pathogenic to shellfish are all thraustochytrids which have been grouped with the lower fungi (slime moulds in the phylum Labyrinthulomycota) and were included in the heterotrophic stramenophiles group by Patterson (2000). The most characteristic feature of thraustochytrids is a unique organelle called the sagenogenetosome (bothrosome or sagenogen).

Morphology and diversity
Although most thraustochytrids are free living and usually associated with organic detrital materials, a few species have been associated with disease in molluscs. In three cases, unidentified thraustochytrids were involved in surface lesions on captive octopus, nudibranchs and squid in the northern hemisphere (Bower 1987a). In addition, *Labyrinthuloides haliotidis* (Fig. 2.3A) was a pathogen of cultured small juvenile abalone (*Haliotis kamtschatkana* and *Haliotis rufescens*) in British Columbia, Canada (Bower 1987a) and an unnamed species, commonly called Quahog Parasite Unknown (QPX), has been associated with mortalities and lesions in hard clams (*Mercenaria mercenaria*) on the eastern seaboard of North America (Whyte *et al.* 1994, Ragone Calvo *et al.* 1998).

In the mollusc host, the vegetative stage (single nucleated organism also called thallus or trophozoite) of parasitic thraustochytrids are usually spheroid and about 2 µm to 10 µm in diameter. In abalone, *L. haliotidis* multiplies by simple binary fission (Bower 1987a). However, in hard clams, QPX develops sporangia from enlarged vegetative cells (10–15 µm in diameter) which undergo endosporulation. Mature sporangia (18–25 µm in diameter) of QPX contain 20 to 40 endospores (immature vegetative cells, 1.5–2 µm in diameter) each with a basophilic cell wall (Smolowitz *et al.* 1998, Ragone Calvo *et al.* 1998). Both pathogens produce biflagellated

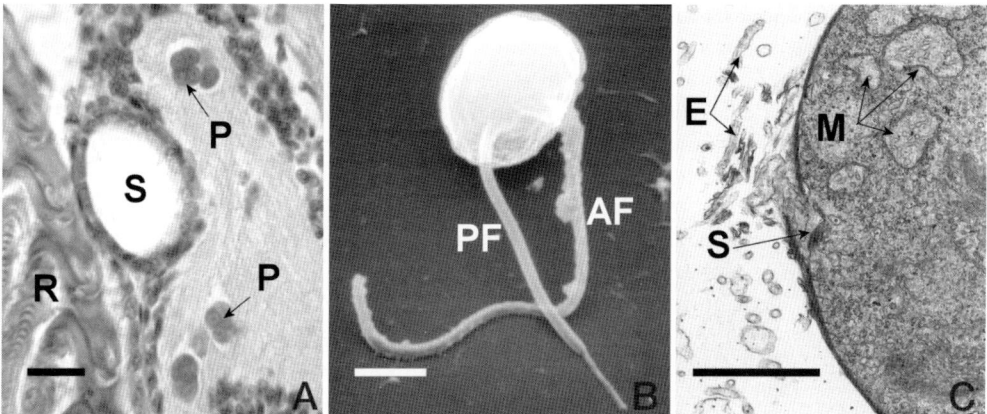

Figure 2.3 *Labyrinthuloides haliotidis*. **A.** Histological section of *L. haliotidis* (P) in the nerve ganglion of *Haliotis kamtschatkana* adjacent to the statocyst (S) and radula (R). Scale bar = 1.5 µm. **B.** Scanning electron micrograph of a zoospore of *L. haliotidis* from sea water. Note the subapical attachment site of the two flagella, the coarse texture of the longer anterior flagellum (AF) where debris has attached to the mastigonemes and the tapered tip of the short glabrous posterior flagellum (PF). Scale bar = 2.5 µm. **C.** Transmission electron micrograph of a *L. haliotidis* from liquid culture media showing the sagenogenetosome (S) on the periphery of a cell that was actively producing an ectoplasm net (E), which consists of a unit membrane with no evidence of cell organelles internally. Mitochondrial profiles (M) containing tubular cristae are evident adjacent to the sagenogenetosome. Uranyl acetate and lead citrate stain. Scale bar = 1 µm.

zoospores (Fig. 2.3B) when transferred to sterile sea water. The zoospores are uninucleate and slightly oval (about 5 µm long and 3.5 µm wide) with two laterally attached flagella. The anterior flagellum (about 12 µm in length) has a brush of mastigonemes along one side and the posterior flagellum (5–10 µm in length) is glabrous and has a tapered tip (Bower 1987a).

The most characteristic feature of thraustochytrids is a unique organelle called the sagenogenetosome (bothrosome or sagenogen) on the cell surface from which arises the ectoplasm net consisting of a unit membrane tube containing no cell organelles (Fig. 2.3C). In the sagenogenetosome an electron-dense plug separates the cell cytoplasm from the ectoplasmic network. Although typical sagenogenetosomes and ectoplasmic nets were absent or very rare in QPX in the clam host and in nutrient culture media where the parasite was usually embedded in a gelatinous matrix or mucofilamentous net (Whyte et al. 1994), Kleinschuster et al. (1998) reported the development of an ectoplasmic net in cultured QPX that had been transferred to sterile sea water. In addition to sagenogenetosomes and ectoplasmic nets, thraustochytrids have scale-like laminated cell walls.

Life cycles

All known parasitic thraustochytrids have direct life cycles. QPX may be an opportunistic facultative parasite not dependent on a parasitic way of life because it appears to be a ubiquitous member of the normal marine and bivalve flora on the east coast of North America. Possibly, *M. mercenaria* disadvantaged in some way (e.g. unfavourable environmental interactions including high planting densities and poor husbandry) may be more susceptible to infection with increased risk of disease development. The same may be true for *L. haliotidis* but nothing is known about its occurrence in the marine environment. Nevertheless, the life cycle of *L. haliotidis* in captive abalone and *in vitro* has been described (Bower 1987a, c) and is similar to that of QPX.

The vegetative cell of *L. haliotidis* removed from a source of nutrients (i.e. placement in sterile sea water), develops by synchronous multiple fission to form a zoosporoblast (6–10 µm in diameter) containing about 10 biflagellate zoospores which escape through a rupture in the zoosporoblast wall. The flagella were shed when the zoospore contacted a hard surface or after about 24 hr of active swimming in sea water. The resulting cell was morphologically similar but slightly smaller than the vegetative stage and survived in sterile sea water at about 5°C for at least two years. Vegetative stages that developed from zoospores were infective to small abalone. Within 4 hr of contacting the host, sagenogenetosomes produced extracellular lytic activity that disrupted the plasmalemma layer of the host epithelial cells adjacent to the parasite, eventually lysing the host cell. By 24 hr post exposure, the ectoplasmic net was well developed, allowing the parasite to move into and within the head and foot tissues of the abalone and dividing forms of the parasite were observed (Bower *et al.* 1989). Division was rapid and tiny abalone were quickly overrun by *L. haliotidis*. As dead abalone decomposed, vegetative stages released from the tissues developed into zoosporoblasts that produced zoospores within about 24 hr to 72 hr. Parasites released from infected abalone were infective to other small abalone on contact. Although alternate hosts have not been described for QPX and *L. haliotidis*, these thraustochytrids can utilise diverse sources of nutrients *in vitro*. Small juvenile Japanese scallops (*Patinopecten yessoensis*) and juvenile Pacific oysters (*Crassostrea gigas*) both less than eight months of age were resistant to infection with *L. haliotidis*. However, juvenile oysters with badly cracked shells became infected suggesting that *L. haliotidis* was capable of utilising oyster tissue as nutrients for growth and multiplication if it was able to gain access to the soft tissues of the oyster (Bower 1987b).

Effects on hosts and ecological importance

Labyrinthuloides haliotidis quickly multiplies in the tissue of it host. Within 10 days after exposure to about 10^4 parasites in 20 mL of sea water, about 90% of the abalone (<4.0 mm shell length and 140 days of age) died with numerous parasites throughout the head and foot (Bower 1987b). Tissues of heavily infected abalone were slightly swollen with a loss of integrity. The prevalence and intensity of infection decreased, and time to death increased as the abalone increased in age and size. Abalone, greater than 1.5 cm in shell length, could not be infected even when about 1.5×10^4 *L. haliotidis* were injected intramuscularly. The mechanism of defence against this parasite is not known.

Although *L. haliotidis* has only been observed in small abalone (<0.5 cm shell length), the high mortalities caused by infection were devastating to the abalone culture facility and this parasite was involved in the demise of an early attempt at abalone culture in British Columbia, Canada (Bower 1987a). Within two weeks of first being detected in a raceway, over 90% of the 100 000 small abalone in that raceway succumbed to infection and the disease quickly spread between raceways. The impact of *L. haliotidis* on wild abalone stocks and its geographical range are unknown because abalone, of the size that are susceptible to infection, are too tiny to be found in the wild.

Quahog Parasite Unknown (QPX) has been associated with mortalities and lesions (swellings and round yellow-tan nodules, 15 mm in diameter) in the mantle, often at the mantle edge, adjacent to the siphon or adductor muscle and gills of hard clams (Whyte *et al.* 1994, Ragone Calvo *et al.* 1998). The mucoid material produced by QPX may prevent phagocytosis by clam haemocytes and thus act as a pathogenic mechanism. However, phagocytic multinucleate giant cells of various sizes containing up to 25 nuclei and haemocyte encapsulation of QPX occur as part of the clam's response to infection (Smolowitz *et al.* 1998). Also, the haemocytic response was often associated with moribund looking QPX (Ragone Calvo *et al.* 1998). Nevertheless, QPX-infected clams grew more slowly and had a lower condition index than uninfected *M. mercenaria* (Smolowitz *et al.* 1998). Observations to date suggest that genetic variability in the host and/or in the QPX pathogen could be responsible for differences in susceptibility toward the infection and in the presentation of the disease. In 1959 QPX was suggested to be the primary cause of significant wild *M. mercenaria* (quahog or hard clam) stock mortalities in New Brunswick and more recently was associated with 80% to 90% mortalities in juvenile *M. mercenaria* (up to 30 mm in shell length) in a nursery and up to 100% in hatchery broodstock on Prince Edward Island (Whyte *et al.* 1994). In addition, QPX has caused severe mortality (80–95% in some instances) in aquacultured and wild stocks of *M. mercenaria* along the north-eastern coast of the United States to at least Virginia (Ford *et al.* 2002). Mortality is usually most severe in the spring and summer months in *M. mercenaria* that are at least one year old. The dynamics of infection and pathogenicity under different holding and handling conditions will require more investigation if QPX proliferation in cultured *M. mercenaria* is to be circumvented (MacCallum and McGladdery 2000).

Important references
Further details, coloured illustrations and a complete list of references are available on the websites of Fisheries and Oceans Canada (2003, 2004).

Haplosporidia (haplosporidians)
Eugene Burreson

Introduction
The Haplosporidia is a small group of parasitic protists consisting of four genera and about 36 species. Molecular phylogenetic analyses support the Haplosporidia as a monophyletic phylum closely related to the phylum Cercozoa. Most species in the phylum have two life history stages: multinucleate plasmodia, and a resistant spore with an orifice at one end that is covered either by an external lid of spore wall material or an internal flange of spore wall material. Most species of Haplosporidia are histozoic in a wide variety of marine invertebrates, although one species occurs in freshwater invertebrates. Species in the genus *Urosporidium* are often hyperparasites.

Morphology and diversity
Historically, the group has been characterised by two life history stages—multinucleate plasmodia, and a resistant spore with an orifice covered either by an external lid of spore wall material (genera *Minchinia* and *Haplosporidium*) or an internal flange of spore wall material (genus *Urosporidium*). Multinucleate plasmodia (Fig. 2.4A) are generally from 5 µm to 20 µm in diameter, but can reach 50 µm or more with over 100 nuclei. Plasmodia of all species are similar and cannot be used to distinguish species. Spores range in length from 4 µm to 12µm depending on the species. The spore stage (Fig. 2.4B) in *Haplosporidium* spp. has ornamentation consisting of tails or wrappings composed of spore wall material. The spore stage of *Minchinia* spp. has ornamentation composed of epispore cytoplasm. Spore ornamentation is an important taxonomic character, although it is usually visible only with scanning electron microscopy. Recent molecular phylogenetic studies have shown that the enigmatic genus *Bonamia* also is a haplosporidian (Carnegie *et al.* 2000, Reece *et al.* 2004). This genus consists of only three species, all of which infect haemocytes of oysters. The most commonly observed cell type in *Bonamia* spp. is a uninucleate 'microcell' from 2 µm to 3 µm in diameter (Fig. 2.4C), although multinucleate plasmodia also have been reported. No spore stage has ever been observed for any *Bonamia* species, although it is possible that the spore stage occurs in some host other than oysters and has not been discovered. If *Bonamia* spp. truly lack spores, then the definition of the Haplosporidia must be modified to include species that have a spore with an orifice as well as those microcell species that infect oyster haemocytes.

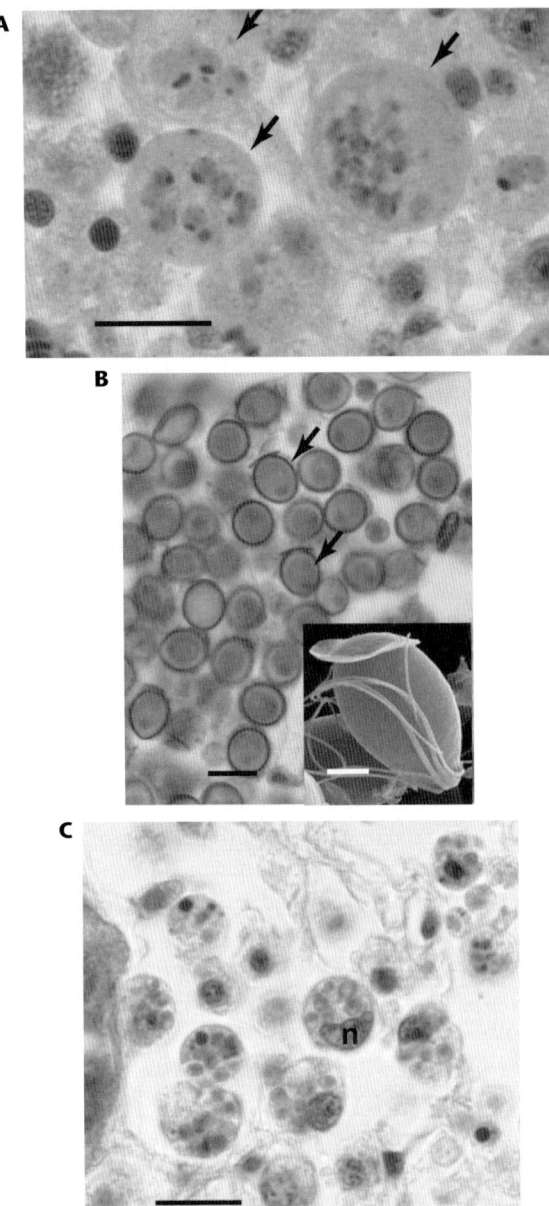

Figure 2.4 **A.** Multinucleate plasmodia (arrows) of *Haplosporidium nelsoni* in the eastern oyster, *Crassostrea virginica*, illustrating the eccentric nucleolus. Plastic section cut at 1 μm and stained with toluidine blue. Scale bar = 10 μm. **B.** Spores of *Haplosporidium louisiana* from the mudcrab, *Panopeus herbstii*. Arrows point to spores demonstrating the typical external lid covering the spore orifice that is present in *Haplosporidium* and *Minchinia*. Scale bar = 10 μm. Inset: Scanning electron micrograph of spore of *H. pickfordi* illustrating the spore lid and spore ornamentation originating from the aboral end. Scale bar = 2 μm. Reproduced from Burreson, EM, Spore ornamentation of *Haplosporidium pickfordi* Barrow, 1961 (Haplosporidia), a parasite of freshwater snails in Michigan, USA, *Journal of Eukaryotic Microbiology* 48(6): 622–626, with permission of the *Journal of Eukaryotic Microbiology*.
C. Microcells of *Bonamia ostreae* infecting haemocytes of the flat oyster, *Ostrea edulis*; n = nucleus of one infected haemocyte. Nuclei of many other infected haemocytes are condensed and necrotic. Scale bar = 8 μm.

Life cycles

In the traditional spore-forming Haplosporidia, multinucleate plasmodia divide by plasmotomy and eventually undergo synchronous sporulation. Resistant spores are released into the environment, usually upon death of the host. The fate of spores is unknown, but they do not seem to be infective to the host in which they are produced. Repeated attempts by many investigators to transmit *Haplosporidium nelsoni* directly from oyster to oyster by cohabitation or injection of plasmodia or spores have been unsuccessful, and most believe that spores are infective to an intermediate host that is a necessary component of the life cycle (Haskin and Andrews 1988). There is one report of direct transmission of *Haplosporidium pickfordi* to freshwater snails using spores (Barrow 1961), but that study has been viewed with skepticism because of infected controls and needs to be repeated. *Bonamia* spp. can be directly transmitted by cohabitation (Elston *et al.* 1986) or injection of microcells and this is evidence that spores are not necessary for transmission and may have been lost in the *Bonamia* lineage (Reece *et al.* 2004).

Effects on hosts and ecological importance

The end result of infections by species of *Urosporidium*, *Minchinia* or *Haplosporidium* is the presence of large numbers of spores, often completely displacing the target tissue. These infections are often fatal; however, most species of spore-forming haplosporidians are rare and are not important pathogens because of their very low prevalence. Some species cause high mortality in commercially important oysters, however. The best studied haplosporidian is *Haplosporidium nelsoni*, causative agent of Multinucleate Sphere X (MSX) disease in oysters along the east coast of the United States of America (USA) and Canada. This parasite causes a general wasting disease resulting from increasing intensity of plasmodia. Epizootic mortality from *H. nelsoni* began in Delaware Bay in 1957 and in Chesapeake Bay in 1959. Over one million bushels of oysters (>35 million L) were killed by the parasite in each bay within a few years (Andrews 1968, Ford and Haskin 1982). Spread of the parasite along the entire east coast of the USA and into Atlantic Canada, and continuing annual mortality has severely impacted the oyster resource and industry in these areas. *Haplosporidium nelsoni* is a natural parasite of the Pacific oyster (*Crassostrea gigas*) in Japan and Korea, and there is strong evidence that the parasite was introduced to the east coast of the USA from the Pacific Ocean (Burreson *et al.* 2001). After nearly 50 years, some natural resistance to the parasite has developed in Delaware Bay oysters.

All species of *Bonamia* are pathogenic to their oyster hosts. *Bonamia ostreae* has decimated populations of the flat oyster (*Ostrea edulis*) in France (Grizel *et al.* 1988), *B. exitiosa* has caused extensive mortality in *Ostrea chilensis* in southern New Zealand (Hine 1996), and *B. roughleyi* is the cause of winter mortality disease in the Sydney rock oyster (*Saccostrea glomerata*), in southeastern Australia (Farley *et al.* 1988). Microcells of *Bonamia* spp. stimulate phagocytosis by host haemocytes and are not killed by the cellular defence mechanisms. Microcells proliferate and eventually lyse the haemocyte, releasing the microcells. The phagocytosis/proliferation/lysis cycle repeats resulting in massive infections of microcells and destruction of host haemocytes leading to death of the oyster.

Important references

The phylum Haplosporidia has been reviewed by Perkins (1990, 2000) and Burreson and Ford (2004). The most recent molecular phylogenetic analysis of the group is by Reece *et al.* (2004). Carnegie and Cochennec-Laureau (2004) reviewed the genus *Bonamia*.

Apicomplexa (sporozoans)
Kálmán Molnár

Introduction
The phylum Apicomplexa is a huge group including rather different protozoan parasites which have a special cell organelle, the apical complex, which facilitates invasion of the host cell. Apicomplexans have three developmental stages during their life cycles: merogony, gamogony and sporogony. Fish apicomplexans are divided into two major groups. Coccidia proper are primarily intestinal parasites and produce resistant oocysts in the host. Adeleid blood parasites (Coccidia *sensu lato*) have the merogonic and some gamogonic stages in fish, while spore formation takes place in parasitic annelids or gnathiid isopods.

Morphology, diversity and development of coccidian apicomplexans
Coccidia proper belonging to the suborder Eimeriorina comprise two families Eimeriidae (including genera *Eimeria*, *Goussia*, *Calyptospora*, *Crystallospora*) and Cryptosporidiidae (with a single genus *Cryptosporidium*). Most of the known coccidians develop in the gut but there are species developing in inner organs (i.e. the spleen, liver, kidney and swimbladder).

Systematics of eimeriid apicomplexans is based on the morphology of the spore, the oocyst. The oocysts of the Eimeriidae (Fig. 2.5) contain four sporocysts and, occasionally, one to two polar granules. Each sporocyst contains two sporozoites and a residual body. The only important difference between genera is in the structure of the sporocyst. Whereas the oocysts of terrestrial animals have resistant and thick oocyst walls, fish coccidia have a thin, sensitive oocyst wall without micropyle. The thickness of the one- or three-layered wall varies between 3 nm and 200 nm. Most fish coccidians have round, less frequently ellipsoidal oocysts. Only a few fish coccidia (e.g. *Eimeria isabellae*) possess a typical *Eimeria* sporocyst (i.e. having a Stieda body). In most species there is only a thickening, plug or cap at one end of the sporocyst. This is where the sporocyst opens and the sporozoites are released in the host or intermediate host. Sporocysts are elliptical, oval or dodecahedral in shape. The sporocyst wall is thin but usually composed of two layers.

Goussia-type sporocysts are composed of two equal-sized, round, elliptical or coffin-shaped valves united by a suture. This suture is hardly discernible using light microscopy but can be seen under an electron microscope. Sometimes the sporocyst is surrounded also by a membraneous veil, which is attached to the sporocyst by special membranes. The sporocysts of *Calyptospora* are characterised by a thickening or projection at the caudal end, by sporopodia arising from the spore surface or from the caudal projection, and a sporocyst veil supported by sporopodia and surrounding the sporocyst. The opening of the sporocyst is a longitudinal suture which extends only to the anterior one-third of the sporocyst. The sporocyst of *Crystallospora crystalloides* is bipyramidal and opens at a suture situated at the foot of the pyramids. *Cryptosporidium* oocysts contain four naked sporozoites and a residual body. There are reports both about thick-walled and thin-walled oocysts. The sporozoites of fish coccidia are banana-, sausage- or comma-shaped. They usually lie in the sporocyst in a head-to-tail presentation. In the Cryptosporidia the sporozoites are side by side and lie in the same direction. The sporozoite has a nucleus easily discernible by light microscopy and is situated in the middle of the body. The conoid apparatus lies at the anterior end, and the posterior end may sometimes be striated. An oocyst residuum exists only in a few species; all sporocysts, however, have a sporocyst residuum. This residuum is granular in the young oocysts and may be compact in the older ones.

Merogonic stages of fish coccidia are intracellular. They have two or three merogonic stages. The meronts develop in the cytoplasm, or occasionally in the nucleus. Usually 8–16 banana-

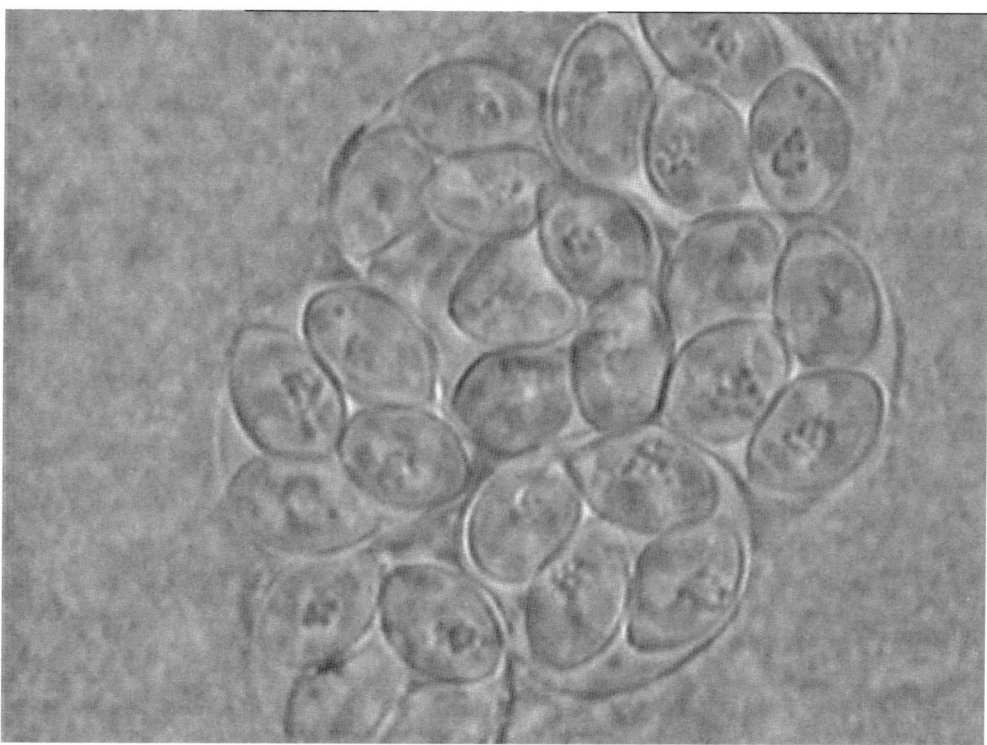

Figure 2.5 Sporulated oocysts of *Eimeria daviesae* from *Gobius kessleri*, a euhaline water fish. The oocysts contain four sporulated sporocysts with sporozoites and sporocyst residuum. × 2000.

shaped merozoites of 8–16 μm in length are formed in the meronts, but in *Goussia cichlidarum*, Landsberg and Paperna (1985) reported meronts containing large numbers of merozoites.

All coccidians develop in a parasitophorous vacuole. In most cases this vacuole is located inside the cytoplasm of the host cells but in case of epicellular development, the parasitophorous vacuole is intracellular but extracytoplasmal and it is covered only by a single unit membrane of the host cell. The merozoites of fish coccidia have a well-discernible nucleus, conoid apparatus, and trimembranous pellicle covering the merozoite. *Cryptosporidium* has incomplete vacuoles. The parasitophorous vacuole formed by enterocyte microvilli surrounds the meronts or gamonts only on the side facing the gut lumen. Between the parasite and the host cell cytoplasm there is a special adhesive zone. All stages of cryptosporidia develop epicellularly.

Gamogonic stages of coccidia comprise male and female developmental stages. Both microgamonts (male) and macrogamont (female) develop intracellularly in a parasitophorous vacuole. Depending on the parasite species the development can be intracytoplasmal, intranuclear or epicellular. Inside the round or ellipsoidal, 10 μm to 20 μm sized microgamonts a large number of microgametes develop after multiple divisions. The macrogamont which develops from the last merozoite generation is always surrounded by a parasitophorous vacuole. Its cytoplasm contains lipid granules and two distinct types of 'wall-forming body'. After being fertilised the macrogamont becomes a young oocyst.

It had been generally accepted that coccidia (among them fish coccidia) develop directly without intermediate or paratenic hosts. Observations made by Landau et al. (1975), however, suggested that vectors might transmit the infection. Experiments made by Molnár (1979), Paterson and Desser (1982) and Steinhagen and Körting (1990) showed that besides direct transmissions, tubifex paratenic hosts served as vectors in infections of cyprinid fishes with enteric coccidia. However, Solangi and Overstreet (1980) and Fournie and Overstreet (1983) reported that *Calyptospora funduli* required a true intermediate host, the grass shrimp (*Palaemonetes pugio*) for its development.

Morphology, diversity and development of adeleid apicomplexans (*Coccidia* sensu lato)

Fish-parasitic Coccidia *sensu lato* have heteroxenous life cycles which involve two hosts, one being the fish and the other a parasitic leech or insect (gnathiid isopods). Levine (1988) classified adeleid apicomplexans into the suborder Adeleiorina, which has two families, Haemogregarinidae and Dactylosomatidae. Three genera, *Haemogregarina*, *Cyrilia* and *Desseria* belong to Haemogregarinidae and two genera, *Dactylosoma* and *Babesiosoma*, belong to Dactylosomatidae.

The development of coccidia *sensu lato* is divided into merogony, gamogony and sporogony. There is also a special association of the two gamonts prior to encystment (syzygy) and this takes place before sporogony.

Davies and Johnston (2000) believe that in the case of the haemogregarinids *Haemogregarina* and *Desseria*, merozoites are transmitted by the invertebrate host, while in that of *Cyrilia* sporozoites are transmitted. In the case of the dactylosomatid *Babesiosoma* merozoites are transmitted, while for *Dactylosoma* the type of transmission has not been demonstrated yet.

Khan (1980) described that sporozoites of *Cyrilia uncinata* were injected into the blood of fish by leeches which entered lymphocytes, monocytes, neutrophils or blast cells and they developed into meronts and formed merozoites. The merogony of *Haemogregarina* and *Cyrilia* spp. occurs in blood cells while in *Desseria* spp. it takes place mainly in the inner organs. Negm-Eldin (1999) described that *Cyrilia nili* had two successive types of merogonic stages in infected fish and the meronts of the second merogonic cycle were destined to form gamonts. The vermiform merozoites enter erythrocytes or leucocytes to form macrogamonts and microgamonts. In some species there is some division of gamonts in the erythrocytes (Fig. 2.6), while in others they directly change into microgamonts and macrogamonts.

Gamonts are taken in by a leech during a blood meal. In the intestine of the leech they are released from blood cells, and the microgamonts and macrogamonts unite in syzygy. During this process they are surrounded by a thin membrane. In *Haemogregarina* and *Cyrilia* the microgamont produces four microgametes, and one of the resulting microgametes fertilises the macrogamont. Oocyst formation takes place, and depending on the parasite species it may occur either within an enterocyte or on the surface in a parasitophorous vacuole. The sporozoites produced migrate toward the salivary gland and the proboscis of the leech. Intraleucocytic meronts of *Haemogregarina* spp. usually harbour two to eight merozoites and do not significantly enlarge the size of the host cell. The length of the banana- or crescent-shaped merozoites and gamonts varies between 4 µm and 5 µm; intraleukocytic meronts, however, may reach 26 µm × 23 µm.

Effects on hosts and ecological importance

Most fish coccidians have relatively low pathogenicity. No mortality was observed even when 85% to 90% of the liver was infected with *Calyptospora funduli* (Solangi and Overstreet 1980). Lethal infections occur primarily in farm ponds, but severe cases have been reported from natural waters as well. MacKenzie (1978) found a species of *Eimeria* sp. which caused 6% to 10% reduction in body mass in blue whiting, *Micromesistius poutassou*. Fiebiger (1913) as well as Odense and Logan (1976) reported mortality in the haddock caused by *Goussia gadi*. Grabda (1983), who investigated *Eimeria jadvigae* infection in *Coryphaenoides holotrachys*, reported thickening of the swimbladder wall. In more severe cases the inner surface of the bladder was

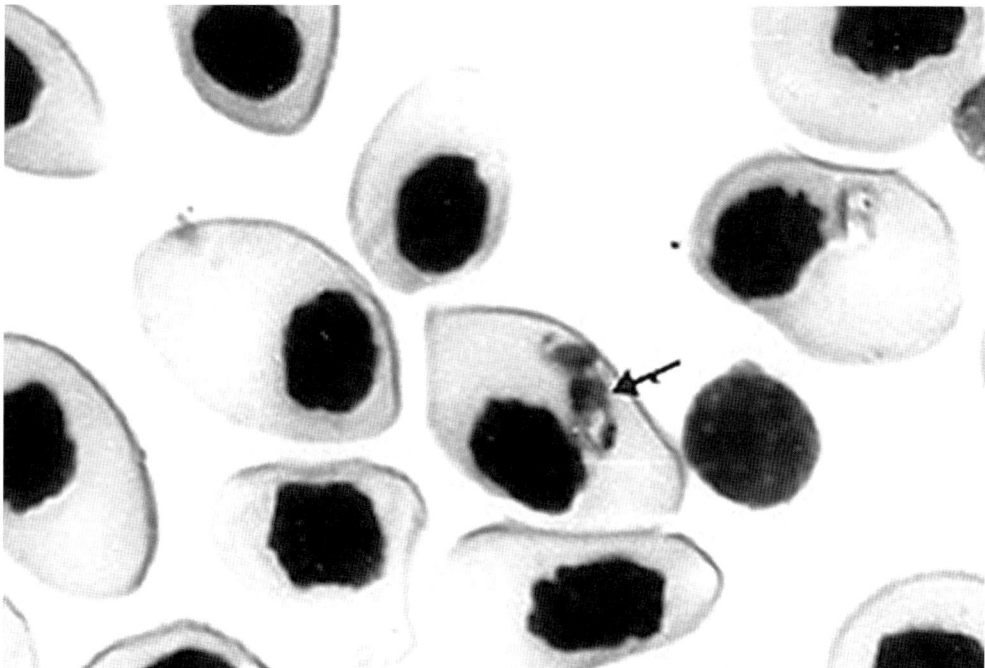

Figure 2.6 Gamont of a *Haemogregarina* sp. (arrow) in one of the Giemsa-stained erythrocytes of a Caspian sturgeon, (*Acipenser persicus*). × 3000. Courtesy of Dr Mahmoud Massoumian.

covered with a white substance, the wall had a spongy texture and the lumen was filled by a mucous exudate containing large numbers of oocysts. Pinto (1956) reported parasitic castration as a result of *G. sardinae* infection. Upton *et al.* (2000) supposed that a heavy infection of the gut with *E. phylloptericis* caused significant morbidity and mortality of the aquarium-cultured sea dragon, but the role of a joint infection with bacterial pathogens could not be excluded.

There is little information on the pathogenic effect of adeleid parasites. In cultured turbot Ferguson and Roberts (1975) reported a proliferative disease of the haematopoietic tissues; Kirmse (1980) found that up to 60% of some populations of *Haemogregarina sachai*-infected turbots were affected with gross tumours in the musculature and viscera. The lesions consisted of necrotic tissue with a caseous centre. Histologically, there was an accumulation of parasitised reticuloendothelial cells, cell debris and pycnotic nuclei. Parasitaemias of up to 36% of all blood cells were observed; most of the infected cells were neutrophils and monocytes. In mackerel caught in certain areas of the Atlantic, 4% of the leucocytes were infected with meronts, and meronts were demonstrated in 100% of impression smears from the spleen. Lesions in the spleen and kidneys contained *Haemogregarina*-like organisms and were often surrounded by a connective tissue capsule (MacLean and Davies 1990).

Important references

Systematics of apicomplexan parasites in this book is mostly based on Levine's (1988) classification, but additional data are available in Duszynski *et al.* (1999) on Coccidia proper. Biology of fish coccidians has been detailed by Davies and Ball (1993). Data on systematics, morphology and development of adeleid apicomplexans are found in Davies (1995) and Davies and Johnston (2000). Useful data on fish apicomplexans in general are summarised by Lom and Dyková (1992) and Molnár (1995).

Microsporidia (microsporans)

Elizabeth Moodie

Introduction

The Microsporidia, also known as Microspora, are a monophyletic phylum of tiny eukaryotic parasites. Growth and reproduction can occur only within host cells. Early research on the ribosomal genes of Microsporidia placed them at the base of the eukaryote evolutionary tree with other 'primitive' amitochondriate protozoans (Keeling and Fast 2002). Subsequent analyses of a variety of genes using more sophisticated methods have indicated that Microsporidia are close to the fungi. Molecular evidence for the relationship between Microsporidia and fungi is supported by biochemical and developmental features of the group (Keeling 2003). Studies on mitochondrial proteins (including Hsp70) have indicated that typical mitochondria have been secondarily lost, although remnants in the form of small membrane-bound organelles have been detected (Williams *et al.* 2002). Parsimonious features of the Microsporidia, including small genome size, prokaryote-like ribosomal genes and the loss of typical eukaryotic organelles are thought to be associated with the highly specialised lifestyle of these parasites (Mathis 2000, Weiss 2000).

Morphology and diversity

The parasites are transmitted between hosts by unicellular spores ranging in length from less than 1 µm to over 30 µm (Larsson 1999). Spores are able to infect new host cells by extruding their contents through the everted polar filament in a manner reminiscent of injection by a hypodermic syringe. In ungerminated spores, the polar filament lies coiled inside the cell (Fig. 2.7). Although the polar filaments of Microsporidia and Myxozoa are superficially similar, they have different origins. Microsporidia, which bear only one polar filament, are single-celled parasites believed to be closely related to Fungi. Myxozoa are of multicellular origin and are thought to be closely related to the Cnidaria. They bear one to four polar filaments, each derived from a different cell.

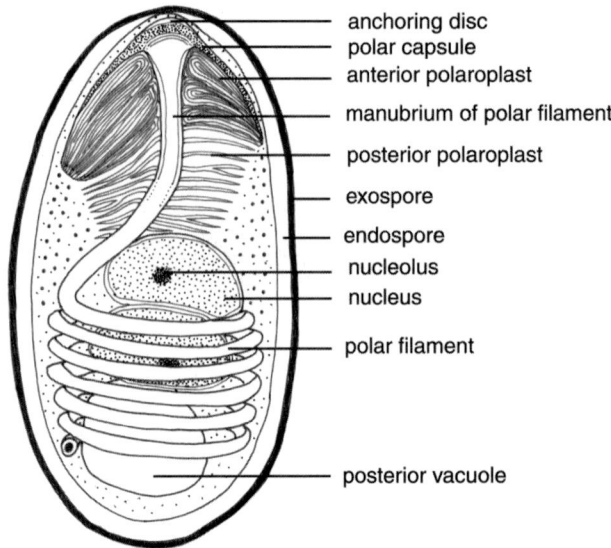

Figure 2.7 Schematic diagram of the ultrastructural features of a binucleate spore.

Table 2.1 Microsporidia genera reported in fish and crustaceans from marine (^{M}) and freshwater (^{FW}) environments

Genera in fish	Genera in crustaceans	
Amazonspora ^{M}	Abelspora ^{M}	Mrazekia ^{FW}
Glugea ^{M}	Agmasoma ^{M}	Nelliemelba ^{FW}
Heterosporis ^{FW}	Alfvenia ^{FW}	Norlevinea ^{FW}
Ichthyosporidium ^{M}	Amblyospora ^{FW}	Nosema ^{M}
Kabatana ^{M}	Ameson ^{M}	Octosporea ^{FW}
Loma ^{M}	Baculea ^{FW}	Ormieresia ^{M}
Microfilum ^{M}	Berwaldia ^{FW}	Orthothelohania ^{M}
Microgemma ^{M}	Chapmanium ^{FW}	Parathelohania ^{FW}
Neonosemoides ^{FW}	Courgourdella ^{FW}	Pleistophora ^{M}
Nosemoides ^{FW}	Duboscqia ^{FW}	Pyrotheca ^{FW}
Nucleospora ^{M}	Glugea ^{M}	Stempellia ^{FW}
Ovipleistophora ^{M}	Gurleya ^{FW}	Thelohania ^{M}
Pleistophora ^{M}	Gurleyides ^{FW}	Tuzetia ^{FW}
Pseudoloma ^{FW}	Holobispora ^{FW}	Vairimorpha ^{FW}
Spraguea ^{M}	Inodosporus ^{M}	Vavraia ^{FW}
Tetramicra ^{M}	Lanatospora ^{FW}	
Microsporidium ^{M}	Marssoniella ^{FW}	

Microsporidia infect most invertebrate phyla and all classes of vertebrates, and to date, more than 1300 species in about 150 genera have been described; the greatest number from fish and arthropods, including marine species (Canning and Vavra 2000). Undoubtedly, many more species exist. Over 158 species in 17 genera are known to infect fish and 34 genera infect crustaceans, as listed in Table 2.1 (Sindermann 1990, Langdon 1991, Undeen 1997, Mathews et al. 2001, Azevedo and Matos 2003, Lom and Nilsen 2003, Moodie et al. 2003). Other aquatic taxa in which Microsporidia have been found include cnidarians, annelids, molluscs and freshwater bryozoans (Nilsen 1999, Clausen 2000, Canning et al. 2002). Microsporidia hyperparasites have been found in gregarines, trematodes and mesozoans that are parasitic in marine invertebrate hosts, and in myxozoan parasites of fish (Canning and Vavra 2000, Lom and Nilsen 2003).

Life cycles

Microsporidia life cycles may be simple or complex. Unfortunately, complete life cycles for most species remain unknown. In those species for which information is available, reviewed most recently by Wittner and Weiss (1999), Petry (2000) and Dunn et al. (2001), transmission may be horizontal or vertical, or a combination of the two. Horizontal transmission by the oral route is common, usually by ingestion of infective spores from the environment. Direct transmission from one definitive host to another or indirect transmission via an intermediate host may occur. Many fish Microsporidia are transmitted directly by ingestion (e.g. *Nucleospora salmonis*, *Glugea* spp. and *Loma* spp.). Small crustaceans may function as paratenic hosts or as intermediate hosts for some species, e.g. *Pleistophora* and *Thelohania* spp., although further studies are required to confirm this (Iversen and Kelly 1976, Shaw and Kent 1999, Lom and Nilsen 2003). Infective spores may be dispersed in the environment or localised in faeces, oral secretions or tissues of infected hosts. Transmission by cannibalism is not unusual (Becnel and Andreadis 1999, Cali and Takvorian 1999). Where vertical transmission occurs, the transovarial route is most commonly used, although venereal transfer has been recorded (Dunn et al. 2001).

Relatively little is known about factors influencing persistence of microsporidian spores in the environment. The subject has been reviewed by Becnel and Andreadis (1999) and Cali and Takvorian (1999) who concluded that survival times are highly variable depending on the parasite species involved, moisture conditions, temperature, exposure to solar radiation, the materials in which spores are deposited and the presence of other microorganisms. Spores of some species stored in sterile refrigerated water can remain viable for several years. Spores utilised for vertical transmission or internal transmission between tissues within a host tend to have thinner walls and shorter polar filaments than spores released to the environment (Dunn et al. 2001).

After infection of a new host, Microsporidia generally undergo a proliferative phase involving binary or multiple fission of vegetative cells called meronts (merogony) that subsequently transform into sporonts. The sporonts undergo a series of divisions by mitosis or meiosis to form sporoblasts (sporogony), each of which matures into an infective spore (sporulation). Meiosis in Microsporidia has been reviewed by Canning (1988) and Flegel and Pasharawipas (1995).

Meronts, sporonts and spores may be uninucleate or binucleate. Sporulation may occur within sporophorous vesicles (SPVs), also known as pansporoblasts in the older literature, or alternatively spores are formed in direct contact with host cell cytoplasm. Sporophorous vesicles (SPVs) are usually spherical or ovoid, and contain a characteristic number of spores depending on species (e.g. 2, 4, 8, 16, 32, 48 or 64). Patterns of cell division, the stage at which the formation of the SPV is initiated, the number of spores within, and ultrastructural features of spores and SPVs are important taxonomic features (Lom and Dyková 1992, Sprague et al. 1992, Larsson 1999, Vavra and Larsson 1999, Canning and Vavra 2000, Lom and Nilsen 2003). Several sporogony pathways may occur in a single species and multiple spore types may be produced.

Effects on hosts

Most Microsporidia that infect fish and many of those that infect crustaceans are detrimental to their hosts, causing either morbidity or mortality. Mass mortalities in wild fisheries and cultured stocks have been reported, for example *Loma salmonae* in salmonid fish, *Agmasoma penaei* in shrimp (Sindermann 1990, Lom and Dyková 1992, Shaw and Kent 1999). Pathogenic effects induced by Microsporidia in host cells include physical disruption of cells due to occupation of intracellular space, host cell hypertrophy, changes to host cell metabolism with destruction, synthesis, or reorganisation of host cell components. Infected host cells often degenerate and die, and if sufficient numbers of cells are affected, tissue function is impaired (Wittner and Weiss 1999, Petry 2000). Different species of Microsporidia vary in their ability to induce severe pathology, depending on the organs invaded and the extent and manner in which parasite proliferation occurs. Direct effects of microsporidiosis include increased mortality and reduced market value of economically important species. Indirect effects include reduced growth, reduced reproductive potential and behavioural changes (e.g. decreased predator avoidance and altered migratory patterns).

Enlarged tumour-like host cells filled with spores (xenomas) are common in fish infected by the genera *Glugea*, *Ichthyosporidium*, *Jirovecia*, *Loma*, *Microfilum*, *Microgemma*, *Nosemoides*, *Spraguea* and *Tetramica*. Diffuse infections without xenoma formation are more common in crustaceans and in fish infected by the genera *Nucleospora*, *Heterosporis*, *Kabatana*, *Pleistophora* and *Thelohania* (Sindermann 1990, Lom and Dyková 1992, Shaw and Kent 1999, Lom and Nilsen 2003). Immune responses are often weak or non-existent where Microsporidia are protected within host cells or encapsulated in xenomas; however, the parasites may be attacked by the host immune system during initial infection or if a host cell or xenoma ruptures.

Suppression of the host inflammatory response and impairment of humoral or cellular responses in association with Microsporidia infection have been reported. Fish that have recovered from infection by *Loma salmonae* show some resistance to reinfection, although the mechanism is unclear. Antibody responses are generally not thought to be protective (Speare *et al.* 1998, Ramsay *et al.* 2002). As yet, no effective vaccine against microsporidiosis has been developed. The ability of a host to resist or contain infection varies between individuals and host taxa, and may be influenced by interactions between the host immune system and environmental factors. Host ranges vary between species. Moderate host specificity is shown by most Microsporidia. Relationships between immune competence and susceptibility to infection by Microsporidia in crustaceans and other marine invertebrates have not yet been explored to any extent.

Detection and treatment of infection

Traditionally, detection of microsporidiosis has depended on examination of tissues for spores under the light microscope or the electron microscope. The sensitivity of microscopic methods is limited by the small size of the parasites and difficulty in detecting early pre-spore stages. polymerase chain reaction (PCR)-based methods are increasingly being used to detect infection and identify the species responsible. They are more sensitive and specific than microscopic methods. Most PCR tests are based on the ribosomal RNA genes and internal transcribed spacer region (Weiss and Vossbrinck 1999).

Antimicrosporidial drugs are usually not cost effective for use in cultured populations of marine organisms, or have limited efficacy. Those that have been used in fish are reviewed by Shaw and Kent (1999). Prevention is a better option than cure; however, preventative measures are hampered by a paucity of information on the complete life cycles of many species (see above).

Ecological importance

Microsporidia may exert a variety of direct effects on their host populations as well as indirect effects on other species and the wider environment. Infection by Microsporidia may result in alteration of sex ratios in host populations, changes to host population demography, changes in host behaviour and alterations to trophic dynamics in communities that include infected hosts.

Vertically transmitted Microsporidia parasites have the potential to alter sex ratios in host populations by male killing or feminisation of the host, as described in the review by Dunn *et al.* (2001). Late male killing, due to the development of large numbers of spores in males but not females, is a strategy which effectively increases horizontal transmission rates. It has been reported in several species that infect mosquito hosts (e.g. *Amblyospora californica*). Feminisation of hosts is a strategy likely to increase transmission success via the transovarial route, and is exhibited by species from at least three genera that infect the marine amphipod *Gammarus duebeni*, including *Thelohania*, *Nosema* and *Octosporea*.

Juvenile hosts are more likely to suffer mortality than adults as a result of infection by some Microsporidia (e.g. *Loma salmonis*) (Shaw and Kent 1999). Where many spores develop in host musculature (e.g. *Thelohania* spp. in shrimp and *Glugea* spp. in fish), there is a greater chance that infected individuals will be caught by predators. Infected hosts may also be unable to engage in normal behaviours such as migratory movements. As yet, no models have been developed that predict the effects of Microsporidia infection on the ecology of wild host populations or their communities, in the marine environment.

Relatively little information is available on the extent to which environmental factors influence host susceptiblity to Microsporidia infection, or whether Microsporidia infection increases the likelihood of coinfection of hosts by other pathogens or vice versa. Chronic exposure to stressors such as pollutants (Barker *et al.* 1994) and high stocking densities in aquaculture are thought to increase prevalence and intensity of infection (Overstreet 1973, Sindermann 1990, Shaw and Kent 1999).

Although relatively little is known about the virulence dynamics of microsporidian infections in marine hosts, elegant experiments by Ebert and his coresearchers on microsporidian parasites of the freshwater crustacean *Daphnia* have contributed substantially to the body of knowledge on factors influencing virulence of microparasites. Increased virulence is associated with the production of larger numbers of spores (Ebert 1994) and may also be a consequence of within-host competition where multiple strains infect the same host (Ebert and Mangin 1997). Manipulative experiments involving infection of clonal populations of *Daphnia* by Microsporidia have shown that parasites may influence microevolution in the host during both asexual and sexual reproduction (Capaul 2003).

Relatively few studies have specifically addressed coevolution of Microsporidia and their marine hosts, although in a recent analysis of fish-infecting species, Lom and Nilsen (2003) found that all but one of the 15 genera they analysed were grouped together on the same branch of the phylogenetic tree, suggesting either coevolution or cospeciation. Further studies on coevolution are warranted.

Important references
Excellent reviews of the Microsporidia have been produced by Wittner and Weiss (1999), Canning and Vavra (2000) and Petry (2000). Important features used in taxonomy prior to the widespread use of molecular techniques are described by Sprague *et al.* (1992) and Larsson (1999). Keeling (2003) reviewed the biology and evolution of the group, including data from recent molecular studies. Life cycles, with an emphasis on transovarial pathways, are reviewed by Dunn *et al.* (2001).

Mikrocytos mackini (microcell)
Susan M Bower

Introduction
Mikrocytos mackini is a protist of unknown taxonomic affiliations, commonly referred to as a microcell. It is characterised by the apparent lack of mitochondria and haplosporosomes.

Morphology and diversity
The tiny size of *M. mackini* (2–4 µm in diameter) and non-descript spheroid shape (Fig. 2.8A) necessitates the examination of specimens by electron microscopy for the observation of relevant features. The three morphological forms identified by Hine *et al.* (2001) include:

1. Quiescent Cells (QC) with a central round to ovoid nucleus, less than seven cisternae of inactive nuclear membrane-bound Golgi, few vesicles and lysosome-like bodies
2. Vesicular Cells (VC) containing many small coated and uncoated vesicles, lacking nuclear membrane-bound Golgi-like arrays and with the nuclear membrane sometimes dilated to form a cisternal chamber
3. Endosomal Cells (EC, Fig. 2.8B) with a dilated nuclear membrane, a well-developed anastomosing endoplasmic reticulum connected the nuclear and plasma membranes and endosomes in the cytoplasm.

There was an overlap in features between QC and VC, between VC and EC and between EC and QC. Few organelles including the apparent lack of mitochondria occurred in all forms of *M. mackini*.

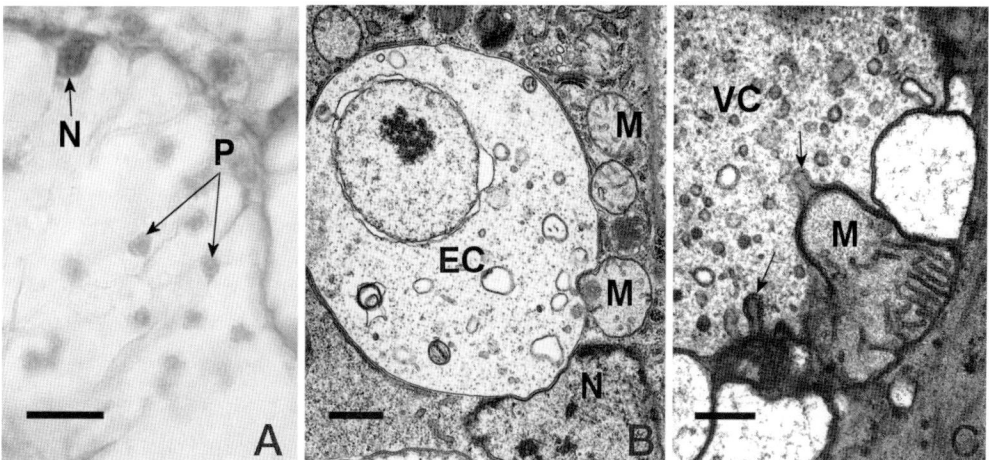

Figure 2.8 *Mikrocytos mackini.* **A.** Histological section of several *M. mackini* (P) in the cytoplasm of vesicular connective tissue cells (N = nuclei of host cells) of the labial palps of *Ostrea edulis*. Scale bar = 10 μm. **B.** Transmission electron micrograph of a *M. mackini* endosomal cell (EC) in close association with mitochondria (M) and the nucleus (N) of a haemocyte of *Crassostrea gigas*. The anastomosing endoplasmic reticulum is not evident in this specimen. Scale bar = 0.5 μm. **C.** Transmission electron micrograph of a *Crassostrea gigas* myocyte mitochondrium (M) with tube-like structures (arrows) extending into the cytoplasm of *M. mackini* vesicular cell (VC). Scale bar = 0.25 μm.

Mikrocytos mackini can de differentiated from other microcells (*Bonamia* spp.) by its location in vesicular connective tissue cells, adductor muscle myocytes and less frequently in haemocytes, and by the apparent lack of mitochondria and haplosporosomes. Also, *M. mackini* seems to have a unique way of obtaining energy from its host cell. Hine et al. (2001) depicted tube-like structures extending into the cytoplasm of *M. mackini* from the mitochondria of its host cell (Fig. 2.8C). Thus, the contents of the host cell mitochondria appeared to pass through a tubular extension into the cytoplasm of the parasite.

The parasite is infective to at least four species of oysters (Pacific oysters; eastern oysters, *Crassostrea virginica*; flat oysters and Olympia oysters, *Ostrea conchaphila*) (Bower et al. 1997). The only other described species of *Mikrocytos*, *M. roughleyi* from the Sydney rock oyster, *Saccostrea glomerata* (=*S. commercialis*) in New South Wales, Australia (Farley et al. 1988), is now believed to be a species of *Bonamia* (Cochennec-Laureau et al. 2003). Although *Bonamia* spp. are also known as microcells that parasitise various species of oysters, they are not related to *M. mackini*. The inability to assign *M. mackini* to a phylum and lack of knowledge on close relatives (Hine et al. 2001, Carnegie et al. 2003), makes it impossible to compensate for information gaps on various biological parameters by extrapolation.

To date, *M. mackini* has been reported only from oysters in the southern part of British Columbia, Canada, and the adjacent state of Washington, USA. However, disease caused by *M. mackini* appears to be restricted to older oysters (over two years) in some locations in British Columbia and mortalities occur in the spring (April and May) after three to four months when temperatures are less than 10°C. The requirement for cool temperatures and the long prepatent period may explain why the disease only occurs during the spring and seems to be confined to oysters cultured in more northerly locations.

Life cycles

The three different morphological forms of *M. mackini* appear to be preferentially located in different tissues and host cells of the oyster. Tissue locations and overlap in features between the

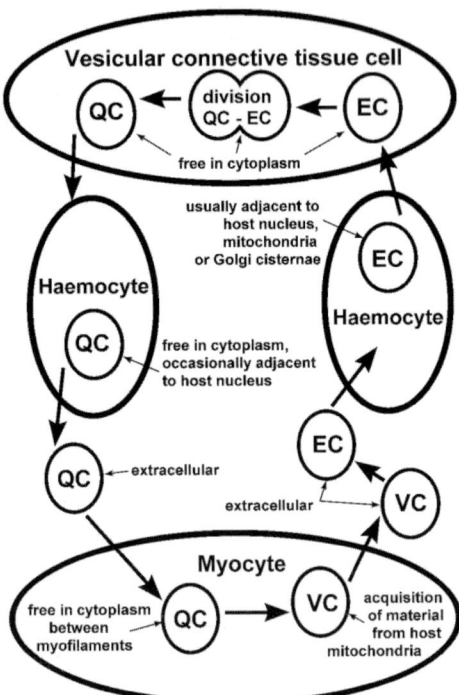

Figure 2.9 Proposed developmental cycle of *Mikrocytos mackini* indicating host cell type and host organelle affiliations for the three recognised morphological forms consisting of quiescent cell (QC), vesicular cell (VC) and endosomal cell (EC). Reprinted with permission from Hine, PM, Bower, SM, Meyer, GR, Cochennec-Laureau, N and Berthe, FCJ 2001. Ultrastructure of *Mikrocytos mackini*, the cause of Denman Island disease in oysters *Crassostrea* spp. and *Ostrea* spp. in British Columbia, Canada. *Diseases of Aquatic Organisms* 45: 215–227.

different morphological forms suggested a developmental cycle of the parasite in its host (Fig. 2.9). Apparently QC, which appear to lack energy reserves and cytoplasmic organelles usually associated with energy production in eukaryotic cells, travel from the vesicular connective tissue cells in haemocytes to the adductor or heart muscles. Once inside the myocytes, the QC changed to VC which appeared to endocytose the contents of myocyte mitochondria, acquiring ATP and other proteins. The VC then changed to EC in the process of leaving the muscle and entering haemocytes. The EC maintain close contact with haemocyte mitochondria and appear to obtain ribonucleoproteins from the haemocyte nucleolus. While travelling to the vesicular connective tissue cells in haemocytes or extracellular, the EC began mitosis by binary fission which was completed in the vesicular connective tissue cells where the daughter generation changed to QC and reinitiated the cycle.

Alternately, *M. mackini* may acquire the morphological form suited to its energy acquisition opportunities available in the host cell. The lack of organelles in *M. mackini*, including mitochondria or their equivalents that are found in most eukaryotic cells, may be due to its obligate parasitism. The utilisation of host cell organelles could have reduced the need for parasite organelles. Possibly, *M. mackini* is a primitive eukaryote that is able to survive because of a parasitic existence or is a highly evolved parasite that has secondarily lost most energy-producing organelles.

In addition to being very cryptic, experimental evidence suggests that most infections of *M. mackini* in oysters are subclinical. Exposed oysters, held at 18°C, can retain *M. mackini* at subclinical levels for at least six months. *Mikrocytos mackini* is only capable of causing disease in oysters held at less than 10°C for at least three months (Hervio et al. 1996). This apparent requirement for long periods at low temperatures for pathogenic expression and the prolonged prepatent period at warm temperatures may explain why *M. mackini* is only detectable in the field during the spring (March–May). Possibly, subclinical infections occur in oysters from enzootic areas throughout the year. *Mikrocytos mackini* can be directly transmitted between oysters but transmission seems to be limited to periods when the infection is active in diseased oysters during the spring. To date, no evident mechanisms of protection for existence outside the oyster host (i.e. no spore-like stage) and no alternate hosts have been detected (Bower 2001).

Effects on hosts and ecological importance

The disease (Denman Island disease or mikrocytosis) associated with *M. mackini* was first detected in the early 1960s among beach cultured oysters on Henry Bay, Denman Island, when it caused high mortalities (about 30%) of oysters at low tide levels (Quayle 1982). From 1960 to 1994, prevalence of infection in Pacific oysters from Henry Bay fluctuated from 11% (1967) to 48% (1988), in mid-March to mid-May (Hervio et al. 1996). In addition to mortalities, active infections of *M. mackini* induce the development of focal green abscess-like lesions (pustules) up to 5 mm in diameter usually within the body wall and adductor muscle or on the surfaces of the labial palps or mantle. The pustules usually contain a central area of tissue necrosis surrounded by haemocyte infiltration. Often, a brown scar occurs on the shell adjacent to a pustule on the mantle surface. Although the lesions caused by infection can persist throughout the summer, *M. mackini* is usually no longer detectable by early summer using routine histological techniques. Experimental evidence suggests that oysters become infected when the disease is active in the early spring but the infection remains cryptic until the following spring. Juvenile oysters (seed) held on affected beaches during the spring will develop the disease the following spring in British Columbia, regardless of subsequent culture techniques. About 10% of infected *C. gigas* appear to recover. *Crassostrea gigas* seems to be more resistant to the disease than the other species of oysters challenged experimentally under laboratory and field conditions (Bower et al. 1997).

Important references

Further details, coloured illustrations and a complete list of references are available on the website (Fisheries and Oceans Canada 2004).

Ciliophora (ciliates)

Jiří Lom

Introduction

Among the protozoans, the monophyletic assemblage of ciliates is perhaps the most numerous with more than 8000 species ranging from about 10 μm to 4.5 mm in size. Most species have a pellicle covered with cilia, although ciliature may be reduced or even absent. They have the most complicated cell structure among protozoans: free-living species inhabit a range of aquatic and terrestrial environments, and other species live in various symbiotic relations with aquatic animals. There is hardly an aquatic animal, be it a fish, invertebrate or even a mammal, without some epibiotic, ectoparasitic or endozoic ciliate; there are also free-living ciliates often behaving like facultative parasites. Although the ciliates in most cases do not inflict any harm, there are some severely pathogenic species (e.g. in fish).

Morphology and diversity

The ciliate cell is covered with a pellicle consisting of an outer cell membrane subtended by a layer of flat pellicular alveoli. The pellicle is covered by cilia, either uniformly, arranged longitudinally in rows or kineties, or some or all cilia grouped together to form compound ciliary organelles like cirri or membranelles. In some groups the ciliature may be reduced or even completely absent. The basal bodies or kinetosomes of cilia are associated with a complex fibrillar network, the infraciliature, composed of microfibrillar ribbons and microtubules. Ciliates exhibit nuclear dualism, there are one to several small, diploid, mitotically dividing micronuclei with a complete genome and one to several large, amitotically dividing ampliploid ('polyploid') macronuclei for vegetative functions of the cell. Ciliates divide essentially by transverse binary fission, rarely by budding or multiple fission. The sexual process is conjugation, in which two individuals undergo a partial and transient fusion and reciprocally fertilise themselves by products of meiotic division of their micronuclei. Feeding takes place by elaborate buccal structures involving either simple somatic ciliature or special ciliary organelles like membranelles and paroral membrane, all these serving for driving the food into the cytostome. In some, suctorial tubes serve for sucking in the prey cytoplasm, and some are secondarily astome. Ciliates overcome unfavourable conditions inside protective cysts; a special kind of cyst is used as shelter for proliferation. Generally, the life cycle is simple, continuous division of trophic stages with insertion of cyst stages.

Evolution of ciliates resulted in diversification that lead to formation of special morphological structures (stalks, loricae and cysts). Essential pressure for evolution of various special morphological and life cycle adaptations is, however, a symbiotic life in different ecological niches. The resulting diversification has led to groups of ciliates that are so different that it is rather difficult to treat them here as a single assemblage.

Free-living ciliates are an important constituent of the food web in marine environments and the same can be said of the symbiotic ones. Symphoriants utilise bacteria, algae and organic particles from the water around them. Commensals feed on substances offered by the hosts (abundant populations of several families of endocommensal scuticociliates (e.g. Entodiscidae, Entorhipidiidae and Cryptochilidiidae live in the intestine of sea urchins). Some of the ciliates, which act as true parasites, feeding at the expense of their hosts, may deplete the host's viability or even reduce the population. This applies to ciliate infection in stressed, cultured fish.

Until the second half of the 20th century, ciliates were classified essentially according to the type of the buccal and also somatic ciliature. This made it possible to determine the taxa using light microscopy. Now, electron microscopy yields characters at a higher level. In the current classification (Lynn 2003), the phylum is divided into two subphyla: Postciliodesmatophora with characteristic microtubular ribbons linking all kinetosomes in a kinety, comprising two classes; and Intramacronucleata, in which macronuclear division involves microtubules that lie inside it. This subphylum comprises nine classes with a total of 19 subclasses. The genera of symbiotic ciliates from the marine environment belong mostly to two classes of this subphylum: Phyllopharyngea with subclasses Phyllopharyngia (*Brooklynella*) and Rhynchodia (*Ancistrocoma*); and Oligohymenophorea with subclasses Scuticociliatia (with holotrichous ciliation, e.g. *Mesanophrys*), Apostomatia (with a few spiralling ciliary rows, primitive buccal structure and a complex life cycle, e.g. *Ascophrys* and *Vampyrophrya*) and Peritrichia (*Trichodina*).

Symbiotic adaptations and effects on hosts

Ciliates display a wide range of adaptations to a symbiotic way of life, ranging from facultative parasites to innocuous symphoriants to true parasites. Species of the genus *Lagenophrys*, peritrichous ciliates like all in this subclass devoid of somatic cilia and only equipped with a buccal ciliary spiral, are essentially innocuous symphoriants. However, on blue crab gills they can grow so exuberantly that its host, if confined in a tank, dies for lack of oxygen (see Hausmann and Bradbury 1996).

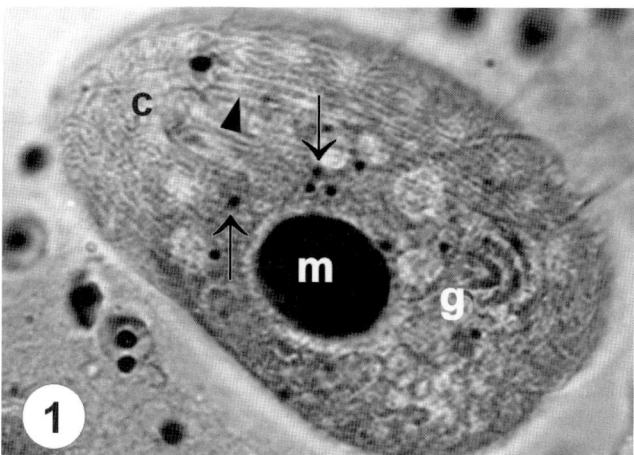

Figure 2.10 *Brooklynella hostilis.* Haematoxylin-stain reveals macronucleus (m) with dot-like micronuclei (arrows) above it, faintly visible ciliary rows on the ventral surface (arrowhead), cytostome (c) and glandular organelle (g). × 850. Reprinted from 'Protozoan Parasites of Fishes'. Developments in Aquaculture and Fisheries Science, Vol. 26, Lom and Dyková (1992), with permission from Elsevier.

The scuticociliate genus *Paralembus* has species which browse innocuously through the surface mucus of sea anemones and feed on trapped bacteria and cellular debris from surface injuries. Free-living ciliates of the genera *Helicostoma* or *Euplotes* may sometimes prove to be pathogenic for debilitated cultured fry of commercial fish, causing lethal skin lesions. Scuticociliates *Miamiensis avidus* or *Uronema marinum* are known as facultative parasites. Especially in fish cultures, they attack the skin and gills first, hampering respiration, and then make their way into the internal body organs, disintegrating them and causing mortalities. In these and other infections, the fish need to be predisposed by stress, environmental or other, for the ciliates to fully develop their pathogenic potential (Lom and Dyková 1992). Like the following *Brooklynella*, they have no host preferences and attack all fish available.

The phylopharyngiid ciliate *Brooklynella hostilis* (Figs 2.10 and 2.11) with flattened cell, only ventrally equipped with rows of cilia, is a scavenger gliding over the surface of fish gills. It feeds on desquamated cells. In stressed captive fish (in cultures, aquaria) it may multiply massively

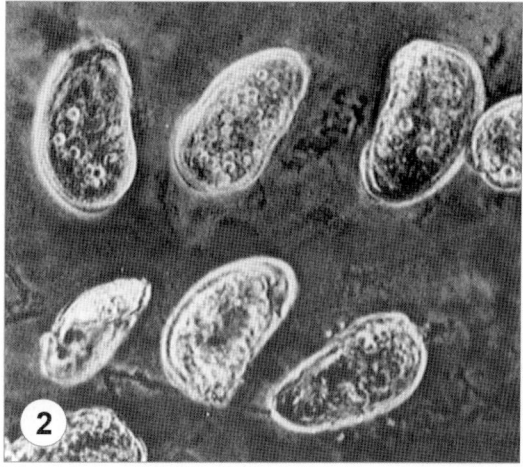

Figure 2.11 *Brooklynella hostilis,* live ciliates. × 330.

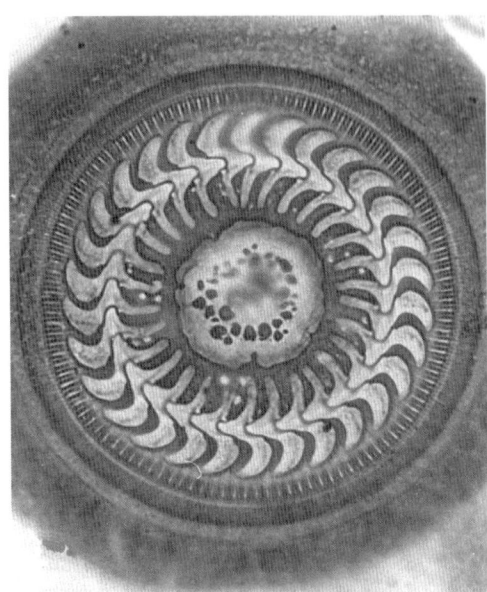

Figure 2.12 Silver impregnated adhesive disc of *Trichodina murmanica*, serving for attachment to fish surface, reveals fine details of the proteinaceous cytoskeleton. × 1000.

and turn into a pathogen. In the lack of cell debris, the extensible cytopharyngeal armature destroys the gill cells and the ciliate feeds on them. Severe gill lesions – the secondary lamellae are sometimes completely denuded of the epithelium – inflict heavy losses in fish stocks.

Species of the peritrichous genus *Trichodina* are common inhabitants of the surface of some freshwater and marine invertebrates and especially of fish (Lom 1995). Their disc-shaped cell has a ciliated spiral on the upper side and a sucker-like, reinforced adhesive disc (Fig. 2.12) on the lower side, by means of which they can attach temporarily to the substrate. In unstressed feral fish populations, they behave like harmless ectocommensals feeding on cell debris and micro-organisms. In cultured fish, they occur in large numbers and damage the surface tissue by the action of the disc and feed on the detached cells.

Species of the scuticociliate genus *Mesanophrys* live as scavengers on the exoskeleton of crabs and lobsters. When by chance they find an animal with a break in its shell, they invade the haemocoel, multiply, destroy the amoebocytes and, feeding on the tissue, they may kill the crustacean. In crustaceans kept in holding tanks or cages where the injuries may be commonplace they may cause many die-offs.

Ciliates of the subclass Apostomatia mostly do their host no harm, feeding on exuvial fluid or secretions of cuticular hairs. Some are, however, pathogenic. Mortalities in cultured *Palaemon serratus* (Deroux et al. 1975) are also due to the apostome ciliate *Ascophrys rodor* causing cuticular lesions. *Vampyrophrya parasitica* has stages (tomites) encysted on the shell of marine copepods. If the host's exoskeleton is breached, the tomites excyst, get through the wound into the body, where they start feeding on the tissues. Eventually, when only an empty shell is left, the grown ciliates encyst and divide into small tomites again. Then they leave the dead host, find another host on the surface of which they encyst again and wait. In other apostomes, like *Synophrya hypertrophica*, the tomites actively bore through the exoskeleton of the gills and feed inside on the blood by osmotrophy.

A severe pathogen of practically all marine teleosts, rare in feral fish but often abounding in pisciculturas and sea aquaria, is *Cryptocaryon irritans* of the class Prostomatea. It has a life cycle

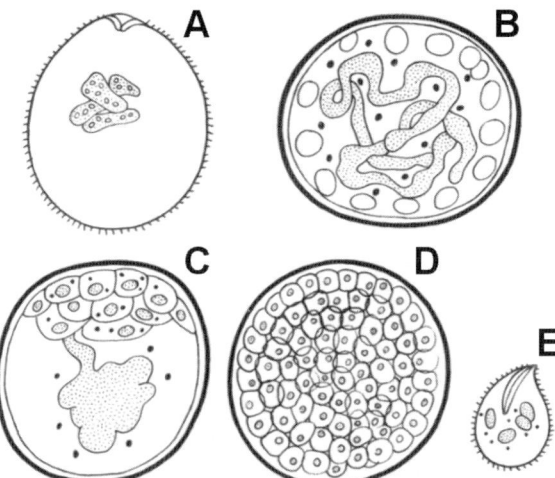

Figure 2.13 *Cryptocaryon irritans*, diagram of the life cycle. **A**. Mature trophont with four macronuclei at the surface of fish. **B**. Large encysted ciliate with macronuclei fused into one ribbon prepares to divide. **C**. Palintomic division within the cyst. **D**. Many new tomites within the cyst. **E**. A tomite transformed into a theront, again with four macronuclei, in search of a new host. Fig. 2.13 modified from 'Protozoan Parasites of Fishes'. Developments in Aquaculture and Fisheries Science, Vol. 26, Lom and Dyková (1992), with permission from Elsevier.

(Fig. 2.13) very similar to its freshwater counterpart *Ichthyophthirius* but is morphologically different (e.g. has quadripartite macronucleus). Small individuals (theronts) burrow into the fish surface, grow there into a trophont up to 450 µm long, which then leaves the host, drops to the substrate as a tomont and encysts. Within the cyst, up to about 200 tomites are produced within six to nine days. Then they emerge from the cyst as theronts and swim in search of another host. The surface tissue may be severely disintegrated and serious epizootics in affected stocks are common.

Marine ciliates offer an almost unfathomable range for studies of adaptations to symbiotic life.

Important references
Comprehensive treatment of ciliates can be found in Corliss (1979), Grassé (1984), Hausmann and Bradbury (1996), Lynn and Small (2000) and de Puytorac (1994).

Myxozoa (myxozoans)
Jiří Lom

Introduction
Myxozoa attract growing attention. Being predominantly parasites of fish, they are a significant and increasing impediment to modern pisciculturesx, which have to cope with steadily emerging new myxozoan pathogens. Presently, more than 2770 species assigned to 61 genera have been described from freshwater and marine fish and many of them have a disease potential. Myxozoa have multicellular spores consisting of specialised cells. Previously assigned to Protozoa, they have now been transferred to the Metazoa. The debate still continues as to what assemblage of metazoans they belong, either to the Radialia or to Bilateria. Thus, they are a group interesting from both practical and theoretical standpoints.

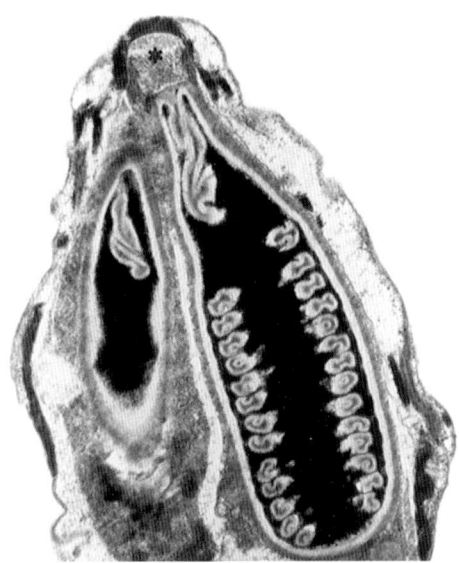

Figure 2.14 The main myxozoan feature, polar capsule with spirally wound ejectible polar filament in the spore of the genus *Henneguya*, as seen in electron microscopical section. × 9.400.

Morphology and diversity

Myxozoa have multicellular spores consisting of specialised cells – valvogenic cells forming the spore shell valves, amoeboid infective germ or sporoplasm (there may be one to many) and capsulogenic cells forming polar capsules with evertible polar filaments (Fig. 2.14). They are virtually identical with cnidarian nematocysts. Although nematocysts are instrumental in food capture, polar capsules serve for attachment to new hosts. The structure of myxozoan spores reflects what is also found in trophic stages (i.e. multicellularity and morphological and functional specialisation of cells). It is now generally accepted that the myxozoan life cycle requires an alternation of vertebrate and invertebrate host.

Spores are produced in different trophic structures. In the class Myxosporea, they originate in sporogonic plasmodia (SP) in the vertebrate phase of their life cycle and in the invertebrate phase of the cycle within a simple envelope of two to several cells, the pansporocyst. In the vertebrate host, coelozoic SP occur in organ cavities (associated with gall and urine secretion, rarely blood vessels or pericardium), histozoic SP occur intercellularly or intracellullarly in almost all tissues. Spores of this phase (myxospores) have mostly two polar capsules, less frequently one, three, four, five, rarely more, and are bilaterally or radially symmetric.

The most common vertebrate hosts are bony fishes, a few species live in elasmobranchs, amphibians and chelonid reptiles, three in agnaths. Most common hosts of the invertebrate phase are oligochaetes, also a few polychaetes and sipunculids; the infection affects mostly intestinal tissue. Spores of this phase (actinospores) are as a rule triradial with three capsules.

In the class Malacosporea, spores originate in a worm-like or sac-like stage living in the body cavity of bryozoans. This happens by differentiation from cells derived from the inner cell sheet or from agglomeration of cells inside the parasite.

Life cycle

Myxosporea: As late as in 1984, Wolf and Markiw discovered that transmission requires a developmental phase in an invertebrate host; since then, about 32 cases of fish–invertebrate life cycles have been discovered. Only one, that of *Ellipsomyxa gobii*, was observed to take place in the marine environment. The universal validity of the two-host cycle, especially for myxosporeans from marine fish, has not yet been safely confirmed; rare cases of direct fish to fish transmission have been reported (Redondo *et al.* 2004).

Myxospores released from the fish (Figures 2.15A–G) can survive in the aquatic environment for more than one year. When ingested by an oligochaete (in freshwater) or by a polychaete

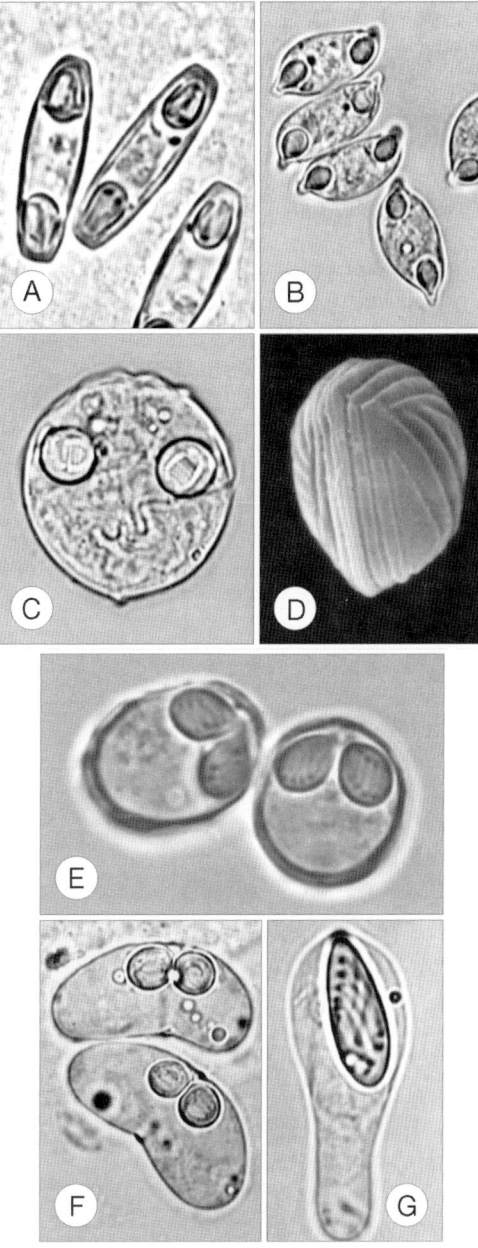

Figure 2.15 Spores of myxosporeans that infect marine fishes. **A.** *Sphaeromyxa magna*, × 2000. **B.** *Myxidium gadi*, × 1900. **C.** *Sinuolinea* sp., × 1300. **D.** *Ortholinea australis* (seen in scanning electron microscope), × 2200. **E.** *Myxobolus spinacurvatura*, × 2400. **F.** *Ceratomyxa macrospora*, × 1300. **G.** *Auerbachia pulchra*, × 2200. Figs A, C, F, G reprinted from 'Protozoan Parasites of Fishes'. Developments in Aquaculture and Fisheries Science, Vol. 26, Lom and Dyková (1992), with permission from Elsevier.

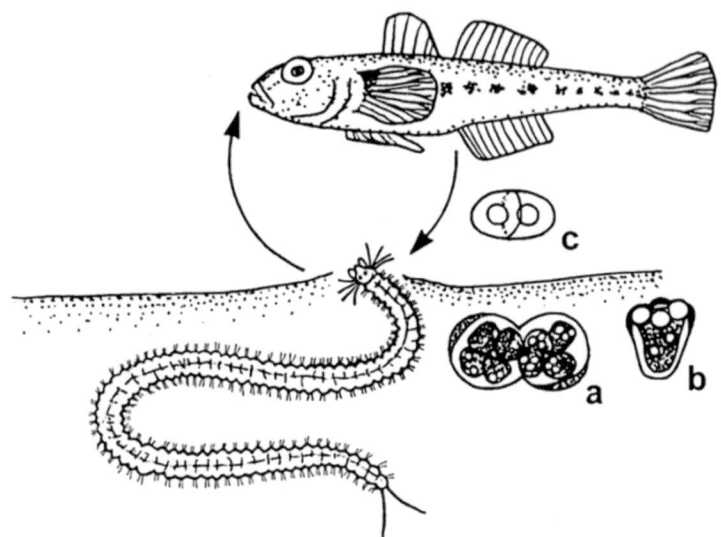

Figure 2.16 Life cycle of *Ellipsomyxa gobii* involves two hosts: it lives as a myxosporean stage (with the spore c) in the gall bladder of the common goby (*Pomatoschistus microps*), and (a) as an actinosporean stage in the polychate *Nereis* sp., producing pansporocysts with (b) spores. From Køie et al. (2004); reproduced with permission from the author, Dr Marianne Køie and the Editor-in-Chief, *Folia Parasitologia*.

(Fig. 2.16) (in the sea; this has been little studied until now – see Køie *et al.* 2004), they extrude their polar filaments, attach to the intestinal wall, the shell valves open, the sporoplasm escapes and invades the worm's intestinal tissue. After a period of proliferation by merogony (El-Matbouli and Hoffmann 1998), there appears a complex of four cells, the outer two will become the envelope cells of the future pansporocysts, the structure in which the spores develop. The inner two cells will divide to produce 16 gametic cells, which fuse to form eight zygotes. This is the only well-proven sexual process of Myxosporea and thus invertebrates must be regarded as definitive hosts. The zygotes develop into triradiate actinospores. Most of them have long, folded caudal projections, which as soon as the spore is released into water extend telescopically to a great length to assure a longer buoyancy in water. The damage done to the annelid hosts has not yet been sufficiently studied; infection of annelids can persist for two years.

When a floating actinospore – which is rather short lived – randomly contacts a fish, the extruded polar filaments attach to the skin or gills, the spore opens, and the sporoplasm penetrates into the surface tissue. From there on, the parasite wanders in a very complicated way, dividing on the way to produce a large number of cells. They may use blood circulation (*Sphaerospora*) or they may make their way as intracellular and intercellular stages through different body organs and tissues (*Myxobolus cerebralis*) before they eventually reach the final site where the sporogonic plasmodium develops (El-Matbouli *et al.* 1995). The stages produced, often in large numbers, in blood or various tissues are called presporogonic stages. They consist of a large mother (primary) cell in which are produced by inner cleavage two to many inner (secondary, and even tertiary) cells. These stages reveal another typical myxosporean feature, the cell-in-cell organisation. Eventually, the mother cell disintegrates and inner cells start the cycle all over again. Ultimately, the final site of infection is reached and the established sporogonic plasmodium starts to produce myxospores.

Sporogonic stages range from simple, small pseudoplasmodia, actually a uninucleate mother cell (about 10 μm) containing generative cells developing into one or two myxospores up to large multinucleate plasmodia (2 cm in *Sphaeromyxa maiyai*) in which a large number of myxospores are produced. Spores originate by division of a single generative cell into the number of specialised cells necessary to constitute the spore, or they arise by union of two cells: a pericyte which envelops a generative cell, which inside this envelope divides to produce the cells necessary to constitute two spores. Mature spores may be released from a living host or are set free after its death when the tissues disintegrate.

Malacosporea are freshwater organisms living as worm-like or sac-like creatures in the body cavity of freshwater bryozoans (Canning and Okamura 2004). Some (*Tetracapsuloides bryosalmonae*) produce spores infective for anadromous salmonids, causing the dangerous Proliferative Kidney Disease (PKD). In kidney tubules they may produce spores of slightly different appearance (Hedrick *et al.* 2004).

Classification and evolution

Phylum Myxozoa has 61 genera in two classes:

1 Malacosporea (one family and two genera – *Buddenbrockia*, *Tetracapsuloides*).
2 Myxosporea, with two orders:
   a Bivalvulida, with three suborders:
      – Sphaeromyxina (one family and the single genus *Sphaeromyxa*)
      – Variisporina (10 families and 38 genera e.g. *Myxidium*, *Ceratomyxa*, *Sphaerospora*)
      – Platysporina (one family and 13 genera e.g. *Myxobolus*, *Henneguya*)
   b Multivalvulida (six families and seven genera e.g. *Unicapsula*, *Kudoa*)

The worm- or sac-like malacosporeans are supposed to be the early steps in the evolution of Myxozoa (Okamura *et al.* 2002); consequently, the simple structure of myxosporean trophic stages (plasmodia) would reflect an extreme simplification due to parasitism. Invertebrates are viewed as the primary and definitive hosts wheras vertebrates were only secondarily acquired to serve for asexual population increase (Canning and Okamura 2004). A series of adaptations has evolved, such as alternation of hosts, mechanisms of invasion, modes of proliferation, migration through body organs and intracellular parasitism.

Effects on hosts and ecological importance

The ecological role of marine myxosporeans lies in their effect on fish populations. Most species live in a balanced state with their hosts in feral fish; striking epizootics have been rather exceptional. However, in high sea fisheries, heavy infections with muscle-invading multivalvulid species (see pp. 378–391) can make a large part of the catch unmarketable because the flesh is unsightly or degraded to mushy substance. In fish in captivity, the pathogenic potential of some species can be fully unravelled. Depending on parasite species and body organs, the pathogenic action may be extremely varied; for example pressure atrophy in tissues exerted by parasite masses (plasmodia of *Myxobolus* species), irritation of epithelia of organ cavities (*Sphaerospora*, *Ceratomyxa*), enzymatic lysis of muscle tissue (species of the order Multivalvulida, like the genera *Hexacapsula*, *Kudoa* or *Unicapsula*), destruction of tissues (*Ceratomyxa shasta* or *Enteromyxum* species) or anaemia (*Tetracapsuloides*). Pathogenicity has been reviewed in Lom and Dyková (1995).

Among the recently emerged myxosporean pathogens in maricultures are the species *Kudoa thyrsites* (Kent *et al.* 2001), a cosmopolitan parasite infecting muscle tissue of many species of fish. It is the cause of 'soft flesh', muscle tissue degradation in *Merlucius productus*. It produces unmarketable flesh quality in pen-reared *Salmo salar*. Species of the genus *Enteromyxum* develop as small sporogonic pseudoplasmodia in the intestinal tissue causing severe enteritis with high rate of mortalities. *Enteromyxum leei* not only infects cultured *Sparus aurata* and some other commercial fish species, it was also found to infect 25 host species from different orders and genera (Padrós *et al.* 2001). *Enteromyxum scophthalmi* is the cause of acute enteritis, starvation and death in cultured *Scopthalmus maximus*. The high virulence and rare development of mature spores in this host suggest that turbot is an accidental host for the parasite (Redondo *et al.* 2004). Cultured *Sparus aurata* also suffers from infections with *Ceratomyxa sparusaurati*. The parasite lives in the gall bladder and causes its swelling, sloughing of the epithelial cells and trickling mortalities (Palenzuela *et al.* 1997).

The challenge for future research of marine myxosporeans is to assess pathogenicity of new intruders into maricultures and especially to discover to what extent marine Myxosporea follow in their life cycle the pattern of alternation of fish and invertebrate host.

Important references
Monographic treatment of Myxozoa can be found in Canning and Okamura (2004), Kent *et al.* (2000, 2001), Kudo (1919), Lom (1990), Lom and Dyková (1992, 1995) and Shulman (1966).

Chapter 3

Helminth parasites

Helminths (parasitic worms) are among the most important groups of marine parasites. They include some turbellarians (most of which are free living), ectoparasitic flukes (Monogenea), endoparasitic flukes (Aspidogastrea and Digenea), the various cestodes including the Amphilinidea, Gyrocotylidea and Eucestoda, nematodes (roundworms) and acanthocephalans (spiny-headed worms). Species numbers of some of these groups are enormous. Only a small proportion of the trematodes and monogeneans, for example, has been described, but total species number of these two groups has been estimated to be many tens of thousands.

Many species have great ecological and economic importance. Monogeneans are known to do much damage to fish in aquaculture, and some cases of mass mortalities in the sea caused by monogeneans have been documented. Monogeneans, jointly with the copepods, are the most diverse group of ectoparasites of fish, as are the digeneans among the endoparasites. Digeneans are by far the most speciose group of metazoan endoparasites of fish and have evolved life cycles ranging from fairly simple to extremely complex. Nematodes are important parasites of many groups of marine animals including fish, marine mammals, birds and invertebrates.

Some marine helminths have considerable theoretical interest. The Aspidogastrea, for example, are believed to be very ancient, hundreds of millions of years old, with many archaic characteristics. They may cast light on the evolution of parasitism in the trematodes. Acanthocephala are useful as pollution monitors: they accumulate certain pollutants at a much higher rate than their hosts. Parasitism in the turbellarians has arisen several times, some living in the intestine, others in the tube feet of echinoderms, and others again on the gills of marine fish. Adaptations to their way of life vary considerably, making them fascinating objects of study in what may be called comparative parasitology. The eucestodes, discussed in a beautifully illustrated section, are important parasites in the digestive tract especially of elasmobranchs, with a whole spectrum of remarkable morphological adaptations of their attachment organs, and some with extremely complex life cycles.

'Turbellaria' (turbellarians)
Lester Cannon

Introduction
Once a taxon (Class Turbellaria) turbellarians are now recognised merely as the lower flatworms (Phylum Platyhelminthes), a diverse, polyphyletic assemblage of worms which are mainly free-living (Cannon 1998). In contrast, the higher flatworms (Neodermata) are exclusively parasitic. Two groups (Acoela and Nemertodermatida) traditionally included in the turbellarian

flatworms, and included in the following discussion, are now recognised to belong to a more 'primitive' group of their own, near the base of the invertebrate animals. Turbellarians have a digestive tract lacking an anus, although a gut may be missing in some species. They lack complicated life cycles involving specialised larvae.

Morphology and diversity

Although mainly free living, the turbellarians contain members in most groups which are known to form associations with other animals. Mostly these associations have not been investigated thoroughly, though many are certainly loose and conveniently termed 'commensal', while a few are examples of true, obligate parasitism. With the notable exception of the Temnocephalida, which associate with (mainly) freshwater crustaceans, most turbellarian associations are marine. Table 3.1 provides details of known symbioses. For brevity, references sometimes are only to a recent source.

Marine parasitic or symbiotic turbellarians, though diverse, share some fundamental characteristics. Most are quite small (1–2 mm long), are generally flat, oval or elongate oval, soft bodied and covered with a ciliated epidermis (i.e. they lack the characteristic integument of the parasitic Neodermata). Furthermore, few have specialised attachment organs and most are hermaphrodites.

Electron Microscope (EM) studies have focused mainly on the epidermis (Tyler and Tyler 1997), glands (Jondelius 1992a), protonephridia (Rohde 2001) and the nature of sperm and spermiogenesis (Watson 1997) particularly with an aim to elucidate phylogeny. Partial loss of ciliation and other epidermal specialisations are the main adaptations to parasitism.

Life cycles

Turbellarians do not undergo metamorphosis, so specialised larvae are not found. As with free-living turbellarians it is general for a sclerotic egg capsule containing one (or more) embryos to be laid often cemented to some substrate. Young emerge from the capsule and develop directly to adults: there are no known cases of the use of intermediate hosts.

From Table 3.1 it is evident that both endo- and ecto-commensalism is found, and that this kind of relatively loose (undefined and frequently unresolved) association is most usual in all but the rhabdocoel turbellarians. This latter group includes those most often considered to have close links to the Neodermata (see Lockyer et al. 2003).

Examples of endocommensalism are to be found with *Meara* (Nemertodermatida) and the acoels (e.g. *Aphanostoma*, and *Avagina*) found in the gut of holothurians which, though little studied, probably feed on food items ingested by their hosts (see Bresslau 1933 for older literature). Similarly, several polyclads are known to be ectocommensal (or intimate predators) with marine hosts: Prudhoe (1985) summarises the older literature. These associations range from those like *Notoplana comes* which may be merely seeking shelter or perhaps food scraps below its host ophiuroid (Cannon and Grygier 1991) to *Discoplana malagasensis* which, according to Doignon et al. (2003), lives only in the brood chamber of *Ophiothrix purpurea*.

Morphological adaptations of ectocommensals to assist in attachment are seen is some triclads. *Bdelloura* spp., for example, have variously modified posterior regions which allow them to attach to the gills and legs of their hosts, the Xiphosuran crabs (e.g. *Limulus* spp.), and *Micropharynx parasitica* has the posterior rolled to provide a grip. It is found in grooves on the upper surface of rays (*Raja* spp.). While the bdellourids appear to feed on food scraps dropped by their hosts, *M. parasitica* may have become a true parasite since it possibly feeds on skin (Ball and Kahn 1976).

Plagiostomum oyense (Prolecithophora) is said to live in the brood chamber of the isopod *Idotea neglecta* where it attaches to the egg capsules (Naylor 1955); and the proseriates *Ectocotyla*

Table 3.1 Marine turbellarian symbioses

Taxon (Order/Family)	Genus	Host group	Symbiosis	Reference
NEMERTODERMATIDA				
Nemertodermatidae	*Meara*	Holothuriodea	Endocommensal?	Westblad (1950)
ACOELA				
Convolutidae	*Aphanostoma*	Holothuriodea	Endocommensal?	Bresslau (1933)
	Avagina	Holothuriodea	Endocommensal?	Hickman (1956)
	Waminoa	Cnidaria	Ectocommensal?	Winsor (1990)
Hallangidae	*Aechmalotus*	Holothuriodea	Endocommensal?	Bresslau (1933)
Otocelidae	*Otocelis*	Holothuriodea	Endocommensal?	Bresslau (1933)
POLYCLADIDA				
Apidioplanidae	*Apidioplana*	Cnidaria	Ectocommensal?	Cannon (1990)
Emprosthiopharyngidae	*Emprosthiopharynx*	Paguridae	Ectocommensal?	Prudhoe (1985)
Hoploplanidae	*Hoploplana*	Gastropoda	Ectocommensal?	Prudhoe (1985)
Leptoplanidae	*Ceratoplana*	Echinoidea	Ectocommensal?	Prudhoe (1985)
	Discoplana	Ophiuroidea	Ectocommensal?	Doignon et al. (2003)
	Notoplana	Ophiuroidea	Ectocommensal?	Cannon and Grygier (1991)
	Stylochoplana	Polyplacophora	Ectocommensal?	Prudhoe (1985)
Prosthiostomidae	*Euprosthiostomum*	Paguridae	Ectocommensal?	Prudhoe (1985)
	Prosthiostomium	Cnidaria	Ectoparasitic?	Prudhoe (1985)
Stylochidae	*Distylochus*	Echinoidea	Ectocommensal?	Prudhoe (1985)
PROLECITHOPHORA				
Plagiostomidae	*Plagiostomum*	Isopoda	Ectocommensal?	Naylor (1955)

Table 3.1 Marine turbellarian symbioses (Continued)

Taxon (Order/Family)	Genus	Host group	Symbiosis	Reference
PROSERIATA				
Monocelididae	*Digenobothrium*	Teleostei	Scavenger?	Palombi (1926)
	Ectocotyla	Decapoda	Ectocommensal?	Fleming et al. (1981)
	Peraclistus	Decapoda	Ectocommensal?	Fleming et al. (1981)
TRICLADIDA				
Bdellouridae	*Bdelloura*	Xiphosura	Ectocommensal?	Sluys (1989)
	Syncoelidium	Xiphosura	Ectocommensal?	Sluys (1989)
Meixneridae	*Jugatovaria*	Decapoda	Ectocommensal?	Sluys and Cannon (1989)
Procerodidae	*Ectoplana*	Xiphosura	Ectocommensal?	Sluys (1989)
	Micropharynx	Elasmobranchia	Ectoparasitic?	Ball and Khan (1976)
RHABDOCOELA				
DALYELLIOIDA				
Acholadidae	*Acholades*	Asteroidea	Parasitic	Jennings (1989)
Fecampiidae	*Fecampia*	Decapoda	Parasitic	Bellon-Humbert (1983)
	Glanduloderma	Myzostomia	Parasitic	Jägersten (1940)
	Kronborgia	Crustacea	Parasitic	Blair and Williams (1987)
Genostomidae	*Genostoma*	Leptostraca	Ectoparasitic?	Hyra (1993)
Graffillidae	*Graffilla*	Mollusca	Parasitic	Schell (1986)
	Paravortex	Teleostei and Bivalvia	Parasitic	Cannon and Lester (1988)
Hypoblepharinidae	*Hypoblepharina*	Crustacea	Ectocommensal?	Karling and Nilsson (1974)
Notenteridae	*Notentera*	Polychaeta	Parasitic	Joffe and Kornakova (1998)
Provorticidae	*Oekiocolax*	Prolecithophora	Parasitic	Reisinger (1930)
Pterastericolidae	*Pterastericola*	Asteroidea	Parasitic	Jondelius (1992b)

Table 3.1 Marine turbellarian symbioses (Continued)

Taxon (Order/Family)	Genus	Host group	Symbiosis	Reference
Umagillidae	Anoplodiera	Holothuriodea	Parasitic	Cannon (1982)
	Anoplodiopsis	Holothuriodea	Parasitic	Cannon (1982)
	Anoplodium	Holothuriodea	Parasitic	Shinn (1985a, b)
	Bicladus	Crinoidea	Parasitic	Cannon (1982)
	Cleistogamia	Holothuriodea	Parasitic	Cannon (1982)
	Collastoma	Sipinculida	Parasitic	Westervelt (1981)
	Desmote	Crinoidea	Parasitic	Shinn (1987)
	Fallacohospes	Crinoidea	Parasitic	Kozloff (1965)
	Macrogynium	Holothuriodea	Parasitic	Cannon (1982)
	Notothrix	Holothuriodea	Parasitic	Hickman (1955)
	Paranotothrix	Holothuriodea	Parasitic	Cannon (1982)
	Parafallacohospes	Crinoidea	Parasitic	Shinn (1987)
	Syndesmis	Echinoidea	Parasitic	Gavaerts et al. (1995)
	Umagilla	Holothuriodea	Parasitic	Cannon (1987)
	Wahlia	Holothuriodea	Parasitic	Shinn (1986)
Urastomidae	'Ichthyophaga'[A]	Teleostei	Parasitic	Cannon and Lester (1988)
	Urastoma	Bivalvia	Parasitic	Robledo et al. (1994)
KALYPTORHYNCHIA				
Schizorhynchidae	Typhlorhynchus	Polychaeta	Ectoparasitic?	Karling (1981)

[A] 'Ichthyophaga' is preoccupied, being a fish-eating eagle.

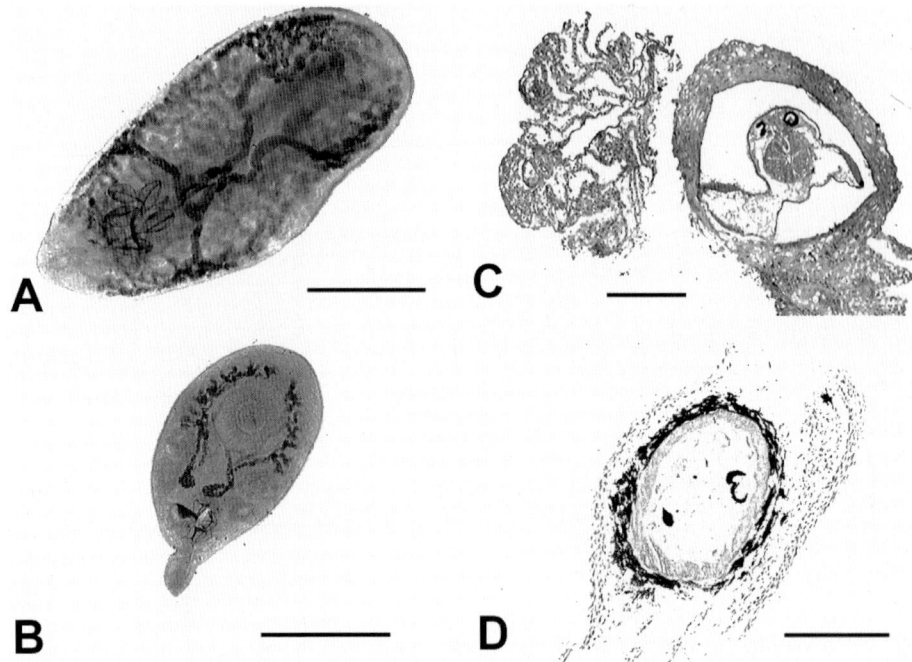

Figure 3.1 A, B. Endoparasitic turbellarians (family Umagillidae) which swim or move freely in the intestine of tropical sea cucumbers from the Great Barrier Reef. **A.** *Wahlia stichopi* from *Stichopus chloronotus*. **B.** *Cleistogamia pyriformis* from *Holothuria impatiens*. Scale bars A, B = 500 µm.
C, D. Endoparasitic turbellarians, sectioned through the eye region, which induce hypertrophied tissue tunnels in fish. **C.** *Paravortex* sp., within a tube of gill tissue. **D.** '*Ichthyophaga*' within a tunnel in the epidermis. Scale bars C, D = 150 µm.

spp. (also with a rear caudal disc) and *Peraclistus oofagus* are known to live on the exterior surfaces of crabs *Chionoecetes opilio* and *Hyas araneus* in the north Atlantic Ocean (Fleming et al. 1981).

Further examples of ecto and endo-commensalism may be found within the rhabdocoels. *Typhlorhynchus nanus* lives on parapodia of the polychaete *Nephthys scolopendrina*, while species of *Genostoma* and *Hypoblepharina* are found on small crustaceans. All are probably ectocommensal although *Genostoma* may feed on eggs and host tissue (Hyra 1993). Some *Paravortex* spp. live in the intestine and stomach of bivalves, but in the intertidal periods migrate to their hosts' digestive glands. Here they feed on partly digested food and also obtain enzymes from the host which work in their own gut to complete the digestion of food (Jennings 1997). Is this endocommensalism or endoparasitism?

True parasitism is known among several families within the rhabdocoels. The greatest diversity is to be found among the endoparasites of (mainly) echinoderms – the Umagillidae and Pterastericolidae. Shinn (1981) analysed the diet of three species of umagillids from echinoids – all consumed intestinal tissue but one supplemented this with the host's ingesta and another with the symbiotic ciliates found in the gut.

Shinn (1985a, b) also elegantly researched the life cycle of *Anoplodium hymanae* (Umagillidae) which lives in the coelomic cavity of the holothurian *Stichopus californicus*. Adults release egg capsules into the coelom where they become incorporated, often in hundreds, in masses of coelomocytes. These masses leave the coelom by way of fine ducts that connect to the exterior near the cloaca, so the eggs pass to the sea without the need for evisceration of the host. Embryos within the capsules are released on later digestion by another host, whereupon the ciliated juveniles penetrate the wall of either the lower gut or the respiratory trees to re-enter the coelom.

Most umagillids and all pterastericolids live in the gut. As far as is known, here they feed on intestinal tissues. Cannon (1982) reported that species may show horizontal segregation along the gut in some hosts, notably in *Stichopus* spp. (Holothuroidea). The torpedo-shaped *Wahlia stichopi* (Fig. 3.1A) can be seen spiralling among the high, ruffled folds of the anterior intestine. *Cleistogamia* spp. are pyriform and flattened ventrally (Fig. 3.1B), they congregate in the midgut where the epithelium is lower, and in the rectum where the rectal epithelium forms a low, flattened pavement. *Paranotothrix queenslandensis*, which is wafer thin, lies adherent to the gut.

Some adaptations to parasitism seen in the Umagillidae and Pterastericolidae are morphological; for example characteristic of many species is the presence of variously developed epithelial regions devoid of cilia (Jondelius 1988, 1991). Physiological adaptations are related to another characteristic of many of these worms, the presence of variously developed red pigmentation which suffuses the body. Haemoglobin is found in modified parenchymal cells distributed around the brain and reproductive organs (Jennings and Cannon 1985, 1987, Jennings and Hick 1990) presumably to assist with oxygen transport for the relatively greater energy needed in these regions. Phillips (1978) similarly reported the presence of haemoglobin in the graffillid *Paravortex scrobiculariae* which lives in the gut of bivalves subject to oxygen stress during low tide.

Effects on hosts and ecological importance

Oekiocolax plagiostomorum living within the parenchyma of *Plagiostomum parasitorum*, a free-living turbellarian, reportedly causes partial castration of its host (Reisinger 1930).

Species of *Graffilla* and *Paravortex* (Graffillidae) are found in molluscs, where (depending on species) they are found in the mantle and gills, in the stomach and digestive gland, in the heart and the kidney. Køie (1969) reported kidney pathology in whelks infected with *G. buccinicola*, namely weak, dilated kidneys, full of mucus. Villalba et al. (1997), however, reported a *Paravortex* sp. in the mussel *Mytilus galloprovincialis* in Spain to cause no pathogenicity, and data from Woods and Hayden (1998) from New Zealand suggest *Paravortex* in scallops, *Pecten novaezelandiae*, is associated more with scavenging than parasitism. In contrast, the unnamed *Paravortex* spp. reported by Kent and Olsen (1986) from Hawaii and Cannon and Lester (1988) from Australia create lesions in the skin and gills of fish (Fig. 3.1C). In Hawaii the young worms are claimed to erode the skin surface with the pharynx and to depart from the fish to mature on the sea floor, whereas in Australia the gravid adult worm lies partially within a 'tube' of host tissue; there is evident hyperplasia, but no necrosis. Two different life cycles seem apparent.

Like the Graffillidae, the Urastomidae has species which infect both molluscs and fish. *Urastoma cyprinae* is a cosmopolitan species infecting bivalves. It was considered to produce little pathogenicity, but recent studies have shown that living in the gills causes disruption and necrosis within gill tissues (Robledo et al. 1994). '*Ichthyophaga*' *subcutanea* is reported from the skin of fish. [More than one species may be involved and the genus is preoccupied by a sea eagle (Lesson 1843).] '*Ichthyophaga*' lives in burrows or cysts in the skin and appears to browse on the epidermal tissue (Fig. 3.1D). The cysts are frequently melanised by the host and are reminiscent of digenean metacercarial cysts.

The remaining true parasites – *Acholades, Notentera*, and the fecampiids, notably *Fecampia* and *Kronborgia* – have lost the gut (Fig. 3.2). *Acholades asterias* lives within the outer layer of the connective sheath of tube feet of the Tasmanian asteroid *Coscinasterias calamaria*, causing no loss of function (Fig. 3.2A). The worm lacks any alimentary structures and feeds directly from all of its modified ciliated epidermis which is more elaborate anteriorly (Jennings 1989). The worm secretes digestive enzymes and partially digests the host's tissue extra-corporeally and

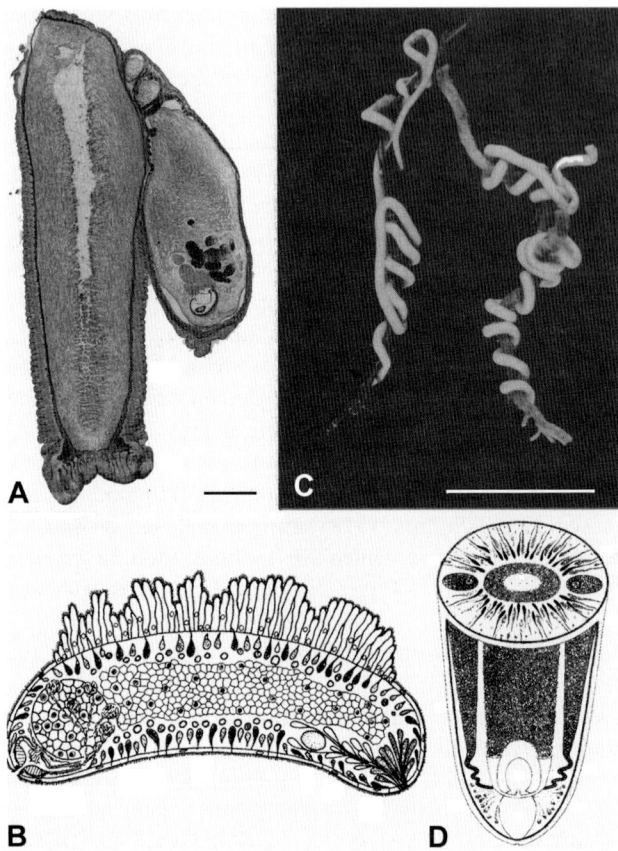

Figure 3.2 **A.** Section through the tube foot of a Tasmanian seastar (*Coscinasterias calamaria*) which shows attached to it the gutless encysted turbellarian worm, *Acholades asterias*. Scale bar = 500 μm. Reproduced with permission from Jennings, JB, Epidermal uptake of nutrients in an unusual turbellarian parasitic in the starfish *Coscinasterias calamaria* in Tasmanian waters. *Biological Bulletin*, 176: 327–336, Fig. 1(A); published 1989. **B.** Freehand drawing of *Notentera ivanovi*, a turbellarian worm with a hypertrophied dorsal epidermis from the gut of a polychaete worm from the Russian arctic. Reproduced from *Hydrobiologia* **383**, Joffe and Kornakova (1998), *Notentera ivanovi* Joffe *et al.*, 1997: a contribution to the question of phylogenetic relationships between 'turbellarians' and the parasitic Platyhelminthes (Neodermata), pages 245–250, Fig. 1, with permission from Kluwer Academic Publishers. **C.** Cocoons attached to seaweed from the Pacific West Coast, USA, containing eggs of the parasitic turbellarian *Kronborgia pugettensis* created after the worms have left their crustacean hosts. Reproduced from *Parasitology* **91,** Shinn and Christensen (1985), *Kronborgia pugettensis* sp. nov. (Neorhabdocoela: Fecampiidae), an endoparasitic turbellarian infesting the shrimp *Heptacarpus kincaidi* (Rathbun), with notes on its life-history, pages 431–447, Fig. 5, with permission from the authors and Cambridge University Press. **D.** Freehand drawing of the posterior end of a female of the parasitic turbellarian *Kronborgia isopodicola* showing terminal parts of the genital ducts. Reproduced from *Journal of Natural History* **21**, Blair and Williams (1987), a new fecampiid of the genus *Kronborgia* (Platyhelminthes: Turbellaria: Neorhabdocoela) parasitic in the intertidal isopod *Exosphaeroma obtusum* (Dana) from New Zealand, pages 1155–1172, Fig. 5 with permission.

then, via pinocytosis, absorbs nutrients to complete digestion intracellularly mimicking the gastrodermis of free-living turbellarians. *Notentera ivanovi* is found in the gut of the polychaete worm *Nephthys ciliata* from the White Sea, Russia (Fig. 3.2B). It too lacks alimentary structures, and like *Acholades* the epidermis is modified, but only dorsally. Furthermore, though the dorsal epithelium looks much like gastrodermis it seems to lack lysosomes or show evidence of pinocytosis (Joffe and Kornakova 1998), so its role in digestion is still uncertain. Nothing is known of the life cycle of either of these worms.

The Fecampiidae include the little known *Glanduloderma*, a hyperparasite from myzostomes parasitic in echinoderms and *Fecampia* and *Kronborgia* from crustaceans. *Kronborgia* spp. are peculiar in having separate sexes (Fig. 3.2D). In *Fecampia* and *Kronborgia* the adults lie in the haemocoel of their crustacean host (crabs, prawns, amphipods, isopods) where their activities may adversely influence organ function (Bellon-Humbert 1983) or even cause sterility or death on exit from the body (Christensen and Kanneworff 1965). There is no alimentary system, but characteristic of the adult worms is the rich and diverse glandular mix found in the body wall. This becomes significant when the mature worm leaves the host for it soon secretes a cocoon about itself, sometimes several centimetres long and attached to a stone or sedentary organism (Fig. 3.2C). As it begins to fill the cocoon with egg capsules the adult worm shrinks. In the case of *Kronborgia* the male is very much smaller than the female and enters the cocoon later to fertilise her. Juveniles (larvae) hatch and penetrate a new host.

In summary, the turbellarians are responsible for negligible economic losses, and most cause little pathology. They occur in all seas, but mainly in invertebrate hosts making them principally of interest to the curious zoologist. They do, however, straddle the divide between predation and parasitism, between independent and dependent lives and therefore present opportunities to explore the nature of symbioses: to ask what constitutes parasitism? What holds associations together?

Important references
The most comprehensive early account of parasitism in the turbellarians is given by Bresslau (1933); Hyman (1951) and de Beauchamp (1961) provide some further data.

Monogenea Polyopisthocotylea (ectoparasitic flukes)
Craig Hayward

Introduction
The Polyopisthocotylea (or Heteronchoinea) is a large group of monogeneans, with about 1000 described species. It is the sister group of the monopisthocotyleans, both groups included in the Class Monogenea (or Monogenoidea). Major characteristics of marine representatives of these two groups are compared in Table 3.2. Typical hosts of both groups are fishes, but they are also known from several other groups of (freshwater) vertebrates. Polyopisthocotyleans have also been reported in a few cases from invertebrate hosts – copepods and isopods (Table 3.2). Members of both groups have direct life cycles (i.e. they infect only a single host). Both groups typically live on the external surfaces of their hosts, and attach using a unique disc-like haptor. Both groups have high host specificity; a few exceptions shed light on the nature of this specificity. Polyopisthocotyleans occur in all seas of the world, from the littoral zone to open oceanic waters, from the poles to the tropics, and from surface waters to the depths of the sea. However, because of their generally high host specificity, the distribution of particular species tends to correspond closely to the distribution and migratory abilities of populations of their particular hosts. Polyopisthocotyleans seem to be less common on tropical reef fishes than in fishes in other warm marine waters, perhaps because they are more often removed there by cleaning organisms.

Morphology and diversity
Although about 1000 species have been described, the true number is probably several times as high. The monophyly of the group is well supported by analyses of morphological, molecular

Table 3.2 Comparison of major characteristics of marine Polyopisthocotylea and Monopisthocotylea

Characteristic	Marine Polyopisthocotylea	Marine Monopisthocotylea
No. of species described	~800	~1000
Microhabitat	Gills, oral cavity	'Skin', fins, gills, oral cavity, cornea, nasal tissue, urogenital system (cloaca, rectal gland, oviducts), body cavity. *Exceptionally*: gut, heart musculature and circulatory system
Major hosts	Chondrichthyes, Teleostei	Agnatha, Chondrichthyes, Teleostei
Minor hosts	Copepods, isopods	Copepods, squid
Diet	Blood	Epithelial cells
Gut structure	Two cell types	Single cell type
Genitointestinal canal	Present	Absent
Attachment structures	Multiple clamps	Hamuli, haptoral suckers, tiny additional sclerites, adhesives
Movement	Most sedentary	Most mobile
Haptoral symmetry	Symmetrical or asymmetrical	Symmetrical

and spermatological characters (Mollaret et al. 2000). The only large non-marine group, the Polystomatoinea, is sister group to all other polyopisthocotyleans, collectively known as the Oligonchoinea (Olsen and Littlewood 2002). The most recent comprehensive treatment of oligonchoineans was over 40 years ago by Yamaguti (1963). The classification in Table 3.3 follows the revision by Lebedev (1995). This system generally agrees with recent phylogenetic analyses based on morphologies and molecules. (An exception among marine groups is rejection of the inclusion of Plectanocotylidae in Mazocraeinea – Jovelin and Justine 2001).

Most of the 800 or so described oligonchoineans are marine. Basal representatives in order Chimaericolidea (Fig. 3.3A) infect holocephalans (chimaeras); and Diclybothridea infect chondrichthyans (sharks and rays) (Fig. 3.3B), catadromous and freshwater acipenseriforms (sturgeons and paddlefishes), and holocephalans (Table 3.3). These form sister groups to representatives that infect various orders of teleosts (ray-finned bony fishes), in Orders Pterinotrematidea (Fig. 3.3C) and Mazocraeidea (Fig. 3.3D–G). Within the latter, a family that is parasitic mainly on (the relatively basal) clupeiform fishes – the Mazocraeidae (Fig. 3.3D) – are thought to be the sister group of all other Mazocraeinea (Mollaret et al. 2000). Among the latter, the Gastrocotylinea (Fig. 3.3F) are sister group of an unresolved group, which includes the Discocotylinea (Fig. 3.3E) and Microcotylinea (Fig. 3.3G) (Jovelin and Justine 2001).

Recent revision of several groups indicates that when characterising the shape and appearance of species (including the number and size of their clamps, and other hard sclerites), worm age and host size have often been neglected. When identifying species, characters based on soft tissues (e.g. total length, and number of testes visible) are of secondary importance, due to the highly variable states of contraction in the body of fixed specimens.

Cryptic species are suspected to exist in some groups. For example, as is common among related species of polyopisthocotyleans, populations of *Hexostoma* (Hexostomatidae) are segregated on different gill arches of individual tunas (*Thunnus* spp.), although reproductive isolation has not yet been confirmed genetically. However, in one exceptional case, a species of

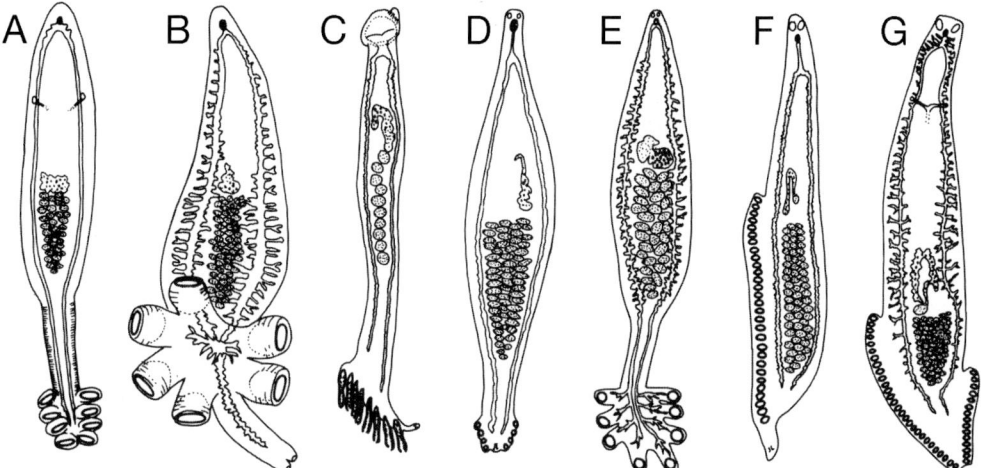

Figure 3.3 Range of body forms among marine Polyopisthocotylea. **A.** *Chimaericola leptogaster* (Chimaericolidae). **B.** *Rajoncocotyle emarginata* (Hexabothriidae). **C.** *Pterinotrema mirabile* (Pterinotrematidae). **D.** *Mazocraes alosae* (Mazocraeidae). **E.** *Neoheterobothrium affine* (Diclidophoridae). **F.** *Gastrocotyle indica* (Gastrocotylidae). **G.** *Zeuxapta seriolae* (Heteraxinidae). Note differences in the numbers and arrangement of clamps. Redrawn and modified from: Systema Helminthum, Volume IV, Monogenea and Aspidocotyla, Yamaguti S, Copyright © 1963, John Wiley & Sons, Inc. This material is used by permission of John Wiley & Sons, Inc.

Table 3.3 Diversity of Oligonchoinea (modified from Lebedev 1995) and their host groups

Order	No. of Families	Species	Host fishes
Chimaericolidea	1	3	Holocephali (chimaeras)
Diclybothridea	2	57	Elasmobranchii (sharks, rays), Acipenseriformes (sturgeons), Holocephali
Pterinotrematidea	1	2	Albuliformes (bonefishes)
Mazocraeidea	30	703	
Superf. Mazocraeinea	4	116	Clupeiformes (herrings), Gadiformes (cods), Myctophiformes (lanternfishes), Perciformes (perch-like fishes), Scorpaeniformes (scorpionfishes)
Superf. Discocotylinea	6	187	Gadiformes, Myctophiformes, Perciformes, Pleuronectiformes (flatfishes), Tetraodontiformes (puffers), Cypriniformes (carps), Salmoniformes (salmons)
Superf. Gastrocotylinea	10	128	Clupeiformes, Gadiformes, Perciformes, Pleuronectiformes, Tetraodontiformes, Anguilliformes (eels), Aulopiformes (grinners), Beloniformes (needlefishes), Syngnathiformes (pipefishes)
Superf. Microcotylinea	10	272	Beloniformes, Perciformes, Pleuronectiformes, Scorpaeniformes, Syngnathiformes, Beryciformes (sawbellies), Gasterosteiformes (sticklebacks), Atheriniformes (silversides)
Total	34	765	

Microcotyle (Microcotylidae) appears to have very low specificity, as it has been recorded from teleosts belonging to five different orders; populations so far tested from two host orders have matching or very closely matching DNA sequences (ITS2 region), and morphological characters also appear to be indistinguishable.

Sexually mature and relaxed marine polyopisthocotyleans range from a few millimetres to 4 cm long in Chimaericolidae. Most species attach exclusively to the gills of their host fishes, and are sedentary. Some Diclidophoridae inhabit surfaces of gill arches and arthropods on the gills (*Choricotyle* spp.), or migrate from gill filaments to embed their haptor in muscles lining the wall of the operculum (*Neoheterobothrium affine*, *Heterobothrium okamotoi*) or within gill arches (*Heterobothrium elongatum*).

Marine Polyopisthocotylea feed on blood of host fish. The mouth is terminal or subterminal, and opens into an oral cavity that may be suctorial (e.g. Hexabothriidae) or contains two buccal organs (Mazocraeidea). The buccal complex has both ciliated and non-ciliated sensory receptors; their function is unknown, but they may aid mating, feeding and site selection. Eyespots are present in larvae but disappear from mature worms (except Diclybothridae). The pharynx is relatively small and muscular (absent in Pterinotrematidae, Fig. 3.3C). The intestine divides into two blind-ending branches (caeca); in some species, these fuse posteriorly. Many species with larger body sizes have smaller lateral branches along the caeca, which increases their surface area; in Hexostomatidae, these fuse to form an extensive reticulated net. The intestine is lined with digestive cells that are separated and supported by a syncytial connective tissue. These intestinal cells digest blood proteins intracellularly. Haematin-rich residues from this process accumulate in vacuoles, giving polyopisthocotyleans their characteristic dark-brown appearance (Kearn 1998). Haematin is released into the gut lumen periodically, and then egested from the mouth. A unique structure, the genitointestinal canal, may permit blood-digesting organisms to pass from the intestine through the oviduct to colonise developing eggs (Kearn 1998).

Clamps for attachment (Fig. 3.4) develop in the haptor, usually soon after larvae settle on host gills. They grip lamellae, or other epithelial surfaces in the gill cavity, over a relatively broad area, and seldom provoke any tissue response. Clamps may be circular, muscular organs that attach by suction to flat host surfaces, or more complex structures containing many supporting skeletal elements that enable jaw-like clasping around three-dimensional surfaces of gill filaments. Worms with complex clamps tend to be more sedentary than those with clamps that act as suckers.

Clamps are supported internally by hard sclerites, which vary in number and articulation (Fig. 3.4). In the two most basal marine families (Chimaericolidae and Hexabothriidae), the clamps are 'open' and act as suckers. In Chimaericolidae, there are three supporting sclerites (Fig. 3.4A). In Hexabothriidae, a single curved sclerite develops in a groove in the roof of each clamp; an extrinsic muscle attached to this sclerite generates suction (Figs 3.4Bi, ii). This sclerite is unique in possessing a sharp point, which appears to aid attachment by piercing gill epithelia. In remaining groups, all from teleosts, clamps are more complex, and contain two opposable, sclerite-supported jaws that usually clamp onto secondary gill lamellae. Clamps of Pterinotrematidae are elongate and contain feather-like spines (Fig. 3.4C); those of Mazocraeidae contain varying numbers of robust sclerites (Fig. 3.4D). Clamps of Discocotylinea contain five rod-like supporting sclerites; these may be modified in some families by thickenings, fusion into plates, or the presence of additional sclerites. In one such family, Diclidophoridae, there are eight or more clamp sclerites (Fig. 3.4Ei); clamping is controlled by a central diaphragm that acts on the contraction of an attached extrinsic muscle (Fig. 3.4Eii). Clamps of Plectanocotylidae close by an external muscle–tendon–fairlead system (Fig. 3.4F). Clamps of the two remaining groups also contain five basic sclerites: in Microcotylinea, clamps (Fig. 3.4G) are numerous; and in

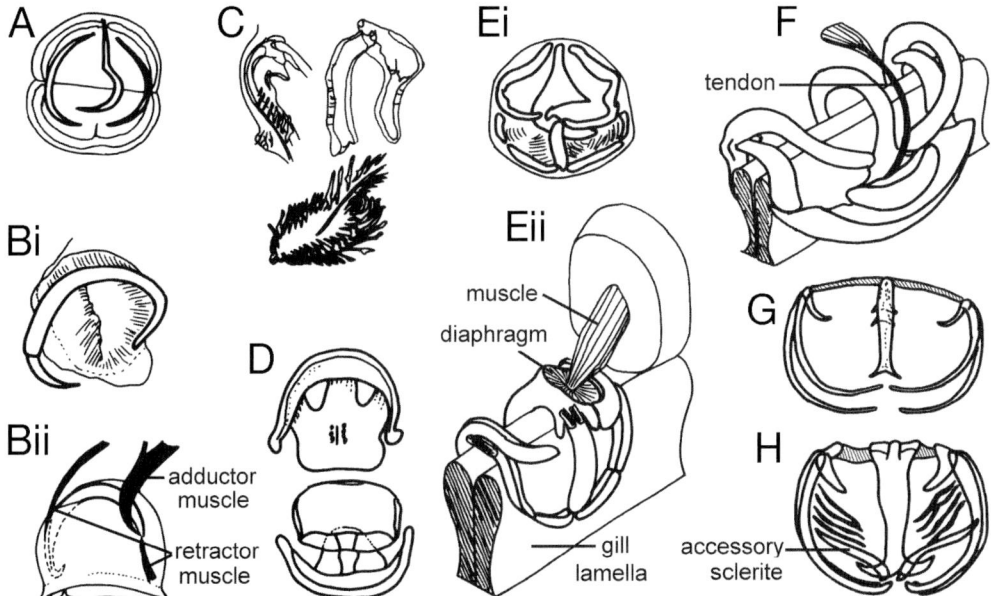

Figure 3.4 Clamp structure and attachment among representative marine Polyopisthocotylea. **A.** *Chimaericola leptogaster* (Chimaericolidae). **Bi.** *Erpocotyle callorhynchi* (Hexabothriidae). **Bii.** Three-dimensional view of a hexabothriid clamp (Hexabothriidae). **C.** *Pterinotrema mirabile* (Pterinotrematidae). **D.** *Mazocraes alosae* (Mazocraeidae). **Ei.** *Neoheterobothrium affine* (Diclidophoridae) **Eii.** Three-dimensional view of clamp of *Diclidophora* (Diclidophoridae) grasping two secondary lamellae. Musculature of clamps not shown. **F.** Three-dimensional view of clamp of *Plectanocotyle* (Plectanocotylidae) grasping two secondary lamellae. **G.** *Heteraxinoides chinensis* (Heteraxinidae). **H.** *Thoracocotyle crocea* (Thoracocotylidae). A, D and Ei redrawn and modified, with permission, from: Bychowsky BE, 1957, Monogenetic Trematodes, their Classification and Phylogeny. Academy of Sciences, USSR, Moscow (in Russian). English translation by Hargis, WJ and Oustinov, PC, 1961 (Copyright, American Institute of Biological Sciences). Bi, C, G and H redrawn and modified from Systema Helminthum, Volume IV, Monogenea and Aspidocotyla, Yamaguti S, Copyright © 1963, John Wiley & Sons, Inc. This material is used by permission of John Wiley & Sons, Inc.; Bii, Eii and F redrawn and modified from *International Journal for Parasitology* **8**, Kearn G, Evolutionary expansion of the Monogenea, pages 1227–1271, Copyright (1994), with permission from Cambridge University Press and Elsevier Science.

Gastrocotylinea, each clamp possesses a pair of accessory sclerites in the distal end (Fig. 3.4H), and other rib-like sclerites are also often present.

The number of haptoral clamps in most groups is fixed at three or four pairs. In a few groups (*Polycliphora* of Diclidophoridae; several families of Gastrocotylinea; and Microcotylinea), clamps are more numerous, ranging from tens of pairs, to over 200 pairs. However, in a given individual of such species, the clamp numbers are plastic, and depend on a combination of several factors, including worm age, water temperature and host size. In at least some cases, the size of the host (and hence of gill lamellae) also determines the maximum size reached by clamps. Such feedback probably increases the likelihood of firm attachment to secondary gill lamellae.

The haptors of most groups are symmetrical, with equal numbers of clamp pairs on both sides. Various groups depart from this basic plan in several ways. Clamps may be present in unequal numbers (e.g. Heteraxinidae), or occur all or mostly on one side (e.g. Axinidae), or be larger on one side than the other [e.g. *Bicotyle* (Axinidae)]. In several groups, some or all clamps are modified; in Anchorophoridae, for example, most or all clamps have been transformed into sclerotised anchor complexes. In a few groups, clamp numbers are reduced; for example, clamps number only two pairs in *Quadrivalvia* (Gastrocotylidae), and have been lost altogether in *Lethacotyle* (Protomicrocotylidae). Clamps may occur on pronounced stalks in some Diclidophoridae (Fig. 3.3E); haptors in Hexabothriidae bear an appendix with a pair of suckers (Fig. 3.3B).

Each adult contains male and female reproductive organs. The testes are usually numerous, and the ovary single and usually tubular. The male copulatory organ may be a muscular penis, a sclerotised tube, or an eversible cirrus (either naked or with spines). Vaginal structures are often present, and may be single or paired, and naked or armed with sclerites. Copulation, which has been seldom observed, transfers sperm by intromission of the male copulatory organ into the vagina; in those species lacking vaginal structures, sperm is injected into the tegument hypodermically.

Marine species can be identified to family and higher-level groups by the number and arrangement of their clamps, and the number and shape of sclerites present within them. In some cases, these characters can also be used to identify worms to species level (e.g. in Gotocotylidae). In families with more or less uniform clamp structure, related species may be distinguished by the shape of reproductive sclerites, particularly the male copulatory organ.

Life cycle and infection mechanisms

Marine polyopisthocotyleans have simple, direct life cycles (i.e. they infect only a single host) (Fig. 3.5A). Two species are exceptional in that they may use optional intermediate hosts: *Gotocotyla acanthura* (Gotocotylidae) and *Pricea multae* (Thoracocotylidae) both mature on Spanish mackerels (*Scomberomorus* spp.), but various small pelagic fishes (belonging to at least eight Orders), which are preyed on by these mackerels, are also infected with immature *G. acanthura* and *P. multae*.

Polyopisthocotylean eggs may be deposited continuously, or at night. Eggs are spheroid or fusiform; the shells of most are rigid and have lids (Kearn 1986). Eggs also usually have filaments at one or both poles. These appendages may be short and needle-like, or long and flexible (Fig. 3.5A); eggs may also form one long continuous string, and may entangle with each other (Kearn 1986). Polyopisthocotylean eggs have been found drifting in plankton, but most probably sink to the sea bottom.

Embryos develop for one to several weeks, and then hatch into small, hook-bearing larvae (less than 300 µm) known as oncomiracidia (Fig. 3.5B). Cues to hatching are variable, and may include time of day (diurnal rhythm), or responses to stimuli such as shadows, mechanical disturbances, and host mucus or tissue (Whittington *et al.* 2000). Oncomiracidia are short-lived, and usually survive less than three days (although some larval Diclidophoridae survive more than nine days). They have many cilia and swim actively, but groups vary in the number and distribution of cilia (Whittington *et al.* 2000). Most swim erratically, and many also rotate. Compared with their hosts, they are also relatively slow swimmers (1–5 mm/s) (Whittington *et al.* 2000). Fast-swimming pelagic hosts are probably infected when schools rest in deeper waters at night.

Cues to larval behaviour leading to settlement on gills vary from species to species, and again depend upon the biology of the target host (Whittington *et al.* 2000). Pigment-shielded eyespots occur in most larvae, usually as one pair (which may be fused together – Fig. 3.5B), rarely as two, or are absent. Other sensory receptors are known (e.g. uniciliated and multiciliated sensilla) but their functions are unknown. Oncomiracidia may respond not only to light (phototaxis), but also to gravity (geotaxis), host compounds (chemotaxis), water currents (rheotaxis), or some combination of these. Some oncomiracidia possess a terminal globule in their haptor (Fig. 3.5B), which contains crystal-like inclusions in large vacuoles. The function of this structure is unknown, but it may detect magnetic stimuli.

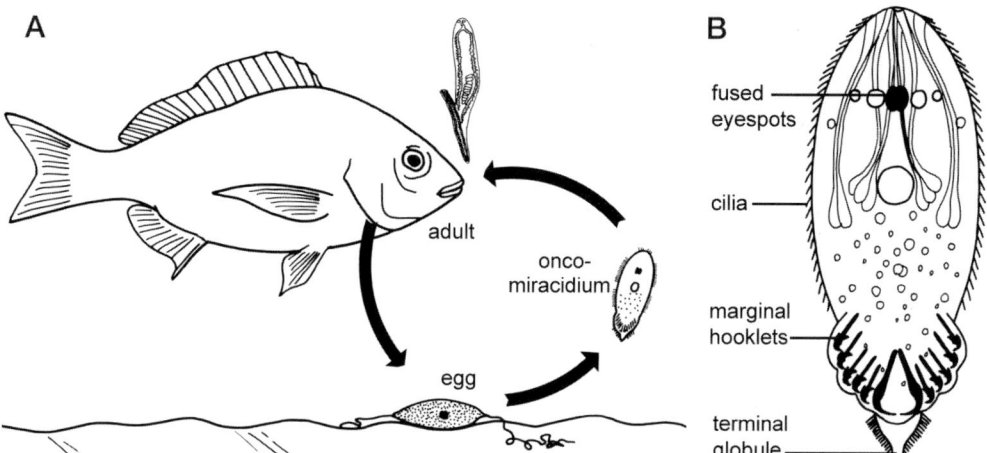

Figure 3.5 **A.** Life cycle of *Polylabroides multispinosus* (Microcotylidae). Adult parasites on gills of bream lay eggs that sink to the sea bottom. At 24°C, eggs hatch after about 7 days. Oncomiracidia use cilia to swim; after finding and attaching to gills of bream, eggs are produced 20 days later. Generation time is about one month at 24°C. Data source: Roubal FR, Diggles BK (1993) The rate of development of *Polylabroides multispinosus* (Monogenea: Microcotylidae) parasitic on the gills of *Acanthopagrus australis* (Pisces: Sparidae). *International Journal for Parasitology* **23**, 871–875.
B. Oncomiracidium of *Plectanocotyle gurnardi*. Redrawn and modified from *Advances in Parasitology*, Volume **44**, Whittington ID, Chisholm LA and Rohde K, The larvae of Monogenea (Platyhelminthes), pages 139–232, Copyright (2000), with permission from Elsevier.

Oncomiracidia enter the gill chamber in the respiratory current of the host. They then settle on gill epithelium using five to eight pairs of tiny (less than 15 µm) marginal hooklets. Other sclerites in the haptors of larvae may also be used for initial attachment; these vary widely, but one or more pairs of hooks, known as hamuli, are usually present. Many larval structures are shed before maturity (cilia, hooklets, hamuli and terminal globule), although hamuli persist in adults of some groups. After initial attachment, juveniles may migrate to particular regions (microhabitats) of the gills or pseudobranchs.

Generation times and longevities of most monogeneans are unknown, or poorly known. The rate of development is inversely related to sea temperature, in those species investigated so far. In one example, the life cycle of *Polylabroides multispinosus* (Microcotylidae) (Fig. 3.5A) can be completed in about one month at 24°C. Individuals reach sexual maturity 20 days after larvae settle on host bream (*Acanthopagrus australis*), and adults probably continue to lay eggs for several months.

Effects on hosts and ecological importance

Polyopisthocotyleans are likely to be of moderate ecological significance in the wild: there are many species, they have high host specificity, but typically they do not cause serious pathological effects.

Most polyopisthocotyleans show strong host specificity for one or a few closely related fishes. In part, this is because the swimming activity of the larva of each species is specialised. Larvae congregate at certain times and positions in the water column in response to environmental cues, and this maximises the chance of encountering a particular fish species. Some larvae may also be attracted to tissues of specific hosts, but results of the few experiments to date are ambiguous (Whittington *et al.* 2000). Instead, a second critical factor in the apparent selectivity of polyopisthocotyleans seems to be fish immunity. The 'wrong' host fishes may produce substances that harm larvae or juveniles. Indirect evidence of the involvement of immunity in host specificity is the phenomenon of enlarged host range in culture conditions. For example, *Polylabris tubicirrus* (Microcotylidae) usually infects several sparid fishes of the genus *Diplodus* in the wild, but has been recorded on another sparid genus (*Sparus aurata*) only when it is held in captivity. During such captivity, assorted stressors compromise host immunity.

In temperate latitudes, the population dynamics of at least some polyopisthocotyleans are clearly seasonal. In most cases, worms are more abundant in summer. This may be because high water temperatures speed up worm metabolism and reproductive rate. At the same time, the immune defenses of many fish species may be overtaxed during prolonged heat. Two Microcotylidae (*Bivagina pagrosomi* and *B. tai*) are exceptional in that they are more common on young snappers (*Pagrus auratus*) in winter months. This unusual phenomenon might be related to the immune status of these snappers, which could be seasonally depressed in their first winter.

In most species of polyopisthocotyleans, worms tend to aggregate on particular individual fishes. The total intensities usually range from 1 to 20 worms, and are probably limited by a combination of: how many oncomiracidia the host individual was exposed to; the immune status of the individual; and whether or not the individual visits cleaning stations.

Polyopisthocotyleans are considered benign to wild hosts. An exception is *Neoheterobothrium affine* (Diclidophoridae), which has been associated with mortalities of olive flounder (*Paralichthys olivaceus*) in the waters around most of Japan since the mid 1990s. This worm was also detected in olive flounder in neighboring Korea, in 2000. This worm may have been introduced to the region in imports of live American flounder (*P. dentatus*). However, the worm was recorded in the Bering Sea over four decades ago on Kamchatka flounder (*Reinhardtius evermanni*), so it seems likely that it spread southwards to Japan naturally on this host.

In contrast with wild populations, farmed marine fishes are usually stressed by conditions under which they are maintained such as being crowded together; being fed diets that contain fewer nutrients than live, natural foods; and low water quality. If stress is ongoing, the resistance of fish to disease soon weakens. At the same time, the most frequently encountered parasites on such confined fishes tend to be those with direct life cycles, and Polyopisthocotylea are a major group of these. Furthermore, the eggs of many polyopisthocotyleans have long filaments that easily entangle on the nets of marine cages. Oncomiracidia hatching from these eggs need travel only a short distance to find a suitable host, and so their populations may build rapidly.

Some farmed fishes are known to respond to polyopisthocotyleans by raising levels of serum antibodies. For example, elevated levels of antibodies have been detected in tiger puffer (*Takifugu rubripes*) infected with *Heterobothrium okamotoi* (Diclidophoridae). Some fish have also been partially immunised against infection, as indicated by significantly reduced worm burdens; for example, *Microcotyle sebastis* (Microcotylidae) in rockfish (*Sebastes schlegeli*).

Blood-feeding by large numbers of polyopisthocotyleans is thought to induce anaemia. For example, *Zeuxapta seriolae* (Heteraxinidae) has been held responsible for anaemia in cultured kingfishes (*Seriola* spp.) for over 40 years. Many of these worms would cause anemia. However, in aquarium conditions, large numbers of polyopisthocotyleans are not necessarily associated with disease. Thus, rockfish (*Sebastes melanops*) heavily infected with *M. sebastis* showed no external signs of stress or pathology in an aquarium. This finding may indicate that the resistance of captive fishes must be compromised on more than one level to induce illness. After the first protective barrier is broken, polyopisthocotyleans may settle and multiply, but it seems that a second level may also need to be breached before fish weaken and die. If the stresses of farming still leave fish susceptible to other diseases, the aim of minimising disease by controlling worm numbers (with chemicals or by vaccination) will ultimately fail.

A few polyopisthocotyleans cause irritation to hosts because their clamps have unusual modes of attachment. The haptor of *Hexostoma* spp. (Hexostomatidae) attaches to the gills of tunas

(*Thunnus* spp.) over secondary lamellae, compressing them strongly; surrounding tissue swells markedly, and this appears to further anchor the haptor in place. *Neoheterobothrium affine* (Diclidophoridae) migrate from gills to the opercular wall in olive flounder (*Paralichthys olivaceus*), and embed their clamps deeply within muscle, leading to necrosis and infiltration of inflammatory cells and fibroblasts. In a final example, the piercing attachment of clamps of *Erpocotyle tiburonis* (Hexabothriidae) leads to gill hyperplasia in captive sharks (*Sphyrna tiburo*).

Important references
The most recent treatise on Polyopisthocotylea was published by Yamaguti (1963). The composition of groups has changed significantly since then (see Lebedev 1995). An important earlier monograph is by Bychowsky (1957, with English translation in 1961).

Monogenea Monopisthocotylea (ectoparasitic flukes)
Ian D Whittington

Introduction
It is unfortunate that even among specialists there is disagreement over higher taxon names for 'Monogenea'. Names in use are Monopisthocotylea (=Polyonchoinea), considered to be a subclass of Class Monogenea (=Monogenoidea) and sister group to the other monogenean subclass, the Polyopisthocotylea (=Heteronchoinea) (see pp. 55–63, and Boeger and Kritsky 2001). Polyopisthocotyleans tend to parasitise teleost gills, are relatively stationary feeding on blood and possess an attachment organ (haptor) bearing several to numerous grasping units in the form of sclerotised clamps (see Table 3.2, Figs 3.3, 3.4). Monopisthocotyleans, in contrast, infect a diversity of microhabitats across many fish groups, can move readily, feed on epithelial cells and possess a haptor forming a single attachment unit (see Table 3.2, Fig. 3.7). About 2500 Monopisthocotylea species have been described. The approximately 1000 marine monopisthocotylean species are classified as belonging to about 170 genera and 16 families, but >10 000 species may exist worldwide.

Morphology and diversity
Monopisthocotyleans can be grouped at the level of Order (e.g. Capsalidea), superfamily (e.g. Capsaloidea) and family (e.g. Capsalidae). Specialist opinions differ, but most consensus occurs at family level. They principally infect 'external' surfaces ('skin', fins, head, gills, oral cavity, see Table 3.2) of fishes including Agnatha (hagfish, Fig. 3.6A), Chondrichthyes (chimaeras, sharks, batoids, Figs 3.6B–D) and teleosts including primitive sturgeons (Figs 3.6E, F). Still considered ectoparasitic, some marine monopisthocotyleans may inhabit *internal* sites that possess external openings including nasal tissue, urogenital system and body cavity of Chondrichthyes parasitised by Monocotylidae and the digestive system of teleosts parasitised by some Dactylogyridae and the Montchadskyellidae (see Table 3.2). *Exceptionally*, heart musculature and circulatory system of batoids can be parasitised by Amphibdellatidae (see Table 3.2). Few marine invertebrates are infected by Monopisthocotylea; for example, *Alloteuthis* squid species (Fig. 3.6G) by Gyrodactylidae (*Isancistrum* species) and parasitic caligid copepods (Fig. 3.6H) by Udonellidae (see Table 3.2). The latter relationship is likely to be phoretic.

Marine Monopisthocotylea range from 300 µm (some Gyrodactylidae, e.g. Fig. 3.7F) to >3 cm [*Capsala martinieri* (Capsalidae, e.g. Figure 3.7C)] long. A unique feature of Monogenea is the *posterior* haptor (Figs 3.7, 3.9A–C, E), the principal organ for attachment. Unlike Polyopisthocotylea, the haptor of Monopisthocotylea comprises a *single*, symmetrical attachment unit (compare Figs 3.3 and 3.7). Most adult monopisthocotyleans (Fig. 3.7) have proteinaceous

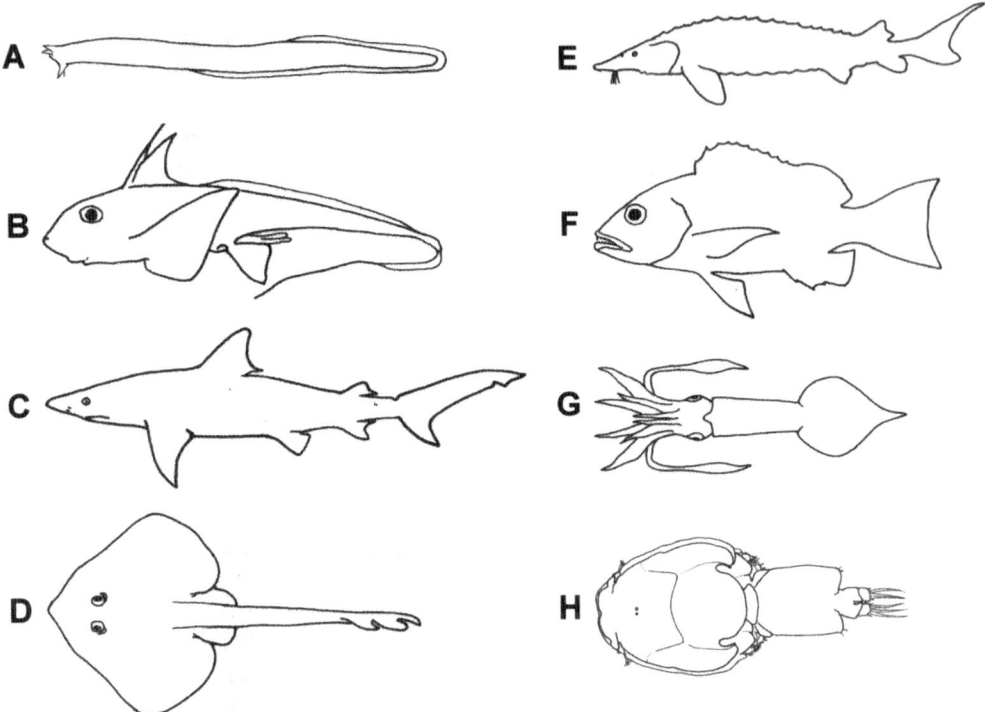

Figure 3.6 Host range of marine Monopisthocotylea Monogenea. **A**. Agnatha (jawless fish). **B–D**. Chondrichthyes. **B**. Chimaeras. **C**. Sharks. **D**. Batoids or rays. **E–F**. Teleosts (ray-finned bony fish) **E**. Sturgeons. **F**. Modern teleosts. Invertebrate hosts (**G–H**) are rare. **G**. Squid. **H**. Caligid copepods. Not drawn to scale.

sclerites in the form of 14 or 16 small (<15 μm) hooklets usually around the haptor periphery for attachment to host epithelial cells. Larvae use haptoral hooklets (e.g. Fig. 3.8D) to attach to a newly invaded host. The adult haptor often also has one or two pairs of larger hooks called hamuli (Figs 3.7A, B, F, H, I, 3.9A, B) for firm attachment to host epithelial cells. Adults in some families may possess additional tiny sclerites resembling teeth, studs and/or spines for extra grip (Figs 3.7B, G, I, 3.9E). Arrangement, morphology and origin of sclerotised haptoral elements are important for classification, systematics and identification (Fig. 3.7). The monopisthocotylean haptor can: form a muscular sucker supplemented by sclerites (e.g. Capsalidae, Fig. 3.9A); be divided into partitions called loculi each able to generate suction (e.g. Monocotylidae, Figs 3.7B, 3.9E; some Capsalidae, Fig. 3.7C); have gland cells that secrete adhesives (e.g. Microbothriidae, Fig. 3.7D; Udonellidae, Fig. 3.7E; Anoplodiscidae); comprise sclerites and adhesives combined (e.g. Dactylogyridae, Fig. 3.7H). Unlike Polyopisthocotylea, Monopisthocotylea have *no* haptoral clamps (see Table 3.2).

Monopisthocotylea also possess *anterior* organs for attachment to hosts. Anterior attachment organs may comprise: different configurations and numbers of gland duct openings that secrete adhesives (Fig. 3.7); a muscular mouth forming an oral sucker-like structure ('pseudosucker'; Fig. 3.7B); saucer-like suckers (Fig. 3.7C); weak muscular grooves (Fig. 3.7D); structures for adhesion and suction (Fig. 3.7B, C). Alternate use of haptor and anterior attachment organs permits leech-like movement over host epithelial cells despite strong shear forces from water flow as hosts swim and respire. Whittington and Cribb (2001) reviewed temporary attachment to wet, slimy epithelial surfaces by adhesion.

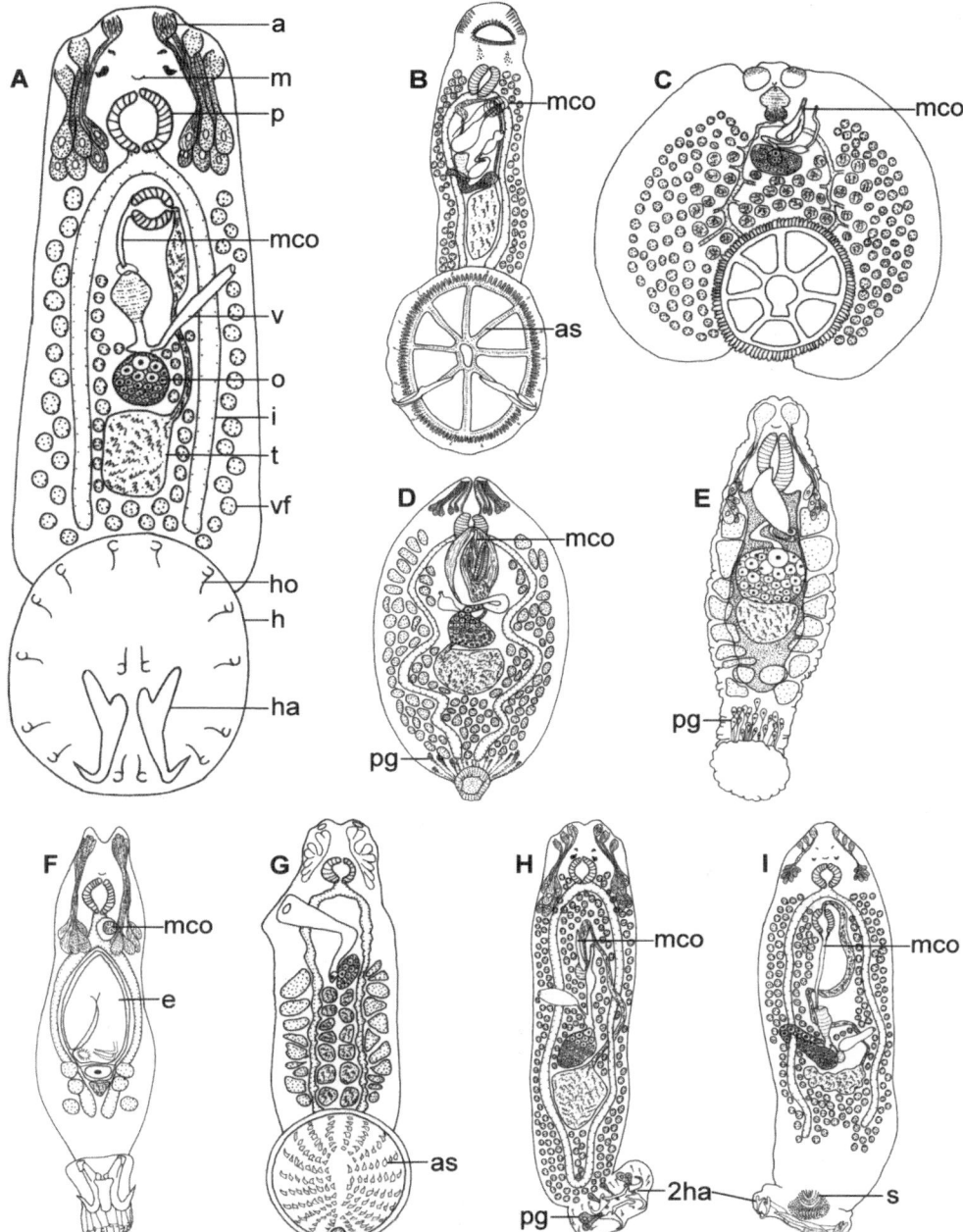

Figure 3.7 Marine Monopisthocotylea Monogenea diversity represented by eight families.
A. Generalised form illustrating major features. **B.** Monocotylidae (based on *Monocotyle* species).
C. Capsalidae (based on *Capsala* species). **D.** Microbothriidae (based on *Leptocotyle* species).
E. Udonellidae (based on *Udonella* species). **F.** Gyrodactylidae (based on *Acanthoplacatus* species).
G. Acanthocotylidae (based on *Acanthocotyle* species). **H.** Dactylogyridae (based on *Haliotrema* species). **I.** Diplectanidae (based on *Diplectanum* species). Not drawn to scale. Abbreviations: a, anterior attachment organ; as, many additional haptoral sclerites; e, developing embryo; h, haptor; ha, hamulus (note two hamulus pairs in H. and I.); ho, hooklet; i, intestine; m, mouth; mco, male copulatory organ; o, ovary; p, pharynx; pg, posterior glands; s, squamodisc; t, testis; v, vagina; vf, vitelline follicle.

The incredible diversity of Monopisthocotylea in form, function and structure of haptor and anterior attachment organs is matched by broad diversity in body plan (Fig. 3.7). An anterior, subterminal mouth connects to a muscular pharynx provided with intrinsic and/or extrinsic gland cells. The intestine usually divides into two blind caeca, one on each side of the body (Figs 3.7A, B, D, F, G, I) although a confluent gut (Dactylogyridae, Fig. 3.7H) and highly branched intestine (most Capsalidae, Fig. 3.7C; a few Monocotylidae) can occur; some have a single, blind caecum (Tetraonchoididae) or a sac-like gut (Udonellidae, Fig. 3.7E). Like Polyopisthocotylea, all Monopisthocotylea are hermaphroditic. Male reproductive organs are a single testis, but multiple testes are present in some Acanthocotylidae (Fig. 3.7G), Capsalidae (Fig. 3.7C), Microbothriidae and Monocotylidae. The male copulatory organ (MCO) may be a penis (muscular, protrusible organ), a cirrus (muscular, eversible organ) or other muscular organ types (spined 'bulb' of some Gyrodactylidae, e.g. *Acanthoplacatus*, Fig. 3.7F). The MCO can also be a sclerotised tube (Monocotylidae, Fig. 3.7B; Microbothriidae, Fig. 3.7D; Dactylogyridae, Fig. 3.7H; Diplectanidae, Fig. 3.7I). Structure, shape and size of the MCO are important for distinguishing species. Functional morphology, copulatory behaviour and insemination methods are known for few species. Female reproductive organs are a single ovary, either compact (most Capsalidae, Fig. 3.7C; Microbothriidae, Fig. 3.7D; Udonellidae, Fig. 3.7E; Acanthocotylidae, Fig. 3.7G; Dactylogyridae, Fig. 3.7H) or elongate, looping around the right gut caecum (e.g. most Monocotylidae, Fig. 3.7B; some Diplectanidae, Fig. 3.7I). A vagina is usually present, sometimes paired, perhaps bearing spines around the pore and can be sclerotised internally. The vitellarium, a large follicular organ, fills the body: fine ducts join the many vitelline follicles which produce vitelline cells that provide eggshell and nourishment to the developing embryo inside eggs.

Anywhere marine fish occur, at any depth from equatorial seas to polar oceans, Monopisthocotylea occur. Monogenea are among the most host specific of all parasites (Whittington *et al.* 2000a; see p. 71). Specificity can be so strict that one Monopisthocotylea species may be restricted to one marine fish species; for example, *Entobdella soleae* (Capsalidae) on *Solea solea* (Soleidae; see Kearn 1998). Strict host specificity means geographic distributions of Monopisthocotylea species may mirror geographic distributions of their specific host species. Geographic range, therefore, can be restricted, but some oceanic fish species with cosmopolitan distributions can be parasitised throughout their range; for example, *Mola mola* (Molidae) by *Capsala martinieri* (Capsalidae). Some Monopisthocotylea are specific at the host generic level, which can result in broad geographic distributions. *Benedenia seriolae* (Capsalidae) infects *Seriola* species (Carangidae) off Japan, Australia, Chile, New Zealand and Mexico and probably occurs wherever *Seriola* species occur.

Life cycle and biology

Life cycle

Marine Monopisthocotylea possess a single definitive, mostly fish, host (see Fig. 3.6 for exceptions) on which obligate adult parasites live. Most species lay eggs (oviparity, Fig. 3.8C). There is great diversity in egg shape (tetrahedral, Figs 3.8C, H, M, 3.9G; ovoid, Figs 3.8G, K, L; spherical, Fig. 3.8I) and size (for review, see Kearn 1986). When eggs hatch, a single, usually ciliated, larva called an oncomiracidium emerges. An exception to oviparity occurs in most Gyrodactylidae, which give birth to live young (viviparity; Fig. 3.8N). Relatively small numbers of tanned eggs relative to other parasitic platyhelminths are laid by fertilised adults and are normally deposited into the sea (Figs 3.8A–C). When eggs are fully embryonated (dependent on environmental variables e.g. temperature, salinity, light periodicity, but hatching may be stimulated by host-generated cues; see p. 67), a generally ciliated swimming larva hatches (Fig. 3.8D). The oncomiracidium must find and attach to the specific definitive host species (Fig. 3.8E) to complete the cycle.

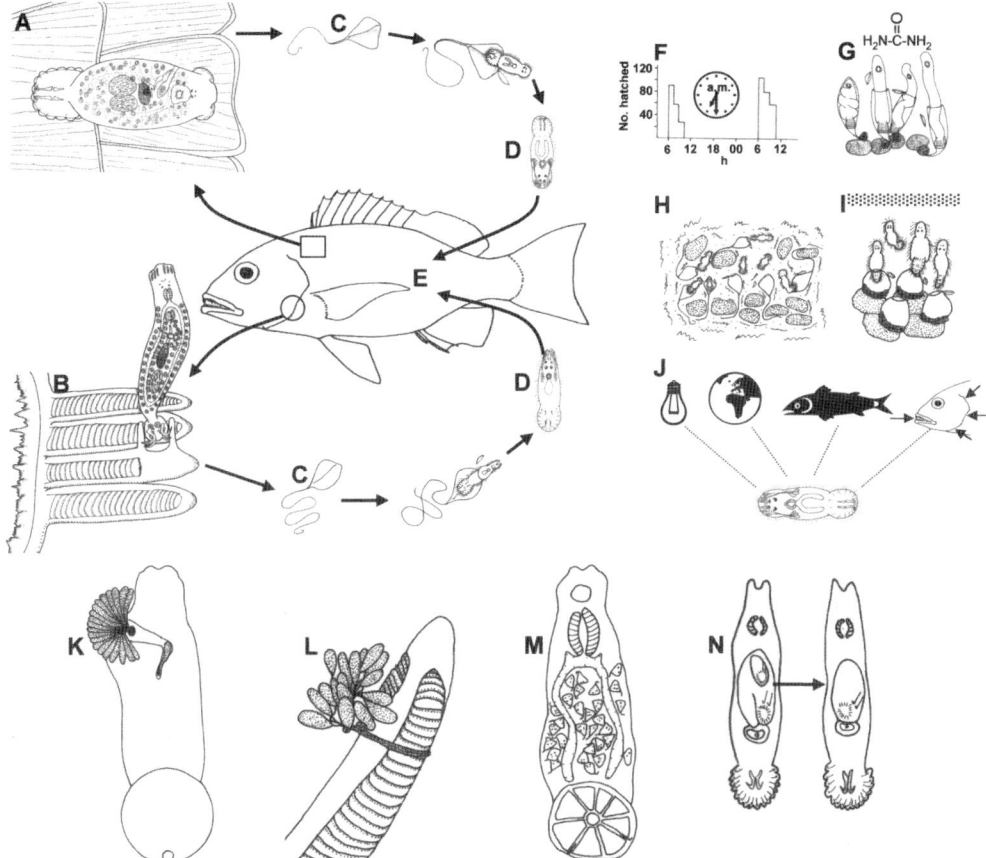

Figure 3.8 Generalised life cycle of marine Monopisthocotylea Monogenea. **A.** Adult parasite on 'skin'. **B.** Adult parasite on gills. Note: diagrammatic tip of primary gill lamella in **B.** shows attachment by a dactylogyrid or diplectanid-like Monopisthocotylea between adjacent secondary lamellae by two pairs of hamuli (see Fig. 3.9D). **C.** Parasites lay eggs deposited freely in the sea. **D.** The ciliated larva (=oncomiracidium) that hatches is the infective stage. **E.** Ciliated larva must locate and attach to the *specific* definitive host fish species. Several adaptations to this simple life cycle can promote infection success. Modifications include **F–N**. **F.** Rhythmical hatching. Stimulated hatching in response to host-generated cues – e.g. **G.** chemicals from skin secretions, **H.** mechanical disturbances and **I.** shadows. Various larval behaviours – e.g. **J.** (from left to right) phototaxis, geotaxis, chemotaxis and rheotaxis. **K.** Egg retention external to parasite. **L.** Eggs attached to host tissue. **M.** Egg storage inside parasite. **N.** Viviparity. Not drawn to scale. For further detail, consult Whittington *et al.* (2000b).

Strict host specificity places considerable significance on the mission of infective larvae. In the vast marine environment, oncomiracidia must find, attach to, and then establish themselves on an individual of their *specific* host species to complete the cycle. However most larval Monopisthocotylea are small (<250 μm), live for <48 h (longevity decreases with increased water temperature) and swim slowly (1–5 mm/s) relative to hosts. The simple life cycle (Figs 3.8A–E) masks many wonderful adaptations to parasitism, especially regarding the time of larval emergence (rhythmical hatching, Fig. 3.8F) or hatching stimulated by host-generated cues (skin secretions, Fig. 3.8G; mechanical disturbances, Fig. 3.8H; shadows, Fig. 3.8I). Furthermore, larval behaviours in response to environmental and host factors can direct oncomiracidia to the habitat of their specific host fish species (Fig. 3.8J). Post-larval migrations may occur after host infection when larvae shed their cilia. Rare exceptions to freely deposited eggs include: egg retention external to parasites (e.g. *Acanthocotyle*, Fig. 3.8K); eggs attached to host tissue (e.g. *Dioncus*, Fig. 3.8L); egg storage inside parasites (e.g. *Monocotyle multiparous*, Fig. 3.8M); no egg laid, but birth of unciliated juveniles (e.g. viviparity in Gyrodactylidae, Fig. 3.8N).

Although hermaphroditic, most Monopisthocotylea must find a partner with which to cross-fertilise, reproduction being sexual, except some Gyrodactylidae. Insemination is mostly mutual occurring by intromission, spermatophore exchange or hypodermic impregnation. Viviparous reproduction in Gyrodactylidae is unique in the Animal Kingdom and involves precocious growth of intra-uterine embryos so that newborn juveniles may contain up to three developing embryos telescoped inside each other (Fig. 3.8N) like toy Russian dolls! Early gyrodactylid embryos may be produced parthenogenetically (i.e. an oocyte may require no fertilisation by sperm). Viviparity in Gyrodactylidae significantly reduces generation time enabling rapid population growth on the same host individual. Gyrodactylids only spread to different host specimens by direct parasite transmission when hosts touch.

Biology
All Monopisthocotylea feed on host epithelial cells (see Table 3.2) except perhaps the poorly studied blood-inhabiting juveniles and recently emerged adults of *Amphibdella* (Amphibdellatidae) after living in the heart and circulatory system of batoids. Epithelial cells are available throughout the range of sites inhabited by Monopisthocotylea, including gills (see Table 3.2). In healthy fish, small wounds from monopisthocotylean attachment and feeding heal rapidly by migration and proliferation of undamaged epithelial cells from the injury edges (but see p. 70). During feeding, the pharynx is either everted through the mouth, then applied to host epithelial cells (e.g. Capsalidae) or tissue is sucked into the oral cavity (e.g. some Monocotylidae). Pharyngeal secretions disassociate and digest epithelial cells. The muscular pharynx then pumps detached epithelial cells and semi-digested 'epithelial soup' into the gut. Intracellular digestion occurs in the gut, which unlike Polyopisthocotylea, is lined by a single cell type. Monopisthocotylean gut contents are usually colourless or grey because undigested epithelial cells leave no pigmented residue unlike dark gut contents (=haematin) of Polyopisthocotylea. Some Monocotylidae (*Dendromonocotyle* species) incorporate host ray 'skin' pigment into their branched gut which may provide camouflage, concealing these skin parasites from cleaner organisms.

Detailed biology for few marine Monopisthocotylea is known, but the best-researched species, *Entobdella soleae* (Capsalidae) from 'skin' of a European soleid flatfish teleost, is increasingly cited in parasitology texts as 'typical' (for details, see Kearn 1998). A close relative, *Benedenia lutjani* (Capsalidae), infects 'skin' and fins of a round-bodied lutjanid species off the Great Barrier Reef (Whittington and Ernst 2002) and is used to highlight briefly the biology of a 'skin'-parasitic Monopisthocotylea. *Benedenia lutjani* eggs hatch only during daylight, larvae perhaps avoiding nocturnal predation by filter feeders. Oncomiracidia cannot feed until they attach to a suitable host. Larvae of *B. lutjani* invade anywhere on the lutjanid's external surfaces and migrate, leech-like, to the pelvic fins. Converging on pelvic fins ensures subadults that have developed a functional male reproductive system by protandry, before the female system is functional, have opportunities to exchange sperm with congeners of similar age. Protandry, common in Monopisthocotylea except Gyrodactylidae, allows sperm exchange until sexual maturity (=egg-laying) is reached. Egg-laying adult *B. lutjani* occur on the branchiostegal membranes between eight and 14 days after larval invasion. Different microhabitats for juveniles and adults may segregate demand for food and space resources (Whittington and Ernst 2002).

Many families of Monopisthocotylea have invaded the gill chamber of teleost fishes with relatively minor modifications to their basic ground plan (e.g. Monocotylidae, Capsalidae, Gyrodactylidae, see Kearn 1994). Families specialised for gill parasitism include Dactylogyridae (e.g.

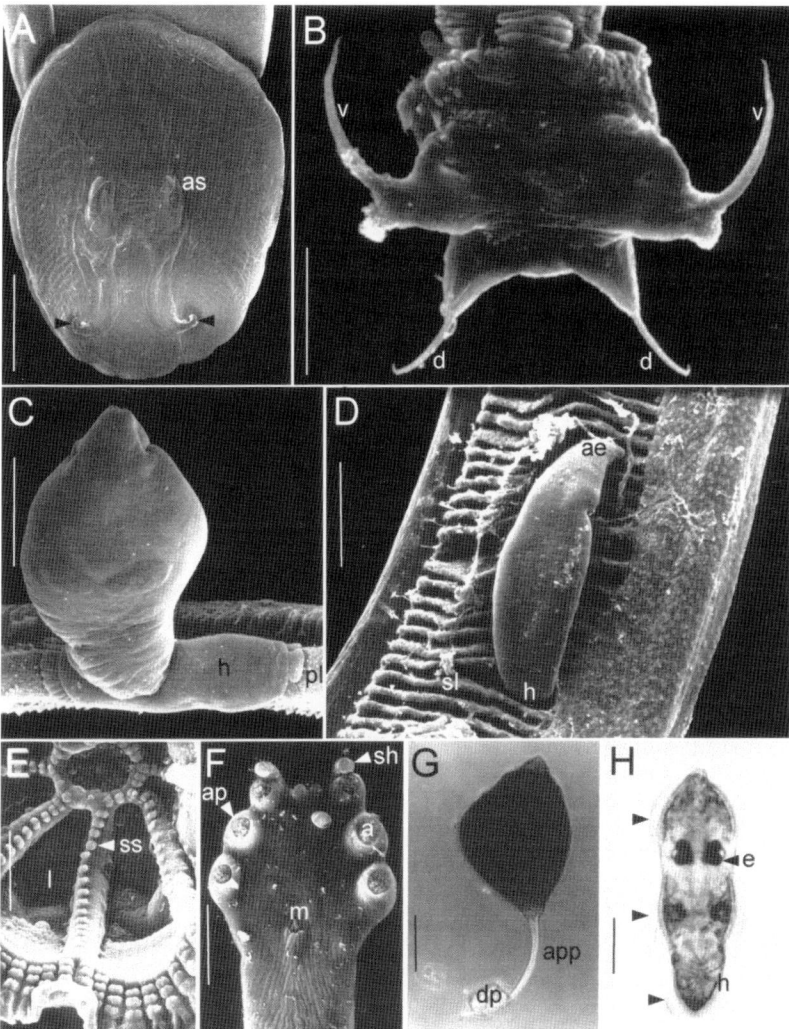

Figure 3.9 Haptors, attached parasites, anterior attachment organs and life cycle stages of some marine Monopisthocotylea Monogenea. **A**. Scanning electron micrograph (SEM) of ventral haptor surface of typical 'skin'-parasitic monopisthocotylean (Capsalidae) from ray skin. Note central accessory sclerites (as) and tips of anterior hamulus pair (arrowheads). Scale bar = 250 μm. **B**. SEM of ventral haptor surface of typical gill-parasitic monopisthocotylean (Dactylogyridae) from teleost gills. Note two pairs of large hamuli directed dorsally (d) and ventrally (v) for secure attachment between secondary gill lamellae (see D). Scale bar = 50 μm. **C**. SEM of *Benedenia rohdei* (Capsalidae) on edge of teleost primary gill lamella (pl) demonstrating ability of capsalid haptor (h) to attach to curved tissue. Scale bar = 250 μm. **D**. SEM of a Dactylogyridae species with haptor (h) attached between teleost secondary gill lamellae (sl) leaving anterior end (ae) free to feed on epithelial cells. Scale bar = 200 μm. **E**. SEM of part of haptor of a *Monocotyle* species (Monocotylidae) showing several loculi (l) capable of suction on host ray gill tissue and many additional stud-like sclerites (ss). Scale bar = 50 μm. **F**. SEM of anterior attachment organs of *Merizocotyle icopae* (Monocotylidae) from ray nasal tissue showing mouth (m), sensory horns (sh) and three conspicuous pairs of apertures (ap) through which adhesive (a) is secreted. Scale bar = 50 μm. **G**. Light micrograph of tetrahedral egg of a Capsalidae species with adhesive droplet (dp) on appendage (app) to maintain eggs on sediment at seabed. Scale bar = 100 μm. **H**. Light micrograph of oncomiracidium of *Clemacotyle australis* (Monocotylidae) showing three bands of cilia (arrowheads), pigmented eyes (e) and haptor (h). Scale bar = 50 μm.

Haliotrema species, Fig. 3.7H) and Diplectanidae (e.g. *Diplectanum* species, Fig. 3.7I). Larval invasion occurs on fish 'skin' followed by migration to gills, perhaps reflecting origins from 'skin'-parasitic ancestors. Invasion routes of larval Monogenea and subsequent migrations and development require more study. Adult dactylogyrids and diplectanids possess two large pairs of hamuli pointing in opposite directions (Figs 3.7H, I, 3.8B, 3.9B) that impale epithelium for secure attachment between adjacent secondary gill lamellae (Fig. 3.9D). Diplectanidae have additional dorsal and ventral structures called squamodiscs comprising many tiny, radially arranged sclerites (Fig. 3.7I) for extra grip. Although the gill chamber is a cramped environment, dactylogyrids and diplectanids are tiny, move actively to feed and their strict site specificity to precise microhabitats may promote opportunities to cross-inseminate (Rohde 1994; see also p. 71).

Effects on hosts and ecological importance

Observations of many Monopisthocotylea species from numerous sites on a diversity of wild marine chondrichthyan and teleost species indicate 'natural' parasite burdens do little or no obvious harm to hosts. Little, however, is known about 'natural' monopisthocotylean population sizes. *Entobdella soleae* (Capsalidae) parasitises 'skin' of 50% to 60% of soles in the North Sea (United Kingdom), each infected fish carrying one to five adults (Kearn 1998), whereas *Neoheterocotyle rhinobatidis* (Monocotylidae) populations on gills of common shovelnose rays off the Great Barrier Reef may typically reach 300–500 parasites. Neither host species appears unduly affected by these parasite populations. Monogenea probably evolved on primitive marine fish >400 million years ago and coevolutionary associations have fine-tuned many parasite–host interactions (e.g. Figs 3.8F–J). 'Natural' monopisthocotylean intensities on healthy hosts limit their attachment and feeding to epithelial cells, rarely provoking tissue responses. Parasite mobility spreads wounds over a broad area reducing local damage. However, infected fish in aquaria or aquaculture cages can succumb to the direct monogenean life cycle (Fig. 3.8) resulting in exponential population increases with severe consequences. 'Moderate' Monopisthocotylea infections on captive fish can be associated with reduced appetite, stunting of growth and emaciation. Captive fish can suffer stress, particularly if host densities are high or if fish are handled regularly (e.g. routine husbandry practices in aquaculture; aquarium touch tanks), exacerbating Monopisthocotylea infections. Management can be achieved using chemicals or drugs as baths (Thoney and Hargis 1991) or in medicated feeds. Freshwater baths can reduce marine 'skin' Monopisthocotylea infections but no treatments protect against larval reinfection, so parasite populations may easily increase again.

Attachment and feeding by large numbers of Monopisthocotylea on fish 'skin' and gills can injure host epithelial cells faster than tissue regenerates. Mechanical irritation by significant 'skin' Monopisthocotylea populations probably occurs because infected fish may rub against hard substrates (tank or cage structures; bottom sediment). Progressive symptoms and damage to 'skin' by heavy infections may include: dark 'skin' patches from parasite feeding activities; excess mucus secretion; compression, erosion or removal of epithelium; loose scales; dermis injuries such as haemorrhages, open sores and ulcers exposing connective and muscle tissue; osmotic problems. On gills, significant Monopisthocotylea burdens can lead to increased mucus production and epithelial inflammation. Where hamuli impale gill tissue, epithelial hyperplasia, oedema and haemorrhages may occur with atrophy of gill capillaries and lamellae. Gill symptoms can affect respiratory function possibly leading to host death. Large Monopisthocotylea populations on 'skin' or gills have been associated with mortalities of captive fish, but damage they inflict is open to secondary infections and perhaps viruses, bacteria and other microorganisms mostly kill infected hosts. Pathology from Monopisthocotylea can be heightened by poor water quality from pollution or high organic loads in intensive aquaculture, but few studies have focused on these mechanisms. Not only mechanical attachment may harm hosts. Adult Microbothriidae species that have no haptoral sclerites but use adhesives (Fig. 3.7D) to stick to hard denticles are associated with skin lesions on wild and captive sharks. The adhesives may chemically aggravate host epithelial cells. Growing economic importance of marine Monopisthocotylea (see pp. 71–72) demands detailed knowledge of biology and epidemiology for pathogenic species in Capsalidae, Microbothriidae, Gyrodactylidae, Dactylogyridae and Diplectanidae for applied purposes.

More fundamentally, Monogenea are renowned in parasitology for strict host and site specificity. These phenomena combined with their direct life cycle makes monogeneans superb natural 'models' to explore basic parasite biology and ecology. Infected fish in aquaria provide a supply of parasites for study. This oversimplifies matters because only considerable effort maintains a Monopisthocotylea–host fish 'model' successfully in a laboratory, but 'models' have contributed significant advances in knowledge about parasite–host interactions, biology and ecology. The nature of what mediates strong specificity remains a mystery but offers exciting avenues for further study with broad significance to parasitology, ecology and biology in general. A chemical(s) from the specific host species may influence oncomiracidial behaviour (e.g. Fig. 3.8J), but physical contact between larva and host may be needed, associated with sensory recognition and/or adhesion compatibility (Whittington et al. 2000a). Other important factors may include larval, juvenile and adult abilities to mechanically attach, feed and persist. Here, the role of the host's immune system is likely critical.

Strong host and site specificity by marine ectoparasitic Monopisthocotylea makes them ideal to examine how and why they are distributed so precisely. Relative uniformity and stability of the marine environment combined with replicate fish samples to map parasite distributions on 'skin' and gills has generated significant data about parasite ecology. Studies indicate that specific microhabitat selection is shaped less by interspecific competition, but that increased mating opportunities are more important (Rohde 1994). Their extreme host specificity and direct life cycle has led to the use of Monogenea as biological tags to discriminate fish stocks, but virtually no data exist about monopisthocotylean life spans, which limit the conclusions that can be drawn.

Baseline data about natural prevalence and intensity of Monogenea, seasonal fluctuations on wild hosts and whether there is any associated pathology are deficient. It is unknown, therefore, whether 'natural' Monopisthocotylea burdens have any role in regulating marine host populations. In this context, the significance of cleaner relationships may be important. These ecological aspects of marine Monopisthocotylea provide stimulating future challenges.

Economic importance

Previous sections outline how life cycle, biology and host-related factors (stress; susceptibility) may combine to promote heavy Monopisthocotylea infections on captive fish. If not quarantined and/or treated appropriately, marine fish introduced into public aquaria can carry their Monopisthocotylea with them. A notorious aquarium pathogen is *Neobenedenia melleni* (Capsalidae). It is especially significant for its apparent atypical broad specificity (>100 teleost species, >30 families, five orders) can infect and kill many fish species rapidly. Outbreaks can force aquarium closure. Surviving fish require treatment. Tanks need cleaning and restocking. These are considerable expenses and are inconvenient for public aquaria whose business is to display live fish. Current research indicates '*N. melleni*' may be a complex of many morphologically indistinguishable species.

Fish farmed in floating sea-cages can be susceptible to Monopisthocotylea living nearby on wild stocks of the same, or related, fish species. Monopisthocotylea eggs tangling on nets or net fouling organisms may promote heavy infections because when they hatch, larvae are close to a large host biomass. Some problematic Monopisthocotylea in aquaculture worldwide include: *Diplectanum aequans* (Diplectanidae) on sea bass gills, *Dicentrarchus labrax*, in the Mediterranean; other diplectanids and dactylogyrids (e.g. *Haliotrema* species) affecting lutjanids and barramundi in south-east Asia; *Neobenedenia* species (Capsalidae) on tilapia, barramundi, carangids, lutjanids, pleuronectids, serranids and tetraodontids off the Caribbean, Hawaii, eastern Australia, Japan, south-east Asia and Israel; *Benedenia seriolae* (Capsalidae) on *Seriola* species off Japan, Australia, Chile, Mexico and New Zealand; *Gyrodactylus salaris* (Gyrodactylidae) on salmonids in Scandinavia.

Some Monopisthocotylea have been accidentally introduced with their 'normal' host species to new environments containing 'new' or naive hosts. Devastation to Aral Sea sturgeons occurred in the 1930s when *Nitzschia sturionis* (Capsalidae) was introduced on Caspian Sea sturgeons. Many Aral Sea sturgeons died due to different susceptibilities between the two sturgeon populations. The fishery took 20 years to recover. A *Neobenedenia* species (Capsalidae) introduced on *Seriola dumerili* (Carangidae) fingerlings imported from Hong Kong and China into Japanese waters proceeded to infect pleuronectids and tetraodontids of commercial value in Japan.

The economic affect of marine Monopisthocotylea in aquaculture is illustrated by *B. seriolae* (Capsalidae) infections of three *Seriola* species (*S. quinqueradiata*, *S. dumerili*, *S. lalandi*; Carangidae) in the Japanese yellowtail industry. Total yearly yellowtail production in Japan is about 150 000 tonnes with an annual market value >US$1 billion, but management of *B. seriolae* may add >20% to production costs. Effects on farmed host health may include reduced growth, increased secondary infections and mortality (see p. 70), but the *economic* impacts of parasite management are also significant, including labour, medication and infrastructure expenses, plus interruptions to stock feeding and heightened host stress. Improved Monopisthocotylea management in sea-cage aquaculture depends on information about the basic biology, ecology and epidemiology of problematic species. These data will extend knowledge about parasite population dynamics on captive hosts and should lead to reduced infections through informed husbandry in concert with enhanced medication treatments and improved strategic delivery (i.e. integrated parasite management).

Important references
Several reviews (Kearn 1986, 1994, 1999) and book chapters (e.g. chapters 4, 5 and part of chapter 6 in Kearn 1998) provide details about Monopisthocotylea, especially chapter 4 of Kearn (1998), a comprehensive account of the biology of *Entobdella soleae* (Capsalidae) from 'skin' of *Solea solea*. Rohde (1994), Whittington *et al.* (2000a, b) and Whittington and Cribb (2001) provide reviews focusing on ecology, oncomiracidia/eggs, host specificity and anterior adhesives, respectively.

Aspidogastrea (aspidogastreans)
Klaus Rohde

Introduction
The Aspidogastrea is a small group of about 80 freshwater and marine species. It is the sister group of the digeneans, both groups included in the Class Trematoda (Littlewood *et al.* 1999). The main characteristics distinguishing the group from the Digenea is the presence of an attachment organ consisting of many suckerlets, rugae or alveoli, and the lack of a complex life cycle involving larvae that multiply.

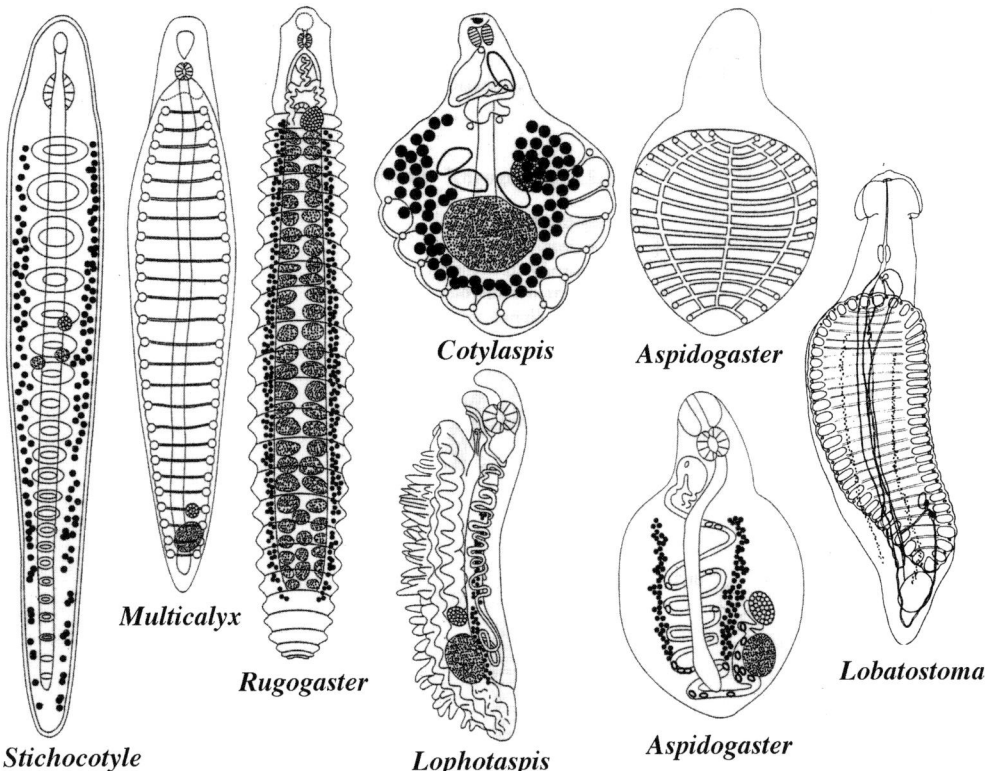

Figure 3.10 The marine genera of Aspidogastrea: *Stichocotyle, Multicalyx, Rugogaster, Lophotaspis, Aspidogaster, Lobatostoma* and *Cotylaspis*. Note the differences in the structure of the adhesive disk. Reprinted with permission from Rohde, K. Subclass Aspidogastrea Faust and Tang, 1936. In: D.I. Gibson, A. Jones and R.A. Bray (eds). Keys to the Trematoda, Vol. 1. CABI Publishing (2002).

Morphology and diversity

Marine species belong to seven genera [i.e. *Stichocotyle, Multicalyx* (=*Macraspis*), *Rugogaster, Lophotaspis, Cotylaspis, Lobatostoma* and *Aspidogaster*) of which the first four are entirely marine (Rohde 2001). Species range in length from less than 2 mm (*Cotylaspis*) to several centimetres (*Stichocotyle, Multicalyx*). Whereas most Digenea have anterior and posterior suckers, the Aspidogastrea have a posterior adhesive disk consisting of a single longitudinal row of suckers (*Stichocotyle*) or rugae (*Rugogaster*), or of numerous suckerlets or alveoli (Fig. 3.10). The species examined in detail have a remarkably complex structure, particularly evident in the nervous system and sensory apparatus. Electron-microscopic studies of the marine species *Lobatostoma manteri* have shown that it has at least eight types of sensory receptors. Examination of (3–5 mm long) specimens under the light, scanning and transition electron microscopes revealed a total number of at least 20 000 to 40 000 surface and subsurface receptors (Rohde 1989).

Aspidogastreans live in the gall bladder and bile ducts of elasmobranchs (*Stichocotyle, Multicalyx*), in the rectal glands of chimaeras (*Rugogaster*) or in the digestive tract of teleost fishes and turtles (all others, and immature *Multicalyx*). Adults of some species live in and on molluscs (Rohde 1972, further references therein). Encapsulated juvenile *Stichocotyle* are known from the intestinal wall of lobsters (Cunningham 1897). All species are hermaphroditic.

Life cycles

The life cycles of aspidogastreans are simple, relative to those of most digenean trematodes, and are of two types. In one type, there are two obligate hosts, an intermediate mollusc host which harbours the sexually immature parasites, and a definitive or final vertebrate host in which sexual maturity is reached. In the second type, the mollusc may serve as intermediate as well as final host, and the vertebrate host is not obligate although vertebrates may become infected by ingesting infected molluscs.

An example of the first type of life cycle is that of *Lobatostoma manteri* (see Rohde 1973) (Fig. 3.11). Adult parasites infect the small intestine of a teleost fish, the snubnosed dart *Trachinotus blochi* on the Great Barrier Reef. They produce eggs which are shed in the faeces of the fish and eaten by prosobranch snails. At Heron Island, Great Barrier Reef, two species were found to be naturally infected, *Cerithium (Clypeomorus) moniliferum* (Cerithiidae) and *Peristernia australiensis* (Fasciolariidae). A third species, *Planaxis sulcatus* (Planaxidae), could be experimentally infected. In other words, host specificity for snails is very low. Larvae hatch in the stomach of snails. They are not ciliated and migrate into the digestive gland of the snails where they slowly develop into the juvenile parasite which has all the characteristics of the adult, even possessing a testis and ovary, but not producing egg cells and sperm. In the smallest snail species, *Cerithium*, usually a single juvenile is present which is coiled up in the digestive gland. In the other two larger snail species several worms may be present. They are commonly found in the stomach, but infect the ducts of the digestive glands as well. But even in *Cerithium*, worms may move between the stomach and the digestive gland. Worms feed on the tissue of the digestive gland. When snails containing infective juveniles are eaten by fish, they move into the small intestine of the fish, where they reach maturity. Although the snails found to act as intermediate hosts have very thick shells, the final host, the snubnosed dart, has very strongly developed pharyngeal plates between which the snails are crushed. A spine-like vomer in the anterior part of the mouth cavity prevents the snails from slipping out of the mouth. Fish without these morphological adaptations cannot become infected.

An example of a life cycle of the second type, in which maturity is reached in a mollusc, is that of *Aspidogaster conchicola*, which infects several genera of freshwater bivalves and prosobranch snails, as well as freshwater tortoises and several genera of freshwater teleost fish. *Aspidogaster conchicola* lives and produces eggs in a variety of organs and tissues and even on the mantle of molluscs. There are contradictory reports on how molluscs become infected, either by eggs containing infective larvae, or by larvae after hatching. Vertebrates acquire the parasites by eating infected molluscs (Bakker and Davids 1973, Huehner and Etges 1977, further references in Rohde 1994). An unidentified species of the genus *Aspidogaster* is also known to infect the pericardium of marine bivalves, and something resembling *Aspidogaster* has even been reported from an ascidian, a very early (von Baer 1827) observation which needs verification.

Not a single life cycle of species infecting elasmobranchs is known. The records of immature *Multicalyx* from the intestine of a teleost, and of juvenile *Stichocotyle* from capsules in the intestinal wall of lobsters, suggest that the life cycles of these species may include additional, perhaps transport, hosts.

Effects on hosts and ecological importance

Little is known about the effects of aspidogastreans on vertebrate hosts. *Lobatostoma manteri* erodes the digestive gland of snails, and little of the digestive gland is left in snails harbouring large juveniles (Rohde and Sandland 1973, Rohde 1975). A study of the population dynamics of snail populations at Heron Island showed that, during low prevalence of infection (i.e. when conditions are presumably unfavourable for transmission), snails with multiple infections with both digenean and aspidogastrean parasites disappeared first. The relative number of egg-

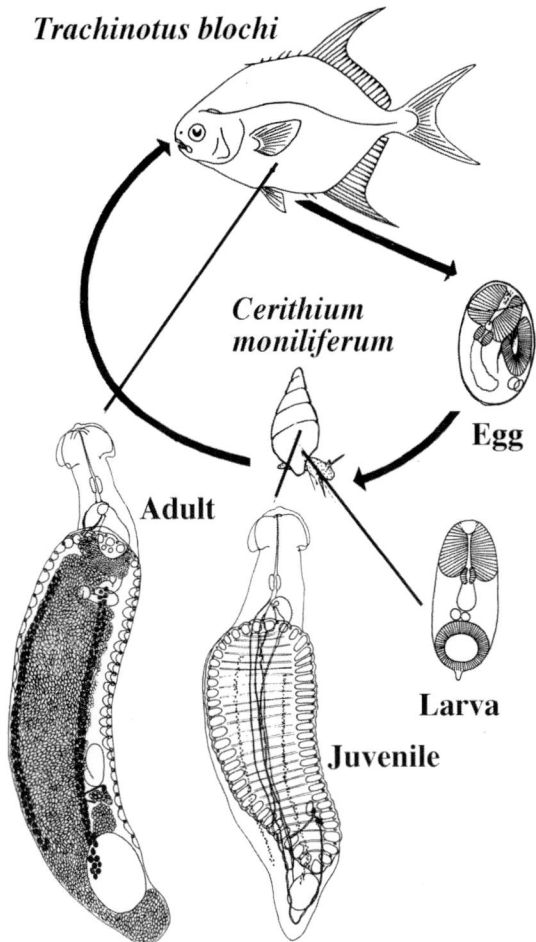

Figure 3.11 Life cycle of *Lobatostoma manteri*. Note: Adult in small intestine of the teleost fish *Trachinotus blochi*, eggs containing non-ciliated larvae are shed in the faeces and eaten by snails. Larvae hatch in the stomach of snails and migrate into the digestive gland. Juveniles develop in the digestive gland and/or the stomach of snails. Fish become infected by eating infected snails. Only fish with pharyngeal plates strong enough to crush the snail's shell can become infected. Based on but strongly modified from Rohde (1999).

producing snails (*Cerithium moniliferum*) was not affected during high prevalence of infection with *L. manteri* (Rohde and Sandland 1973).

Overall, the ecological significance of aspidogastreans is likely to be small, considering the small number of species and the lack of obvious pathological effects on the vertebrate hosts. Nevertheless, the group has attracted considerable attention for various reasons. Species of the group may be living fossils, sharing many characteristics that are likely to be archaic. They can survive for long periods (up to weeks) in simple media (water, saline) outside a host and continue producing eggs, suggesting poor adaptation to a parasitic way of life (references in Rohde 1972). They have low host specificity and a remarkably complex structure, again more suggestive of free living than parasitic animals (references in Rohde 1972, 1994). The adhesive organ and aspects of their internal structure could be called 'pseudo-segmented', which may suggest that the Aspidogastrea (and therefore all Platyhelminthes) are related to segmented taxa such as annelids (Rohde 2001). The complex life cycles of digenean trematodes (see pp. 80–84) may have evolved from the simple life cycle of aspidogastreans (Rohde 1972).

Important references
The group has been reviewed by Rohde (1972, 1994, 1999, 2001). Illustrated discussions of the morphology of larvae and adults, life cycles, early development, a key to the families and genera, and an extensive list of references are available on a webpage (Rohde 1999).

Digenea (endoparasitic flukes)
Thomas H Cribb

Introduction
The Digenea, jointly with the Aspidogastrea, form the Class Trematoda (Platyhelminthes). Digeneans, in contrast to the aspidogatreans, have complex life cycles involving several larval stages some of which multiply in the intermediate hosts. Sexual adults of most species have an anterior and ventral (posterior) sucker, which – however – have been lost in some species. Sexual adults infect all classes of marine vertebrates, asexual reproduction occurs in molluscs, and metacercariae occur in many groups of invertebrates and vertebrates.

Morphology and diversity
Digeneans are primarily parasites of the gut, but in fish also occur under the scales, on the gills, in the swim bladder, body cavity, urinary bladder, gall bladder, flesh, ovary and circulatory system. In tetrapods, they also occur in extra-intestinal sites including the circulatory system, lungs, air sacs, oesophagus, urinary bladder, liver, eye and ovary.

Figure 3.12 shows sexual adults of many of the families referred to in this Section. Digeneans range from tiny species no more than 250 µm in length to giants such as *Hirudinella ventricosa*, a parasite of the stomach of fishes that may be over 10 cm long. Some thread-like didymozoid are exceptionally long; *Nematobibothrioides histoidii* reaches over 12 m in length in the muscle of the ocean sunfish. Despite these extremes, most digeneans are 0.5 mm to 5 mm in length. They have a living tegument as their external surface, smooth or armed with spines.

Most digeneans have an unremarkable shape that gives little clue to their identity. A few are distinctive; Transversotrematidae are transversely elongate, Hemiuridae have a telescoping ecsoma at their posterior end and Didymozoidae may be helical, threadlike or almost spherical. Most digeneans have an oral sucker that opens into the gut, and a ventral sucker used only for attachment. The ventral sucker is never complex like those of aspidogastreans. The presence of two suckers does not characterise the Digenea as clearly as is often thought because in several groups one or both suckers are absent. The gut usually comprises a short tubular prepharynx, a muscular pharynx, an oesophagus and a pair of blind caeca (occasionally single). Occasionally the caeca form a complete loop (cyclocoel), separate or common ani or open into the excretory vesicle (uroproct). The excretory system comprises flame cells and ducts that lead to the excretory vesicle which opens through the excretory pore at the posterior end of the body.

The reproductive system usually fills most of the body (Fig. 3.13). Digeneans are hermaphrodites with the exceptions of all Schistosomatidae (blood flukes of tetrapods) and some Didymozoidae (tissue parasites of fishes). The male system typically has two testes (one or many also occur) (Fig. 3.14). The sperm passes via ducts to the cirrus-sac. The cirrus-sac is a muscular structure that has an internal seminal vesicle and an eversible ejaculatory duct that, when everted, forms the cirrus that is the intromittent organ. Although this arrangement is common, many digeneans lack cirrus-sacs. Whatever form it takes, the male system eventually opens

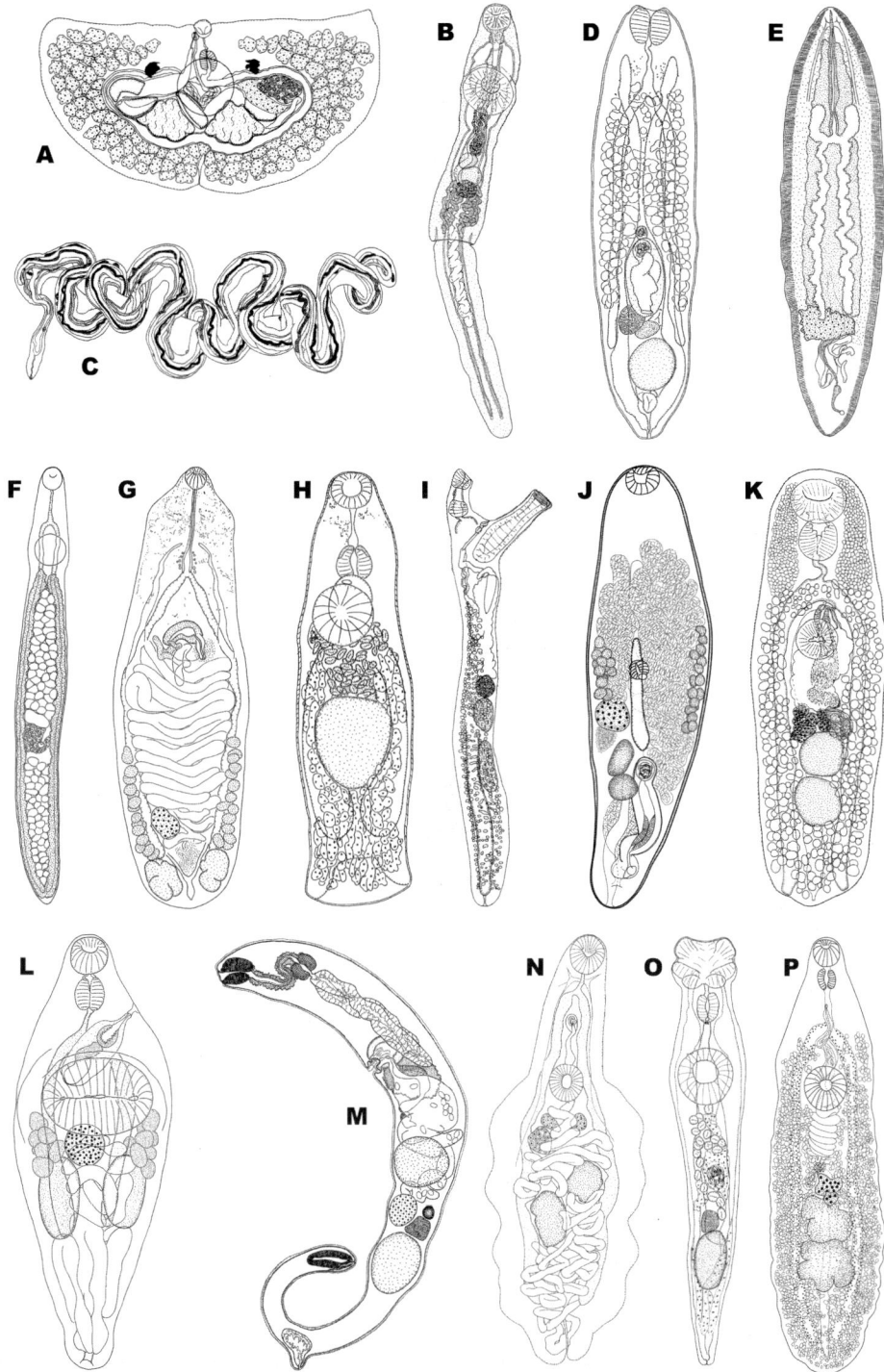

Figure 3.12 Representatives of sexual adults of digenean families. **A**. Transversotrematidae. **B**. Hemiuridae. **C**. Didymozoidae. **D**. Bivesiculidae. **E**. Sanguinicolidae. **F**. Spirorchiidae. **G**. Pronocephalidae. **H**. Haploporidae. **I**. Acanthocolpidae. **J**. Bucephalidae. **K**. Lepocreadiidae. **L**. Zoogonidae. **M**. Gyliauchenidae. **N**. Gorgoderidae. **O**. Haplosplanchnidae. **P**. Opecoelidae.

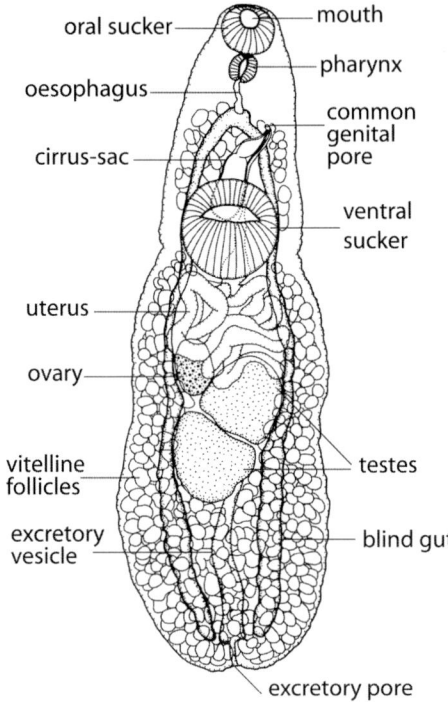

Figure 3.13 Typical morphology of a digenean sexual adult.

through the male genital pore. Usually the male and female pores open next to each other in a small genital atrium which itself opens to the exterior through a common genital pore.

The female reproductive system produces eggs that pass to the exterior via the uterus (Fig. 3.14). The ovary is usually a single mass but it is occasionally follicular. Oocytes are produced in the ovary and eggs are assembled one at a time in a special chamber of the oviduct called the ootype where (or near where) the oocyte becomes fertilised by incoming sperm. Vitelline cells surround the zygote in the ootype. The ootype is itself surrounded by gland cells called Mehlis' glands that stimulate the release of eggshell precursors from the vitelline cells. The eggshell material surrounds the zygote and vitelline cells and the ootype molds the form of the egg. There is often a preformed line of weakness that becomes the operculum (opening) of the egg. Eggs range in size from <20 μm to >100 μm in length. Most eggs have a simple oval outline but some have filaments. Vitelline follicles, which are often scattered throughout much of the body, produce the vitelline cells. Laurer's canal is a duct sometimes present that leads from the egg-forming complex to the dorsal surface. It is probably usually a drain for waste reproductive products but some digeneans may be fertilised via its opening. Sperm (incoming) may be stored in a special chamber known as a canalicular seminal receptacle that branches off the Laurer's canal, or in masses in the uterus (uterine seminal receptacle).

Host groups
Chondrichthyans are the least richly infected group of marine vertebrates. They have only a few families of digeneans and most of these are rare. The only moderately common families are the Azygiidae, Gorgoderidae and Syncoeliidae. It is striking that the spiral valve of elasmobranchs almost never has digeneans but is usually full of cestodes. There is no completely satisfying explanation for this disparity.

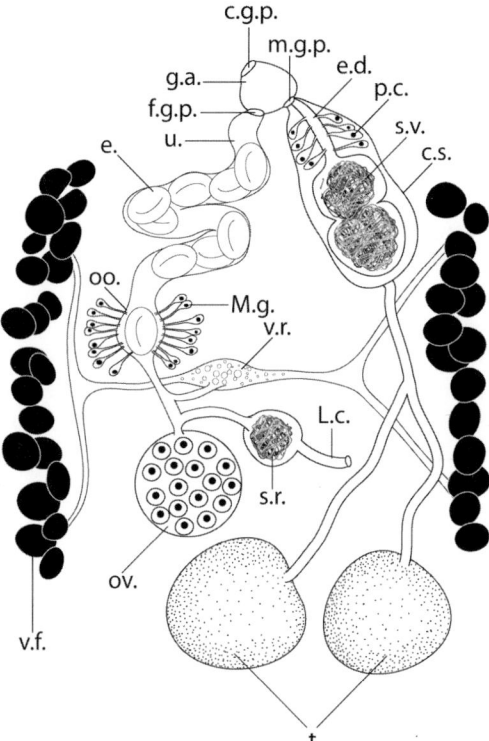

Figure 3.14 Male and female reproductive systems. c.g.p., Common genital pore; c.s., cirrus-sac; e., egg; e.d., ejaculatory duct; f.g.p., female genital pore; L.c., Laurer's canal; M.g., Mehlis' gland; m.g.p., male genital pore; oo, ootype; ov., ovary; s.r., seminal receptacle; s.v., seminal vesicle; t., testes; u., uterus; v.f., vitelline follicles; v.r., vitelline reservoir.

About 70 families of digeneans occur in teleost fishes and there are well over 5000 species known from all fishes (including freshwater species). Despite the tremendous diversity of digeneans in fishes, a few groups dominate the fauna. Ten families (Acanthocolpidae, Bucephalidae, Cryptogonimidae, Derogenidae, Didymozoidae, Fellodistomidae, Hemiuridae, Lecithasteridae, Lepocreadiidae and Opecoelidae) account for about two-thirds of records of digeneans in fishes.

Marine reptiles (turtles, sea snakes, semi-marine crocodilians and marine iguanas) all are susceptible to infection with digeneans. Marine turtles have an exceptionally rich fauna of Pronocephalidae and Spirorchiidae. Marine crocodilians are rich in cryptogonimids. Many groups of birds have developed strong associations with the sea. These include albatrosses, gannets, herons, pelicans, penguins, procellariids, and many others. Important components of their parasite fauna are the Heterophyidae, Renicolidae and Schistosomatidae. Marine mammals (cetaceans, sirenians, seals and various lesser groups) also all carry digeneans. Important families of digeneans include Brachycladiidae, Heterophyidae and Notocotylidae.

Classification
The classification of Digenea is difficult because the group is so large. Few parasitologists would claim to understand the whole group. The first problem is that the classification of the group above the superfamily level is not obvious (unlike the orders of cestodes). A new classification recently proposed for the Digenea by Olson *et al.* (2003) recognised just two orders: Diplostomida and Plagiorchiida. The members of the two orders are not immediately recognisable except

that the Diplostomida have genital pores towards the posterior end of the body whereas the Plagiorchiida have them towards the anterior end of the body. Even this distinction is imperfect as the Bucephalidae (Plagiorchiida) have posterior genital pores. The two orders have most value and significance in a phylogenetic context. The new classification also recognised 14 suborders of which just one does not occur in marine animals. Although some of the suborders are practically useful in that all their members are clearly related (e.g. Hemiurata and Opisthorchiata) they generally have less utility than superfamilies. The classification recognised 22 superfamilies, which is probably the best level of generalisation for the combination of uniformity of morphology of the sexual adult and the form of the life cycle. However, 22 superfamilies are far too many to be learnt by almost anyone. The best approach to classification is therefore to appreciate that the group is large, to understand that there is no obvious higher classification, and to be familiar with some of the most important groups at either the family or the superfamily level.

Identification of digeneans is challenging. Fortunately, a three volume series, *Keys to the Trematoda*, is being published (volume 1, Gibson *et al.* 2002). This contains keys to superfamilies, families and genera of all trematodes (Digeneaand Aspidogastrea). It will be the starting point for digenean identification for many years. These volumes do not go to the level of species; for that, it is necessary to go to the primary literature or to consult experts.

Life cycle

The digenean life cycle is usually complex (involving at least two hosts) and involves both free-living and parasitic stages. It always incorporates both asexual and sexual reproduction (Fig. 3.15). Almost all digeneans alternate between a mollusc first intermediate host (the exceptions are some sanguinicolids that infect polychaetes) and a vertebrate definitive host (exceptionally the life cycle is abbreviated and an invertebrate is the definitive host).

Sexual adults produce eggs that pass to the environment. Eggs typically hatch to release a motile, short-lived, non-feeding, ciliated larva, the miracidium. The miracidium swims, usually for no more than a few hours, and penetrates a molluscan first intermediate host. Most digeneans infect gastropods but about 10 families can infect bivalves. Just a handful infect scaphopods (tusk shells) or polychaete annelids (some Sanguinicolidae). As the miracidium penetrates the mollusc, it sheds its ciliated epithelial cells and develops into a mother sporocyst. The eggs of Hemiuroidea, Opisthorchioidea and Pronocephaloidea are very small, do not hatch until the mollusc eats them, and then penetrate the mollusc through its gut.

The mother sporocyst is a simple sac that lacks any trace of feeding structures or gonads; it absorbs all its food directly across the tegument. Although it has so few features, the mother sporocyst is actually an adult and it is misleading to refer to it as a 'larval' stage; it is better to refer to it as the first intramolluscan generation. This is why the form from the vertebrate is called the 'sexual adult' here. The mother sporocyst produces a second intramolluscan generation asexually. The nature of the reproduction in the intramolluscan generations is controversial but involves a 'germ cell lineage' of cells that remain undifferentiated from the zygote formed in the sexual adult. The generation produced by the mother sporocyst comprises either multiple daughter sporocysts or multiple rediae. A daughter sporocyst usually resembles the mother sporocyst, whereas a redia has a mouth, pharynx and short saccular gut. Apart from the distinction in how they feed, there is no important difference between daughter sporocysts and rediae. Daughter sporocysts and rediae both reproduce asexually in the same way as the mother sporocyst. Their progeny are cercariae, the larvae of the sexual adult. This can be something of a simplification. Daughter sporocysts and rediae do not always produce cercariae; sometimes they reproduce themselves; if a cercaria-producing rediae is transplanted from an infected to an uninfected mollusc, it will stop producing cercariae and instead produce more rediae. This capacity reflects the ability of these parasites to exploit the habitat provided by the mollusc to its limit (see pp. 85–86). All the intramolluscan stages live in the tissues (haemocoel) of the mollusc.

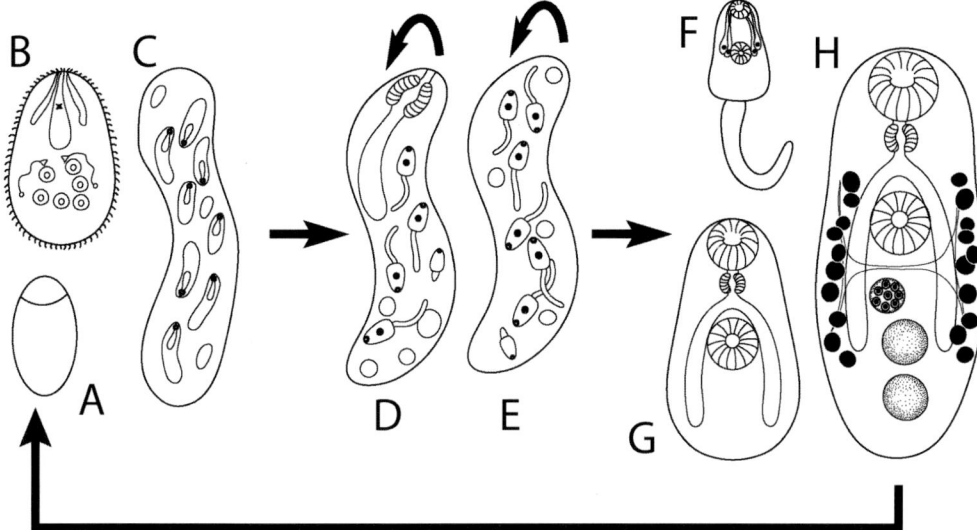

Figure 3.15 Digenean life cycle. There are usually three distinct generations: 1. – **A**. generation of egg, **B**. miracidium and **C**. mother sporocyst; 2. – **D**. generation of rediae or **E**. daughter sporocysts (these may reproduce themselves); 3. – **F**. generation of cercaria, **G**. metacercaria and **H**. sexual adult.

Asexual reproduction in the first intermediate host is one of the great evolutionary 'inventions' of the Digenea. In comparison to the Aspidogastrea, which have no asexual reproduction, it probably explains the difference in the success (=species-richness) of the two groups. Its importance is that one egg (and sexual adults often produce thousands of eggs) may infect a mollusc and lead to the production of hundreds, thousands or perhaps even millions of cercariae over the life of the mollusc. A recent report of mass emergence of 250 000 opecoelid cercariae from an abalone may be the largest known marine output. Cercarial output relates to the size of the mollusc and the cercaria and, doubtless, other aspects of the mollusc–parasite interaction. It is undoubtedly the reproductive firepower conferred by asexual reproduction that makes complex digenean life cycles possible.

Cercariae are highly variable and have morphology related to their infection behaviour (Fig. 3.16). The cercarial body sometimes looks like the adult worm (e.g. Transversotrematidae, Fig. 3.16C) but more often there is substantial development required and it may bear no resemblance to the adult worm. The cercaria almost always has a tail that it uses for swimming, usually tail-first. In a few cases the tail takes on a specialised secondary function. In most cycles the cercaria emerges actively from the mollusc, probably most frequently in the gill chamber. The process of emergence is often called 'shedding', but this is misleading as it wrongly implies that the mollusc is in charge of the process. Cercariae frequently have a distinct periodicity to their emergence. Periodicity is thought to relate to some advantage for the transmission of the parasite (e.g. cercariae available when the appropriate host is active), but a clear connection has rarely been demonstrated for marine digeneans. After it emerges, the cercaria behaves in one of several distinct ways that ultimately leads to active or passive infection of the vertebrate definitive host. Six distinct behaviours are important for marine digeneans:

1. The cercaria may penetrate the definitive host directly (Fig. 3.16A). This behaviour occurs only in (and in all of) the blood flukes (Schistosomatoidea – Schistosomatidae of birds and mammals, Spirorchiidae of turtles and Sanguinicolidae of fishes). The cercaria swims to its definitive host, attaches to the skin, drops its tail, penetrates, enters the

circulatory system and begins to develop directly to the sexual adult. Cercariae of blood flukes have abundant penetration glands that produce secretions that assist in penetration of the skin. Cercariae of bird schistosomes often attempt to infect humans and produce transitory irritation and inflammation at the site of penetration before they die. Such 'swimmer's itch' occurs worldwide.

2. The cercaria may be eaten directly by the definitive host and develop directly to the adult in the vertebrate gut (Fig. 3.16B). This behaviour occurs in Bivesiculidae, Azygiidae, Fellodistomidae and Tandanicolidae (but not in all species of these families), which are all parasites of fishes. The cercariae in these life cycles are large. A recent analysis suggests that this may be the most primitive route of infection in the Digenea.

3. The cercaria may attach directly to the body surface of the definitive host and develop directly to the adult there. This occurs only in the Transversotrematidae (Fig. 3.16C) which are parasites of teleost fishes. The cercaria is large and has unique processes at the base of its tail, which it uses to recognise its host. There are only about 10 species in this family, presumably an indication that this life cycle strategy is not particularly effective.

4. The cercaria may emerge from the mollusc, encyst in the open on potential food of the definitive host, and wait there to be eaten (Fig. 3.16D). The cercaria has cystogenous glands in its parenchyma which secrete a protective cyst around the cercaria. This encysted stage is called a metacercaria. (A metacercaria is a stage intermediate between the cercaria and the sexual adult. In a two-host life cycle, it differs from the cercaria only by having lost its tail and encysted. In three-host life cycles, where the metacercaria occurs inside another host, there is often substantial development towards the adult form.) In this two-host cycle, the metacercaria is infective to the definitive host as soon as it has encysted. Several important groups of digeneans of marine animals have this behaviour (Paramphistomoidea, Pronocephaloidea, Haplosplanchnoidea and Haploporidae). Encystment confers an important advantage relative to the life cycles described above. This is because an encysted cercaria (now a metacercaria) expends almost no energy as it waits to be eaten by a definitive host. It thus survives longer than a free-swimming cercaria. In this cycle, the parasite waits for the definitive host to find it, rather than *vice versa*.

The two remaining life cycle types both involve a second intermediate host. A high proportion of all digeneans have such cycles. There are probably three related reasons why this type of life cycle is so common. First, the metacercaria can grow in an intermediate host which allows the cercaria to be small and thus to be produced in larger numbers than bigger cercariae that must be able to infect the definitive host directly. Second, the intermediate host may act as an ecological link between otherwise unlikely molluscan and definitive host combinations (e.g. tuna and bivalves). Finally, the definitive host now actively seeks out the parasites as it seeks its prey.

5. In the first of the three-host life cycles, the second intermediate host eats the cercaria (Fig. 3.16E). The cercarial body may penetrate the gut of its second host and enter the body cavity or it may remain in the gut. The life cycle is completed when the definitive host eats the second intermediate host. This kind of life cycle occurs in all Hemiuroidea and in some Azygiidae, Fellodistomidae and Gorgoderidae. The Hemiuroidea is a large superfamily and its cercaria, the cystophorous cercaria, withdraws its body into its tail, which becomes a specialised infection device. Small crustaceans eat the cercariae and when this occurs, a delivery tube everts, penetrates the wall of the gut of the crustacean and injects the cercarial body into the body cavity of the crustacean where it develops to a metacercaria. In several cases in the Hemiuroidea, the life cycle extends to four hosts. Many hemiuroids occur in large marine fishes that do not eat the tiny crustaceans. These

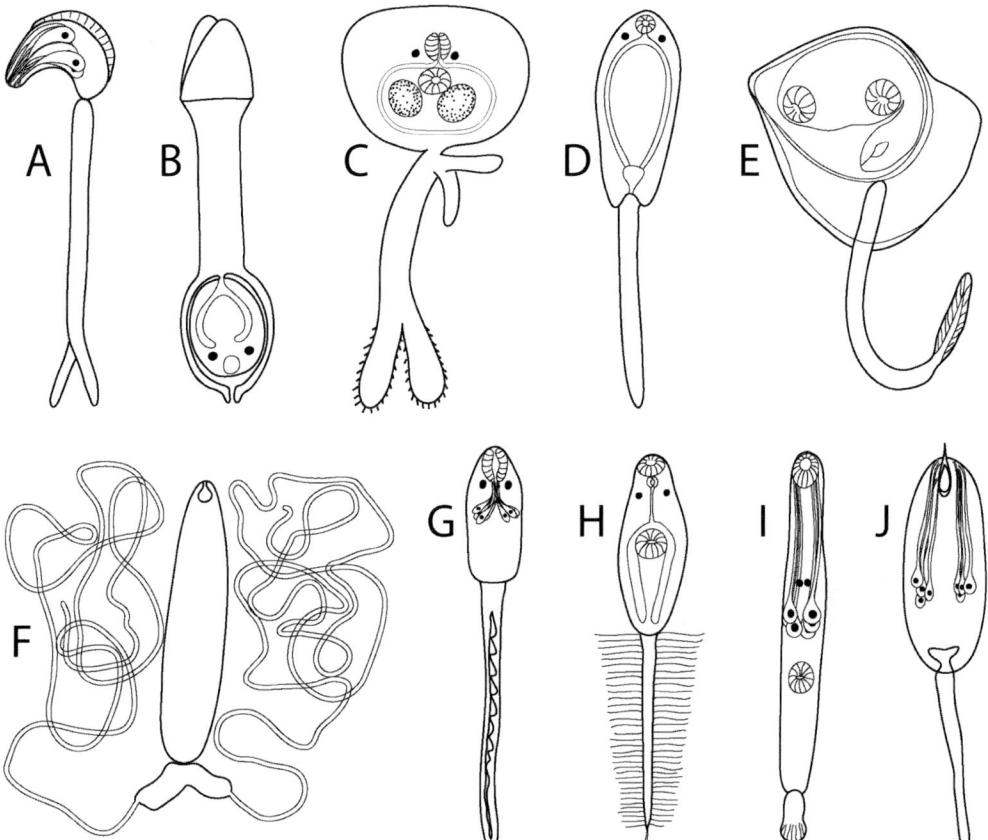

Figure 3.16 Digenean cercariae. **A**. Cercaria penetrates definitive host directly – Sanguinicolidae. **B**. Cercaria ingested directly by definitive host – Bivesiculidae. **C**. Cercaria attaches directly to definitive host – Transversotrematidae. **D**. Cercaria encysts externally on host food – Notocotylidae. **E**. Cercaria eaten by second intermediate host – Hemiuroidea. After Køie (1991). Cercaria penetrates second intermediate host externally (F–J). **F**. Bucephalidae. **G**. Cryptogonimidae. **H**. Lepocreadiidae. **I**. Opecoelidae. **J**. Microphallidae.

large fish become infected by eating any of a huge range of fishes, crustaceans, cnidarians and molluscs (third intermediate or, in some cases perhaps, paratenic hosts) that eat the small crustaceans.

6   The second form of three-host life cycle involves the external penetration of a second intermediate host (Figs 3.16F–J). Cercariae that penetrate their hosts in this way have well-developed penetration glands and often have cystogenous glands. This life cycle form is the most common of all the life cycles in the Digenea. It occurs in the Bucephaloidea, Diplostomoidea, Echinostomatoidea, Lepocreadioidea and Opisthorchioidea and in the Xiphidiata (Allocreadioidea, Gorgoderoidea, Microphalloidea and Plagiorchioidea). Cercariae of the Xiphidiata usually have an extra structure, a stylet (Fig. 3.16J), that helps penetrate their intermediate hosts. The stylet appears to have been a particularly important evolutionary development allowing infection of arthropods.

Sometimes digeneans have life cycles shorter than those described in the previous six points. Most authors think these are secondarily abbreviated. The two most common abbreviations are where metacercariae produce eggs (progenesis) and the normal definitive host is absent, and where the cercaria fails to emerge from the mollusc and the definitive host eats the mollusc. Rarely the life cycle is abbreviated to just a single host; all life cycle stages, including sexual adults, of the hemiurid *Parahemiurus bennettae* can be found within the gastropod first intermediate host.

The longevity of stages of the digenean life cycle is highly variable and in several aspects surprisingly poorly known. Development is generally temperature dependent. Eggs typically take a few days to a week to embryonate and hatch. Eggs that must be eaten probably survive much longer. The free-swimming miracidium is always short lived; survival for as long as 24 hours is probably exceptional. Development of the intramolluscan generations from penetration of the miracidium to emergence of the first cercariae typically requires at least 28 days. Free-swimming cercariae are usually also short lived and life spans of over 48 hours are exceptional; most survive less than one day. Little is known about the longevity of metacercariae but they probably survive for at least weeks and possibly for months. There is also comparatively little known about the longevity of sexual adult digeneans. Many digeneans show some seasonality in abundance, which suggests life spans of less than a year. Sanguinicolids may live no more than 70 days and Transversotrematidae live no more than three months. Probably larger digeneans live longer than smaller ones. Some of the larger digeneans of medical importance live for years and it is likely that this will be the case for larger marine digeneans too.

Two unique digenean life cycles deserve brief mention. Some species of Cyathocotylidae that have sexual adults in fish-eating birds have sporocysts that produce miracidia (as well as cercariae and more sporocysts) that emerge from the snail, apparently to infect other snails. This behaviour is not known for marine species. How this behaviour evolved and why it is rare is unknown. Gymnophallids are parasites of the intestines of shore birds as sexual adults and have their asexual reproduction in bivalves and metacercariae in molluscs and annelids. In a few species, a completely new series of asexually reproducing generations has appeared in the second intermediate host. This provides a direct parallel to the asexual development in the first intermediate host and presumably with the same effect, increasing the reproductive output of the parasite. Both life cycles demonstrate the tremendous plasticity of digenean life cycles.

Marine digenean life cycles are generally far less well known than are freshwater and terrestrial cycles. Whereas we may not know the details of particular cycles, often we can predict their outline confidently. Thus, if we find a bucephalid digenean then we can be confident that it will have sporocysts in a bivalve, a fork-tailed cercaria, and that its metacercaria will occur in the tissues of fishes. What we do not know is the detail of the identity of the hosts and the specific ecology and physiology of the infections. There are still some life cycles about which we know nothing (e.g. Brachycladiidae, Enenteridae, Gorgocephalidae and Gyliauchenidae).

Host-specificity

The host-specificity of digeneans is highly variable. Every system needs careful analysis because predictions about host specificity are difficult. In general, digenean host specificity for vertebrates is probably not as strict as that of the monogeneans and many cestode groups.

Host specificity for major vertebrate host groups (classes) can be first divided into groups that parasitise tetrapods (reptiles, birds and mammals) and those in fishes. Few families occur in both groups of hosts. Reptiles and fishes share a few families (e.g. Angiodictyidae, Pronocephalidae and Gorgoderidae) but generally each parasite group dominates in one host group and there are relatively few in the other. Few species of the families found in fishes also occur in birds or mammals.

Individual families of digeneans show a range of host specificity. A few are restricted to just one small host group. Thus, the Opisthotrematidae and Rhabdiopoeidae occur only in sirenians (dugongs and manatees) and the Opistholebetidae occur only in the closely related Tetraodontidae and Diodontidae (pufferfish and porcupinefish). This is unusual. Most larger families of digeneans occur in a range of ecologically related hosts. Diet is usually the link that defines the distribution of digenean families. For example, the Bucephalidae occur in dozens of families of piscivorous fishes. In some cases, the host range extends to hundreds of families of fishes (e.g. Opecoelidae). Just two groups of digeneans, the blood-fluke families and the Transversotrematidae, are transmitted entirely independent of host diet; both have wide host distributions.

Individual species of digeneans have host specificities ranging from oioxenous (strict – a single host species), to stenoxenous (phylogenetically related hosts) to euryxenous (ecologically related hosts). All three patterns are common. Determination of the reality of host specificity of individual species is difficult because of uncertainty about whether sampling is sufficient to identify the full host range (probably for most species it has not) and the increasing probability that cryptic species, perhaps with unrecognised strict host specificity, are present. The records of Thomas Cribb suggest that about 80% of all fish digeneans (including those of freshwater fishes) occur in just a single family of fishes. This varies between major taxa, however. Species of Hemiuridae occur in nearly three families per species on average whereas the Haploporidae average only about 1.14 host families per species.

The host specificity of digenean species to marine molluscs has not been studied much but it seems that the range rarely extends beyond a few closely related species.

Effects on hosts and ecological importance

First intermediate hosts

Digeneans typically have a dramatic effect on the health of their first intermediate host. Intramolluscan stages live within the haemocoel and typically occupy either digestive gland or gonad space (or both). Because of the asexual reproduction, molluscs are typically infected with many sporocysts or rediae and developing cercariae. These stages may occupy over half the normal volume of the digestive gland and replace the gonad entirely. Typically, infected molluscs are either completely castrated or their reproductive output is reduced dramatically. Parasitic castration implies that the animal is biologically dead in that it will no longer produce descendants. Indeed, by continuing to live and feed it may also compete with uninfected individuals. Typically, molluscs stay infected for life. Sometimes infection leads to parasitic gigantism where the mollusc grows larger than uninfected individuals.

The importance of the infection of molluscs can be considered at two levels, the individual and the population. For the individual the outcome is unambiguously bad – the animal is typically biologically dead, or nearly so. For the population, the significance of infection depends on its prevalence. Prevalences vary dramatically from well over 50% (prevalences of 100% are known for cohorts of some digenean/mollusc combinations) down to well under 1%. Where a parasite castrates over half the population of an animal, the parasite is a major driver of the ecology of the species. Equally, where only a tiny proportion is infected, the importance is low. The populations of molluscs that suffer the highest prevalences of infection are often those that occur in huge numbers in intertidal zones. In most such cases, the parasites are digeneans of shore birds. The combination of high concentrations of birds, high concentrations of molluscs and, perhaps, periodic absence of water (or low volumes) are probably the factors that allow high prevalences. Much fascinating ecological work is done on such systems. For example, the snail *Cerithidea californica* matures at smaller sizes where parasitism is high than it does in areas where it is low.

A final implication of intramolluscan infections is that the infection may interfere with the performance of the mollusc (e.g. burrowing ability) so that it is more susceptible to predation or death from physiological stress. This is an unwelcome outcome for the parasite because it dies too, but it may be more common than is realised.

Second intermediate hosts
When a second intermediate host is infected (many digenean groups do not have a second intermediate host), it will have encysted or unencysted metacercariae in its tissues. Most metacercarial infections appear to be relatively benign. The growth at the expense of the host is relatively modest (no reproduction is involved) and usually it is by absorption through a cyst wall rather than by direct browsing on host tissue. It is probable, therefore, that a few 'typical' metacercariae in an intermediate host will have only marginal pathogenicity.

Metacercariae can be significantly pathogenic in two ways. First, if the infection is sufficiently heavy then the combined effect may be damaging. Massive infections may severely affect or even kill animals trapped in small bodies of water where the concentration of cercariae is high. Harm may also take the form of reduced performance expressed, ultimately, as reduced reproductive output or more directly and immediately as an increased chance of predation of the infected host. By example, a fish with 1000 metacercariae in its muscle is more likely to be eaten by a predator than an uninfected individual. In this context, even a relatively small difference in performance may mean the difference between life and death. A system where the infection of the intermediate host can increase the chance of its predation by the definitive host is susceptible to exploitation by the parasite by 'intentional' (=behaviour promoted by natural selection) interference with the behaviour of the host. Such promotion of transmission is best known for the Phylum Acanthocephala [spiny or thorny-headed worms – see pp. 116–121, Acanthocephala (spiny-headed worms)] and is reported surprisingly infrequently for the Digenea. Microphallid metacercariae may encyst in the protocerebrum of gammarid amphipods, however, and produce aberrant escape behaviour that increases their susceptibility to predation by birds. Such effects may be more common than we realise.

Definitive hosts
Most digeneans are parasites of the digestive tract where they feed on mucus, epithelial cells and, sometimes, blood. Some groups (e.g. Paramphistomidae, Gyliauchenidae) have their ventral suckers close to the posterior end of the body and probably eat host gut contents. There are few reports of significant pathogenesis and perhaps none of mortalities associated with such infections. This is probably because the worms are tiny relative to the size of their hosts, they are mobile so do not produce permanent damage at a single site, and their diet (except perhaps for the few blood feeders) is innocuous. Tiny digeneans of the human gut, however, can be pathogenic, so intestinal digeneans should not be assumed to be entirely harmless, but overall they do not appear to be significant pathogens.

The digeneans that do cause significant problems for their hosts as sexual adults are mainly those that occur in non-gut sites. Pre-eminent among these are the blood flukes. Blood flukes occur in the blood vessels of marine birds (Schistosomatidae), marine turtles (Spirorchiidae) and fishes (Sanguinicolidae). In all cases there is potential for harm associated with the presence of the adults (interference with blood flow, erosion of blood vessel walls) and with the eggs. The problem with the eggs is that there is no direct route for their exit to the water so that they must pass across host tissue. In doing this, the eggs may cause major inflammatory responses. Not infrequently, they become trapped in tissues that do not lead to the exterior. Schistosomatids and spirorchiid eggs typically cross the gut wall but they cause inflammatory responses in other sites too (liver, brain). Eggs of sanguinicolids typically cross the gills and may cause blockages that ultimately lead to suffocation of the fish. Sanguinicolids have been associated with mass mortality of fish in aquaculture.

Digeneans also cause serious pathology in toothed whales. Species of the genus *Nasitrema* (Brachycladiidae) live as adults in sinuses of the head (tympanic cavities, pterygoid sinuses). By processes that are not entirely understood (perhaps the large adult worms wander into the surrounding tissues), the eggs of these digeneans end up in nervous tissue including the brain and the eighth cranial nerve where they are associated with severe degeneration. The animals in which these infections and lesions occur are typically from strandings and it is now thought that the infections affect neurological performance and that they are a significant cause of mass strandings of toothed whales.

Important references
The literature on the Digenea, including those in marine animals, is now huge. The most recent analysis of the classification of the Digenea is that by Olson *et al.* (2003). To identify digeneans the three volume series (Gibson *et al.* 2002) is the best starting point. The compilation of Yamaguti (1971) is still useful for lists of species. Huge numbers of species of digeneans are still being described; Jousson and Bartoli (2002) is a good example of high-quality taxonomy that incorporates a molecular approach and explores issues of cryptic species. Important explorations of digenean evolution include Pearson (1972), Brooks *et al.* (1985) and Cribb *et al.* (2003). There is no recent summary of digenean life cycles and that of Yamaguti (1975) is still the most comprehensive account available. Many life cycles are still being described: Køie (1979) is a good example of a thorough life cycle study. Poulin and Cribb (2002) explore the abbreviation of digenean life cycles. Many studies of the ecology and significance of digenean infections in molluscs include for example: Lively (1987) and Lafferty (1993). There are many interesting accounts of pathology associated with marine digeneans (e.g. Morimitsu *et al.* 1987, Ogawa and Fukudome 1994, Wegeberg and Jensen 1999, and Helluy and Thomas 2003).

Amphilinidea (unsegmented tapeworms)
Klaus Rohde

Introduction
The Amphilinidea is a small group of Platyhelminthes comprising eight species (Dubinina 1982). *Gigantolina magna* is a parasite of marine teleosts, all other species infect freshwater teleosts and turtles. However, one of these, *Amphilina foliacea*, a parasite of sturgeon, is acquired in fresh water but its host migrates into the sea.

Morphology and diversity
The amphilinids are several centimetres long with a flattened leaf-like body that is not divided into proglottids – in contrast to the related true cestodes (Littlewood *et al.* 1999) (Fig. 3.17). The marine species *G. magna* from the Indo-Pacific Ocean grows to 17 cm long, 1.5 cm wide and 1.5 mm to 2 mm thick. Worms live in the body cavity of their hosts and, lacking a gut, absorb food through the surface tegument. Ten peculiarly shaped larval hooks are retained at the posterior end by the adult (Fig. 3.17). The species are hermaphroditic. The small follicular testes are scattered through large parts of the body, and the single ovary is located at the posterior end. The uterus extends from the ovary to the anterior end, back again to the posterior end, and forward to open at the anterior end. Arrangement of the uterine coils is used for defining genera.

Life cycles
The life cycle of *Gigantolina magna* from marine teleosts, *Plectorhinchus* and *Diagramma* (Haemulidae), is not known. However, larvae of three species for which the intermediate hosts

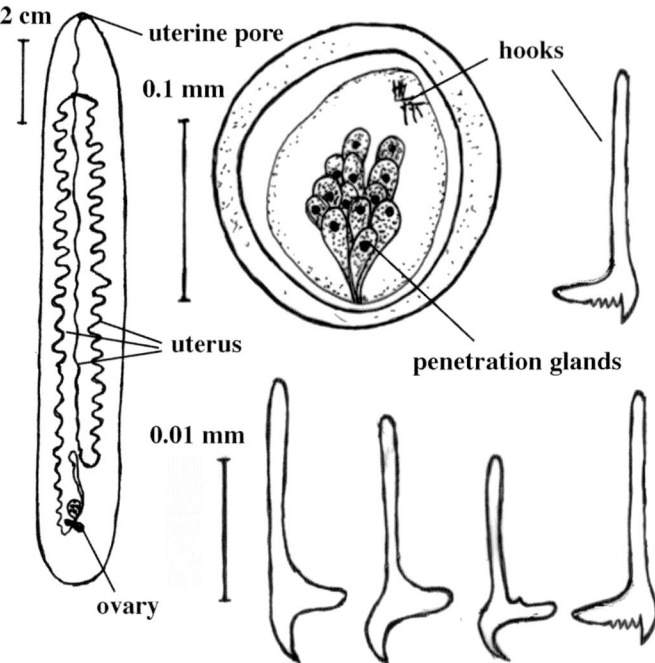

Figure 3.17 *Gigantolina magna*. Adult, egg capsule containing a larva, and larval hooks. Based on but strongly modified from several illustrations in Dubinina (1982). Note: larval hooks retained in the adult but are too small to be visible in the illustration.

are known use amphipod or decapod crustaceans. Best known is the life cycle of *Gigantolina elongata* (=*Austramphilina elongata*) which is closely related to the marine species (Rohde and Georgi 1983). It infects the body cavity of freshwater turtles in Australia. Eggs containing ciliated larvae are shed into fresh water. Larvae hatch, swim around and penetrate through the exoskeleton of freshwater crayfish and prawns, shedding the ciliated epidermis during penetration. The hooks, some of which are serrate, help in penetration, sawing through the cuticle. Secretions of anteriorly opening glands also contribute to penetration. Larvae migrate into the gill and muscle tissue and slowly grow, becoming infective to turtles after several months. Large juveniles (a few millimetres long) have been found only in the abdominal muscles. Turtles become infected by eating infected crayfish. In experimentally infected prawns, larvae stopped growing at an early stage. Juveniles penetrate through the oesophageal wall of turtles and migrate along the trachea and oesophagus into the body cavity, where they mature. Adults are sometimes also found in the lungs. It is possible that eggs are laid in the lungs, carried in the trachea to the mouth cavity, and expelled from the body either in the sputum or swallowed and shed in the faeces. Experimental evidence for these assumptions is, however, not available.

Amphilina foliacea in European sturgeon, *Acipenser*, lay eggs which escape through the coelomic pore which connects the body cavity to the outside. Eggs containing infective larvae are ingested by amphipods. The shell is broken by chewing movements of the crustacean mouth parts. The larva escapes and penetrates into the intermediate host. Final hosts become infected by eating infected amphipods (Janicki, references in Dubinina 1982) (Fig. 3.18).

Effects on hosts, economic and ecological importance

Sturgeon have considerable economic importance, particularly as producers of caviar. Russian workers have studied the effects of *Amphilina foliacea* on the fish (Andreev and Markov 1971,

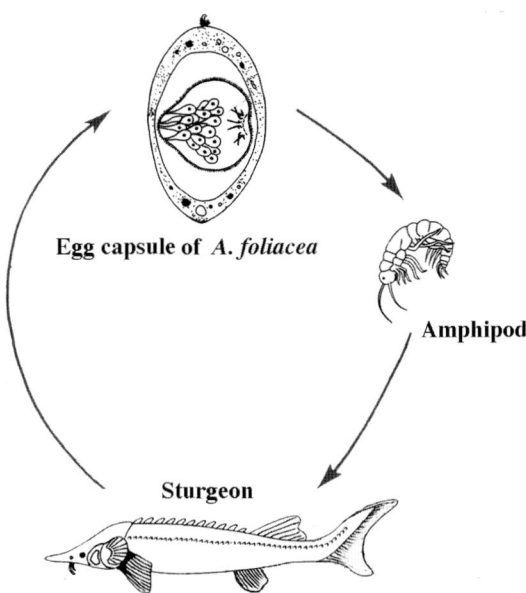

Figure 3.18 Life cycle of *Amphilina foliacea*. Note: egg capsule in water, larva develops to juvenile in intermediate amphipod host, and adult matures in body cavity of sturgeon. Egg leaves body cavity through coelomic pore of fish. Based on but strongly modified from several illustrations in Dubinina (1982).

Popova and Davydov 1988). Worms in the body cavity penetrate the liver, gonads and muscles and cause hyperaemia and haemorrhages followed by extensive inflammation. Capsules are formed around the worms. Nevertheless, the worms continue to grow and may leave the capsules. Also affected are, for example, haemoglobin levels, red blood cell counts and fat contents. The ecological importance of amphilinids is likely to be small, considering the few species.

Important references
A detailed and well-illustrated review of all early work on the Amphilinidea was given by Dubinina (1982, in Russian). A recent review is by Rohde (1994). A webpage (Rohde 1999) contains descriptions of all species, a key to the species, illustrated discussions of morphology of larvae and adults, life cycles, effects on hosts and a complete list of recent references (not included in Dubinina 1982).

Gyrocotylidea (unsegmented tapeworms)
Willi Xylander

Introduction
The Gyrocotylidea is a small group of about 10 known species in two genera all of which are intestinal parasites of the entirely marine Holocephali. Unlike the true tapeworms but like the amphilinids, gyrocotylids are not 'segmented'. They have a posterior attachment organ, a so-called rosette organ.

Morphology and diversity
The genera are *Gyrocotyle* (comprising at least nine valid species, see Table 3.4) and *Gyrocotyloides* with one species (i.e. *G. nybelini* in the European hagfish *Chimaera monstrosa*).

Table 3.4 Host species and parasites belonging to the two different subgroups of the genus *Gyrocotyle* accordings to van der Land and Dienske (1968)

Host species	Species of the 'urna' group	Species of the 'confusa' group
Chimaera monstrosa	*G. urna*	*G. confusa*
Hydrolagus colliei	*G. fimbriata*	*G. parvispinosa*
Callorhynchus millii	*G. rugosa*	*G. maxima*
Chimaera obilgyi		*G. nigrosetosa*
Hydrolagus affinis	*G. major*	*G. abyssicola*

Gyrocotyloides nybelini differs significantly from species of *Gyrocotyle* in its morphology and has been considered to be a stage of contraction rather than a valid species. However, recent studies support the validity of its species status. Some described species of Gyrocotylidea are considered to be synonyms (e.g. *Gyrocotyle meandrica* is considered a synonym of *G. maxima*) although they occur in different host species, *Callorhinus callorhynchus* and *C. millii*, respectively.

Most species of holocephalian investigated so far harbour two species of *Gyrocotyle* (Table 3.4): One morphotype (called 'urna' group) is more frequent, has a posterior rosette organ with very elaborate folds, and many lateral undulations of the body margins. The second type (called 'confusa' group) is less frequent (in most hosts), has a smaller rosette organ with fewer folds, a more elongated body and less elaborate lateral undulations. Body size in gyrocotylids ranges from about 2 cm to 30 cm.

The Gyrocotylidea is the sister group of all other cestodes (see Xylander 2001) although, superficially, in their external morphology they resemble the Monogenea (Fig. 3.19), sharing with them a single set of genital organs, a posterior sucker, paired anteriorly located nephridiopores, and no body 'segmentation'. However, several apomorphic characters, for example the brushborder neodermis, the lack of an intestine in all stages, the protonephridial system and the

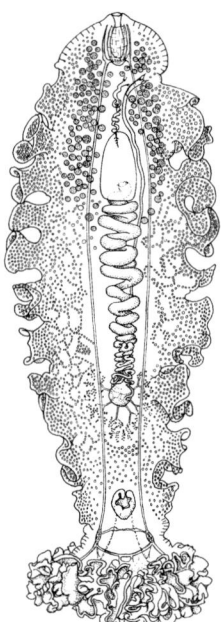

Figure 3.19 Adult *Gyrocotyle fimbriata*.

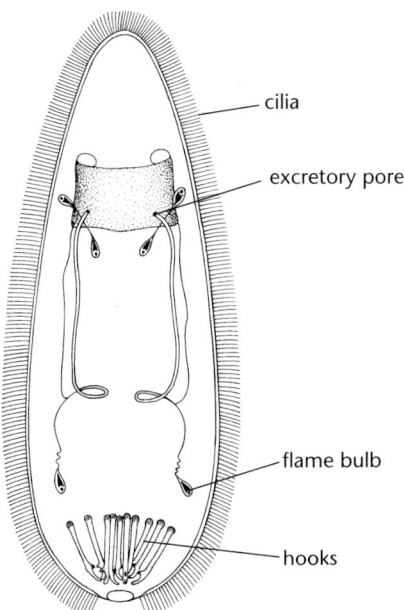

Figure 3.20 Lycophora larve of *Gyrocotyle*.

morphology of the larvae clearly show that they are cestodes. At the anterior tip of the body is a cavity which is completely covered by neodermis. The body margins show undulations and a posterior holdfast (the 'rosette organ') used for attachment to the intestinal wall of their hosts. This sucker grows during development by sequentially enclosing more intestinal microvilli to form the complex rosette organ. In contrast, in *Gyrocotyloides* there is only a small cup-like sucker located at the tip of a caudal 'stalk'.

The internal morphology is rather homogeneous throughout the whole group. The testes follicles are located in groups mostly at the anterior body end. The vitellarium consists of hundreds of lateral follicles from the anterior to the posterior body end. The germarium ('ovary') is located in the posterior third of the body and may be folliculate. The uterus is sack-shaped in most species; only in *G. rugosa* is it branched. The protonephridial system is reticulate as in other cestodes but has two separate nephridiopores at the anterior end and there is non-terminal ciliation in the nephridioducts (which is lacking in all other groups of cestodes). The nervous and sensory systems are well developed (e.g. about 10 different sensory structures have been described from the anterior end of postlarval stages).

The primary ciliated larva of Gyrocotylidea, the 'lycophora' is about 0.3 mm to 0.5 mm long (Fig. 3.20). These larvae are homologous with the lycophores of Amphilinidea and the coracidia of higher cestodes. They bear 10 hooks at the posterior end, six terminal cells of the protonephridial system, a well-developed central nervous system ('brain or cerebral ganglion') and various sensory cells.

Life cycle

The life cycle of Gyrocotylidea is still unknown. The only known stages of these helminths are the intestinal stages from the spiral valve of Holocephali (which have a neodermis) and the ciliated larvae which hatch from eggs set free from the uterus of mature specimens in the spiral valve. No intermediate host has yet been detected and several authors therefore considered this group to have a direct life cycle as found in the Monogenea. However, no observations on infection and no experiments (or other indications) support this view. Arguments for an indirect life cycle have been presented by Xylander (1989):

- All – even the youngest – stages found in the holocephalan hosts have a cup-shaped organ at the anterior end. The anterior cavity of Gyrocotylidea resembles the anterior pit of neodermal Amphilinidea, procercoids and plerocercoids. In Cestoidea and the Amphilinidea such pits develop in the intermediate host.
- Young holocephalians, which still feed on their yolk reserves, are never infected whereas specimens which have started to feed (mainly on amphipod, isopod and decapod crustaceans) show high prevalence and intensity of infection. This can only be explained by an oral infection via an infected intermediate host.
- Young chimaeras feed mainly on small benthic crustaceans (e.g. amphipods and isopods); later their diets shift to other groups of prey. Amphipods, however, have been shown to be the intermediate hosts of Amphilinidea and many other cestodes with aquatic hosts. Therefore, small crustaceans may also be the intermediate hosts of Gyrocotylidea.

In most host individuals only one gyrocotylid species occurs but mixed infections have also been observed. The various species seem to attach to different parts of the spiral valve of their hosts, one settles anteriorly closer to the pylorus and the other is found more distally.

In larger hosts usually not more than two parasites are found per host, whereas in younger holocephalans larger numbers are common. Halvorsen and Williams (1968) speculated that the parenchymal postlarvae might explain the reduction in intensity. Parenchymal postlarvae are fully developed specimens of gyrocotylids which are embedded in the tissue of a larger (sometimes slightly larger) specimen of the same species. Parenchymal postlarvae have a neodermis with typical cestode microvilli; the postlarvae seem to disintegrate after some time due to lysis by the 'hosts'. The fully developed neodermis implies that the parenchymal postlarvae are not a transitional stage of asexual multiplication as are the sporocysts and rediae of trematodes.

Effects on hosts and ecological importance
As chimaeroids are of no or very little importance, the ecological/economic significance of Gyrocotyloidea is very low. Except in high intensity infections, no pathogenic effects have been noticed.

Important references
The most recent reviews of the morphology, systematic position and life cycle of the Gyrocotylidea were given by Xylander (2001, 2003) and Rohde (1994). Bristow (1992) clarified the status of some species of *Gyrocotyle*.

Eucestoda (true tapeworms)
Janine N Caira and Florian B Reyda

Introduction
There are well over 5000 recorded species of cestodes. Adults inhabit the digestive tract, or occasionally its associated organs, in the vertebrate definitive host. Cestodes lack a digestive system, absorbing nutrients through a specialised outer layer of the body, the neodermis. Two major subgroups are recognised: the Cestodaria, consisting of the orders Gyrocotylidea and Amphilinidea, and the Eucestoda (tapeworms), consisting of the remaining 11 cestode orders. The eucestodes possess a distinct anterior holdfast organ, the scolex, which varies conspicuously in

shape among orders (Fig. 3.21). The ribbon-like body, typical of most eucestodes, is called the strobila, and is subdivided into a linear series of compartments (proglottids), each of which houses one or more sets of reproductive organs (Fig. 3.21). Eucestodes are thus said to be polyzoic. In contrast, Cestodaria lack a distinct scolex and proglottisation, and are monozoic, possessing only a single set of male and female reproductive organs per individual.

The marine environment figures prominently in cestode biology (Fig. 3.22). Over 1400 species of cestodes are known to occur in marine habitats. Three of the 11 orders of eucestodes are exclusively marine: the Diphyllidea (36 species), Lecanicephalidea (about 70 species), and Tetrabothriidea (about 70 species). Three of the most speciose orders are predominantly marine: Pseudophyllidea (about 50% of approximately 300 species), Trypanorhyncha (95% of about 350 species), and Tetraphyllidea (95% of about 540 species). In addition, six of the seven known species of Spathebothriidea occur in euryhaline or marine habitats. With over 3000 species, the Cyclophyllidea is the most speciose of the cestode orders. Several hundred species in seven of the 15 families in this order have been reported from euryhaline or marine environments. The three remaining cestode orders are predominantly non-marine. The Proteocephalidea (about 400 species) are essentially restricted to fresh water and terrestrial environments; the occasional reports from marine or euryhaline hosts are considered to be incidental infections. To date, the Nippotaeniidea (six species) and Caryophyllidea (137 species) are known only from freshwater teleosts. Readers are referred to Schmidt (1986) for preliminary lists of species in these orders. Khalil *et al.* (1994) is a useful source of information on the morphology and host associations beyond the brief summaries provided below for marine cestode orders. Joyeux and Baer (1961) remains a classic work on cestode morphology and biology.

Marine eucestode orders

Spathebothriidea (seven species, six of which are euryhaline or marine)
The scolex of spathebothriideans is either undifferentiated or consists of a weakly muscular cup-like apical organ (Fig. 3.21A). The spathebothriideans are polyzoic but the strobila is not subdivided into proglottids. Adult spathebothriideans are generally small organisms ranging in length from about 10 mm to 4 cm. Six species in five genera parasitise the digestive tracts of euryhaline or marine fishes, generally of the orders Acipenseriformes, Perciformes, Salmoniformes and Pleuronectiformes.

Pseudophyllidea (about 300 species, approximately 50% marine)
The scolex of pseudophyllideans consists of a pair of weakly muscular bothria (Fig. 3.21B). Adults range in size from 5 mm to about 30 m; this order includes the largest known tapeworms. Five of the seven pseudophyllidean families include marine species; two of these (Philobythiidae and Echinophallidae) are exclusively marine. Species in four families (Bothriocephalidae, Philobythiidae, Echinophallidae and Trianophoridae) parasitise the digestive system of marine teleosts. The remaining family, Diphyllobothriidae, includes species that parasitise the digestive system of a diversity of piscivorous birds (most commonly, Charadriiformes, Gaviiformes, Podicipediformes and Pelecaniformes), marine carnivores (e.g. pinnipeds, sea otters), cetaceans, and, not infrequently, humans. The degree of specificity for the definitive host varies widely among pseudophyllideans; many species are known from only a single host species, but several, such as *Ligula intestinalis*, parasitise tens of host species.

Diphyllidea (36 species; all marine)
The diphyllideans typically possess a scolex bearing a dorsal and ventral set of apical hooks and two bothria supported by a cephalic peduncle (Fig. 3.21C). Most species possess eight columns

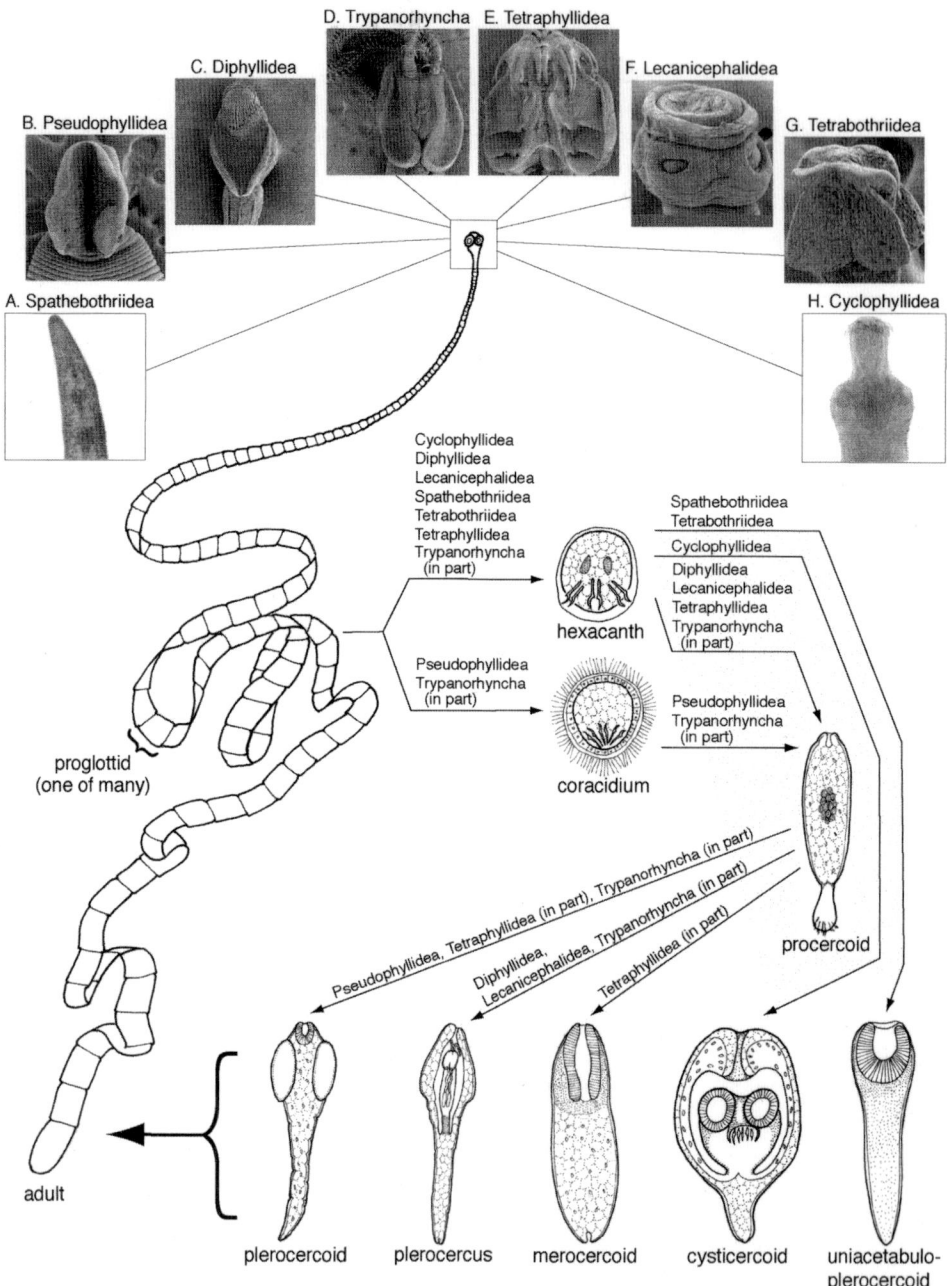

Figure 3.21 A–H. Adult scolex morphologies, and larval forms and their sequences in the life cycles of marine cestode orders.

Helminth parasites 95

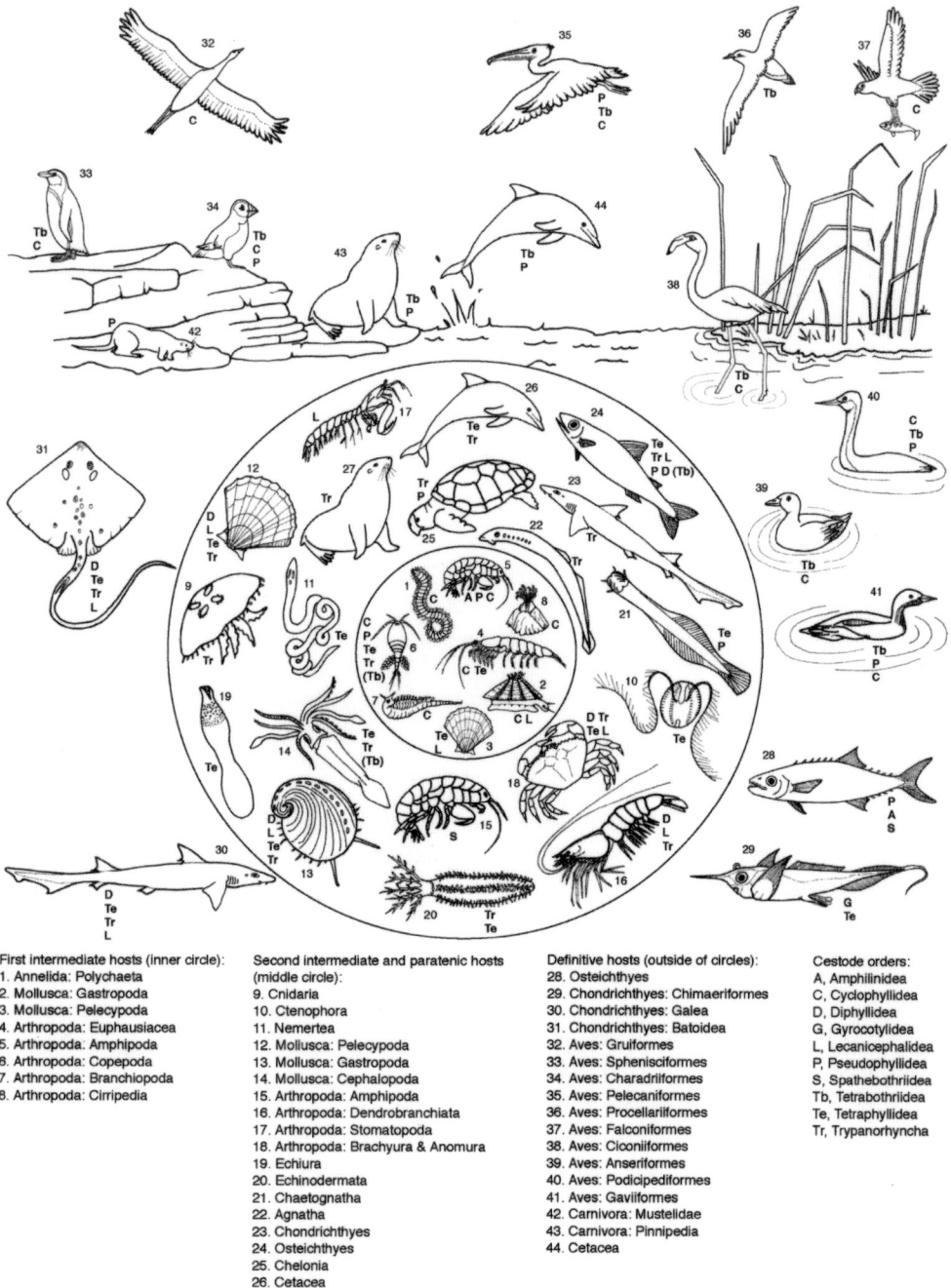

First intermediate hosts (inner circle):
1. Annelida: Polychaeta
2. Mollusca: Gastropoda
3. Mollusca: Pelecypoda
4. Arthropoda: Euphausiacea
5. Arthropoda: Amphipoda
6. Arthropoda: Copepoda
7. Arthropoda: Branchiopoda
8. Arthropoda: Cirripedia

Second intermediate and paratenic hosts (middle circle):
9. Cnidaria
10. Ctenophora
11. Nemertea
12. Mollusca: Pelecypoda
13. Mollusca: Gastropoda
14. Mollusca: Cephalopoda
15. Arthropoda: Amphipoda
16. Arthropoda: Dendrobranchiata
17. Arthropoda: Stomatopoda
18. Arthropoda: Brachyura & Anomura
19. Echiura
20. Echinodermata
21. Chaetognatha
22. Agnatha
23. Chondrichthyes
24. Osteichthyes
25. Chelonia
26. Cetacea
27. Carnivora: Pinnipedia

Definitive hosts (outside of circles):
28. Osteichthyes
29. Chondrichthyes: Chimaeriformes
30. Chondrichthyes: Galea
31. Chondrichthyes: Batoidea
32. Aves: Gruiformes
33. Aves: Sphenisciformes
34. Aves: Charadriiformes
35. Aves: Pelecaniformes
36. Aves: Procellariiformes
37. Aves: Falconiformes
38. Aves: Ciconiiformes
39. Aves: Anseriformes
40. Aves: Podicipediformes
41. Aves: Gaviiformes
42. Carnivora: Mustelidae
43. Carnivora: Pinnipedia
44. Cetacea

Cestode orders:
A, Amphilinidea
C, Cyclophyllidea
D, Diphyllidea
G, Gyrocotylidea
L, Lecanicephalidea
P, Pseudophyllidea
S, Spathebothriidea
Tb, Tetrabothriidea
Te, Tetraphyllidea
Tr, Trypanorhyncha

Figure 3.22 Marine animals serving as first intermediate hosts (inner circle), second intermediate and/or paratenic hosts (middle circle), and definitive hosts (outside of circles) for marine cestodes.

of spines extending throughout the length of the cephalic peduncle. Adult diphyllideans are relatively small, ranging in size from about 1 mm to 10 cm. Most parasitise the spiral intestine of elasmobranch fishes. They are known primarily from batoids (Rajidae, Rhinopteridae, Platyrhinidae, Rhinobatidae, Taeniuridae, Urobatidae, Dasyatidae and Myliobatidae), but occur in the shark families Triakidae and Scyliorhinidae as well. Diphyllidean species generally each parasitise only a single species of definitive host (i.e. they are oioxenous). This order was revised by Tyler (2001).

Trypanorhyncha (~350 species, about 95% marine)
Most trypanorhynchs are conspicuous in their possession of a scolex bearing four retractable hooked tentacles at its apex, and two or four bothria (Fig. 3.21D). Adult trypanorhynchs range in size from about 1 mm to 50 cm. They are known to parasitise the spiral intestine of essentially all orders of elasmobranchs (sharks, rays). Adults of the families Tentaculariidae and Lacistorhynchidae are also known from the stomach and gall bladder. Trypanorhynchs are the least host specific of the four orders of tapeworms that parasitise elasmobranchs. While some trypanorhynch species are known from only a single host species, many have been reported from several host species.

Tetraphyllidea (about 540 species, 95% marine)
Tetraphyllideans possess a scolex that bears four muscular bothridia (Fig. 3.21E) that vary impressively in form among genera. Adult tetraphyllideans range in size from about 1 mm to 100 cm. They parasitise the spiral intestine and occasionally the stomach of chondrichthyan (cartilaginous) fishes. Two species have been reported from ratfish (order Holocephali). The remaining species parasitise essentially all orders of sharks and rays (elasmobranchs). Most tetraphyllidean species each parasitise only a single species of definitive host (i.e. they are oioxenous). In general, rays host more diverse tetraphyllidean assemblages than sharks. *Cathetocephalus* is included in the Tetraphyllidea here as suggested by Khalil *et al.* (1994). But evidence is mounting to suggest it is more appropriately placed in its own order, Cathetocephalidea. The phylogenetic relationships among tetraphyllidean genera were explored by Caira *et al.* (2001).

Lecanicephalidea (about 70 species; all described taxa marine)
The lecanicephalideans have a scolex bearing four suckers or bothridia. Most also possess an apical modification of the scolex (Fig. 3.21F) that varies widely in form among genera. The Lecanicephalidea rivals the Cyclophyllidea in including some of the smallest marine tapeworms known. Lecanicephalideans generally range in size from about 300 μm to 3 cm. Adults parasitise a subset of elasmobranchs, primarily batoids of the families Myliobatidae, Narcinidae, Rhinopteridae, Mobulidae, Dasyatidae, Rhynchobatidae, Urolophidae and Rhinobatidae. However, three distantly related families of sharks, Megachasmidae, Hemiscylliidae and Squatinidae, also host lecanicephalideans. As adults, lecanicephalideans are restricted to the spiral intestine of their hosts. Most species are oioxenous. Jensen (2001) provided a detailed treatment of this order.

Tetrabothriidea (about 70 species; all marine)
The tetrabothriidean scolex usually consists of four muscular bothridia that bear anterolateral extensions known as auricles (Fig. 3.21G). Adult tetrabothriideans range in size from about 6 mm to 60 cm. The definitive hosts of tetrabothriideans represent an eclectic suite of marine homeotherms, specifically pinnipeds (Phocidae, Otariidae), cetaceans (Balaenidae, Balaenopteridae, Delphinidae, Eschrichtiidae, Phocaenidae, Physeteridae and Ziphiidae) and marine

birds (Gaviiformes, Podicipediformes, Sphenisciformes, Procellariiformes, Pelecaniformes, Charadriiformes and rarely Anseriformes). Five of the six genera are restricted to one of these three major vertebrate taxa. *Anophryocephalus* parasitises only pinnipeds, *Chaetophallus* only birds, whereas *Priapocephalus*, *Strobilocephalus* and *Trigonocotyle* are exclusively parasitic in cetaceans. Most of the 58 species of *Tetrabothrius* parasitise marine birds; the remaining eight occur in cetaceans. Individual species generally parasitise one or at most, several vertebrate species. However, exceptions occur. For example, *Tetrabothrius cylindraceus* has been reported from tens of species of marine birds. The evolutionary history and the host associations of the tetrabothriideans have been addressed in detail by Hoberg (e.g. 1995, 1996).

Cyclophyllidea (about 3000 species, several hundred marine)
The cyclophyllidean scolex bears four muscular suckers. An apical extension, or rostellum, which often bears hooks, is also found in many species (Fig. 3.21H). Marine cyclophyllideans are generally small, ranging in length from slightly less than 1 mm to 2 cm. Despite the overall diversity of the Cyclophyllidea, several cestode orders (Tetraphyllidea, Trypanorhyncha and Pseudophyllidea) exceed it in their diversity in the marine environment. Seven of the 15 cyclophyllidean families include marine species: Progynotaeniidae, Acoleidae, Dioecocestidae, Amabaliidae, Davaineidae, Dilepididae and Hymenolepididae. While none of these families is exclusively marine, there exist genera that are (e.g. the dilepidid genus *Alcataenia*). Adult marine cyclophyllideans are restricted to birds of the orders: Charadriiformes, Gruiformes, Ciconiiformes, Podicipediformes, Pelecaniformes, Falconiformes and Sphenisciformes. They generally parasitise the digestive systems of these hosts. The phylogenetic relationships among cyclophyllidean families were investigated by Hoberg *et al.* (1999).

Origins and distribution of marine cestodes

Origins
Interpretations of the habitats of origin of the various marine cestode taxa depend heavily on the availability of a robust phylogenetic hypothesis for the cestodes. Although consensus regarding cestode interrelationships has not yet been achieved (e.g. see Olson *et al.* 2001), several generalities are beginning to emerge. The Cestodaria are basal with respect to the Eucestoda. The Tetraphyllidea are not monophyletic; the Lecanicephalidea and Proteocephalidea are each likely to be monophyletic, but both group among tetraphyllidean taxa. The tetrabothriideans and nippotaeniids are either basal to, or members of, the Cyclophyllidea. Bothriate orders such as the Pseudophyllidea, Diphyllidea, and possibly the Trypanorhyncha, appear to be basal to the tetrafossate (bothridiate) groups.

On the basis of these relationships, the following observations are offered regarding the origins of marine eucestodes. Association with the marine environment appears to be the ancestral condition for cestodes. The spathebothriideans, diphyllideans, pseudophyllideans, trypanorhynchs, lecanicephalideans and tetraphyllideans are all likely to have originated in the marine environment. The few members of these orders that currently occupy hosts in freshwater situations, such as the tetraphyllideans and trypanorhynchs of South American freshwater stingrays, are widely believed to have resulted from colonisation of these hosts and/or habitats by marine ancestors. These freshwater incursions probably occurred independently in these two orders. The habitat of origin of the tetrabothriideans is difficult to assess. If tetrabothriideans are determined to be basal to the nippotaeniids and cyclophyllideans, a relatively strong case could be made for a marine origin for the group. However, if, as has been suggested by trees resulting from some analyses (e.g. Olson *et al.* 2001), the tetrabothriideans represent a clade within the Cyclophyllidea, a case could be made that the ancestral tetrabothriideans occurred in terrestrial and/or freshwater environments, in which case their current existence in marine habitats is the result of one or more colonisation events (Hoberg 1996, 2002). This interpretation would be consistent with the fact that tetrabothriideans parasitise a subset of species in three relatively distantly related groups of marine vertebrates, including birds, pinnipeds and cetaceans. In fact, the data suggesting that cyclophyllideans occur in marine environments as a result of colonisation are rather compelling. According to the phylogenetic hypotheses of cyclophyllidean interfamilial relationships generated by Hoberg et al. (1999), six of the seven families that include marine species belong to a derived clade within the order. The seventh family (Davaineidae) appears to be a close relative of that clade. Given that all of the more basal cyclophyllidean families parasitise freshwater or terrestrial hosts, it is likely that the marine cyclophyllideans originated from terrestrial or freshwater ancestors that colonised the marine environment. Since all seven families also include freshwater and/or terrestrial species, cyclophyllideans probably have colonised the marine environment multiple times.

Geographic distribution
Temperature is among the most important factors structuring the distributions of marine cestodes globally. This influence is manifested in both latitude and depth. Temperature can directly affect the cestodes themselves. For example, the eggs of many cestode species successfully embryonate only under a very restricted range of water temperatures. Temperature can also have indirect effects. Most conspicuously, cestodes that parasitise hosts with specific temperature preferences are themselves restricted to those temperatures. For example, species of *Amabalia*, which parasitise flamingos, are unknown from polar areas; species of *Parorchites*, which occur in penguins, are known only from Antarctica.

In general, most orders of marine cestodes can be characterised as either cooler or warmer water taxa. Three eucestode orders (Spathebothriidea, Pseudophyllidea and Tetrabothriidea) are commonly encountered in cooler water habitats and the spathebothriideans in temperate or subarctic regions in the northern hemisphere. Most marine pseudophyllideans occur in temperate or higher latitudes. Some of the pseudophyllideans known from subtropical regions (e.g. *Tetrapapillocephalus*) occur at depth and thus are effectively inhabitants of cooler waters, but others, such as *Pseudeubothroides* in Hawaii, are legitimate inhabitants of warmer waters. Despite the migratory natures of many of their hosts, most tetrabothriideans have been reported from higher latitudes. Many species are restricted to either the northern or southern hemisphere. For example, at least 15 species of tetrabothriideans occur exclusively in Antarctica, and 10 or more species occur exclusively at latitudes north of 55°N. A small number do, however, occur in subtropical seas including the waters around several Caribbean islands, Angola, Sri Lanka and New Caledonia.

Records of the four cestode orders that parasitise elasmobranchs come predominantly from temperate or warmer latitudes, with an emphasis on, but not limited to, the IndoPacific. This is expected given the diversity of elasmobranchs in these regions. However, some tetraphyllideans, trypanorhynchs and diphyllideans can survive much colder temperatures. For example, species in the former two orders are known from Antarctica, and several species in the latter order have been collected in the North Sea. The lecanicephalideans exhibit the most restricted distribution of these four orders. Jensen (2001) observed that all existing records come from hosts collected at latitudes between 45°N and 35°S. However, data on the cestodes of higher latitude elasmobranchs, and in particular skates and scyliorhinid sharks, are limited.

The marine cyclophyllideans also occur frequently in temperate or warmer latitudes. The geographic distributions of the birds hosting marine cyclophyllideans obviously constrain the geography of these cestodes. Essentially all of the marine cyclophyllideans known are restricted to shallower coastal regions.

The data on deep-sea cestodes are limited. However, Klimpel *et al.* (2001) provided a compilation of records from deep-sea fishes from which the following observations can be made. Cestodes are not the most predominant parasite group, but occur in about 20% of the 421 fish species found below 200 m that have been examined. The deep-sea cestode fauna is dominated by tetraphyllideans, followed by trypanorhynchs and then pseudophyllideans. Spathebothriideans and diphyllideans comprise only a small component of the fauna. Larval cestodes, particularly trypanorhynchs, account for a large proportion of the records. Campbell (1983) offered the following observations and insights into the cestodes of deep-sea organisms. Deep-water cestode faunas are generally less diverse than shallow-water faunas, but prevalence and intensity of infection with individual species may rival those seen in shallower waters. Some cestode taxa are restricted to deep-waters (e.g. the pseudophyllidean family Philobythiidae). Such taxa are generally more closely related to deep-water taxa in other areas than to shallower-water taxa in the same geographic region. In general, the occurrence of cestodes in benthic species is more pronounced than in mesopelagic or bathypelagic species. This may be because life cycles proceed through food chains comprised of benthic communities (i.e. horizontally), rather than vertically through the water column. As a consequence, eggs and larval stages are concentrated rather than diluted. Abyssal cestode assemblages appear to be the least diverse deep-water assemblages, but they are also the most poorly studied.

Development and life cycles

Host associations

Cestodes are more intimately tied to the biology of their hosts than are most other parasite groups. To quote Mackiewicz (1988), the life cycles of cestodes have become 'interpolated into host biology'. With a few exceptions, the life cycles of tapeworms lack free-living stages and rely instead on the natural dynamics of the food web to affect transfer between hosts. Unfortunately, the life cycles of marine cestodes are poorly known (Beveridge 2001). Although fragmentary data exist, a complete life cycle has yet to be established for any species of Spathebothriidea, Diphyllidea, Tetraphyllidea, Lecanicephalidea or Tetrabothriidea. Complete cycles have been determined for only a handful of species in the remaining orders (i.e. Pseudophyllidea, Trypanorhyncha and Cyclophyllidea). With these limitations in mind, the following observations regarding the life cycles of marine cestodes can be made.

Vertebrates serve as definitive hosts for all marine cestodes. During their lives, most marine cestodes also parasitise a minimum of one intermediate host in which some development occurs. The Cyclophyllidea utilise a single intermediate host. The life cycles of Spathebothriidea, some Trypanorhyncha, and possibly the Diphyllidea, Lecanicephalidea, Tetrabothriidea and some Tetraphyllidea, appear to involve two intermediate hosts. Three or four hosts, in addition to the definitive host, may occur in the life cycles of pseudophyllideans, trypanorhynchs and tetraphyllideans in which the phenomenon of paratenesis occurs. In such cases two of the hosts usually serve as intermediate hosts and the remainder as paratenic hosts. The number of paratenic hosts depends on the difference in trophic level between the second intermediate and definitive hosts.

Figure 3.22 provides a summary of the major groups of marine animals known to host the various orders of marine cestodes (e.g. Dollfus 1976 and references therein). Conspicuous differences exist in diversity among hosts serving the various roles in these life cycles. A remarkably limited slate of only three phyla serve as first intermediate hosts. In contrast, at least eight phyla of invertebrates and a range of vertebrates, in particular teleosts, serve as second intermediate or paratenic hosts. With the exception of marine reptiles (sea snakes, sea turtles, marine iguanas and saltwater crocodiles), and the Sirenia (manatees, dugongs), all classes and orders of marine vertebrates have been found to host adult cestodes.

Larval stages
Marine tapeworms exhibit a relatively diverse array of larval forms. The extensive and complex literature on cestode larval terminology was organised by Chervy (2002), whose terminology is followed here. All eucestodes begin life as a hexacanth embryo. In pseudophyllideans and some trypanorhynchs, this embryo is surrounded by a ciliated membrane. It hatches from the egg and exists for a short time as a free-swimming stage known as a coracidium. The hexacanth embryos of most marine eucestodes lack cilia and remain within the egg, awaiting consumption by an appropriate intermediate host. The developmental trajectory beyond the hexacanth stage varies among, and in some cases, within cestode orders. These are summarised in Figure 3.21.

Larvae attributed to spathebothriideans have been reported from a range of amphipods. Some of these plerocercoids possess reproductive primordia, suggesting that the amphipod serves as the final intermediate host. However, attempts to directly infect amphipods by feeding them macerated spathebothriidean bodies or eggs have failed. This has led to speculation that the spathebothriidean life cycle may involve a phase prior to the plerocercoid in the amphipod.

Following the coracidium, pseudophyllideans pass through two parasitic larval stages, the procercoid and plerocercoid. Procercoids have been found in crustaceans such as amphipods and a wide array of copepods. Plerocercoids generally parasitise vertebrates, although the occasional record of parasitism in a chaetognath exists. Teleosts of a range of types and sizes commonly host plerocercoids; this stage has also occasionally been found in sea turtles. In general, small planktivorous teleosts serve as intermediate hosts whereas larger, piscivorous teleosts serve as paratenic hosts.

The fragmentary developmental information available for diphyllideans and lecanicephalideans suggests that both orders lack a coracidium and both possess a plerocercus as their terminal larval stage. Plerocerci attributed to diphyllideans and lecanicephalideans have been reported from bivalve and gastropod molluscs, as well as a range of crustaceans and, on occasion, teleosts. A larval stage prior to the plerocercus may exist in both orders. In fact, a procercoid attributed to the lecanicephalidean genus *Tylocephalum* has been reported from pelecypod and gastropod molluscs (e.g. Cake 1978).

At least two distinct life cycle strategies are found among trypanorhynchs (Mattis 1986). These differ in the number of hosts and larval types involved. In species utilising three hosts, the larval sequence is hexacanth embryo, procercoid and plerocercoid. In species utilising four hosts, the sequence is coracidium, procercoid and plerocercus; the plerocercus stage persists in the paratenic host. Procercoid development generally occurs in a copepod. Plerocercoids and plerocerci are known from four phyla of invertebrates as well as many vertebrates. However, most records by far of these larval stages come from teleosts (see Bates 1990). As in the pseudophyllidean four-host life cycle, larger teleosts serve to bridge gaps in the food chain between smaller teleosts and the apex predators of the system, in this case elasmobranchs. Trypanorhynch plerocerci and plerocercoids generally have a scolex bearing fully formed tentacular armature like that seen in the adult stage. As a consequence, some trypanorhynch species are known only from larval stages, and an extensive literature exists on the host associations of the larval stages of a diversity of trypanorhynch species.

Existing data suggest that the basic tetraphyllidean life cycle involves three hosts, but as many as five may be involved if paratenesis occurs (Euzet 1959). A procercoid develops when an egg bearing a hexacanth embryo is consumed by a copepod or euphausid shrimp, or occasionally a pelecypod mollusc (Cake 1978). This is followed by either a merocercoid or a plerocercoid depending on the tetraphyllidean taxon. Tetraphyllidean merocercoids, categorically referred to as *Phyllobothrium delphini* and *Monorygma grimaldi* regardless of their true generic affinities, are commonly found in the blubber and digestive system of cetaceans. Tetraphyllidean plerocercoids, traditionally referred to as *Scolex polymorphus* or *Scolex pleuronectis*, have been reported from seven phyla of invertebrates and tens of species of teleosts. Some tetraphyllideans pass through a uniacetabulo-plerocercoid stage prior to the bothridiate plerocercoid stage, but it is unclear how widespread this phenomenon is within the order. It appears that both plerocercoids and merocercoids can survive passage through one or more paratenic hosts. These stages transform into the adult when consumed by an appropriate elasmobranch.

The tetrabothriidean life cycle might involve two intermediate hosts: perhaps a crustacean, and then either a cephalopod or teleost. However, few larvae attributed to tetrabothriideans are known from natural infections. Hoberg (1987) provided evidence that the stage infective to the definitive host may be a uniacetabulo-plerocercoid, in which case at least some of the records of larvae of this form, such as *Scolex bothriosimplex*, may represent tetrabothriids.

Complete life cycle data are available for one or more marine species in the cyclophyllidean families: Progynotaeniidae, Davaineidae, Dilepididae and Hymenolepididae. In all of these cases, development involves two hosts and a larval stage known as a cysticercoid, which follows the hexacanth. A diversity of crustaceans, and occasionally gastropods and polychaete annelids, serve as intermediate hosts for marine cysticercoids.

One of the primary handicaps of marine cestode life cycle work is that, with the exception of the final larval stages of some of the hook-bearing cestode groups (e.g. trypanorhynchs, most diphyllideans and cyclophyllideans) and a few specialised lecanicephalideans (e.g. *Polypocephalus*), larval forms are notoriously difficult to identify. Most larvae are undifferentiated and thus possess none of the morphological features of the adult on which the taxonomy of species is normally based. Molecular sequence data have recently begun to be applied to questions of larval cestode identification with great success. This approach is likely to continue to greatly facilitate work on cestode life cycles.

Transmission

The greatest challenge facing a marine cestode is not successful movement between hosts, for this is a passive process on the part of the cestode. The primary challenge falls to the hexacanth embryo, either in the egg or as a coracidium larva, for this stage must locate and successfully parasitise an intermediate host in the vastness of the marine environment before it can progress to the definitive host and sexually reproduce. Marine cestodes use both strategies recognised by Mackiewicz (1988) as effective for increasing the probability of transmission to the first intermediate host. For example, the cocoons or egg packets of many lecanicephalideans, tetraphyllideans and cyclophyllideans bear filaments or inflation devices. These elaborations affect position in the water column so that cocoons float, or sink and become entangled in marine vegetation, depending on whether the target host is a pelagic or benthic invertebrate. Other species (e.g. some spathebothriideans, tetraphyllideans, cyclophyllideans, trypanorhynchs and lecanicephalideans) produce eggs with shells that bear similar elaborations and likely serve a similar purpose. Jarecka (1961) provided an illuminating account of egg morphologies and their implications for transmission. Particularly intriguing are the cestodes that produce eggs that appear to mimic the food items of their hosts. Cocoons and packets, such as the paruterine organs of cyclophyllideans, also serve to ensure that each encounter exposes a host to multiple individuals of the parasite.

Substantial fecundity is a hallmark of marine cestodes. Most species are polyzoic and possess multiple sets of reproductive organs, usually one set per proglottid but sometimes more. An extreme case is the cetacean-dwelling *Hexagonoporus physeteris*, which has up to 14 sets of reproductive organs per proglottid and over 40 000 proglottids on a strobila up to 30 m long. Another strategy, although uncommon, is multistrobilation as exemplified by the shark-dwelling adults *Cathetocephalus thatcheri*.

There is a marked dichotomy among marine eucestodes with respect to strategies for releasing eggs from proglottids. Anapolytic taxa, such as the pseudophyllideans and tetrabothriids, release eggs through pre-formed uterine pores. In such taxa proglottids remain attached to the strobila actively shedding eggs throughout the life of the adult worm. Apolytic taxa, such as the diphyllideans, lecanicephalideans, trypanorhynchs, and tetraphyllideans, lack uterine pores. Adults of these taxa regularly shed immature, mature or gravid proglottids, depending on whether a species is hyperapolytic, euapolytic, or fully apolytic, respectively. Detached proglottids are extremely mobile. Those of hyperapolytic and euapolytic species live in the digestive tract of the host maturing, locating and mating with one another, ultimately producing eggs. Mating is facilitated by a behaviour that has been observed in many species whereby free proglottids actively migrate and congregate in the anterior chambers of the spiral intestine (e.g. Williams et al. 1970). Gravid proglottids and/or eggs ultimately pass into the marine environment with the faeces of the host. Contact with sea water causes the body wall of the proglottid to break open at one or more areas of weakness and eggs are released.

Ecological importance and effects on hosts

Ecology

The paucity of information available on the ecology of marine cestodes was noted over 20 years ago by both Campbell et al. (1980) and Esch (1983). Unfortunately, this situation has changed little. Several of the more basic aspects of marine cestode ecology are addressed here.

Some cestodes take advantage of the spatial aspects of the ecology of host systems to assist them in overcoming the potentially overwhelming problem of encountering an appropriate host in the vast marine environment. For example, the life cycles of many marine cestodes, particularly cyclophyllideans, tetraphyllideans, diphyllideans, lecanicephalideans and some tetrabothriids, proceed in the shallower waters of the neritic zone in which eggs and potential intermediate and definitive hosts are concentrated. Vertebrates whose biologies are not conducive to this strategy, such as pelagic seabirds and deep-sea pelagic fishes, often harbour depauperate cestode faunas relative to those of their neritic counterparts (e.g. Campbell 1983, Hoberg 1996).

Some cestodes take advantage of the temporal aspects of the system. For example, a marked increase in the prevalence of certain avian cestodes has been observed during breeding season when birds are concentrated in large numbers (e.g. Williams and Harris 1965). This is likely to be the case in pinnipeds as well.

It is not known if cestodes of pelagic hosts (e.g. cetaceans, deep-sea teleosts) take similar advantage of the spatial and temporal aspects of the ecologies of their hosts. Interestingly, most cestodes found as adults in pelagic hosts (i.e. pseudophyllideans, tetrabothriids) are anapolytic and thus have the potential to produce and release individual eggs continuously over long periods of time.

Parasite-induced alteration of host behaviour has been documented in the freshwater relatives of many marine cestode groups, but this phenomenon is poorly studied in marine systems. Data from freshwater systems suggest pseudophyllidean plerocercoids and cyclophyllidean cysticercoids as potentially productive targets for future work.

Effects on hosts

Work conducted on the pathogenicity of marine cestodes is limited and has tended to emphasise fish hosts (e.g. Williams and Jones 1994). The effects adult cestodes have on their hosts range from negligible to, in exceptional cases, death. Pathogenesis associated with adult cestode infections is often correlated with scolex morphology and the size and number of worms present. The scolex can cause changes in the mucosa and underlying layers of the intestinal wall, including fibrosis or loss of epithelium at or near the site of attachment, focal pressure necrosis, hyperplasia and metaplasia of the mucosa, and haemorrhage. Infections with adult pseudophyllideans have been known to decrease growth rate and diminish the ability to adapt to sea water in sockeye salmon. Such infections also render fish more susceptible to other environmental stresses, such as pollutants (see Williams and Jones 1994). Some of these same effects have been observed in vertebrates other than fish. For example, hyperplasia of connective tissue at the site of attachment is known to occur in blue whales infected with adults of the tetrabothriidean *Tetrabothrius affinis* (see Rees 1967). Some of the larger pseudophyllidean species have been reported to cause acute intestinal obstruction in pinnipeds. It is likely they serve as agents of this phenomenon in cetaceans and possibly also in teleosts, particularly when infection intensities are high. It is, important to note, however, that it is not uncommon to see thousands of lecanicephalideans, tetraphyllideans and/or trypanorhynchs parasitising the spiral intestine of stingrays which otherwise appear perfectly healthy.

Larval cestode infections have also been known to compromise the health of vertebrate and invertebrate hosts; in many instances pathogenicity is associated with the intensity of infection. For example, butterfish with heavy infections of trypanorhynch larvae in their musculature have been found to weigh less than those with lower burdens (see Williams and Jones 1994); heavy infections of trypanorhynch larvae in the guts of pelecypod molluscs have been observed to restrict passage of food. Blood changes, suggestive of anaemia, have been observed in cutlassfish infected with tetraphyllidean larvae of the *Scolex pleuronectis* form. Heavy infections of larval lecanicephalideans may result in physiological stress, and affect growth and reproduction of pelecypod molluscs (Cake 1978); reduced vigour has been observed in shrimp hosting heavy infections of cestode plerocercoids.

Economic and medical significance

Although it is difficult to quantify the economic significance of marine cestodes, their effects on the fishing industry are diverse and potentially extensive, both in natural and aquaculture situations. The general deterioration of condition, weight loss and reduced growth rate that can result from infection with larval or adult cestodes directly affect the productivity of teleost, shrimp and pelecypod harvests (see pp. 402–403). Cestode infections can occur in aquaculture facilities that maintain organisms in natural enclosures or artificial enclosures that involve unfiltered sea water. In such situations captive fish and shellfish are exposed to infection from natural intermediate hosts, and also eggs and coracidia. Infections beyond these sources can be avoided using the standard protocol of freezing food items that may serve as second intermediate or paratenic hosts for cestodes.

Larval cestodes residing in the musculature of shellfish and teleosts are often of sufficient size to be visible to the naked eye and can significantly reduce the marketability of these hosts. This effect is particularly marked in heavy infections of trypanorhynchs, tetraphyllideans and lecanicephalideans. Plerocercoids of pseudophyllideans are often so large that the palatability of their teleost hosts is reduced even in very low intensity infections.

Considering the diversity of marine cestode species, and the ubiquity of their larval stages in the marine environment, it is remarkable that humans do not acquire infections with cestodes more frequently. One explanation is that several of the more diverse, predominantly marine cestode orders that utilise 'seafood' as intermediate or paratenic hosts (e.g. Tetraphyllidea, Trypanorhyncha), normally parasitise elasmobranchs as adults. Humans are unlikely to provide an environment suitable to sustain these cestodes. Nonetheless, several cases of trypanorhynch larvae infecting humans are known (e.g. Bates 1990); most are thought to have occurred as a result of the consumption of uncooked squid. The etiological agents of most human infections with marine cestodes are pseudophyllideans of the family Diphyllobothriidae, most commonly species of *Diphyllobothrium* and *Diplogonoporus*. Such infections are often the result of humans having inserted themselves into the life cycles of marine pseudophyllideans, which normally parasitise piscivorous marine carnivores or cetaceans, by consuming teleosts infected with plerocercoids. Although standard cooking practices are sufficient to kill the larvae, such practices are not universally employed. Individuals in countries in which it is traditional to consume dishes consisting of raw or undercooked fish, such as Peru, Finland and Japan, are particularly vulnerable to infection (Rees 1967). The number of cases of Diphyllobothriasis is estimated as being in the millions globally. Most infections result in minor symptoms, including abdominal discomfort, diarrhoea and nausea. But in a few instances, pernicious anaemia can result as a consequence of the particularly high affinity for vitamin B_{12} exhibited by these cestodes (see pp. 430–431).

Important references
Contributions in Arme and Pappas (1983) discuss various aspects of the biology of eucestodes. A key to the cestodes of vertebrates was provided by Khalil *et al.* (1994), and Olson *et al.* (2001) analysed the interrelationships and evolution of cestodes.

Nematoda (roundworms)
Gary McClelland

Introduction
The Nematoda, comprised of 256 families and more than 40 000 species, is one of the largest and most successful groups in the animal kingdom (Williams and Jones 1994, Anderson 2000). Although most are free living, numerous species parasitise plants or animals. Soil-dwelling nematodes of the rhabditean and dorylaimian lines are believed to have given rise to the 125 families of zooparasitic nematodes, including species that exploit freshwater and marine hosts.

Nematodes are bilaterally symmetrical and generally take the form of an elongate cylinder, tapered at each extremity and varying from less than 1 mm to more than 1 m in length at maturity (Roberts and Janovy 2005). They have a fluid-filled pseudocoele and complete digestive system, with the mouth at the anterior extremity and the anus found at or near the posterior end. The body is covered by a non-cellular cuticle that is secreted by the underlying hypodermis and shed four times during maturation. The excretory system consists of lateral canals and/or ventral glands and opens through a ventral excretory pore near the anterior end of the body. Most nematodes are dioecious, sexually dimorphic and oviparous, but some taxa are ovoviviparous and hermaphrodism and parthogenesis also occur. The ventral excretory pore and the spicules, part of the male copulatory apparatus, distinguish the Nematoda from other pseudocoelomate groups.

Morphology and diversity
The phylum Nematoda is comprised of two classes: the Enoplea (=Adenophorea; =Aphasmidia) and the Rhabditea (Roberts and Janovy 2005). Parasitic species are found in the enoplean subclass Dorylaimia and the rhabditean subclass Rhabditia (=Phasmidea; =Sercenentea). Dorylaimians typically have five or more oesophageal glands and a buccal stylet present at some stage during their life cycles. Rhabditians have an excretory system with one or two lateral canals, but lack caudal and hypodermal glands. With regard to surface sensory structures, rhabditians possess poorly developed amphids on or near the lips, while deirids (cervical papillae) and phasmids (often lacking in enopleans) are commonly found bilaterally, in the oesophageal and caudal regions, respectively. Marine parasitic nematodes are found in three dorylaimian orders, the Trichurida, Dioctophymatida and Mermithida, and four rhabditian orders, the Strongylata, Ascaridida, Oxyurida and Spirurida. The status of two additional orders of exclusively marine parasitic nematodes, the Marimermithida and Benthimermithida, remains uncertain. As many

vertebrates found in marine settings also feed on land or in fresh water, it is probable that species of terrestrial or freshwater origin have been included among nematode taxa surveyed below. This would be particularly true of species reported from anadromous (salmonid, clupeid and perciform) and catadromous (anguillid) fish, coastal birds and saltwater crocodiles (*Crocodylus porosus*).

Class Enoplea, Subclass Dorylaimia

Order Trichurida

In trichuridans (Figs 3.23A–D), the anterior portion of the body is usually more slender than the posterior portion, and the oesophagus is capillary-like and embedded within one or more rows of large glandular cells, the stichocytes; spicules are either single or lacking in males, and the eggs have a polar plug (opercula). Marine species belong to four families: Capillaridae, Cystoopsidae, Trichinellidae and Trichosomoididae (Moravec 2001). Capillarids (*Capillaria, Capillostrongyloides, Gessyella, Hepatocola, Paracapillaria, Piscicapillaria, Pseudocapillaria* and *Schulmanela* spp.) are particularly widespread and abundant in the alimentary canal of Elasmobranchs and ray-finned fish (Actinopterygii). Other capillarids reported from marine hosts include *Capillaria* and *Eucolius* spp. from the digestive tract of coastal birds (Torres *et al.* 1991, Bosch *et al.* 2000), *Capillaria delamurei* from the intestine of the Caspian seal, *Pusa caspica*

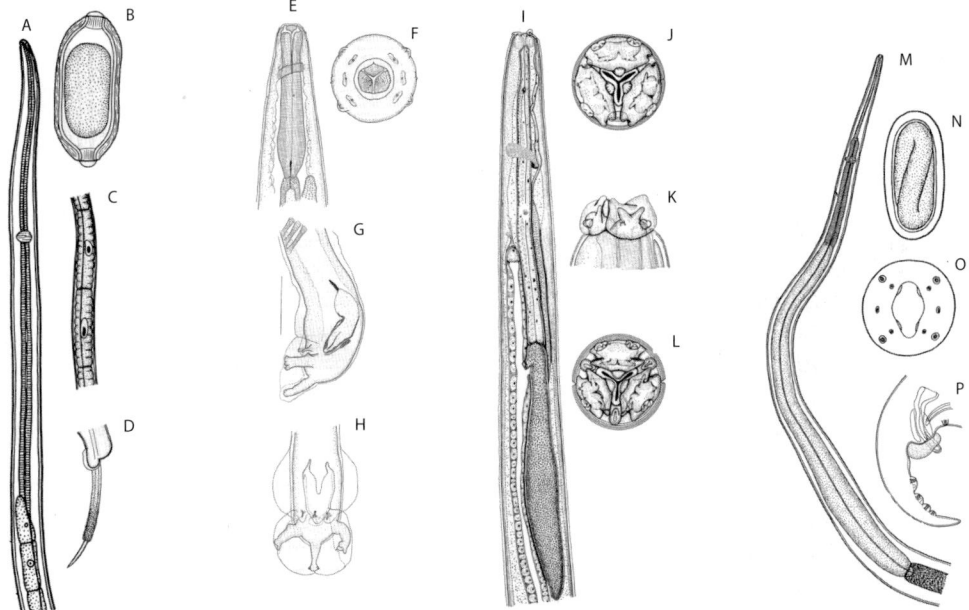

Figure 3.23 Examples of marine parasitic nematodes. **A–D.** *Capillaria gracilis* (Trichurida, Capillaridae) from marine fish, after Moravec (2001). **A.** Anterior end of female. **B.** Operculate egg. **C.** Stichosome region. **D.** Caudal end of male with protruding spicule sheath and spicule.
E–H. Lungworm (*Stenurus minor*) (Metastrongyloidea, Pseudaliidae) from harbour porpoise (*Phocoena phocoena*). After Arnold and Gaskin (1975). **E.** anterior end of female. **F.** En face view.
G. Lateral view of male caudal region. **H.** Ventral view of male caudal region. **I–L.** *Contracaecum osculatum* (Ascarida, Anisakidae) from seals. **I.** Anterior end of third-stage larva. **J.** En face view of fourth-stage larvae. **K.** Lateral view of cephalic end of fourth-stage larva. **L.** En face view of adult.
M–P. *Pancreatonema americanum* (Spirurida, Rhabdochonidae) from pancreatic ducts of dogfish. After Moravec *et al.* (2001). **M.** Anterior portion of worm. **N.** Egg. **O.** En face view. **P.** Male caudal region.

(Zablotzky 1997) and *Crocodylocapillaria* and *Paratrichosoma* spp. from the intestine and skin, respectively, of the saltwater crocodile (Moravec 2001). Eggs of *Huffmanela* spp. (Trichosomoididae) have been found in the skin, musculature and viscera of a variety of marine elasmobranchs and ray-finned fish, although adult nematodes remain to be found. *Cystoopsis scomber* (i.e. species *inquirenda*), described from the gill chamber and mouth of Black Sea mackerel, *Scomber scomber* (Moravec 2001), and *Trichinella nativa* (Trichinellidae), which mature in the intestines of marine mammals (Dailey 2001), remain the only marine records for their respective families.

Order Dioctophymatida
Dioctophymatidans are large, stout worms with highly developed, multinucleate oesophageal glands, a terminal anus in both sexes, and a bell-shaped copulatory bursa and single spicule in males. Species of the family Eustrongylidae mature in the crop of fish-eating birds, and are transmitted primarily by freshwater fish. Larvae of *Eustrongylides* spp., parasites of waders (herons and egrets), however, are also transmitted by estuarine and marine fish (US Food and Drug Administration/Center for Food Safety and Applied Nutrition 1991).

Order Mermithida
Mermithidans, recognised by the presence of six or eight hypodermal cords in cross section, are primarily found in the haemocoeles of terrestrial and freshwater invertebrates (especially insects) as larvae, but are usually free living as adults. Marine species include *Thaumamermis zealandica* (Mermithidae) from intertidal amphipods (Poinar et al. 2002), *Nematinermis enoplivora* (Tetradonematidae) from littoral and sublittoral free-living nematodes (Tchesunov and Spirodonov 1993) and *Echinomermella grayi* and *E. matsi* (Echinomermellidae) from sea urchins (Jangoux 1990). Atypically, *N. enoplivora* and *Echinomermella* spp. reach maturity in their invertebrate hosts. Free-living adults of *Thallasomermis megamphis* (Mermithidae) have been found in abyssal sediments, but hosts of immature worms are not known (Tchesunov and Hope 1997).

The status of many mermithid-like marine nematodes remains uncertrain. Roberts and Janovy (2005) place the Marimermithidae among families of the Mermithida, and do not include the benthimermithids. Based on molecular evidence, however, De Ley and Blaxter (2002) give the marimermithids order status (Marimermithida) among the dorylaimians, while assigning benthimermithids to the Benthimermithida, an order of uncertain taxonomic placement. Marimermithidans include *Ananus asteroideus* (Rubstov 1977) and *Marimermis kergelensis* (Rubstov and Platonova 1974) of seastars, *M. maritima* (Tchesunov 1996) of sea urchins, and *Australonema eulagiscae* in polychaetes (Tchesunov and Spiridonov 1985). Hosts of larval stages remain unknown for *Marimermis* and *Trophomera* spp. described as free-living adults from marine sediments (Rubstov and Platonova 1974). Benthimermithidans are parasites of invertebrates from abyssal sediments. *Adenodelphis eurythenes* has been described from the amphipods (Petter 1983), but only the free-living adults of *Benthimermis* spp. are known. Taxonomic status remains to be assigned for mermithid-like nematodes described from free-living deep-sea nematodes by Petter and Gourbault (1985) and those found in tubicifid oligochaetes by Hallett et al (2001).

Class Rhabditea, Subclass Rhabditia

Order Strongylida
Strongylidans are generally long, slender worms with oesophagi that are swollen at the posterior end, but lack a definite bulb; males have a well-developed copulatory bursa supported by

sensory rays (Figs 3.23E–H). Although they parasitise most classes of vertebrates, they seldom occur in fish. Marine species are confined to six families found in avian and mammalian hosts (Borgsteede 1997, Dailey 2001). Hookworms, (Ancylostomatoidea, Ancylostomatidae) mature in the intestines of juvenile land-breeding seals, with *Uncinaria lucasi* occurring in northern fur seals, *Callorhinus ursinus*, in the Northwest Pacific, and *U. hamiltoni* in *Otaria flavescens* and *Mirounga leonina* in the South Atlantic and Subantarctic. The remaining marine mammal strongylidans are lungworms (Metastrongyloidea). 'Small lungworm', *Parafilaroides* (Filarioididae) spp. occupy the bronchioli and alveoli of both eared seals (Otariidae) and true seals (Phocidae), and 'large lungworm', *Otostrongylus* spp. (Crenosomatidae), the bronchi of phocid seals only. Dailey (1985) lists eight genera of cetacean lungworm (Pseudaliidae): *Delamurella*, *Halocercus*, *Otophocaernus*, *Pharurus* (syn. *Torynurus*), *Pseudalius*, *Pseudostenurus*, *Skrjabinalius* and *Stenurus*. Infections with *Halocercus* spp. are usually confined to the lungs, but other pseudaliids (e.g. *Stenurus* spp.) invade the pulmonary and mesenteric arteries, cranial sinuses, brain, middle and inner ears, eustachian tube and oral cavity. *Amidostomum* spp. (Trichostrongyloidea, Amidostomatidae) are reported in the gizzard, and *Cyathostoma* and *Syngamus* spp. (Strongyloidea, Syngamidae), in the respiratory tract of various coastal waterfowl, but are probably transmitted terrestrially.

Order Ascaridida

Ascarididans (Figs 3.23I–L) are characterised by the presence of three prominent lips, external labial papillae and numerous caudal papillae. They are parasites of the gastrointestinal tract of all classes of vertebrates including marine fish, reptiles, birds and mammals. Among marine species, they are outranked in diversity only by the spirurids, and together with spirurids they comprise most of the nematodes described from marine fauna. There are marine representatives in three ascarididan superfamilies, the Ascaridoidea, Cosmocercoidea and Seuratoidea.

In Ascaridoids, the oesophagus is typically divided into a muscular preventriculus and a posterior ventriculus, which may be muscular or glandular (Fig. 3.23I). Marine species are found in four families, the Acanthocheilidae, Anisakidae, Ascaridae and Heterocheilidae, with anisakids being the most numerous and diversified members of the group (Rohde 1984, Lauckner 1985a–d, Nadler and Hudspeth 2000, Nadler *et al.* 2000, Dailey 2001). The anisakid subfamily Anisakinae includes *Pulchascaris* and *Terranova* spp. of elasmobranchs (Bruce and Cannon 1990), *Parheterotyphlum* spp. of sea snakes (Hydrophiidae), *Sulcascaris sulcata* of marine turtles (Chelonia) (Lauckner 1985a), *Terranova* spp. of saltwater crocodiles (Machida *et al.* 1992), and *Anisakis*, *Contracaecum*, *Pseudoterranova* and *Phocascaris* spp. of cetaceans, pinnipeds and sea otter *Enhydra lutris* (Mustelidae) (Dailey 2001). *Contracaecum* spp. also infect various piscivorous birds such as cormorants, pelicans (Nadler *et al.* 2000) and penguins (Ranum and Wharton 1996), but *Porrocaecum* spp. found in coastal waterfowl are probably of terrestrial origin (Anderson *et al.* 1996). *Goezia* spp. (Goeziinae) are parasites of marine ray-finned fish (Petter and Cabaret 1995) and, in the case of *G. holmesi*, sea snakes (Lauckner 1985a), while marine actinopterygians and, occasionally, elasmobranchs, cephalospidomorphs (lampreys) and holocephalans (chimaeras) are parasitised by species of the anisakid subfamily Raphidascaridinae (*Aliascaris*, *Heterotyphlum*, *Hysterothylacium*, *Icthyascaris*, *Iheringascaris*, *Lappetascaris*, *Maricostula*, *Mawsonascaris*, *Paranisakiopsis*, *Paranisakis*, *Raphidascaris*, *Raphidascaroides* and *Rostellascaris*) (Rohde 1984, Sprent 1990, Petter and Cabaret 1995, Rocka and Stephanski 2002). Marine representatives of the family Ascarididae include *Dujardinascaris* and *Multicaecum* spp. of saltwater crocodiles (Machida *et al.* 1992, Sprent *et al.* 1998), and *Paradujardina halicoris* of sirenians (Raga 1997). The Acanthocheilidae (*Acanthocheilus*, *Metanisakis* and *Pseudanisakis* spp.) are confined to elasmobranchs (Bruce and Cannon 1990, Moravec and Nagasawa 2000,

Sanmartin *et al.* 2000), while the only heterocheilid described from marine fauna is *Heterocheilus tunicatus* from manatee (Raga 1997).

Cosmocercoids have oesophagi with a well-defined corpus, isthmus and bulb with valve, and typically occur in the lower intestines of vertebrates. Seuratoids, an artificial group of vertebrate parasites, have a club-shaped oesophagus with no swellings. A cosmocercoid *Kathlania leptura* (Kathlaniidae) and a seuratoid, *Cucullanus carettae* (Cucullanidae) have been reported from loggerhead turtles (*Caretta caretta*) (Lauckner 1985a), while numerous cucullanids (*Cucullanus*, *Dichelyne* and *Truttaedacnitis* spp.) are found in marine actinopterygians (McDonald and Margolis 1995, Køie 2000b, 2001b, Sanmartin *et al.* 2000).

Order Oxyspirurida
In oxyurids, commonly known as pinworms, the oesophagus has a prominent posterior bulb and valve, and the tail is sharply pointed. These small- to medium-sized worms occur primarily in the lower intestine and rectum of terrestrial invertebrates and vertebrates. Of the nearly 500 species described from vertebrates, only 19 species are from fish, and *Laurotravassoxyuris travossi* (Pharyngodonidae), which occurs in *Holocanthus tricolor*, an inhabitant of coral reefs, is the only marine representative (Anderson and Lim 1996, Moravec 2000). Larval oxyuroids have been reported in gill washings from Pacific herring (*Clupea pallasi*) (McDonald and Margolis 1995), but the marine turtle parasite, *Kathlania leptura*, referred to as an oxyuroid by Lauckner (1985a), has been assigned, with other kathlaniids, to the Cosmocercoidea (Ascaridida) (Chabaud 1978).

Order Spirurida
In spiruridans (Figs 3.23M–P), the mouth may be surrounded by six lips, two lateral pseudolabia, or lips may be lacking altogether. Usually, there is a well-developed buccal capsule, and the oesophagus is divided into anterior muscular and posterior glandular portions. Spiruridans occur enterically, subcutaneously and in deeper tissues and organs of all classes of vertebrates. They are the most diversified group among marine parasitic nematodes, with representation in nine superfamilies and 16 families, and the greatest number of marine species of any nematode order.

The spiruridan suborder Camallanina is partioned into two superfamilies, the Camallanoidea and the Dracunculoidea, with marine species being found in both. In camallanoids, lips are lacking, but the buccal capsule is well developed, often in the form of two opposing valves. In dracunculoids, the buccal capsule is reduced and the oesophugus is not partitioned into distinct muscular and glandular portions. The marine camallanoids, *Onchophora*, *Procamallanus* and *Spirocamallanus* spp. (Camallanidae) occur primarily in the stomach, pyloric caecae and intestines of ray-finned fish (Rigby and Adamson 1997, Moravec *et al.* 1999), although they have also been identified from the spiral valve of elasmobranchs (Knoff *et al.* 2001). Marine dracunculoids assigned to the Philometridae (*Clavanema*, *Ichthyofilaria*, *Margolisianum*, *Paraphilometroides*, but mainly *Philometra* spp.) infect a broad spectrum of ray-finned fish, where they may be found subcutaneously in the fins and opercula, in the musculature, or in the body cavity on the liver, gonads and mesenteries (Moravec *et al.* 1998, 2002). Other marine dracunculoids include *Pseudodelphis oligocotti* (Guyanidae) from the intestinal mesenteries of sculpins (Adamson and Roth 1990), *Phlyctainophora squali* (Phlyctainophoridae) from the skin and buccal cavity of dogfish (Adamson *et al.* 1987) and, although it is transmitted in fresh water, *Anguillicola crassus* (Anguillicolidae) from the swim bladder of sea-going eels (*Anguilla* spp.) (Kirk *et al.* 2002).

Marine acuariids (Acuarioidea) parasitise a variety of coastal birds including gulls and waders, cormorants and ducks, but some (*Seuratia* spp.) infect pelagic birds such as shearwater (Hoberg and Ryan 1989) and albatross (Langston and Hillgarth 1995). With the exception of

Paracuaria spp., which occur in the intestines (Roca *et al.* 1999), acuariids (*Ancryocanthopsis, Chevreuxia, Cosmocephalus, Schistorophus, Sciadiocara, Skrjabinocerca, Skrjabinoclava, Streptocara, Syncuaria, Viktorocara* and *Voguracuaria* spp.) occupy the oesophagus, crop or gizzard (Wong and Anderson 1993, Anderson *et al.* 1996, Latham and Poulin 2002). Coastal (herons, egrets and cormorants) and pelagic birds (albatross) are also infected with *Desmidocerca, Desmidocercella* and *Diomedenema* spp. (Aproctoidea, Desmicercidae) which invade the air sacs (Anderson and Bain 1976).Marine species of the suborder Spirurina are distributed among seven superfamilies and 11 families. Species which mature in marine fish belong to four families, the Gnathostomatidae and Metaleptidae (Gnathostomatoidea), the Physalopteridae (Physalopteroidea) and the Rhabdochonidae (Thelazioidea). Gnathostomatids (*Echinocephalus* spp.) and metaleptids (*Metaleptus* spp.) occur in the stomachs and intestines of elasmobranchs and chimaerids (Beveridge 1987, Hoberg *et al.* 1998, Moravec and Nagasawa 2000). Among physalopterids, *Proleptus* spp., occupy either the stomach or spiral valve of elasmobranchs (Heupel and Bennet 1998), and *Bulbocephalus* spp., the intestines of ray-finned fish (Hoa *et al.* 1972). Marine rhabdochonids include *Fellicola, Hepatinema, Heptochona* and *Johnstonmawsonia* from the intestines, body cavity, liver, bile duct, gall bladder and pancreatic duct of actinopterygians (Chabaud 1975, Petter and Køie 1993, Tanzola and Gigola 2002), and *Pancreatonema* spp. from the pancreatic duct of elasmobranchs (Borucinska and Frasca 2002).

Marine spirurinans distributed within four families (Cystidicolidae, Crassicaudidae, Habronematidae and Tetrameridae) of the Habronematoidea are parasites of fish, birds and cetaceans. Cystidicolids (*Ascarophis, Caballeronema, Capillospirura, Cristitectus, Cystidicola, Metabronema, Pseudoproleptus, Spinitectoides* and *Spinitectus* spp.) usually occupy the stomach, pyloric caecae or intestines of ray-finned fish (Køie 1993, McDonald and Margolis 1995, Sanmartin et al 2000, Rocka and Stefanski 2002) although *Parascarophis sphyrnae* occur in the spiral valve of elasmobranchs (Knoff *et al.* 2001). Habronematoids of marine birds include *Stellacaronema* spp. (Habronematidae) from the gizzard of shore birds (Wong and Anderson 1993), and *Tetrameres* spp. (Tetrameridae) from the crop and gizzard of coastal (Anderson *et al.* 1996) and pelagic birds (Bourgeois and Threlfall 1979). Crassicaudids are parasites of cetaceans, with *Crassicauda* spp. occurring in deep tissues of the mammary glands, in cranial sinuses, the inner ear, and the urogenital system, and *Placentonema gigantisma*, in the uterus and placenta of sperm whale (*Physeter macrocephalus*) (Dailey 2001). *P. gigantisma*, which reaches 9 m in length and 2.5 cm in width, is the largest nematode known (Dailey 1985).

Acanthocheilonema (*Dipetalonema*) *spirocauda* and *A. odendhali* spp. (Filarioidea, Onchocercidae) from the heart, aorta and intermuscular fascia of phocids and otariids, respectively, appear to be only spirurids identified from seals (Measures *et al.* 1997). Together with *Oswaldofilaria kanbaya* from saltwater crocodile (Manzanell 1986), however, they also seem to be the only filarioid worms described from marine or estuarine hosts.

Life cycles

According to Anderson's (1984, 1996) hypothesis, parasitic nematodes first evolved in terrestrial hosts and were able to invade aquatic environments, only after the development of heteroxeny (use of intermediate hosts) and paratenesis (use of transport hosts). Intermediate hosts support larval growth and development to a stage where the nematode is capable of infecting its definitive host, and both intermediate and paratenic hosts participate in the temporal and spatial dispersal of the parasite, thereby increasing the likelihood that it will be ingested by the final host.

While heteroxeny and paratenesis are common in the (known) life cycles of marine parasitic nematodes, there are notable exceptions. The Mermithida are monoxenic (single host) nematodes which are common in aquatic hosts, and in the case of the mermithoid-like Marimermithida and Benthimermithida, exclusively marine. Mermithidans may have exploited aquatic invertebrates prior to the development of terrestrial forms (Adamson 1986). Their frequency in aquatic hosts stands in stark contrast to the dearth of freshwater and marine examples of oxyurids ('oxyspirurida' above), monoxenous parasites believed to have evolved in terrestrial invertebrates (Anderson and Lim 1996). As a rule, mermithidans complete larval development in an invertebrate host and are free living as adults, although for some (e.g. the sea urchin parasites *Echinomermella* spp.) maturation and reproduction also occur within the host (Jangoux 1990). The success of mermithidans in the marine environment is probably attributable to the hosts being primarily sedentary, benthic organisms (annelids, molluscs, echinoderms and crustaceans) as opposed to the more active and wider ranging fish and higher vertebrates which host nematodes with heteroxenic life cycles. With the exception of ascarididans (*Hysterothylacium* spp.) and spiruridans (*Ascarophis* spp.) which occasionally undergo neotenic development to adults in crustacean intermediate hosts (Anderson 2000), mermithidans are the only nematodes which reach maturity in marine invertebrates.

Trichinella nativa (Trichurida) is a monoxenous parasite that matures in the intestine of carnivorous mammals, and produces numerous first stage (L1) larvae which migrate to the musculature (Roberts and Janovy 2000). Marine mammals become infected by ingesting flesh that contains encapsulated L1s while scavenging on mammal remains (e.g. polar bears, seals, walruses and sled dogs) or meat discarded by hunters (Kapel *et al.* 2003). Experimental evidence, however, suggests that *T. nativa* may also be transmitted by sarcophagous crustaceans and fish acting as transfer hosts. Direct transmission of eggs or larvae, transplacentally, in breast milk, or in urine contaminated breast milk has been proposed for pseudaliids (Strongylida, Metastrongyloidea) and crassicaudids (Spirurida, Habronematoidea) of cetaceans, but again, there is speculation that small crustaceans may be involved in transmission (Dailey 1985, Dailey *et al.* 1991). Direct transmission has also been proposed for the sea cow parasite, *Paradujardina halicoris* (Ascarida, Ascarididae) (Lauckner 1985d). Although hookworms (Strongylida, Ancylostomidae) are essentially monoxenous, it can be argued that *Uncinaria lucasi* may use the northern fur seal as either a definitive or transfer host (Raga 1997). Hatching and two moults to the third larval stage (L3) occur terrestrially in host breeding colonies. The L3s penetrate the skin of the flippers of juvenile seals and migrate to the blubber where they persist indefinitely until host hormonal activity stimulates movement to the mammary glands of mature female. Larvae ingested with beast milk complete the third and fourth moults, mature and reproduce in the intestines of pups.

Heteroxenic life cycles have been reported for diverse taxa of enopleans and rhabditeans. Among capillarids (Trichurida) of fish, transmission is either direct or indirect with oligochaetes usually serving as intermediate hosts (Moravec 2001). Experimental transmissions have shown, however, that oligochaetes are transfer hosts, and small fish obligate intermediate hosts for *Capillaria gracilis*, a rectal parasite of Atlantic cod (*Gadus morhua*) (Køie 2001a). Similarly, *Eustrongylides* spp. (Dioctophymatida), which mature in the crop of wading birds, use oligochates or small fish as intermediate hosts, and larger fish as paratenic hosts (Coyner *et al.* 2001). Seal lungworms (Metastrongyloidea), *Filaroides* (*Parafilaroides*) spp. (Filaroididae) of otariids and phocids, and *Otostrongylus circumlitus* (Crenosomatidae) of phocids, develop to the infective L3 in the intestinal wall of fish, bypassing the molluscan intermediate host used in transmission of terrestrial metastrongyloids (Bergeron *et al.* 1997).

Heteroxeny has frequently been described in marine ascarididans, with particular attention having been paid to the medically and economically important anisakines. In the life cycle of the sealworm *Pseudoterranova decipiens* (*sensu stricto*) (Fig. 3.24) (McClelland 2002), for example, partially embryonated eggs passed in seal faeces settle to the sea bed where they complete development to the L3 and hatch. Newly hatched larvae are 0.2 mm in length, ensheathed in the cuticle of the previous larval stage and caudally attached to the substrate. When ingested by benthic

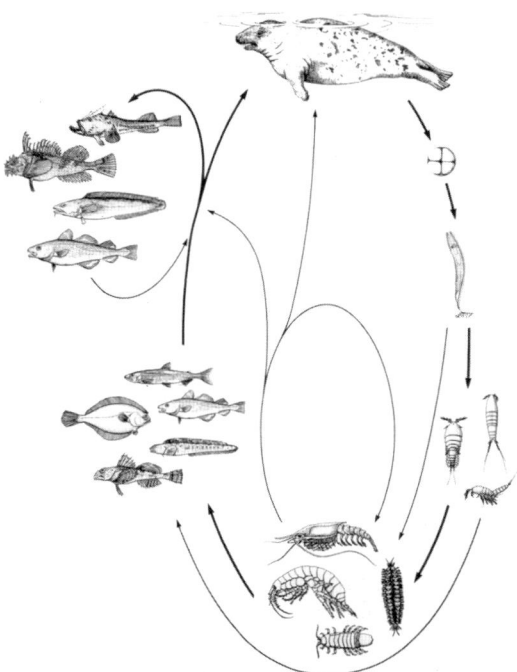

Figure 3.24 Life cycle of *Pseudoterranova decipiens* (*sensu stricto*) (Ascaridida, Anisakidae). See text for explanation.

crustaceans (harpacticoid and cyclopoid copepods, and juvenile amphipods and mysids), they exsheathe, penetrate to the haemocoele and begin to grow. Copepods, evidently, serve only to enhance transmission to a larger array of benthic macroinvertebrates (e.g. mature amphipods and mysids, polychaetes). In macroinvertebrates, larval sealworm reach 2 mm in length, at which point they are infective to fish, and may ultimately exceed 8 mm in length, whereupon they are also infective to seals. The invertebrate hosts, however, are usually ingested by primary fish hosts, which are typically small benthic consumers, including juveniles of larger demersal species. The larvae penetrate the gut wall of the fish host and usually establish themselves in the musculature where they continue to grow to 30 mm to 60 mm in length. Large, demersal piscivores may serve as a second or third fish host, but probably represent a cul de sac for larval sealworm, as they are seldom exploited by seals. Following ingestion by the definitive seal host, infective sealworm L3s escape from the bodies of fish or invertebrate hosts, embed their anterior extremities in the gastric mucosa, and, after completing the final two moults in the life cycle, mature and reproduce. Whaleworms (*Anisakis* spp.) seem to have equally complex life cycles involving pelagic invertebrates (e.g. ephausiids, squid) and fish (Raga 1997). They are also transmitted to demersal fish by euphausiids, shrimp and small fish (capelin, sandlance) that are found near bottom in daytime but feed pelagically at night (McClelland *et al.* 1990). *Sulcascaris sulcata*, an anisakine parasite of marine turtles, however, has a relatively simple life cycle with marine bivalves (scallops, oysters) serving as intermediate hosts (Lauckner 1983).

Squid are not only important in the transmission of whaleworms but also host larvae of the raphidascarines, *Hysterothylacium* spp. (Hochberg 1990) and *Lappetascaris* sp. (Nagasawa and Moravec 2002), which mature in piscivorous marine fish. *Hysterothylacium aduncum*, a cosmopolitan species with invertebrates as obligate intermediate hosts and both invertebrates and fish as paratenic hosts, has been reported from dozens of host species in seven invertebrate phyla

(Coelenterata, Ctenophora, Mollusca, Annelida, Arthropoda, Echinodermata and Chaetognatha) (Rohde 1984). Other ascaridid taxa with indirect transmission include *Goezia holmesi* (Goeziinae) of sea snakes, which uses copepods as intermediate hosts and fish as paratenic hosts (Lauckner 1985a). Acanthocheilids of elasmobranchs similarly employ decapods as intermediate hosts and fish as paratenic hosts (Rohde 1984), while cucullanids (Seuratoidea) of flatfish and cod reach the infective stage in polychaetes (Køie 2000b, 2001b) or small fish (Køie 2000a, c).

Heteroxenous transmission is also frequently reported for marine spiruridans. Acarioids of coastal and pelagic waterfowl use crustaceans (amphipods, decapods) as intermediate hosts, with fish serving as paratenic hosts for species that mature in piscivorous birds (Langton and Hillgarth 1995, Anderson et al. 1996, Jackson et al. 1997, Køie 1999). Amphipods and decapods also act as intermediate hosts for camallanids (*Spirocamallanus* sp.) (Bashirullah and Aguada 1993), physolopterids (*Proleptus* spp.) (George-Nascimento et al. 1994) and cystidicolids (e.g. *Ascarophis* spp.) (Matrorelli et al. 2000) of elasmobranches and ray-finned fish, and for tetramerids (e.g. *Tetrameres* spp.) of shore birds (Jackson et al. 1997). *Echinocephalus* spp. (Gnathostomatidae) of elasmobranchs and chaemeriform fish use pectinid molluscs (e.g. oysters, abelone) (Lauckner 1983) and sea urchins (Jangoux 1990), while ectoparasitic seal lice (*Echinopthyrius horridus*) are suspected of transmitting phocid heartworm, *Acanthocheilonema spirocauda* (Filaroidea, Oncocercidae) (Dailey 2001).

Effects on hosts

Although the impact of parasitic nematodes on marine fauna can generally be described as benign or, perhaps, in need of further study, some marine enopleans and rhabditians have proven highly pathogenic and even lethal to their hosts. Loss of appetite, emaciation and mortality have been attributed to *Capillaria* spp. (Trichurida) infections in fish (Williams and Jones 1994), while lethagry, inappetitia and weight loss were observed in seals with experimental *Trichinella nativa* infections (Fig. 3.25A) (Kappel et al. 2003). *Eustrongylides* spp. (Dioctophymatida) produce fibrous lesions in the proventriculus of piscivorous wading birds and have been associated with retarded growth and mortality in their nestlings (Coyner et al. 2001). In many instances, invertebrate hosts of mermithidans die as the nematodes leave the host to reproduce in the external environment (Poinar et al. 2002). In green sea urchins, the consequences of infection with *Echinomemella matsi* include retarded growth, functional castration and an attenuated life span (Stein 1999).

The deleterious effects of hookworms (Ancylostomatoidea) and lungworms (Metastrongyloidea) in marine mammals have been reviewed frequently (Lauckner 1985c, Raga 1997, Dailey 2001, Measures 2001). Hookworms (*Uncinaria* spp.) cause haemorrhagic enteritis and anaemia and are often fatal to young seals, while respiratory obstruction (Fig. 3.25B), pneumonia and death are frequently attributed to the lungworms, *Filaroides* (*Parafilaroides*) spp. of phocids and otariids and *Otostrongylus circumlitus* of phocids. Cetacean lungworms (Pseudaliidae) are associated variously with respiratory obstruction, pathology of the small bronchi, granulomatous or necrotic fibrous lesions, pneumonia and mortality of juvenile hosts. Those (e.g. *Stenurus* spp.) that invade the cranial sinus system of odontocetes are believed to interfere with hearing and navigation, and, hence, may be implicated in strandings. While records of strongylidan pathology in marine birds are lacking, gizzard worms (Trichostrongyloidea, *Amidostomum* spp.) and gapeworms (Strongyloidea, *Cyathostoma* and *Syngamus* spp.), known pathogens of waterfowl, have been found in gulls and other coastal birds (Borgsteede 1997).

Pathogenesis of adult ascarididans is generally confined to the gut of marine vertebrate hosts. Lesions or haemorrhaging in the gastric or intestinal mucosa have been found in association with acanthocheilids and anisakines (*Terranova* spp.) of elasmobranchs and raphidascarines

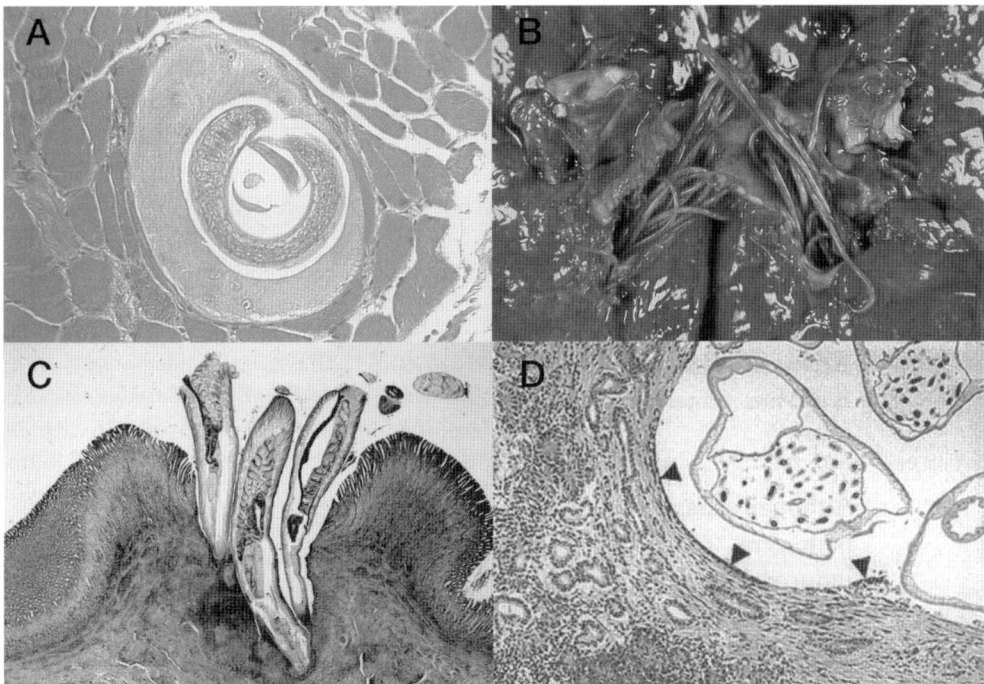

Figure 3.25 Pathology associated with nematode infection in marine fauna. **A.** First stage larva of *Trichinella nativa* (Trichurida) in musculature of grey seal (*Halichoerus grypus*). Photograph courtesy of Lena Measures, Maurice Lamontagne Institute, Fisheries and Oceans Canada. **B.** *Otostrongylus circumlitus* (Metastrongyloidea, Crenosomatidae) in bronchi of harbour seal (*Phoca vitulina*). Photograph courtesy of Lena Measures. **C.** *Pseudoterranova decipiens* (Anisakidae) associated with granuloma in stomach of harbour seal. **D.** Transverse sections of *Pancreatonema americanum* (Spirurida, Rhabdochonidae) in the dilated pancreatic duct of dogfish; arrowheads indicate the flattened and eroded ductal epithelium (after Borucinska and Frasca 2002).

(*Hysterothylacium*, *Lappetascaris* and *Thynnascaris* spp.), goeziines and cucullanids of ray-finned fish (Rohde 1984, Williams and Jones 1994). The attachment sites of *Cucullanus carettae* in the upper intestine of loggerhead turtles were marked by haemorrhaging (Lauckner 1985a), while ulcerative gastritis in cormorants (Kuiken *et al*. 1999) and emaciation and death of brown pelicans (Dyer *et al*. 2002) were attributed to *Contracaecum* spp. (Anisakinae) infections, albeit of freshwater origin. Gastritis or ulcers (Fig. 3.25C) are often found in association with aggregations of L3, L4 and adult anisakines (*Anisakis*, *Contracaecum*, *Pseudoterranova* and *Phocascaris* spp.) in the stomachs and upper intestine of pinnipeds, cetaceans and sea otter (Dailey 2001). Symptoms of heavy infection often include diarrhoea, dehydration and anaemia. Intestinal perforations leading to peritonitis and death have been attributed to *Contracaecum* infections in sea lions and *Pseudoterranova* infections in sea otter.

Although adult spiruridans, like ascaridans, are often associated with lesions in the gastrointestinal tract of their definitive vertebrate hosts, they may also be pathogenic in subcutaneous tissues, and deep tissues and organs (Rohde 1984, Williams and Jones 1994). *Echinocephalus* (Gnathostomatidae) and *Proleptus* spp. (Physalopteridae) are found in association with nodules, ulcers and connective tissue proliferation in the lining of the stomach, spiral valve or intestine of elasmobranchs (Rohde 1984, Heupel and Bennet 1998). Rohde (1984) notes that, although reports on pathology of *Spirocamallanus* spp. (Camallanidae) in the gut of marine fish are lacking, infection with related nematodes (*Camallanus* spp.) have proven lethal in aquarium fish. Lesions and enteritis in the intestinal wall, and ulcers in the swim bladder wall of ray-finned fish have been attributed to *Cystidicola* and *Spinitectus* spp. (Cystidicolidae), respectively (Williams and Jones 1994), and pancreatitis in spiny dog fish (*Squalus acanthius*), to the presence of *Pancreatonema americanum* (Rhabdochonidae) in the pancreatic ducts (Fig. 3.25D) (Borucinska and Frasca 2002). Among dracunculoids, *Philometra* spp. may cause visceral edema, granulomata, adhesions, and atrophy of the ovaries (Rohde 1984, Hesp et al. 2002), while visceral adhesions resulting from *Philonema* spp. infections in salmonids may prevent spawning (Nagasawa 1987) or compromise the hosts ability to forage (Moles 2003). The dracunculoid, *Phlyctainophora squali*, has been found in association with ulcers and cysts in the skin and buccal cavity of dogfish (Adamson et al. 1987). But for evidence that infection with *Seuratia* spp. may result in delay of moulting and fewer wing primaries in albatross (Langston and Hillgarth 1995), the effect of acuarioid and habronematoid nematodes in the foregut of marine birds remains largely unknown. Crassicaudids in cetaceans have been implicated in the erosion of cranial bone, renal lesions and obstruction of the urethra, and lesions in the mammary glands, which possibly result in reduced milk production and reproductive success (Dailey 1985, 2001). Fetal death has been attributed to the crassicaudid, *Placentonema gigantisma*, in the uterus and placenta of sperm whales, and anorexia, dyspnea and erratic breathing to the heartworm, *Acanthocheilonema spirocauda* (Oncocercidae), in the right ventricle of pinnipeds.

Nematode parasites of marine vertebrates may also be pathogenic in their intermediate hosts. Loss of tone in the abductor muscle, and discolouration of both the abductor muscle and gonads of scallops have been attributed to infection with larvae of the turtle parasite, *Sulcascaris sulcata* (Anisakidae), while the host reaction to encapsulated larvae of elasmobranch parasites, *Echinocephalus* spp. (Gnathostomatidae) may lead to obstruction of the genital ducts in oysters (Lauckner 1983), and interference with gametogenesis in sea urchins (Jangoux 1990). Larval anisakines (*Pseudoterranova decipiens*) caused erratic behaviour and death in experimentally infected marine crustaceans (McClelland 1990), and raphidascarine larvae, *Hysterothylacium* spp., are believed to interfere with gametogenesis in naturally infected chaetognaths (Pierrot-Bults 1990). Various larval anisakids (*Anisakis, Contracaecum, Pseudoterranova, Hysterothylacium* spp.) have been implicated in mechanical compression or necrosis of the liver, lesions in the gut wall, viscera and musculature, depletion of lipids and mortality in heavily infected marine fish (Rohde 1984, Williams and Jones 1994). Erratic behavior and death of fish with heavy *Pseudoterranova decipiens* infections has been attributed to mechanical damage of vital organs and tissues resulting from migration and feeding of the larval nematodes (McClelland 2002). Further, the anaesthetic effect of volatile ketones, which larval sealworm produce as metabolic byproducts, may impair the intermediates host's ability to forage or avoid predators, thereby promoting transmission to a subsequent (fish or seal) host. Similarly, *Skrjabinoclava morrisoni* (Acuariidae) larvae induce their intermediate host, a tube-dwelling amphipod (*Corophium volutator*) of intertidal mud flats, to increase surface activity during daylight hours, when there is greater likelihood of ingestion by the parasite's final host, the sandpiper *Calidris pusilla* (McCurdy et al. 1999).

Medical, economic and ecological importance

Nematodes of medical significance are found among trichinellids, dioctophymatids and anisakids of marine mammals and birds. Outbreaks of human trichinellosis in the arctic and subarctic have attributed to consumption of *Trichinella nativa* larvae in walrus meat (for a detailed account, see pp. 436–439). While adult *Trichinella* may cause gastrointestinal or respiratory distress and fever, a plethora of conditions, including pneumonia, pleurosy, encephalitis, meningitis, peritonitis, fatal myocarditis, and severe muscle pain and fatigue have been associated with larval migration and penetration of muscle fibres (Roberts and Janovy 2005). Evidently,

T. nativa larvae in mammalian flesh are freeze tolerant for up to four years, and can survive traditional methods of food preparation in northern communities (Forbes *et al.* 2003). Marine mammal anisakids (*Anisakis*, *Contracaecum*, *Phocascaris* and *Pseudoterranova* spp.) served in raw, undercooked and lightly marinated seafood are also potential human pathogens (Smith 1999). Archetypal cases of anisakiasis or anisakiosis (Couture *et al.* 2003), involving penetration of the alimentary tract and associated organs and clinical symptoms (nausea, severe epigastric pain and vomiting), are reported largely from Japan where seafood is an important component of the diet and traditionally served raw (for a detailed account, see pp. xx–xx). *Anisakis* spp. larvae have been diagnosed as the pathogen in most cases, the remainder being attributed to infection with *Pseudoterranova* spp. larvae. Most cases of *Pseudoterranova* spp. infection documented in Europe and the Americas have been largely asymptomatic, being diagnosed after expulsion of the nematodes by coughing, vomiting or defaecation (Smith 1999). Although cases are relatively rare, *Eustrongylides* spp. larvae, which normally mature in fish-eating birds, may produce severe abdominal pain in humans when consumed in improperly prepared marine fish (US Food and Drug Administration/Center for Food Safety and Applied Nutrition 1991). Experimental evidence suggests that larvae of the elasmobranch parasite, *Echinocephalus sinensis* (Gnathostomatidae), and the rat lungworm, *Angiostrongylus cantonensis* (Metastrongyloidea), potential and known human pathogens, respectively, may be transmitted by oysters or marine clams (Lauckner 1983). Case histories which implicate marine bivalves as the source of human gnathostomiasis or *Angiostrongylus* infection, however, are lacking.

Aside from their medical significance, larval anisakids also represent a chronic and costly cosmetic problem for seafood processors (McClelland 2002). Because of their large size and reddish-brown colour, larval sealworms (*P. decipiens* species complex) are particularly conspicuous in the white flesh of demersal fish. Sealworm infections have cost processors in the North Atlantic and Northwest Pacific tens of millions of dollars annually in increased processing costs related to detection and removal of the parasite, and downgrading and discard of product. Infection with larvae of the sea turtle parasite, *Sulcascaris sulcata*, which causes discolouration and loss of tone in the abductor muscle have caused similar increases in the cost of processing scallops (Lauckner 1983). Surf clams were withheld from commercial trade along the eastern seaboard of the United States of America during the mid 1970s, when *S. sulcata* larvae, which are normally lightly coloured and had previously gone unnoticed in surf clams, became a more conspicuous brownish-black as a consequence of pseudocoelomic infection with haplosporidian hyperparasites.

As evident above, parasitic nematodes are often highly pathogenic in marine hosts, but, generally, their effects on host abundances, and ultimately on ecosystems, remain to be described or quantified. The mermithidan, *Echinomermella matsi*, however, serves as a marine example of a keystone parasite which, in controlling the abundance of its sea urchin host in Norwegian waters, regulates the entire ecosystem (Marcogliese 2002). Left unchecked, sea urchins decimate kelp beds creating urchin-dominated barrens, but outbreaks of *E. matsi* infection reduce sea urchin populations and promote redevelopment of species-rich kelp forests.

Important references
References on marine parasitic nematodes are scattered throughout the literature, but few texts offer comprehensive reviews. Well-illustrated sections on nematode parasites of marine fauna are found in a book by Williams and Jones (1994), and volumes II (1983), III (1990), IV Part 1 (1984) and IV Part 2 (1985) of 'Diseases of Marine Animals' (Ed. O Kinne). Chabaud (1974) provides keys to the subclasses, orders and superfamilies of nematodes, while Anderson (2000) discusses nematode life cycles and transmission. Nematode systematics and morphology described in this section primarily follow Roberts and Janovy (2005).

Acanthocephala (thorny or spiny-headed worms)
Horst Taraschewski

Introduction
Acanthocephalans, the 'thorny headed worms', comprise at least 1000 species. Vertebrates are final hosts, although recent findings suggest that certain marine ichthyoparasitic species can mature in squids. About half of the described species parasitise fish, in which the tubiform worms may reach a length of a few centimetres. Many parasitologists still treat the Acanthocephala as a phylum of its own, others consider them and the Rotatoria as one taxon, the 'Syndermata' (Herlyn et al. 2003).

Morphology and development
The body of a typical acanthocephalan consists of the metasoma (trunk) lying inside the intestinal lumen of the host, and the presoma (proboscis and neck) inserted into the intestinal wall (Figs 3.26, 3.27) This holdfast organ resembles the 'introvert' of Rotatoria, Priapulida, Kinorhyncha and Nemamorpha larvae. *In situ*, the proboscis is usually kept semi-invaginated (Fig. 3.26). A fully evaginated proboscis as shown in textbooks and taxonomy papers is a post-mortem condition (Fig. 3.28). For taxonomic investigation the proboscis hooks must be counted and measured in worms relaxed in chilled tapwater. Acanthocephala do not have an intestine. They absorb their nutrients through densely packed crypts of the outer membrane of the metasoma from the intestinal lumen, and from the intestinal wall through crypts of the presoma. The proboscis cavity formed by a semi-invaginated proboscis might represent a remnant of the former gut, functioning as a collecting funnel for nutrients from the necrotic host tissue

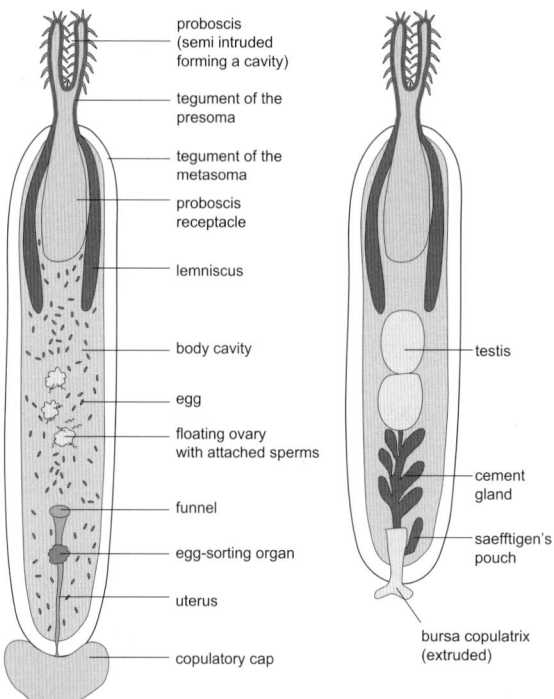

Figure 3.26 Schematic drawings of a typical female and male acanthocephalan (short-necked echinorhynchid palaeacanthocephalan parasitising fish). According to Taraschewski (2000).

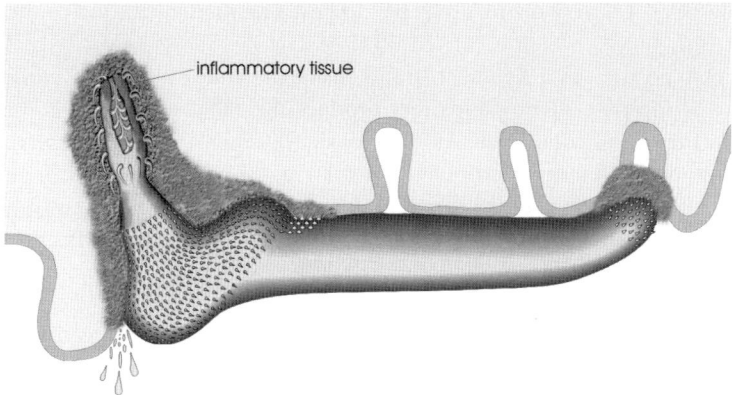

Figure 3.27 Schematic drawing of a male *Corynosoma* sp. attached to the mucosa of a seal's intestinal wall. According to Supper, Hammill and Taraschewski unpublished.

at the point of attachment. In Palaeacanthocephala and Eoacanthocephala the pointed proboscis hooks are covered by a thin layer of sclerotised tegument and, thus, can take up nutrients from pierced host tissue. They also can discharge substances into it. Furthermore, Polymorphida possess trunk spines which function as an additional holdfast organ (Taraschewski 2000).

The tegument of the Acanthocephala is a syncytium. The nuclei of the presoma-tegument lie within the lemnisci (i.e. sack-shaped outgrowths of the presoma-tegument projecting into the body cavity). The subtegumental musculature consists of longitudinal and circular muscles, whose fine structure resembles that of nematodes as consisting of an outer myogenic and an inner cytoplasmic section. In mature females, which are larger than the males, ovaries, eggs (and sperm from a copulation) float within the voluminous body cavity. Embryonated eggs are sorted out by an 'egg-sorting-organ' and are released into the host's intestinal lumen via the

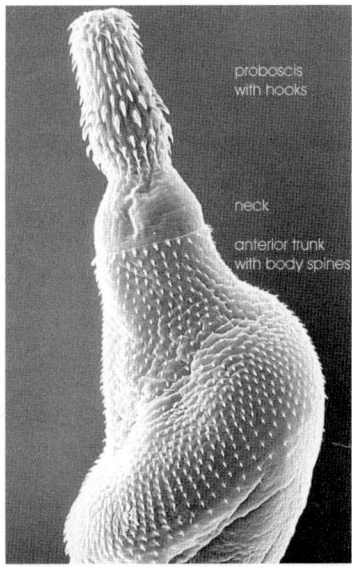

Figure 3.28 Scanning electron micrograph of the anterior end of a species of *Corynosoma* with post-mortem evagination of entire proboscis. According to Supper, Hammill and Taraschewski unpublished.

vagina. Females after sex carry a copulatory cap, imposed on them by the males after injection of sperm ('the selfish gene'!), which is carried till they start discharging eggs. In males the sexual organs are kept in position by a 'ligament strand' inside the body cavity. They consist of two testicles and one or several cement glands that produce the substance used for 'locking up' the female body at the end of copulation. Among Eoacanthocephala a cement reservoir is present. Infectious acanthocephalan eggs harbour the first larva, the acanthor, which is surrounded by four (Neoechinorhynchidae: five) envelopes interspersed with electron-lucent interstices. Each envelope has a specific chemical composition and is used, for instance, in attaching the egg to a food substrate used by the intermediate host, or in activating and hatching of the acanthor. Once the spiny acanthor has penetrated into the haemocoel of the intermediate host, it discharges a thick, spongy membraneous envelope, which first provides protection from the phenoloxidases and peroxidases of the host´s haemocytes, and then becomes compressed, detaches from the larva's surface and has camouflaguing properties (provided the infected intermediate host is a susceptible species). In such a host, the further development via the acanthella towards the cystacanth, the stage infectious to the definitive host, proceeds. The name cystacanth implies that the larva is encysted. However, this is only the case if it belongs to an acanthocephalan species using final hosts that grind their food in the anterior part of their digestive tract, such as ducks. If the final host is a fish which simply swallows the infected crustacean, the cystacanth usually is sausage-shaped with little encystment. The stage persisting in the viscera of paratenic hosts does not have a specific name. It resembles a 'cystacanth' (Taraschewski 2000).

Diversity and origin of marine acanthocephalans

The Acanthocephala comprise four subtaxa: Palaeacanthocephala, Eoacanthocephala, Polyacanthocephala and Archiacanthocephala (Garcia-Varela *et al.* 2002) which differ in morphology, life cycles and several ecological features. Apparently, the Acanthocephala have not evolved in the sea. Ancient marine fish like elasmobranchs are not known to be final hosts of any acanthocephalan and not even single cases of a host switch from teleosts to sharks or rays have been described. Among extant Acanthocephala (except for the terrestrial Archiacanthocephala) certain genera comprise or include marine species. Often only one or a few species of one genus are marine, like *Neoechinorhynchus agilis* (Table 3.5). A few genera such as *Rhadinorhynchus* or *Corynosoma* (Table 3.5) seem to have undergone considerable speciation in the marine environment. Table 3.5 lists important examples.

Concerning the marine *Echinorhynchus gadi* (Table 3.5) as well as the limnic species *Pomphorynchus laevis*, speciation appears to be continuing in brackish water. In the Baltic Sea *E. gadi* can be found in fish of areas with rather low salinities where it is associated with limnic gut dwellers such as the nematode *Raphidascaris acus*.

Acanthocephalans are very abundant in deep sea fish (even *E. gadi* frequently occurs in greater depths) (Klimpel *et al.* 2001), and in Antarctic fish (Zdzitowiecki and White 1996), mammals and birds. Generally, marine acanthocephalans have wider distributional ranges than freshwater ones, probably because of less effective barriers to genetic exchange in the sea compared with fresh water. *Neoechinorhynchus agilis* has a worldwide distribution in temperate waters, like its major final host *Mugil cephalus* (Table 3.5).

Life cycles in and adaptations to the marine environment

Life cycles of Acanthocephala inhabiting the sea (see Fig. 3.29) do not principally differ from species transmitted in fresh water, however, the involvement of paratenic hosts is more common among marine species. Most whales and pinnipeds do not feed on crustaceans, so extraintestinally positioned parasite stages in fish are necessary to close the ecological gap in the transmis-

Table 3.5 Abundant and well-known Acanthocephala with marine life cycles

Species, systematic affiliation	Definitive, intermediate, paratenic hosts	Geographical distribution
Palaeacanthocephala		
Echinorhynchida	Gadidae and other fish, amphipods	North Atlantic (including Western und Central Baltic Sea)
Echinorhynchus gadi		
Aspersentis megarhynchus	Various teleost fish, amphipods	Antarctic waters
Polymorphida		
Corynosoma cetaceum	Cetacea (whales, dolphins), amphipods, fish	Cosmopolitan species
C. strumosum	Pinnipeds (seals), amphipods, fish	Northern hemisphere (Palaearctis)
C. australe	Cetaceans and pinnipeds, amphipods, fish	Southern hemisphere (not Antarctis)
C. enhydri	sea otters (*Enhydra lutris*), amphipods	West coast of North America, Aleutean Islands
Profilicollis botulis	eider ducks (*Somateria mollissima*), green crab (*Carcinus maenas*)	North Atlantic
P. antarcticus	Shore-birds, Brachyura (crabs)	Subantarctic waters (New Zealand South Island, Chile etc.)
Eoacanthocephala		
Neoechinorhynchida		
Neoechinorhynchus agilis	Mugilid fish, probably ostracods	Cosmopolitan species (in temperate waters)

sion cycle. Palaeacanthocephan life cycles (most marine acanthocephalans belong to this taxon) permit the following generalisations:

- Intermediate-host specificity is less pronounced than in freshwater species: crustaceans belonging to different families may transmit a species. Final-host specificity may be narrow or wide, as in freshwater acanthocephalans.
- In most Echinorhynchida amphipods serve as intermediate hosts and fish as final hosts. Paratenic hosts do not seem to be necessary, but it is probable that postcyclic transmission (shown for *Echinorhynchus gadi*, and known from many freshwater acanthocephalans) commonly occurs (i.e. juvenile or mature worms located inside the gut of a host are transmitted to another host where they reattain an intraintestinal position).
- Species of the Polymorphida use mammals and/or birds as final hosts. If seals or whales are the preferred hosts, amphipods serve as intermediate and various fish as paratenic hosts. The longevity and host–parasite relationships of (extraintestinal) acanthocephalans in marine paratenic hosts have not yet been investigated.

  If shore birds serve as final hosts, amphibic crabs usually act as intermediate hosts. The cycles less commonly include amphipods and fish. Birds may function as an ecological sink of seal parasites, that is, many individuals of an acanthocephalan's

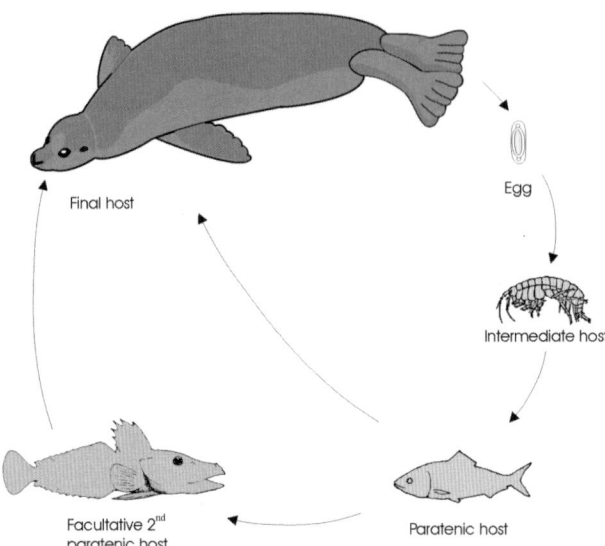

Figure 3.29 Marine acanthocephalan life cycle involving amphipods, fishes and seals. Based on information from various authors.

population 'end up' in hosts where they cannot reproduce, negatively affecting the parasite population. Alternatively, seals act as sinks of bird parasites. Ice foxes picking up fishes from the beach and feeding on fish wastes derived from human activities reflect the occurrence of certain polymorphid species in the nearby sea, which are, however, probably not transmitted by the foxes.

Effects on individual hosts and host populations

Parasites are known to regulate host populations by affecting fitness parameters of infected individuals, resulting in a reduced net reproduction of the population, provided the abundance of the parasite is high enough. Also, sporadically or periodically occurring parasite-induced mortality may be an important regulatory factor. Concerning intermediate hosts, the phenomenon of 'favourisation' has been best studied in freshwater and terrestrial acanthocephalans: aspects of the phenotype like the colour of the carapace as well as various behavioural parameters (e.g. light response, cryptic and evasive behaviour) of infected host individuals are manipulated to make hosts significantly more vulnerable to predation (Taraschewski 2000). Pairing and reproductive success are affected in intermediate hosts as well, ranging from total castration to reduced fecundity. Only a few marine intermediate hosts of acanthocephalans, however, have been investigated in this respect. Crabs infected with *Profilicollis* spp. reveal alterations in cryptic (burrowing) behaviour and show elevated mortality rates, but effects on the fecundity of the crabs have not yet been demonstrated (Lathan and Poulin 2002).

In the final host the pathogenicity of Acanthocephala is related to the mode of attachment of the parasite species and to the systematic affiliation of the host. Long necked acanthocephalans which deeply penetrate into or through the intestinal wall with their presoma, are more pathogenic than species with a short neck and a shallow mode of anchorage (Taraschewski 2000). In fish even dense infrapopulations of acanthocephalans with either mode of attachment usually do not cause mortality. In warm-blooded hosts, species that do not perforate the intestinal wall have not been diagnosed as causing mortality either. *Corynosoma* species infecting seals or whales (Fig. 3.26) belong to this category. Because of their small size (a few millimetres) their relatively short neck and the thickness of the hosts' intestinal wall, the worms do not project deeply into the intestinal wall and do not reach into the peritoneal cavity. Also, the trunk spination of these worms functions as an auxiliary holdfast organ (Fig. 3.27), so that a deep penetration of the presoma may not be necessary for preventing displacement towards to the anus. In smaller homoeothermic hosts like sea otters or eider ducks, however, the polymorphid acanthocephalans that parasitise these hosts (Table 3.5) have the potential to project into the body cavity causing peritonitis and eventually mortality. As to sea otters, there is no doubt about the high pathogenic potential of their acanthocephalans (Mayer *et al*. 2003), but in shore birds, co-stress factors seem to be necessary to trigger mortalities. Accordingly, if mass mortality occurs in populations of eider ducks, the predominant causative agent remains uncertain.

Ecological, economic and medical importance

Thus far, marine acanthocephalans have been little studied at the level of infrapopulations or component communities. Concerning microhabitats, *Corynosoma* species may prefer the stomach of their hosts (*C. cetaceum*, Table 3.5) or may at least occur in this part of the gastrointestinal tract, which is unusual among Acanthocephala. Marine acanthocephalans are often very abundant in host populations. In individuals of the final hosts they form considerable biomass which, in addition to their high accumulation capacity for metals, makes them useful for biomonitoring studies (see pp. 421–425). In intermediate hosts like amphibic crabs, the prevalence may be 90% or 100% (Latham and Poulin 2002). In contrast, non-littoral amphipods are much less frequently infected, similar to infection rates of intermediate hosts in fresh water. In grey seals (*Halichoerus grypus*) off the Irish coast the prevalence of *Corynosoma strumosum* was 100% with an average intensity of 416 (range: 80–846) worms (O'Neill and Whelan 2002). In areas with dense and growing populations of seals the abundance of extraintestinal acanthocephalans in local fishes can also increase. Protection programs for seals are therefore controversial. However, since the encapsulated juvenile acanthocephalans usually remain restricted to the viscera of fish, high infection rates do not seem to have major economic implications (Zdzitowiecki and White 1996, O'Neill and Whelan 2002).

Nature conservation authorities are well aware of acanthocephalans in their efforts to protect several endangered vertebrates such as sea otters or eider ducks. It appears that the acanthocephalans of these key species (Table 3.5) are major factors regulating their population dynamics by causing morbidity and mortality (Mayer *et al*. 2003). Human infections by marine acanthocephalans very rarely occur and are harmless, in contrast to *Macracanthorhynchus hirudinaceus* (with terrestrial transmission via insects) which is of medical importance in China and south-east Asia (Taraschewski 2000).

Important references

Herlyn *et al*. (2003) discuss the phylogenetic position of the Acanthocephala, and Garcia-Varela *et al*. (2002) that of the Palaeacanthocephala. Important references on marine acanthocephalans are by Zdzitowiecki and White (1996), Klimpel *et al*. (2001), Latham and Poulin (2002), O'Neill and Whelan (2002) and Mayer *et al*. (2003). Taraschewski (2000) gave a thorough review of host–parasite interactions in the group.

Chapter 4

Crustacean parasites

Of all the metazoan groups discussed in this book, the crustaceans are the most diverse and ubiquitous. Among them, the copepods are dominant. They, jointly with the monogeneans, are the most speciose group of metazoan ectoparasites of marine fishes; in addition, they infect a wide range of marine invertebrates. Thousands of species are already known, but many potential host groups have not been examined, and for this reason even approximate estimates of species numbers are impossible. Reflecting the diversity of hosts, copepods show an amazing variety of adaptations which secure infection of and survival on the hosts. Many copepods have great economic importance as agents of disease in wild and aquacultured fish populations. Isopods are primarily found in warm waters, they infect fish but also other crustaceans. Larval isopod parasites of the family Gnathiidae are abundant on the gills of tropical marine fish and represent a primary source of food for cleaner fish. Most branchiurans occur in fresh water, but a few species of the genus *Argulus* are ectoparasites on the skin of marine fish. The tiny tantulocarids are ectoparasites of other crustaceans. To date only 28 species have been described, and little is known about their biology. Thoracica and Rhizocephala are included in the Cirripedia. Few species of the Thoracica are parasitic (on dogfish and polychaetes), whereas the rhizocephalans parasitise other crustaceans. The latter are particularly fascinating because of their extreme sexual dimorphism, the extreme reduction of morphological complexity in the parasitic female, and their ability to change the behaviour of host crabs which benefits the parasite. The Ascothoracida infect various echinoderms and cnidarians. Amphipoda use many groups of marine animals as hosts, including medusae, siphonophores, ctenophores, and thaliaceans. Others (including the whale-lice) infect various marine mammals. When occurring in large numbers, whale-lice may even damage very large humpback whales.

Copepoda (copepods)
Geoff Boxshall

Introduction
Copepods are typically small and inconspicuous aquatic crustaceans but they are extremely abundant. They outnumber even the insects since free-living copepods dominate the zooplankton community in the open pelagic water column, the largest biome on the planet. About 11 500 valid species are known, about half of which live in symbiotic associations. Most of these are probably parasitic but the precise nature of the relationship with the host has yet to be elucidated for the majority. Because of this uncertainty, such forms are typically referred to using the neutral term 'associates' in the copepod literature. Parasitic copepods utilise an extraordinary

range of hosts, occurring on virtually every available phylum in the marine environment from the sponges and cnidarians up to the echinoderms and chordates, including sea squirts, fishes and even mammals. They occupy a similarly wide range of microhabitats on their hosts, both as ectoparasites and as endoparasites. Most research has been conducted on the fish parasites and the biology of the parasites of marine invertebrate hosts is relatively less well known, with the exception of a few species found on commercially important invertebrates such as mussels or scallops. The classification of copepods is in a labile state. Numerous lineages have moved independently into parasitism as a mode of life and the old concept of Copepoda Parasitica as a taxon has no validity (Kabata 1979). Ten orders were recognised by Huys and Boxshall (1991) on the basis of morphological characters, but only nine were adopted by Boxshall and Halsey (2004), four of which are either wholly parasitic (order Monstrilloida), largely parasitic (orders Siphonostomatoida and Cyclopoida) or contain some parasitic forms (order Harpacticoida). Molecular methods based on DNA sequence data are of immense power in the analysis of phylogenetic relationships, but they have yet to fully impact copepod systematics; a period of profound change is to be expected.

Morphology

Basic structure

Copepods exhibit two body plans: the gymnoplean plan in which the body is divided into two tagmata, an anterior prosome and posterior urosome, at the articulation between fifth pedigerous (leg-bearing) and genital segments (referred to as somites by convention), and the podoplean plan in which the prosome and urosome articulate one somite nearer to the head, between the fourth and fifth pedigerous somites (Fig. 4.1). All the parasites conform to the latter type and, though many are profoundly modified, all can be derived from the basic cyclopiform body plan. Cyclopiform copepods are so-called because they resemble the free-living copepod *Cyclops* in possessing well-defined body segmentation, clear tagmosis and the entire set of limbs. The prosome comprises a cephalosome made up of the five cephalic somites typical of all crustaceans plus the incorporated maxilliped-bearing (first thoracic) somite, and the first to fourth pedigerous somites. The urosome typically comprises the fifth pedigerous, genital and four free abdominal somites. In adult males all these somites are separate but in most females the genital and first abdominal somites secondarily fuse at the final moult, to form a genital double-somite. Podoplean copepods typically carry their eggs in paired egg sacs, which are extruded from the paired genital apertures and carried by the female until ready to hatch. The presence of paired egg sacs is a useful clue to the identity of very transformed copepod parasites that lack any other morphological characteristics.

The basic set of appendages comprises five cephalic and seven thoracic limbs, plus the paired caudal rami located on the anal somite. In order from the front the limbs are: antennules, antennae, mandibles, maxillules, maxillae, maxillipeds and first to sixth swimming legs. The ancestral segmentation and setation patterns were hypothesised for all appendages by Huys and Boxshall (1991), who noted that the dominant evolutionary trend in copepods is oligomerisation (fusion of body somites and reduction and loss of appendage segments and setal elements). Oligomerisation typically results from progressive reduction and loss, culminating in the extreme simplification exhibited by the terminal branches of several parasitic lineages within the copepods.

The reader is referred to Kabata (1979), Huys and Boxshall (1991) and Boxshall and Halsey (2004) for detailed illustrated accounts of copepod morphology.

Parasites of fishes

Copepods have been parasitic on fishes at least since the Lower Cretaceous, about 110 to 120 million years ago. Nearly 30 families of copepods contain parasites that utilise fishes as hosts

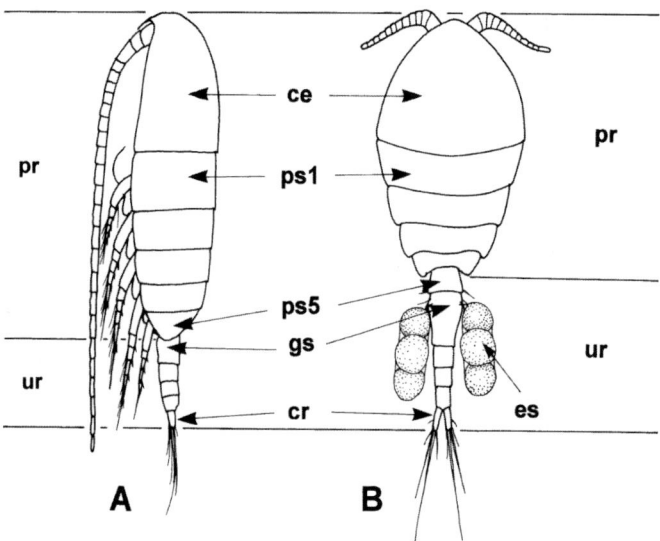

Figure 4.1 Body plans in the Copepoda. **A.** Gymnoplean plan, showing division between prosome and urosome located posterior to the fifth leg-bearing somite. **B.** Podoplean plan, showing division between prosome and urosome located anterior to the fifth leg-bearing somite. Abbreviations: ce, cephalosome; cr, caudal rami; es, egg sac; gs, genital double-somite; pr, prosome; ps1–5, pedigerous somites 1 to 5; ur, urosome.

(Table 4.1) and most are found exclusively on fishes. The body form of fish parasites varies from cyclopiform through to highly metamorphic (rather amorphous bodies that lack expressed segmentation and either have reduced limbs or may lack limbs altogether). Accompanying this trend towards a more transformed morphology is a tendency towards larger body size. Parasites with cyclopiform bodies, such as the Bomolochidae (Fig. 4.2A) and Taeniacanthidae, are typically small (about 1.0–2.0 mm in body length) whereas members of strongly metamorphic families often attain larger body sizes, in the range of 5 mm to 20 mm or greater. There are a few closely related (i.e. bomolochid-like) families, such as the Tegobomolochidae, Telsidae and Tuccidae, that have more transformed, swollen bodies and larger body size. Similar trends can be noted within particular families. The Ergasilidae, for example, contains many estuarine and coastal forms (as well as freshwater species) and exhibits a range from cyclopiform to vermiform, metamorphic bodies. Cyclopiform ergasilids rarely exceed 1.0 mm in body length whereas metamorphic forms commonly attain larger body sizes.

Table 4.1 Copepod families parasitic on marine fishes

Common	Intermediate abundance	Rare
Caligidae	Cecropidae	Lernaeosoleidae
Bomolochidae	Dichelesthiidae	Hyponeoidae
Chondracanthidae	Dissonidae	Tanypleuridae
Ergasilidae	Eudactylinidae	Anthessiidae[A]
Hatschekiidae	Pseudocycnidae	Macrochironidae[A]
Pandaridae	Shiinoidae	Tisbidae[A]
Pennellidae	Sphyriidae	
Lernaeopodidae	Tegobomolochidae	
Lernanthropidae	Telsidae	
Philichthyidae	Tuccidae	
Taeniacanthidae		
[A] Large families with only one species on fish host.		

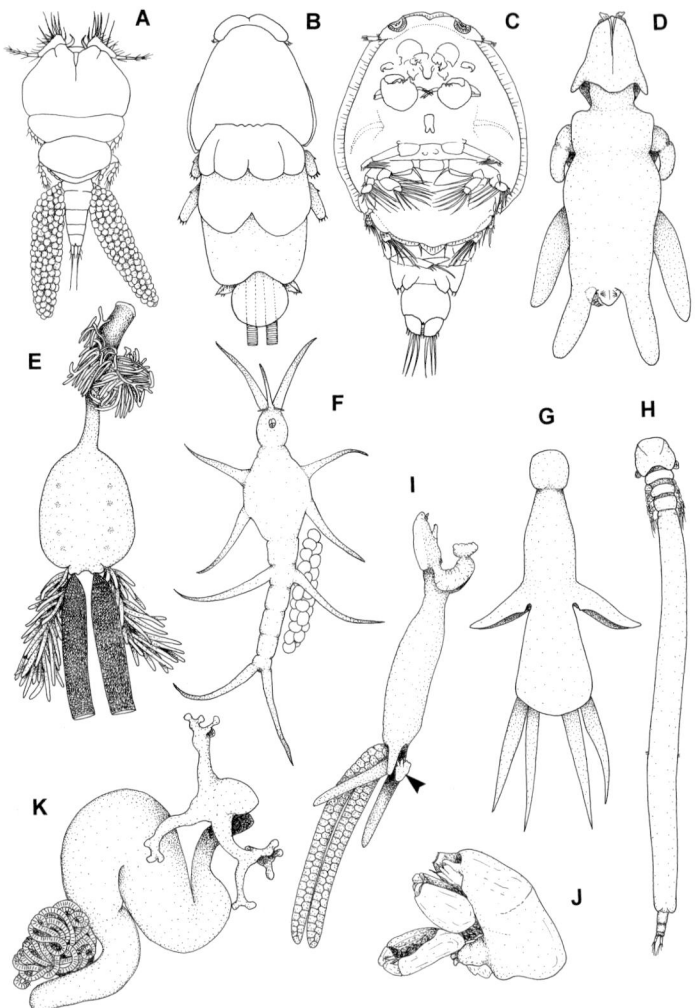

Figure 4.2 Range of gross morphology of copepod families parasitic on fishes. **A.** Bomolochidae. **B.** Pandaridae. **C.** Caligidae. **D.** Chondracanthidae female with male attached. **E.** Sphyriidae. **F.** Philichthyidae. **G.** Lernanthropidae. **H.** Kroyeriidae. **I.** Lernaeopodidae female, showing maxillary arms holding bulla (b) at tip, and with male attached (arrowhead); **J.** Male. **K.** Pennellidae.

Cyclopiform families frequently inhabit relatively sheltered microhabitats on their hosts including the gill chambers and nostrils, although some species may occur on the outer body surface, on the fins or around the eyes. They usually attach using clawed antennae, but some also display modifications of the ventral body surface and limbs that allow them to generate suction onto the surface of the host. In the Bomolochidae the antennules and first swimming legs are flattened and armed with swollen setae, and they function as part of the sucker rim that forms the seal against the mucous-covered skin of the host. Primary attachment is by claws: the suckers provide a secondary attachment mechanism, as does the embedding of the body in the more metamorphic ergasilids such as *Mugilicola*.

The caligiform families within the order Siphonostomatoida are characterised by dorso-ventrally flattened bodies divided into an anterior cephalothorax and a post-cephalothoracic genital trunk. These families can be viewed in sequence of increasing modification of body form, particularly in the number of leg-bearing somites incorporated into the cephalothorax and fused beneath the dorsal cephalothoracic shield. In the Dissonidae and Pandaridae (Fig. 4.2B) only the first pedigerous somite is incorporated under the dorsal shield, in the Trebiidae the first and second are incorporated, while in the Caligidae the first to third are all incorporated. The Caligidae (Fig. 4.2C) is the most speciose family of fish parasitic copepods, comprising over 465 species, and includes the sea lice, which can cause severe economic losses to fin-fish aquaculture, particularly of salmonids (Boxshall and Defaye 1993, Pike and Wadsworth 1999). Caligids and related caligiform families typically attach to the host using a combination of claws and suction: the entire cephalothorax is rimmed with marginal membrane and modified limbs, and forms an effective seal against the host surface. On the ventral surface, within the cephalothoracic sucker, are the clawed antennae and maxillipeds, which serve primarily for attachment by hooking into the skin of the fish. In some caligids attachment is further enhanced by additional paired suckers, the lunules, located ventrally on the frontal plates, as well as by the various spines and processes which enhance friction between parasite and host. Despite these modifications most caligid adults are freely motile over the surface of their hosts and adults of several species are routinely taken in plankton samples. In addition to using clawed appendages to attach to the skin of their elasmobranch hosts most pandarids (Fig. 4.2B) also have adhesion pads located on the limbs or ventral cephalothoracic surface.

The remaining families of siphonostomatoids that use fish as hosts can be divided into two categories: those exhibiting an intermediate level of transformation in body morphology (often elongate bodies typically retaining indications of external segmentation and attaching by clawed appendages) and those that have highly derived body forms (typically lacking expressed segmentation in adults, lacking functional swimming legs and often attaching by embedding of specialised anchor-like structures). The former group comprises families such as the Eudactylinidae, Kroyeriidae (Fig. 4.2H), Dichelesthiidae, Hatschekiidae, Pseudocycnidae and Lernanthropidae (Fig. 4.2G), and can be found on a wide range of elasmobranch and actinopterygian fishes. Most species within these families inhabit the branchial chambers of their hosts, usually the gills, where they attach using clawed antennae, maxillipeds (if present) and/or maxillae. Adults are never found away from their hosts. The group of profoundly modified families includes the Lernaeopodidae, Pennellidae (Fig. 4.2K) and Sphyriidae (Fig. 4.2E). Lernaeopodids typically have a large, fleshy body and attach to the host by means of the bulla (b, Fig. 4.2I), a small chitinous plug which is inserted into the host epidermis and held by the maxillary arms. Adult females are permanently anchored to the host by the bulla whereas males (Fig. 4.2J) are relatively tiny and hold on to the females using their strong antennae. The lernaeopodid *Naobranchia* has no bulla, attaching using ribbon-like maxillary arms to encircle the gill filaments of the host. Pennellids and sphyriids are large parasites, with the former family including the largest of all copepods, *Pennella balaenopterae*, a parasite of baleen whales that can attain a body length in excess of 16 cm. These parasites are anchored to the host by a cephalic holdfast, formed by the entire head, which develops more or less complex processes and is deeply embedded in the host tissues. This form of attachment is secure for these large-bodied copepods.

There are several important fish parasitic families belonging to the order Cyclopoida, which now incorporates the Poecilostomatoida, formerly treated as a distinct order by Kabata (1979) and by Huys and Boxshall (1991). These have a flat, plate-like upper lip (labrum) and short, blade-like or toothed mandibles, and are readily distinguished from the siphonostomatoids which are characterised by possession of an oral tube, formed by the labrum and the fused paragnaths (Boxshall 1990a), containing the stylet-like mandibles. The more highly modified cyclopoids, such as the Chondracanthidae (Fig. 4.2D), Shiinoidae and Philichthyidae (Fig. 4.2F), typically have bodies lacking in expressed segmentation and lacking at least some of the swimming legs. The first two of these families are most frequently found in the branchial chamber of the host where they attach by means of robust antennal claws. Security of attachment is often enhanced by the hyperplastic tissue response of the host, in which the epidermis overgrows the head of the parasite, effectively creating an anchor process. Most members of the Philichthyidae inhabit the subcutaneous spaces associated with the sensory canals of the lateral line and skull bones of their hosts. In both the Chondracanthidae and Philichthyidae the body is typically fleshy and may be provided with elaborate processes, the function of which is uncertain.

Parasites of marine invertebrates

The body form of copepods parasitic on marine invertebrates is astonishingly varied: many retain the basic cyclopiform body but some of the most extreme examples of secondary reduction in body segmentation combined with the loss of paired limbs are found in copepods from invertebrate hosts. Sponges, for example, act as host to a wide variety of copepods (particularly from the Siphonostomatoida) many of which, such as the Asterocheridae (Fig. 4.3A) and Dinopontiidae, are basically cyclopiform even though they inhabit the internal canals of the sponge. The Entomolepididae are more modified, having dorsoventrally flattened and rather scale-like bodies, but they are recognisably derived from a cyclopiform plan. Families such as the endoparasitic Spongiocnizontidae and Sponginticolidae have a highly derived morphology, with unsegmented, vermiform bodies and almost complete loss of limbs in the latter.

Copepods can be found as parasites across the entire range of cnidarian groups from the hard corals and sea anemones to the medusae and siphonophores (Humes 1985). Many common families, such as the Anchimolgidae, Asterocheridae and Rhynchomolgidae, are ectoparasitic, typically retaining cyclopiform morphology and attaching to their hosts by clawed antennae. Other families are more derived. Members of the Xarifiidae and Corallovexiidae (Fig. 4.3G) for example, which inhabit the gastrovascular cavities of scleractinian corals, have vermiform, unsegmented bodies and reduced limbs. More extreme are the Lamippidae which are endoparasitic in octocorals, particularly the alcyonacean soft corals. Some lamippids have unsegmented bodies equipped with enormous processes that give them a stellate appearance.

Relatively few species of copepods make use of host groups such as the Nemertea, Platyhelminthes, Bryozoa, Phoronida, Echiura, Brachiopoda, Enteropneusta, Hemichordata, Vestimentifera and Sipuncula (Boxshall and Halsey 2004). Those that parasitise these taxa mostly retain cyclopiform bodies and attach by clawed cephalic limbs. The exceptions are the Echiurophilidae, which can have elongate body processes, some Catiniidae which attach to their sipunculan hosts using antennal suckers, and the *Akessonia*- and *Ive*-groups that are endoparasitic in sipunculans and acorn worms, respectively, and have highly transformed vermiform or lobate bodies, lacking several limbs.

Eleven families of copepods are recorded exclusively from polychaete worms, but most are rarely encountered. Few retain cyclopiform morphology, such as members of the Eunicicolidae and Clausiidae. Even some clausiids are modified by reduction of expressed body segmentation and increase in body size, as are members of the Anomoclausiidae, Entobiidae, Spiophanicolidae and Serpulidicolidae. All of these families attach by means of clawed cephalic appendages. The Herpyllobiidae, found on polynoid worms, contains some of the most extremely modified parasitic copepods known. Adult females are sac-like (Fig. 4.3J), lacking any segmentation and all limbs, and can only be identified as copepods by reference to their paired egg sacs and their early larval stages. They are mesoparasites – living with the anterior portion, the endosoma, embedded in the host and with the posterior portion, the ectosoma, sticking out from the host's surface and carrying the egg sacs. Dwarf males are found attached to the genital region of the female. Four other families, also mesoparasites of polychaetes, exhibit similar extreme oligomerisation, these are the Bradophilidae, Phyllodicolidae, Saccopsidae and Xenocoelomatidae.

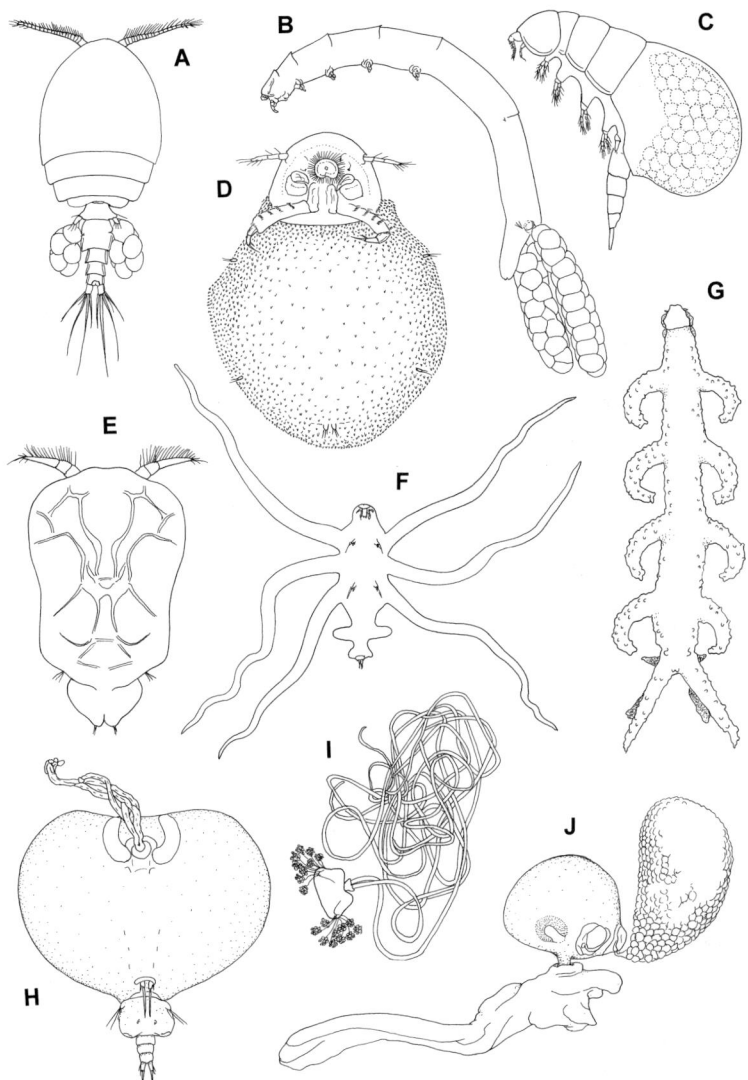

Figure 4.3 Range of gross morphology of copepod families parasitic on invertebrate hosts. **A.** Asterocheridae. **B.** Mytilicolidae. **C.** Notodelphyidae. **D.** Nicothoidae. **E.** Stellicomitidae. **F.** Splanchnotrophidae. **G.** Corallovexiidae. **H.** Nicothoidae (*Nicorhiza*, showing rootlet system derived from oral tube). **I.** Chitonophilidae, with rootlets. **J.** Herpyllobiidae, showing body divided into endosoma and ectosoma, bearing egg sac (other sac removed).

Copepods parasitise most molluscan groups, from aplacophorans and polyplacophorans to the cephalopods. Most have bodies that are cyclopiform or only slightly modified and this applies particularly to ectoparasitic forms found on the gills of bivalves or in the mantle cavity of gastropods. Important families include the Anthessiidae and Lichomolgidae, species of which attach using clawed antennae. More modified forms with elongate bodies, often with reduced segmentation and appendages, include the Mytilicolidae (Fig. 4.3B) and Myicolidae, many of which inhabit the intestinal tracts of their hosts. Some of these are major pests in commercial bivalve culture. The Splanchnotrophidae (Fig. 4.3F), which live in the tissue sinuses and hepatic diverticulae of their nudibranch and sacoglossan hosts, have highly transformed bodies equipped with elongate processes. Most highly metamorphic of all are the mesoparasitic Chitonophilidae (Fig. 4.3I), which are reduced to a sac-like or vermiform body plus a rootlet-system that penetrates the tissues of their chiton and gastropod hosts.

One family, the Nicothoidae, parasitises other crustaceans. Many of these inhabit the brood chamber or marsupium of their hosts and have a globular body form (Fig. 4.3D) that mimics the eggs of the host in size as well as shape. They apparently feed on the host's eggs. The most highly derived forms with the family, however, are mesoparasitic with the adult females reduced to an external sac-like body anchored by a rootlet system (Fig. 4.3H), which penetrates extensively though the host tissues.

Thirteen families of copepods are known to occur exclusively on echinoderm hosts, each of which is specific to a particular host class (Boxshall and Halsey 2004). Other families, such as the Asterocheridae, Taeniacanthidae, Rhynchomolgidae and Pseudanthessiidae, parasitise echinoderms, but not exclusively. These generalist families are typically cyclopiform and attach to the outer surface of their hosts by clawed antennae and maxillipeds. They can be extremely abundant: Humes (1973) reported a total of 27 209 specimens of the asterocherid *Collocherides astroboae* and the rhynchomolgid *Doridicola micropus* from just three individual basket stars of the genus *Astroboa*. Most specialist echinoderm parasites are characterised by small body size and, although there is some reduction and loss of external segmentation, the bodies are relatively weakly transformed. These tiny parasites (Fig. 4.3E) can also reach enormous population densities: Humes (1971) reported 1420 individuals of a stellicomitid, *Stellicomes supplicans*, on two individual seastars. Members of the Calverocheridae and Pionodesmotidae form cysts in their sea urchin hosts and have swollen, rather globular bodies. The most highly transformed morphology is found in the Cucumaricolidae, the species of which have lobate bodies and inhabit the coelom of sea cucumbers, and the Chordeumiidae which live in cysts or within the genital bursae of their brittle star hosts.

Copepods are very commonly associated with both solitary and colonial tunicates, and can be found in all parts of the zooids and in the matrix of colonial forms. Among these, the Notodelphyidae and Botryllophilidae are widely distributed and are most frequently reported, perhaps because their large body size renders them readily visible through the body tunic of their hosts. Within a single family, such as the Notodelphyidae (Fig. 4.3C), morphology can vary from cyclopiform to vermiform, lobate or stellate body shapes.

Sexual dimorphism
Most parasitic copepods are sexually dimorphic in body form and in appendage structure. Females are typically larger than males, have more transformed bodies and may have more robust attachment mechanisms, whereas males typically have at least one pair of limbs (either the antennules or maxillipeds) adapted for grasping the female during copulation. Sensory systems associated with mate detection and mate recognition behaviour are also dimorphic, typically being better developed in males. In families such as the Lernaeopodidae (Figs 4.2I, J) and Chondracanthidae (Fig. 4.2D), the size differential is very pronounced and males are often referred to as dwarves, although comparison with free-living relatives suggests that in such cases it is equally appropriate to interpret the females as giants. In the Chondracanthidae the tiny males attach to special glandular organs on the females, the nuptial organs, which may directly provide them with nutrients. The most extreme form of sexual dimorphism within the copepods is cryptogonochorism. In the Xenocoelomatidae and in the unrelated genus *Gonophysema* males are tiny. They penetrate the female and move into a special receptacle, the receptaculum masculinum, where they undergo a metamorphic reduction to become what is effectively a functional 'testis', resulting in a pseudohermaphrodite condition (Bresciani and Lützen 1961).

In certain Notodelphyidae and Myicolidae, parasites of invertebrate hosts, there are two forms of adult males: a sedentary or typical form and a swimming or atypical form. Isolated atypical males are capable of moving to alternative host individuals containing females. In the notodelphyid *Pachypygus*, atypical males have a unique sensory organ, the cephalic pleural organ, which may act as a chemosensor involved in the detection of pheromones produced by the female, or of metabolites produced by the female/host complex (Hippeau-Jacquotte 1987).

Life cycles

Basic life cycle pattern

The basic life cycle of copepods comprises two phases, naupliar and copepodid. The egg typically hatches into a nauplius larva defined by its small, unsegmented body and the possession of only three pairs of functional appendages, antennules, antennae and mandibles. There is a maximum of six naupliar stages (designated NI–NVI) and all six are retained in most free-living copepods and in some parasites. Nauplii may be planktotrophic, feeding on other planktonic organisms, or lecithotrophic, relying on yolk stores for nutrients. Parasitic copepods typically have lecithotrophic nauplii characterised by reduced setation on the three limb pairs and by the absence of the so-called naupliar feeding process on the coxa of the antenna. The final nauplius stage, primitively NVI, undergoes a metamorphic moult to the first copepodid, which has a segmented body, a full adult set of cephalic appendages and the first and second swimming legs. In free-living copepods there is a maximum of five copepodid stages (designated CoI–CoV) and one body somite is added at each moult through this phase. In both sexes the fifth copepodid stage moults into the adult. This is a definitive or final moult and the female becomes sexually receptive on moulting.

Mating takes place soon after the female becomes sexually receptive and adult males may engage in pre-copulatory mate guarding, holding pre-adult females until the final moult (Boxshall 1990b). The sequence of mating behaviours consists of mate detection, mate recognition and mate capture, culminating in copulation during which sperm-containing spermatophore(s) are transferred to the female. There is strong evidence that mate detection and recognition behaviours are chemically mediated, with males using an array of chemosensory aesthetascs on their antennules to detect pheromones produced by females. Spermatophores typically discharge via copulatory pores, into seminal receptacle(s) located internally within the genital region of the female and sperm are stored until required for fertilisation. Fertilisation occurs as egg batches are laid and a single female may produce several batches of eggs during the course of her reproductive life. Most parasitic copepods extrude their eggs into paired egg sacs or uniseriate egg strings, although some, for example some Notodelphyidae (Fig. 4.3C) and the genus *Pectenophilus*, store eggs internally.

The full life cycle comprising six naupliar stages and five copepodid stages preceding the adult is retained in many families of parasitic copepods, especially those utilising invertebrates as hosts (Fig. 4.4). Examples include the Asterocheridae and Cancerillidae among the Siphonostomatoida, and the Myicolidae and Notodelphyidae among the Cyclopoida. Rarely is the full number of stages retained in fish parasites, the only example being the Ergasilidae, which is unusual in that the naupliar stages are planktotrophic (Fig. 4.5A), feeding on unicellular algae. In parasitic copepods the infective larva is, with rare exceptions, the first copepodid and life cycles are direct, involving only a single host. One of the exceptions is the Notodelphyidae, in which it is the second copepodid that serves as the infective larva. The infective copepodid larva provides the transition between the preceding dispersal phase and the parasitic phase.

Modification of the nauplius phase

In many parasites the naupliar phase of the life cycle is more or less abbreviated, occasionally it is lost altogether. In fish parasites, particularly in siphonostomatoids, the nauplius phase is reduced to two stages (NI and NII) and these are lecithotrophic (Fig. 4.5B). Most siphonostomatoids on

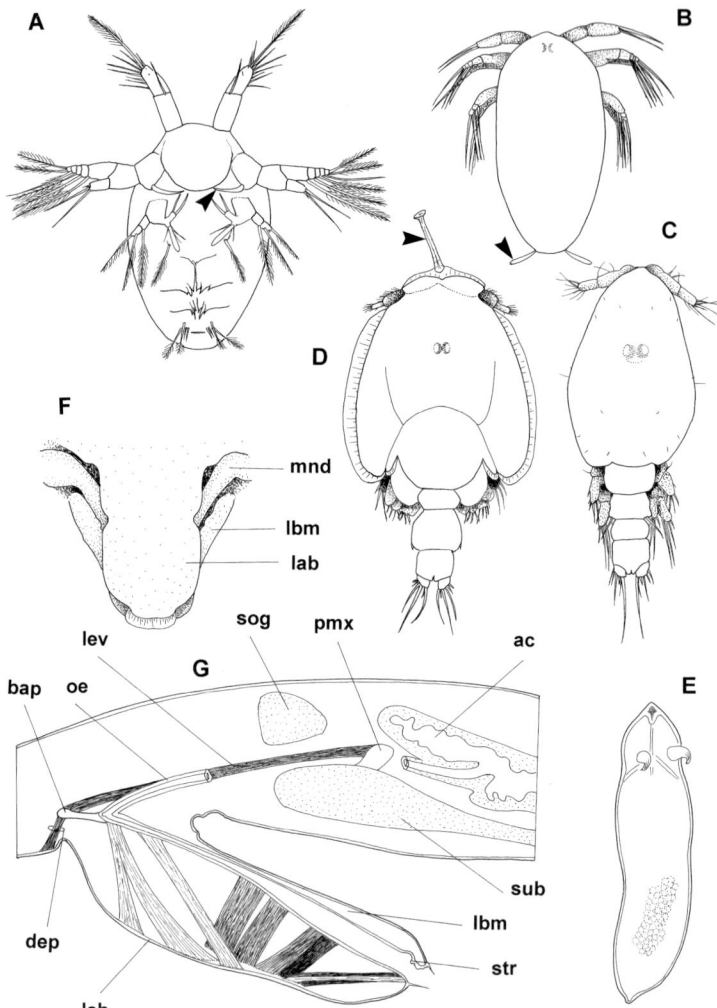

Figure 4.5 Larval stages. **A.** Planktotrophic nauplius (NVI) of Ergasilidae, showing feeding process (arrowed) on coxa of antenna. **B.** Lecithotrophic nauplius of *Lepeophtheirus* (Caligidae), showing paired balancers (arrowed). **C.** First copepodid larva of *Lepeophtheirus*. **D.** Chalimus larva of *Lepeophtheirus*, showing frontal filament (arrowed). **E.** Onychopodid larva of *Gonophysema*. Oral tube of Caligidae. **F.** Ventral view of oral tube, showing mandible entering tube via lateral slit between labrum and labium. **G.** Median longitudinal section through oral tube showing musculature. Abbreviations: ac, anterior midgut caecum; bap, buccal apodeme; dep, oral tube depressor muscle; lab, labrum; lbm, labium; lev, oral tube levator muscle; mnd, mandible; pmx, postmaxillulary apodeme; oe, oesophagus; sog, supraoesophageal ganglion; str, strigil; sub, suboesophageal ganglion.

fishes have uniseriate egg strings in which disc-shaped eggs are closely packed into a single row extending the length of the string. Nauplii hatching from such egg strings have a single pair of modified caudal setae known as balancers, although their function is unknown. Related families that have multiseriate egg strings, namely Lernaeopodidae, Sphyriidae and two genera of Hatschekiidae, have nauplii that lack balancers. The NI moults quickly, usually within six hours to 24 hours, and the NII moult into the first copepodid is equally rapid, so from hatching to the infective stage usually takes one to two days depending on temperature. In some

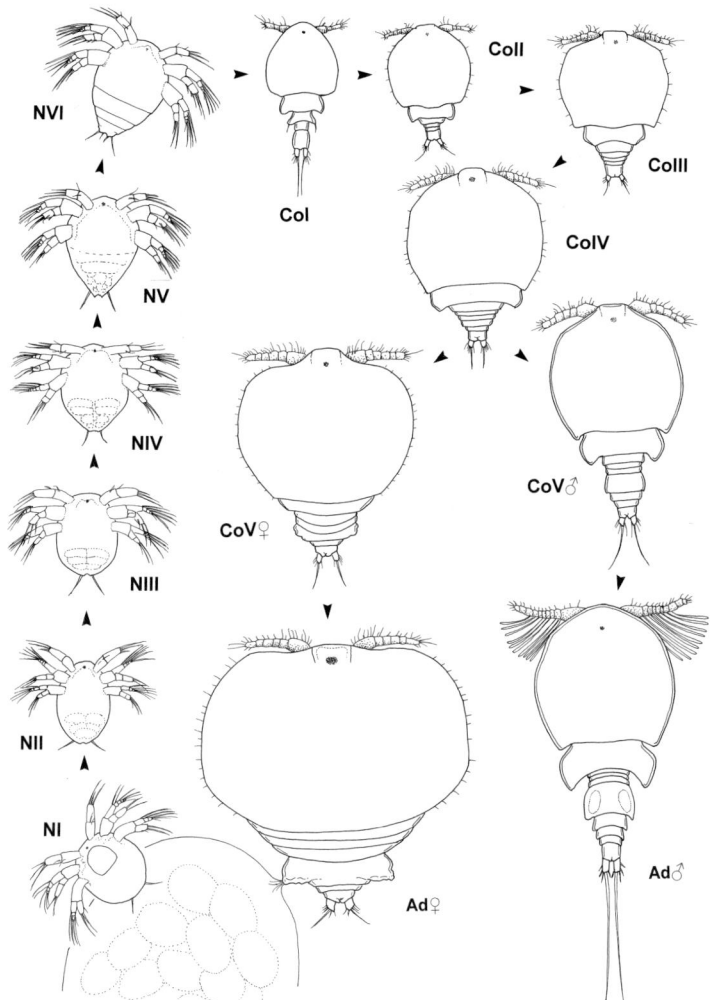

Figure 4.4 Life cycle of *Cancerilla tubulata* comprising six nauplius stages and five copepodid stages plus the adults of both sexes. Abbreviations: NI – NVI, nauplius stages 1 to 6; CoI – CoV, copepodid stage 1 to 5; Ad, adult.

Lernaeopodidae the nauplius phase is reduced to a single stage, as in *Allela*, *Clavella* and *Nectobranchia*. In other Lernaeopodidae and some Pennellidae, such as *Salmincola*, *Cardiodectes* and *Peroderma*, it is lost completely and developing eggs hatch directly into the infective first copepodid.

Abbreviation of the life cycle is not just a feature of fish parasites, some parasites of invertebrates also have abbreviated cycles but the information available is so fragmentary that no definitive pattern can be identified. The Herpyllobiidae and the mytilicolid *Trochicola* have only two naupliar stages and only a single nauplius is reported for *Gonophysema* and for some genera of Nicothoidae. In other genera of Nicothoidae, as well as in some Chordeumiidae and the Cucumaricolidae, the entire nauplius phase is lost.

The most spectacular modification in biology of the nauplius phase is the order Monstrilloida (Fig. 4.6). All monstrilloids have endoparasitic naupliar stages and free-swimming, non-feeding adults. Females carry their eggs on long ovigerous spines. These eggs hatch into infective

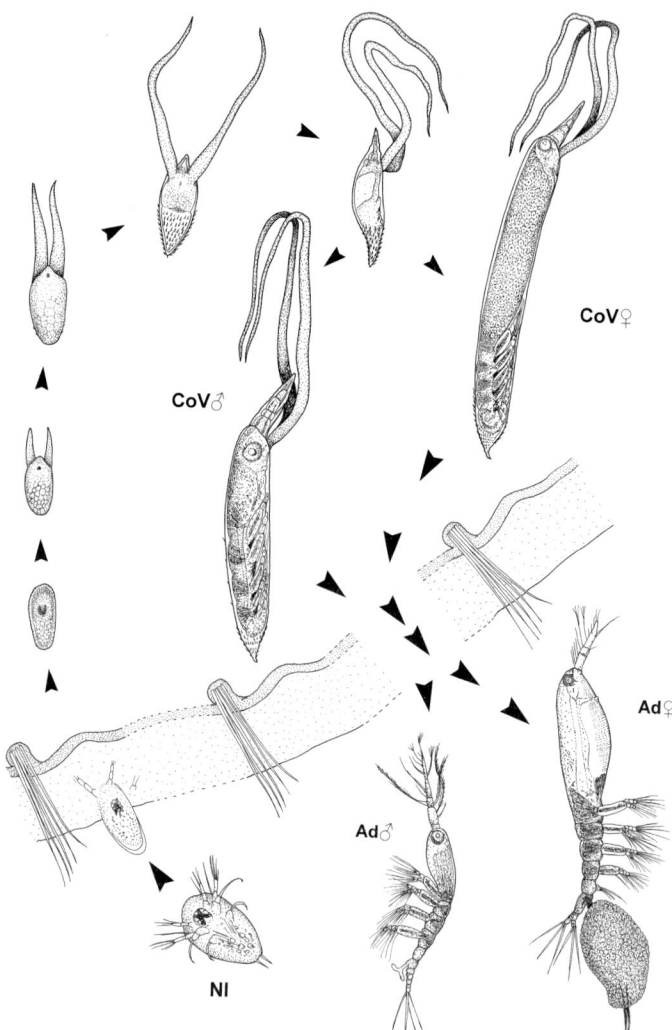

Figure 4.6 Life cycle of the Monstrilloida, comprising a free-living phase consisting of the non-feeding adults and the first nauplius stage released by the female, and an endoparasitic phase during which development from infective nauplius to fifth copepodid stage takes place. Abbreviations: NI, infective first nauplius stage; CoV, copepodid stage 5; Ad, adult; host, host body wall.

nauplii that locate a host, either a polychaete or a mollusc, and burrow into its tissues. Once within the host's blood system the nauplii transform into a sac-like body bearing paired, root-like anterior processes. Development through the copepodid phase, up to the ?fifth stage takes place within the host. Once development is complete the monstrilloid leaves its host as the final copepodid stage and undertakes a single moult into the adult. Adults of both sexes live in the marine plankton community but lack mouthparts, presumably serving as a dispersal and reproductive phase in the life cycle. A family of cyclopoids, the Thaumatopsyllidae (once classified as monstrilloids), has a similar life cycle, with parasitic nauplii inhabiting the gut of brittle stars and non-feeding adults living in the plankton. The copepodid phase in *Thaumatopsyllus paradoxus* comprises the full five stages preceding the adult, and the entire phase from final nauplius to adult is completed without further food intake.

Modifications of the copepodid phase

The first copepodid (Fig. 4.5C) is relatively uniform in structure and is characterised by the presence of two functional pairs of biramous swimming legs, each with one-segmented rami. In general, the copepodid phase provides a gradual transition from the copepodid body form, which is common to all podopleans, to adult morphology, however transformed. In more derived families successive copepodid stages show increasing levels of modification in body form and limb structure, as compared to their free-living relatives. In almost all copepod parasites the first copepodid is a free-swimming stage, the few exceptions include the monstrilloids and the brittle star parasite *Parachordeumium amphiurae*, which hatches directly as an infective second copepodid, having passed through the first within the egg.

A common modification to the copepodid phase is the specialisation of the later stages as particular larval forms. In most siphonostomatoid fish parasites, for example, the first copepodid produces a chitinous frontal filament soon after settlement on the host. This filament is secreted by an anteriorly located gland and anchors the developing larva securely to its host. These are chalimus larvae (Fig. 4.5D) and the basic life cycle of the caligid sea lice contains four chalimus stages followed by either one or two pre-adult stages. Pre-adults secrete a frontal filament during moulting, but it is temporary and after moulting caligid pre-adults detach and are motile. Frontal filaments and chalimus larvae are present in most families of fish parasitic siphonostomatoids for which the larvae are known. However, none is present in the Lernanthropidae (Raibaut 1985). Elsewhere in the Copepoda a similar chitinous filament is found in the Nicothoidae, attaching the developing larva to the exoskeleton of its crustacean host (Hansen 1897).

In the tunicate parasite *Gonophysema* the infective copepodid larva settles on the host then undergoes a metamorphic moult into an onychopodid larva (Fig. 4.5E). This larva is reduced to a simple elongate sac-like body provided with grasping antennae that are used to attach to the host. The onychopodid then penetrates the skin of the tunicate and begins a transformation into the amorphous, lobate adult (Bresciani and Lützen 1961).

The basic copepodid phase, as primitively retained in siphonostomatoids (e.g. *Cancerilla*), comprises five stages plus the adult. The presence of pre-adults, as true moult stages, is a feature of many fish parasites. Either one pre-adult (e.g. in *Caligus clemensi*) or two pre-adult stages (e.g. in *Lepeophtheirus salmonis*) may be interpolated into the basic life cycle, and represent an addition to the ancestral copepod life cycle. The addition of a stage or stages is remarkable, given the general pattern of abbreviation and loss of stages exhibited across the parasitic copepods. It appears to be shared by only certain families of fish parasitic siphonostomatoids, but its presence in related families has yet to be confirmed. Reports of additional copepodid stages or of pre-adults in cyclopoid families are doubtful and are more likely to represent growth stages than moult stages.

The infection process

The infective copepodid larva is typically small, in the order of 0.3 mm to 0.7 mm body length, but the oceans are enormous, potential hosts are patchily distributed and many of them are highly mobile. Clearly, overcoming the problems of locating a host and successfully infecting it are critical to the completion of the life cycle in parasitic copepods. Infection biology has been studied in a few fish parasites, especially sea lice. The probability of encounter between a copepodid and a potential host is increased by aggregation of larvae at the appropriate part of the environment. For example, copepodids of *Lepeophtheirus pectoralis*, a parasite of flatfish, become negatively phototactic after some hours and tend to sink to the sea bed where they would be more likely to encounter hosts. In contrast, copepodids of *Lepeophtheirus salmonis*, a salmonid parasite, tend to aggregate around the subsurface halocline in coastal waters, a depth horizon at which salmonids are known to cruise. Infective larvae in close proximity to fish hosts have been shown to respond to mechanical signals generated by the locomotory or respiratory motions of the potential host. The primary sensory interface between the copepodid and its environment is the paired antennules and their array of chemosensory, mechanosensory and bimodal setation elements. In addition, *L. salmonis* larvae also respond to visual signals, exhibiting bursts of high velocity swimming in a modified shadow reflex. Caligid copepodids have a nauplius eye consisting of three ocelli, the two dorsal ocelli being provided with a spherical lens as well as a reflective tapetal layer. The optical properties of such an eye probably allow the formation of a simple image. Final attachment to the host appears to be a chemosensorily-mediated behaviour and this may form the basis for host recognition, but the details of this mechanism are unresolved.

Many marine invertebrate hosts provide sessile targets for infective stages, although copepods also utilise motile invertebrates from scyphozoans to cephalopod molluscs. As in the fish parasites, the infection process has been studied in relatively few species. Infective stages of the *Sabelliphilus sarsi* locate their sessile polychaete host by responding to chemical cues emanating from the host. On this basis larvae are able to distinguish between different polychaete species.

In shallow coastal seas host densities may be relatively high and the environment exhibits a complex physico-chemical structure that may be exploited by the aggregation behaviour of the infective larvae. In the deep ocean the more structured waters lie close to the bottom and only three families of parasitic copepods are routinely found, the Chondracanthidae, Lernaeopodidae and Sphyriidae, primarily on bottom-living rather than pelagic host fishes. In the vast open pelagic biome away from the effects of the coast and the sea bed, the likelihood of an encounter with potential hosts is lowest of all. Only two species of copepods, *Cardiodectes medusaeus* and *Sarcotretes scopeli*, are commonly found in the pelagic. Both species exhibit low host specificity, occurring on several different host families (Boxshall 1998) and this may be a factor in their success.

Two-host life cycle
Some pennellids, such as *Sarcotretes scopeli*, retain a direct life cycle, while others have an unusual life cycle involving two different hosts. After a brief planktonic phase, primitively comprising two nauplius stages, the infective copepodid larva locates the first host. This may be a fish, as in the case of *Lernaeocera* (Fig. 4.2K), or a gastropod mollusc, as in the case of *Cardiodectes*. Development through the attached chalimus stages to the sexually mature adults takes place on the gills or in the mantle cavity of the first host. Mating takes place on the first host, after which the mated adult female, which is still basically cyclopiform in shape, leaves the first host and finds a second, usually a fish but occasionally a marine mammal. Once attached to the second host the female embeds and commences a profound metamorphosis involving considerable increase in body volume. The post-metamorphic female then produces egg strings while on the second host, using stored sperm to fertilise the eggs as they are extruded. The advantages of the two-host life cycle have not been fully analysed but must presumably outweigh the apparent disadvantage of requiring two separate host infection events to complete the cycle.

Effects on hosts

Attachment
Parasitic copepods damage their hosts directly by their attachment mechanisms and by their feeding activities. Attachment by means of clawed limbs is typical for ectoparasites and penetration of the skin by the claws causes local lesions, the pathology of which varies according to site and other factors. On the skin of fishes attachment can cause pressure necrosis and epidermal erosion and the host tissue responses can include swelling, hyperplasia, proliferation of fibro-

blasts, fibre production and cellular infiltration. Any surface lesion may also render the host susceptible to secondary infections. Attachment to gill filaments typically results in hypertrophy of the gills and fusion of secondary lamellae, with consequent loss of respiratory surface area. For some fish parasites, such as *Ommatokoita elongata*, the preferred site on the shark host is the cornea of the eye. This species and other eye parasites, such as *Lernaeenicus sprattae* (the eye maggot of sprat), cause blindness and this may have a profound impact on the host, given that 84% of the Greenland shark, *Somniosus microcephalus*, carried *Ommatokoita elongata* in both eyes (cf. Kabata 1979).

Feeding

Most parasitic copepods feed by rasping at the surface of the host using their mandibles. Dislodged fragments of host tissue are taken directly into the mouth or oral tube and carried into the gut for digestion. In siphonostomatoids such as the entomolepidids, the distal part of the oral tube forms a long, narrow siphon which is used for piercing host tissues or cells to withdraw fluids (Boxshall 1990a). In caligids the oral tube is movable, being erected by special muscles during feeding (Fig. 4.5F–G) and held flat against the body when not feeding. Once erected the mandibles contained within the oral tube extend through the distal opening and scrape at the host's epidermis in concert with the strigil, a toothed ridge on the labium. Feeding activity produces surface lesions where the epidermis has been removed and, in cases of heavy or prolonged infestation, deeper lesions result, with damage to the dermis and blood vessels. Deeper lesions often result in haemorrhaging and can cause anaemia.

In many pennellids and sphyriids the entire head is permanently embedded in a fixed position in the host and feeding appears to be mainly on host fluids. Blood feeders, such as *Cardiodectes*, have evolved systems for sequestering excess iron derived from feeding. *Cardiodectes medusaeus* synthesises ferritin crystals and stores them as large aggregates within the cephalic holdfast. Fish infected with large blood-feeding parasites commonly show signs of anaemia, such as reduced packed cell volume, reduced haemoglobin levels, the presence of numerous immature erythrocytes and altered white blood cell counts.

A surprising variety of copepods possess a system of rootlets which penetrate host tissues. In the case of chitonophilids, herpyllobiids and their relatives, and the nicothoids *Rhizorhina* and *Nicorhiza*, these rootlets appear to be absorptive in function, extracting nutrients from the host and transporting them to the ectosoma where the reproductive organs are located. The rootlet system in the siphonostomatoid *Nicorhiza* is derived from the modified oral tube, but in the other families the homology of the rootlets is unknown. The elaborate body processes found in some endoparasites, such as *Echiurophilus fizei*, splanchnotrophids, antheacherids and some genera of Chordeumiidae and Lamippidae, may also function as nutrient absorbing systems. The structure of the body integument provides evidence of modification to facilitate absorption of nutrients through the body surface. For example, in *Antheacheres* and *Linaresia* the thin integument consists of only epicuticle, the procuticle having been lost. Vesicles may be present within the cuticle and the underlying epidermal cells may have modified apical surface provided with microvilli (Bresciani 1986).

Economic importance

Infestation by any parasitic copepod may result in loss of condition of the host. For example, *Pectenophilus* can attain prevalence rates of 100% on its host the Japanese scallop, *Patinopecten yessoensis*, in aquaculture facilities. It causes significant loss of condition in cases of heavy infestation. This in turn may result in reduced growth rates, reduced reproductive effort and greater mortality. Caligid sea lice, especially species of the genera *Lepeophtheirus* and *Caligus*, are serious pests in fin-fish aquaculture: in salmonid farming in both northern and southern hemispheres and in brackish water tilapia culture facilities in southern Asia (Boxshall and Defaye 1993, Pike and Wadsworth 1999) (see pp. 378–391). Caligid sea lice can cause severe economic losses due to the reduced growth of infected fish, to the costs of chemotherapy and to mortality. The economic effects of parasitisation of fishes by copepods include reduced marketability. Farmed fish with unsightly lesions have reduced value. Integrated parasite management, which may include routine inspections for sea lice, chemotherapy, separation of year classes, fallowing and use of cleaner fish, is widely employed but sea lice levels remain problematic in some areas and have been implicated in raising infestation levels in adjacent populations of wild salmonids. Effects on the host are generally less pronounced in wild fish, although the redfish, *Sebastes*, with large and obvious parasites like *Sphyrion lumpi* embedded in its flanks is problematic for filleting and processing, and has reduced market value.

Important references
Some important books on parasitic copepods are available. They include Kabata (1979) who gives a very detailed and beautifully illustrated account of copepods infecting British fishes, and Boxshall and Halsey (2004), *An Introduction to Copepod Diversity*.

The volume edited by Boxshall and Defaye (1993) focuses on the biology and control of sea lice on wild and farmed fish.

Isopoda (isopods)
RJG Lester

Introduction
Parasitic isopods are typically marine, and usually inhabit the warmer seas. Their body form (Fig. 4.7) varies from an easily recognisable isopod to a relatively amorphous sac recognised as an isopod only from the less modified male found within the folds. Though free-living isopods tend to be detritivores, parasitic forms feed on host blood or host haemolymph. Their mouthparts form a cone with maxillipeds that tear at the flesh and tiny pointed mandibles that pierce into the tissue to penetrate blood vessels or blood sinuses. The gut, particularly the hind gut, is quickly filled, often swelling the body, then the contents are slowly transferred to the midgut glands for digestion. Thus the parasites tend to be intermittent feeders. They can be a major drain on the host, frequently affecting reproductive performance and sometimes affecting growth rate.

Most parasitic isopods are ectoparasites. There are three major groups: cymothoids, epicaridians and gnathiids. Cymothoids are parasites of fish, both as immature forms and adults. Epicaridians are parasites of Crustacea, again as immatures and adults. Gnathiids are larval parasites of fish, the adults being free living and non-feeding. Genetically, the cymothoids and epicaridians appear to be closely related whereas the gnathiids appear to have evolved from a different isopod line.

Cymothoidae

Morphology and diversity
These are the isopods commonly seen on teleosts in tropical and subtropical waters, attached to the body surface, in the mouth or on the gills (Brusca 1981, Bunkley-Williams and Williams 1998; Lester and Hayward in press; Fig. 4.7A). They resemble free-living isopods except for their hook-like legs. The stages normally found are the non-swimming, permanently attached mature females, often with a small male nearby.

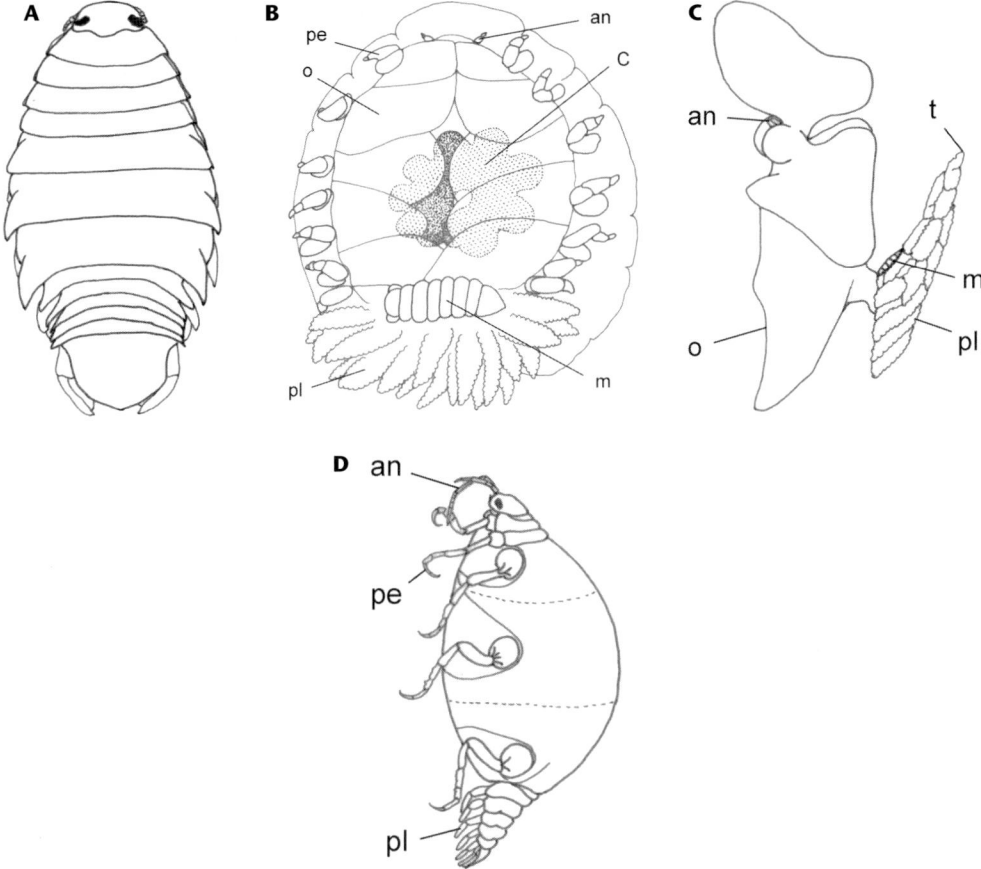

Figure 4.7 Parasitic Isopoda – variation in body form of adult females. **A.** A cymothoid, *Nerocila orbignyi*. **B.** A bopyrid, *Epipenaeon ingens*, a parasite of prawns, itself parasitised by the cryptoniscid *Cabirops orbionei*. **C.** An entoniscid, *Pinnotherion vermiforme* from within a pea crab, *Pinnotheres* sp., itself parasitising a mussel, *Modiolus modiolus*. **D.** A gnathiid, *Paragnathia formica*. Abbreviations: an, antenna; C, cryptoniscid; m, male; o, oostegite; pe, pereopod; pl, pleopod; t, tip of abdomen. Figures: redrawn from Bruce (1987); drawn from unpublished photograph taken by L Owens; modified from Giard and Bonnier (1893); modified from Monod (1926), respectively.

Though most adult isopods on fish belong to the Cymothoidae, there are parasitic forms in other families. The Aegidae, distinguished from Cymothoidae by having less modified pereopods, includes the notorious *Alitropus typus*, which parasitises fishes in India and south-east Asia in fresh and brackish waters. The parasites attack fish to feed but retain their free-swimming capability as adults. Unlike cymothoids they do not appear to be protandrous hermaphrodites.

Tridentellid and corallanid isopods are mostly free living but have a few representatives that are parasites of fish, such as the corallanid *Argathona macronema* which is common in the nasal passages of serranids and lutjanids on the Great Barrier Reef. Some corallanids are parasitic on Crustacea. Those belonging to the genus, *Tachaea*, are parasites of freshwater shrimps in Asia and Australia, and are usually found attached to the outside of the cephalothorax.

Life cycle
Gravid females release eggs into a brood pouch or 'marsupium' formed from their ventral oostegites. Here the eggs embryonate, hatch and undergo two or more moults to form the 'manca' or

'pullus II' stage. These are released from the brood pouch, sometimes more or less simultaneously as a result of contractions from the parent. The parent then moults, feeds, digests the meal and eventually produces the next batch of eggs. Several batches may be produced during her life span. The mancae have only six pairs of legs (compared to seven in juveniles and adults), large compound eyes and heavily setose pleopods with which they swim extremely rapidly. After a short free-swimming period they are parasitic and need to find a fish to take their first meal within one to two days or they will die. In genera such as *Anilocra* and *Nerocila* the mancae then leave the fish, moult, reattach to another fish, and so on until they complete their juvenile moults, the number of which has not been determined for any species, and approach adulthood. In other species such as the gill-inhabiting forms (e.g. *Mothocya*, some *Livoneca* spp.), the 'tongue biters' (e.g. *Ceratothoa* spp.) and the tissue dwellers (e.g. *Ourozeuktes* spp.) the mancae or an early juvenile stage move to the preferred site and remain attached to the fish.

Cymothoids are protandrous hermaphrodites. The first male to parasitise a fish changes into a female. Males attaching to the same fish remain as males. It seems likely that a pheromone or neurohormone is released by the female which inhibits further development of the males. Whether this is through the host's blood as suggested by Raibaut and Trilles (1993), through the host's mucus as suggested by Trilles and Hipeau-Jacquotte (1996) or through the water is not clear. Egg development apparently depends on the presence of a male, for each batch. Fertilisations occur immediately after the female has moulted in some species. In some skin-inhabiting forms males are rarely found. They presumably remain free-swimming and stay with females only long enough for fertilisation to occur. In others the small males are permanently attached alongside females and have lost the ability to swim. In gill-inhabiting species such as *Enispa convexa*, non-swimming males are found on the same fish as females, though not necessarily in the same gill cavity, and move back and forth to fertilise the females, presumably again in response to a female pheromone. In 'tongue biters' such as *Ceratothoa imbricata*, juveniles and one or more males occur on the gills and the adult female in the mouth. In tissue-inhabiting forms such as *Ourozeuktes*, small males are found in the pouch with the female. Male cymothoids are usually narrower than females and the ratio of length to width has been used as a measure of femininity, the Montalenti Index (Montalenti 1948). There is usually a strong correlation between parasite length and host length, in some cases because the fish are parasitised when small and the parasites live for many years and in other cases where the parasite apparently grows to fill the available space and then stops.

Effects on hosts

Cymothoids harm the fish in several ways. Mancae feed voraciously and easily kill fry and fingerlings through the tissue damage they cause. Permanently attached adults stunt the growth of fish and retard or inhibit reproduction, probably because of the nutritional drain though more subtle mechanisms such as through hormonal changes have not been ruled out. Those in the gill chamber are usually associated with stunted gills, partly from pressure atrophy and partly from damage associated with feeding and attachment. They have also been frequently associated with anaemia. Those in the mouth affect the development of oral structures and may completely replace the tongue, as in *Ceratothoa oestroides*. Menhaden infested by the buccal parasite, *Olencira praegustator*, school separately from uninfested fish of the same age. The tissue-inhabiting forms such as *Ourozeuktes* spp., which form a pouch from a depression in the skin, cause pressure atrophy of adjacent muscle and visceral organs. Though cymothoids penetrate the skin with their pereopods and mouthparts, and the tissue-inhabiting forms maintain a small opening to the outside, little secondary infection occurs.

In Mediterranean mariculture, infections of *Nerocila orbignyi* (Fig. 4.7A) on the gills of sea-caged bass, *Dicentrarchus labrax*, and bream, *Sparus aurata*, are associated with poor growth.

Bragoni et al. (1984) recommended using fine mesh nets near aquaculture cages to keep out mullet believed to be the source of the parasite. To avoid the parasite *Emetha audouini*, cages containing sea bass were moved away from the shore into deeper water with a stronger current (Papapanagiotou et al. 1999).

Salmon farms in Chile and Australia have been plagued by tongue biters, *Ceratothoa* spp. In Chile the problem became less acute when the numbers of the normal host *Trachurus murphyi* declined. In Australia, infections disappeared when the fish were treated with fresh water to control amoebic gill disease.

Epicaridea

Morphology and diversity

Though there has been relatively little morphological diversity among the Cymothoidae, isopods with *Tachaea*-like ancestors have radiated as Epicaridea (the 'Bopyridae' *sensu latu* of Dreyer and Wagele 2001), which are parasites of Crustacea. Indeed, Kuris (1974) estimated that 3% of all crustacean species were parasites of other Crustacea. The Epicaridea contain the bopyrids, dajids, entoniscids and cryptoniscids. All known life cycles involve two hosts, both of which are Crustacea.

Bopyridae conform to the isopod pattern of distinct segmentation, seven pairs of pereopods, and with a brood pouch formed from oostegites (Fig. 4.7B). Adult females occur in the gill chamber (e.g. subfamily Pseudioninae), or less commonly attached to the pleon (e.g. Athelginae), of shrimps and crabs. Adult males are much smaller than the females and are usually found attached between her pleopods. Females feed on host haemolymph by piercing a blood sinus usually on the inside wall of the gill cover or 'branchiostegite'. The parasite may take up to 25% of the shrimp's haemolymph in one day though presumably as in cymothoids this rate of uptake is not continuous. The male apparently has no contact with the shrimp. Whether males are hyperparasites on the female, or do not feed at all, is not known.

Life cycle

Eggs released into the brood pouch, embryonate and hatch into an 'epicaridium' larva with styliform suctorial mouthparts and six pairs of clawed pereopods. They swim rapidly, do not feed, and last for one to two weeks while they seek their first host, a copepod, frequently a calanoid. Once attached to the side of the copepod they pierce the exoskeleton, feed and within a few days moult to the 'microniscus' stage (Fig. 4.8). Unlike the biphasic ecdysis of most isopods (posterior first), epicaridia apparently moult in one piece. The microniscoi remain attached to the copepod for several weeks during which they enlarge to 10 times their original size. They are frequently found on copepods in fresh zooplankton samples but quickly drop off when the samples are preserved. Microniscoi transform into free-swimming larvae called 'cryptonisci' (not to be confused with the family Cryptoniscidae, see pp. 143–144). The change apparently occurs without a moult by expanding folds in the cuticle (Anderson and Dale 1981). The cryptonisci then leave the copepod and seek a definitive host.

In *Leidya distorta*, a parasite of the fiddler crab, *Uca uraguayensisa*, the cryptoniscus settles between the gill lamellae and after moulting one or more times migrates to the roof of the branchial chamber and matures into a female. Other cryptonisci and males were found attached to various parts of the maturing females (Roccatagliata and Jorda 2002). In *Probopyrus pandicola*, a parasite of shrimp, female cryptonisci may penetrate into the tissues and become endoparasites for up to two weeks, during which they can cause host mortality, before appearing in the gill chamber. Males are not endoparasitic and are attracted straight to the female (Anderson 1990). Gender appears to be environmentally determined in *Epipenaeon* species, the first

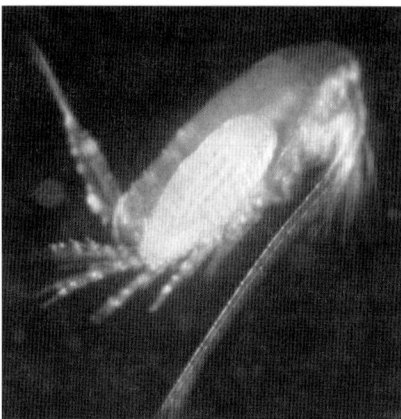

Figure 4.8 A microniscus (larval epicaridean) attached to a copepod at Heron Island, Australia. L Newman, photo.

cryptoniscus to attach to a shrimp becomes a female, subsequent ones become males. In *Parapenaeon* species gender seems to be genetically determined as about equal numbers of male and female cryptonisci settled on adult females whereas in *Epipenaeon ingens*, an environmentally determined species, virtually all the settled cryptonisci were males (Owens and Glazebrook 1985). In many species it appears that the definitive host is infected when a juvenile and remains infected for life. When the host moults, the parasites crawl through a split in the carapace and immediately reattach at the same location. In many species, the release of epicaridea and the moult of the parent is synchronised with the moult of the shrimp and will occur a few days prior. After the shrimp has moulted the male moves into the female brood pouch, presumably for insemination, and within several hours, a new batch of eggs is laid (Cash and Bauer 1993).

A female in the gill chamber of a shrimp or crab causes the chamber to be greatly enlarged. In those bopyrids that do not survive as long as the host, signs of the branchial enlargement frequently remain. The swelling caused by bopyrids is a problem in commercial shrimp fisheries where mechanical sorters select small parasitised shrimps with much larger unparasitised shrimps, and as a consequence staff have to be employed to remove the parasitised shrimps from the premium grades (Owens 1993).

Effects on hosts
Bopyrids may cause a slight decrease in host growth (Somers and Kirkwood 1991), or in the case of male shrimp, a slight increase, and may cause small changes in the host's secondary sexual characters but the most dramatic changes are in the host's reproductive capability. Both partial and total sterilisation of parasitised female shrimps and crabs have been reported by many authors. Males are less affected. When the parasites are removed, females recover, sometimes partially and often totally. The parasites cause a major energy drain. *Probopyrus* on *Palaemonetes* takes 10% of host's energy intake and reduces egg production by 50% (Anderson 1977).

The metabolic activity of parasitised copepods and shrimps is reduced and their activity decreases (Anderson 1975a,b, Bergey *et al.* 2002). This affects the rate at which parasitised shrimps capture food (Bass and Weis 1999) and limits their migration capability (Somers and Kirkwood 1991). Infected shrimp showed a reduced tolerance to salinity stress (Moles and Pela 1984). The gills of infected crabs and shrimps are flattened and sometimes deformed from pressure atrophy. The efficiency of oxygen uptake is reduced because of hydrodynamic changes in the gill chamber (Schuldt and Rodrigues-Capitulo 1987).

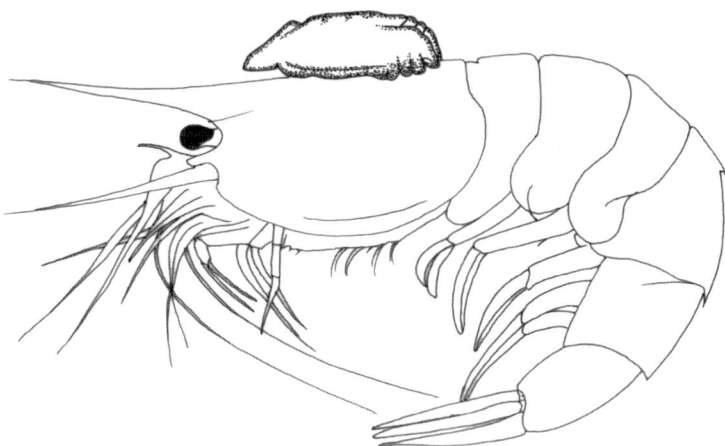

Figure 4.9 An adult dajid, *Holophryxus acanthephyrae*, on its prawn host. Figure drawn from Jones and Smaldon (1986).

Thus bopyrids partially or totally remove the reproductive capability of females without seriously affecting host moulting and growth. Most theories suggest that this is the result of either excessive blood loss or a direct effect of the parasite on the host's hormones. For the latter to occur, bioactive compounds from the parasite must pass to the host. Adult female bopyrids attract many cryptonisci, possibly using pheromones and it is conceivable that these also inhibit the sexual maturation of the host. Alternatively, the parasite could inject material into the host to prevent clotting of the haemolymph and this could include compounds active against ovarian development. However, there are few reports of inflammation or melanisation associated with the feeding sites, suggesting little if any foreign material is injected. Cymothoid isopods on small fish also inhibit sexual maturation, apparently solely through nutritional drain, and perhaps this is the most likely explanation here.

Dajids have only five pairs of pereopods and females have a brood pouch formed from oostegites. They occur on mysids, euphausiids and prawns often on the cephalothorax facing to the rear with their mouthparts penetrating directly into the shrimp's pericardium (Brandt and Hanssen 1994, Fig. 4.9). How they cope with the host's moult is not known. Their epicaridian larvae are found on copepods.

Adult female entoniscids have lost most signs of pereopods but do retain oostegites that form the brood pouch (Fig. 4.7C). They are internal parasites of decapods, generally enclosed by a host sheath in the host's visceral cavity. Males occur within the pleopods of the female parasite or elsewhere within the crab. Epicaridean larvae are released to the outside through a small opening in the host's branchial chamber which is kept open by the tip of the parasite's abdomen. The larvae apparently pass through a microniscus stage on a copepod before transforming to a cryptoniscus (Trilles 1999). The cryptoniscus seeks a decapod, may penetrate through a weak point in the gills and after an internal phase a developing female makes the small opening to the outside. An entoniscid may consitute up to 12% of the weight of an infected crab and generally causes sterility. Some are encapsulated and killed by the crab's internal defence mechanisms.

Adult female cryptoniscids are typically without pereopods or oostegites (see C in Fig. 4.7B). They parasitise barnacles (including Rhizocephala), ostracods and other isopods, often occurring as hyperparasites. *Hemioniscus balani* is a worldwide parasite of barnacles in temperate waters. The female attaches to the barnacle's ovaries and removes ovarian fluid thus sterilising the ovary though not affecting sperm production. Epicaridian larvae develop within the female parasite which eventually ruptures and frees the larvae (Blower and Roughgarden 1989). In *Liriopsis* sp. the anterior lobe of the mature female is buried in the host and the parasite presumably feeds on the barnacle's haemolymph. *Clypeoniscus hanseni* lies within the brood pouch of the isopod *Idotea pelagica* and feeds on its eggs after they have been laid (Sheader 1977). Like other cryptoniscids, *Cabirops orbionei* sterilises its host, the bopyrid parasite, *Epipenaeon ingens* (Fig. 4.7B). As a result it has been proposed as a possible biological agent to control the bopyrid (Owens 1993). Cryptoniscids are thought to use copepods as first intermediate hosts.

Gnathiidae

Morphology and life cycle

Gnathiids are a small, relatively homogeneous group of isopods that are parasitic as juveniles on teleosts and elasmobranchs (see pp. 266–278; Lester and Hayward, in press). Adult males develop formidable jaws, adult females (Fig. 4.7D) resemble juveniles. The adults do not feed. They are found in small groups in marine cavities such as in mud banks, in dead barnacles or coral, or in sponges. Typically a cavity will contain a male and a group of females. Young females appear to find the cavity in response to a pheromone produced by the male. Other males that enter the cavity are either inhibited from maturation, as in *Gnathia calva*, or ejected by the resident male (Wagele 1988). Females mature in the cavity. Eggs are brooded in the ventral brood pouch and they hatch to produce a 'zuphea' or unfed juvenile. These leave the cavity, swimming rapidly using their setose pleopods and seek a fish to which they attach using their hooked pereopods. They tear their way into the tissue, eventually pierce a blood vessel and engorge themselves on the host's blood. In doing so, the hind gut becomes greatly distended and is accommodated by the folded carapace of their three mid segments expanding rather like a concertina so the parasites, now called 'praniza' become dilated in the mid section and may appear reddish. After several hours or days depending on species and temperature, the praniza leave the host and enter a cavity in the substrate where they digest the meal and eventually moult to the second zuphea stage. This is repeated twice more until the third stage praniza leaves the fish to seek a cavity in which to mature. In their behaviour on fish and their free-living, egg-laying stage, they are reminiscent of terrestrial ticks.

On teleosts, praniza may remain for only a few hours while they feed. Praniza on elasmobranchs, however, may remain for weeks. A good way to identify praniza is to keep them in clean sea water until one moults into an adult male, which can be identified. If an adult male and several females are kept together, the females will eventually become gravid and produce a new generation of zuphea.

The prevalence of gnathiids is often underestimated because many species leave their host immediately the host is captured, and others only feed on fish at night. However, gnathiids may be so abundant on coral reefs as to form the main component of the diet of cleaner fish (see pp. 266–278).

Important references

Trilles and Hipeau-Jacquotte (1996) presented a comprehensive account of parasitism in the crustaceans. Lester and Hayward (in press) reviewed fish parasitic isopods. A monograph on the Gnathiidae is by Monod (1926), and one on the Epicaridea is by Trilles (1999). Bunkley-Williams and Williams (1998) wrote a synopsis of isopods associated with fishes. Raibaut and Trilles (1993) reviewed the sexuality of parasitic crustaceans. Among interesting papers dealing with specialised aspects is Dreyer and Wagele (2001), who used molecular and morphological evidence to postulate that bopyrids which parasitise crustaceans evolved from fish parasites.

Branchiura (fish lice)
Geoff Boxshall

Introduction
The Branchiura comprises about 175 species classified in four genera placed in a single family, the Argulidae, but only the genus *Argulus* occurs in the marine environment. Branchiuran fish lice range in length from a few millimetres to about 30 mm and they have strongly flattened bodies, with a low profile when attached to their hosts.

Morphology
The body (Fig. 4.10A) comprises a head of five limb-bearing segments and a trunk, divided into a thoracic region carrying four pairs of strong swimming legs, and a short abdomen. Fish lice have paired compound eyes located in a blood sinus below the cuticle in the anterolateral part of the head. The head has well-developed carapace lobes, which form as posterior extensions of the dorsal head shield, and they typically cover the legs on either side of the body. In some species they may extend further to cover the abdomen. These carapace lobes contain highly branched gut caeca and have two specialised areas ventrally, which have traditionally been referred to as 'respiratory areas' although they appear to be involved in regulating the internal body fluids (Haase 1975).

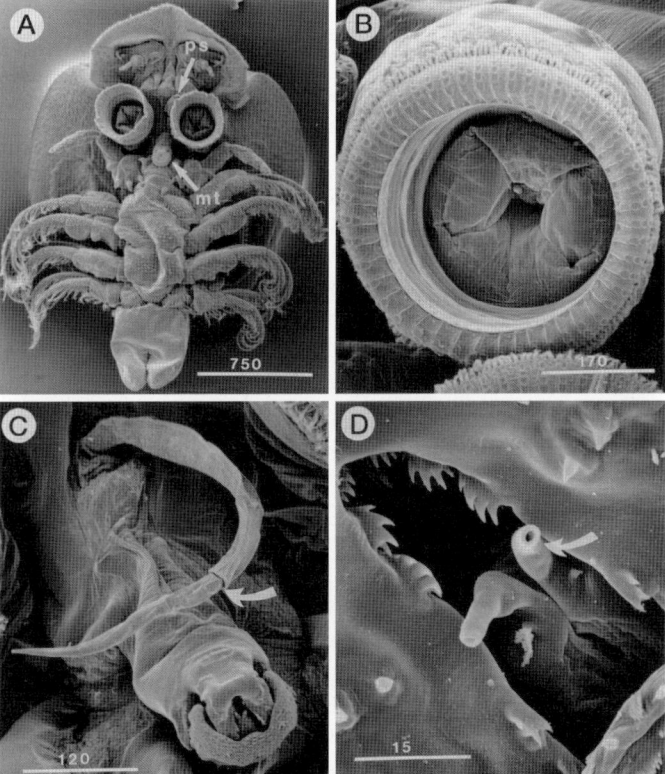

Figure 4.10 Scanning electron micrographs of *Argulus*. **A.** Ventral view of adult, with mouth tube (mt) and pre-oral poison spine (ps) arrowed. **B.** Sucker derived from modified maxillule. **C.** Mouth tube and everted poison spine, with groove separating distal spine from proximal sheath arrowed. **D.** Mouth opening showing paired labial spines (arrowed). After Gresty *et al.* 1993.

Branchiurans have nine pairs of limbs in total. Anteriorly on the ventral surface of the head lie the short antennules and antennae. Both are provided with claws and function as organs of attachment to the host. The claw of the antennule is located proximally and the cylindrical distal segments carry arrays of short setae which are probably sensory. Branchiurans have a tubular sucking mouth equipped with rasping mandibles located at the tip of the mouth tube. In *Argulus* there is a retractable poison stylet located just in front of the mouth tube (Fig. 4.10A, C). This stylet is absent in two of the freshwater genera, *Chonopeltis* and *Dipteropeltis*. The maxillules are developed into powerful muscular suckers (Fig. 4.10B) in the adults except in the freshwater genus *Dolops*, which retains clawed maxillules into the adult phase. The maxillae are uniramous limbs with spinous processes on the basal segments and small claws at the tip. The four pairs of thoracic swimming legs are biramous and directed laterally. Dorsally, the first and second legs commonly carry an additional process, the flagellum, originating near the base of the exopod. The third and fourth legs are usually modified in the male and are used for transferring sperm to the female during mating. The abdomen contains the paired testes in the male and the paired seminal receptacles, where sperm are stored until needed to fertilise eggs, in the female. The abdomen terminates in paired abdominal lobes separated by the median anal cleft, in which lies the anus and the minute, paired caudal rami.

Life cycles

The sexes are separate and in most branchiurans males transfer sperm directly to the females using a variety of modified structures on the third and fourth thoracic legs. The sperm are elongate and filiform in structure, and they are motile. The fine structure of the sperm has been used to link the Branchiura with the Pentastomida (Wingstrand 1972), a close phylogenetic relationship that is also supported by molecular sequence data (Abele et al. 1989). Only the life cycle of *Argulus* is well known; little is known of the other genera or of marine species. In freshwater species, after taking a meal, a mature female *Argulus* will leave its host and begin to lay eggs in rows on any hard, submerged surface. Up to 1200 eggs are laid at any one time and cemented to the substrate. The abandoned eggs hatch after between 12 days and 80 days according to species, but development time is also very dependent on temperature. Eggs within a string tend to hatch within a week. These eggs hatch into free-swimming larvae equipped with setose swimming antennae and mandibles, plus rudiments of the maxillules, maxillae and first two pairs of swimming legs. These larvae function as a dispersal phase and moult into the second stage, in which strong claws have replaced the setae on the antenna and the setose palp of the mandible is lost (Gresty et al. 1993). The first larval stage lasts about six days and moults occur at intervals until maturity. Branchiurans are parasitic from the second larval stage onwards but appear to leave the host and then find a new host at intervals throughout development. Changes during the larval phase are gradual, mainly involving the development of the thoracic legs and reproductive organs, except for the maxillule, which undergoes a profound metamorphosis around the fifth larval stage (Rushton Mellor and Boxshall 1994) changing from a long limb bearing a powerful distal claw into a short but powerful circular sucker. This is one of the most remarkable transformations known for any arthropod limb.

Effects on hosts and ecological importance

Branchiurans are primarily ectoparasites of fishes, but have occasionally been reported from the tadpoles of amphibians. They live mainly in freshwater habitats, both running and static water, and may occur at high density in artificial water bodies such as reservoirs, ornamental fish ponds and fish farms. A few species of *Argulus* infest estuarine and coastal marine fishes but they do not occur in oceanic waters. Infestations with *Argulus* have been reported from marine fish-farming facilities in Chile and Canada and can cause mortality in farmed salmonid stocks.

Branchiurans attach to the skin of their fish hosts and feed on its blood and external tissues. They have rasping mandibles, which scrape tissues into the opening at the tip of the tubular sucking mouth. In *Argulus* the poison stylet is used to inject a secretion into the host. The secretion may contain digestive enzymes to begin to break up host tissues before ingestion. Paired labial stylets lying within the opening of the mouth tube (Fig 4.10D) are also secretory and may produce secretion with a similar pre-digestive function. Host blood is also taken and is digested within paired, lobate gut caeca that lie within the carapace lobes (Overstreet *et al.* 1992).

Important references
Overstreet *et al.* (1992) reviewed the fine structure of Branchiura, and Gresty *et al.* (1993) the fine structure and function of the cephalic appendages of *Argulus japonicus*. Rushton-Mellor and Boxshall (1994) described the developmental sequence of *Argulus*.

Tantulocarida (tantulocarids)
Geoff Boxshall

Introduction
Tantulocaridans are tiny ectoparasitic crustaceans that spend most of their lives attached to the external surface of their hosts, a wide range of other crustaceans. All tantulocaridans are marine and they occur at all depths, from shallow coastal waters to the deep ocean, and in all temperature regimes from polar to tropical. The Tantulocarida is a small group comprising 28 species placed in 20 genera and four families, but its true diversity is undoubtedly underestimated as these parasites are often overlooked because of their minute size. It is classified in the class Maxillopoda and is regarded as most closely related to the Thecostraca (barnacles and relatives) with which it shares a similar body plan and the positions of the genital openings in both sexes (Huys *et al.* 1993).

Morphology
The adult asexual female consists of a minute head, a neck of varying length, and a sac-like trunk full of eggs or developing tantulus larvae, and it may attain lengths of up to 2 mm. It attaches to the exoskeleton of its host by means of a tiny oral sucker, only 12 µm to 15 µm in diameter. This stage has no limbs at all and no genital apertures, and it appears to release mature larvae by rupturing of the wall of the trunk sac. The sexual female, where known, is smaller, less than 0.5 mm in body length, and consists of a large cephalothorax and a five-segmented post-cephalic trunk. The cephalothorax carries a pair of sensory antennules but no mouthparts. A few large eggs lie within the cephalothorax which also carries a conspicuous median genital opening, interpreted as a copulatory pore (Huys *et al.* 1993). The first two of the trunk segments each carry a pair of biramous thoracic legs, which appear to be used for grasping, and the fifth segment bears the elongate caudal rami. The adult male resembles the sexual female in size and basic body plan, with a large cephalothorax and six-segmented trunk, but it has more limbs: vestigial sensory antennules, six pairs of biramous swimming legs, a well-developed median penis and caudal rami. Adults of both sexes develop within posteriorly located, sac-like expansions of the trunk of the still-attached preceding tantulus larva.

Knowledge of tantulocaridan biology is fragmentary. They spend most of their lives attached to their hosts, which include isopod, tanaid, amphipod, cumacean, ostracod and copepod crustaceans. The tantulus larva functions as the infective stage in the life cycle and has been found living free in marine sediments (Huys 1991). The sexual adults have never been collected away from the host, but probably inhabit the hyperbenthic zone, just above the sea bed.

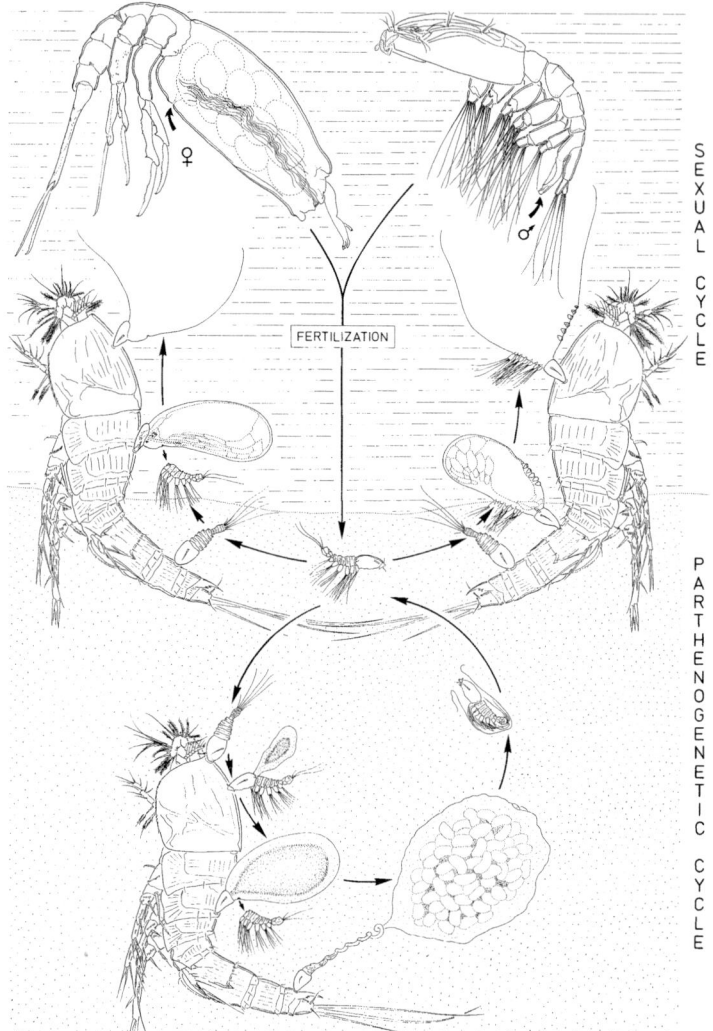

Figure 4.11 The double life cycle of the Tantulocarida. Composite Figure based largely on stages of *Itoitantulus misophricola* known from copepod hosts. The sexual cycle comprises the free-living, non-feeding sexual adults, which mate and the female then releases larvae. The parthenogenetic cycle contains only the tantulus larva and the sac-like, limbless asexual female. From Huys *et al.* (1993); reprinted with permission from The Crustacean Society.

Life cycles

Tantulocaridans have a bizarre double life cycle (Fig. 4.11), with a sexual phase and an asexual phase. The asexual phase is common and the sexual phase is rare. Sac-like asexual females release fully formed tantulus larvae (Boxshall and Lincoln 1987), which are capable of infecting a new host and developing directly into another asexual female, without mating and without even moulting. This tantulus larva, ranging from 85 µm to about 180 µm in length, comprises a head, which has an oral disc but lacks any cephalic limbs, and a trunk of eight segments. The first six trunk segments carry biramous swimming legs equipped with reduced endites. After release from the mother, infective larvae spend time in the sediment before encountering a suitable benthic or hyperbenthic host. Host location and attachment mechanisms are poorly understood as these forms lack eyes and antennae, the main sensory interfaces of other crustaceans. After successfully locating a host the larva punctures the host cuticle using its oral stylet. The cephalic musculature that operates the stylet then degenerates. The tantulus larva develops into an asexual female and the post-cephalic trunk of the larva is shed, so the female remains attached to the host by the adhesive oral disc of the preceding larval stage. The trunk of the female expands to accommodate the growing larvae until they are released.

In the sexual phase, the cycle again begins with the infective tantulus larva attaching to its host by its oral disc. A sac-like expansion forms on the trunk, within which either a sexual adult male or adult female then develops. The precise location of this expansion varies according to family. Developing adults are supplied with nutrients from the host, transported via an umbilical cord originating in the still-attached larval head. Fully formed adults develop within the sac, which is attached to the host only by the oral disc of the larva. On reaching maturity these sexual adults are released by rupturing of the sac wall. They have never been observed alive, but it is assumed that the male, which has well-developed swimming legs and paired clusters of chemosensory aesthetascs representing the antennules, actively searches out and locates the receptive female. The male has a large penis and presumably inseminates the female via the mid-ventral copulatory pore. The fertilised eggs develop within the expandable cephalothorax of the female until ready to hatch as a fully formed tantulus or other larva.

Effects on hosts and ecological importance
Tantulocaridans exhibit varying degrees of host specificity: for example, members of the family Deoterthridae occur on cumacean, isopod, tanaid, amphipod, ostracod and copepod hosts, members of the Microdajidae on tanaid hosts only, members of the Doryphallophoridae on isopods only, and members of the Basipodellidae on copepods only. They attach to the host's outer surface by their adhesive oral disc but they do not appear to suppress the host's moulting (Boxshall and Lincoln 1987). Nutrients are obtained from the host via the puncture in the cuticle, made by means of the oral stylet, which is protruded through a pore in the centre of the disc. There is evidence of an absorptive rootlet system extending from the oral disc of the tantulocaridan and penetrating through the tissues of the host (Boxshall and Lincoln 1983, 1987), but this awaits confirmation.

Important references
The group was established and the main asexual cycle elucidated by Boxshall and Lincoln (1983, 1987). The free-living larval phase was studied in most detail by Huys (1991) who also reviewed available information on the group. The life cycle was completed by the discovery of the sexual female (Huys *et al.* 1993).

Ascothoracida (ascothoracids)
Mark J Grygier and Jens T Høeg

Introduction
The Ascothoracida comprise about 100 species of parasitic crustaceans divided among six families in two orders. All Ascothoracida are marine parasites and occur from the intertidal to the deep sea (5000 m). Hosts are various echinoderms (excluding regular urchins and sea cucumbers) and cnidarians (i.e. gorgonians, zoanthids, scleractinian corals, antipatharians). Their body is enclosed by a carapace that is fundamentally bivalved, but often modified and enlarged for brooding and possibly food absorption in females. The most speciose genera are *Dendrogaster* (about 30 species in seastars) and *Baccalaureus* (about 11 species in zoanthids).

Grygier (1996b) provides a comprehensive entry to the literature. Many largely inaccessible Russian works were summarised by Wagin (1976) in the only book-length review devoted to this group.

Phylogenetic relationships of the Ascothoracida and the evolution of parasitism

The monophylum Thecostraca comprises the Cirripedia (with the orders Acrothoracica, Rhizocephala and Thoracica), the Ascothoracida and the Facetotecta. The Ascothoracida and Rhizocephala (see pp. 154–165) are parasitic. The Facetotecta occur throughout the marine waters of the world as y-nauplii and y-cyprids; adults are completely unknown, but they are also believed to be parasitic (Grygier 1996a). The likely sister group to the Thecostraca is the Tantulocarida (see pp. 147–149), parasitic on Crustacea (Høeg and Kolbasov 2002, Pérez-Losada et al. 2002). Thus a major clade of Crustacea consists of an assortment of parasitic taxa, along with the filter-feeding barnacles (the Acrothoracica and the speciose Thoracica). Morphology and phylogeny indicate that parasitism evolved convergently in the Tantulocarida, Ascothoracida and Rhizocephala. The Ascothoracida, which in contrast to the Cirripedia are a fundamentally non-sessile group (e.g. *Synagoga*), appear to be a relict group that has survived through adoption of a parasitic mode of life.

No ascothoracidans are known as fossils, but several trace fossils (excavations and galls) dating back to the Cretaceous on echinoderm and anthozoan hosts have with considerable confidence been attributed to the Ascothoracida. Although none is yet recorded, fossil galls formed by Petrarcidae in scleractinian corals should be easy to recognise.

Morphology and diversity

Compared to the parasitic Cirripedia Rhizocephala, the Ascothoracida are surprisingly diverse in morphology, biology and host range. The basic body plan comprises a bivalved carapace, with diverticula of the gut and gonads, which encloses the main body. The cephalon has grasping antennules and an oral cone and the 11-segmented trunk has six pairs of biramous thoracic legs, male genitalia and caudal rami (Fig. 4.12). This body plan remains evident in adult females of the family Synagogidae (especially *Synagoga* and *Waginella*). In other families females tend to

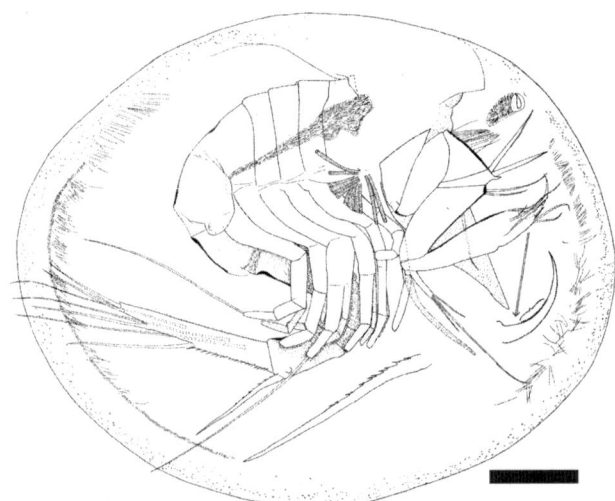

Figure 4.12 *Synagoga paucisetosa* (from Grygier 1990); this ectoparasite deviates but little from the free swimming a-cyprid (ascothoracid larva). Scale = 500 μm.

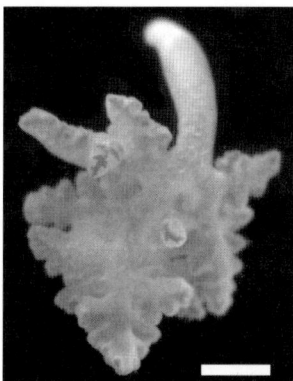

Figure 4.13 *Dendrogaster ludwigi*, a parasite in seastars, with the carapace modified into a heavily ramified mantle enclosing the brood chamber. Scale = 1 mm.

have reduced trunk appendages and abdomen and a large brood chamber formed from the carapace, which may sometimes be lobed or branched (Fig. 4.13). Size ranges from about 1.5 mm in some species of Petrarcidae (in galls in coral polyps) to over 16 cm in some species of *Dendrogaster* (filling the inside of the seastar host). Most studies have dealt with taxonomy and morphology. Thorough studies on life cycles and on demography and ecology are very few, notably Wagin (1947), Brattström (1947, 1948), and Grygier (1991).

Host relations

Some ascothoracidans are ectoparasites, and *Synagoga* may even be capable of swimming from host to host. Many are mesoparasites, either located in a cavity of the host that is connected to the outside (e.g. bursae of brittlestars) or in a cyst or gall with an opening to the outside, as in *Gorgonolaureus* and allies on gorgonians and Ctenosculidae on seastars. True endoparasites seem to include Petrarcidae in galls in scleractinian corals and *Dendrogaster* and allies in seastars. *Ulophysema* is initially an endoparasite in its heart urchin host, but by inducing a hole in the test becomes a mesoparasite (Figs 4.14, 4.15). Parasites of cnidarians may occupy individual polyps, or nodules formed from several polyps; in corals, so-called external galls involve a proliferation of the coenenchyme in general.

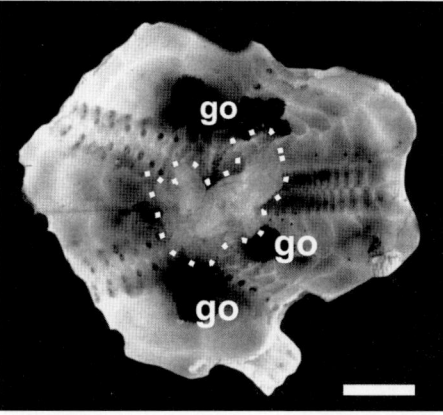

Figure 4.14 *Ulophysema oeresundense*, adult female (outlined), attached to the inside of the host echinoid test (broken open and viewed from the ventral side). Scale = 1 mm; go, gonad.

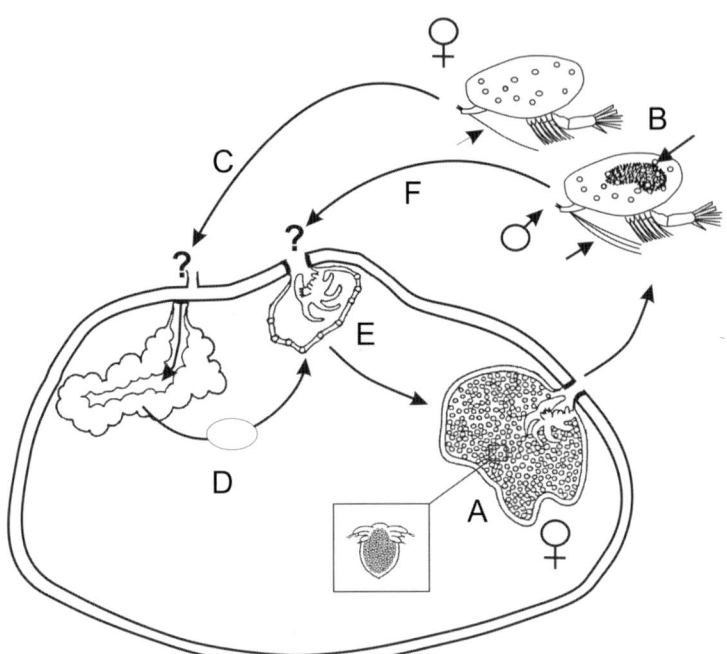

Figure 4.15 Life cycle (in part tentative) of *Ulophysema oeresundense* infesting the heart urchin *Echinocardium cordatum*. After Michael Hartmann, MSc Thesis, Univ. Copenhagen. The adult parasite is internal. **A.** The larvae are brooded in the mantle cavity until the second a-cyprid instar is released into the plankton through the parasite-induced pore in the test of the host. **B.** Male and female larvae differ morphologically, male a-cyprids having three (females one) antennular aesthetascs and testes with mature sperm (arrows). **C.** Females possibly enter the host through a gonopore, eventually breaking into the host's body cavity. **D.** In the host's body cavity the females are free but covered in ciliated host cells. **E.** The female eventually attaches by its apertural region to the host's test and produces a hole to the exterior, which could allow male a-cyprids to fertilise the female. **F.** Male a-cyprids fertilising the female. Young females without such a hole for the entry of males never contain developing embryos or larvae.

Many of the parasites of echinoderms (*Ascothorax*, *Ulophysema* and *Dendrogaster*) have been implicated in host castration. Little is known of feeding. Primitively, they undoubtedly feed by the piercing-sucking mouth apparatus. The ultrastructure of the cuticle of *Ulophysema* is suggestive of absorptive feeding in the manner of rhizocephalan barnacles, but whether this is its primary mode of food intake and occurs in other ascothoracidans remains unknown.

Life cycle and sexual system

Sexes are separate except for the hermaphroditic Petrarcidae. Sex determination seems to be genetic, and in some species male and female larvae differ morphologically (e.g. in the armature of chemosensory aesthetascs, just as in most Cirripedia Rhizocephala). Primitively, larval development comprises six naupliar instars followed by one or two cypris-like stages (a-cypris), homologous to the single cypris instar of the Cirripedia and the y-cypris of the Facetotecta. Lecithotrophy is common, but some species have planktotrophic nauplii. The a-cypris, also known as the ascothoracid-larva, attaches by its grasping antennules rather than by glandular secretions as in cirripede cyprids (Figs 4.16, 4.17). In species with two a-cyprid instars, only the second is the actual settlement stage. Many species have an abbreviated ontogeny, and the entire naupliar phase is sometimes brooded, or even embryonised.

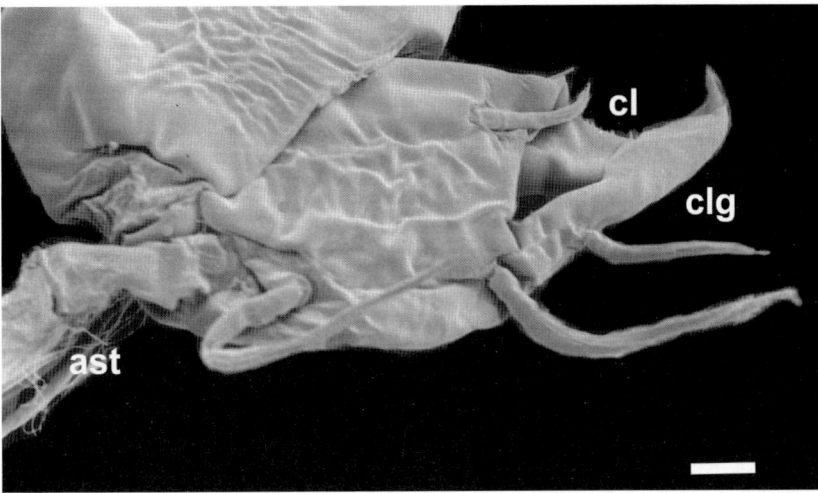

Figure 4.16 Terminal part of antennule of a-cyprid of *Ulophysema oeresundense* with chemosensory aesthetascs (ast); attachment is purely mechanical using the claw (cl), which is retractable into the claw guard (clg). Scale = 10 μm.

No complete life cycle has been worked out for any ascothoracidan, and neither host infection nor copulation has ever been witnessed directly (seminal receptacles are in the female's legs, but absent in Dendrogastridae). The two best-known species are *Ulophysema oeresundense* (see Brattström 1948) and *Baccalaureus falsiramus* (see Itô and Grygier 1990). *Ulophysema oeresundense* (Fig. 4.15) has two brooded naupliar instars and two a-cyprid stages, the latter being released into the plankton. *Baccalaureus falsiramus* has six naupliar stages and a single a-cyprid stage. Its manner of host infestation (in *Zoanthus* sp.) is unknown, but like its congeners the female becomes a mesoparasite, with the carapace aperture open to the outside. The dwarf male

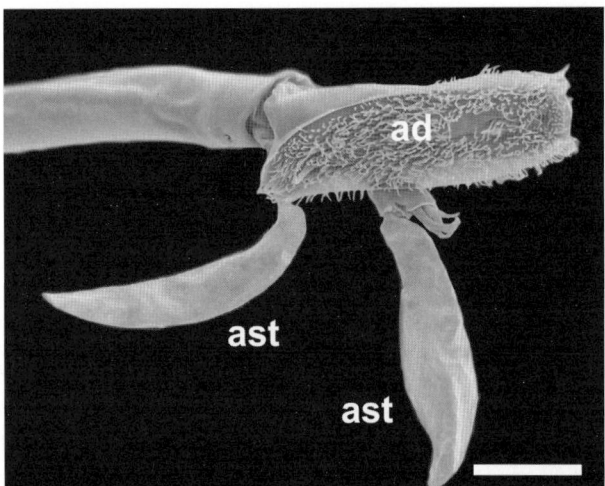

Figure 4.17 *Briarosaccus callosus* (Cirripedia Rhizocephala), male cyprid, terminal part of antennule; attachment is by a cement secretion from the attachment disc (ad). Both the ascothoracidan (a-cyprid in Fig. 4.16) and rhizocephalan (cyprid) antennules carry large chemosensory aesthetascs (ast), assumedly to assist the location of a host (females) and a female parasite (males). Scale = 10 μm.

sits close to the aperture, and larvae presumably escape this way. Dwarf males of *Dendrogaster* live within the mantle cavity of the female, presumably feeding upon it; how they arrive there is unknown, but self-production of an initial brood of males by the female has been suggested. Females of *U. oeresundense* are not permanently associated with males, but male a-cyprids have testes with mature sperm, and they possibly fertilise the females through the pore induced by the latter in the test of the host echinoid (Fig. 4.15). In agreement with this hypothesis, females of *U. oeresundense* that are free in the host's body cavity never contain developing embryos or larvae. In no highly modified species has metamorphosis from the last larva to juvenile been observed, although the transformation from bivalved to sac-like carapace form has been supposed to take place in one moult in *Ulophysema* and *Dendrogaster*. In *Parascothorax* a 'post-larval' stage of both the female and male has been distinguished and second-stage a-cyprids ready to moult to the male stage have been observed.

Ecology and epidemiology
Few species have been thought to be abundant enough for meaningful ecological and particularly demographic study. There are two main exceptions. One is *Ulophysema oeresundense*, parasitising Scandinavian heart urchins (Brattström 1947), including up to 20% of *Echinocardium cordatum* in the Sound connecting the Baltic Sea and the Kattegat. The other is *Parascothorax synagogoides*, parasitising the brittlestar *Ophiophthalmus normani* in the bathyal basins off California, with prevalence varying among basins (0.5–9.0%), multiple infestations being more frequent than chance would predict, and brood sizes varying from 3 to 183 (Grygier 1991). Male a-cyprids apparently join females before the cysts induced by the latter become closed; as many as five males may accompany a female. Grygier (1991) is perhaps the only demographic study ever published of a deep-sea invertebrate parasite; this paper also makes a comparison with known ecological information on two species of confamilial *Ascothorax*.

Hyperparasitism of ascothoracidans by cryptoniscid isopods is not uncommon, occurring in four of the six families; parasitic castration of the ascothoracidan was noted in the case of *P. synagogoides*.

Important references
Comprehensive accounts of the Ascothoracida are by Wagin (1976, in Russian) and Grygier (1996b). Itô and Grygier (1990) described the larval development of an ascothoracid. Grygier (1996a) gave a detailed account of the Facetotecta. The phylogenetic relationship of the Ascothoracida and Cirripedia and the phylogenetic position of the Facetotecta were analysed using morphology by Høeg and Kolbasov (2002) and on the basis of 18S rDNA sequences by Pérez-Losada et al. (2002).

Cirripedia Thoracica and Rhizocephala (barnacles)
Jens T Høeg, Henrik Glenner and Jeffrey D Shields

Introduction
The Rhizocephala comprise about 250 species, which is about one-quarter of all cirriped species. Rhizocephalans are parasites of other Crustacea, principally Decapoda. Adult Rhizocephala are sessile, in contrast to the related Ascothoracida, discussed in the previous Section (see pp. 149–154), which are fundamentally (primitively) mobile. Adult females have lost all obvious crustacean morphological traits and consist of a sac-like part attached to the crab host (Fig. 4.18) with outgrowths of tissue (rootlets) extending cancer-like into the host's body. The males are reduced to dwarfs and live within the parasitic female. Only the larvae are free living and

they closely resemble those of other barnacles. Rhizocephalans are particularly fascinating because they induce marked behavioural changes in their hosts that can benefit host survival. They also have considerable economic importance causing mortality and castration, thus reducing the profitability of crustacean fisheries.

Thoracica

Nearly all parasitic Cirripedia belong to the Rhizocephala, which parasitise Crustacea. The few parasitic cirriped species that do not belong to the rhizocephalans possess rudimentary thoracopods and are tentatively included in the Thoracica. These are *Rhizolepas*, with a few species infecting polychaetes, and *Anelasma squalicola*, which infects dogfish. Neither of these genera is related to the Rhizocephala. Both are ectoparasites in which a body with rudimentary thoracopods (cirri) is attached by means of a peduncle that sends nutrient-absorbing roots penetrating into the host. *Anelasma* is hermaphroditic and close in morphology to a filter-feeding barnacle, retaining both mouth appendages and a gut, albeit these structures may not be functional. The dogfish host seems always to carry at least two *Anelasma* together, attached at the dorsal fin. More recently additional thoracican barnacles have proven to be ectoparasites including the family Microlepadidae on diadematid echinoids, *Koleolepas* on sea anemones inhabiting hermit crab shells and Hoekiini barnacles on corals. Unlike *Rhizolepas*, *Anelasma* and the rhizocephalans, all of these species feed on the host as micropredators, using their appendages and mouth for food intake (Grygier and Newman 1991, Ross and Newman 1995, Yusa and Yamato 1999, Yusa *et al.* 2001). Whale barnacles (Thoracica: Coronuloidea), although deeply anchored within the dermis of their cetacean hosts, feed exclusively by filtration and do not absorb nutrients from the host.

Effects on hosts

Most epizootic Thoracica are better known as fouling agents rather than parasites because they do not penetrate into their host or otherwise cause physical damage. However, at sufficiently high densities, infestations of balanid and pedunculated barnacles may become physiologically deleterious. *Chelonibia patula* is an acorn barnacle symbiotic on turtles, whales and other crustaceans. Heavy infestations on crabs can burden the host by hampering movement and increasing its vulnerability to predation (e.g. Overstreet 1983). The pedunculated *Octolasmis* spp. infest the gills of decapod hosts. Oxygen uptake, lactate levels, pH and other blood parameters do not differ between infested and uninfested blue crabs infected with *O. muelleri*, but infested hosts show an elevation in heart rate and ventilation rate as if in a constant state of exercise (Gannon and Wheatly 1992, 1995). Heavily fouled crabs may indicate the presence of other disease agents, senescence or anecdysis, and they probably do not survive long in nature.

Economic importance

The thoracican barnacles are not known to cause significant economic impacts to crustacean or other fisheries. However, heavily fouled hosts are not typically selected for high grading and may end up in lesser valued processed foods rather than as whole product. Heavily fouled hosts may indicate the presence of other disease agents or senescence and, thus, they are often culled from further use.

Rhizocephala

The Rhizocephala are a bizarre, highly modified group of barnacles that are barely recognisable as Crustacea. While the morphology and life cycle of these intriguing parasites is now well described, their ecology and coevolution with their hosts is only just beginning to be understood.

Most rhizocephalans parasitise brachyuran and anomuran crabs. A few are parasites on carideon shrimp, stomatopods, peracarids and even other barnacles. None are found on palinuran lobsters or astacidean crayfish, which is surprising given the broad range of infections in disparate host taxa.

Most infect more than a single host, and many of these do not appear to be cryptic species (Murphy and Goggin 2000). The well-studied *Sacculina carcini* infects more than 10 crab species from several genera (Øksnebjerg 2000). Larval female cyprids can settle on species that never harbour adult parasites, with far reaching implications for the colonisation of new hosts (i.e. habitats) (Thresher et al. 2000). However, a few rhizocephalans are specialists, using only a single host species or genus.

Rhizocephalans are found from the deep sea to the pelagic and into high intertidal habitats. They extend into brackish waters, but only a few species occur on truly freshwater or semiterrestrial crustaceans, leading to marked changes in life history (Andersen et al. 1990).

Several reviews have dealt with the taxonomy, morphology or life cycle of the Rhizocephala (Høeg and Lützen 1985, Høeg 1992, 1995, Høeg and Lützen 1995, 1996, Øksnebjerg 2000, Walker 2001). The ecology of only a few species has been studied. Høeg (1995) and Høeg and Lützen (1995) reviewed the literature until about 1992. Later important references are Hochberg et al. (1992), Alvarez et al. (1995), Hines et al. (1997) and Tindle et al. (2004) on species of *Loxothylacus*, Alvarez et al. (2001) on *Lernaeodiscus porcellanae* and Werner (2001) on *Sacculina carcini*.

Systematics and evolution

The Rhizocephala consists of two monophyletic suborders, the Kentrogonida with more than 230 species and the Akentrogonida with 30–40 even more highly specialised species. More than 120 species belong to the genus *Sacculina*, including the 'classic' *S. carcini*, but molecular evidence indicates that the genus is polyphyletic. Species level systematics is very problematic, but scanning electron microscopy promises to offer characters for comparatively easy identification (Rybakov and Høeg 2002).

These parasites must have evolved from sessile filter-feeding cirripedes, but the details of this process, including the intercalation into the life cycle of an internal migratory phase, the vermigon, remains largely unexplained (Glenner and Høeg 2002). While other barnacles are well represented in the fossil record, there are no fossils of Rhizocephala (Conway Morris 1982); thus molecular techniques will be needed to further study their evolutionary and coevolutionary patterns.

Morphology, life cycle and reproductive ecology

The adult parasite has lost virtually every arthropod trait, including segmentation, appendages, sense organs, excretory organs and gut. It consists of an external reproductive body, the externa, connected to a nutrient-absorbing system of rootlets infiltrating the host, the interna. Specialisation lies mainly in their sexual systems and the advanced level of behavioural control exercised by these parasites upon their hosts.

A typical rhizocephalan life cycle includes several life history stages (Figure 4.19).

Larval development and dispersal

Despite the parasitic life style, larval development of Rhizocephala is virtually identical to that of other Cirripedia (Walossek et al. 1996). There are several naupliar instars and a terminal cyprid, the settlement stage unique to all cirripedes. Larval development is entirely lecithotrophic and normally of short duration, lasting two to three days in warm waters, about two weeks at 10°C, but up to 30 days in the arctic. As a probable adaptation to enhance survival or dispersal, several

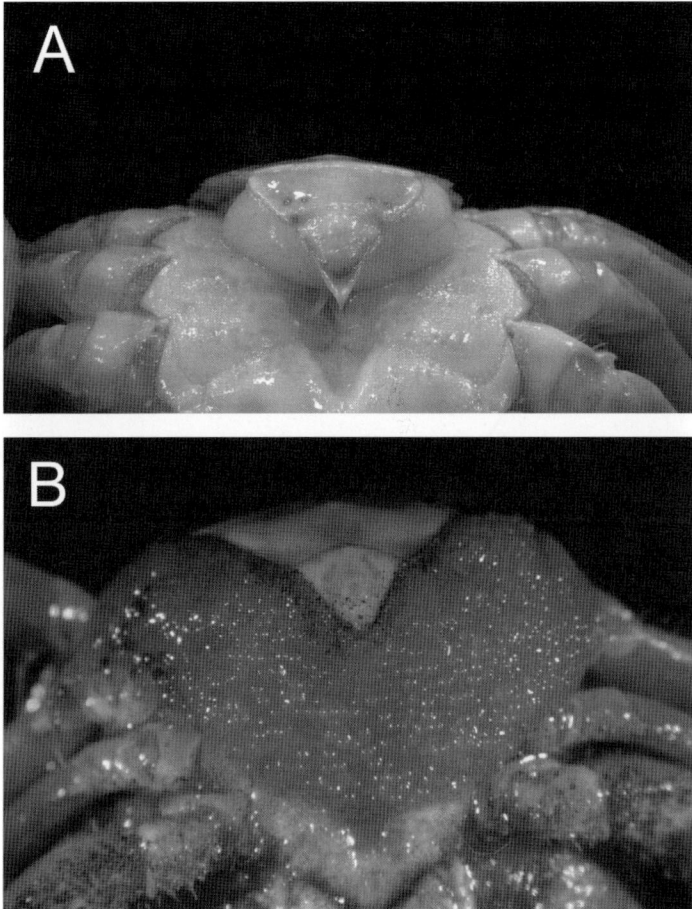

Figure 4.18 **A**. *Sacculina carcini* infesting the shore crabs *Carcinus maenas* and *C. mediterraneus* in western European and Mediterranean waters. The external reproductive body sits under the abdomen where **B**. healthy females normally carry their brood, but the host in **A**. is a feminised male crab.

rhizocephalans release their larvae as cyprids, including most deep sea species, all freshwater and semi-terrestrial species and all members of the specialised suborder Akentrogonida.

Female parasite strategies
Female cyprids probably rely on odour trails in the water column to locate the host (Pasternak *et al.* 2004a,b). At settlement, the cyprids recognise the host by means of carbohydrate or glycoprotein cues in the epicuticle (Boone *et al.* 2003, 2004). The host's behavioural defences can be circumvented by three means: efficient host finding; avoidance of detachment before penetration; and avoidance of the host's immune response coupled with a rapid takeover of host control once injected by the cyprid larva. In general, the female cyprid can locate recently moulted crabs by chemosensory ability. Rhizocephalans settle at specific sites on the host, but always in areas with a thin or soft cuticle to facilitate penetration. The well-studied example of *Lernaeodiscus porcellanae* and its host, the porcelain crab *Petrolisthes cabrilloi*, illustrates the effectiveness of host cleaning defences and parasite countermeasures (Fig. 4.20). Female cyprids settle in the narrow confines between the gill filaments in the branchial chamber to avoid being groomed away by

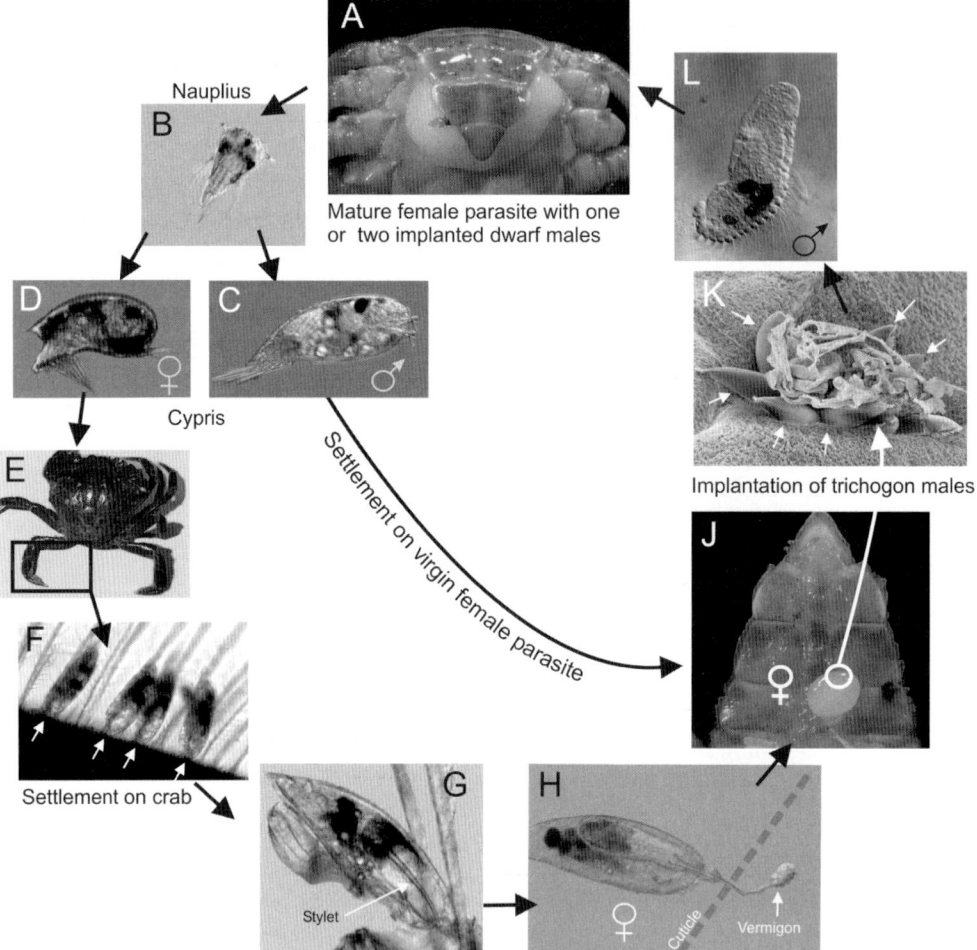

Figure 4.19 Life cycle of *Sacculina carcini* (Kentrogonida) parasitising the brachyuran crab, *Carcinus maenas*. **A.** The parasite is externally manifested by a reproductive body (externa) located on the abdomen and connected with a nutrient-absorbing system of rootlets (interna) inside the host. The parasite is female, but is host to and nourishes two highly reduced dwarf males in a pair of receptacles. A series of broods are released and fertilised by sperm produced cyclically by the resident males. **B.** Larvae are released as nauplii that develop lecithotrophically into cyprids. Female cyprids (D) settle on potential host crabs (E, F). They form a kentrogon (G), which uses a hollow cuticular stylet (G, arrow) to penetrate the host cuticle and inject a vermigon stage into the hemocoel; the stippled line in (H), indicates where would have been the crab cuticle in this *in vitro* preparation. After an internal growth period, the parasite emerges externally as a small, virgin female (J). These are the targets for male cyprids (C), which accumulate around and fight for access to the narrow opening into the brood chamber. At most two of them will successfully inject a trichogon stage into the chamber and become established in a receptacle as highly reduced sperm producing dwarf males (See also Fig. 4.21). Implantation of one or two males immediately induces the female to grow into sexual maturity. The males produce sperm in synchrony with the cycles of egg production in the ovary of the female. They are nourished by the female and remain with her for the duration of her lifetime. (Based on Delage 1884, Lützen 1984, Høeg 1984, 1987, Walker 1988, Glenner 2001 and previously unpublished micrographs.)

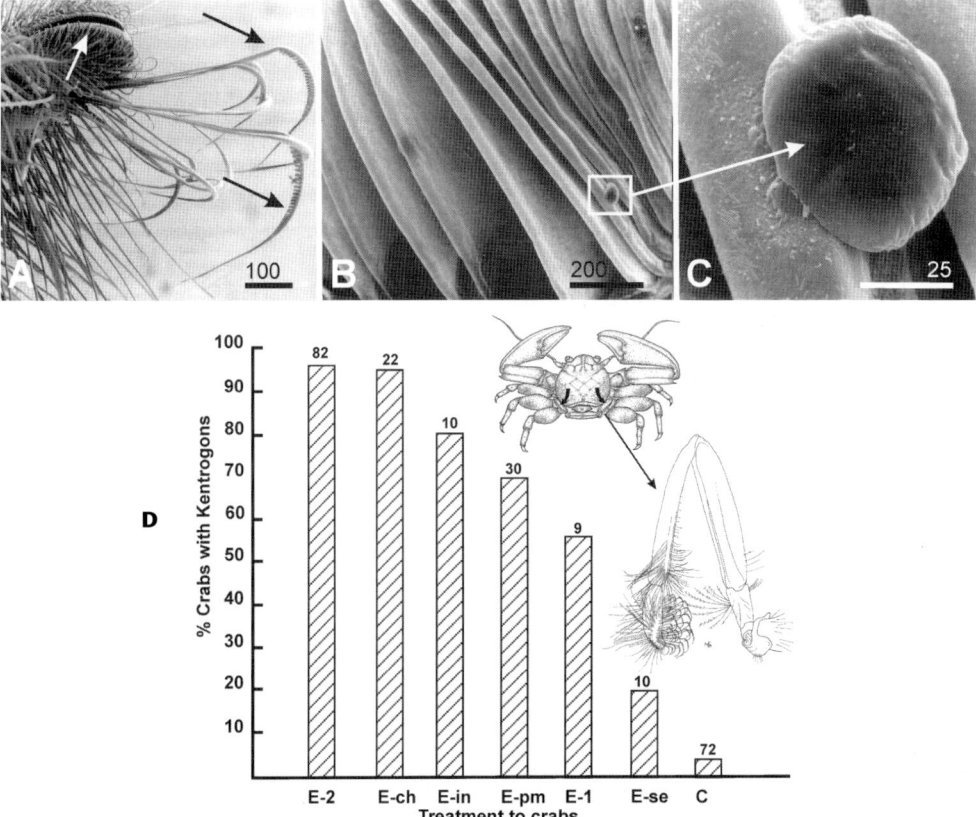

Figure 4.20 Host cleaning defences and parasite countermeasures in the porcelain crab, *Petrolisthes cabrilloi*, and the rhizocephalan, *Lernaeodiscus porcellanae*. The line drawings depict a crab and an enlarged grooming limb with clusters of setae at the tip. SEM micrographs show the following. **A**. The tip of a grooming limb; the sickle-shaped setae (black arrows) remove parasite cyprids and general debris on the gills, but only the cheliped (white arrow) can remove the scale-shaped and more firmly attached kentrogons. **B**. Several gill filaments with an attached kentrogon (square). **C**. Kentrogon enlarged. All scales in micrometers. **D**. Experiments on gill cleaning efficiency (see details in text). Healthy control crabs can normally groom away all parasite larvae before they manage to invade, while crabs with missing or damaged grooming limbs (E-2, E-ch, E-1, E-se), injured crabs (E-in) and immediate post-moult crabs (E-pm) risk infection because they retain substantial numbers of infective kentrogons after exposure to parasite cyprids. Numbers above bars indicate how many crabs were used for each experiment.

the crab before they succeed to invade. Using the specialised 5th pair of thoracopods (Fig. 4.20D), the crabs can normally groom away most of the settled cyprids before formation of the infective kentrogon (Fig. 4.20C). Removal of cyprids is by means of the sickle-shaped setae at the tip of the grooming limbs (Fig. 4.20A, black arrows). The smaller, firmly attached and low-profiled kentrogons (Fig. 4.20B, C) can only be removed using the chelipeds at the tip of the grooming limbs (Fig. 4.20A, white arrow). As a parasite countermeasure, formation of the better protected kentrogon is very rapid (<1 hour after settlement), thus minimising the period when the larvae are critically exposed as cyprids. Experimental crabs in Figure 4.20D were exposed to female cyprids for 30 minutes and thereafter isolated for 90 hours, followed by counting the numbers of kentrogons on their gills to estimate the efficiency of host grooming. Crabs with both grooming limbs ablated (E-2) cannot remove parasite larvae and are thus very susceptible to infection. Decreasing numbers of kentrogons due to increasing grooming efficiency is seen in crabs with both grooming chelipeds crushed using forceps (E-ch), crabs injured with a needle prick (E-in), immediate post-moult crabs with soft cuticles (E-pm), crabs with only one grooming limb removed (E-1), and crabs with the sickle-shaped setae removed (E-se). Control crabs (C) can almost completely remove parasite larvae, even at the unnaturally high densities in this laboratory experiment.

In *Sacculina carcini*, settlement at the base of a seta may offer a similar protection against host grooming. Injection of the vermigon (the migratory internal stage) happens within one to three days after settlement and the parasite immediately begins to take over control of the host, one of the first effects being to damage the nervous system by penetration of the parasite's rootlets and by bringing hormonal slavery. How host control is achieved and maintained is little understood and is wide open to a modern molecular approach.

Male parasite strategies
Unlike the female cyprid, which at settlement meets the hostile environment of the potential host, the male cyprid must locate and inseminate the virgin externa of the female via implantation. However, the male also faces several hurdles mainly due to competition with other males (Fig. 4.21). The male cyprid must not only locate a host crab, but also one already carrying an established virgin parasite. Furthermore, no more than two males can normally be implanted in any parasite, and once a single male is acquired, the female externa immediately begins to mature sexually; the accompanying morphological and physiological changes soon prevent the insemination by other males. Where prevalence is high there is a surplus of male larvae, so males must compete to find the rare virgin. Many male cyprids can land on a female vying for insemination, so only the fastest and strongest ones will successfully inseminate the female (Høeg 1987, 1991). At low prevalences, male larvae are fewer and virgin parasites may remain receptive over extended time periods (Høeg and Ritchie 1985).

When a virgin is found, the cyprid can penetrate no farther than the narrow entrance to the brood chamber, which contains the paired receptacles that will eventually host the one or two successful males (Fig. 4.21B). Only those settled closest to the orifice can inject a trichogon stage into the brood chamber, and once inside (Fig. 4.21C), they must be the first to reach an empty receptacle. Having achieved this, the trichogon sheds its spiny cuticle, thereby blocking access to any subsequently arriving competitors (Fig. 4.21D). This spatial barrier into the brood chamber may serve both to protect against multiple matings and to insure that only the most vigorous males gain access. Male cyprids have more chemosensory organs than females (Fig. 4.17), and the efficiency of these organs is highlighted by the whole process, from locating the externa to entering a receptacle, being able to be completed within minutes. Once established, the male undergoes spermatogenesis and is nourished by the female parasite for the duration of its lifetime (cryptogonochorism). However, the male may still face competition with a potential partner in the other receptacle (Fig. 4.21E). It is, therefore, of benefit to the male that the female commences sexual maturation immediately after insemination by the first male. Which sex controls this process is not known. Apparently a single male suffices to fertilise all of the ova in the broods produced by a female. The presence of two receptacles (i.e. two possible males) may therefore benefit the female either by increasing genetic variability or by insuring continued reproduction should one male accidentally perish.

Sexual biology and reproductive strategies
Unlike most other cirripedes, which are hermaphroditic, rhizocephalans have separate sexes, but the detailed mechanism of the genetically based system for sex determination remains unknown (Yanagimachi 1961). For unknown reasons, male larvae are generally larger than females, but it

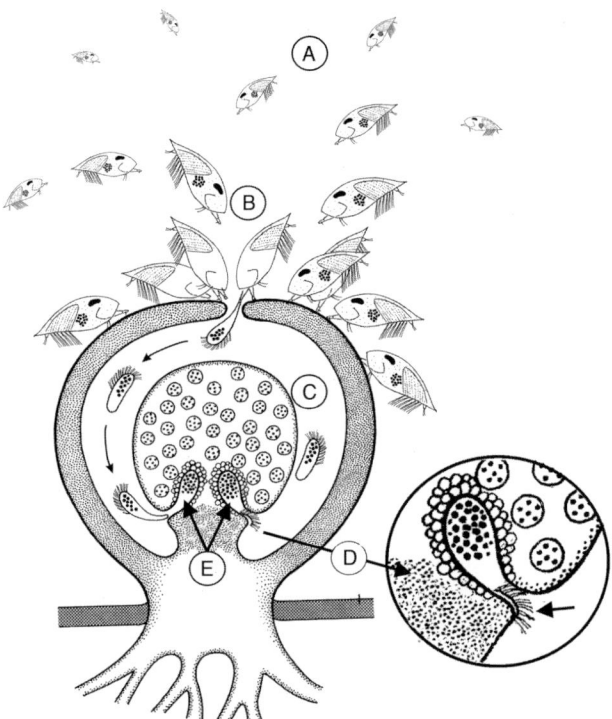

Figure 4.21 The battlefield of the virgin externa: The male larvae faces many obstacles in achieving reproduction. **A**. A cyprid locates a host carrying a virgin parasite externa. **B**. Competition for space around the narrow confines of the aperture and implantation of a trichogon stage into the brood chamber. **C**. The race against time to migrate to the opening of one of two male receptacles. **D**. Blockage of one of two male receptacles (arrow) by the shed cuticle of a previously arrived male. **E**. Competition of an established dwarf male with its male partner in the other receptacle in fertilising the eggs of the female. In this virgin externa, several males have settled close to the aperture and injected trichogons; one is being successfully implanted in the left side receptacle (curved arrows); the right side receptacle is already blocked by a previously arrived male (D) and the trichogons arriving there will eventually perish. Based on Høeg and Ritchie (1985) and Høeg (1987).

may allow them longer time to search for virgin females. The brooded eggs can have any sex ratio, although it is normally wholly or predominantly biased towards one sex. Surprisingly, the sex ratio varies through successive broods for a single externa and follows the yearly cycle of the population. In *S. carcini*, broods are heavily biased to the male sex when virgin females are most frequent (Table 4.2). Adding to the complexity, the mean sizes of both male and female larvae varies with season, again for reasons entirely unknown. In the Akentrogonida, sex determination and larval ecology remains largely unknown.

The number of males hosted by a single female parasite varies. Kentrogonids with colonial externae, each with two receptacles, can host many males, thereby increasing genetic variability. An extreme case is seen in the akentrogonid and colonial Thompsoniidae, where the female consists of hundreds or thousands of externae, each of which can be fertilised by males.

The adult rhizocephalans have two strategies for host use dictated by their own longevity. Peltogastrid and lernaeodiscid species that infest anomuran hosts remain on the host for a term approaching the life expectancy of a healthy crab. During this period they regularly release broods of larvae every two to four weeks. These species also allow their hosts to continue moulting and sometimes also to increase in size; they probably do not extract more resources than

Table 4.2 The percentage of juvenile (virgin) females, adult females and male cyprid larvae of *Sacculina carcini* infesting *Carcinus maenas* along the French coast of the English Channel

	Juvenile female parasites (%)	Adult female parasites (%)	Male cyprid larvae (%)
Spring	51	49	86
Late summer	37	63	14
Data combined for 1982, 1983, 1986.			

would otherwise be used in the host's own reproduction. Conversely, sacculinids that infest brachyuran crabs produce a limited series of large broods over a rather short period, one to two reproductive periods, after which they die. The crab host is normally prevented from moulting and therefore becomes deleteriously fouled. Some akentrogonids produce only a single brood during the life time of the parasite and thus display semelparity, an extreme form of this short-term infection pattern.

A more complicated form of the long-term infection pattern is seen in the species with colonial externae. In some species each such externa reproduces only once, but is replaced by regeneration from the root system. The advantages of this system are not well understood, but it spreads the risk of lethal injury over several externae. This is especially advantageous in species, such as thompsoniids, where the externae protrude anywhere on the body of the host and are thus prone to being lost. But colonial externae may also serve as a means to optimise reproductive output. This is directly correlated to externa volume, which is smaller in a solitary externa than the total volume of multiple externae from a similarly sized host (Wardle and Tirpak 1991, Galil and Lützen 1995).

Epidemiology
Single infections are the rule in most species. Multiple (colonial) externae connected to a common root system are frequent or obligatory in some species, but true multiple infections by separate cyprids can occur. Prevalence varies extensively in the Rhizocephala, both within and between species, but few studies have examined more than a single season or differences between host populations. Prevalence ranges from well below 1% to nearly 100% in some populations of *Sacculina carcini* from Mediterranean lagoons. Prevalence often shows wide local variations over short scales. For example, Werner (2001) followed *S. carcini* on *C. maenas* on the west coast of Sweden and found prevalence to consistently vary by an order of magnitude between stations separated by only a few kilometres, the most sheltered sites having the highest abundance. At Roscoff on the French channel coast, prevalence varies from almost nil to 20% over a few hundred metres. The reasons for such variation are often obscured, but host factors such as sex, size, moulting frequency, moult stage and migratory behaviour show correlations with prevalence. For example, blue crabs, *Callinectes sapidus*, have a peak in moulting activity during spring, and most infections occur at this time because the parasite larvae have a preference for post-moult crabs (Tindle et al. 2004). Female blue crabs appear more susceptible to infection at this time because they prefer high salinity waters and thus come into contact with the parasite *Loxothylacus texanus* which cannot tolerate low salinities (Boone et al. 2004). Environmental factors such as hydrography, salinity, depth and turbulence often show strong correlations with prevalence in several species (e.g. Reisser and Forward 1991, Walker and Lester 1998, Boone et al. 2004). In *L. porcellanae* parasitising intertidal porcelain crabs, hosts inhabiting high wave energy environments have more damage to their grooming appendages making them prone to infection (Fig. 4.20), while those in sheltered habitats such as mussel beds suffer less damage and have lower prevalences. In *Loxothylacus panopaei*, prevalence is highest in hosts

living at salinities of 10 parts per thousand (ppt) to 15 ppt and zero at salinities below 10 ppt (Reisser and Forward 1991). Hydrographic features have been associated with increased prevalence of *Briarosaccus callosus* on king crabs, presumably because fjords with shallow sills entrain the water mass thereby retaining high levels of infectious larval stages (Sloan 1984, 1985). Even when prevalence is low, these parasites can linger for extended periods such as in a hermit crab infested by *Clistosaccus paguri* (1–2.5% over three years, Høeg 1982). However, rare host species may limit the spread of their parasites simply as a result of low population densities.

Effects on hosts
Rhizocephalans change the morphology, physiology and behaviour of their hosts so that hosts become nothing but automatons serving only the purpose of the parasite. Almost all Rhizocephala sterilise their host through hormonal or biochemical alteration of the gonads. In most cases, infections cause feminisation of castrated males through the destruction of the androgenic gland, the male sex-determining organ in crustaceans. Castrated hosts are essentially taken over by the parasites and in most cases remain altered for life, becoming sterile competitors in the host population. This is probably the worst competitor any species can face. Shields and Wood (1993) found that male crabs will even attempt to mate with sterile, infected females and feminised males! Because of such sterile matings, the effect of the rhizocephalan to the host population may be even greater than that predicted by prevalence alone, particularly when considering the lengthy pre-copulatory and post-copulatory embraces elicited by receptive crabs.

Parasitic castration also channels metabolic energy otherwise used in host reproduction to the parasite. This is notable because the sterile, infected host reacts to its parasite by cleaning and caring for it as if it were its own offspring. This effect also extends to male hosts which are feminised behaviourally as above, but also morphologically, often acquiring secondary sexual characters of the females such as a broader abdomen used for 'brood protection' in female crabs and the development of female pleopods (Ritchie and Høeg 1981, Innocenti *et al.* 1998). This elaborate 'brood mimicry' is possible because the reproductive body, or externa, of the parasite generally protrudes from within the brood chamber of the host. Some host species will even assist the parasite by exposing the abdomen when the virgin parasite must acquire males and by ventilating it when the mature female releases its broods of nauplii. Additionally, the parasite can arrest the moult cycle of the host (i.e. cause anecdysis), but, notably, this is not universally so. Species infesting anomuran crabs allow their hosts to continue moulting, thus enabling it to groom and often also to grow. Due to the effect on the moult cycle, infestation by rhizocephalans can affect the size distribution of the host population, normally resulting in large numbers of stunted 'females' in the population, but the pattern may vary both between and within host-rhizocephalan systems (O'Brien and Van Wyk 1985).

Host strategies
Ecologically, the effect of castration is the loss of the reproductive fitness of an infected host. A successful infection immediately begins to take over control of the hormones and behaviour of the host. Host defences may operate at three levels: 1, behaviourally, by evasion of the female cyprids; 2, physically and behaviourally, by preventing the larvae from attaching and penetrating through the carapace; and 3, immunologically, by fighting the internal parasite before it becomes established. Female cyprids probably rely on odour trails in the water column to locate the host and this may be one reason why they prefer settlement on recently moulted crabs, which are presumably either more odiferous or exude a different odour than intermoult crabs (Glenner and Werner 1998). There is no evidence for behavioural evasion by crabs, but host grooming is important in preventing infections. Porcellanid crabs are excellent groomers and normally remove all parasite larvae before they penetrate (about 48 hours after settlement) and

thus avoid becoming infested (Fig. 4.20). Susceptible individuals are those that for several reasons (damage, disease or moulting) are deficient in their grooming effectiveness. Porcellanids react within seconds to the presence of female cyprids by initiating fervent grooming of their gills (the settlement site for their rhizocephalan), yet in the presence of harmless nauplii or male cyprids, no grooming behaviour is elicited and the crabs continue to filter feed. Defensive gill grooming must be a time-consuming behaviour and may perhaps be relaxed in populations where the prevalence of the parasite is low; however, such critical population-level studies on coevolution have not been performed (Nuismer et al. 2003). Another question is whether hosts other than porcellanids use grooming to fight rhizocephalan infection.

Immunologically, the host defences must attack and kill the parasite before it takes over hormonal control. In abnormal hosts there is evidence of cellular reactions to the parasite. Melanised internae can be observed in such hosts, and occur in the few normal hosts that survive the loss and death of the parasitic externa (hence, death of the parasite) (Sparks and Morado 1986). Susceptible hosts often show little indication of a defensive response, so the parasite somehow avoids recognition by the host's defences. There are also indications that if parasitism does not succeed, the host may simply die. The same can be seen in abnormal hosts that are artificially infested (i.e. the parasite may be more lethal to them than to their regular host) but experimental studies have used high exposure levels that may have swamped the host's defences leading to host death. Given that rhizocephalan larvae seem to be opportunistic in their choice of settlement targets, survival in an otherwise abnormal host may have important consequences for their ability to spread into a new geographical area.

Economic importance

Rhizocephalans cause castration, anecdysis, stunting and increased mortality to their crab hosts, and thus, can have direct and indirect effects on crustacean fisheries. Prevalence can reach extraordinary levels in focal outbreaks and the resulting stunting and increased mortality of infected hosts can result in direct losses to a fishery because stunted and moribund animals cannot be marketed. Indirect effects include castration, sterile matings, loss of fecundity and competition with phenotypically identical parasites (Shields and Wood 1993). Fortunately, few commercial fisheries are plagued by rhizocephalans, but they do damage segments of lithodid and portunid crab fisheries. *Briarosaccus callosus* infects numerous lithodid king crabs from the Antarctic to the Arctic. In the golden king crab, *Lithodes aequispina*, prevalences reached 48% in the fjords of British Columbia (Sloan 1984). In the blue king crab, *P. platypus*, prevalences reached 76% in the fjords of south-eastern Alaska (Hawkes et al. 1986). Interestingly, the red king crab, *Paralithodes camchaticus*, hosts *B. callosus* in its native range, but the population introduced into the Barents Sea and now spreading rapidly southwards along the Norwegian coast is without this parasite.

Two important portunid fisheries are also impacted by rhizocephalans. *Sacculina granifera* infects *Portunus pelagicus* with prevalences reaching 40% (Shields and Wood 1993). Infected crabs are frequently stunted and their hepatopancreas turns a vivid green that can impart a poor flavour to the flesh (Bishop and Cannon 1979). *Loxothylacus texanus* on *Callinectes sapidus* can reach prevalences of 30% to 70% in the lagoons around the Gulf of Mexico. All of these rhizocephalans stunt their hosts, producing individuals too small for sale which are left to accumulate in the fishing grounds as 'shorts', serving as foci for transmission to new hosts (Meyers 1990). Human activities such as culling stunted, parasitised crabs back into the water or moving animals between locations (culling while underway) may contribute to the spread of these parasites.

Recent models have suggested that parasitic castrators can regulate populations of crustaceans (Blower and Roughgarden 1989a,b, Kuris and Lafferty 1992). Such systems may be considered hydrodynamically or demographically closed for purposes of immigration (e.g. Sloan 1984, Cowen et al. 2000). Thus, the potential impacts of rhizocephalans on the connectivity between fecundity, larval dispersal and juvenile recruitment represent an exciting area for productive research. As an example, the portunid crab *Charybdis longicollis* has recently spread through the Suez into the eastern Mediterranean and created an important fishery. It was later accompanied by its natural rhizocephalan and the dynamics and spread of this host-parasite system is now being monitored (Galil and Lützen 1998). (For commercial effects see also p. 364.)

Important references

Walker (2001) wrote a brief Introduction to the Rhizocephala. Important comprehensive contributions on the group are given by Bresciani and Høeg (2001, root system), Høeg (1992, microscopic anatomy; 1995, biology and life cycle), Høeg and Lützen (1995, life cycle and reproduction), Høeg and Lützen (1996, general account and taxonomy) and Glenner (2001, up-to-date life cycle). O'Brien and Van Wyk (1985) gave an account of the effects of parasitic castrators (isopods and rhizocephalans) on growth of crustacean hosts. Høeg and Lützen (1985) reviewed Scandinavian rhizocephalans and Øksnebjerg (2000) the taxonomy, biogeography and ecology of rhizocephalans of the Mediterranean and Black seas. Walker (1988) described the larval development of *Sacculina carcini* in some detail. An early but thorough discussion of parasitism by rhizocephalans is that by Veillet (1945). Important papers on parasitic Thoracica are by Grygier and Newman (1991), Yusa and Yamato (1999), Yusa et al. (2001) and Ross and Newman (1995).

Amphipoda (amphipods)

Jørgen Lützen

Introduction

Two of the four suborders of Amphipoda, the Hyperiidea and the Caprellidea, contain species with a parasitoid or parasitic life style. Species of the first suborder typically have a large cephalothorax (head and first thorax segment) with very large eyes. The second suborder contains the whale lice which are dorso-ventrally flattened and whose pleon (most posterior part of the body) is strongly rudimentary. These two suborders are discussed separately in the following.

Hyperiidea

Morphology and diversity

The suborder Hyperiidea includes a group of marine pelagic parasitoids that are associated with gelatinous zooplankters such as medusae, siphonophores, ctenophores and thaliaceans. The current view based on a comparison of hyperiid and gammarid amphipods is that hyperiids represent a polyphyletic group of descendents of different lineages of benthic gammaridean ancestors and that the structural similarities uniting recent hyperiids have arisen as a result of their association with planktonic hosts. The structure of hyperiids is poorly adapted to a pelagic life style and differs only slightly from gammarid amphipods (Fig. 4.22A). The maxillipeds have no palps, and in many, but far from all, species the compound eyes are large to huge, sometimes covering the entire cephalon. Species with large eyes may select their hosts visually. The body ranges from almost spherical to an elongate shape. The eggs are brooded in a female brood pouch as in other amphipods. The total number of species amounts to 254 (Thurston 2000) in 22 families.

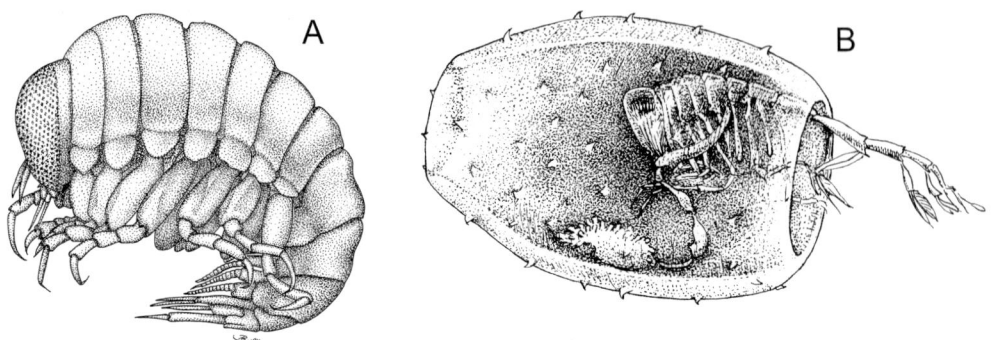

Figure 4.22 Hyperiidea from the North Atlantic. **A**. Lateral view of female *Hyperia galba*, an associate of the scyphozoan *Aurelia aurita*. **B**. *Phronima sedentaria*, female nursing its young in a pyrosome barrel.

Most species occur mainly in the open ocean, and occur most abundantly within the tropical or warm temperate zones, less so in coastal waters. Some species occupy panoceanic distributions within their preferred temperature ranges, while others apparently are limited to one ocean, or even part of an ocean. Several species extend their distribution into the Polar regions or penetrate into the mesopelagic, bathypelagic and abyssopelagic layers. Benthic representatives are completely lacking.

Hosts
Some genera or families are almost exclusively restricted to a particular type of host, such as *Lestrigonus* and *Hyperia*. A few families are associated with siphonophores. Members of the family Lycaeidae are less specific, as their hosts include medusae, siphonophores, pteropods, heteropods and salps. In several genera and families a parasitic habit has not yet been established, but because these hyperiids do not in any essential respect differ from the rest, they are nonetheless believed to have a similar life style.

At the onset of their existence, hyperiids are obligatorily associated with gelatinous zooplankton which is used as a nursery, shelter and food reservoir for the developing young. Later in life the association becomes looser, but the adults still depend on zooplankton as the principal food source.

Adult hyperiids are only partly pelagic and between bursts of swimming seek out gelatinous zooplankton on which to prey or to rest, or using it as a shelter and a means of transport. Laboratory or field observations have shown that some adult hyperiids are able to consume zooplankton within a short time. Undigested nematocysts are frequently found in the gut contents or faecal pellets of species known to associate with cnidarians. *Hyperia galba* has been observed to feed upon and even sometimes destruct the gonads of its scyphomedusan host although it is believed that it mainly consumes the planktonic matter collected by its host and merely occasionally eats the tissue. Adults of *Vibilia* feed exclusively on the food strand of the salp hosts, while those of *Lycaea* feed on salp tissues (Madin and Harbison 1977).

Life cycles
Following mating, which has only been observed rarely and takes place on a host plankton, adult males, which are better swimmers than females, most likely maintain a free-swimming predatory life style, although still preying on zooplankton. The ovigerous females probably stay permanently upon the host while brooding the eggs. The brood size of hyperiids ranges from half a

hundred to several hundred eggs. Compared to gammarids of similar size the egg number is often greater by an order of magnitude. This is attained by reducing the size of the eggs.

The early part of the life cycle comprises one or a few specialised stages that are passed while still in the brood pouch. Compared to gammarids, the eggs hatch at a precocious stage. The earliest stage is the 'pantochelis' larva which is provided with four cheliform pereopods and has unsegmented and limbless metasome and urosome. The pantochelis stage metamorphoses into a 'protopleon' larva (often divided into three substages) having a segmented metasome and imperfect pleopods. In many species there is no pantochelis stage and the egg hatches directly into a first protopleon stage. The last (or only) protopleon stage gives rise to the first juvenile stage, which resembles a miniature adult, and corresponds to the hatching stage of gammarids. The metamorphosis of the last protopleon stage marks the 'demarsupiation', or the deposition by the female of the larvae (in rare cases the juveniles) into a host.

During demarsupiation the gravid female swims from host to host to deposit a single or a few larvae upon each host specimen, the number being obviously limited by the host size and capacity as a food source for the developing juveniles. Examples of how the larvae are transferred to the host, often involving a complex and stereotyped behaviour of the female, are given by Laval (1980). In *Lestrigonus schizogeneios* the female penetrates the host's subumbrella (a hydromedusa), splits a gonad with its mouth parts and inserts the larvae deeply into the organ. As they grow older, the juveniles abandon the gonad and start to feed upon the prey trapped by the host. Ovigerous females of *Vibilia* deposit the pantochelis larvae on the surface of salps with the specialised 7th pereopods. When the larvae moult a few hours later, the ensuing protopleon larvae enter the branchial cavity and start to eat its wall or to feed on the collected suspended matter. If the host cannot support the number of young as they grow up, they probably leave it to invade another salp in the chain.

Species of the family Phronimidae excavate solitary salps or pyrosomes into 'barrels' open at both ends and in which they hide and which they use as a nursery for their young. The female *Phronima sedentaria* (Fig. 4.22B) enters a salp and consumes or removes all internal organs and finally scrapes the internal wall of the resulting barrel smooth with the mouthparts and pereopods 1 and 2. The larvae are demarsupiated into the barrel where they soon bunch together into a tight cluster which slowly moves around on the inner barrel wall. The female exhibits maternal care, as it stays with the barrel and prevents the young from passing to the outer surface. At intervals she makes short excursions into the water and returns with prey to feed the brood. Both sexes of *P. sedentaria* live in barrels. The pereopods are used to maintain the position of the animal within the barrel and beating of the pleopods propels the combined barrel–hyperiid through the water.

Effects on hosts and ecological importance

Juveniles and adults of many hyperiids produce visible damage to their hosts as they eat the tissue and vital organs such as the gonads. In chain salps, the defence against this is their rapid growth and faculty of regeneration. Even so, by physically damaging the host or depriving it of its food, hyperiids most likely have long-term effects on growth, longevity and reproduction of salps. In medusae, the harm is roughly proportional to the size of the hyperiid in relation to that of the host (Madin and Harbison 1977). The growth of medusae of *Phialidium* becomes seriously impeded as the subadult hyperiids steal their food, and when the medusa is not adequately fed, it is eventually consumed itself.

Members of the Hyperiidae seem to be immune to the stings of the host medusae, but the involved mechanism has not been studied. Species of the globularly shaped Platyscelidae and Parascelidae shield themselves from the nematocysts of their siphonophore host by rolling up into a tight ball. Some hyperiids mimic in colour or shape the body of the host, or some of its organs (e.g. the nucleus of salps). It is likely that this helps them to escape the attention of predators such as fishes which detect their prey visually.

Among the marine crustacean zooplankton, hyperiid amphipods rank third in abundance following copepods and euphausiids, but quantitatively they make out only a small part of the zooplankton biomass, and their ecological significance is accordingly relatively limited. However, in Polar seas they may contribute significantly to the food for fishes, diving birds, whales and seals.

Vinogradov et al. (1996) has written a book about the world hyperiidean fauna. Laval (1980) has surveyed many aspects of hyperiid biology such as host associations, feeding, reproduction and demarsupiation.

Caprellidea

Morphology and diversity

Among the 300 or so species of this suborder, the Cyamidae (whale lice) (Fig. 4.23) form a very homogeneous family with about 30 species. At every stage of the life cycle all species are obligatory parasites on whales, dolphins and porpoises. One species parasitised the now extinct sirenian, the North Pacific Steller's sea cow, *Hydrodamalis stelleri*. The body is short and broad, dorsoventrally flattened, the cephalothorax anteriorly pointed and usually coalescent with the first pereon segment, while the abdomen is reduced to a minute process. Two pairs of the pereopods (3 and 4) are absent and merely represented by four clavate or corkscrew-shaped gills. The pereopods 1, 2, and 5–7 are broad, flattened prehensile appendages with which the whale lice cling firmly to the skin of the hosts. The mandibles, maxillules and maxillae are furnished with

Figure 4.23 Caprellidea, Cyamidae, whale lice. A heavy infection by *Cyamus ovalis* on skin of the Northern right whale, *Eubalaena glacialis*, from East Iceland.

setae and short spines by which the parasites excavate shallow pits in the skin, and ingest small pieces of the tissue. Some species are stationary and restricted to, for instance, the crevices of the flippers and lips, the surroundings of the nostrils and the genital and anal openings, or to the edges of wounds. Other slowly crawl around on the whale's body from one place to another.

The adult length ranges from 8 mm to 27 mm. The females, which are generally broader but shorter than the males, have a brood pouch made up of four oostegites ventrally on segments 3 and 4. The number of eggs produced over a single spawning season varies from 50 to a few hundred, but may reach more than 1000 in the larger species.

Life cycle
The life cycle has been studied in *Cyamus scammoni*, parasitic on the gray whale, *Eschrichtius robustus*, which spends the summer in high latitudes of the Pacific Ocean, but winters in lagoons in Baja California. The eggs are laid over an extended period during summer, and egg-laying has been completed and the eggs are fertilised before the whales reach the winter destinations. The juveniles are released from the brood pouch as miniature adults with clawed pereopods and immediately attach themselves firmly to the host's skin. When the whales return homewards in February the young female *C. scammoni* reach 6 mm to 8 mm, the males 8 mm to 10 mm. In March both sexes have attained maturity at a length of 10 mm to 12 mm and 14 mm to 18 mm, respectively, and the females start egg laying during early summer. It is estimated that eight to nine months are required to complete the life cycle.

Whale lice are unable to swim and a planktonic phase in their life cycle is totally absent. It is, therefore, believed that new hosts become infected only when they come into intraspecific bodily contact with an already parasitised individual, for instance during sexual intercourse and social interactions, at birth or when a female nurtures her young. Some whale lice parasitise several species of whales (e.g. 12 species by *Isocyamus delphini*).The care-giving (epimeletic) behaviour exhibited by many whales towards individuals of their own species, if extended to those of other species, may help to explain how whale lice are able to spread between host species.

Effects on hosts
Infestations by whale lice are usually not very heavy except in the humpback whale, *Megaptera novaeangliae*, and the gray whale. When abundantly present on the skin of these species, they may greatly damage even the largest of the whales.

Martin and Heyning (1999) have published a key to the genera of the whale lice and a checklist of the species and their hosts. The systematics of the family has been reviewed by Gruner (1975). Studies on biology of individual species were presented by Leung (1976) and Balbuena and Raga (1991).

Important references
Vinogradov *et al.* (1996) published an extensive monograph of hyperiid amphipods of the world oceans, and Gruner (1975) provided a catalogue of whale lice. A discussion of the taxonomy of pelagic amphipods including parasitic ones is by Thurston (2000). Madin and Harbison (1977) discussed the associations of hyperiids with salps, and Martin and Heyning (1999) presented a key to the genera of Cyamidae, as well as a checklist of cyamids and their hosts. Laval (1980) reviewed hyperiid amphipods as crustacean parasitoids associated with gelatinous plankton. Important papers dealing with more specific aspects are by Balbuena and Raga (1991), who discussed the ecology and host relationships of the whale louse *Isocyamus delphini* parasitising long-finned pilot whales, and by Leung (1976), who described the life cycle of *Cyamus scammoni*.

Chapter 5

Minor groups and fossils

This chapter consists of many sections dealing with a great variety of taxa. Some taxa contain few parasites that have been little studied, others contain parasites with well-known and fascinating adaptations to a parasitic way of life. The approach chosen by different authors reflects this diversity: some sections out of necessity contain brief descriptions of morphological characters, others go into great detail in discussing behavioural adaptations, complex life cycles or epidemiological patterns.

Parasitism is a very ancient way of life, but few parasite fossils are known because most parasites are small and soft-bodied and do not easily fossilise. Most evidence is indirect, based on galls, cysts and drill holes interpreted as resulting from parasite activity. Nevertheless, some parasites have fossilised, including larval pentastomids and crustaceans. The section on fossil parasites in this Chapter covers the literature up to 2003.

By far most parasitic species belong to a few phyla, such as various protistans, Myxozoa, Crustacea, Platyhelminthes and Nematoda. However, in addition, some small phyla consist entirely of parasites, and many phyla contain at least some parasite species. Such groups are discussed in this Chapter. Small groups consisting entirely of parasites at least during part of their life cycle are the Mesozoa, Myzostomida, Nematomorpha, Pycnogonida (possibly with a few non-parasitic species), and Pentastomida. Of these, the mesozoans, myzostomids and pycnogonids are entirely marine, whereas the nematomorphs and pentastomids are predominantly terrestrial/freshwater parasites, only a few being found in the marine environment. Groups including at least some parasitic species are the sponges, cnidarians, ctenophores, polychaetes, leeches, nemerteans, rotifers (and *Seison*, long thought to be a rotifer), mites and ticks, insects, tardigrades, molluscs, echiurans, echinoderms and vertebrates. A fascinating phylum described only recently is the Cycliophora. Cycliophorans are probably ectocommensals rather than genuine parasites, living on the lip of lobsters. Like many genuine parasites restricted to specific hosts and microhabitats, they have an extremely complicated life cycle that ensures production of a huge number of offspring, necessary to guarantee infection of the very narrow microhabitat on one particular host species.

Many of the minor groups discussed in this Chapter, because of the few species involved and low prevalences and intensities of infection, are neither ecologically nor economically important. However, some include species of some significance. Parasitic sponges damage cultured oysters, and may be the most destructive organisms responsible for bioerosion of coral reefs. The parasitic hydrozoan *Polypodium* is the only intracellular metazoan parasite, it has a negative impact on caviar and the reproductive capacity of sturgeon. Some polychaetes are important pests in mollusc cultures, and marine leeches may be vectors of blood protistans transmitted to fish. Nemerteans may be important parasites (or perhaps predators) feeding on the eggs of crabs, which makes them useful for controlling crab pests; and although there is no evidence that mites inflict damage on their invertebrate hosts, mites and ticks may cause severe damage including death to birds, either directly or, in the case of ticks, by transmitting disease agents. Larval pycnogonids may destroy bryozoan and hydrozoan colonies partly or entirely. Chewing lice may transmit parasites to seals, and it seems likely that fleas may be important in causing nesting mortality in marine birds.

Other groups are of great biological interest because of their unique and often mysterious adaptations. Thus, mesozoans have unique morphology and their life cycles are largely unknown; the best known of the few parasitic marine tardigrades has two types of males, including one of dwarf size. Transmission of the only marine pentastomid has not been clarified; parasitic molluscs have an extraordinary reduction in morphological complexity, some resembling worms rather than molluscs. Echiurans have extreme sexual dimorphism: the dwarf male parasitising the much larger female. The reader may be surprised to find that quite a few fish are genuine parasites. Parasitic fish are often ignored in discussions of parasitism, but they have ingenious adaptations to their way of life, which makes the section dealing with them so fascinating.

Fossil parasites
Greg W Rouse

Introduction
The broad range of Metazoa was clearly established by the Cambrian period, more than 500 million years ago. Given the present-day assortment of parasites across this assemblage, it is reasonable to assume a great diversity of parasites, similar to that of today, has been present for hundreds of millions of years. While this inference may be justifiable, and is well supported by the fossil record of Metazoa, direct fossil evidence of parasitism is poor and is likely to remain so. Most contemporary parasites are soft-bodied and they are furthermore associated with soft tissues that are unlikely to fossilise (Conway Morris 1990), and this is likely to have been the situation for most parasites. Therefore, the best chance of detecting parasitism via the fossil record is through indirect evidence such as cysts, galls or drill holes left on the host's skeleton or shell. Thus, there is a bias, with a few taxonomic groups (e.g. echinoderms) preserving parasite activity better than others. There are only a few exceptional cases of marine parasites where there has been direct fossilisation. The paucity of data on fossil parasites means that few studies have dealt with evolution of parasites and parasite–host interactions, and the ecological impact of parasitism on fossil communities. However, Gahn and Baumiller (2003) and Littlewood and Donovan (2003) point to new developments in these areas.

Indirect fossil evidence
Although there may be little direct evidence of parasites, the existence of those that had hosts with hard skeletons can often be inferred. This may be in the responses of the host's skeleton and there is clear evidence of such parasitism on groups such as bivalves, crustaceans and echinoderms. Many papers concerning traces of fossil parasite activities are rather speculative, however, particularly when concerning the parasite's identity. A thorough and critical review of the existing literature is needed.

Bivalves are often the intermediate host for digeneans and cestodes, and the metacercariae of the former and metacestodes of the latter have been implicated in causing pearl formation. Littlewood and Donovan (2003) surveyed current bivalve diversity and the incidence of pearl formation and contemporary digenean parasitism. Their study suggests that the occurrence of fossil pearls back to the Triassic period may reflect a long history of parasitism on bivalves by Digenea, even though there is little direct evidence for this. Drill holes in the valves of bivalves

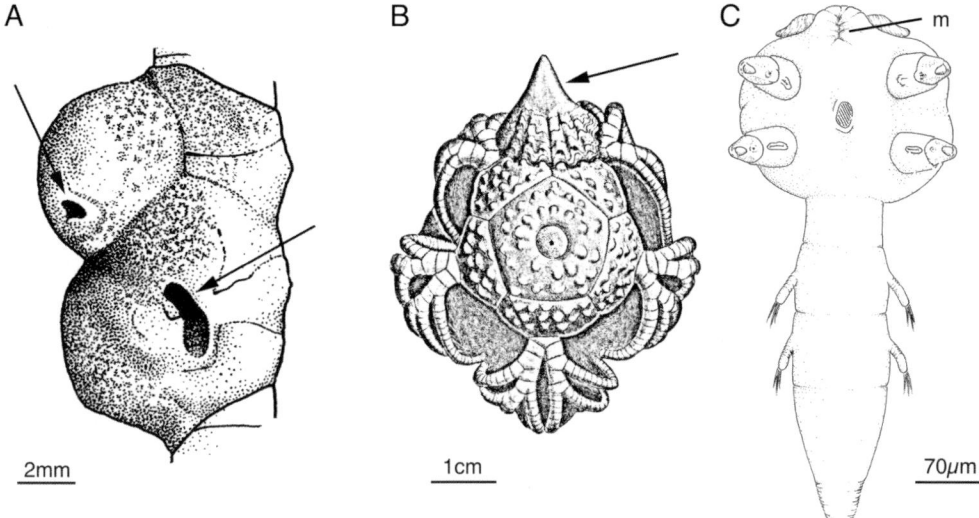

Figure 5.1 **A.** Galls of putative myzostomids on the arm of a fossilised crinoid. From Hess *et al.* (1999), 'Fossil crinoids'; Reprinted with the permission of Cambridge University Press). **B.** Ventral view of Paleozoic stalked crinoid *Platycrinites hemisphaericus* fossilised with a specimen of a *Platyceras* gastropod lying over the anal opening. Reproduced with permission from Wachsmuth C and Springer F (1897) *Harvard College Museum of Comparative Zoology*, Memoir **20, 21**: 1–897; published by the Museum of Comparative Zoology and Harvard University. **C.** Ventral view of a generalised 'round-headed' larval pentastomid. Reprinted with permission from Waloszek D and Müller KJ, Pentastomid parasites from the Lower Palaeozoic of Sweden, *Transactions of the Royal Society of Edinburgh: Earth Sciences*, 85: 1–37. Copyright (1994) The Royal Society of Edinburgh, D Waloszek and KJ Müller.

and brachiopods have been documented from at least the Permian period and while many of these represent predation, it is possible that some are parasitic (Kowalewski *et al.* 2000). Some fossilised crabs from the Jurassic period have swellings evident in the branchial region that have been inferred to represent parasitism by epicarid isopods (Radwanski 1972). Feminisation of fossil xanthid crabs from the Miocene epoch has been found by Feldmann (1998) who argued that this represents parasitic castration by rhizocephalan barnacles. Feminised crabs dating even further back, to the Cretaceous period, indicate that rhizocephalans may have evolved during the Mesozoic era (Feldmann 1998). Ascothoracican barnacle parasitism on fossil echinoid echinoderms from the Cretaceous period has been inferred from characteristic elongate boreholes in the test of the host (see Baumiller and Gahn 2002).

Indirect evidence of parasitism has in general been well studied in fossil echinoderms, particularly crinoids dating back to the Ordovician period. Various kinds of galls have been described on the arms and stalks of Palaeozoic crinoids and many of these would appear to have been caused by Myzostomida (Fig. 5.1A). Other galls dating through the Palaeozoic appear to have been caused by now extinct organisms called *Phosphannulus* (Hyolothelminthes). The galls had a tube that penetrated into the crinoid stem (Welch 1976), suggesting the parasite directly consumed the crinoid. Organisms such as ophiuroid echinoderms have also been found on the fossilised remains of crinoids (Hess *et al.* 1999). These date back to the Devonian period and similar suspension-feeding ophiuroids use crinoids as perches for feeding today (see pp. 248–250).

Direct fossil evidence

Platyceratidae, an extinct group of gastropods, were associated with crinoids, as well as the now-extinct blastoids and cystoids, throughout the Ordovician and Triassic periods. The snails are generally found around the anal area of the echinoderms (Fig. 5.1B) and this has been used to argue that the group was coprophagous. Others have suggested that the snails were suspension feeders, while others have argued that the snails were drilling parasites on the echinoderms (Gahn and Baumiller 2003). Gahn and Baumiller (2003) argue that the parasitic interpretation is correct and show that crinoids infested with the platyceratids were significantly smaller than those that were not. Fossilised siphonostome copepods have been found in the gills chambers of Cretaceous fish (see Conway Morris 1981), but the most remarkable recent fossil discovery that concerns parasites is that of larval Penstastomida fossils. Contemporary Penastomida are endoparasitic and feed on blood or other fluids in the respiratory tract of tetrapods ranging across amphibians, turtles, crocodiles, birds and mammals (see pp. 235–240). There are no known pentastomids living in marine organisms and no fossils of the adults are known. This makes the finding of fossilised pentastomid larvae dating back to the Ordovician and upper Cambrian periods, initially by Andres (1989) and further documented by Walossek and colleagues (e.g. Walossek and Müller 1994) absolutely stunning. Several different species of pentastomids have been described from the larvae, preserved three-dimensionally via secondary phosphatisation of the cuticle ('Orsten' preservation). They show extraordinary details (Fig. 5.1C) and two basic kinds have been recognised, 'hammer-headed' and 'round-headed' larvae. These details have allowed important reinterpretations of recent pentastomid anatomy. Obviously, the hosts of these ancient larvae are unknown, but Walossek and Müller (1994) postulate that they may have still been chordates since the fossil record of this group does extend back that far. They also suggest that the pentastomids were originally ectoparasitic forms.

Important references
There have been some reviews on fossil parasites including those of Conway Morris (1981, 1990) and Littlewood and Donovan (2003), the latter from a helminthological perspective. Baumiller and Gahn (2002) provide the most recent tabulation of the occurrence of parasitism in the fossil record. Boucot (1990) edited an extensive book on organismal interactions and the fossil record.

Porifera (sponges)
John NA Hooper

Introduction
The Porifera (sponges) is a large group of presumed monophyletic aquatic invertebrates at the base of the Metazoa, with about 15 000 species worldwide, predominantly marine, living from the abyssal zone to ephemeral aquatic habitats. They are united by the unique possession of choanocytes (flagellated or collar cells that actively beat to produce a water current), lining simple or complex chambers, and connected to the external water column by a system of inhalant and exhalant canals with external pores, together forming a highly efficient aquiferous system that maintains basic metabolism and contributes significantly to reef filtration. They lack a tissue grade of construction but have a highly mobile population of cells capable of totipotency, and possess siliceous or calcitic spicules in many (but not all) species. Sponges have free-swimming or creeping sexual larvae, so far with no known pelagic phase (thus potentially limiting their ability to disperse over long ranges, unlike e.g. corals), although most groups have considerable means of asexual propagation, and all have extensive regenerative powers that appear to be vital for sustaining local populations. There are three distinct classes (Hexactinellida, Demospongiae and Calcarea), with the extinct class Archaeocyatha having suspected affinities with Demospongiae. An overview of the phylum, including a revision of nearly 700 genera, has been undertaken (Hooper and Van Soest 2002).

Parasitic sponges

Sponges have long been known as a host to a plethora of prokaryote and eukaryote commensals (e.g. Carter 1871, Humes 1996), including other sponges (e.g. Annandale 1915) – hence the term 'sponge hotels' (Wilkinson 1984) – and in some cases these may greatly exceed the native sponge cell biomass (e.g. Price *et al.* 1984, Wilkinson 1992). These commensal relationships span the continuum from obligatory symbionts to parasites (e.g. Uriz *et al.* 1992). It is less well known, however, that some sponges themselves are obligatory parasites on other life forms, for part of their life history at least. These predominantly consist of the excavating (or 'boring' or bioeroding) sponges belonging to the Hadromerida demosponge families Clionaidae, Alectonidae and Spirastrellidae – although a parasitism has been independently acquired in other sponge groups too, such as the poecilosclerid *Paracornulum* and *Zyzzya*, and the haplosclerid *Aka* – but together consisting of only relatively few species (perhaps several hundreds). These parasitic sponges excavate chambers in calcium carbonate substrata, contributing significantly to reef carbonate recycling, but also damaging living tissue and sometimes killing the host (e.g. Wesche *et al.* 1997, Schönberg and Wilkinson 2001). They have been responsible for extensive damage to commercial shellfish (e.g. Rosique *et al.* 1996), other molluscan hosts (e.g. Lauckner 1983), hard corals and octocorals (e.g. Schönberg and Wilkinson 2001, Rützler 2002b), and they have reportedly caused extensive damage to community structure and physical stability of shallow reefs (Rützler 2002a).

Effects of sponges on hosts

The most visibly destructive cases of bioeroding sponge infections can be seen in edible oyster cultures infected with clionaid species (Fig. 5.2A), producing 'spice bread disease' (Thomas 1981), with reported infection rates of up to 50% in some commercial oyster leases (Wesche *et al.* 1997). Their ability to infect both living and dead calcitic substrates, their reported tolerance to low salinities (20 parts NaCl per thousand, ppt – g of solute per kg of sea water) and ability to survive when the host is exposed to the air (Hartman 1958), their ability to infect several species of host (Thomas 1981), and the common practice of translocating young oyster spat between commercial oyster beds has produced nearly cosmopolitan distributions for some species (e.g. *Pione vastifica*). At larger spatial scales excavating sponges seem the most destructive internal

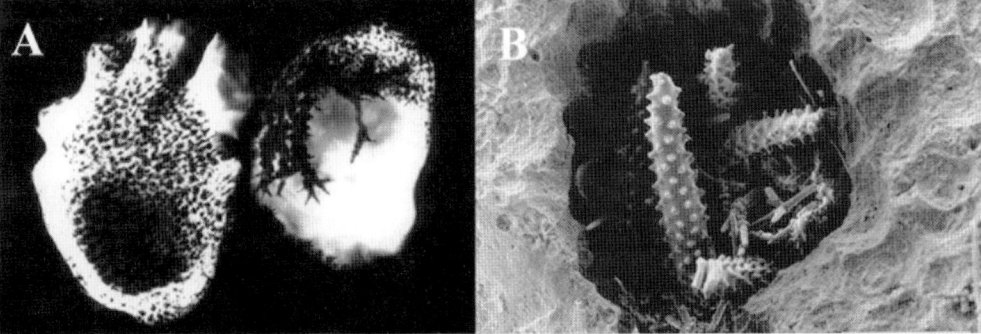

Figure 5.2 Sponges parasitic in mollusc and coral hosts. **A.** Radiograph of an oyster shell (*Saccostrea glomerata*) infected with *Pione vastifica*. **B.** Parasitic sponge (*Alectona millari*) in an excavated coralline chamber, showing spined acanthoxea megasclere spicules (centre) and pits on the chamber wall. Scale bar = 100 µm. Figure 5.2A is reprinted from *Aquaculture*, Volume 157, Wesche, Adlard and Hooper, 'The first incidence of clionid sponges (Porifera) from the Sydney rock oyster *Saccostrea commercialis* (Iredale and Roughley 1933)', pp. 173–180, Copyright (1997), with permission from the authors and Elsevier. Figure 5.2B is reproduced from Rützler (2002b), with permission from the author and Kluwer Academic/Plenum Publishers.

bioeroding organisms of coral reefs both in terms of effects (i.e. weakening coral platforms, production of dead coral rubble) and rates of destruction (up to 15 kg/m² per year), and much of the damage caused to corals during storms has been attributed to weakening of basal structures by bioerosion (Wilkinson 1983). Moreover, there is evidence of a significant increase in the prevalence of bioeroding sponges over the past 20 years within the Caribbean, which has been attributed to increased seawater temperatures (either from local or global mechanisms) and probably other biotic, climatic and anthropogenic sources of stress on coral reefs (Rützler 2002a). Although clionaid sponges have the ability to invade living coral tissue and to survive direct contact with coral polyps (Schönberg and Wilkinson 2001), their ecological success may be largely due to their ability to undermine and erode the coral skeletal base, thus avoiding contact with the coral polyp defensive mucus and nematocysts (Rützler 2002a).

Adaptations to a parasitic existence

The families Clionaidae (*Cliona*, *Cliothosa*, *Pione*, *Thoosa* and *Volzia*) and Alectonidae (*Alectona*, *Delectona*, *Dotona*, *Neamphius* and *Spiroxya*) are capable of bioeroding in the alpha, beta and gamma growth stages (Fig. 5.3), while Spirastrellidae (*Diplastrella*, *Spirastrella*) lacks an adult endolithic habit and excavates only in the alpha stage (Rützler 2002b). In clionaid sponges excavation is a two-step cellular process, via special etching cells in the sponge larvae (and/or in mature sponges) in contact with the calcitic substrate (Rützler and Rieger 1973). The pseudopodial processes of these cells produce a carbonic anhydrase-regulated acid phosphatase secretion at the periphery of the filopodial sheet, and together with the action of a lysosomal enzyme system which dissolves organic matter, limestone chips are created and physically liberated into the sea water via the sponge exhalant canal (Pomponi 1979). Etching initially produces a cavity, then a series of connected chambers, some extending up to 80 mm through the substratum, depending on the porosity of the host (MacGeachy 1977), and eventually consuming the entire calcitic substratum (in the case of corals at least). As the sponge grows it fills the excavated chambers with its cells and siliceous spicules (megascleres and microscleres) (Fig. 5.2B), and maintains contact with the external water column by growing papillae bearing inhalant (ostia) and exhalant pores (oscula). Three growth stages are defined for excavating sponges: in the alpha stage sponges are predominantly confined within the excavated chambers with only the papillae protruding outside the host (Figs 5.2B, 5.3A). There are differentiated inhalant and exhalant papillae to minimise incoming water contamination, the former with numerous ostia and the latter with only a single terminal osculum. In the beta stage the papillae fuse to form a continuous sponge crust covering the external surface of the host (Fig. 5.3B), and in the gamma stage sponges become massive (often burrowing inside the calcitic substratum) (Fig. 5.3C), with

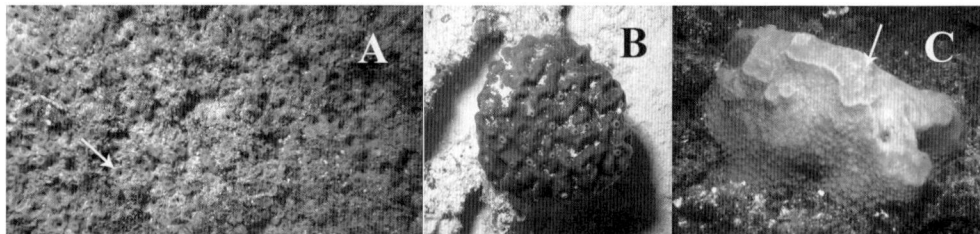

Figure 5.3 Growth stages of parasitic sponges. **A.** Alpha stage, *Cliona* sp. #2670 boring on dead coral substrate, orange alive (arrow indicates an individual pit). **B.** Beta stage, *Cliona orientalis* overgrowing a faviid coral, olive green alive. **C.** Gamma stage, massive *Cliona montiformis* on top of a coral head, yellow alive (arrow indicates pore sieve plate on the apex of the sponge). All photos JNA Hooper.

the exhalant papilla (or fistule) sometimes also developing a special pore sieve at its end to prevent smothering from surrounding sediments. Gamma stage growth forms are known only for a few of the excavating sponges, and are thought to occur as the sponge becomes massive with age, or after the original substrate has been completely outgrown and the sponge becomes free living (Rützler 2002b).

Estimating the effects of parasitic sponges
During the processes of parasitism etching pits and scars are left inside the host, with the geometry of scarring differing according to the species of sponge, and these are thus a useful diagnostic tool for both Recent and fossil faunas (Calcinai et al. 2003). Similarly, the extent of sponge bioeroding activity in time and space can be measured by the abundance of etched limestone chips in surrounding sediments, with one estimate indicating that 20% to 40% of all silt on patch reefs in Belize are excavated chips (Halley et al. 1977), and some of which are subsequently lithified to form consolidated reef rock (Wilkinson 1983). Clearly, therefore, parasitic sponges – although not widely recognised as such – have contributed significantly to modern day coral reef structure and composition (and have existed at least since the Mesozoic as evidenced from endolithic scar patterns in the fossil coral record; Perry and Bertling 2000), as well as continuing to be a major pest of commercial molluscan fisheries.

Important references
Important references on the effects of boring sponges on coral reefs are by MacGeachy (1977), Wilkinson (1983) and Rützler (2002a). Thomas (1981) discussed boring sponges destructive to economically important molluscan beds and coral reefs in Indian seas. The chapter by Lauckner (1983) on diseases of molluscs contains a section on sponges. Rosique et al. (1996) evaluated the influence of the sponge *Cliothosa hancocki* on the European flat oyster bed, *Ostrea edulis*. The fine structure of *Cliona lampa* penetrating calcareous substrata was described by Rützler and Rieger (1973), and the ultrastructure and cytochemistry of the etching area of boring sponges by Pomponi (1979).

Cnidaria and Ctenophora (cnidarians and comb jellies)
Ferdinando Boero and Jean Bouillon

Introduction
Cnidaria and Ctenophora (see Grassé and Doumenc, 1987, 1993 for a detailed account) are known both as benthic organisms (polyps of the Cnidaria) and as gelatinous plankton (medusae of Cnidaria, and Ctenophora). The parasitic forms are usually neglected, with the exception of the hydrozoan *Polypodium hydriforme*, parasite of sturgeon eggs, one of the few metazoan intracellular parasites. Within the two phyla, the Hydrozoa show the greatest diversity in parasitic life. Many species are epibiotic and not parasitic, using the surface of other organisms as a settling substrate. Two forms of association are here considered as parasitic:

1. at least part of the guest body is embedded in the host tissues
2. the guest lives inside the host (e.g. in the flagellate chambers of sponges, the mantle cavity of mollusc bivalves or the atrial cavity of tunicates).

A third but rare case is that of predators that feed only on parts of their prey, without killing it. Since most hosts have almost no economic importance, the impact of these associations is unstudied and, in most cases, it is not clear if they cause any harm. Relatively few species have been directly observed while feeding on their hosts. In some cases, the host might receive some compensation from the guest (e.g. encrusting bryozoans inhabited by hydroids whose hydrorhizae grow under the host skeleton have an increased competitive ability). Such cases might suggest a long history of coexistence, beginning as mutualism and ending as parasitism (Piraino et al. 1992).

In almost all species, host infection presumably occurs via the planula stage. The asexual reproduction so common in cnidarian polyps is a way of increasing the number of zooids deriving from a single zygote and resembles a common life history pattern of many parasites: larval amplification by asexual reproduction. Due to the paucity of records, the geographic distribution of these associations seems to be scattered. Proper observation on potential hosts at any marine location might reveal the presence of parasitic Cnidaria, especially Hydrozoa.

Parasitic Cnidaria
The parasitic relationships of Cnidaria are described by considering selected examples of each parasitised phylum. A list of known species with similar habits is also given.

- **Sponges**. Many polyps live embedded in sponges (26 Hydrozoa and two Scyphozoa), projecting distally from their surface. Some species live in the flagellated chambers. Penetration of tissues by the guests should cause some harm to the hosts and, thus, the association can be considered as parasitic, although Uriz et al. (1992) suggested that the sponge profits from the association by receiving physical support from the guests, thus saving energy in producing a skeleton of its own. The polyps may use the currents produced by the sponge to obtain food, and receive protection by being partially embedded in its body. Species: *Dipurena halterata, D. simulans* (Fig. 5.4A), *D. spongicola, D. strangulata, Sarsia bella, Bicorona elegans, Sphaerocoryne agassizi, S. bedoti, Heterocoryne caribbensis* (Fig. 5.4B), *Cladonema* sp., *Hybocodon cryptus, H. prolifer, H. unicus, Zyzzyzus calderi, Z. floridanus, Z. robustus, Z. spongicolus, Z. warreni* (Fig. 5.4C), *Ectopleura exxonia, Tubularia ceratogyne, Bibrachium euplectellae, Cytaeis spongicola, C. abyssicola, C. nuda, Gastroblasta* sp., *Nausithoe punctata, N. racemosa*.
- **Cnidaria**. The polyps of 12 hydrozoans live embedded in the bodies of colonial Hydrozoa, Alcyonacea, Pennatulacea, Gorgonacea and Scleractinia. They use the host for support, but might also feed on its tissues. Species: *Hebella dispolians* (Fig. 5.4D), *Hebella furax, Pteroclava krempfi, Ralpharia coccinea, R. gorgoniae, R. magnifica, R. neira, R. parasitica, R. sanctisebastiani, Hydrichthella epigorgia* (Fig. 5.4E), *Ptilocodium repens, Zanclea gilii* (Fig. 5.4F).

    The larval stages of several Narcomedusae (seven species, but probably many more) are parasitic in the gastric pouches or the manubrium of other medusae, using their food. Once grown, they leave their hosts and live freely. Tentaculate Actinaria larvae are parasitic in the gastrovascular system of medusae, using food particles of their host by means of siphonoglyph currents. Species: *Cunina becki, C. octonaria, C. peregrina* (Fig. 5.4G), *C. proboscidea, Pegantha clara, P. rubiginosa, P. triloba, Peachia* spp., *Halcampa* spp.
- **Ctenophora**. Young *Edwardsia* (Actiniaria, Anthozoa) are parasites of the ctenophore *Mnemiopsis leidyi*.
- **Annelida**. The Narcomedusa (Hydrozoa) *Cunina globosa* parasitises the coelomic cavities of the pelagic polychaete *Tomopteris*.
- **Mollusca**. Cnidaria living on gastropod shells containing living molluscs or hermit crabs are either simple epibionts or even mutualists.

    Polyps of three hydrozoan species live on the shells of pteropods, feeding on their epithelium and on their embryos (i.e. they are real parasites). Species: *Kinetocodium danae* (Fig. 5.4H), *Perigonella sulphurea, Pandea conica*.

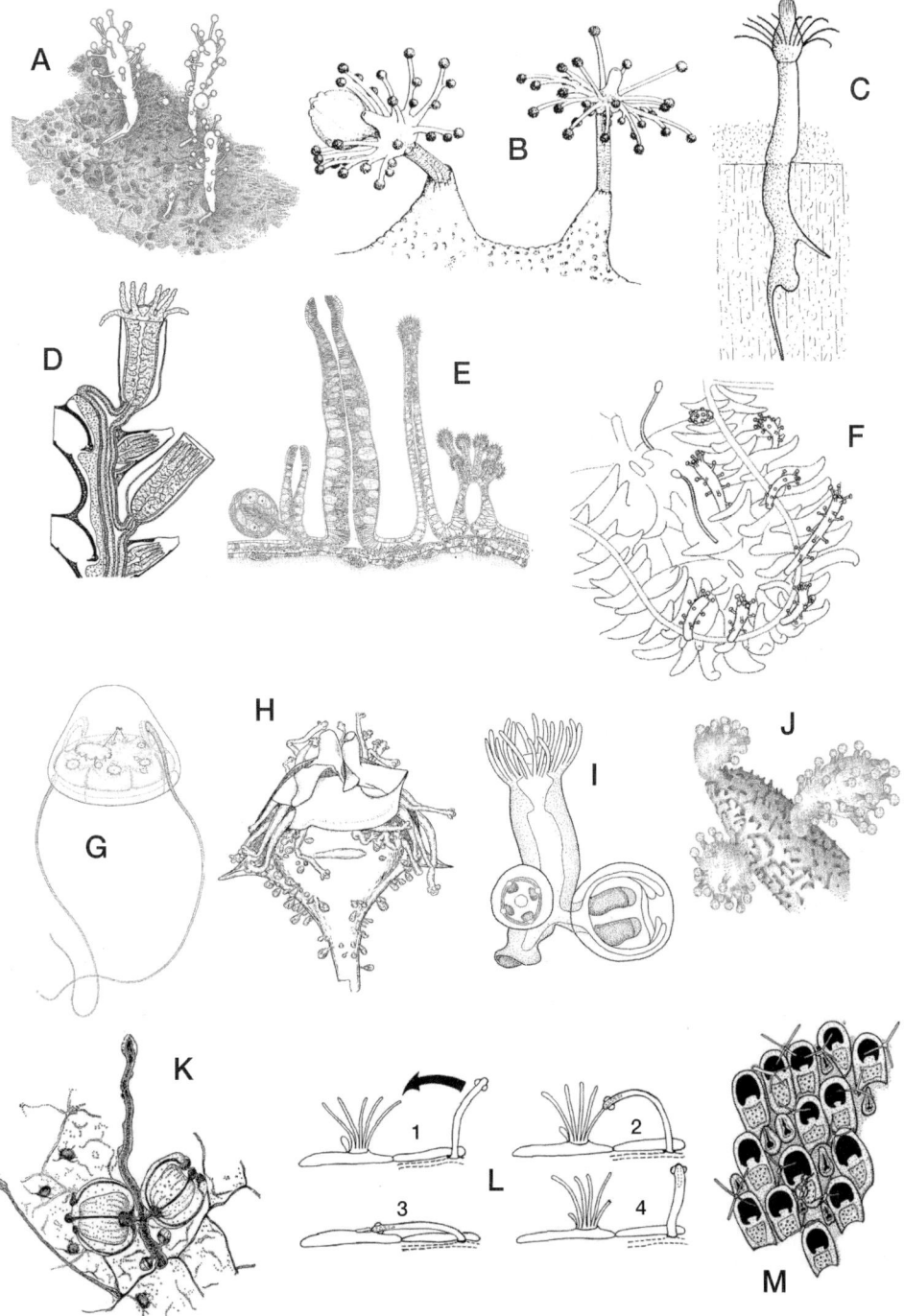

Figure 5.4 **A**. *Dipurena simulans*. After Bouillon (1971). **B**. *Heterocoryne caribbensis*. After Wedler and Larson (1986). **C**. *Zyzzyzus warreni*. After Millard (1975). **D**. *Hebella dispolians*. After Boero et al. (1997). **E**. *Hydrichthella epigorgia*. After Bouillon (1967). **F**. *Zanclea gilii*. After Boero et al. (2000). **G**. *Cunina peregrina*. After Bouillon (1987). **H**. *Kinetocodium danae*. After Kramp (1957). **I**. *Eugymnanthea inquilina*. After Palombi (1935). **J**. *Rosalinda incrustans*. After Vervoort (1966). **K**. *Halocoryne epizoica* (J Bouillon unpublished). **L**. *Halocoryne epizoica*, four stages of feeding on bryozoan tentacle. After Piraino *et al.* (1992). **M**. *Octotiara russelli*. After Boero and Bouillon (1989).

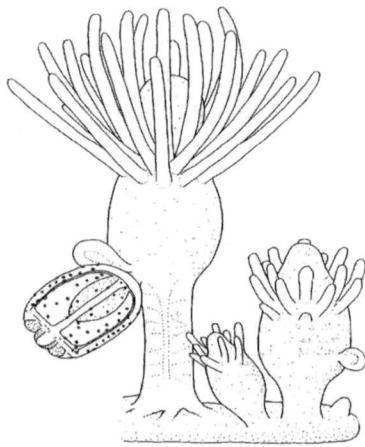

Figure 5.5 *Bythotiara parasitica*. After Schuchert (1996, modified).

Polyps of five hydrozoan species live in the mantle cavity of bivalve molluscs, attached to the tissues of the mantle cavity by stolonal sucker-like structures or by hydrorhizae penetrating into the host tissues. They use food collected by the ciliary movements of the bivalve gills and the labial palps but their exact parasitic relationships are not known. Some have been observed feeding on the larvae of other parasites of their hosts. Species: *Eugymnanthea inquilina* (Fig. 5.4I), *E. japonica*, *Eutima commensalis*, ?*E. ostrearum*, *E. sapinhoa*.

- **Crustacea**. Some Crustacea Decapoda use Cnidaria as defence, actively putting them on their bodies; only two species of Hydrozoa live on the exoskeleton of crustacean hosts: *Rosalinda incrustans* (Fig. 5.4J) on decapods, *Hydrichthys sarcotretis* on parasitic copepods.
- **Bryozoa**. Hydrorhizae of the hydrozoan guest grow in spaces among bryozoan zooids and are covered by host skeleton. Polyps project from holes in the host skeleton. The host receives protection by the nematocysts of the guest which, in turn, receives protection by the calcareous skeleton of the bryozoan and takes advantage of the currents created by the lophophore. One species feeds on the lophophore arms of the bryozoan and this habit might be common in other species as well. Species: *Zanclea sessilis*, *Z. protecta*, *Z. bomala*, *Z. divergens*, *Z. giancarloi*, *Z. retractilis*, *Z. polymorpha*, *Z. hirohitoi*, *Z. exposita*, *Zanclella bryozoophila*, *Z. diabolica*, *Z. glomboides*, *Halocoryne epizoica* (Figs 5.4K, 5.4L), *H. frasca*, *H. pirainoid*, *Octotiara russelli* (Fig. 5.4M), *Cytaeis schneideri*, *Hydranthea margarica*.
- **Urochordata**. The Hydrozoa *Bythotiara parasitica* (Fig. 5.5) and *B. hunstmanni* live in the prebranchial cavities of ascidians.

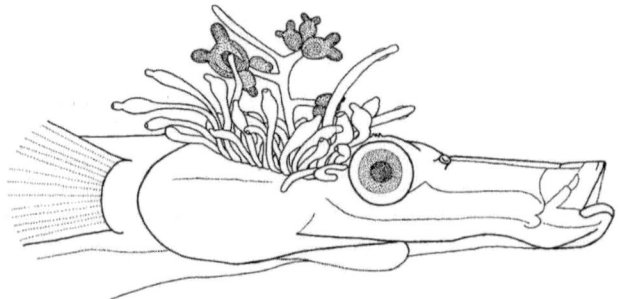

Figure 5.6 *Hydrichthys mirus*. After Boero *et al.* (1991).

- **Chordata**. Seven Hydrozoa live on the body surface of fish, with a hydrorhiza forming a naked encrusting plate that can erode the host tissues with haustorium-like structures, the tentacle-less polyps bend over and feed by applying their open mouth to the injured parts, sucking in blood and tissues. Species: *Hydrichthys mirus* (Fig. 5.6), *H. boycei*, *H. cyclothonis*, *H. monocanthi*, *H. pacifica* and, *H. pietschi*, *Larsonia pterophylla*.

  Polypodium hydriforme (Fig. 5.7), the single representative of the hydrozoan subclass Polypodiozoa, is one of the few known metazoans adapted to intracellular parasitism. It

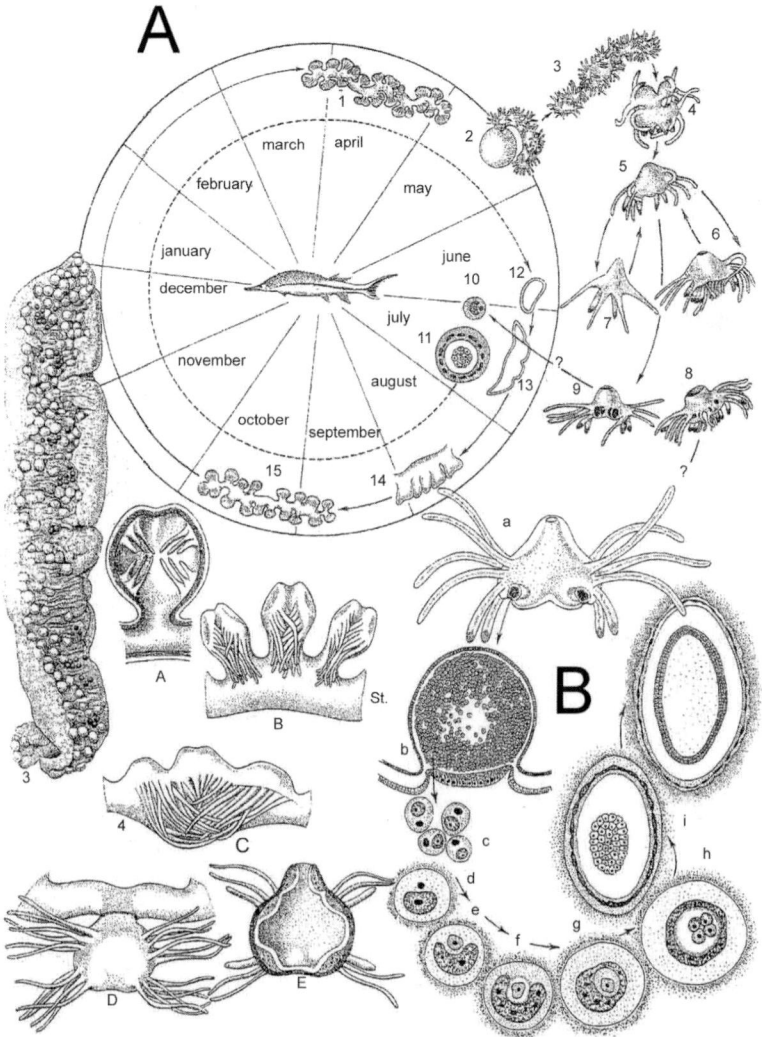

Figure 5.7 *Polypodium hydriforme*. After Raikova (1973). **A**. General cycle: 1, reversed stolon prolifer; 2, stolon prolifer living in infected Acipenserid egg; 3, stolon prolifer in water; 4, fragment of stolon; 5, polyp with 12 tentacles; 6, polyp with 24 tentacles; 7, polyp with six tentacles; 8, female polyp; 9, male polyp; 10, binucleate cell parasite of an Acipenserid oocyte; 11, encapsulated morula; 12, planula; 13, budding planula; 14, stolon prolifer without tentacles; 15, reversed stolon prolifer with internal tentacles. **B**. Diagram of formation of the medusae. **a**, sexual medusa; **b**, gonad with binucleated cells; **c**, binucleated cells; **d–g**, binucleated cells in a fish oocyte; **h**, segmentation of embryo; **i**, morula stage; **j**, planula stage; **3**. ovary of *Acipenser*, large eggs are parasitised, small ones are not. **4**. A–C, reversed buds with internal tentacles on a stolon prolifer (St.); **D**, stolon prolifer with external tentacles; **E**, medusae liberated from an infected fish egg.

has a unique life cycle, with a succession of a free-living stage and an intracellular parasitic stage in some Acipenseridae and Polyodontidae fish eggs. The species decreases the quality of caviar and the reproductive potential of the host.

Parasitic Ctenophora
Gastrodes parasiticum is an internal parasite in the tunic of *Salpa fusiformis*. Some authors believe that it is the juvenile cydippoid larva of *Lampea pancerina* that, as adult, is usually attached to the body of salps with its pharynx.

Haeckelia rubra eats the tentacles of the narcomedusa *Aegina citrea*, and retains the prey's unfired cnidocysts for its own defence (cleptocnidae).

Important references
Comprehensive reviews of the Cnidaria can be found in Millard's (1975) monograph on the Hydroida of southern Africa, and in Grassé and Doumenc (1987, 1993). Relevant papers on Cnidaria are by Palombi (1935), Kramp (1957), Vervoort (1966), Bouillon (1967, 1971, 1987), Wedler and Larson (1986), Boero and Bouillon (1989), Boero *et al.* (1991, 1997, 2000), Piraino *et al.* (1992), and Schuchert (1996). Raikova (1973) gave a detailed account of the life cycle and systematic position of *Polypodium hydriforme*. Uriz *et al.* (1992) examined the question of whether parasitism, comensalism or mutualism is found in the Scyphozoa (Coronatae).

Mesozoa Orthonectida (orthonectids)
Hidetaka Furuya

Introduction
Orthonectids (Phylum Orthonectida) belong to a small group of mesozoans with about 20 described species. Orthonectids have been reported from temperate and cold waters of coastal regions of the English Channel, the Dover Strait, the Strait of Kattegat, the Barents Sea, the White Sea, the North-western Pacific Ocean (Japan), and the North-eastern Pacific Ocean, United States of America (USA). One family, the Rhopaluridae, contains all four genera: *Rhopalura*, *Intoshia*, *Ciliocincta*, and *Stoecharthrum* (Kozloff 1992).

Morphology and diversity
The genera, *Rhopalura*, *Intoshia* and *Ciliocincta* are dieocious and sexually dimorphic. The genus *Stoecharthrum* is characterised by hermaphroditism. Adults range in length from 0.05 mm to 0.8 mm. The body of the adult consists of a jacket of ciliated and unciliated epidermal cells arranged in rings around an internal axial mass (Fig. 5.8). The pattern of ciliation is characteristic of the genus. In *Rhopalura* some rings of epidermal cells are completely ciliated, whereas others lack cilia. In *Intoshia* the ciliated rings are mostly covered with cilia. In *Ciliocincta* and *Stoecharthrum* the cilia are restricted to the anterior or posterior margins, or both, of certain rings. Contractile muscle cells differentiate to pack the gonad with longitudinal, circular and oblique orientations. The oocytes in females of *Intoshia*, *Ciliocincta* and *Stoecharthrum* form a single row of cells within the axial mass. In *Rhopalura* the oocytes are packed into the central mass (Fig. 5.8).

Orthonectids live in tissues of a wide variety of marine invertebrate phyla, Platyhelminthes, Nemertea, Annelida, Mollusca, Echinodermata, Bryozoa and Urochordata. In *Rhopalura* all species are associated with the gonads of brittle stars, various bivalves and gastropods. The remaining genera, *Intoshia*, *Ciliocincta* and *Stoecharthrum*, are found in the parenchyma or tissues of the body wall, rather than the gonads.

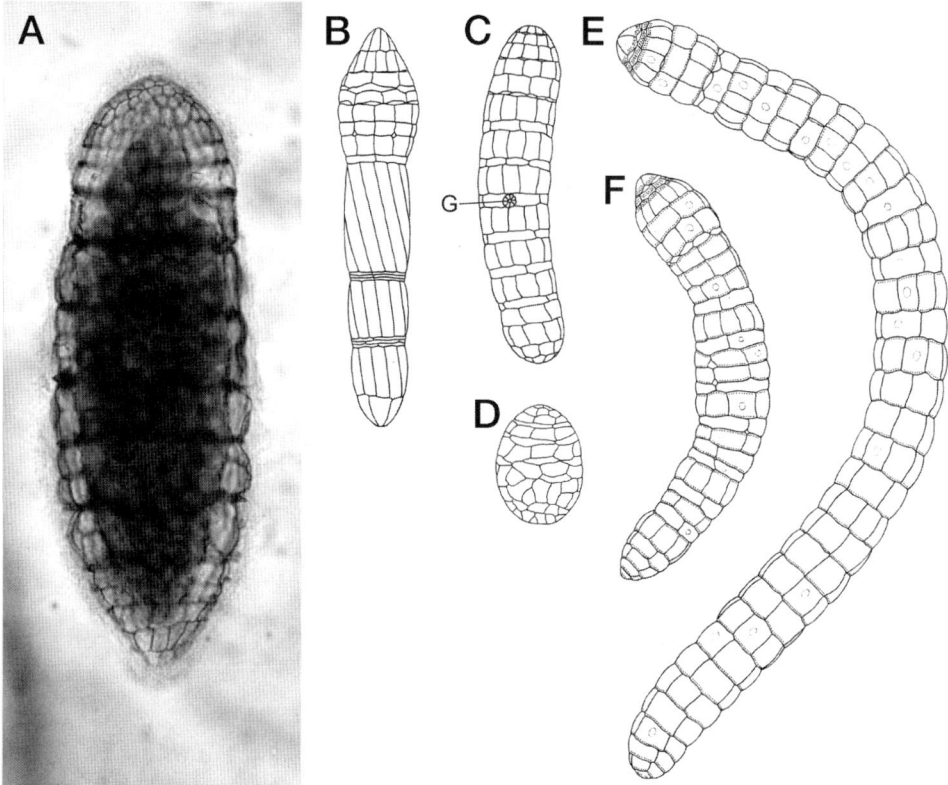

Figure 5.8 Morphology and arrangement of jacket cells in orthonectids. **A**. light micrograph of *Rhopalura ophiocomae* (female). **B**. *R. ophiocomae* (male). **C**. *Intoshia linei* (female). **D**. *I. linei* (male). **E**. *Ciliocincta sabellariae* (female). **F**. *C. sabellariae* (male). Cilia are omitted in Figs **B–F**. Abbreviations: G, genital pore. **A**. From the collection of Santa Barbara Museum of Natural History. **B–F**. after Kozloff (1992). Reprinted from *Cahiers de Biologie Marine* 33 (1992) pp. 377–406. Kozloff: The genera of the phylum Orthonectida, Figs 2, 12, 14, 24, 26. With permission from the Director, *Cahiers de Biologie Marine* and the author.

Life cycles

The life cycle of orthonectids is complex (Fig. 5.9). In *Rhopalura ophiocomae* the infection takes place through a genital bursa or the gut of the host *Amphipholis squamata* (Kozloff 1969, 1971). The infective germinal cell penetrates the epithelium and enters a contractile cell, in which embryos develop from the germinal cell. This structure is called plasmodium; however, it belongs not to the orthonectid, but to the host contractile cell (Kozloff 1994). Thus, the plasmodium consists of modified host tissues. The germinal cells give rise to morula or berry-like clusters that form sexual individuals in the plasmodium (Metschnikoff 1881). Adults escape through the genital slits of the host. In *R. ophiocomae* males are smaller than females. During mating males make brief contact with females when sperm are released. Fertilisation is inseide the female bodies. In this species maturation of the egg is not completed until after entrance of the sperm (Caullery and Lavallée 1908). Embryos are formed around 22 hours after the first cleavage of eggs. When the embryos are fully developed, the female ruptures and dies, releasing ciliated dispersal larvae *in vitro*.

The method of infection is unknown. In *R. ophiocomae*, Caullery and Mesnil (1901) considered that larvae enter the genital slits of the ophiuroid host. Mature larvae have not been studied

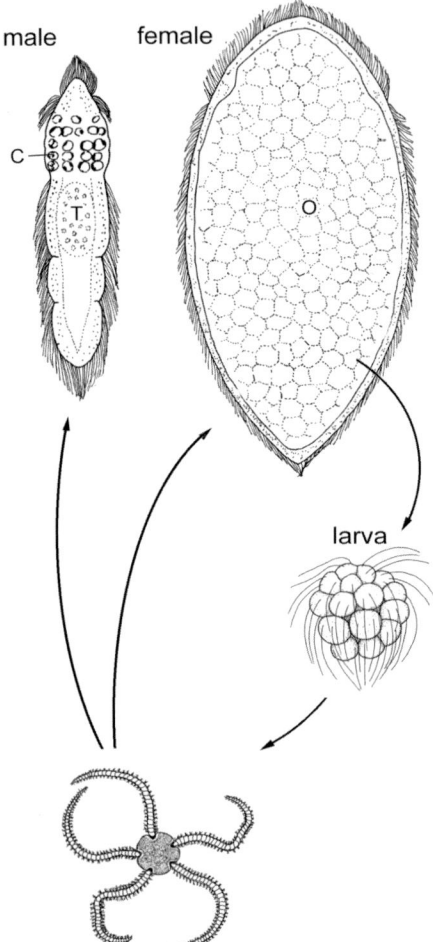

Figure 5.9 Life cycle of *Rhopalura ophiocomae*. Note: Sexually mature adults appear to be released synchronously from the host *Amphipholis squamata*. During mating, males make brief contact with females, at which time sperm are released. Fertilisation is internal. Males die shortly after mating, whereas females live for a few days following mating while brooding the developing embryos. When the embryos are fully developed the female ruptures, releasing ciliated dispersal larvae that enter a new host. Abbreviations: C, crystalin inclusion; O, oocyte; T, testis.

in detail but appear to differ morphologically between species (Caullery and Lavallée 1908, 1912, Atkins 1933). Differences in larval form are possibly related to host differences (Furuya 2001). In contrast, host species of dicyemids are specific to benthic cephalopods. Larvae of dicyemids demonstrate less interspecific variation than larvae of orthonectids.

Effects on hosts and ecological importance

Heavy infections have some effects on hosts. In *R. ophiocomae* young plasmodia are located in ophiuroid hosts in the ventral part of the disk, the interradial spaces, and near the gonads and genital ducts. As embryos grow in size the plasmodia enlarge, eventually destroying the host gonad and even invading the host coelom. Ophiuroids infested with older plasmodia of *Rhopalura ophiocomae* are easily recognised. The ventral surface is grayish instead of orange. They are more flaccid than normal because they are sterile and the genital pouches are not distended with embryos (Kozloff 1969). *Ciliocincta sabellariae* infects the muscle tissue of the dorsal body wall and cirri of the polychaete *Sabellaria cementarium*. Heavily infested individuals can be recognised by the presence of grayish patches on the dorsal surface (Kozloff 1971).

Except for a brief free-living phase during which mating and the incubation of developing embryos take place, the entire life cycle is confined to the gonads and other organs and tissues of their invertebrate hosts. The incidence of infection ranges from about 1% to 15% but typically is less than 5%. Infections are often highly localised and in some study sites may be restricted to areas of only a few hundred square metres (Caullery and Mesnil 1901). The ecological importance of orthonectids is likely to be small because of the few species and the low prevalence.

Important references
The group was reviewed by Kozloff (1990, 1992). A key to the genera was given by Kozloff (1992). The morphology, life cycles and early development were studied by Giard (1880), Caullery and coworkers (1901, 1908, 1912), Julin (1882) and Metschnikoff (1881).

Mesozoa Dicyemida (dicyemids)
Hidetaka Furuya

Introduction
Dicyemid mesozoans (Phylum Dicyemida) are the commonest and most characteristic parasites of benthic cephalopod molluscs. About 120 species have been reported in at least 50 species of benthic cephalopods distributed in many geographical localities: Sea of Okhotsk, Japan Sea, Western and Eastern North Pacific Ocean, New Zealand, North Indian Ocean, Mediterranean, Western North and Eastern Atlantic Ocean, Gulf of Mexico and the Antarctic Ocean.

Morphology and diversity
Species of dicyemids range in length from 0.1 mm to 5 mm. Dicyemids have neither body cavities nor differentiated organs. Their bodies consist of a central cylindrical cell called the axial cell and a single layer of 8 to 30 ciliated external cells called the peripheral cells (Fig. 5.10). The phylum Dicyemida includes three families, Conocyemidae, Dicyemidae and Kantharellae. The number of peripheral cells is species specific and constant in the families Conocyemidae and Dicyemidae. The family Kantharellidae contains only one species that is characterised by variable number of peripheral cells. At the anterior region of dicyemids, 2 to 10 peripheral cells form the calotte. Genera are characterised by the number and orientation of cells in each tier of the calotte (Hochberg 1990). The calotte shape varies, depending on the species, and might be an adaptation for attachment to the various regions of host renal tissues (Fig. 5.11). The family Conocyemidae it is characterised by an irregular shape of the adult form and includes two genera, *Conocyema* and *Microcyema*. . The head of *Conocyema polymorpha* looks like a balloon (Fig. 5.12A). In *Microcyema vespa*, peripheral cells are fused into a syncytium (Fig. 5.12B).

Dicyemids generally are found in benthic cephalopods, namely octopuses and cuttlefishes. But a few species of dicyemids were reported from squids, *Sepioteuthis lessoniana* (Nouvel 1947) and *Loligo* sp. (Kalavati and Narasimhamurti 1980). Such cases have been considered to be exceptional. Recently, two undescribed dicyemid species have been found in two species of squid, *S. lessoniana* and *Todarodes pacificus* (Furuya and Tsuneki 2003). Host species of dicyemids might not be necessarily restricted to the benthic cephalopods. Dicyemids are found primarily in the fluid-filled renal sacs or kidneys. In decapod cephalopods, squids, sepiolids and teuthoids they have been found also in the reno-pancreatic coelom and occasionally in the pericardium (Hochberg 1990).

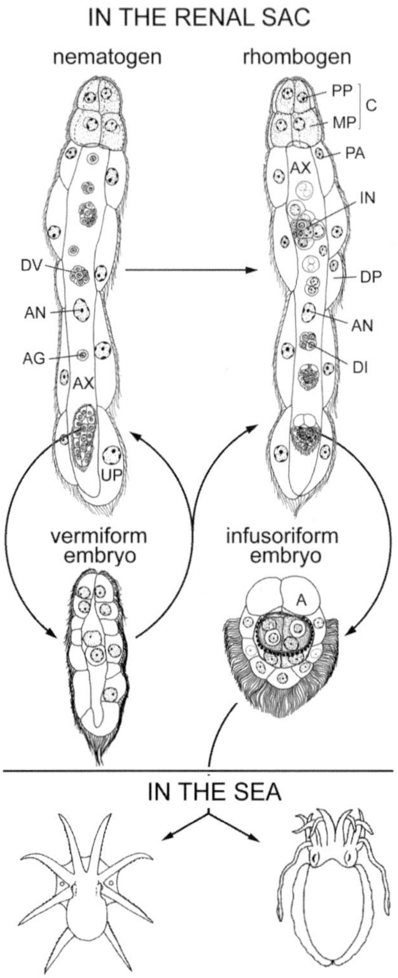

Figure 5.10 Life cycle of dicyemids. The process involved in the infection of a new cephalopod and the development into the adult are not known. In vermiforms (nematogen, rhombogen, vermiform embryo), a large cylinderical axial cell is surrounded by peripheral cells. Four to 10 anterior peripheral cells (propolars and metapolars) form a calotte. The other peripheral cells are diapolars. Two posterior diapolars are somewhat specialised as uropolars. The development of infusorigens, gametogenesis around the infusorigen and development of two types of embryo all proceed within the axial cell cytoplasm. Abbreviations: A, apical cell; AG, agamete; AN, axial cell nucleus; AX, axial cell; C, calotte; DP, diapolar cell; DV, developing vermiform embryo; MP, metapolar cell; PP, propolar cell; UP, uropolar cell. After Furuya and Tsuneki (2003).

Life cycles

The life cycle of dicyemids consists of two phases with different body organisation (Fig. 5.10): (1) the vermiform stages, in which the dicyemid exists as a vermiform embryo formed asexually from an agamete, and as a final form, the nematogen or rhombogen, and (2) the infusoriform embryo which develops from a fertilised egg produced around the hermaphroditic gonad called the infusorigen. The infusorigen itself is formed from an agamete. The name 'dicyemids' is derived from the existence of two types of embryo in the life cycle. A high population density in the cephalopod kidney may cause the shift from an asexual mode to a sexual mode of reproduction (Lapan and Morowitz 1975). Vermiform stages are restricted to the renal sac of cephalopods, whereas the

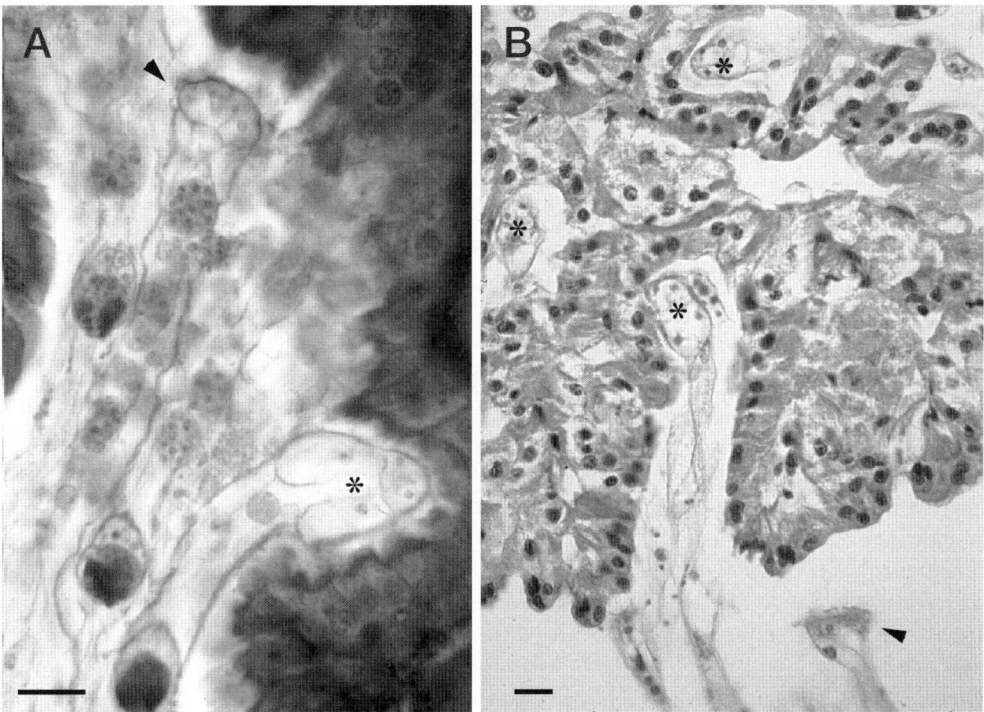

Figure 5.11 Light micrographs of stained sections through the renal organ. **A.** *Pinnoctopus cordiformis*. **B.** *Octopus vulgaris*. Note that the cone-shaped calotte and the anterior part of the body of *Dicyema knoxi* (Fig. 5.11A) and *D. misakiense* (Fig. 5.11B) are inserted into folds in the renal tissue (asterisks). In contrast, *D. maorum* (Fig. 5.11A) and *D. japonicum* (Fig. 5.11B), with a slightly inflated, disc-shaped calotte, are attached to the surface of the renal appendages (arrow heads). The specimen of *D. japonicum* that is shown is slightly detached due to an artifact. Scale bars = 20 μm. After Furuya *et al.* (2003b).

infusoriform embryos escape from the host into the sea to search for a new host. However, it remains to be understood how infusoriform larvae develop into vermiform stages in the new host.

In dicyemids, the population may develop from a small number of individuals (one or few) at the initiation of the infection of the renal sac because success of infecting new hosts is apparently low. Dicyemids are occasionally found in only one of the two renal sacs in a host octopus. Two different dicyemid species are occasionally detected, one each in the right and left renal sacs of the same host individual. These cases suggest that only a few propagules may infect an individual host. Subsequently asexual multiplication leads to a large population in the renal sac. In such conditions, cross-fertilisation is of little advantage and self-fertilisation via a hermaphroditic gonad has developed.

Infusoriform larvae actively swim close to the bottom for only a few days *in vitro* (McConnaughey 1951), an adaptation to the close proximity of cephalopod host individuals. At the anterior region of the embryo, there is a pair of unique cells called the apical cells (Fig. 5.10). Each contains a refringent body composed of a hydrated magnesium salt of inositol hexaphosphate (Lapan and Morowitz 1975). Its high specific gravity imparts a negative buoyancy to the dispersal larvae. This suggests the role of refringent bodies is to help the larvae to remain near the sea bottom where larvae can encounter another host. Dicyemids eventually enter the kidney and apparently do not move once attached. A very short larval stage in the plankton also is typical in colonial benthic animals. The analogy between colonial animal and dicyemids can be attributed to their sedentary life styles (Furuya et al. 2003a).

Effects on hosts and ecological importance

The renal sac of cephalopods is a unique environment providing living space for the dicyemids. The fluid-filled renal coelom provides an ideal habitat for the establishment and maintenance of dicyemids (Hochberg 1982). More than one species of dicyemids are usually found in each cephalopod species or individual. Most species of dicyemids are host specific (Furuya 1999). Generally, dicyemids occur at high prevalences (Furuya et al. 2003b). Dicyemids are usually found to be infecting renal organs heavily, but damages have never been observed. Lapan (1975) suggested that dicyemids facilitate host excretion of ammonia, contributing to acidification of the urine. In addition to the contraction of renal appendages, ciliary activity of dicyemids makes urine flow constant and, as the result, it assists in removal of urine. Thus, dicyemids are in a symbiotic, rather than a parasitic, relationship with cephalopods.

There is an interesting relationship between the calotte configuration and the co-occurence pattern in hosts. The calotte shapes are distinctly different when dicyemid species co-occur (Furuya et al. 2003b). Four basic types of calotte shapes are recognised. A conical calotte (Fig. 5.12F) is by far the most typical configuration. Dicyemids with a discoidal calotte (Figs 5.12C, D) frequently are found together with species having cone-shaped calottes. Cap-shaped calottes (Fig. 5.12E) appear to be intermediate in shape between the conical and discoidal type, and tend to occur in the cephalopods when more than two species of dicyemids are present. Dicyemids with irregularly shaped bodies and calottes (Figs 5.12A, B) occur when more than three species coexist. When more than two dicyemid species were present in a single host

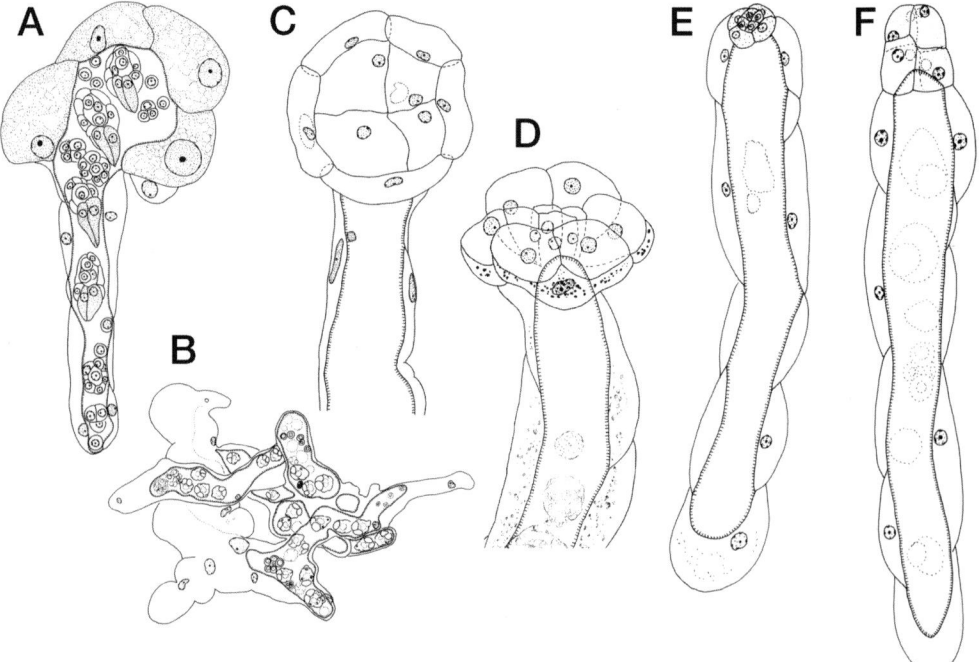

Figure 5.12 Diversity of dicyemid species. **A.** *Conocyema polymorpha*. **B.** *Microcyema vespa*. **C.** *Dicyemennea trichocephalum* (anterior part). **D.** *Dicyemodeca anthinocephalum* (anterior part). **E.** *Pseudicyema nakaoi*. **F.** *Dicyema acuticephalum*.

individual, calotte shapes were dissimilar as a rule. It is a common phenomenon that calotte shapes in dicyemid species from different host species more closely resemble each other than those of dicyemids observed within the same host species. Species of dicyemids that possess similar calotte shapes are very rarely found together in a single host individual, and in all such cases one species is dominant. In these cases, the species best adapted to the habitat possibly becomes a dominant species. In a host individual, interspecific competition between dicyemids may result when they have similar calotte shapes, and therefore, these dicyemids tend to infect different host individuals.

Species of dicyemids with different calotte shapes, for example, *D. misakiense* (conical) and *D. japonicum* (discoidal) inhabit different regions of the renal organs (Fig. 5.11). In general, dicyemids with conical calottes insert the anterior region of the body into crypts or folds in the renal appendages. In contrast, dicyemids with cap-shaped or disc-shaped calottes attach to the broad, flat or gently rounded surfaces of the renal appendages. Interspecific competition is most likely avoided by the habitat segregation in dicyemids that possess different calotte shapes.

Important references
The group was reviewed by Nouvel (1947, 1948), McConnaughey (1951), Lapan and Morowitz (1975), Hochberg (1990), and Furuya and Tsuneki (2003). A key to the families Conocyemidae and Dicyemidae was given by Hochberg (1990). The family Kantharellidae was discussed by Czaker (1994). A review of the development of two types of larvae and hermaphroditic gonad is by Furuya et al. (1996).

Myzostomida (myzostomids)
Greg W Rouse

Introduction
Myzostomida is a small taxon comprising around 150 described nominal species. There is debate on the overall position of the group, with authors variously favouring a position among polychaetes (Rouse and Pleijel 2001) or outside annelids and closer to platyhelminths (Eeckhaut et al. 2000) or rotifers (Zrzavy et al. 2001). Myzostomida are all parasitic on echinoderms, mainly crinoids, but also a few asteroids and ophiuroids. There are also records of them from sponges and from inside antipatharian corals (Grygier 2000).

Diversity and geographical distribution
Most myzostomes are mobile, reaching up to 1 cm in length, and roam over the host stealing food from it (e.g. *Myzostoma, Hypomyzostoma*), but others are sessile near a convenient 'feeding' site or even in the host's mouth. Some induce the host to form hard galls or soft cysts (Fig. 5.13A–D) around them (e.g. *Notopharyngoides, Endomyzostoma*), while others are endoparasitic (Fig. 5.13E), reaching 3 cm or more in length and live in the gut lumen, coelom or gonads of their host (e.g. *Mesomyzostoma*). There is a wide range of body forms in Myzostomida, with flattened oval or disc-like forms being predominant. However, many are elongate (Fig. 5.14E, F), ridged, or have extensions of the body that allow them, for example, to resemble the pinnules of their crinoid hosts (Fig. 5.14B, C). Mobile, free-living Myzostomida show a wonderful diversity of colour patterns that generally matches those of their hosts. Gall-forming and internal parasitic forms (Fig. 5.13) tend to be pale and unpigmented. Most Myzostomida are placed within *Myzostoma* (Fig. 5.14A–D) and are external parasites. The diversity of sessile and internal parasitic forms is much lower.

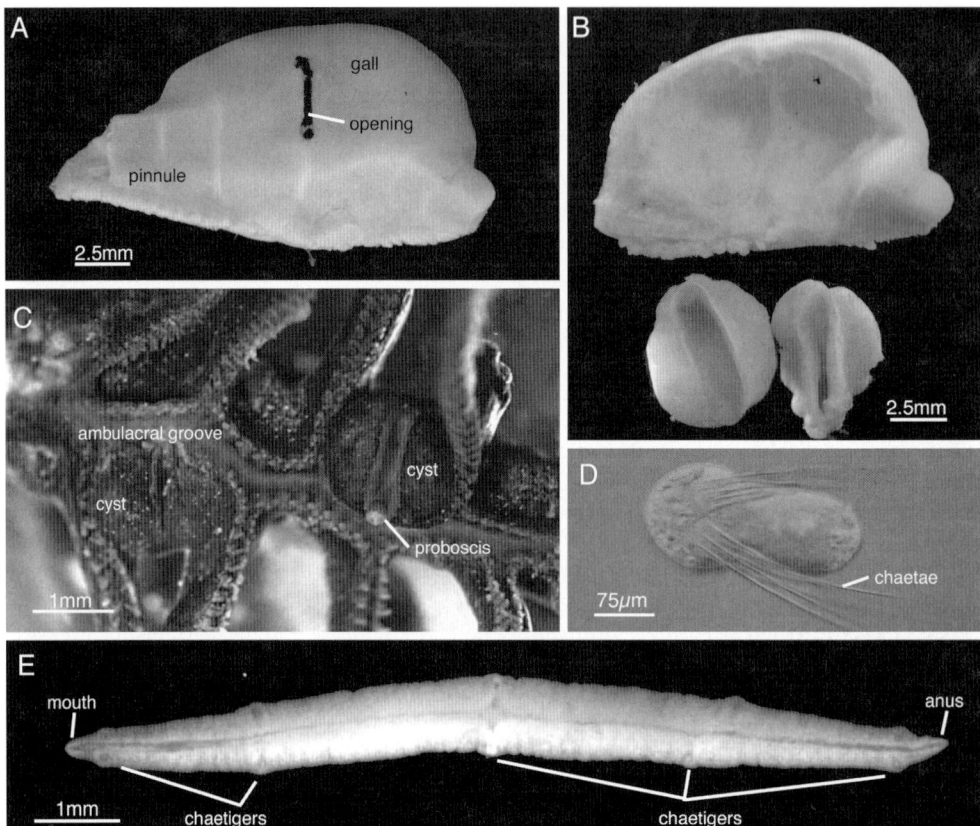

Figure 5.13 Gall, cyst and endoparasitic myzostomes. Larval myzostome stage. **A.** *Endomyzostoma deformator* gall on a pinnule of the stalked crinoid *Endoxocrinus alternicirrus* (from deep waters off Japan). **B.** Gall of *Endomyzostoma deformator* dissected open and the two occupants removed. **C.** Two cysts of *Contramyzostoma sphaera* adjacent to the central ambulacral groove of host crinoid *Comatella nigra* (from Lizard Island, Australia). **D.** Swimming larval stage of mobile myzostome *Myzostoma furcatum* with one pair of chaetal bundles. **E.** Endoparasitic *Mesomyzostoma katoi* from gonad of host *Oxycomanthus japonicus* (from Japan). All photos by G Rouse.

Myzostomida have been recorded from most marine areas, wherever their hosts are found, down to depths of more than 2000 m. Not surprisingly, myzostome diversity is highest where crinoid diversity is greatest, namely coral reef environments, and more than 100 described species are known from the Indo-Pacific (see Grygier 1990, 2000). Myzostome diversity is low in colder waters, with, for instance, only a few known from the well-studied European waters. Interestingly none have been recorded from the eastern tropical and south-eastern Pacific (Grygier 2000). The most comprehensive regional reviews on the occurrence of Myzostomida have been for the Indo-Pacific and around Australia by Grygier (1990, 2000). Some groups have relatively few records. *Protomyzostomum*, living in ophiuroid basket stars, have been found only in temperate (Japan) or cold seas (Arctic and Antarctic), as have those ectoparasitic (on asteroids) taxa in *Asteromyzostomum* (Arctic and Antarctic). *Asteriomyzostomum*, another taxon containing endoparasites (of asteroids), has taxa described from the Mediterranean and from off California. *Mesomyzostoma* (Fig. 5.13E) contains two endoparasitic taxa (of crinoids) described from Indonesia and Japan (Grygier 2000). Presumably, further investigation of asteroids, crinoids and ophiuroids will yield many more Myzostomida, particularly endoparasites. Host-

Figure 5.14 Mobile forms of myzostomes, mainly from crinoids at Lizard Island (Australia). **A.** Dorsal view of *Myzostoma furcatum* from host *Amphimetra tesselata*. **B.** Dorsal view of *Myzostoma cuniculus* from host *Comanthus mirabilis*. Posterior extensions are held up to look like crinoid pinnules. **C.** Dorsal view of *Myzostoma laigense* from host *Stephanometra oxyacantha*. Posterior extensions are held up to look like crinoid pinnules. **D.** Ventral view (SEM) of *Myzostoma australe* from crinoid host *Ptilometra macronema* from southern Australia. **E.** Lateral view of an undescribed *Hypomyzostoma* sitting on the arm of host *Colobometra perspinosa*. **F.** Ventral view (SEM) of the undescribed *Hypomyzostoma*. All photos by G Rouse.

specificity of Myzostomida varies considerably. Some will only infest a single kind of crinoid, but others live on three or four different crinoid taxa. In addition, some crinoid taxa may be hosts to several different myzostomes (Grygier 1990, 2000).

Morphology

The head of myzostomes is not a distinct structure, there are no eyes and the buccal opening is a ventral or terminal structure. The body is segmented and in most myzostomes there are five pairs of appendages (Fig. 5.14D, F), arguably similar to the parapodia of polychaetes, each bearing an emergent hook-shaped chaeta and an internal 'acicula'. In internal parasitic forms such as *Mesomyzostoma* the appendages are limited to the emergent chaetae only (Fig. 5.13E). The lateral margin of many myzostomes has cirri, usually 10 pairs, though they maybe absent or far more numerous than this. The ventral body surface generally has four pairs of 'lateral organs' that may be eversible and are probably sensory. They generally alternate with the parapodia.

Myzostoma, *Hypomyzostoma* and *Notopharyngoides* have a protrusible proboscis that Rouse and Fauchald (1997) regarded as homologous with the muscular axial proboscis found in some polychaetes. The everted tip of the proboscis is the mouth and this may be fringed with 4 to 30 papillae. In all others the proboscis, if present, is of a different organisation and represents the posterior region of the buccal cavity that is everted through the mouth. In some taxa, such as internal parasites like *Mesomyzostoma* a proboscis is absent (Fig. 5.13E). Behind the pharynx in most myzostomes is a large 'stomach' with several pairs of diverticula that branch extensively. From the stomach there is a straight intestine leading to the terminal anus. No circulatory system has been described for any Myzostomida and the stomach diverticula are thought to transport nutrients and wastes from around the body (Grygier 2000). Five pairs of protonephridia have been described in *Myzostoma cirriferum* by Pietsch and Westheide (1987). The pairs of nephridia lie anterior to each pair of parapodia. A pair of ciliated ducts that usually connect the uterus to the intestine have commonly been referred to as metanephridia. They are thought to serve an excretory function in removing waste gametes but may not actually represent metanephridia (Grygier 2000). Apart from the lateral organs and some sensory patches of cilia scattered over the body, no other sensory organs have been reported in Myzostomida.

Reproduction and life cycles

Myzostomida are usually protandric hermaphrodites and pass through a functional male stage to be simultaneous hermaphrodites at maturity (Grygier 2000). The male system is paired with diffuse testes lying ventral to the gut. Seminal vesicles are connected to the testes via vasa deferentia and exit on each side of the body next to the third parapodia. A pair of protrusible penises is present in taxa such as *Myzostoma* and *Notopharyngoides*, but are absent in the endoparasitic forms. Spermatogenesis is unusual and sperm develop in vacuoles inside spermiocysts. Spermatophores are placed on the surface of recipient worms and the spermiocysts pass into the body of the worm before migrating to the uterus (Grygier 2000) The female system has one or two ovaries dorsal to the gut that lead into uterine diverticula that fuse into a median uterus that exits, via an oviduct, near the anus. A pair of ciliated ducts (referred to above as possible metanephridia) also lead from the diverticula to the intestine.

Fertilisation is internal and probably occurs in the uterus. Larval development has been described for several *Myzostoma* taxa (see Eeckhaut *et al.* 2003) and it would appear that all fertilised eggs are spawned into the surrounding water, though Grygier (2000) reports brooding in some *Endomyzostoma*. The fertilised eggs of *Myzostoma* are very small (50 µm in diameter or less) and give rise to a planktonic, non-feeding trochophore larva (Fig. 5.13D) that is ready to settle on a host after five to eight days. Larvae develop two bunches of very long chaetae, similar to those seen in the larvae of some polychaetes. After settling, the larvae pass through further development in a cyst before becoming free living. Reproduction in endoparasitic and permanent cyst-dwellers has yet to be documented.

The only study on population dynamics of Myzostomida has been on *Myzostoma cirriferum*, living on the crinoid *Antedon bifida* from Europe (Eeckhaut and Jangoux 1997). They showed that juveniles live encysted for two months after settling and then live freely for four to five months. Population density increases through the spring and reaches an average maximum of around 70 per host.

Effects on hosts and ecological importance

Most Myzostomida associated with crinoids (*Contramyzostoma*, *Endomyzostoma*, *Hypomyzostoma*, *Myzostoma*, *Notopharyngoides* and *Pulvinomyzostomum*) would appear to feed by taking food from their host's ambulacral grooves, or from the oesophagus or gut before it is digested. The exceptions are the endoparasitic *Mesomyzostoma*, which presumably eats its crinoid host's gonad tissue and *Mycomyzostoma* that lives in cysts on the stalk of its sea lily host and so has no possibility of stealing food. *Asteromyzostomum* taxa living in the ambulacral grooves of asteroids insert their mouth through the skin of their host. *Asteriomyzostomum* and *Protomyzostomum* presumably live off their asteroids or ophiuroid host's internal tissues. No studies have been done on any possible detrimental effect of myzostomes, but the high densities of up to 3200

Myzostoma cirriferum per host (Woodham 1992) would presumably have some effect on the crinoid's food-gathering ability. High infestation of the endoparasitic *Mesomyzostoma* can also occur (up to 10 or more per host, GW Rouse pers. obs.) and these may have the potential to castrate their host.

Important references
The most important publications on the taxonomy of myzostomes are by Graff who described more than half the known species (e.g. Graff 1877, 1884, 1887). Recent reviews on myzostomes include a comprehensive account by Grygier (2000) and a briefer one in Rouse and Pleijel (2001). Eeckhaut and colleagues have published excellent papers on the larval development, phylogenetic placement and taxonomy of myzostomes (Eeckhaut 1998, Eeckhaut *et al.* 1998, Eeckhaut *et al.* 2000, Eeckhaut *et al.* 2003).

Polychaeta (bristle worms)
Greg W Rouse

Introduction
Polychaetes comprise many of the annelids and occur in most marine habitats from the intertidal to the deepest sediments. Some polychaetes dominate hydrothermal vent communities, while others comprise a major component of the interstitial fauna. Most polychaetes are cryptic, living under rocks or burying themselves in sediment. Some dig continuously through the sediment, while others make permanent burrows or tubes that they secrete or construct from gathered materials. Although most polychaetes are free living, it is not surprising that parasitism in the group has evolved several times. There are estimated to be around 9000 valid species (Rouse and Pleijel 2001), although there are many more to be described.

Morphology and diversity
The best known parasitic Annelida, the leeches (Hirudinea, Clitellata) are blood-feeders on animals (including humans). However, following the broad definition of parasitism used in this volume, namely that the parasite depends on the host and derives some benefit from it, there are many polychaete groups that can be characterised as parasitic. Parasitism among the polychaetes has evolved often and occurs in a great variety of forms. Most parasitic polychaetes are placed in the clade Aciculata, which comprises more than half of described polychaetes and contains two major groups, Eunicida and Phyllodocida. Parasitic Eunicida include a few Dorvilleidae, all Histriobdellidae and many Oenonidae. Phyllodocida has a wide range of parasitic forms including *Ichthyotomus*, Nautiliniellidae, *Antonbrunnia*, Syllidae and many scaleworms (Polynoidae). *Spinther* is another parasitic genus in Aciculata, but its affinities are uncertain. Outside the Aciculata, the incidence of parasitism is more sporadic, with the exception of Spionidae, though a few Sabellidae are found in association with molluscs.

While most Dorvilleidae are free living, some *Ophryotrocha* and all *Iphitime* are parasites on decapods, where they graze with their jaws on material drawn into, or growing in, the branchial chamber. They are best known from European waters. *Iphitime* and *Ophryotrocha* were placed in Iphitimidae until analysis by Eibye-Jacobsen and Kristensen (1994) established that they are dorvilleids. Histriobdellidae (Figs 5.15A, B) are also only found in association with crustaceans. *Histriobdella homari* occurs on North Atlantic marine lobsters such as *Homarus gammarus* and grazes on bacteria and blue-green algae on the gills and branchial chamber (Jennings and Gelder 1976). *Stratiodrilus* live on the gills of freshwater crustaceans in Australia (Fig. 5.15B), Madagascar and South America. An exceptional form, *Stratiodrilus cirolanae* (Fig. 5.15A), is marine and

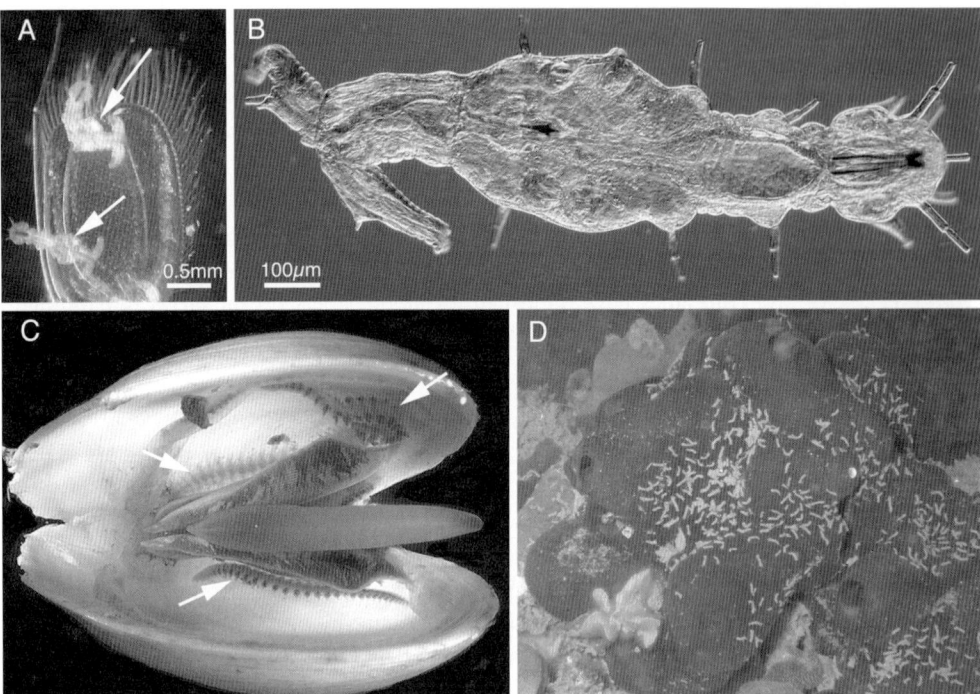

Figure 5.15 Parasitic polychaetes. **A**. Arrows indicate *Stratiodrilus cirolanae* (Histriobdellidae) on pleopod of host isopod *Cirolana venusticauda*. **B**. Male of *Stratiodrilus tasmaniae*. The head has appendages and jaws similar to other Eunicida. **C**. Deep-sea vesicomyid clam containing three specimens (arrows) of an undescribed nautilienellid. **D**. Sediment tubes of an undescribed *Polydorella* (Spionidae) visible on the surface of a sponge in South Australia. A, B, and D by G Rouse; C courtesy of J Dreyer.

found on the pleopods of the intertidal isopod *Cirolana venusticauda* in South Africa. Many Oenonidae live at least part of their lives in the coelomic cavity of other polychaetes (Fig. 5.16F) or bivalves. *Notocirrus* parasitises other polychaetes with infection intensities of up to 50; it ranges in size from a few to hundreds of segments. The group is found worldwide.

Within Phyllodicida, Nautiliniellidae represents a recent discovery from deep-sea hydrothermal vents and cold seeps, where they are parasites of bivalves. Adults of Nautiliniellidae reach a length of 10 cm and have hundreds of segments. Several species may inhabit the same host (Fig. 5.15C). Nautiliniellidae have been found off southern Japan, California, western Mexico and the western North Atlantic. *Antonbruunia viridis* was described by Hartman and Boss (1965) from lucinid bivalves from Madagascar. All were collected on a single occasion and have not been reported since. Most hosts contained one female (up to 14 mm long) and one dwarf male, about half the size. Eisig (1906) described *Ichthyotomus sanguinarius* and found large numbers attached to the fins of several different Mediterranean eels. *Ichthyotomus sanguinarius* was recorded once, it reaches 10 mm in length with about 90 segments and has separate sexes. Eisig (1906) showed that it pierces the host's skin with its jaws to suck blood. Syllidae is a diverse group with numerous parasitic forms. Syllid parasites have been mainly reported from sponge hosts, the best known are *Haplosyllis spongicola* (thousands of worms can occur on a single sponge) and *Myrianida prolifer* (Fig. 5.16A–D), a parasite of hydroids. *Calamyzas amphictenicola* and *Asetocalamyzas laonicola* are also thought to be syllids, found in European waters and the White Sea, respectively; they parasitise polychaetes, penetrating the body wall or gills with their pharynx. *Asetocalamyzas laonicola*, which lives on a spionid polychaete, may actually be a

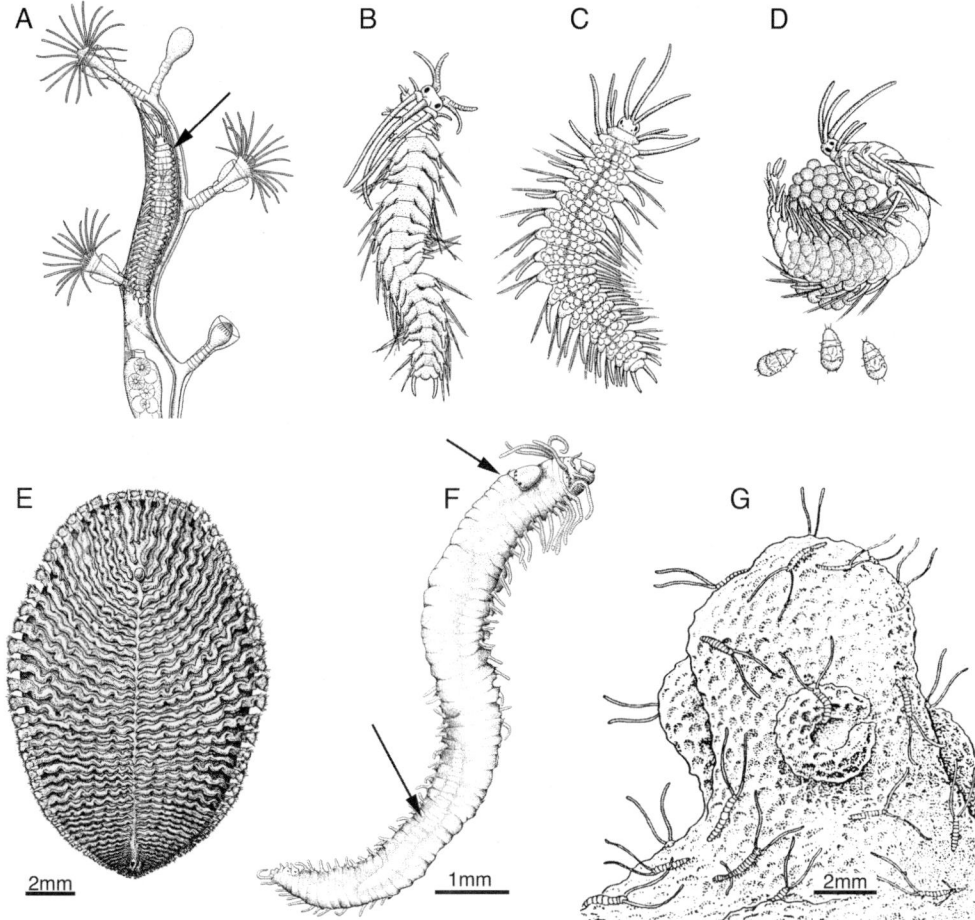

Figure 5.16 Parasitic polychaetes. **A–D.** Life cycle summary of *Myrianida prolifer* (Reproduced with permission from Fischer A , Reproductive and developmental phenomena in annelids: a source of explanatory research problems, *Hydrobiologia* 402: 1–20; copyright (1999) Albrecht Fischer). **A.** Adult specimen on host hydroid. This will bud off free-swimming epitokes. **B.** Male epitoke with large eyes and palps to detect females. **C.** Female epitoke. **D.** After mating the female broods the embryos on her ventral surface until they swim away as trochophore larvae to find a host hydroid. **E.** Dorsal view of *Spinther alaskensis* taken from a sponge. Reproduced with permission from Hartman O, The polychaetous annelids of Alaska, *Pacific Science* 2: 3–58; published by the University of Hawai'i Press 1948. **F.** Lateral view of a syllid, *Haplosyllis spongicola*, that parasitises sponges. In this case the syllid itself is host to two specimens of the oenonid *Labrorastatus luteus*. The arrow near the head of the host points to the emerging head of one parasite. The other arrow points to the stipples outline of another *L. luteus* lying in the coleom of the host. Modified from Uebelacker (1978) *Journal of Parasitology*, 64: 151–154, with permission of the *Journal of Parasitology*. **G.** The bodies and palps *Polydorella smurovi* (Spionidae) are visible on the surface of a sponge. Reproduced with permission from Tzetlin AB and Britayev TA, A new species of the Spionidae (Polychaeta) with asexual reproduction associated with sponges. *Zoologica Scripta* 14: 177–181; published by Blackwell Publishing 1985.

parasitic dwarf male and is the topic of further research (AB Tzetlin, pers. comm. July 2004.) Many Polynoidae (scaleworms) are commensal, predominantly with echinoderms or burrowing animals such as other polychaetes and decapods. Some appear to be parasitic, especially those found on crinoid echinoderms. *Spinther* is a small group, all species are parasites on sponges. The body consists of up to 50 segments and may reach several centimetres in length. They are flattened and rounded to oval in outline (Fig. 5.16E), and occur on the surfaces of sponges from shallow to moderate depths, with bright colouration that may be cryptic.

Outside Aciculata, parasites are much less diverse. The 'polydorid' group of Spionidae is well known for boring into the shells of molluscs and is found worldwide. Other spionids, *Polydorella*, are found only on sponges (Tzetlin and Britayev 1985) and are easily recognised by their sediment tubes on the surface sponges (Figs 5.15D, 5.16G). *Terebrasabella heterouncinata* is a recently described sabellid that forms burrows in the shells of molluscs and is a pest of abalone (Fitzhugh and Rouse 1999). *Caobangia* is another genus of sabellids, recorded from rivers in Asia, where they infest freshwater snail shells (Jones 1974).

Reproduction and life cycles
Reproductive mechanisms of most parasitic polychaetes are unknown. One phenomenon that is common among Aciculata is epitoky, and this also occurs in some parasitic forms. It is illustrated in the life cycle of the syllid *Myrianida prolifer* (formerly *Autolytus*), which is parasitic on hydroids (Fig. 5.16A). In *M. prolifer*, the sexes are separate and each individual buds off epitokes from the posterior end (Fig. 5.16B, C). These are filled with either eggs or sperm and swim to near the surface where they meet other epitokes. The females retain the eggs, after they are fertilised by the male epitoke (which dies), until they develop into trochophore larvae (Fig. 5.16D) before dying themselves. Meanwhile 'stock' animals remain on the hydroid and bud off further epitokes. Spionids show a range of reproductive mechanisms, but most are brooders of larvae within the parental tube, including parasitic forms such as *Polydora* species. *Polydorella* species, all parasites on the surface of sponges, are unusual in showing asexual reproduction and this allows them to reach high densities on their host. Histriobdellidae are known to have separate sexes and lay embryos attached to the base of the gill filament, the directly developing young then hatching out onto the host (GW Rouse, pers. obs.). In the simultaneous hermaphrodite sabellids *Caobangia* and *Terebrasabella heterouncinata* the larvae are also brooded on the tube of the parent and have a brief dispersal phase (Fitzhugh and Rouse 1999).

Effects on hosts and ecological importance
Whereas Aciculata contains most parasitic forms, the most important economically are Spionidae, which can have an effect on commercial bivalve fisheries, such as oysters. *Polydora* spp. use acid secretions and chaetae to create a burrow in the shell, and sediment is carried in from outside to create an inner tube. When the burrow reaches the inner shell layers, the mollusc reacts by secreting more shell material. This can result in damaging 'mud blisters'. Sometimes the mud-blisters may be so numerous that they lead to the host's death (Martin and Britayev 1998).

Important references
Symbiotic polychaetes have been comprehensively reviewed by Martin and Britayev (1998). Older reviews are by Paris (1955) and Clark (1956). For a general overview of polychaete diversity that places this review in context see Rouse and Pleijel (2001).

Hirudinea (leeches)
Fredric R Govedich, Bonnie A Bain and Ronald W Davies

Introduction
Leeches are a diverse group of animals with many species living in marine environments, including intertidal regions and the deep ocean, with other species living in lakes and rivers on every continent (except Antarctica), and in moist terrestrial environments of Australasia and Oceania.

They can be predators or temporary ectoparasites (sanguivory) and feed on a range of invertebrate and vertebrate prey (Sawyer 1986a,b, Davies and Govedich 2001, Govedich 2001, Kutschera and Wirtz 2001). This diversity makes them an interesting group of animals to study.

Morphology and diversity

The Annelid subclass Hirudinea includes the Acanthobdellida, which consists of only two species of salmonid ectoparasites in the genus *Acanthobdella*, and the Euhirudinea or 'true' leeches (Siddall and Burreson 1995, Davies and Govedich 2001, Govedich 2001, Kutschera and Wirtz 2001). Leeches (Euhirudinea) are divided into two major groups or orders: The Rhynchobdellida (leeches with a protrusible proboscis and true vascular system) which include the marine and freshwater members of the Piscicolidae, Ozobranchidae and Glossiphoniidae, and the Arhynchobdellida (leeches with a non-protrusible muscular pharynx, either with or without jaws, and a haemocoelomic system) which consists of the freshwater and terrestrial members of the Hirudinidae, Haemadipsidae and Erpobdellidae (Sawyer 1986b, Siddall and Burreson 1995, Davies and Govedich 2001, Govedich 2001, Kutschera and Wirtz 2001).

Leeches have sensory structures including simple eyes and oculiform spots, papillae and sensilla that allow them to find prey or hosts and interact with their environment. Eyes are typically arranged on the dorsal surface of the 'head' and are found either along the margins or near the midline (Fig. 5.17). The eyes in some species are very close together or even fused into lobed composite eyes and the original number of eyes can only be determined by counting the lobes. Oculiform spots or eyespots can also be found along the margins of the body and on the

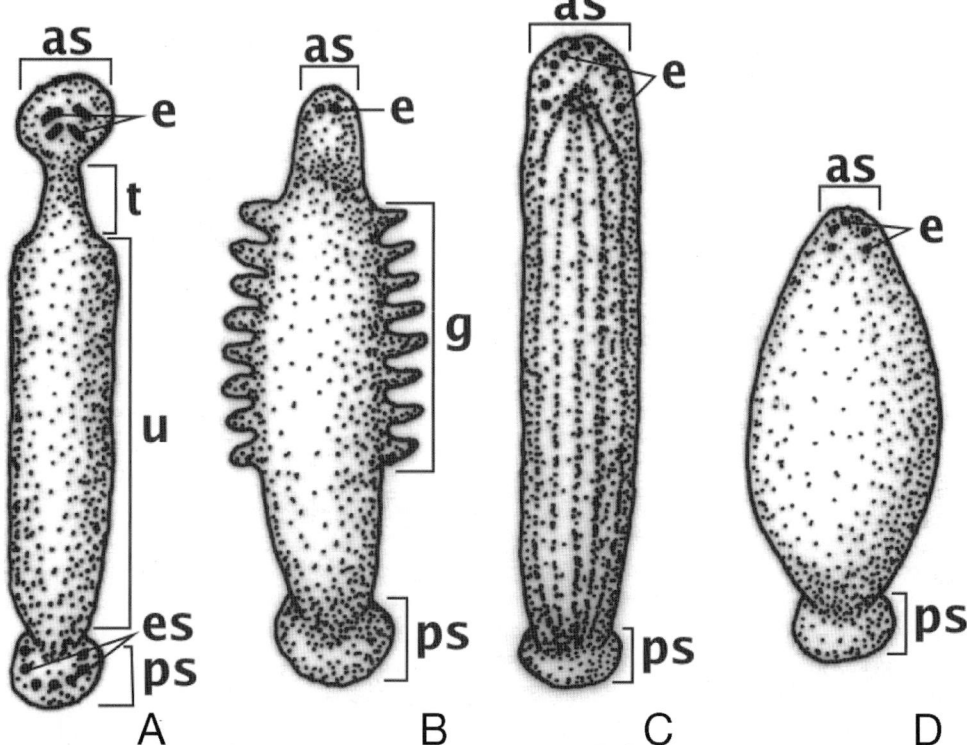

Figure 5.17 Dorsal view showing the body outline and eye position in four families of leeches (Euhirudinea). **A**. Piscicolidae. **B**. Ozobranchidae. **C**. Hirudineae. **D**. Glossiphoniidae. Figure labels: as, anterior sucker; ps, posterior sucker; e, eye; es, eyespot; t, trachelosome; u, urosome; g, gills.

posterior suckers of some piscicolids. Sensilla act either as mechanoreceptors or chemoreceptors and may be located on papillae and tubercles. Papillae are typically small protrusible sense organs and tubercles are large fleshy protrusions that consist of some dermal tissues and muscles. Both types can be arranged in rows or scattered on the dorsal and ventral surfaces of the animal (Sawyer 1986b, Davies and Govedich 2001, Govedich 2001).

Leeches lack chaetae (bristles) and have both an anterior or oral sucker which contains the mouth and a wider ventrally directed posterior or caudal sucker (Figs 5.17, 5.18). Suckers are often used to attach to substrates, prey, and hosts, and are used for locomotion by some aquatic and most terrestrial leech species. Crawling typically involves a looping motion consisting of body elongation and shortening and the alternation of attachment by the anterior and posterior suckers. In addition to crawling, leeches swim by using dorsoventral undulations of their flattened body. Many aquatic leeches are very good swimmers (particularly the erpobdellids and hirudinids), but only a few piscicolids and glossiphoniids are able to swim well (Sawyer 1986a,b, Davies and Govedich 2001, Govedich 2001, Kutschera and Wirtz 2001).

Bodies of most leeches are not externally divided into distinct regions. However, most species in the primarily marine family Piscicolidae (except for the genera *Myzobdella* and *Piscicolaria*) have a body that is divided into a narrow 'neck' or trachelosome and longer and wider 'body' or urosome (Fig. 5.17). Leech bodies consist of two preoral, non-metameric segments called the prostomium and peristomium and 32 postoral somites (metameres) labeled I through XXXIV (Sawyer 1986a,b, Davies and Govedich 2001, Govedich 2001). Each of the postoral somites is externally subdivided into annuli with mid-body (complete) somites having the full number (2–16) of annuli. Within each somite, individual annuli are numbered based on the three primary annuli, a1, a2, and a3, counting from the anterior. The a2 (middle) or neural annulus contains the ventral nerve cord ganglion and is typically delineated externally by a transverse row of papillae or sensilla. Some species have fewer than three annuli due to the loss or fusion of the primary annuli (biannulate condition) and others have more than three annuli resulting from the repeated bisection of the primary annuli. These additional annuli give the more complex annulation patterns observed in many species and are labeled depending on the original annulus that was bisected. For example, if the a1 annulus is bisected, the anterior annulus becomes b1 and the second becomes b2. With further bisection, b1 becomes c1 and c2 and so on. The resulting numbering system can become complex with some annuli bisected and others remaining in the primary condition. For example, a five annulate leech may have the a1 and a3 annuli bisected, but still have the primary a2, giving the formula b1, b2, a2, b5, b6 for the somite (Moore 1900, Sawyer 1986a,b, Davies and Govedich 2001, Govedich 2001).

Leeches primarily respire through their epidermis; however, many piscicolids and ozobranchids may also have paired pulsatile vesicles or in some species 'gills' on the neural (a2) annuli of the urosome. Pulsatile vesicles and gills aid in gas exchange and are connected to the circulatory system via special coelomic passages (Sawyer 1986a,b, Davies and Govedich 2001). The coelomic system of leeches is highly modified and reduced with intersegmental septa absent in adults (although remnants are found in some glossiphoniids). Piscicolids, ozobranchids and glossiphoniids have a coelom that has been modified to enclose the blood vascular system with a ventral lacuna surrounding the ventral nerve cord and blood vessel, and a dorsal lacuna surrounding the dorsal blood vessel. Dorsal, ventral, and lateral lacunae are connected via transverse communicating lacunae and many piscicolids have pulsatile vesicles that are modified coelomic chambers connected into the coelomic network via additional lacunae (Sawyer 1986a,b, Davies and Govedich 2001, Govedich 2001). Under hypoxic or low oxygen conditions, many species compensate behaviourally by using dorsoventral undulations of the body to move water across their epidermis. Most piscicolids and ozobranchids do not ventilate. Instead they use their gills or pulsatile vesicles to move oxygenated coelomic fluid through the body (Sawyer

1986a, Davies and Govedich 2001). Some species are also able to compensate for low oxygen levels metabolically and use alternative energy reserves including amino acids, lipids and glycogen, producing succinate and alanine rather than lactic acid as metabolic wastes (Reddy and Davies 1993, Davies and Govedich 2001).

The excretory system of leeches includes up to 17 pairs of modified metanephridia that are connected to small external pores located in somites VII to XXII with a reduced number in the gonad-containing somites and at the anterior end of the animal. In addition, some marine species have an extensive network of finely branched tubes connected to the nephridia, forming a modified plectonephridium. This network is located under the longitudinal muscle layer and may be an adaptation to life in the ocean as it is found only in marine or closely related freshwater species that may have reinvaded freshwater ecosystems. In all leeches, the metanephridia excrete nitrogenous wastes such as ammonia and maintain salt and water concentrations within the body (Sawyer 1986a,b, Davies and Govedich 2001). Specialised botryoidal (Arhynchobdellida) and chloragogen (Rhynchobdellida) tissues function in a manner similar to the liver and kidneys of other animals. These special tissues are located in coelomic chambers around the gut or in clusters between longitudinal muscle layers and are important in metabolic processes such as lipid metabolism, carbohydrate catabolism, oxidation and detoxification. They also function in the storage of lipids, phospholipids and pigments resulting from the breakdown of blood in sanguivorous species (Sawyer 1986a, Govedich 2001).

Leeches are typically simultaneous hermaphrodites, although some may have a brief protandrous stage, and all have internal fertilisation (Sawyer 1986a,b, Davies and Govedich 2001, Govedich 2001, Kutschera and Wirtz 2001). Typically both partners exchange gametes; however, this is not always the case. Glossiphoniids mate through hypodermic implantation of a spermatophore and some individuals (called 'sneakers') have been observed mating with individuals that were unable to implant a reciprocal spermatophore either because they were feeding or otherwise occupied. Though rare, self-fertilisation has also been observed in some glossiphoniids (FR Govedich unpublished data).

Gonopores are located (Fig. 5.18) near the midline on the ventral surface of somites XI and XII, respectively, and are visible in reproductively mature individuals. The male gonopore is anterior to the female gonopore with the two gonopores separated by a number of annuli that are often species-specific. The male gonopore is typically larger, more obvious and may be raised or surrounded by papillae. The female gonopore is posterior to the male gonopore and is smaller in size, sometimes making it difficult to locate (Sawyer 1986a,b, Davies and Govedich 2001, Govedich 2001).

Leeches do not have true testes or ovaries, instead they have paired testisacs and ovisacs. These thin-walled sacs are derived from coelomic sacs that are lined by a special germinal epithelial layer that produce either spermatozoa (testisacs) or ova (ovisacs). Spermatozoa and ova remain within the liquid-filled sacs where they develop and mature (Anderson 1973; Sawyer 1986a; Davies and Govedich 2001; Govedich 2001). Multiple pairs of testisacs (Fig. 5.18) are usually found in the somites posterior to XI and may be discrete spherical structures (Piscicolidae, Ozobranchidae, Glossiphoniidae and Hirudinidae) or multifollicular columns (resembling bunches of grapes) located next to the crop on either side of the ventral nerve cord (Erpobdellidae) (Davies and Govedich 2001, Govedich 2001). The testisacs are connected to the vasa deferentia via a short vasa efferentia. The vasa deferentia run along each side of the body, and form large coiled epidymes or sperm vesicles at their anterior end. Ejaculatory ducts run from the epidymes through the atrial cornua and unite to form the male atrium that in some species may be modified into a protrusible penis. A single pair of ovisacs (Fig. 5.18) are usually found in the somites posterior to XII and may be small spherical organs or elongate tubes that can be straight, coiled or recurved back on themselves. The ovisacs are connected by a pair of short

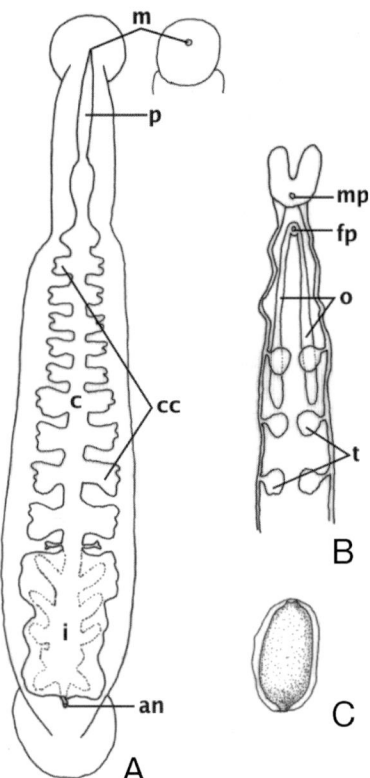

Figure 5.18 Internal features of a marine piscicolid leech. **A.** Digestive system (dorsal view with ventral view of mouth). **B.** Reproductive system (ventral view). **C.** Cocoon. Figure labels: m, mouth; p, proboscis; c, crop; cc, crop caeca; i, intestine; an, anus; mp, male gonopore; fp, female gonopore; o, ovisac; t, testisac.

oviducts that converge to form a common oviduct that leads either to a vagina or directly to the female gonopore (Sawyer 1986a, b, Davies and Govedich 2001, Govedich 2001).

The digestive system consists of a mouth (Fig. 5.18) that in some leeches is a small pore located in the center or anterior part of the oral sucker, and in others the sucker may be almost completely replaced by the mouth. Leeches with a small mouth pore (piscicolids, ozobranchids and glossiphoniids) typically have an eversible muscular proboscis (Fig. 5.18), and those with a large mouth which fills the oral sucker have a buccal cavity that may lack jaws (erpobdellids) or have two or three sets of jaws that resemble half circular saw blades (hirudinids and haemadipsids) (Sawyer 1986a,b, Davies and Govedich 2001, Govedich 2001, Kutschera and Wirtz 2001).

The mouth leads to the pharynx that is connected to the salivary glands which produce chemicals used for the processing and breakdown of food. The saliva of predacious and sanguivorous (blood feeding) leeches typically serve different functions. The salivary glands of predacious leeches secrete digestive enzymes that break down the tissues and body fluids of their prey. Salivary glands in most sanguivorous leeches secrete compounds that are used for bloodsucking rather than digestion. Functions of these compounds include mucus to lubricate the mouth parts, a hyaluronidase or spreading factor to make the host's skin more permeable, a histamine-like secretion used for vasodilation (blood vessel dilation), an anticoagulant that may prevent or break down blood clots, and possibly an anaesthetic-like compound, though this has yet to be verified (Sawyer 1986a,b, Davies and Govedich 2001, Govedich 2001).

The pharynx leads to a crop that is adapted for the storage of blood in sanguivorous leeches and food storage, digestion and absorption in many predacious leeches. The crop may be either acaecate or contain 1–11 pairs of lateral caeca (Fig. 5.18) but the number and arrangement of caeca is highly variable between genera. The crop is connected to the intestine (Fig. 5.18) where digestion and absorption of food and nutrients occurs in both predacious and sanguivorous leeches. Predacious leeches are able to digest their food within a few days, but sanguivorous ones take much longer (weeks or even months) due to their reliance on endosymbiotic bacteria that produce the enzymes needed to break down blood (Sawyer 1986a,b, Graf 1999, Davies and Govedich 2001, Govedich 2001). The anus usually opens dorsally just anterior to the posterior sucker in or near somite XXVII. The anus in a few species including *Branchellion torpedinis*, *Actinobdella peduculata* and *Marsupiobdella africana* is displaced further anteriorly (Sawyer 1986b, Davies and Govedich 2001, Govedich 2001).

Life cycle
Leeches are hermaphrodites with functional male and female reproductive systems and internal fertilisation. Once individuals reach maturity, typically three to four months after hatching under ideal conditions, individuals will mate and produce several cocoons (Fig. 5.18). In most leeches, including marine species, these cocoons are large fluid-filled chambers that contain the eggs and are provisioned with a nutrient fluid. This fluid supplies the eggs and developing young with all of the energy and nutrition they will require for growth and development. Following cocoon production, most leeches abandon their cocoons after they have been attached to vegetation, solid substrates or even future hosts (Sawyer 1986a,b, Kutschera and Wirtz 2001). In contrast glossiphoniids have extended parental care with eggs brooded in an external nest, on the ventral surface of the parent, or, in a few species, in a special brood pouch. Once glossiphoniid eggs hatch, the young are brooded and cared for by the parent (Sawyer 1986a,b, Kutschera and Wirtz 2001).

Adult marine leeches are often found attached within or near gill chambers or at fin bases of many fish, including salmon, grouper, sharks, skates and rays although a few leeches prefer to live on turtles and crocodiles. Most species must leave their host to reproduce and free-living adults are often found in estuaries where they mate and then produce their cocoons. The cocoons are abandoned and often the adults die soon after their production. Once the juveniles hatch they find their way to potential hosts where they begin feeding and growing. The hosts will often move out of estuaries after a time carrying the leeches with them (Sawyer 1986a, b).

Economic and medical importance
Freshwater and terrestrial leeches are recognised for their medicinal properties and have been used as a component of traditional medicines for many centuries. Leeches are used in the treatment of a variety of circulatory diseases and for reconstructive plastic surgery (Sawyer 1986a,b, Govedich 2001, Kutschera and Wirtz 2001). The saliva of leeches is also economically important with spreading factors such as hyaluronidase, and anticoagulants such as hirudin, hementin, hemenerin and destabilase synthesised and sold as medications for the treatment of blood-clotting disorders and other forms of heart disease (Sawyer 1986a,b, Govedich 2001, Kutschera and Wirtz 2001). Salivary compounds (including anticoagulants and vasodilators) of marine species have not been extensively studied and there may be many new medicinal compounds yet to be discovered.

Marine leeches have been known to reduce the value of fish catches, particularly when they have reached high densities on economically important fish species (Cruz-Lacierda *et al*. 2000). The economic value of these losses is not known due to inadequate records and a lack of reporting (FR Govedich pers. obs).

Effects on hosts and ecological importance

Leeches are distributed worldwide and can be found living in marine, estuarine, moist terrestrial and freshwater ecosystems. Leeches can be an integral component of benthic and occasionally pelagic communities with some species feeding on invertebrates, and others, including many piscicolid, glossiphoniid, hirudinid and haemadipsid species acting as temporary ectoparasites (sanguivory) on fish, amphibians, reptiles (turtles and crocodiles), birds and mammals. Sanguivorous species are typically not host-specific and will feed on a range of available hosts (Sawyer 1986a,b, Davies and Govedich 2001, Govedich 2001, Kutschera and Wirtz 2001).

Sanguivorous leeches typically do not permanently harm their hosts and act only as temporary ectoparasites, leaving their host following a blood meal. This is not always the case with some marine leeches spending most of their life on a host, leaving only to reproduce and lay cocoons. Typically leeches are only a minor irritant; however, there have been cases where hosts have become infested with a large number of individuals causing them stress and in extreme cases, death (Cruz-Lacierda et al. 2000). Also leeches can serve as intermediate hosts of haematozoa, particularly trypanosomes, and can act as vectors for the infection of fish (Sawyer 1986a, b, Negm-Eldin 1997).

Important references

Comprehensive reviews of many aspects of the morphology, taxonomy and biology of leeches are by Sawyer (1986a,b). Davies and Govedich (2001) discuss North American leeches, and Govedich (2001) those of Australasia and Oceania. Siddall and Burreson's (1995) account of the phylogeny of the Euhirudinea addresses the question of the independent evolution of blood feeding by leeches.

Cycliophora (wheel wearers)

Iben Heiner and Reinhardt Møbjerg Kristensen

Introduction

The marine phylum Cycliophora was described by Funch and Kristensen (1995), the third phylum described in the 20th century. It consists presently of a single species *Symbion pandora*, two other species are under description. All occur in the Northern hemisphere. *Symbion pandora* is a microscopical animal found on the mouthparts (setae) of the lobster *Nephrops norvegicus* (Figs 5.19, 5.20A and B), the two other species are from the lobsters *Homarus gammarus* and *H. americanus* (Figs 5.20C and D). The Cape lobster *H. capensis* from the southern hemisphere has not been found in recent years and has not yet been examined. Other decapods from the Southern hemisphere seem to lack cycliophorans (RM Kristensen, pers. obs).

Cycliophora have a very complicated reproduction, involving asexual and sexual reproduction. The sessile and feeding forms of Cycliophora are epibionts (i.e. they live on another living organism, a lobster). Cycliophorans filter the water for bacteria and algae (Funch and Kristensen 1997), and therefore are commensals rather than parasites. However, they face the same problem as do parasites (i.e. they have to locate a specific host and a very small microhabitat on it), a problem they have solved in an ingenious way, having adopted a complicated life cycle that leads to the production of a great number of offspring. In addition to the feeding sessile stage, there are three non-feeding larval stages, the Pandora larva (asexual larva), the Chordoid larva (dispersal stage) and the Prometheus larva (primary male), and two mature sexual stages, the dwarf males and the females.

The phylogenetic position of Cycliophora has been much debated since their discovery (Kristensen 2002b). The general opinion of morphologists is that they are a sister group to either

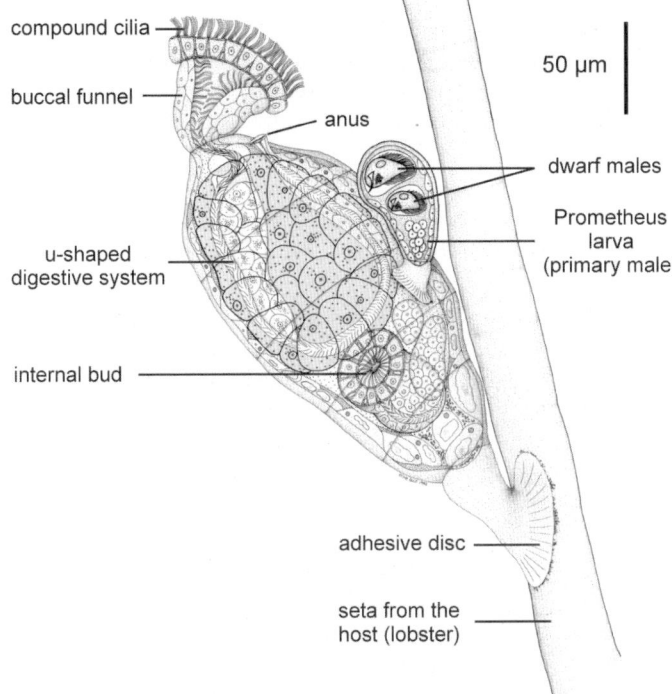

Figure 5.19 The holotype of *Symbion pandora* from the Norway lobster *Nephrops norvegicus* (collected close to Frederikshavn, Denmark). The sessile feeding individual has an attached Prometheus larva outside and a new internal bud inside. The Prometheus larva has already developed two dwarf males. Illustration drawn by Stine B Elle.

Entoprocta or to the clade Entoprocta/Ectoprocta (Funch 1996, Funch and Kristensen 1995, 1997). However, the first molecular analysis using 18S rRNA of Cycliophora showed a sister group relationship with Rotifera and Acanthocephala (Winnepennincks *et al*. 1998), and investigations of the phylogenetic position of Micrognathozoa using four molecular loci (Giribet *et al*. 2004) suggest a relationship of Cycliophora with Micrognathozoa and Syndermata (Rotifera and Acanthocephala).

Morphology and life cycle

Cycliophorans are bilateral-symmetric and have a well-differentiated cuticle. The sessile and feeding stage of *Symbion pandora* is about 0.33 mm long. The body of the animal consists of a ciliated buccal funnel, an ovoid trunk with a U-shaped digestive system, and a short stalk with an adhesive disc that attaches the animal to its host. The U-shaped digestive system ends in an anus located near the buccal funnel and the upper part of the trunk (Fig. 5.19).

In the asexual reproductive cycle the sessile animal grows by producing a single internal bud with a new buccal funnel and intestine, which later replaces the old feeding structures. The old intestine is reabsorbed and the old buccal funnel is cast off. This form of growth, inner budding, is repeated several times in each animal (Kristensen 2002a).

At some point, when the sessile animal is large enough, it produces a mobile Pandora larva inside a brooding chamber, together with the formation of an internal bud. This larva is produced asexually and while in the brooding chamber the larva already develops a new 'head' (buccal funnel), which is later used in the feeding stage (Kristensen and Funch 2002, fig. 11.2).

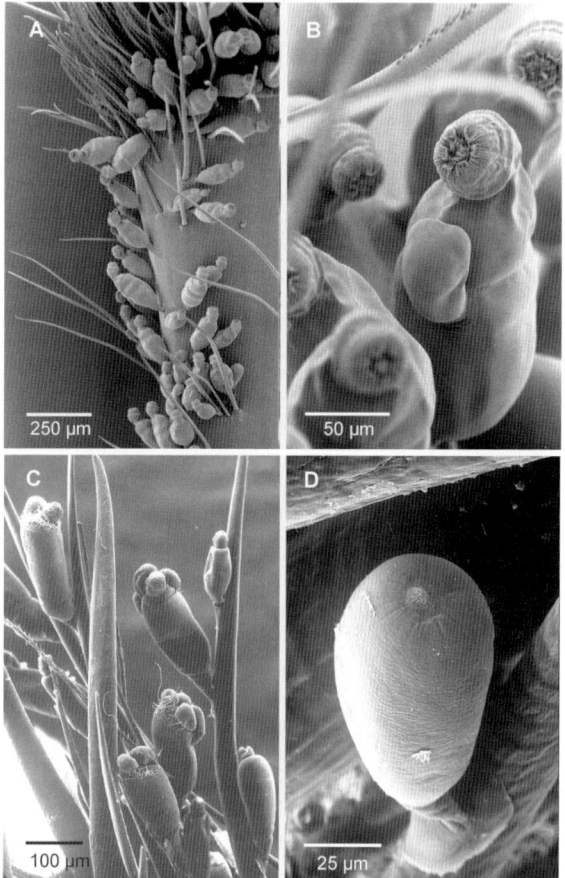

Figure 5.20 Scanning electron micrographs of cycliophorans. **A.** *Symbion pandora* from Kaldbak Fjord, Faroe Islands, collected from a Norway lobster. The colony consists of young feeding stage individuals. **B.** Sessile feeding stage individual of *S. pandora* with attached Prometheus larva. The lobster was the same as in Fig. 5.20A. **C.** An undescribed species of *Symbion* from an American lobster (Maine, USA). The colony consists of old feeding stage individuals with several attached Prometheus larvae. **D.** A newly settled Pandora larva of *Symbion* from an American lobster (Maine, USA).

The Pandora larva leaves the maternal animal when the old buccal funnel is cast off. Consequently, the maternal individual can successively produce more larvae and thereby produce many offspring (Funch and Kristensen 1999). The larva moves around somewhat before settling close to the maternal individual with its head down: the new buccal funnel being formed at the rear end (Fig. 5.20D). The Pandora larva is now sessile and grows in the same way as described for the mother animal. It can then produce its own Pandora larva and thereby the asexual cycle can continue and produce more sessile feeding animals.

The sexual reproductive cycle (Kristensen 2002a) begins at the end of the lobster's moulting cycle. The cycle starts with the sessile animal developing a Prometheus larva or a female in the same way as the Pandora larva (Obst and Funch 2003). The female resembles a Pandora larva, which has become sexually mature (neoteny) containing a large oocyte (Funch and Kristensen 1997, 1999). The Prometheus larva leaves the brooding chamber and settles on a sessile feeding individual, which starts producing a female inside its brooding chamber. The Prometheus larva develops one to several dwarf males by internal budding (Obst and Funch 2003). These dwarf males consist mainly of a large brain, a cuticular penis and a cluster of spermatozoa. Fertilisation of the female is internal: the penis from one of the dwarf males penetrates the chamber of the sessile individual, where the female is located.

The fertilised female escapes the sessile feeding animal and may settle on the same seta as the maternal individual or further away on the palp of the mouth limb. The female also settles head down by secreting a kind of cement from numerous glands on the head forming an adhesive disc. The fertilised egg develops into a new type of larva, the Chordoid larva, eating the female, which becomes cyst-like. The Chordoid larva hatches from the cyst of the female. The larva has well-developed locomotory cilia, which it uses to swim away, dispersing to a new host. It settles on a new host, a lobster, and metamorphoses into a sessile feeding individual.

Important references
Funch, and Kristensen (1995) established the new phylum, Funch and Kristensen (1997, 1999), Kristensen (2002a,b), and Kristensen and Funch (2002) reviewed the knowledge of the group. Investigations on the phylogenetic position of the Cycliophora are by Winnepenninckx et al. (1998) and Giribet et al. (2004). Obst and Funch (2003) described the dwarf male of *Symbion pandora*. Funch (1996) showed that the chordoid larva of *Symbion pandora* is a modified trochophore.

Nemertea (ribbon worms)
Kirsten Jensen and Patricia S Sadeghian

Introduction
Ribbon worms are characterised by a ciliated epidermis and an eversible proboscis. The phylum contains more than 1100 species and about 250 genera (Gibson 1995, Thollesson and Norenburg 2003). Most species are found in benthic and pelagic marine habitats, a few species in freshwater or on land. Traditionally, the Nemertea have been divided into four subclasses: Palaeonemertea, Heteronemertea, Hoplonemertea and the Bdellonemertea. Based on molecular sequence data, Thollesson and Norenburg (2003) showed the monogeneric Bdellonemertea to be hoplonemerteans and the palaeonemerteans to be paraphyletic. To date, at least 40 species have been reported to live in symbiotic relationships with other organisms. Most of these symbiotic nemertean species are hoplonemerteans; only two records exist of symbiotic heteronemerteans. In all cases the associations are with marine invertebrates. Reports of up to ten additional undescribed symbiotic species exist in the literature (see e.g. Wickham and Kuris 1985, Sadeghian and Kuris 2001). The true nature of the symbiotic association in many cases is unknown. In early descriptions, most symbiotic nemertean taxa were considered to be true parasites (Coe 1902, Humes 1942), but most are now considered to be commensals, parasites or specialised egg predators (Berg and Gibson 1996).

Hoplonemertea
Within the Hoplonemertea, most symbiotic species belong to the families Malacobdellidae, Carcinonemertidae, Tetrastemmatidae and Emplectonematidae (see Thollesson and Norenburg 2003). However, the familial affiliations of several taxa are unclear.

Malacobdellidae
The Malacobdellidae contains a single genus, *Malacobdella*, with six valid species (Gibson 1995, Ivanov et al. 2002) (Table 5.1). Species of *Malacobdella* are generally considered to be endocommensals, although Sundet and Jobling (1985) argue for a parasitic life style. They inhabit the mantle cavity of bivalves predominantly of the subclass Heterodonta, but also Protobranchia

and Pteriomorphia. Five species have been reported from a single host species, whereas the type species, *Malacobdella grossa*, has been reported from at least 27 species of bivalves (in 20 genera, 13 families, 3 orders, 2 subclasses). The family has a global distribution occurring in the Eastern and Western Atlantic, and the Eastern Pacific Ocean, north of 30°N and south of 30°S latitude.

Malacobdellans possess a short, ventral, flattened, leech-like body. Their anterior end is notched and the posterior end bears a single, ventral, large muscular sucker. They possess a pharynx with pharyngeal papillae. The proboscis is unarmed with the stylet armature lacking. Malacobdellans are dioecious (Gibson 1968). Species of *Malacobdella* filter feed through a ciliary feeding mechanism with pharyngeal papillae trapping the food (Gibson and Jennings 1969). They ingest microscopic organisms (e.g. small algae, bacteria, protozoans) and possibly larger organisms (e.g. small copepods, nauplii) (Gibson and Jennings 1969). Typically only one adult worm is present per host clam (Gibson 1982), and, at least for *M. grossa*, a longevity of up to 20 years has been suggested (Sundet and Jobling 1985).

Carcinonemertidae

Carcinonemertids are symbiotic on decapod crustaceans. Historically, four genera (*Carcinonemertes, Ovicides, Alaxinus* and *Pseudocarcinonemertes*) have been considered to belong to this family. However, Uhazy *et al.* (1985) and Campbell *et al.* (1989) have suggested that *Pseudocarcinonemertes* does not belong in this family, and the taxonomic position of *Alaxinus* remains uncertain. Species in the genera *Carcinonemertes* and *Ovicides* are considered ectocommensal specialised egg predators of decapod crustaceans. Adult worms are primarily found on the eggs of ovigerous females, whereas, depending on the species, juveniles can be found on the exoskeleton and gills of non-ovigerous females or the exoskeleton and gills of male crabs (Shields and Kuris 1990, Kuris 1993).

Twelve species of *Carcinonemertes* are recognised (Sadeghian 2003) (Table 5.1). Of these, seven (i.e. *C. australiensis, C. coei, C. errans, C. humesi, C. pinnotheridophila, C.* sp. A and *C. wickhami*) have each been reported from a single host species, and one, *C. regicides*, occurs on two host species, a lithodid and a majid crab; *C. mitsukurii* has been reported from five species of brachyuran crabs (in 3 genera, 2 families); *C. carcinophila imminuta* (awaiting elevation to species level) from 17 species of brachyuran crabs (in 11 genera, 5 families); *C. epialti* from 11 species of brachyuran crabs (in 6 genera, 5 families); *C. carcinophila*, the type species, from 17 species of brachyuran crabs (in 11 genera, 6 families) and from one species of anomuran crab. At least six undescribed species have been reported (Wickham and Kuris 1985, Sadeghian and Kuris 2001), including reports from new host families, Hippidae (see Subramonian 1970 in Wickham and Kuris 1985) and Atelecyclidae (see Wickham and Kuris 1985). *Ovicides* is monotypic, with *O. juliae* found on the brachyuran crabs *Chorodiella nigra* and *C. xishaensis* (Xanthidae).

Carcinonemertids are dioecious, rarely hermaphroditic, and are capable of parthenogenesis (Roe 1986, Shields *et al.* 1989). They feed on the eggs of their host by piercing the egg coat with their stylet and subsequently ingesting yolk. Life cycles of carcinonemertids range from simple (e.g. male and female worms of *C. regicides* occur only on the egg mass of ovigerous female crabs throughout the year) (Fig. 5.21), to complex (e.g. mature male and female worms of *C. epialti* live on the egg mass of ovigerous female crabs, while juvenile worms (Fig. 5.22) ensheath on the exoskeleton of juvenile and non-ovigerous females and on juvenile and adult male crabs) (see Shields and Kuris 1990, Kuris 1993). In all cases, however, male and female worms require a reproductive female crab for nutrition, maturation and reproduction.

Tetrastemmatidae

Four genera of tetrastemmatids have been reported to live in association with other invertebrates: *Tetrastemma, Asteronemertes,* and the two monotypic genera *Amphinemertes* and

Table 5.1   Symbiotic monostiliferan Hoplonemertea (Nemertea)

Nemertean taxon	Host taxon
Monostilifera	
Malacobdellidae	
Malacobdella (**endocommensal or parasite**) (see e.g. Gibson 1968, Gibson & Jennings 1969, Sundet & Jobling 1985)	
M. arrokeana	Bivalvia: Heterodonta: Hiatellidae
M. grossa	Bivalvia: Heterodonta & Pteromorphia
M. japonica	Bivalvia: Heterodonta: Mactridae
M. macomae	Bivalvia: Heterodonta: Tellinidae
M. minuta	Bivalvia: Protobranchia: Yoldiidae
M. siliquae	Bivalvia: Heterodonta: Pharidae
Carcinonemertidae	
Carcinonemertes (**ectocommensal/egg predator**) (see e.g. Kuris 1993)	
C. australiensis	Decapoda: Palinura: Palinuridae
C. carcinophila	Decapoda: Brachyura & Anomura
C. coei	Decapoda: Brachyura: Portunidae
C. epialti	Decapoda: Brachyura
C. errans	Decapoda: Brachyura: Cancridae
C. humesi	Decapoda: Brachyura: Majidae
C. carcinophila imminuta	Decapoda: Brachyura
C. mitsukurii	Decapoda: Brachyura
C. pinnotheridophila	Decapoda: Brachyura: Pinnotheridae
C. regicides	Decapoda: Brachyura & Anomura
C. wickhami	Decapoda: Palinura: Palinuridae
C. sp. A (see Sadeghian & Kuris 2001, Sadeghian 2003)	Decapoda: Brachyura: Leucosiidae
Ovicides (**ectocommensal/egg predator**) (see e.g. Sadeghian 2003)	
O. juliae	Decapoda: Brachyura: Xanthidae
Tetrastemmatidae	
Amphinemertes (**commensal/parasite ?**) (see Roe 1988, Gibson 1995)	
A. caeca	Urochordata: Ascidiacea
Asteronemertes (**commensal**) (see e.g. Gibson 1995)	
A. commensalus	Echinodermata: Asteroidea: Solasteridae
A. gibsoni	Echinodermata: Asteroidea: Solasteridae
Pseudocarcinonemertes (**ectocommensal/egg predator**) (see e.g. Kuris 1993)	
P. homari	Decapoda: Stenipodidea: Nephropidae
Tetrastemma (**symbionts**) (see e.g. Gibson 1995)	
T. flavidum	Urochordata: Ascidiacea
T. fozensis	Bivalvia: Heterodonta: Semelidae
T. kefersteinii	Urochordata: Ascidiacea
T. marionis	Urochordata: Ascidiacea
T. suhmi	Crustacea: Brachyura: Grapsidae
T. vittigerum	Urochordata: Ascidiacea

Table 5.1 Symbiotic monostiliferan Hoplonemertea (Nemertea) (Continued)

Nemertean taxon	Host taxon
Emplectonematidae	
***Coenemertes* (parasite ?)**	
C. caravela	Decapoda: Anomura: Callianassidae
(*Dichonemertes hartmanae*) **(free-living ?)**	Decapoda: Anomura: Upogebiidae
(*Emplectonema kandai*) **(free-living ?)**	Urochordata: Ascidicea
***Nemertopsis* (endocommensal/parasite ?)** (see e.g. Roe 1988)	
N. quadripunctatus	Cirripedia: Thoracica: Pollicipedidae & Lepadidiae
N. tetraclitophila	Cirripedia: Thoracica: Tetraclitidae
Uncertain familial status	
***Alaxinus* (ectocommensal/egg predator)** (see e.g. Gibson et al. 1990)	
A. oclairi	Decapoda: Anomura: Lithodidae
***Cryptonemertes* (ectocommensal)** (see e.g. Gibson 1986)	
C. actinophila	Anthozoa: Actiniaria: Actinostolidae & Hormathiidae
***Gononemertes* (endocommensal)** (see e.g. Gibson 1974)	
G. australiensis	Urochordata: Ascidicaea: Pyuridae
G. parasita	Urochordata: Ascidiacea: Ascidiidae
***Oerstedia* (symbionts ?)**	
O. rustica	Urochordata: Ascidiacea

Pseudocarcinonemertes. The genus *Tetrastemma* is comprised of 107 species (Gibson 1995), six of which have been reported to live in association with another invertebrate (symbionts in Table 5.1): *Tetrastemma flavidum*, *T. kefersteinii*, *T. marionis* and *T. vittigerum* live in the branchial or mantle cavity of tunicates (Roe 1988); *T. fozensis* lives in the mantle cavity of the bivalve *Scrobicularia plana*; and *T. suhmi* occurs on the abdomen of the grapsid crab *Planesminutus* in the Sargasso Sea (Gibson 1995). The two species of *Asteronemertes* have been reported from the ambulacral grooves of seastars, that is *A. commensalus*, from *Crossogaster papposus* from the Sea of Okhotsk, and *A. gibsoni* from *Solaster pacificus* from the Pacific coast of Russia (Gibson 1995).

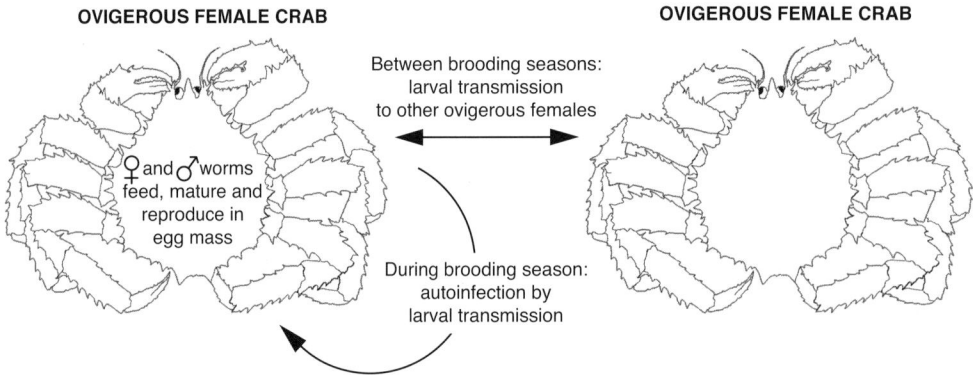

Figure 5.21 Life cycle of *Carcinonemertes regicides* on the red king crab, *Paralithodes camtschatica*. Note, arrows indicate movement of worms. After Kuris (1993).

Figure 5.22 Photograph of immature specimen of *Carcinonemertes australiensis*. Abbreviations: R, rhynchodeum; E, eye; C, cerebral ganglion. Scale bar = 1 mm. Courtesy of R Gibson.

Amphinemertes lives in association with tunicates dredged from Kodiak Island, Alaska (Gibson 1995). Like *Carcinonemertes*, *Pseudocarcinonemertes* is an ectocommensal egg predator. It is exclusively found on the eggs and gills of the American lobster *Homarus americanus* (Nephrodidae), on the Atlantic coast of Canada (Fleming and Gibson 1981). The life cycle of *P. homari* is similar to that of carcinonemertids.

Emplectonematidae
Two genera in this family are considered to be symbionts. The monotypic genus *Coenemertes* has been reported 'gliding among the thoraxic legs … , near the branchial chamber' (Corrêa 1966, p. 366) of the Ghost shrimp *Callianassa* sp. Of the six or seven valid species of *Nemertopsis*, two species have been reported from the mantle cavity, rarely the outer surfaces, of barnacles from the Banda Sea, Hong Kong and Japan (Gibson 1995). Roe (1988) suggested that *N. quadripunctatus* might be a true commensal or parasite. Two additional species have been reported in loose association with another invertebrate (*Dichonemertes hartmanae* associated with burrows of mud shrimp and *Emplectonema kandai* on tunicates).

Genera of uncertain familial status
Five species in four genera are considered to live in association with other invertebrates: species in the monotypic genera *Alaxinus* and *Cryptonemertes*, both species of *Gononemertes* and one of the 19 species of *Oerstedia*. The familial placement of these genera is currently unclear.

Alaxinus oclairi is an ectocommensal egg predator on the eggs of the anomuran crab *Paralithodes camtschatica* (Lithodidae). *Cryptonemertes actinophila* occurs as a potential commensal

Table 5.2 Symbiotic Heteronemertea (Nemertea)

Nemertean taxon	Host taxon
***Nemertoscolex* (endoparasite ?)** (Berg & Gibson 1996)	
N. parasiticus	Echiura: Echiuridae
***Uchidana* (endoparasite ?)** (Roe 1988)	
U. parasita	Bivalvia: Heterodonta: Macridae

beneath the pedal disk of five different species of sea anemones (in three genera, three families) (Gibson 1986). Both species of *Gononemertes* are endocommensals within ascidians. Specifically, *G. australiensis* has been reported from the liver or anus, in the atrium under the pharynx, or among liver and gonad tissue (Gibson 1974). Each species has been reported from two host species, in two different families of ascidians (Table 5.1). Finally, one of 19 species of *Oerstedia* recognised by Gibson (1995) is found 'with ascidians' (p. 524).

Heteronemertea

Only two heteronemertean worms have been documented in association with other invertebrates (Table 5.2). *Nemertoscolex parasiticus* was reported from the coelomic fluid of the echiuran *Echiurus echiurus* collected on the west coast of Sweden (Berg and Gibson 1996). It may be a true endoparasite. *Uchidana parasita* was described from the mantle cavity, but also from 'the spaces between the shell and mantle' (p. 135) of the bivalve *Mactra sulcataria* from the Mie Prefecture in Japan (Iwata 1967). *Uchidana parasita* may be a true ectoparasite, since the species appears to ingest host gill tissue (Iwata 1967).

Economic importance

Nemertean egg predators (i.e. species of *Carcinonemertes, Ovicides, Alaxinus* and *Pseudocarcinonemertes*) have the potential for great ecological or economic impact on their host populations. Egg mortality arising from these infections have been documented from several host populations including the Dungeness crab, *Cancer magister*, and the Tanner crab, *Chionoecetes bairdi* (see e.g. Kuris and Wickham 1987), the Yellow shore crab, *Hemigrapsus oregonensis* (see e.g. Shields and Kuris 1988), the American lobster, *Homarus americanus* (see e.g. Fleming and Gibson 1981) and the Red king crab, *Paralithodes camtschatica* (see e.g. Kuris et al. 1991). Widespread outbreaks of *C. errans* and *C. regicides* occurred in the 1970s and 1980s resulting in significant damage to both the Dungeness and Red king crab fisheries (Wickham 1980, Kuris et al. 1991). Due to the massive presence of these egg predators (mean intensities up to 1000 worms per pleopod), many crab hosts were effectively castrated, losing all of their eggs (Kuris et al. 1991). Kuris (1993) suggested significant egg loss on a population level due to egg predation and/or increased grooming behaviour of hosts as a result of infection. At least for *C. regicides*, outbreaks were associated with hydrographic features that contributed to the retention of the infectious larvae in the system (Kuris et al. 1991).

Important references

An older account of nemertean parasites of crabs is by Coe (1902). Comprehensive more recent taxonomic accounts are those by Gibson (1982, 1995). Humes (1942) reviewed the morphology, taxonomy and bionomics of the nemertean genus *Carcinonemertes*. Important contributions to nemertean life cycles are by Kuris (1993), and to nemerteans as egg predators and their ecology by Wickham and Kuris (1985), Kuris and Wickham (1987), Roe (1988) and Kuris et al. (1991). Thollesson and Norenburg (2003) discussed phylogenetic relationships.

Rotifera and *Seison* (rotifers)
Wilko H. Ahlrichs

Introduction
Most rotifers occur in freshwater, and fewer than 5% of the about 2000 species are marine. Very few are parasitic. The genus *Seison* was formerly included in the Rotifera. However, recent ultrastructural studies have shown that it should not be included in that group (Ahlrichs 1995).

Rotifera

Morphology and diversity
Females are about 100 µm to 200 µm long, males of most species are not known. Those that are known are generally much smaller than females and usually lack a digestive system and some other organs. The body of the female is elongated and consists of three regions: head including the rotatory organ, body or trunk, and foot (Fig. 5.23). Each region shows characteristic folds, the so-called pseudosegments. These folds function like joints, but genuine segmentation is not present.

Life cycle
In Bdelloidea no males are present at all. Parthenogenesis is common in rotifers; females produce diploid amictic ('without mixing') eggs by mitosis that develop into diploid females. During unfavourable periods, females may produce haploid mictic eggs that *if not* fertilised, develop into males. If these mictic eggs *are* fertilised, they become resistant eggs that can remain dormant.

A parasitic life has been suggested for *Zelinkiella synaptae*, *Albertia crystallina*, *Albertia naidis*, *Proales paguri*, and *Proales gonothyraeae*.

Zelinkiella synaptae (Bdelloidea) (150–200 µm long) has a rotatory organ that has two separate trochoal discs. It is viviparous and lives on the body surface (tentacles) mainly of the sea cucumbers *Synapta digitata* and *S. inhaerens* (Holothuria, Echinodermata) but is also found on *Amphithrite* (Annelida). It is not known if *Z. synaptae* harms its hosts, it may be commensal (Remane 1929b).

Albertia naidis (Monogononta) (Syn. *A. intrusor*; *A. soyeri*) (94–340 µm long) is oviparous and cosmopolitan, and parasitic in freshwater oligochaetes. Levander (1894) found this species in *Stylaria lacustris* in brackish water in the far eastern Baltic Sea. It has also been found in *Nais elinguis* from brackish water but has not been reported from saltwater habitats (Remane 1929b, De Smed 1996).

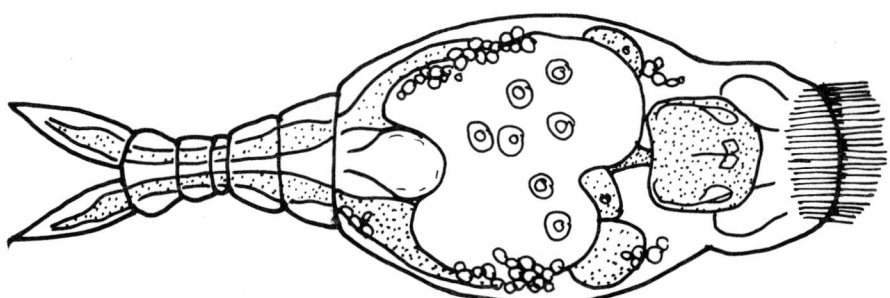

Figure 5.23 *Proales paguri*, female, dorsal view. Redrawn and modified after Thane-Fenchel (1966).

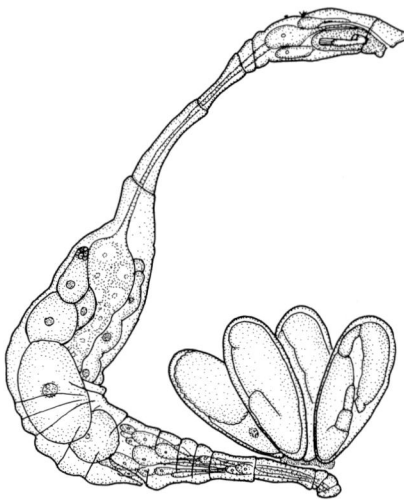

Figure 5.24 *Seison annulatus*, female. Redrawn and modified after Koste (1975).

Albertia crystallina (220 µm long) lives attached to the intestinal epithelium of brackish or marine *Paranais litoralis* (Oligochaeta, Annelida) in the Baltic Sea (?). It has not been found again since its first description in 1851. The description is insufficient (Remane 1929b, De Smed 1996).

Proales paguri (Monogononta) (199–213 µm long) (Fig. 5.23) lives on the gills of the hermit crab, *Eupagurus bernhardus,* and apparently feeds on the gill epithelium of its host (i.e. it is a genuine parasite) (Thane-Fenchel 1968, De Smed 1996).

Proales gonothyraeae, (250–300 µm long), has an almost ventral rotatory organ. It lives within the theca of the hydroid polyp, *Laomedea loveni. Proales. gonothyraeae* may be ectoparasitic, but whether it harms the host is unclear (Remane 1929a, De Smed 1996).

Seison

Morphology and diversity

The taxon *Seison* consists of two dioecious species, *Seison nebaliae* (2000 µm long) and *S. annulatus* (1100 µm long). Both species have a head, a long and narrow telescopic neck, a ventrally bent trunk, and a foot with several pseudosegments and ending in an adhesion disc (Fig. 5.24). They live on the surface of leptostracan crustaceans (e.g. *Nebalia bipes*). It seems that only older hosts with a carapax larger than 6 mm are chosen. *Seison. nebaliae* lives on the entire surface of the crustaceans, whereas *S. annulatus* seems to prefer the gills of the thoracopods below the carapax. Ultrastructural findings indicate that *S. annulatus* feeds on haemolymph of the host, whereas *S. nebaliae* feeds on bacteria (Ahlrichs 1995).

Life cycle

Males and females are of equal size. Parthenogenetic reproduction, common in rotifers, is absent. The eggs are glued to the body surface of the host (Koste 1975, Ahlrichs 1995, 2003).

Important references

Ahlrichs' (1995) electron microscopic studies suggest that *Seison* is not a rotifer. Koste (1975) described *Seison annulatus* in some detail. Remane (1929a) described *Proales gonothyraeae*, a parasitic rotifer on hydroied polyps. Comprehensive accounts of rotifers are by Remane (1929b) and De Smet (1996).

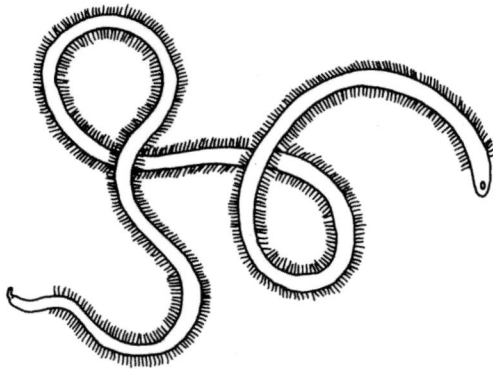

Figure 5.25 General organisation of a male *Nectonema agile*. The posterior end is bent 90° in relation to the remaining body. Note natatory bristles along the body. After Fewkes (1883).

Nematomorpha (horse-hair worms)
Andreas Schmidt-Rhaesa

Introduction
The Nematomorpha have a superficial resemblance to nematodes. They are parasites of arthropods which complete their development within hosts but emerge from these for reproduction. About 300 species have been described. Most (taxon Gordiida) reproduce in freshwater and parasitise terrestrial hosts, but the genus *Nectonema* with five species is marine.

Morphology and diversity
Species of *Nectonema* show sexual dimorphism. Females are generally larger than males and have a round posterior end, whereas the posterior end of males is tapering and curved at an angle of 90° towards the ventral side (Figs 5.25, 5.26). The smallest body length reported is 10 mm (male of *N. melanocephalum*), the maximum length is 960 mm (female of *N. munidae*). Average lengths are 100 mm to 150 mm for males and around 400 mm for females. Males are almost transparent and generally much more mobile than females, while females are opaquely white in colour. Some species have darker pigments in some body regions. Characteristic for *Nectonema* is a ventral and a dorsal row of cuticular bristles (Figs 5.25, 5.26, 5.28, 5.29). On close view, it consists of two parallel rows. The function is likely to support a pelagic lifestyle, therefore they are named natatory bristles. The body is covered with a cuticle which is moulted once

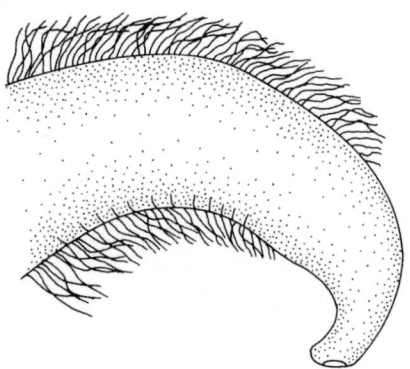

Figure 5.26 Male posterior end of *Nectonema agile*. After Beattie (1987).

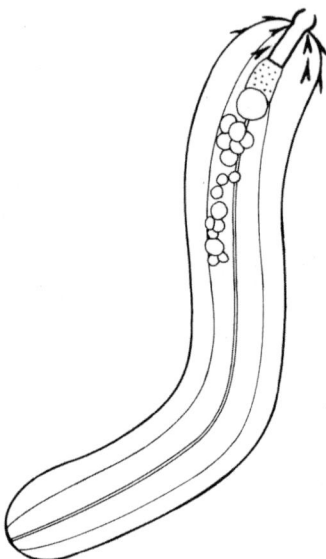

Figure 5.27 Larva of *Nectonema munidae*. After Huus (1932).

during the parasitic phase. Internally, the reproductive system is dominant and adult worms are almost completely filled with gametes. The head region is separated from the remaining body by a solid muscular septum. Within the head is a cavity which includes up to four giant cells with unknown function. Ultrastructural studies have shown a connection to the nervous system and the cells might be sensory in function (Schmidt-Rhaesa 1996a). Nutrients are consumed during the parasitic phase through the integument and probably additionally through the intestinal system (Skaling and MacKinnon 1988). In adults, the intestine is non-functional and ends blindly.

Five species and some unidentified specimens have been reported from several locations. The most abundant species is *Nectonema agile*. It was described and reported several times from the northwest Atlantic, mainly from the region around the marine station of Woods Hole (Massachussetts, USA) and from the Bay of Fundy, New Brunswick, Canada. Additionally, specimens

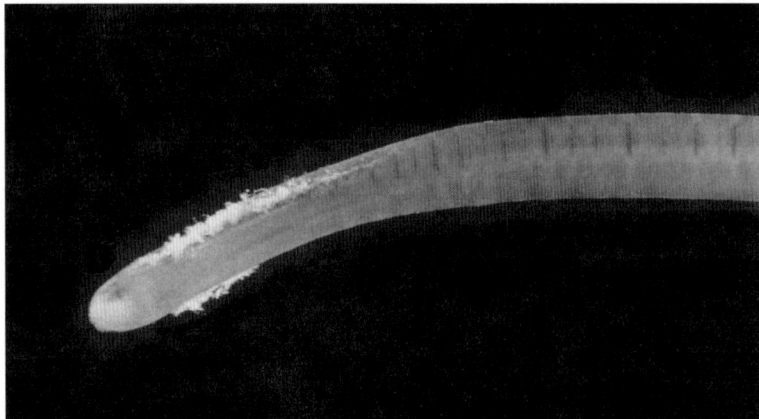

Figure 5.28 Anterior end from a specimen of *Nectonema agile* from the Mediterranean Sea, close to Naples. From museum specimen, Museum für Naturkunde, Berlin, accession no. 5284).

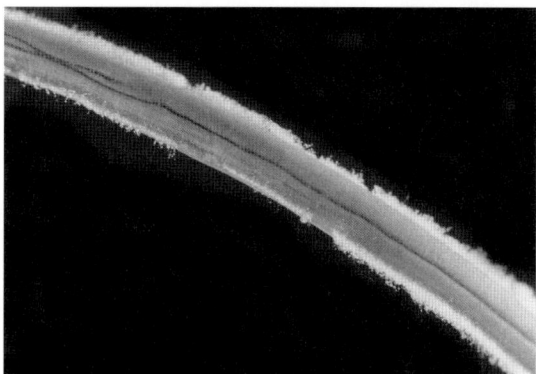

Figure 5.29 Midbody region of same specimen as Figure 5.28. Note natatory bristles along dorsal and ventral line.

from the northeast Atlantic (near Bretagne, France, and in the Mediterranean Sea) and the Black Sea are also assigned to *N. agile*. Regionally restricted, but collected several times is *N. munidae* from a few fjords near Bergen, Norway. Three species have been found only once: *N. melanocephalum* from near the Indonesian islands (Nierstrasz 1907), *N. svensksundi* from Svalbard (Bock 1913) and *N. zealandica* from the South Island of New Zealand (Poinar and Brockerhoff 2001). Additional undetermined records come from Western Greenland (Nouvel and Nouvel 1938), Northern Norway (Bakke 1975) and Japan (Oku et al. 1993).

Species resemble each other in general but differ in size, pigmentation and the presence or absence of septum and giant cells. Because septum and giant cells appear late in the development of *N. munidae* (see Schmidt-Rhaesa 1996a), their partial or complete lack in species such as *N. svensksundi* and *N. melanocephalum* might be due to immature specimens.

Life cycles

Life cycles are incompletely known. *Nectonema* specimens are most often found in the body cavity of their host as late, worm-like juveniles. Rarely have they been caught in the sea where they were found close to the surface. In two cases, specimens were kept in the laboratory and some details on reproduction and early larval development were observed (Huus 1932 and a Masters thesis by Beattie 1987, some details from this thesis can be found in Schmidt-Rhaesa 1999). Sperm transfer is direct, the male inserts its complete posterior end into the female genital opening. From eggs hatch microscopic larvae with two rings containing six spines each and two central hooks (Fig. 5.27). This roughly resembles the larvae of freshwater Gordiida which have three rings of hooks and three central stylets. These structures are used for host infection which can also be assumed for the larva of *Nectonema*. Whereas in Gordiida, a paratenic host is often present, there is no such evidence in *Nectonema* and the final host may be infected directly. Life cycle is most mysterious in *N. munidae*, because reproduction takes place close to the water surface, but hosts (mostly *Munida* species) live benthic in the more than 700 m deep fjords. Probably larval, planktonic stages of the crustaceans are already infected by larvae of *N. munidae*.

All hosts of *Nectonema* species are decapod crustaceans. A summary has been given by Poinar and Brockerhoff (2001). A variety of decapods are parasitised, ranging from shrimp-like forms to galatheid crabs, hermit crabs and brachyuran crabs. Rates of parasitisation seem to be generally low. For *N. munidae*, a prevalence of less than 10% has been reported, in *N. agile* it ranges from 0.5% to 50% (summary in Arvy 1963), and in *N. zealandica* 12.7% of the investigated crabs are parasitised (Poinar and Brockerhoff 2001). Parasitisation is periodic and varies throughout the year (Poinar and Brockerhoff 2001). Usually one single specimen of *N. zealandica* was found per crab, but multiple parasitations (up to four specimens) also occur (Poinar and Brockerhoff 2001).

Important references
There are only about 40 publications dealing with *Nectonema*. Early morphological investigations are by Bürger (1891), Ward (1892), Nierstrasz (1907), Bock (1913) and Feyel (1936). Ultrastructural details have been reported by Bresciani (1975, 1991), Skaling and MacKinnon (1988) and Schmidt-Rhaesa (1996a,b, 1997, 1998). Summaries of hosts and data on parasitisation rates can be found in Arvy (1963), Nielsen (1969), Leslie et al. (1981), Brattey et al. (1985) and Poinar and Brockerhoff (2001).

Acari (mites and ticks)
Jacek Dabert

Introduction
Mites and ticks (subclass Acari) are members of the class Arachnida (Evans 1992). Forty thousand species have been described, but there may be more than one million (Walter and Proctor 1999). Therefore, mites may be among the most species-rich groups of animals, after insects and possibly nematodes. The main and most important evolutionary novelty of Acari – an extreme body miniaturisation – allows them to exploit many niches that most other arthropods cannot use. The group is primarily terrestrial, but many mites live in fresh or salt water.

All ticks are blood-sucking parasites of vertebrates, attacking mammals, birds, reptiles and occasionally amphibians. In contrast, mites utilise many food resources as predators, phytophages, fungivores, saprophages, ectoparasites and, more rarely, endoparasites of invertebrate and vertebrate hosts, including many marine animals.

Systematics and morphology
It is not certain that the two main lineages of the class Acari, superorders Actinotrichida (Acariformes) and Anactinotriochida (Opilioacariformes + Parasitiformes) are monophyletic. Parasitic mites of marine animals belong to both lineages (i.e to the four orders: Astigmata and Prostigmata) (Acariformes), and Ixodida and Mesostigmata (Parasitiformes). Members of the clade of mites called 'ticks' (Ixodida, Metastigmata) are much larger than other mites (to over 1 cm), and because of their leathery, stretchy integument may reach 3 cm when engorged (Evans 1992). The movable gnathosoma is adapted for blood sucking and has characteristic anteriorly projecting saw-like hypostome and chelicerae with outwardly directed teeth modified for cutting skin. A complex chemosensory apparatus, Haller's organ, is situated dorsally on the tarsus of the first leg pair. These sense organs together with various mechanoreceptory and olfactory setae on the palps enable ticks to locate hosts and attach to them.

Mites of the Mesostigmata (=Gamasida) are a large and diverse group adapted to many habitats and life strategies, including ectoparasitism and endoparasitism. They range in size from 0.2 cm to almost 2.5 cm. Modifications for parasitism concern mainly the chelicerae, in particular the degree of development of the fixed digit, dentation of both digits, and the relative lengths of the first and second segments (Evans 1992). Body modifications may also be pronounced (e.g. body elongation in some Halarachnidae parasitising pinnipeds) but many parasitic forms have a body plan similar to the predatory ones (e.g. haematophagous dermanyssids or macronyssids of birds).

Prostigmata (=Actinedida) is the most enigmatic taxon and needs fundamental revision (Evans 1992). It is the most heterogeneous mite order, comprising very diverse forms that use many habitats and food resources. About 70% of the Prostigmata are parasites or parasitoids of various animals includimg other mites (Kethley 1982). The biggest prostigmatid mites are water and velvet mites, which may be 0.5 cm to more than 1 cm long, but most are smaller than 1 mm. The body of parasitic forms may be greatly modified (e.g. in worm-like Demodecidae in hair follicles, or cigar-like Syringophilidae in feather quills). The chelicerae may be dentate-chelate but often, especially in parasitic forms, they are modified to long stilettos. Legs, especially in parasitic forms, may be greatly modified and one or more pairs may even be lost.

Most taxa within the Astigmata (=Acaridida) are associated with vertebrate and invertebrate animals and relatively few are exclusively free living. However, the most characteristic feature of astigmatid mites is the phoretic, highly modified deuteronymph (hypopus) that occurs in many free-living groups and is lost in more derived parasitic forms. The body size range is generally about 0.5 mm to 1 mm long. Astigmata are usually soft-bodied but the hypopi and many commensals and parasites (e.g. most pterolichoid and analgoid feather mites) are covered by strongly sclerotised shields. The primarily ovate body of free-living species may be strongly modified in parasitic ones. The most bizarre body shapes are displayed by feather mites (Fig. 5.30A–D) with, for example, hypertrophied particular body parts and/or legs, variously shaped opisthosomal lobes, leaf-like setae or fancily sculptured cuticular shields; males of some species are asymmetric. Chelicerae are generally not modified, chelate-dentate with exception of Histiostomatoidea in which chelicerae are modified to filter-feeding structures. In some parasites chelicerae and/or palpi may be hypertrophied. The pretarsi in most parasitic and commensal forms are modified into a membranous ambulacrum. Legs of parasitic forms may be variously modified (e.g. to clasping organs, Myocoptidae) or to hypertrophied legs with various apophyses in males of many feather mites.

Diversity of parasitic Acari of marine animals

Most parasitic Acari are associated with terrestrial or freshwater hosts. Examples of parasitic acarids of marine animals are briefly discussed below.

Parasites of invertebrates
Almost all parasitic mites of marine invertebrates belong to the prostigmatid family Halacaridae (Bartsch 1987) and are associated with hosts of several phyla. *Spongihalacarus longiscutus*, a halacarid mite with completely reduced palps, is a probable parasite of the sponge *Haliclona cymaeformis* (Otto 2000). Members of the genus *Bradyagaue* live on colonial sublittoral and bathyal hydrozoans (Cnidaria) and have hind legs adapted for grasping the stolons (Bartsch 1989). However, hypotheses about parasitic lifestyles of these mites (Walter and Proctor 1999) are not fully supported by recent investigations (I Bartsch, pers. comm. 24 Feb. 2004). *Parhalixodectes travei* is probably a sublitoral ectoparasite on *Cerebratulus hepaticus* (Nemertini) (Walter and Proctor 1999). *Halixodes chitonis* (Halixodinae) is a parasite of gills and mantle cavity of chitons (Mollusca) in New Zealand (Walter and Proctor 1999). Another mollusc, *Mytilus* (Bivalvia), is a host to two mite genera: the halacarid *Copidognathus* and astigmatid *Hyadesia* (Cáceres-Martímez et al. 2000). Halacarids are relatively common parasites of decapods (Crustacea). Two *Copidognathus* species are parasitic in the gill chambers of the slipper lobster *Parribacus antarcticus* and on the egg packets of the spider crab *Maja squinado* (Walter and Proctor 1999). The gill chambers of a marine decapod (*Peltarium spinulosum*) are infested by *Veladeadcarus gasconi*. *Enterohalacarus minutipalpis* (Enterohalacarinae), a mite with remarkably reduced palps, has been recorded as an internal parasite in digestive tracts of deep-sea urchins *Plesioderma indicum* (Echinodermata) in the Pacific (Bartsch 1987).

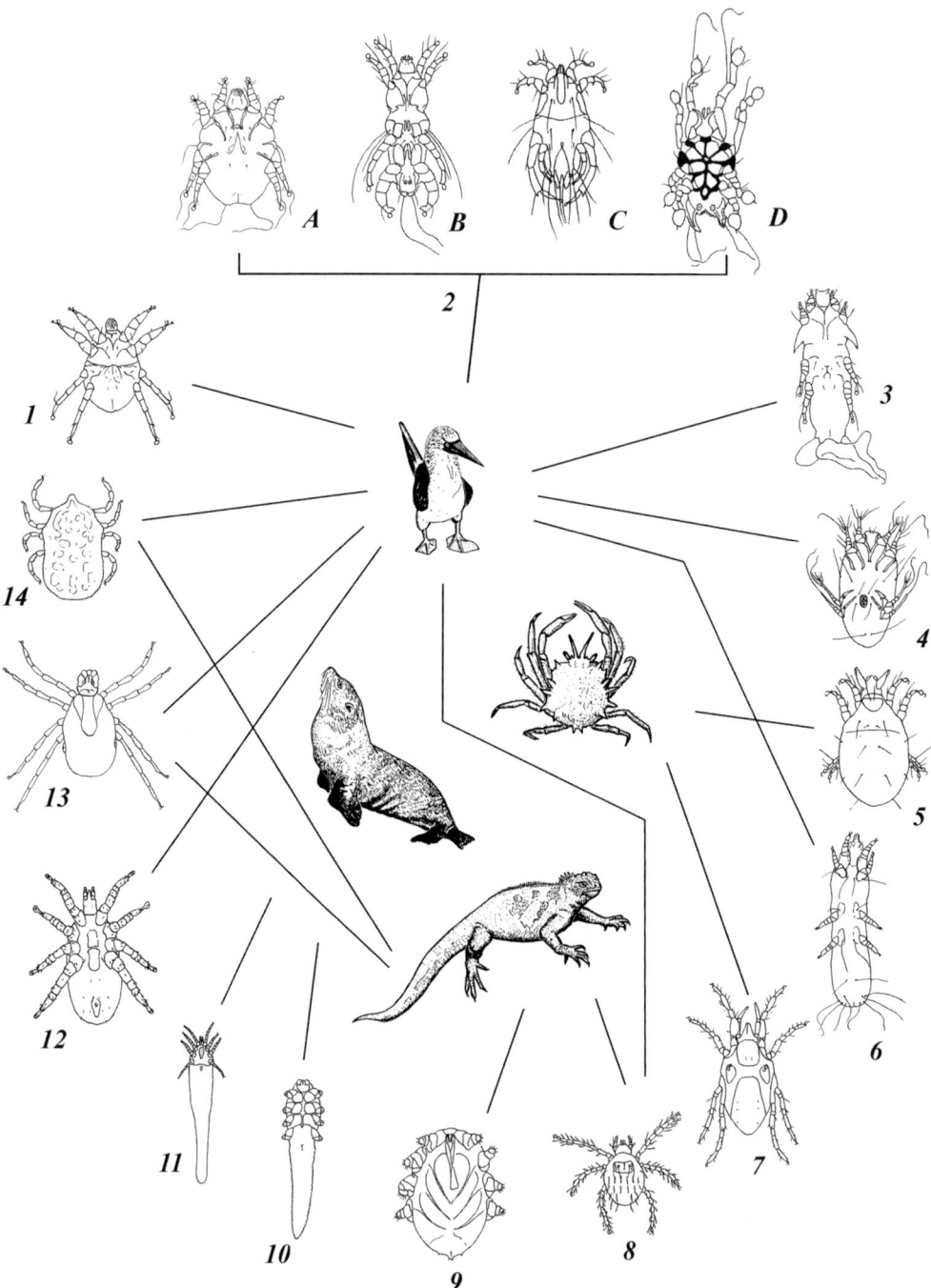

Figure 5.30 The biodiversity of parasitic Acari inhabiting marine invertebrates and vertebrates (mammals, birds and reptiles). Astigmata (1–5): **1.** Turbinoptidae. **2.** Feather mite superfamilies Analgoidea and Pterolichoidea represented by four main morphotypes inhabiting **A.** skin, **B.** quills, **C.** down feathers, and **D.** vane surface of contour feathers. **3.** Laminosiptidae (Fainocoptinae). **4.** Hypoderatidae. **5.** Hyadesidae. **6.–10.** Prostigmata: **6.** Syringophilidae. **7.** Halacaridae. **8.** Trombiculidae. **9.** Cloacaridae. **10.** Demodecidae. **11–12.** Mesostigmata: **11.** Halarachnidae. **12.** Rhinonyssidae. **13–14.** Ixodida. **13.** Ixodidae. **14.** Argasidae.

Parasites of vertebrates
Marine mammals, birds and reptiles are hosts of several unrelated mite groups: Halarachnidae and Rhinonyssidae (Mesostigmata); Ixodidae and Argasidae (Ixodida); Demodecidae, Syringophilidae, Trombiculidae and Cloacaridae (Prostigmata); Hypoderatidae and numerous families of feather mites (Astigmata). Records of mite infestation in fishes concern exclusively freshwater ones (Walter and Proctor 1999, Halliday and Collins 2002). Mites of hosts that spend at least part of their life in the open ocean or sea are discussed below.

Reptilia. Marine iguanas *Amblyrhynchus cristatus* (Iguanidae) are infected by hard ticks *Amblyomma* (Ixodidae) and soft ticks *Ornithodoros* (Argasidae) (Wikelski 1999). Trombiculid mites of the genus *Vatacarus* live in their nasal fossae, tracheal passages and lungs (Wikelski 1999). *Vatacarus* undergoes remarkable neosomy (cuticular growth within a stage) and feeds only as a larva (Krantz 1978, Walter and Proctor 1999). Another member of this genus (*V. ipoides*) is found in the same locations in sea snakes (Krantz 1978). Sea snakes are also hosts of other trombiculids (*Eutrombicula poppi*) (Krantz 1978). *Chelonacarus elongatus* (Cloacaridae) is found in the cloacal tissue of the sea turtle *Chelonia mydas* from the Atlantic coast (Pence and Scott 1998).

Aves. Terrestrial and aquatic birds are hosts of many acarids (Proctor and Owens 2000). Ticks belong to the most abundant and harmful ectoparasites. *Ixodes uriae* (Ixodidae) is the most common parasite of many sea birds in the northern and southern circumpolar regions and has one of the widest distributions of any tick species (McCoy and Tirard 2002). In the northeast Atlantic it infects dense colonies of cliff-nesting sea birds, (e.g. the common guillemot, *Uria aalge*, and the black-legged kittiwake, *Rissa tridactyla*) (Danchin 1992, Barton et al. 1995). It commonly infects also many penguin species (Sphenisciformes), especially chicks. *Haemaphysalis leporispalustris* is another hard tick infesting marine birds (Laridae) (Hyland et al. 2000). Soft ticks of the genus *Ornithodoros* (Argasidae) are largely restricted to the Western Hemisphere and some are truly cosmopolitan, parasitising many birds, including marine species (Keirans et al. 1980, Wikelski 1999). Rhinonyssidae (Mesostigmata) are other blood-sucking ectoparasites that inhabit the nares of many birds, including marine ones (*Larinyssus* on gulls, *Rhinonyssus* on marine ducks) (Butenko 1975, Butenko and Staniukovich 2001).

Bird plumage, skin and subcutaneous tissue are infested by numerous mite parasites of the orders Astigmata and Prostigmata. The deuteronymphs of Hypoderatidae live under the skin of many terrestrial birds and are also found in marine birds (many Pelecaniformes and Charadriiformes) (OConnor 1985, Pence and Hoberg 1991, Pence and Cole 1995). The prostigmatid Syringophilidae inhabit feather quills of many bird taxa and feed on soft tissue fluids by piercing the quill wall with long chelicerae (Kethley 1971). Marine birds are parasitised by the genera *Creagonycha*, *Philoxanthornea* (on Lari), *Syringonomus* (on Procellariiformes) and *Stibarokris* (Pelecaniformes). The prostigmatid, *Eutrombicula orlovensis*, is known from gulls (Kudryashova 1998).

Two astigmatid superfamilies, Analgoidea and Pterolichoidea, are the most diverse mite groups inhabiting plumage and skin of all avian orders, excluding penguins (Gaud and Atyeo 1996). Feather vanes of large contour feathers are inhabited by analgoid Alloptidae, Avenzoariidae, Psoroptoididae and pterolichoid Freyanidae, Kramerellidae, Ptiloxenidae (on Gaviiformes, Procellariiformes, Pelecaniformes, Charadriiformes, Podicipediformes and Anseriformes). In down feather live mites of the analgoid family Xolalgidae (on Procellariiformes, Pelecaniformes, Charadriiformes, Podicipediformes and Anseriformes). Feather quill are parasitised by three families: analgoid Apionacaridae, Dermoglyphidae and pterolichoid Syringobiidae (on Charadriiformes and Pelecaniformes). Skin parasites belong to the analgoid families Dermationidae and Epidermoptidae (on Procellariiformes, Charadriiformes, Pelecaniformes and Anseriformes). Some mites of the family Laminosioptidae (Fainocoptinae) that are sometimes included into the Analgoidea are parasites of the follicles of developing feathers in Podicipediformes. Nasal parasites of the pyroglyphoid family Turbinoptidae are known from Charadriiformes (Fain 1977).

Mammalia. Endoparasitic mites of pinnipeds (seals, walruses and sea lions) and sea otters belong to the mesostigmatan Halarachnidae (lung mites), and include the viviparous genera *Halarachne* and *Orthohalarachne* (Kenyon et al. 1965, Konishi and Shimazaki 1998). Halarachnids commonly occur on the surface of mucous membranes of the nasal passages, trachea, bronchi, and lungs and feed on the blood and mucosal tissue (Kim 1985). These marine mammals are infected by two species of minute (0.1 mm) mites of the genus *Demodex* (*D. zalophi* and *D. phocidi*, Demodecidae) (Dailey and Nutting 1980, Desch et al. 2003). *Demodex* mites are observed in hair follicles and sebaceous glands around the face, over the genitalia, flippers and ventral body.

Life cycles

The primary number of stages in the life cycle of Acari is seven: egg, prelarva, larva, protonymph, deuteronymph, tritonymph and adult (Evans 1992). This full scheme is shortened by omitting particular stages in most recent Acari (Walter and Proctor 1999). Selected examples of life cycles of main parasitic mite groups from marine animals are briefly discussed below.

Hard ticks have only three active stages: larva, nymph (probably protonymph) and adult; in soft ticks there are two to eight nymphal moults (Hoogstraal 1973). Most ixodid ticks have different hosts for each stage, but some species are one-host or two-host parasites (Oliver 1989). The single-host tick *Ixodes uriae* shows a highly seasonal pattern of feeding which coincides with the main breeding period of its sea bird hosts taking its blood meal during the few months in which the sea birds return to land to breed (Barton et al. 1995). The tick cycle lasts for at least three years. Each developmental stage (with exception of adult males) takes one blood meal during the bird's breeding season. The duration of the blood meal is at least six days for each tick stage (Murray and Vestjens 1967, Barton et al. 1995). Ticks are on the hosts only during the blood meal, spending most of their lives in a limited area surrounding the sea bird's nests on the ground, mainly under stones (Frenot et al. 2001).

Mesostigmatans have reduced ontogeny to a larva, two nymphs and adults by supression of the tritonymph. In parasitic Halarachnidae the larva is an active, feeding and dispersing stage. Short-lived proto- and deuteronymph are non-feeding stages. Adults are feeding stages, sometimes with highly modified worm-like body (*Orthohalarachne attenuata*). The suppression of nymphal stages is also a common phenomenon in other mites (e.g. Halacaridae). The larva after a short immobile period moults into various numbers of nymphal stages. All three nymphs are present only in one freshwater species, in most other species the tritonymph is suppressed or only protonymph is retained as in marine species.

Adult Hypoderatidae are nidicolous (living in a nest) and free-living forms. Females are ovoviviparous. The larvae moult in the nest into protonymphs. The next stage is the small, heteromorphic deuteronymph (hypopus) that penetrate actively through the skin of the nestlings. In gannets, *Sula*, the hypopus greatly grow and live to five years waiting for the host maturity (Fain and Clark 1994). Because of hormone activity of hosts the tissular hypopi are rejected into the nest where they moult to the tritonymph or directly to the adult.

The complete life cycle of feather mites takes place on the body of bird hosts. As in all Psoroptidia the deuteronymph is absent. The gravid female is most often the dispersal stage. Generally the transfer is vertical from adult birds to fledglings; exceptionally horizontal by means of parasitic insects (Epidermoptidae). Eggs are stuck to the bases of feather barbs or in the feather quills; ovoviviparity sometimes occurs (e.g. in some Syringobiidae). The life cycle may be synchronised with those of the hosts (Mironov and Malyshev 2002); however, no data on marine birds exist.

Effects on hosts

There is no evidence of detrimental effects of parasitic halacarid mites on marine invertebrates. However, there are some data indicating that some Acari harm their marine bird and mammal hosts, both directly and as vectors of diseases. Mites and ticks do not invariably cause severe disease, but environmental conditions may sometimes be suitable for heavy infestations resulting in debilitation and death of affected animals (Kerry et al. 2000).

Ticks may cause damage by their consumption of blood (anaemia) and dermatological illnesses caused by mechanical injuries and toxins injected (dermatosis and paralysis) (Sonenshine 1991, Roberts and Janovy 1996). But much more dangerous are pathogens transmitted by ticks. Ticks transmit more infectious agents than any other blood-feeding arthropods (e.g. Lyme disease, relapsing fever, tick fever, and boutenneuse fever) (Nuttall 1984). Lyme disease spirochetes, *Borelia burgdorferi*, have been found in both sea birds nesting on sub-Antarctic islands and their tick *Ixodes uriae* (Gauthier-Clerc et al. 1999, Gylfe et al. 1999). Also viruses are carried by ticks and can infect the birds on which ticks feed. Antibodies to flaviviruses have been found in several sub-Antarctic penguin species, although associated disease has not been observed (Morgan et al. 1985).

Gauthier-Clerc et al. (1998) reported that the high infestation of king penguins by *Ixodes uriae* may induce some deaths, and several dozen birds were observed in poor condition. The blood loss on a dead hyperinfested adult was estimated at 100 mL to 300 mL, which corresponds to more than 10% of its total blood volume. Penguins with ticks had lower success in rearing chicks than uninfested parents. Tick infestation leads to large featherless areas in king penguins, which could increase heat loss and affect diving efficiency (Mangin et al. 2003).

The effect of halarachnid infection on pinnipedes or sea otters depends on the intensity of infection. These mites cause copious amounts of mucus in the upper respiratory tract and nose, nasal discharge, dyspnea, and coughing. Severe infections may lead to an impairment of respiration followed by lesions in the lungs. Intensive infection predisposes the host to more serious diseases, or even kills the host. Transmission from animal to humans with subsequent ocular discharge has been reported in one case (Dahme and Popp 1963, Dunlap et al. 1976, Fay and Furman 1982, Raga 1992).

Follicle mites (Demodecidae) are generally considered harmless. However *Demodex zalophi* from sea lions has been implicated in the development of alopecia and thickening of the skin over host genitalia, flippers and belly. Sea otters have also been observed to have *Demodex* mites but no serious dermatologic conditions have been observed. They cause only a mild follicular ectasia and minimal associated folliculitis (Moeller 1997).

Examples of detrimental parasitism among astigmatid feather mites of marine birds are rare. Some of them are undoubtedly detrimental causing skin lesions (Epidermoptidae) or destroying the feathers by chewing out the medula (some Syringobiidae). However, other feather mites are paraphages constantly removing preen oil from the feathers and perhaps forcing birds into energetically costly supplementary production of this substance. There are experimental data supporting the hypothesis that feather mites negatively affect the fitness of hosts (J Dabert, unpubl. data). Another astigmatid group, Hypoderatidae, comprises 'real' parasites; their deuteronymphs live under the skin destroying the subcutaneous tissue.

As a final note, crested auklets, *Aethia cristatella*, nest in dense colonies that smell like tangerines from a long distance. A provocative hypothesis is that the birds may produce this unique scent, among other things, to repel ectoparasites (H Douglas, Institute of Marine Science, University of Alaska pers. comm.).

Important references
Monographs on Acari are by Krantz (1978), Oliver (1989), van der Hammen (1989), Sonenshine (1991), Evans (1992), and Walter and Proctor (1999). Further comprehensive accounts are by Hoogstraal (1973) and Kethley (1982). Bartsch (1989) published a geographical and ecological survey of marine mites, and Proctor and Owens (2000) reviewed diversity, parasitism and coevolution of mites and birds. Raga (1992) discussed parasitism including that by Acari in Pinnipedia. Other discussions of parasitism in marine wildlife including relevant information are by Moeller (1997) and Kerry et al. (2000).

Pycnogonida (pycnogonids)
David Staples

Introduction
Over 1200 species belonging to about 80 genera are recognised worldwide. Records of endoparasitic or ectoparasitic relationships are so common that probably a parasitic stage is obligatory in the life cycle of all species. Associations are often with colonial animals such as hydroids, bryozoans and corals, colonies of which are made up of individual intercommunicating hydranths, zooids and polyps. Observations of pycnogonids feeding on these groups could simply be interpreted as predation, but because feeding does not necessarily lead to the demise of the colony they are regarded here as parasitic. Movement of pycnogonids is typically sluggish and tentative; consequently recorded associations are commonly with sessile or slow-moving animals such as sea anemones, bryozoans, hydroids, algae and sponges.

Morphology and diversity
The relationship of pycnogonids to other groups of arthropods has been a contentious issue resulting in varying classifications of the group. The rank of subphylum is common in the literature; however, current thinking based primarily on morphological characters places them as a class of the subphylum Cheliceriformes. Possession of multiple gonopores, a well-developed proboscis, ovigers (specialised grooming and egg-bearing appendages) and a much reduced abdomen, highlight the distinctiveness of the group, tempting some authors to go so far as to treat the taxon Pycnogonida as an infraphylum or phylum. Current researchers using DNA sequencing to establish relationships between arthropods are also divided on the standing of the group. It is anticipated that the growing body of molecular work will ultimately provide a greater understanding of these unresolved issues.

Only two diverse families lack records of a parasitic stage but both are difficult to observe *in situ*; the Colossendeidae primarily because it inhabits great depth and the Rhynchothoracidae because of its tiny size and interstitial life style. The remaining eight families are the Nymphonidae, Ammotheidae, Callipallenidae, Pallenopsidae, Phoxichilidiidae, Endeidae, Pycnogonidae and Austrodecidae. All families are accommodated in the only extant order, the Pantopoda.

The adult body consists of an anterior proboscis with a mouth at its tip, a trunk, usually consisting of four (but seven species with five and two with six) segments that support the four, five or six pairs of walking legs and a reduced one-segmented abdomen with terminal anus. Each body segment has a pair of lateral processes with which the legs articulate. The anterior-most segment (cephalon) supports several additional appendages and bears a dorsal ocular tubercle. A reduced one-segmented abdomen is attached to the posterior-most segment. Classification is primarily dependent on the presence, absence or degree of development of the appendages attached to the cephalon. These are the chelifores, originating above the proboscis, the palps, carried either side of the proboscis, and the ventrally situated ovigers posterior to the proboscis.

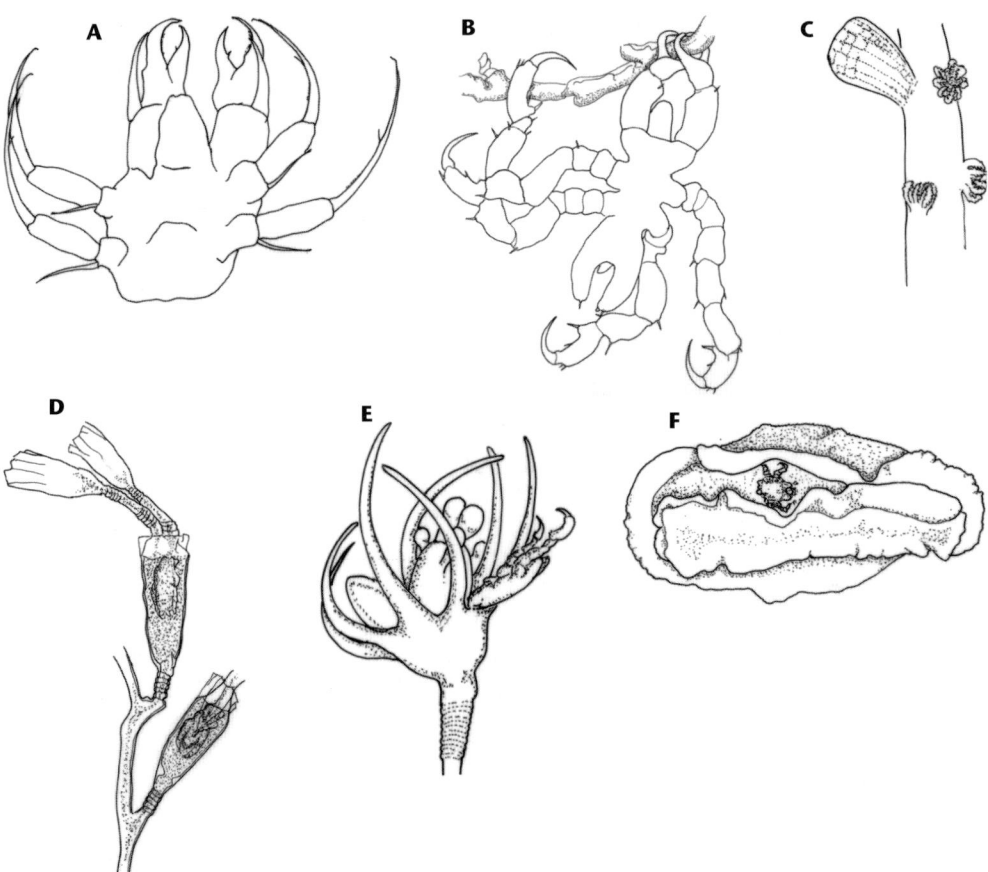

Figure 5.31 **A.** Protonymphon *Ammothea australiensis* removed from adult male. Blanket Bay, Vic., Australia. **B.** Unidentified protonymphon with two pair of adult legs developed, grasping stolon of hydroid *Plumularia australis*. Penguin I., WA. **C.** Juvenile *Ascorhynchus*? inside gall on gorgonia *Chrysogorgia papillosa*. After Stock (1953). **D.** Unidentified protonymphon encapsulated in hydrotheca of hydroid *Obelia bidentata* from Papua New Guinea. **E.** Larva of *Tanystylum* sp. attached to the hydranth of hydroid *Pennaria wilsoni*. **F.** Unidentified protonymphon attached to the mantle fold of nudibranch *Jorunna* sp. (ventral view) from Bass Strait, SA.

Life cycles

With few exceptions the sexes are distinct and separate. The externally fertilised eggs are passed from the female and gathered by the male onto the ovigers where they are cemented individually into balls or bracelets or embedded into a gelatinous ribbon-like mass wrapped around both ovigers. In some species, ovigers are entirely absent in which case the eggs are cemented directly onto the ventral surface of the trunk of the male.

The larval form is called the protonymphon in which stage a parasitic phase is most common (Fig. 5.31A). The extraordinary numbers of transitional forms that separate adult pycnogonids are not apparent in the protonymphon, which makes it difficult to identify specimens to species level. Larval development is variable but typically they hatch from the egg with an ovoid body bearing an anterior proboscis and functional chelifores (present in all protonymphon) and two pair of 'larval legs'. Spinous or tendril-like processes that possibly provide additional means of attachment and entanglement around a host often accompany these larval legs. In some cases a long thread, presumably serving a similar function, is also secreted from a gland at the base of each chelifore. Walking legs are added sequentially with subsequent moults.

The chelae function as the primary attachment to the host, each bearing one fixed and one moveable finger. The fingers are gaping, pointed and pincer-like, capable of penetrating the host tissue. In some instances the chelae serve the sole purpose of grasping the host during the early life stages after which they are shed or atrophy to non-functional knobs with the last larval moult.

The protonymphon leaves the male at differing stages of development, in some cases shortly after emerging from the egg to invade cnidarians, especially hydroids (Fig. 5.31B). In other instances the protonymphon develop to a more advanced stage before commencing a free-living existence on the same host substrate as the adult, usually a hydroid or bryozoan. Species associated with sand, some living interstitially, may deposit protonymphon among the grains, which could account for their parasitic association with soft-bodied invertebrates found in these environments such as opisthobranchs, holothurians and bivalve molluscs. The diversity of hosts and infrequency of repeated observations suggests that many of these associations are a consequence of a chance encounter. In other instances a greater degree of host specificity is demonstrated.

Cnidarians are particularly well represented in parasitic associations within diverse pycnogonid families; the Pycnogonidae, Endeidae, Phoxichilidiidae, Nymphonidae, Ammotheidae, Pallenopsidae and Callipallenidae. Three families are recorded living parasitically with anemones. *Pycnogonum* has a particular relationship with several anemone species; the adults feeding ectoparasitically on the outside wall of the host and the juveniles endoparasitically in the gastrovascular cavity. The Ammotheid *Ammothella biunguiculata* (Dohrn 1881) has also been recorded in the gastrovascular cavity of an anemone; in one instance 67 juveniles infested a single host (Miyazaki 2002).

In their natural environment, individuals appear able to feed without obvious ill-effect on the anemone, survival possibly being dependent on the pycnogonid moving on in search of a new host. Stock (1953) figured gall-like swellings on the gorgonian *Chrysogorgia papillosa* (Fig. 5.31C). Each gall was slightly larger than a retracted polyp and contained a tightly folded pycnogonid. There are numerous records of associations between pycnogonids and scleractinian corals. Child (1988) recorded six pycnogonid families being associated with a single species of coral. Similarly five pycnogonid families have been recorded in association with zoanthids.

The most frequently reported endoparasitic association in pycnogonid literature is between larval pycnogonids and hydroids, where pycnogonids are either encapsulated in gall-like vesicles or attached to a hydranth of the host (Staples and Watson 1987) (Figs 5.31D, E). Researchers noted pyriform sacs for some time before it was realised that a solitary larva was tucked inside an aborted hydrotheca or gonotheca. Most records of larval encystment are in athecate hydroids and involve larvae predominantly of the family Phoxichilidiidae. These larvae have also been observed living free in the gastral cavity of Tubularian hydroids without any apparent ill effects on the host (Loman 1907). The Nymphonidae also have a strong association with hydroids, adults and juveniles being frequently recorded feeding ectoparasitically on the same hydroid colony.

One of the most puzzling associations is that of larval pycnogonids and hydroid medusae. As many as 30 individuals of the larva stages of *Ammothella alaskensis* have been found attached to Anthomedusae dredged from a muddy sea floor. The larvae were on the wall of the manubrium under the exumbrella with their proboscides piercing the soft tissue of the host (Okuda 1940). Other records indicated similar infestation of medusae belonging to several hydroid genera but it is still unknown how the protonymphon become attached or if this association eventually leads to the demise of the medusae. Arango (2001) recorded a species of *Endeis* feeding in large numbers on the Hydrozoan (hydrocoral) *Millepora exaesa*. The only endoparasitic record with a hydrocoral is of encapsulated pycnogonid larvae in the partly aborted gastrozooids of a deep-sea *Stylasterid* (Moseley 1879).

Predation on bryozoans by pycnogonids is well documented. Adult and juvenile pycnogonids feed by consuming the contents of individual zooids either by crushing the zooid before feeding or by sucking out the contents directly through the frontal pore. Such an association has been observed in southern Australian waters where species of *Pseudopallene* have an obligate association with catenicellid bryozoans, namely, Orthoscuticellid spp. Wyer and King (1973) record the use of palps by *Achelia echinata* to prevent closure of the operculum on the frontal pore of bryozoan zooids allowing insertion of the proboscis to suck out the contents. An encystment stage has not been recorded in bryozoans.

Juvenile pycnogonids of the Ammotheid genus *Nymphonella* have been reported living parasitically in several species of infaunal and epifaunal bivalve molluscs in Japanese waters. From one to 21 young and adults of *Achelia chelata* were also found infesting the mussel *Mytilus californianus*, partly destroying the tissue, in some instances to the point of atrophy (Benson and Chivers 1960).

Opisthobranchs are a diverse group of molluscs that include the bubble shells, the sea hares and the nudibranchs, all of which have a documented parasitic association with pycnogonids. Young pycnogonids are most often found attached to the foot and mantle of the host by means of their chelifores and with their proboscides piercing the host tissue (Fig. 5.31F). Forty young stages of *Ammothea* were recorded as ectoparasites on the nudibranch *Arminia variolosa* (Ohshima 1933).

In a collection of ophiuroids from the Seychelle Islands, three species of the genus *Ophiocoma* were found infested with 1 to 29 specimens of *Anoplodactylus ophiurophilus*. All three species of ophiuroid feed by mucus entrapment of particles suggesting that the presence of pycnogonids on the ventral side of the arms, near the mouth and along the ambulacra placed them in the best position to encounter the mucus-impregnated food being transported towards the mouth of the host. Fewer pycnogonids were, however, observed with podia firmly grasped in their chelae, possibly feeding on the soft tissue of the host (Sloan 1979).

In the few ectoparasitic associations with asteroids, all specimens have been taken from the oral surface and ambulacral grooves.

There are also few recorded associations with sea urchins. In one case 19 juvenile specimens of *Pycnosomia strongylocentroti* fixed by their chelae were found among the spines of *Strongylocentrotus* sp. (Losina-Losinsky 1933). This could suggest a parasitic association involving the protruding soft parts of the echinoid such as the podia.

Ohshima (1927) gives an account of up to 30 individuals of *Ammothea hilgendorfi* attached to the holothurian *Holothuria lubrica*. Parasitic associations involving both errant and sedentary polychaetes have been recorded. Juveniles of what were thought to be *Ammothella spinifera* have been found as apparent ectoparasites on the polychaete *Sabella melanostigma*. Several juveniles living inside the Sabellid tube were attached by their chelifores, depressions in the body wall of the Sabellid suggesting that they were feeding on the host tissue (Salazar-Vallejo and Stock 1987).

Effects on hosts and ecological importance

It is difficult to assess the importance of pycnogonids in marine food chains. They are active predators of hydroids and bryozoans and there is no doubt that when present in large numbers their impact on individual colonies can be significant. Entire colonies of the hydroid *Halocordyle wilsoni* can be stripped completely by a species of *Tanystylum* found in southern Australian waters and significant portions of bryozoan colonies are destroyed as a consequence of predation by *Pseudopallene* spp. Ryland (1976) is of the opinion that pycnogonids are the most important consumers of bryozoans.

Important references
For comprehensive reviews of the group, readers are referred to Arnaud and Bamber (1987) and Cadien (1997).

Insecta (insects)
Kirsten Jensen and Ricardo L Palma

Introduction
The Insecta is the most diverse group of metazoan animals on land, possibly including over one million species (most of them not yet described). Few free-living species have invaded marine habitats, but many are parasitic on marine mammals and birds, often causing considerable harm to their hosts. Only five of the over 12 000 species of caddisflies are marine, and only one of these is known to be a larval parasite of starfish.

Phthiraptera
Phthiraptera (lice) are wingless, dorso-ventrally flattened, ectoparasitic insects with single host life cycles, completing their life cycle in or on the feathers, hair or skin of vertebrates. They are hemimetabolous and their developmental stages consist of the egg, three nymphal stages and adult. Four suborders are recognised in the order (Barker *et al.* 2003), three of which are known to parasitise marine vertebrates: the Amblycera (Fig. 5.32A) and Ischnocera (Fig. 5.32B), both known as chewing lice, and the Anoplura (Fig. 5.32C), known as sucking lice. Amblycera and Ischnocera are ectoparasites of birds and mammals. The Anoplura exclusively parasitise mammals.

Chewing lice of marine birds (Amblycera and Ischnocera)
Over 4460 valid species of amblycerans and ischnocerans are recognised (Price *et al.* 2003), 88% parasitising birds, and the rest mammals. In marine ecosystems, amblycerans and ischnocerans are restricted to birds. This section is focused on amblyceran and ischnoceran lice that are associated with bird families that consist entirely or largely of marine species (e.g. Diomedeidae and Laridae, respectively).

Amblycera and Ischnocera possess mandibulate chewing mouthparts and a head that is as wide as or wider than the prothorax. Two characters used to distinguish between these groups are the presence or absence of maxillary palps and form of the antennae. Amblycerans possess maxillary papillae and have four-segmented antennae concealed in lateral grooves. Ischnocerans lack maxillary papillae and possess filiform antennae that are fully visible (Johnson and Clayton 2003). Amblycerans feed mainly on feathers, host skin products and blood, as well as on eggs and moulting nymphs of conspecifics or lice of other species. In contrast, most ischnocerans feed on feathers and dead skin (see Murray 1976, Johnson and Clayton 2003). In general, amblycerans are less site-specific than are ischnocerans (see Marshall 1981), but some degree of site specificity is exhibited by most taxa. For example, *Saemundssonia* species parasitise the heads of their hosts (see Marshall 1981), *Naubates* primarily the wing feathers (see Palma and Pilgrim 2002), and adults of *Piagetiella* the throat pouch (see Marshall 1981). The major mode of transmission of these lice between hosts is via direct contact of individual birds (Johnson and Clayton 2003).

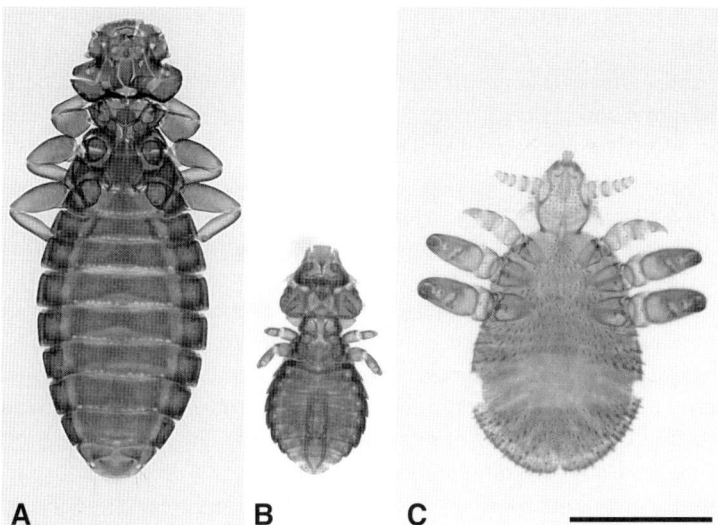

Figure 5.32 Examples of Phthiraptera known to parasitise marine vertebrates. **A.** Female *Actornithophilus piceus lari* (Phthiraptera: Amblycera: Menoponidae) from the California gull, *Larus californicus*. **B.** Male *Saemundssonia* (*Saemundssonia*) *lari* (Phthiraptera: Ischnocera: Philopteridae) from the glaucous-winged gull, *Larus glaucescens*. **C.** Nymph (pharate) *Antarctophthirus trichechi* (Anoplura: Echinophthiriidae) from the walrus, *Odobenus rosmarus*. Scale bar = 1 mm. Figures A.–C. are depicted to same scale.

Lice of marine birds, in general, do not possess adaptations to cope with the marine environment since they essentially lead a terrestrial life style, including lice parasitising penguins, which live in a pocket of air among the water repellent feathers, without contact with the marine environment.

Sucking lice of marine mammals (Anoplura)
Over 532 species of anoplurans in 15 families are known (Durden and Musser 1994). Echinophthiriidae is the only anopluran family reported from marine mammals. Anoplurans possess a head that is narrower than the prothorax, and piercing–sucking mouthparts. They feed on host blood by inserting their mouthparts directly into the host blood vessels.

Lice of hair seals (Phocidae) live exposed to the marine environment among sparse, scale-like hair on regions of the body used in heat dissipation (e.g. flippers, tail, genital and anal orifices), and thus are required to spend prolonged periods of time submersed (see Marshall 1981). These lice require adaptations to cold-water temperatures, and submersion. Lice of fur seals (Otariidae) either parasitise the pelage and therefore live in the air space created by the fur, not exposed to the marine environment, or, like lice of hair seals, live on the exposed/naked areas of the body (e.g. fins, nostrils and eyelids). Adaptations include unusually well-developed musculature and valves of the atria of the thoracic and abdominal spiracles, presumably to prevent water from entering during prolonged dives (Marshall 1981). In addition, modified scale-like setae on the abdomen are thought to trap air to allow for gaseous diffusion through the cuticle when submersed (see Murray 1976, Mehlhorn *et al.* 2002).

Host associations and specificity
Eight genera of Amblycera have been reported from marine bird families (Table 5.3), but five of them also include species parasitic on non-marine birds: species of *Colpocephalum* and

Austromenopon parasitise birds in 11 and four avian orders, respectively; species of *Actornithophilus* parasitise birds in 11 families of Charadriiformes, but only three of them are exclusively marine; and *Eidmanniella* and *Piagetiella* include a few species parasitic on freshwater species of cormorants and pelicans. Therefore, only three amblyceran genera are restricted to exclusively marine bird families: *Ancistrona* and *Longimenopon* with species parasitic on petrels, and *Fregatiella* unique to the frigatebirds.

Based on the index of host specificity as defined by Caira et al. (2003) and the data from (Price et al. 2003), specificity index values were calculated for 22 species of amblycerans in five genera (i.e. genera parasitising bird groups in which all or most of the species are marine). Five of the six species of *Longimenopon* are oioxenous (only found in one species of host), and one species is euryxenous (parasitising hosts in more than one family). All nine species of *Piagetiella* are either oioxenous or mesostenoxenous (restricted to a number of species in one genus). Four of the five species of *Eidmanniella* are mesostenoxenous, and one species is metastenoxenous (parasitising hosts in one family, but more than one genus). The single species of *Fregatiella* is mesostenoxenous. *Ancistrona vagelli* is euryxenous.

Ischnoceran lice of marine birds are more diverse than amblycerans. Eighteen genera have been reported from marine birds (Table 5.3). Fifteen of these are restricted to exclusively marine birds, whereas most species of *Quadraceps* and *Saemundssonia*, but only a few species of *Pectinopygus*, are found also on non-marine birds. Specificity of these lice at the generic level is pronounced. Eleven of the 18 genera are each unique to a single bird family. These are: *Austrogoniodes* and *Nesiotinus* on penguins (although see Mey et al. 2002), *Bedfordiella*, *Pseudonirmus*, *Naubates* and *Trabeculus* on petrels, *Episbates* and *Harrisoniella* on albatrosses, *Haffneria* on skuas and jaegers, *Pelmatocerandra* on diving petrels and *Philoceanus* on storm petrels. Five genera (*Pectinopygus*, *Docophoroides*, *Paraclisis*, *Halipeurus* and *Perineus*) are each reported from two or more families within one order.

Host specificity index values (*sensu* Caira et al. 2003) based on data from Price et al. (2003) and Banks and Palma (2003) were calculated for 140 species of ischnocerans in 16 genera which parasitise only marine birds. By this analysis, 66 species are oioxenous, 59 are mesostenoxenous, 14 are metastenoxenous and one is euryxenous.

Because of their narrow specificity (89% of the 162 amblyceran and ischnoceran species analysed parasitise one or more species within a single genus), lice of marine birds have been used in cophylogenetic studies, mainly comparing louse and host phylogenies using a variety of methods (Paterson et al. 2000, Paterson and Banks 2001, Banks and Paterson 2004, Page et al. 2004). Evidence of cospeciation is not consistently unequivocal.

Marine mammals serving as hosts for sucking lice are pinnipeds, belonging to the carnivore families Odobenidae (walruses), Otariidae (eared seals, fur seals and sea lions) and Phocidae (true, earless or hair seals). The sucking louse family Echinophthiriidae consists of five genera, four of which (*Antarctophthirus*, *Echinophthirius*, *Lepidophthirus* and *Proechinophthirus*) are restricted to pinnipeds. While *Antarctophthirus* parasitises all three pinniped families, *Echinophthirius* and *Lepidophthirus* parasitise only phocids, and *Proechinophthirus* is restricted to otariids (Durden and Musser 1994). A total of 11 species of sucking lice have been reported from 21 of the 33 species of pinnipeds. Seven of the 11 species are oioxenous (also Kim 1985). Among the species that parasitise more than one species of host, none parasitise hosts from more than one family (i.e. are mesostenoxenous or metastenoxenous).

Effects on hosts
Studies on the effect of chewing lice on marine birds are few (see Johnson and Clayton 2003) and little is known overall.

Table 5.3 Distribution of louse genera in families of birds which include marine species exclusively or as a majority

	PHTHIRAPTERA	
Avian family	**Amblycera**	**Ischnocera**
Spheniscidae	–	*Austrogoniodes; Nesiotinus*
Diomedeidae	*Austromenopon*[A]	*Docophoroides; Episbates; Harrisoniella; Paraclisis; Perineus; Saemundssonia*[A]
Procellariidae	*Ancistrona; Austromenopon; Longimenopon*	*Bedfordiella; Docophoroides; Halipeurus; Naubates; Paraclisis; Perineus; Pseudonirmus; Saemundssonia; Trabeculus*
Hydrobatidae	*Ancistrona; Austromenopon; Longimenopon*	*Halipeurus; Philoceanus; Saemundssonia*
Pelecanoididae	*Austromenopon*	*Halipeurus; Pelmatocerandra*
Fregatidae	*Colpocephalum*[A]; *Fregatiella*	*Pectinopygus*
Phaethontidae	*Austromenopon*	*Saemundssonia*
Pelecanidae	*Colpocephalum; Piagetiella*	*Pectinopygus*
Sulidae	*Eidmanniella*	*Pectinopygus*
Phalacrocoracidae	*Eidmanniella; Piagetiella*	*Pectinopygus*
Stercorariidae	*Austromenopon*	*Saemundssonia; Haffneria; Quadraceps*[A]
Laridae	*Actornithophilus*[A]; *Austromenopon*	*Saemundssonia; Quadraceps*
Chionididae	*Actornithophilus*	*Saemundssonia; Quadraceps*
Alcidae	*Austromenopon*	*Saemundssonia; Quadraceps*

[A] Specificity index values were not calculated for species in these genera since most species parasitise non-marine birds.

Direct and indirect effects on the health of their marine mammal hosts by anopluran lice have been suggested. During times of decreased food availability, high burdens of the louse *Echinophthirius horridus* on harbour seals, *Phoca vitulina*, may ultimately contribute to a reduction in the survival of young seals (Thompson *et al.* 1998). In addition, *E. horridus* is the intermediate host of the seal heartworm, *Dipetalonema spirocauda*. Infection by this nematode impairs the host's blood circulation (Geraci and Lounsbury 2002).

Siphonaptera

Siphonaptera (fleas) are wingless, bilaterally flattened, ectoparasitic insects on warm-blooded vertebrates. Fleas are holometabolous and their developmental stages consist of the egg, usually three larval stages, pupa and adult. With two exceptions, the adult is the only parasitic stage. Adult fleas possess piercing–sucking mouth parts and feed exclusively on host blood. When the adults reproduce and oviposit on the host, the eggs are not attached and drop off, commonly into the nest of their hosts or the surroundings. Consequently, development from egg to pupae occurs away from the host, with the larvae feeding on debris, including the faeces of the adult fleas.

The order Siphonaptera consists of almost 2000 species (not including subspecies) in 15 families (Lewis 1998), most of which parasitise mammals; however, some species are found parasitising birds. There are no records of fleas from marine mammals. According to Lewis (1998), 106 species of fleas in 16 genera parasitise birds. At least 30 species of fleas in 10 genera (i.e. *Actenopsylla, Ceratophyllus, Dasypsyllus, Glaciopsyllus, Listronius, Megabothris, Mioctenopsylla, Notiopsylla, Parapsyllus* and *Xenopsylla*) belonging to four families (Ceratophyllidae, Pulicidae, Pygiopsyllidae and Rhopalopsyllidae), parasitise marine birds as their primary host (Johnson 1957, Smit 1979, Hoberg and Wehle 1982, Traub *et al.* 1983, Smit 1984, Holland 1985). A marine bird is considered the primary host if the bird shows the highest prevalence of infection with this flea as compared to other non-marine birds, and if the flea breeds in the marine bird's nest. One species of flea with a most unusual life history is *Glaciopsyllus antarcticus* (Ceratophyllidae), a species parasitising two petrel species breeding in Antarctica. This species is one of only two flea species in which the larvae are also parasitic on the host, feeding on blood, and pupating on the host, instead of in the nest (Bell *et al.* 1988).

While there are few studies in which the effects of fleas on marine bird populations have been measured (see e.g. Merino *et al.* 1999), effects similar to those of ticks and mites (i.e. nestling mortality and nest abandonment) can be speculated with high intensities of infestation (Merino *et al.* 1999).

Trichoptera
Though poorly known, one species of caddisfly is intriguing because of its apparent parasitic association with starfish. While all of the about 12 000 caddisfly species (order Trichoptera) develop in freshwater, only the five species in the family Chathamiidae are known to have marine larvae. One of these five species (*Philanisus plebeius*) is unique among caddisflies in that females may oviposit individual, small clusters or strings of eggs in the coelomic cavity of starfish of the genus *Patiriella* (Asteroidea, Asterinidae), where they hatch. The larvae subsequently leave the starfish and become free living in the intertidal zone. *Philanisus plebeius* occurs in coastal south-eastern Australia and New Zealand. Anderson *et al.* (1976) reported caddisfly eggs from the starfish *Patiriella exigua* in Australia, while Winterbourn and Anderson (1980) reported them from *Patiriella regularis* in New Zealand. Eggs of *Philanisus plebeius* are not surrounded by a protective gelatinous or cement-like matrix common among other trichopterans (Anderson and Lawson-Kerr 1977). Unfortunately, oviposition and escape of larvae from the starfish have not been observed in nature.

Important references
An important contribution on insect parasites of marine birds and mammals is by Murray (1976). Extensive accounts of parasitic insects were given by Marshall (1981), who published a monograph, and Durden and Musser (1994) and Price et al. (2003), who published checklists of parasitic insects. Johnson and Clayton (2003) reviewed the biology, ecology and evolution of chewing lice; Holland (1985) gave an account of the fleas of Canada, Alaska and Greenland, and Smit (1979) of the fleas of New Zealand. Geraci and Lounsbury (2002) discussed the impact of parsitic insects on marine mammal health. Studies on a marine parasitic caddisfly are by Anderson *et al.* (1976) and Anderson and Lawton-Kerr (1977).

Tardigrada (water bears)
Reinhardt Møbjerg Kristensen and Jesper Guldberg Hansen

Introduction
The phylum Tardigrada (water bears) consists of microscopic, multicellular coelomates with four pairs of segmented legs. Tardigrades belong to the Panarthropoda group, together with fossil lobopodians, onychophorans (velvet worms) and arthropods (Nielsen 2001). Recent

molecular studies support the tardigrade–arthropod relationship (Giribet *et al.* 1996, Garey *et al.* 1999). With an adult size of 0.08 mm to 1.2 mm (most marine species are only 0.1 mm to 0.3 mm) they are among the smallest Metazoa. The herbivorous tardigrades move with a slow, bear-like gait, hence their common name 'water bears', whereas the carnivorous tardigrades are some of the fastest moving animals in the world (relative to their size). About 930 species have been described worldwide today, but taxonomists expect that at least 10 000 species exist. About 770 species are semiterrestrial and the rest (160 species) are true marine species. Tardigrades are found from highest elevations in the Himalayas to deep trenches in the deep sea, and from hot springs to the ice cathedrals inside the ice cap of Greenland. However, they are all aquatic animals in the active state, whether from the dry desert or tropical rainforest. Only three extant ectoparasitic tardigrades have been described; however, a recently recorded fossil from Siberian limestone (mid Cambrian, approximate 520 million years ago) indicates that parasitism in heterotardigrades may have developed very early (Maas and Waloszek 2001). The claws and the whole body configuration are similar to the extant tardigrade *Tetrakentron synaptae* living on holothurians (Kristensen 1980). True heterotardigrades and eutardigrades are found in Cretaceous amber from North America (Cooper 1964, Bertolani and Grimaldi 2000) (i.e. the modern lineages of heterotardigrades and eutardigrades have existed for at least 90 million years).

Tardigrades have been classified as belonging to the 'Aschelminthes' (especially nematodes, loriciferans and kinorhynchs) and the Arthropoda. According to the Ecdysozoa theory (Aguinaldo *et al.* 1997), aschelminth phyla with a body skeleton that is moulted are closely related to the Arthropoda. Hence, the Tardigrada can now be seen as a very close relative of the Arthropoda that retains many primitive (plesiomorphic) ecdysozoan traits. The phylum comprises three classes: Heterotardigrada, Mesotardigrada and Eutardigrada (Ramazzotti and Maucci 1983).

Morphology
The body is bilaterally symmetrical with five distinct body segments including a cephalic segment and four trunk segments each bearing a pair of segmented legs. The terrestrial and limnic forms have reduced the segmentation in their stumpy legs with two to four claws, while marine forms may have telescopically segmental legs with up to 13 claws or four to six toes with complex claws. Other marine tardigrades have rod-shaped adhesive discs or round suction discs also inserted on the foot via toes.

The digestive system consists of three major parts: the foregut (ectodermal origin), the midgut (mesodermal origin) and the hindgut (ectodermal origin). The foregut is a very complex feeding structure and consists of a mouth cavity, buccal tube, pharyngeal apparatus with placoids and oesophagus all lined with cuticle. Furthermore, associated with the buccal tube is a stylet apparatus, which in the plesiomorphic condition consists of a calcium carbonate-encrusted stylets and stylet supports. The stylets and the stylet supports may be strongly reduced mouth limbs (mandibles and maxillae).

The nervous system is distinctly metameric consisting of the three-lobed brain, the subpharyngeal ganglion and the four ventral trunk ganglia. Characteristic for all tardigrades is the paired commissure going from the protocerebrum to the first ventral trunk ganglion. Both heterotardigrades and eutardigrades may have eyespots, usually located inside the protocerebrum. The sensory structures found on the head, trunk and legs seems to be homologous to that of single arthropod sensillae. Tardigrades lack respiratory organs or gas exchange structures. Gaseous exchange takes place trough the epidermal cells and the complex cuticle.

Diversity and life styles
Hermaphrodism has been reported to occur in representatives of most families of eutardigrades; however, only one marine heterotardigrade species is hermaphroditic, all others are dioecious.

Males are unknown in many species of the heterotardigrade genus *Echiniscus*. Sexes are not always easily distinguished externally but males can be smaller, and in heterotardigrades the chemoreceptors (clavae) can be longer. Internal fertilisation by copulation is present in nearly all marine species, preceded by complex courtship behaviour. Eggs are round or oval and possess a smooth or ornamented shell. Ornamented eggs are generally laid free and are found in many eutardigrade genera and in the heterotardigrade genus *Oreella*. All other known heterotardigrades lay smooth eggs either free or in the exuvium.

The life span of tardigrades varies according to temperature; arctic species can live more than one year, temperate–tropic species only a few months. As a reaction to desiccation or low temperature the life cycle can be lengthened by cryptobiosis, a state without any metabolism. In this state the tardigrade can survive many years.

Most heterotardigrades are herbivorous, feeding by piercing plants cells with the two stylets, and sucking the contents by the strong muscular pharyngeal apparatus with armature (placoids). Almost all true carnivorous species are terrestrial eutardigrades. Bacterivorous tardigrades are perhaps more common than we know today, and many small soil tardigrades feed on bacteria. Symbiotic bacteria in the head are found in many arthrotardigrades of the subfamily Florarctinae.

Parasitism in the tardigrades

Tardigrades are found in association with several other invertebrates, for example *Actinarctus doryphorus* with the echinoderm *Echinocyamus pusillus* (see Schultz 1935, Grell 1937) and *Echiniscoides sigismundi* in *Mytilus edulis* (see Green 1950) or on *Semibalanus balanoides* (see Kristensen and Hallas 1980). However, in all these cases the tardigrades have been found later to be free living, interstitial or living on algae. True ectoparasitism on animals is only found in three marine species:

1. *Tetrakentron synaptae* (subfamily Styraconyxinae, Arthrotardigrada) is an obligate ectoparasite on the holothurian *Leptosynapta galliennei*
2. *Pleocola limnoriae* (subfamily Styraconyxinae, Arthrotardigrada) is a facultative ectoparasite on the pleotelson and pleopods of the isopod *Limnoria lignorum*
3. *Echiniscoides hoepneri* (family Echiniscoididae, Echiniscoidea) is an obligate ectoparasite on the cirriped *Semibalanus balanoides*.

Pleocola limnoriae was only collected once on *Limnona lignorum* at Roscoff, France (Cantacuzène 1951). Only one to two tardigrades were found on the isopod, and about 3% of the isopods were infected. There is no sign of adaptation to parasitism in the morphology of the tardigrade. It has never been found again, in spite of thorough studies. LW Pollock has claimed that he has observed the species free living in the interstitial environment, but his observation could be on a closely related species.

The tidal tardigrade *Echiniscoides hoepneri* was only found on the cirriped *Semibalanus balanoides* at three localities in Greenland. It lives together with its sister species *Echiniscoides sigismundi* (see Kristensen and Hallas 1980); however, whereas *E. sigismundi* sucks out green algae on the barnacle, *E. hoepneri* sucks out the embryos in the brood chamber of its host. The species was only found on the basal membrane, when the barnacle was released from the substrate. The species is adapted to ectoparasitism. The stylets are extremely long compared with the stylets of other species of free-living *Echiniscoides*. Furthermore, the stylets grow longer after each moult, while the length of the stylets in other species of *Echiniscoides* usually stop growing after the last instar of juveniles (Kristensen and Hallas 1980, Fig. 49). Elsewhere, the morphology of *E. hoepneri* does not show adaptation to parasitism; however, the species has lost the ability to anhydrobiosis, a physiological character exhibited by the sister species *E. sigismundi*.

The ectoparasite *Tetrakentron synaptae* on the holothurian *Leptosynapta galliennei* has been well investigated since its discovery in Roscoff, France (Cuénot 1892). The species has never been collected outside Brittany and it has not been found on other holothurians (Marcus 1929, van der Land 1975, Kristensen 1980). Cuénot (1932) recorded only one to three specimens on each host's tentacles; however, more than 100 may be present on the trunk of the host. The incidence of infection may be as high as 25% to 50% in the Bay of Roscoff depending of how old the host is. The tardigrade is strongly adapted to parasitism and was for the first time observed to suck out epidermal cells of the holothurian in the investigation of Kristensen (1980, Fig 7a). New data of the morphology and life cycle of this aberrant tardigrade are included in the following paragraph. *Tetrakentron synaptae* specimens were investigated in a period from 1974 to 2004. Adults were found on the sea cucumber from April to October, juveniles only in June and July. The sediment of the slimy tube of the sea cucumber was investigated. Two species of the tardigrade genus *Batillipes* were found, but not one specimen of *Tetrakentron synaptae*, indicating that this species really is closely associated with its host.

The species, *Tetrakentron synaptae*, is closely associated with the sea cucumber *Leptosynapta galliennei*. It does not occur on *L. inhaerens*, which was also investigated for the tardigrade at Roscoff, Brittany. *Tetrakentron synaptae* adheres strongly to the sea cucumber; however, the tardigrades can be detached by freshwater treatment of the host. The tardigrades survive freshwater, and especially the large females (182–238 µm) will survive for several days (Fig. 5.33A). In the laboratory the female lays free eggs in clusters (4–8), thereafter the female moults. Hence, the eggs are not deposited in the exuvium as seen in some other heterotardigrades. One week after the eggs are laid, embryos are observed inside the eggs (Fig. 5.33B). The exuvium of the female was investigated, and the very complex seminal receptacles were observed for the first time. Later the two cuticular receptacles were also found in the adult female (rs, Fig. 5.33A). The female is strongly dorsoventrally flattened and has an enlarged epicuticle without pillars. The lack of pillars makes the outer epicuticle very loose, and this part of the cuticle breaks off often (arrows, Fig. 5.34). The anus is located almost dorsally and the gonopore caudally. Two different types of males were collected. One type of male (127–198 µm) is less modified than the female (Fig. 5.33D) and is more slender. This male is strongly mobile and virile and was seen moving around on the sea cucumber. Mating was not observed; however, females had many spermatozoans in the seminal receptacles. The secondary clavae (chemoreceptors) are dome-shaped (sc, Fig. 5.33D) and larger than in a sessile female. This type of male actively locates the female on the sea cucumber. The other type of male is a dwarf male (75–83 µm). It is as dorsoventrally flattened as the female (Fig. 5.33E), and this type was always observed close to the females. The dwarf male was observed only when the population of tardigrades was large (more than 200 animals on one sea cucumber). Both type of males have mature spermatozoa in the testis (te, Fig. 5.33D, E). The egg measured 45 µm, and the smallest juvenile (probably the first instar), 55 µm in length. However, this juvenile had already four sets of claws (Fig. 5.33C), unique for an arthrotardigrade, whose first instar larva usually has two sets of claws at hatching.

Adaptation to parasitism is also observed in the enlarged claws and stylet glands, and the reduction in all sensory structures except for the cirrus E (cE) and secondary clava. Furthermore all stages have four sets of three-pointed claws on each leg.

Important references

For comprehensive accounts of Tardigrada the reader is referred to the monographs by Marcus (1929), Cuénot (1932) and Ramazzotti and Maucci (1983). Important papers on the best known species, the parasitic marine tardigrade *Tetrakentron synapta*, are by Van der Land (1975) and Kristensen (1980).

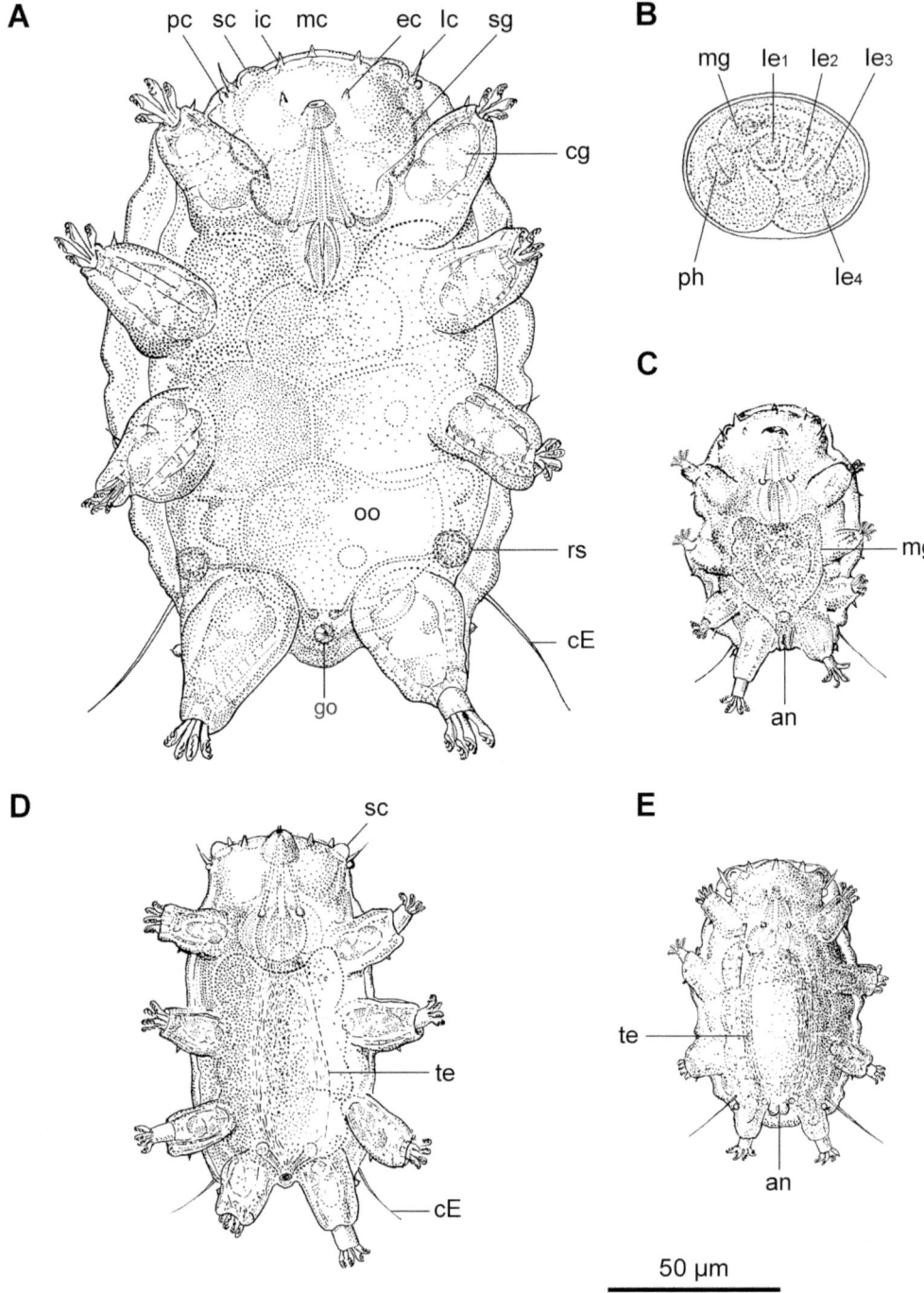

Figure 5.33 Life cycle studies of the arthrotardigrade *Tetrakentron synaptae*. **A.** Ventral view of an adult female. **B.** Lateral view of embryo inside the free-laid egg. **C.** Ventral view of juvenile (female?). **D.** Ventral view of mobile male. **E.** Dorsal view of nearly sessile dwarf male. Abbreviations: an, anus; cE, cirrus E; cg, enlarged claw gland; ec, external cirrus; go, gonopore; ic, internal cirrus; lc, lateral cirrus; le1–le4, leg anlage; mc, median cirrus; mg, midgut; oo, oocyte; pc, primary clava; ph, pharyngeal bulb; sg, enlarged stylet gland; te, testis.

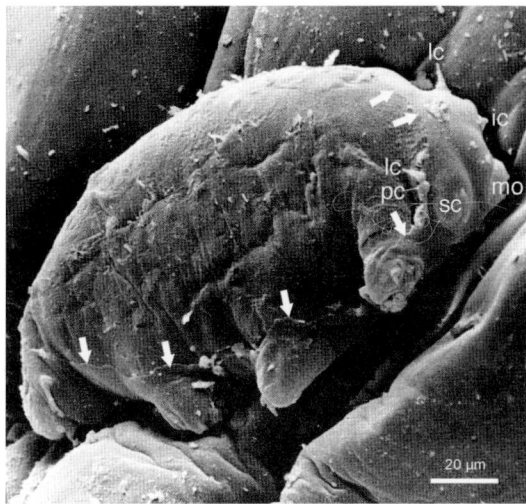

Figure 5.34 Scanning electron micrograph of *Tetrakentron synaptae* on the sea cucumber *Leptosynapta galliennei*. Arrows, broken epicuticle. Abbreviations: ic, internal cirrus; lc, lateral cirrus; mo, mouth cone; pc, primary clava.

Pentastomida (tongue worms)
Wolfgang Böckeler

Introduction
The pentastomids (tongue worms) include about 120 species. It is still uncertain whether they are a separate phylum closely related to the arthropods (Storch and Boeckeler 1979, Böckeler 1984a, Storch 1993) or whether they should be included within the branchiuran Crustacea (Wingstrand 1972, Abele 1989, Storch and Jamieson 1992, Zrzavi 2001).

Pentastomids have two pairs of hooks (Fig. 5.35) at the distal end of more or less pronounced appendages. The two orders of pentastomids (Storch 1993) can be distinguished by the position of the hooks. In the approximately 40 species of Cephalobaenida, they lie behind each other, whereas in the Porocephalida they are aligned in one row besides the mouth opening. In addition to the genera *Cephalobaena* and *Raillietiella*, the 'lower' Cephalobaenida also contain the marine *Reighardia*. All other genera are included within the *Porocephalida*.

The free-living ancestors of pentastomids were microscopic (several hundred micrometres) with seven to eight metameres, and can be traced back to the Cambrium (about 500 million years ago, mya). They were discovered as part of the marine fauna among the upper Cambrian 'Orsten' fossils of Sweden (Waloszek and Möller 1994, Waloszek *et al.* 1994). They are thought to have adapted to parasitism by enlarging their body size (especially of females) and adapting their digestive apparatus to blood meals (Böckeler 1984a).

Ontogenetic studies on *Reighardia sternae* show clearly that the expansion of body size was due to a sac-like enlargement of the last original segment caudal to the genital pore (Böckeler 1984a).

Morphology and diversity
Pentastomids are obligate parasites at all stages of their life cycle. The definitive hosts of Pentastomida are usually reptiles, about 70% of which are snakes. Intermediate hosts are found among invertebrates and vertebrates of all classes, except birds. Within the intermediate host, larvae

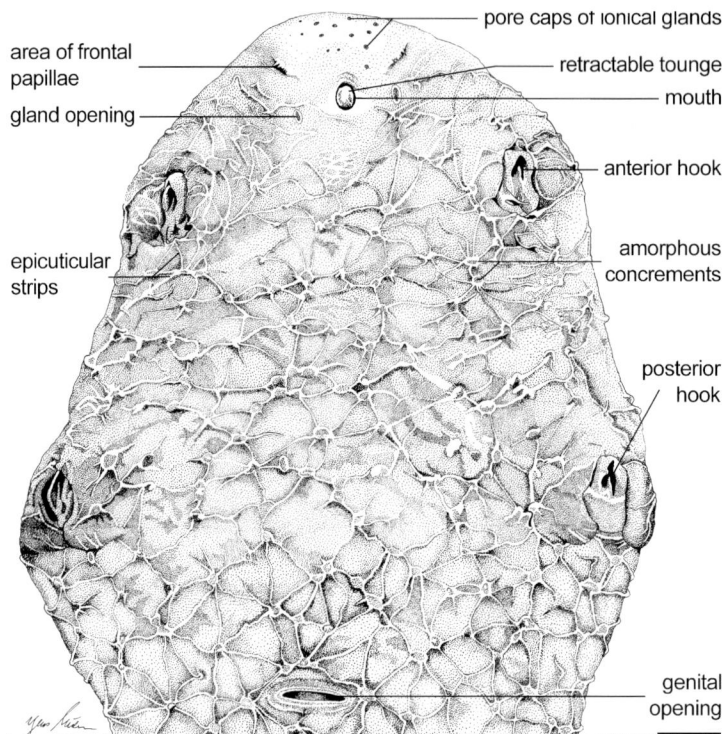

Figure 5.35 Anterior end of a female of *Reighardia sternae*. Scale bar = 0.5 mm.

become encapsulated and develop to the 5th larval stage by moulting. Species of the genus *Linguatula* have mammals as final and intermediate hosts.

All recent species are vermiform; they are bloodsuckers (respiratory system and air sacs of sea birds and reptiles) or feed on mucus and sloughed cells (nasopharyngeal cavity of mammals) and attain maturity in the respiratory tract of their final hosts. The length of the females ranges from 20 mm to 160 mm and thus they belong to the largest endoparasites within the arthropods. Males reach a size between 10 mm and 20 mm. Ten different types of glands have been described from *Raillietiella* sp. (Stender-Seidel *et al.* 2000). Penetration by histolytic processes, evasion of the host's immune response (Riley *et al.* 1979, Riley 1992), pheromones, migration within the body cavity of the host and antimicrobial avoidance in the lungs of the host require high specialisations of the glandular system. Their migration over a long distance in the final hosts also demands a surprising number of different kinds of sensilla (Storch and Böckeler 1979). However, a vascular system and respiratory organs are lacking. Special cells, the ionocytes may have replaced an excretory system. Strong sexual dimorphism is known for all species. The species *Reighardia sternae* parasitising sea birds differs in its morphology from that of other species: the exocuticle exhibits tubercular structures, the intestine ends blindly and the eggs develop synchronously in the uterus.

The genus Reighardia

Although the genus *Reighardia* occurs and is transmitted within the marine environment, it is not a typical marine parasite since it only infects marine birds. *Reighardia. sternae* has been recovered from 17 species of sea birds worldwide (Laridae; Sternidae; Alcidae). Dyck (1975) has described a second species, *R. lomvia*, from *Uria aalge*. Transmission is direct and no intermediate hosts (e.g. fish or invertebrates) are involved.

Table 5.4 Incidence of *Reighardia sternae* in *Larus argentatus* off the German coasts

Nestling			One-year-old 'juveniles'[C]			Two-year-old			Three-year-old			Four-year-old adult			Summary		
n	+	%	n	+	%	n	+	%	n	+	%	n	+	%	n	+	%
37[A]	-	-	475	74	15.6	187	44	23.5	86	3	3.5	380	7	1.8	1165	128	11.1
49[B]	-	-	251	25	10.0	162	27	16.7	74	11	14.9	248	4	1.6	784	67	8.5
			129[C]	32	24.8				31	1	0.3	161	2	1.2	321	35	10.9

[A]Böckeler (1984b), [B]Vauk-Hentzelt (1984), [C]Riley (1972).

The biology of Reighardia sternae

Incidence of infection

The occurrence of *R. sternae*, which to date has only been reported in sea birds, makes it the only pentastomid species occurring in a marine environment (beside *R. lomvia*). *Reighardia sternae* is euryxenous (i.e. a parasite with a wide host range), since it has been isolated from 17 different species of sea birds worldwide (Laridae; Sternidae; Alcidae).

Collections of seagulls, mainly *Larus argentatus*, from British coasts (Halton Deans, Calverly Bridge, Scarborough and at other locations on the east coast of Yorkshire) and from German coasts (Helgoland, Cuxhaven, Lübeck and Kiel) followed by laboratory experiments (Banaja *et al.* 1975, Böckeler 1984b, Riley 1972, Vauk-Hentzelt 1984, von Haffner and Rack 1965) strengthen the opinion that *R. sternae* is homoxenous (i.e. occurs only) in sea birds. The questions of the time and mode of transmission and the quest for a facultative intermediate host are fascinating and, until now, have only been solved indirectly.

Investigations of seagulls from Helgoland, Cuxhaven, Lübeck and Kiel have elucidated the epidemiological situation of *R. sternae* (Böckeler 1984b, Vauk-Hentzelt 1984) on German coasts and on the east coast of Yorkshire (Riley 1972). The investigations were based on 2270 *Larus argentatus* collected to clarify extensive and intensive infections by *R. sternae*.

Studies by different authors at different places (Table 5.4) show remarkably similar results: the lowest incidence of *R. sternae* is observed in adults. All authors have reported no findings in nestlings. Similar outcomes have been obtained in *L. marinus* (Böckeler and Vauk-Hentzelt 1979) and *L. ridibundus* (Riley 1972, Böckeler 1984b).

Seasonal occurrence

The distribution of *R. sternae* in the host population is heterogeneous throughout the year. In October–November on German coasts, more seagulls than usual are found to be infected. Not only have recent infections by *Reighardia* larvae been found, but also seagulls with transmittable gravid females. A corresponding biometrical analysis has suggested that, in the autumn and winter, subspecies populations of *L. argentatus* move from the north Atlantic to the German coasts. This implies that infected seagulls migrate to the 'warmer' regions, since feeding resources in the north Atlantic become scarce. Another (lower) peak has been shown for March and April. Between May and September, the prevalence of *R. sternae* is very low, since the infected gulls return to their original biotope. In May–June, adult *L. argentatus* are feeding and raising their nestlings. As adults are rarely infected and nestlings never contain *R. sternae* (Table 5.4), transmission to the nestlings cannot take place during the spring or by feeding.

Intensity of infection

The results of different authors are both similar and remarkable: the intensity of *R. sternae* infections is strikingly low. Out of 185 gulls, 96 individuals were inhabited by only one or two pentastomids; three or four *R. sternae* individuals were isolated from 40 gulls; five or six from 26; and seven to eight from 16, and nine or 10 from only seven gulls. A strong, intensive infection (more than 10 *R. sternae* specimens per seagull) appeared seldom (only in 23 cases of 208). Only twice were heavy infections (more than 50 parasites) observed. Similarly low infections have been observed in *Larus marinus, L. canus, and L. ridibundus* (Böckeler 1984b). Riley (1972) has reported an average of three *Reighardia* females per bird.

Epidemiology and life cycle
The results given above indicate a homoxenous, but euryxenous, life cycle for *R. sternae* (Fig. 5.36): intensive infections and egg production are relatively low. *Reighardia* embryos grow until the transmissible (5th) infective stage for final hosts inside the uterus of the female and not as in other pentastomids (e.g. *Raillietiella* sp.) (Thomas 1995), encapsulated inside an intermediate host, where the 5th stage remains until an oral transmission to the final host takes place. Investigations on freshly shot and dissected seagulls, laboratory experiments and the study of the complete ontogenesis (Böckeler 1984b) strengthen the opinion of direct transmission in *R. sternae* (Banaja *et al.* 1975, 1976, Riley 1972, Böckeler 1984b). An intermediate host has not yet been detected, either naturally or experimentally. Some marine organisms (Amphipoda, Decapoda, and the flatfish *Pleuronecta platessa*) have been studied for their possible role as intermediate hosts for *R. sternae* (Vauk-Hentzelt and Schumann 1980), but without success.

Seagulls (e.g. *L. argentatus*) come into contact with the eggs or total gravid females of *R. sternae* containing the 5th stage larvae. Depending on abiotic factors (ground/temperature/humidity), the females or eggs stay alive for different time periods. Experiments (Vauk-Hentzelt 1984) show the longest survival of 84 hours for vomited eggs or females, when kept in original ocean water near Helgoland at a temperature of 10.7°C.

The expelling of pentastomids via the nasopharynx by infected final hosts is quite normal in *Linguatula* sp. and has been observed by the author in *Raillietiella* sp. (coughed out by the gecko *Hemidactylus* sp.) and in an unknown pentastomid by a snake regurgitating complete conglutinated egg masses and destroyed females.

Possible paths of transmission
On leaving the interclavicular air sac and migrating via the trachea, *R. sternae* females reach the buccal cavity. The parasite might be passed here directly or via the external environment to the next sea bird:

1. when, for example, seagulls hunt each other and strangle their prey containing *Reighardia* eggs or females, the parasite will be taken up with the prey by the superior gull, thereby facilitating direct transmission
2. because they irritate the tracheal and pharyngeal mucous membranes, the parasites cause the bird to cough and vomit complete pentastomids or aggregated egg masses into the environment
3. when spit balls are regurgitated, the complete pentastomids or aggregated egg masses will be passed into the environment
4. when *R. sternae* stays in the buccal cavity of the original host, the worms or eggs will be swallowed; since an autoreinfection has been reported, the beginning of a new life cycle in this way cannot be excluded – deposition of the eggs or parts of the female(s) by defaecation into the environment where they are ingested by a not yet immune host, is, however, likely.

Figure 5.36 *Reighardia sternae*, life cycle.

Behaviour inside the host
Within three days of ingestion, digestion causes the rupture of egg shells; the larvae hatch, penetrate the gut wall and remain outside the duodenum within the body cavity. There they feed on ruptured intestinal capillaries. They grow to about 10 mm. Sexual differentiation and mating take place within four weeks. Females become precociously mature. Copulation takes place when both sexes have a length of about 10 mm and the uterus of the female is not yet filled with eggs. Spermatozoa are stored in two seminal receptacles.

Males then migrate and can be found in all part of the gull's interior where they die, whereas females start their migration by penetration into the abdominal air sac. Within the air sac system, they move via two thoracic air sacs (anterior, posterior) to the clavicular air sac, feeding on the capillaries of the air sac layer. This phase is characterised by an elongation of the slender body of the pentastomid to up to about 50 mm, and high activity of the ovary (oogenesis). The eggs are shed via the oviduct and along the seminal receptacle towards the uterus, only when the females arrive at the clavicular air sac, and start synchronous development. Females grow to their final size (65 mm) and store a total of about 6000 eggs. In about six weeks, the embryos reach the 5th larval stage with two pairs of limbs.

After a prepatent time of about six months, the second migration of the female leads from the clavicular air sac via the bifurcatio tracheae and trachea to the pharynx. There, the whole animal or part of it can be coughed out or swallowed. *Reighardia* eggs, hundreds of which are embedded as a conglomerate in a sticky mucus (produced by the dorsal organ), or whole females are expelled into the environment orally or via faeces.

Effects on hosts and ecological importance
Pentastomids rarely have pathological effects on their hosts. *Reighardia .sternae* has a low incidence of infection. According to biometrical analysis (size, body weight and plumage), seagulls infected with *R. sternae* are without anomalies. Repeated infections are reduced by the development of an immune response (Riley 1979, 1992, Böckeler 1984b). As known to date, pentastomids have no significant ecological importance.

Important references
A general account of the Pentastomida is by Storch (1993). Banaja *et al.* (1976), Riley (1972), Riley (1975), Banaja *et al.* (1976) and Böckeler (1984b) made important contributions to our understanding of the life cycle of *Reighardia sternae*.

The drawings of Jens Müller, Kiel (Germany) are gratefully acknowledged.

Mollusca (molluscs)
Felix Lorenz

Introduction
With more than 50 000 living species, the Mollusca are the second largest phylum and among the most important components of marine ecosystems. The first molluscan fossils appear in the early Cambrium. The phylum comprises seven classes, the Aplacophora (worm snails, marine, 250 species), Polyplacophora (chitons, marine, 600 species), Monoplacophora (marine, 10 species), Bivalvia (mussels, marine and limnic, 7500 species), Scaphopoda (tusk shells, marine, 350 species), Gastropoda (snails, marine, limnic and terrestrial, >40 000 species) and Cephalopoda (octopuses and squids, marine, 600 species). Among the bivalves, certain larval stages parasitise the gills of freshwater fish, but only gastropods have become parasitic in and on marine animals.

Morphology of molluscs

Molluscs are bilateral Spiralia in which the pericard, gonocoel and nephrocoel are the only remnants of a reduced coelom. The body of molluscs is usually composed of four functional sections:

1. The *head* carries optical and sensory organs and the proboscis, a snout or tube containing the radula, the universal feeding-organ of the Mollusca – it is usually a tape-like structure covered with tiny calcareous teeth that enable the snail to rasp food particles off a hard substrate; on either side of the radula there are 'jaws' which may be modified or absent in some groups. The Bivalvia are lacking a head.
2. The *foot* is a muscular organ of locomotion. It is densely ciliate and rich in mucous glands. In many gastropods it has a horny or calcareous operculum.
3. The *visceral sac* contains stomach and interstine, digestive gland, gonads, heart and excretory organs.
4. The *mantle* carries respiratory organs as well as sensory organs (ctenidium). It forms an ectodermal skeleton, the shell, consisting of inorganic (calcium carbonate, aragonite) and organic (conchin) components. The shell may cover the visceral sac, protect the head and the mantle cavity housing the respiratory organs, or as in the Bivalvia and the Scaphopoda, the entire animal. The shell may consist of eight plates (Polyplacophora), two valves (Bivalvia) or is single. It characterises the classes and has taxonomical importance.

Morphology and development of gastropods

Only the Gastropoda have a significant number of parasitic forms in marine ecosystems, although some limnic Bivalvia have parasitic phases in their larval development: their glochidian larvae settle in the gills of various species of fish. There are no significant parasitic Bivalvia known from marine ecosystems.

Most gastropods are characterised by a torsion of the visceral sac and a subsequent coiling of the univalve shell, and a highly modifiable radula. These characteristics have probably led to the evolutionary success of the gastropods. The protective shell may be modified for several purposes: shelter, camouflage, brood-case, attachment platform, tool for opening bivalve shells, and the radula, used for rasping off food, may show great variability. In some groups (Toxoglossa) the radula is modified to poison-loaded barbs which are shot into the prey. In parasitic groups, it may be reduced or absent, whereas the jaws may be stiletto-like for penetration of host tissue.

Gastropods have a characteristic veliger larva with an operculum and ciliate sail-like protrusions on either side of a simple, coiled larval shell. Many families disperse with a planktonic veliger phase, others are intracapsular developers that may use sponges and other marine organisms as vehicles for their brood and hence for dispersal. This strategy might have lead to some forms of parasitism in the Gastropoda.

Parasitism in the gastropods

Marine gastropods can be ectoparasites or endoparasites of many slow moving or sessile marine invertebrates, with a variety of transitional stages (Table 5.5). In the following parts of this subsection, the origins and types of parasitism are discussed, as well as sexual dimorphism and hermaphroditism in some species, and finally, the transition from a commensal life to specialised parasitism.

The way to ectoparasitism

The transition from a free-living mollusc that feeds on debris, sponges and soft corals to a temporary ectoparasite of a particular host can be observed in closely related species. Among the cowries,

Table 5.5 Parasitic gastropods (modified after Warén 1983, Rosenberg 1993)

Family	Parasitic species	Size (mm)	Hosts	Type of parasitism
Eulimidae	1500	2–140	Echinoderms	Ecto-, gall-, endo-
Pyramidellidae	>500	2–25	Molluscs, Annelids	Ecto-, suction
Ovulidae	400	5–80	Coelenterates	Ecto-
Epitoniidae	200	2–60	Coelenterates	Ecto-
Triviidae	100	3–20	Ascidians	Brood-
Architectonicidae	30	3–50	Coelenterates	Ecto-
Colubrariidae	30	15–70	Fish	Ecto-, suction
Velutinidae	30	5–25	Ascidians	Brood-
Cypraeidae	<20	8–150	Sponges	Ecto-
Cancellariidae	>150	5–40	Fish	Ecto-, suction
Marginellidae	>300	2–90	Fish	Ecto-, suction

family Cypraeidae, *Cypraeovula algoensis* from the Atlantic coast of South Africa feeds on small sponges and soft corals. On the Indian Ocean side of the Cape, *Cypraeovula mikeharti* is adapted to parasitising black sponges *Tetrapocillon* sp. and *Guitarra* sp. into which it eats holes and chambers that serve as hiding places and for the deposition of egg clusters. Similar transitions are observed in other cypraeid genera which are ectoparasitic on large sponges. Some of these species may return to preying on various sponges in the absence of their preferential host sponges. This potential of becoming parasitic is found mainly in cowries adapted to areas with extreme conditions and high competition with closely related species. Species of *Zoila* eat deep cavities into sponges in which they hide and deposit their spawn. When torn off the ground, the sponge may serve as a vehicle for juveniles developing in capsules (Lorenz and Hubert 1993, Lorenz 2001).

Obligatory ectoparasitism
The large family of egg-cowries, Ovulidae, comprises obligatory ectoparasites on gorgonians, alcyonarians and antipatharians. The species have spindle-shaped to cup-shaped shells with a concave base that protects the foot of the snail which firmly attaches to the stem or the branches of the host. Most species have separate sexes. The mantle lobes usually cover the shell and have the same colour as the host. They often carry papillae which resemble the host's polyps, camouflaging the parasite perfectly. Ovulids usually spend their whole lives on their host coelenterates, with differing degrees of host specificity. They feed on polyps or body fluids, but retain the capability of moving rather quickly about the host, sometimes causing obvious damage to the colony (Liltved and Gosliner 1983, Liltved 1989, Fretter and Graham 1997).

Brood parasitism
As a form of parental care, some species place their spawn into pockets eaten into the tissue of compound tunicates with the aid of an ovipositor (Fretter 1946). Adults of the family Triviidae live on or in close association with the tunicates, feeding on them and using them as hiding places. Some Triviidae have developed mantle patterns that camouflage them perfectly among their hosts (Liltved and Gosliner 1987).

Parasitism or symbiosis?
In the derived ovulid genus *Pedicularia*, a transition from a parasitic to a symbiotic life can be observed. The female stage of these protandric hemaphrodites becomes sessile and firmly

attaches the shell to stems of sylasterine corals. The proboscis is prolonged for feeding on the mucus that the sylasterine corals secrete (Liltved 1989).

The path between symbiosis and parasitism is narrow in many families of gastropods. Examples for this phenomenon are species of the large family Epitoniidae, the often bizarre-shelled Coralliophilidae as well as the Architectonicidae. Young specimens of these families may crawl about freely, but at a certain size become attached to a host coelenterate. For example, many species of *Epitonium* and *Coralliophila* attach firmly to soft corals of the genus *Palythoa* and subsequently become overgrown (Robertson 1970, 1981). Some of them feed on the host's tissue and hence are parasitic, others feed exclusively on the mucus that the coelenterates secrete. Colonies of discosomatid anemones were observed to grow denser and larger when inhabited by *Coralliophila* which keeps the coelenterate colony clean of secretions (Lorenz 1996).

Temporary ectoparasitism
Those ectoparasitic species that feed by suction usually have reduced radulae and a prolonged proboscis, a suction pump is formed by the buccal mass. The jaws may be modified to stylets that aid in the penetration of host tissue. Temporary ectoparasitism by sucking body liquids or blood is found in many families of gastropods (e.g. Eulimidae and Pyramidellidae) (Vaney 1913, Fretter and Graham 1949, Fretter 1951, Morton 1979, Robertson and Mau-Lastrovicka 1979, Wise 1993). The hosts may be coelenterates, molluscs, annelids, echinoderms, but also fish. Certain species of the families Marginellidae, Cancellariidae and most Colubrariidae (Fig. 5.37A) are known to approach sleeping rays, parrot fish and others, and insert their extremely prolonged proboscis in body openings such as the mouth to reach thin, well-blooded tissues (e.g. of the gills) (O'Sullivan et al. 1987, Bouchet 1989, Johnson and Jazwinski 1995, Bouchet and Perrine 1996).

Sexual dimorphism, hermaphrodites
The large family Eulimidae represents all levels of parasitism, commonly on all classes of Echinodermata (Humphreys and Lützen 1972, Lützen 1972a,b, Warén and Crossland 1975 and Warén 1981, 1983). The members differ from other gastropods in a high degree of sexual dimorphism. The limpet-shaped Indo-Pacific *Thyca cristallina* is a famous ectoparasite living on the starfish *Linckia laevigata*. Subadult specimens crawl across the arms of their host, female adults attach firmly to the oral side of the arms, to the right of the ambulacral groove, facing the oral opening. Dwarfed males live in the mantle cavity of the females. *Thyca* burrows its proboscis deeply into the host tissue, feeding on body fluids (Elder 1979, Egloff et al. 1988).

Echineulima (Fig. 5.37B) parasitises the echinoid families Diadematidae and Echinometridae. They are protandric hemaphrodites. The presence of a female suppresses other males to become females. These parasitic snails attach themselves to the host's test with a disk-like 'snout' (opening of the proboscis) that forms microvilli-like protuberances attaching it firmly to the host tissue. Through this 'anchor', the long proboscis reaches deep into the perivisceral cavity of the host (Lützen and Nielsen 1975).

Gall formation and the way to endoparasitism
The infection with some parasitic eulimids causes the formation of galls within which the parasitic snail is embedded. *Sabinella* inhabits the spines of sea urchins and forms galls in which female, male and egg capsules find shelter (Fig. 5.37C). The proboscis of the adults reaches into the host tissue through an opening at the base of the spine (Warén 1983).

Stilifer burrows deep into the tissue of its host asteroid, forming galls. The shell is entirely covered by a pseudopallium. Typical organs of a snail (e.g. tentacles) are absent in this genus as an adaptation to parasitic life. In the related genus *Gasterosiphon* (Fig. 5.37D) only a long canal leaves an opening from the internalised parasite through which the larvae can escape from the

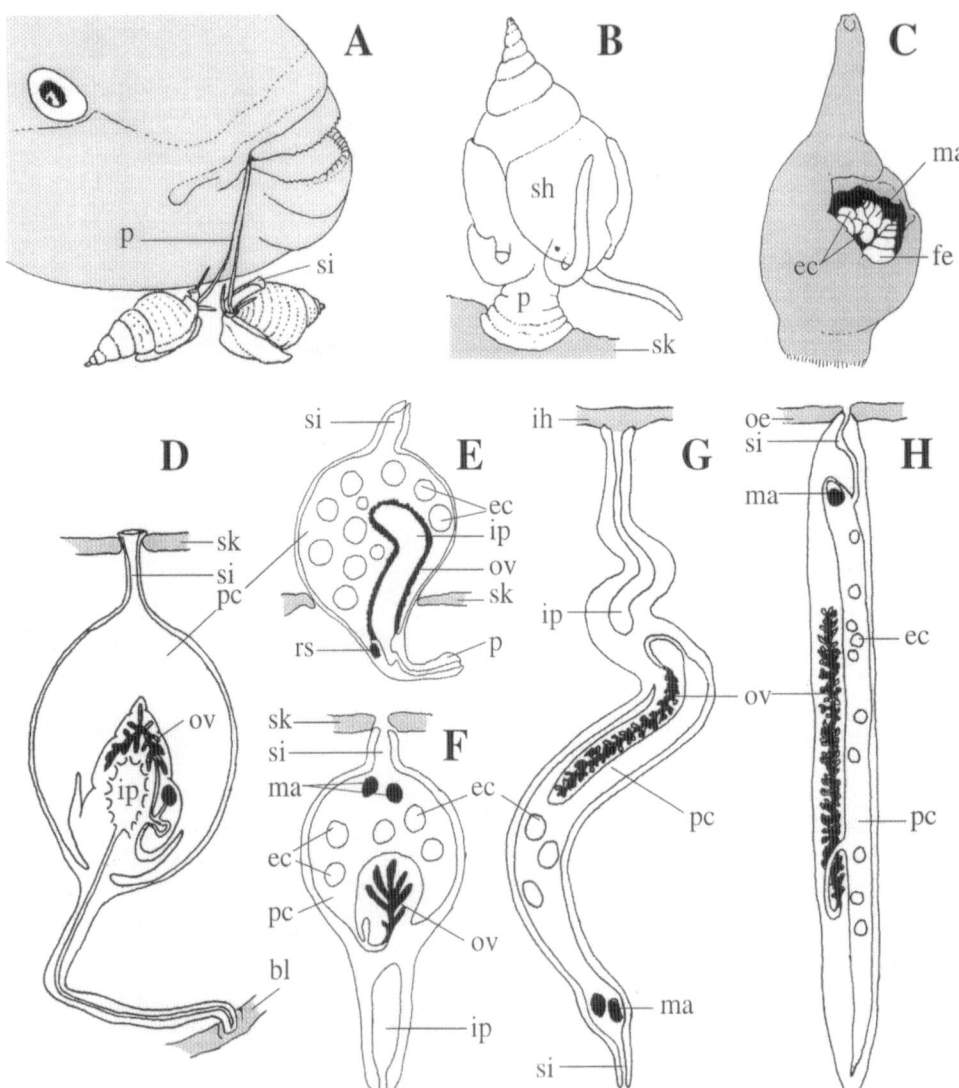

Figure 5.37 **A.** Temporary ectoparasitism. A parrot-fish, *Scarus sordidus*, parasitised by two *Colubraria obscura*. **B.** Ectoparasitism: *Echineulima* sp. parasitising an echinoid. **C.** Gall-formation: male and female *Sabinella* with egg capsules in a gall of a spine of *Eucidaris*. **D.–H.** Transition to endoparasitism. **D.** *Gasterosiphon* in a holothurian. **E.** Female *Diacolax* in host holothurian. **F.** *Entocolax*. **G.** *Entoconcha*. **H.** *Enteroxenos*. Abbreviations: bl, blood-lacuna of host; ec, egg capsule; fe, female; ih, intestine of host; ip, intestine of parasite; ma, male; oe, oesophagus of host; ov, ovary; p, proboscis; pc, pseudopallial chamber; rs, receptaculum seminis; sh, shell; si, sipho; sk, skin of host. **A.** After a photo by M. Strickland. **B.** and **C.** after Warén 1983. **D.–H.** After Warén (1983).

host holothurian. These genera demonstrate transitions to endoparasitism (Warén 1983, Fretter and Graham 1997).

Endoparasitism
Endoparasitism is an exception in the Mollusca. Some genera of the large family Eulimidae have become mostly worm-like, highly modified endoparasitic snails lacking head, radula,

blood system, nervous system and most other organs (*Diacolax*, *Entocolax*, *Entoconcha*, *Thyonicola* and *Enteroxenos*), their host is usually a holothurian. The female of *Diacolax* (Fig. 5.37E) has a rostrum (probably a modified proboscis) deeply inserted in the body cavity of its holothurian host *Cucumaria*. Its intestine forms a blind sac at whose outer side a massive ovary releases its products into a large brood pouch for eggs and larvae formed by the pseudopallium. It covers the entire animal, at whose terminal end there is an opening ('sipho') to release the larvae (Voigt 1901, Koehler and Vaney 1903, Mandahl-Barth 1941, 1946, Tikasingh 1961, Lützen 1979, Warén 1983).

Entocolax (Fig. 5.37F), *Entoconcha* (Fig. 5.37G), *Thyonicola* and *Enteroxenos* (Fig. 5.37H) are genera with endoparasitic species in holothurians. They show gradual morphological transitions to highly derived endoparasites without sensory and locomotory organs, large ovaries, sexual dimorphism, occurrence of dwarf males to absorption of the male, very short larval periods outside their host, and finally, complete loss of intestines (Vaney 1913, Warén 1983).

A well-studied example is *Enteroxenos*, parasitic in aspidochirote holothurians. It has lost most of its organs, including its mouth and an alimentary canal. The larva enters the host through the oral opening, sheds its larval shell and the operculum and becomes a female, up to 14 cm long, a worm-like appendage to the host's viscera. It is covered by a peritoneum of the host, and has a ciliated tubule, a canal communicating with the host's oesophageal lumen on one side, opening into a central body cavity on the other side. An ovarial ridge protrudes into this cavity. The male enters the female through the ciliated tubule from the host's oesophagus, enters the female's body cavity, sheds the shell and operculum and attaches to the 'receptaculum masculinum', an epithelial protrusion of the female's body cavity at the terminal end of the tubule. The male grows to an irregular-shaped vesicle – basically the male becomes nothing but a testis. After oviposition, the female detaches from the host's viscera. Larvae and parasites are released during the holothurian's annual evisceration. The veligers are suspected to have a very short phase outside the host, which probably gets reinfected by swallowing larvae with the debris they feed on (Lützen 1979).

Effects on hosts

The effect of parasitism on the host, with few exceptions, is relatively low. Coelenterate colonies usually seem to suffer very little damage from gastropod parasitism (Robertson 1970). One exception is *Ovula ovum* (Ovulidae), an ectoparasite of the soft coral *Sarcophyton*. A single individual of this 5 cm to 8 cm long snail can destroy a 30 cm diameter colony of polyps within one day (F Lorenz, pers. obs.). Ectoparasitic and endoparasitic eulimids seem to have little effect on their hosts (Warén 1983). Infected holothurians do not show any effects of the parasite on their growth or fertility (Lützen 1979).

Important references

An early account of adaptations of gastropods to parasitism is by Vaney (1913). Humphreys and Lützen (1972), Lützen and Nielsen (1975), and Lützen (1972a, b, 1979) made important contributions to gastropods infecting echinoderms. Papers by Bouchet (1989), Johnson and Jazwinski (1995) and Bouchet and Perrine (1996) contain information on gastropods parasitising sleeping fishes. Robertson (1970) reviewed the predators and parasites of stony corals, with special reference to symbiotic prosobranch gastropods, and Robertson (1981) discussed two prosobranch gastropods symbiotic with Indo-Pacific zoanthids. A detailed study of the anatomy and functional morphology of the feeding structures of an ectoparasitic pyramidellid gastropod is by Wise (1993).

Echiura (spoon worms)

Kirsten Jensen

Introduction

The phylum Echiura (spoon worms) comprises slightly more than 150 species (Hessling and Westheide 2002). Echiurans are bilaterally symmetrical coelomate worms that are considered closely related to annelids (see Edmonds 2000, Hessling and Westheide 2002). In general, echiurans are marine. They are gonochoric and possess a free-swimming, planktotrophic (i.e. feeding) ciliated larva, known as a trochophore. Typically, adult echiurans have a body consisting of two parts, an anterior proboscis and a posterior trunk; and they possess a complete digestive system, with a mouth at the base of the proboscis, and an anus terminally on the trunk. The main organs of excretion are a pair of anal vesicles. In both females and males, metanephridia function as gonoducts, opening via pores on the ventral side of the trunk. Mature female or male gametes are released from the gonads into the coelom, filtered out of the coelomic fluid by the nephrostome and subsequently stored in the gonoduct.

Four families of echiurans (Echiuridae, Bonelliidae, Urechidae and Ikedaidae) are recognised (Murina 1998, Brusca and Brusca 2003). Species in the family Bonelliidae are unique within the phylum in that they exhibit pronounced sexual dimorphism. Female bonellids with everted proboscis can reach a length of over 1 m, whereas males rarely exceed a length of 3 mm (Baltzer 1931a). Murina (1998) recognised 70 bonellid species in 28 genera. Bonellids, however, are most unusual in that dwarf males live in a symbiotic ('parasitic') relationship with females. In general, the morphology of a female bonellid is like that of a typical echiuran, while the morphology of the dwarf male is rather atypical (Fig. 5.38). Males possess a ciliated epidermis, and lack a proboscis. They possess a rudimentary digestive system that lacks both mouth and anus. Instead of gonoducts, they possess a sperm sac consisting of three functional regions: the vas deferens, the sperm reservoir and the terminal funnel with its inlet canal (Schuchert and Rieger 1990). Individual mature male gametes are filtered out of the coelom into the sperm sac, and released via the vas deferens. In addition, whereas time to maturation for females is estimated to be at least two years, males reach sexual maturity within one to two weeks. Depending on the species, bonellid males have been reported from either the body wall, the gonoducts, the anterior regions of the digestive system or the coelomic cavity of the female (Dawydoff 1959). Most information on life cycles and sex determination in the bonellids has resulted from work on *Bonellia viridis*.

Sex determination and life cycle of Bonellia viridis

Sex determination in *B. viridis* is highly unusual, in that for most trochophore larvae sex is not determined at fertilisation (i.e. larvae are metagamic), but in response to an environmental factor: trochophores usually develop into females in the absence of adult females, but develop into males when they come into contact with adult females. The main factor inducing masculinisation is the green pigment bonellin produced by adult females (e.g. Baltzer 1931a,b). Bonellin is produced in the tegument of the proboscis, and to a lesser degree in the trunk (Jaccarini *et al.* 1983, Edmonds 2000). Jaccarini *et al.* (1983) demonstrated that in addition to metagamic larvae, a small percentage of larvae (about 17%) are syngamic (i.e. direction of sexual differentiation (female, male or intersex) is determined genetically at the time of fertilisation).

In the life cycle of *B. viridis*, fertilised eggs are released by the female during annual spawning periods (Baltzer 1931a). Eggs of *B. viridis* are 70–80 µm in diameter. Trophophore larvae of *B. viridis* are unusual among echiurans in that they are lecithotrophic (i.e. non-feeding), with large yolk supply (i.e. with a large yolk supply on which they feed), pigmented green and possess

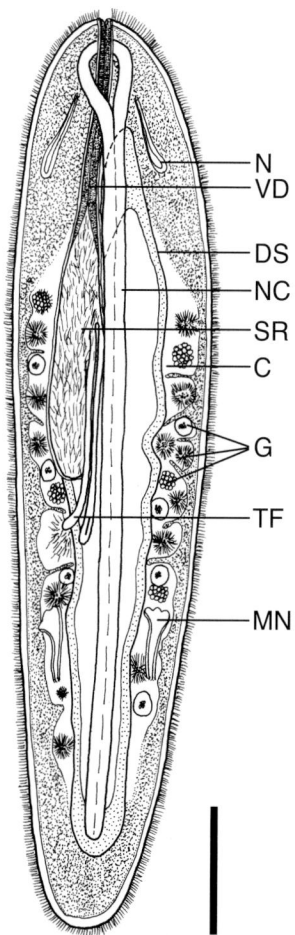

Figure 5.38 Male *Bonellia viridis*. Redrawn and modified from Baltzer (1931a). Scale bar = 250 μm. Abbreviations: C, coelom; DS, digestive system; G, gonads (with sperm in different stages of development; MN, metanephridium; N, nephridium; NC, nerve cord; SR, sperm reservoir; TF, terminal funnel; VD, vas deferens.

eyespots (Dawydoff 1959) (Fig. 5.39). Trochophore larvae first settle on the proboscis of the adult female and attach using their anterior ciliary band, the prototroch. Development into the male body form occurs within about 100 hours of settlement. During this metamorphosis, the anterior end of the larvae is reduced in size, the green pigmentation and the eyespots are lost, the coelom develops, spermatogenesis initiates in the gonads and precursors to the sperm sac develop. At day four to eight post-settlement, males migrate towards the mouth of the female, settling in the pharynx and foregut. Up to 85 males have been found in the foregut of a single female (Baltzer 1931a). A subset of males subsequently migrates from the digestive system to the gonoduct. Males release their gametes directly into the gonoduct of the female. Consequently, whereas fertilisation in most echiurans is external, fertilisation in bonellids is internal.

Although the morphology of dwarf males has been intensively studied, details on the association between male and female bonellids are sparse (Baltzer 1931a). Dawydoff (1959), for example, referred to the association between male and female bonellids as parasitic, whereas Baltzer (1931a) suggested a more symbiotic relationship of the male with the female. Benefits for the

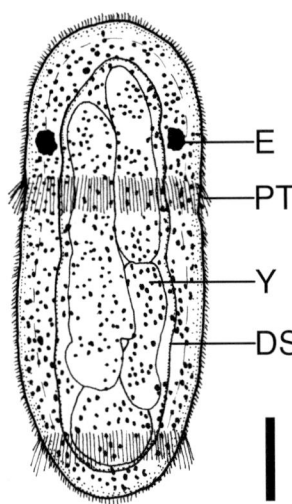

Figure 5.39 Lecithotrophic trochophore larva of *Bonellia viridis*. Redrawn and modified from Baltzer (1931a). Scale bar = 200 µm. Abbreviations: E, eyespot; DS, digestive system; PT, prototroch; Y, yolk.

male include shelter and unchanging environmental conditions inside the female. It has been suggested that after settlement on females, the large yolk deposit of *Bonellia* larvae (Dawydoff 1959, Edmonds 2000) is slowly consumed (Jaccarini *et al.* 1983), and might not last the male over its seven to eight month life span (Baltzer 1931a). The male may be able to obtain nutrition from its location living in the foregut of the female where digestion in the females is initiated. In addition, the gonoducts are lined with secretory cells which have been suggested to produce substances of nutritive value to the male (Baltzer 1931a).

Important references
General accounts of the Echiurida are by Baltzer (1931a), Dawydoff (1959), Murina (1998), Edmonds (2000) and Brusca and Brusca (2003). Baltzer (1931b) described the development of *Bonellia viridis*. Hessling and Westheide (2002) conducted an immunohistochemical analysis of the nervous system in the development of *Bonellia viridis* and examined the question of whether echiurids are derived from a segmented ancestor. The ultrastructure of the dwarf male of *Bonellia viridis* was examined by Schuchert and Rieger (1990).

Echinodermata (echinoderms)
Greg W Rouse

Introduction
With around 7000 described extant species and a fossil record dating back more than 500 million years (Rowe and Gates 1995), it is somewhat surprising that echinoderms have very few incidences of parasitism, even in a broad sense. All extant echinoderms are marine, and most are relatively large. This large size, combined with their hard calcitic skeleton, may explain the scarcity of parasitic forms. Echinodermata contains five major clades, usually given the rank of class: Asteroidea (seastars), Ophiuroidea (brittlestars and basketstars), Echinoidea (sea urchins), Holothuroidea (sea cucumbers) and Crinoidea (featherstars). Concentricycloidea, a group containing only two described species and given class rank by Baker *et al.* (1986) is now regarded as

Figure 5.40 Parasitic echinoderms. **A**. *Rynkatorpa pawsoni*. Lateral view of entire specimen. Reproduced with permission from Martin, WE, 1969. *Biol. Bull.* 137: 332–337, Figure 1. **B**. Arrows indicate *Colobometra perspinosa* (Colobometridae: Comatulida: Crinoidea) perched on a gorgonian. The crinoids, reaching up to 25 cm arm length, are usually found on sea whips or branching gorgonians. Photo G Rouse.

derived Asteroidea (Janies 2001). Of these groups, only Holothuroidea and Ophiuroidea contain taxa that can be regarded as parasitic.

Parasitic echinoderms

Rynkatorpa pawsoni (Synaptidae: Apodida) was referred to by Martin as the first ectocommensal sea cucumber (Martin 1969). Four specimens of this extraordinary holothurian were found, attached by their posterior ends to the host anglerfish *Giganactic macronema*. The fish had been trawled from between 1000 m and 2000 m depth off southern California and the four holothurians were found in a group on one side of the host. *Rynkatorpa pawsoni* (Fig. 5.40A) reaches around 7 mm in length and 2.5 mm in width. It lacks many typical holothurian features, including tube feet and a respiratory tree. It also appeared to lack skeletal elements such as anchors and plates, though these may have been destroyed in the treatment of the specimens (Martin 1969). It does have 12 digitiform retractable tentacles and typical echinoderm features such as a water vascular system. Martin (1969) found that, although the *R. pawsoni* were firmly attached to the host's skin, there was no evidence of invasion. Benefit to the sea cucumbers from the association is unknown.

Ophiuroids are commonly found using other organisms as substrates. Members of Euryalinida (snake-snakestars and basket-stars) are found perched on sponges and colonial cnidarians such as gorgonians, soft corals and black corals. Hyman (1955) provides a summary of known associations between ophiuroids and other animals and many snakestars seem to be using the 'hosts' in order to be higher in the water column in order to capture zooplankton. A few arms will be coiled around the host with other arms waving freely in the water column. Basketstars such as *Gorgoncephalus* are known to attach to soft corals while very young and once their arms have started to branch will move onto adults of their own species. Panikkar and Prasad (1955) reported that several hundred specimens of the ophiuroid *Ophiocnemis marmorata* could be found on the scyphozoan *Rhopilema hispidum*. Marsh (1998) reported *Ophiocnemis marmorata*, which is well known as a benthic ophiuroid, 'hitch-hiking' on another jellyfish, *Cephea cephea*. It appears the ophiuroids settle on the oral arms of the jellyfish as larvae and stay there. Their food source is not known. Some comatulid crinoids (featherstars) also appear to have favoured 'hosts' on which to perch for favourable feeding. These are usually soft corals or gorgonians. One example is *Colobometra perspinosa* (Fig. 5.40B) that is usually found clinging, often in large numbers, to gorgonians.

A commonly seen association is that between ophiuroids using their close relatives crinoids, or even echinoids (sand dollars) as hosts. Numerous ophiuroids are known to be associated with comatulid crinoids and wrap their arms around the cirri and arm bases of their hosts (see Hyman 1955). Many would appear to be obligate associates of crinoids (e.g. *Ophiomaza*) and lie with their mouths close to the crinoids oral disc or amubulacral grooves of the arms. Presumably they steal food from the crinoid before it reaches its mouth (GW Rouse, pers. obs.). Other ophiuroids such as *Nannophiura lagani* and *Amphilycus androphorus* are intimately associated with echinoids and lie attached to the undersides of their hosts. The benefits they gain from this association are not known.

Important references
Hyman (1955) has reviewed echinoderm parasitism. Ophiuroids (Euryalinida) in New Zealand that are often found associated with other animals were recently reviewed by McKnight (2000). Martin (1969) described *Rynkatorpa pawsoni*, and Marsh (1998) discussed 'hitch-hiking' ophiuroids.

Parasitic marine fishes
David Woodland

Introduction
Apart from kleptoparasitic birds such as skuas and frigate birds (*Fregata* spp.) which harass other sea birds, causing them to regurgitate their prey of fish, squid or prawns, and the kelp gull, *Larus dominicanus*, which parasitises whales by feeding on their flesh, all other vertebrate marine parasites are fishes. Even among the fishes relatively few species are parasitic. One group, the pearlfishes, are endoparasites of sea cucumbers, but the rest, that is the lampreys, anglerfishes, cleanerfish mimics, fangblennies, scale feeders, cookiecutters and various browsers, are ectoparasites, especially of other fishes. Of the ectoparasites, some spend extended times attached to a single host; others strike, detach a piece of flesh, and depart. We will treat the ectoparasites first, beginning with the most primitive, the lampreys.

Lampreys
The lampreys, along with the hagfishes, are the only surviving representatives of that ancient group of jawless fishes, the *Agnatha*. Hagfishes are scavengers; but the adults of some lampreys, *Petromyzontiformes*, are parasitic: 15 of the 34 species in the northern hemisphere Petromyzontidae, and three of the four species in the southern hemisphere Mordaciidae and Geotriidae (Gill *et al.* 2003). Of these 18 parasitic species, nine spend their adult life in the sea (Potter and Gill 2003). These sea lampreys occur off the coasts of countries bordering the North Atlantic and North Pacific, along the coast of northern Eurasia, in the western Mediterranean and the Caspian Sea, and off the coasts of South America, southern Australia and New Zealand (Hubbs and Potter 1971).

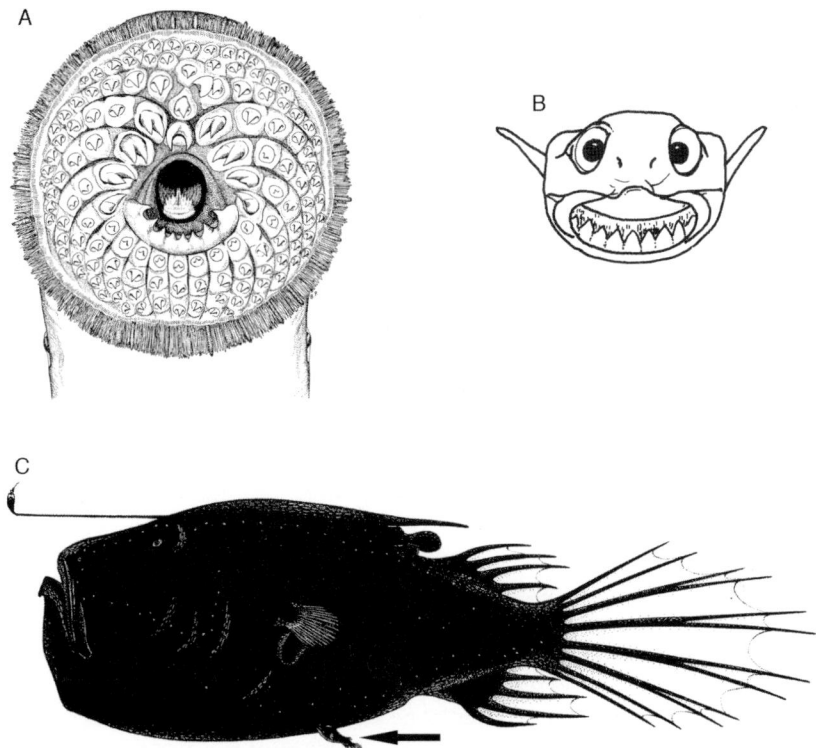

Figure 5.41 A. Suctorial disc of the adult parasitic sea lamprey *Petromyzon marinus*; besides the extensive dentition on the disc, the three rasping plates on the tip of the piston are visible within the oral opening. B. The cookiecutter shark, *Isistius plutodus,* displaying its suctorial lips and dentition; a small shark (length 40 cm), but its lower teeth are relatively the largest of any living shark. C. Female deepsea anglerfish, *Cryptopsaras couesii*, 45 cm total length, with fully grown parasitic male attached ventrally (arrowhead). A. Reprinted From Hardisty MW and Potter IC (1971) *The Biology of Lampreys.* Vol. 1, fig 5, p 22. Copyright (1971), Academic Press, with permission from Elsevier. B. From: Compagno LJV (1984) Food and Agriculture Organization of the United Nations Species Catalogue. Vol. 4 Sharks of the world. [SQUAL Isist 2 'frontal view of head', p.95]. FAO Fisheries Synopsis No. 125, Vol. 4, Part 1. Copyright (1984), with permission from FAO of the UN, Rome. C. Copied from Pietsch TW (1986) Systematics and distribution of bathypelagic anglerfishes of the family Ceratiidae (Order Lophiiformes). *Copeia* **1986**, 479–493; but this figure is reproduced from Bertelsen E and Pietsch TW 1983 The ceratioid anglerfishes of Australia. *Records of the Australian Museum* 35: 77–99, and is redrawn after Tanaka 1911 Copyright (1983), The Australian Museum, Sydney. Drawing by R Nielsen. With permission from The Australian Museum.

Lampreys spend much of their life as larvae; they live in burrows in soft bottoms of rivers and streams, feeding on detritus, diatoms and bacteria (Rogers *et al.* 1980, Potter *et al.* 1986). Depending on the species, larval life span varies between three and seven years, after which a radical metamorphosis occurs which includes the development of a suctorial disc (Fig. 5.41A). Following metamorphosis the adolescents migrate downstream to larger bodies of fresh water or to the sea. The anadromous species spend anywhere between a few months to a few years in the open sea (the exact duration is not known for some species) where they feed mainly on bony fishes. They then re-enter freshwater and cease to feed. Over the next several months their gonads ripen, they migrate to the headwaters of rivers and streams, spawn and die.

The shift from microphagy to parasitism requires radical changes in anatomy and physiology. These include the development of buccal glands whose secretion has both anticoagulant and cytolytic properties (Baxter 1956) and the development of a tooth-bearing suctorial disc that serves for attachment to the body of the host. (The disc is present even in non-parasitic species where it is used, as in all lampreys, for grasping the bottom and for moving stones when constructing the nest used for spawning). Attachment to a host depends on two things, the embedding of disc teeth into its integument and the creation of a negative pressure within the buccal funnel. A fringe of mucus-secreting fimbriae around the circumference of the disc assists in maintaining the suction pressure (Lethbridge and Potter 1979).

The dentition on the disc of the adult lamprey consists of many circumoral teeth. Their principal role is to lock the disc in place on the host, but in tissue feeders the supraoral tooth plate can be involved in the excision of tissue. The processes of gouging out or of abrading tissue are carried out largely by the backward and forward thrusting of 'the piston' that sits on the floor of the oral cavity. On the tip of the piston there are three horny plates (laminae) – one transversely placed, two longitudinally placed (Fig. 5.41A). As the piston moves forward these plates ride up and over its tip. In tissue feeders, the transverse lamina is large and toothed; it closes against the supraoral plate, gouging out tissue; the longitudinal laminae have opposed, serrated edges that snip off bits of teased tissue. In blood feeders, the laminae are small and finely serrated; they abrade the surface of the wound to ensure a continuous supply of blood (Potter and Hilliard 1987).

All nine anadromous species feed on bony fishes. Some feed exclusively on blood (*Petromyzon marinus*, *Mordacia lapicida* and *M. mordax*); others feed primarily on tissues (*Geotria australis* and *Lampetra* spp.) and these may even invade the internal organs leading to the death of the host. Many different species serve as hosts. In laboratory tests involving seven fish species some species were preferred over others, but none was immune from attack (Farmer and Beamish in Hardisty and Potter 1971, p. 148). In the field, attachment to a host is probably opportunistic. At least two species (*Petromyzon marinus* and *Lampetra tridentata*) attach to whales and porpoises (Pike 1951, van Utrecht 1959). As most *P. marinus* attachment sites on right whales, *Eubalaena glacialis*, occur on the highly vascularised flukes, it seems likely that attachment is for feeding rather than for 'hitchhiking' (Nichols and Hamilton 2004).

While most sources classify adult lampreys as parasitic, in some situations they act more like predators. Mortality is likely to be highest where the lamprey is a tissue feeder and the host small. Nevertheless, the frequency of healed scars (40%) on commercial catches of a small species like the smelt, *Osmerus eperlanus*, due to attacks by the tissue feeding *Lampetra japonica* (Nikol'skii 1956 in Hardisty and Potter 1971, p. 150) demonstrates that individuals often survive. Clearly, in many cases lampreys are indeed parasites.

Water flow for respiration in lamprey larvae is unidirectional, entering via the mouth and exiting via the gill openings. But when an adult lamprey is attached to a host this method of water circulation is not possible; instead, water flow is tidal, entering and leaving through all seven pairs of openings into the branchial chambers (Roberts 1950, Randall 1972). While jawless fishes have solved the problem of respiring while attached to a host, simultaneous host attachment and respiration is a far greater evolutionary obstacle for jawed fishes. Of the marine species, only the male ceratioid anglerfishes have risen to the challenge.

Ceratioidei
Apart from their larvae, which live in warm surface waters down to about 200 m, deepsea anglerfishes (with the exception of the genus *Thaumatichthys* which is benthic) are pelagic at a depth of 300 m to 2000 m (Pietsch 1999). This is one of the most impoverished zones in the sea; here communities depend on dead and decaying mesopelagic zooplankton and faecal material raining down from above to support their food web. Despite this, the number of bathypelagic fishes is respectably large (Marshall 1979, p. 407). However, with its members dispersed in a

three dimensional world – unlike those of surface and bottom dwelling species – the distances between nearest neighbours of any single species is likely to be large. The ceratioids face two problems then: how to get enough to eat and how to locate a mate at spawning time.

Their solution to the first problem is extreme sexual dimorphism in size. Females grow to 30 cm to 100 cm; they are equipped with a luminescent lure at the end of a moveable rod (except Neoceratiidae) and a large gape to accommodate, hopefully, the largest prey they might attract (Fig. 5.41C). However, post larval males lack a lure, have very small mouths, and most free-living individuals attain a length of less than 3 cm during their lifetime (Pietsch 2005). Thus, by leaving food acquisition largely to the females, a population might increase its density almost twofold. Even so, there is still the problem of being able to locate a mate. Adolescent males have large, involuted olfactory organs; in females they are only small papillae. It is presumed females produce a pheromone that can be traced to its source by the olfactory sensitive males. Further, males have hooked denticles on their chin and usually on their snout that enable them to attach themselves to a female. In the caulophrynids, ceratiids, neoceratiids, linophrynids and the oneirodid *Leptacanthichthys gracilispinis* – the so-called 'parasitic' species – attachment may be permanent and intimate. Typically, at any one time only one male is attached to a female although as many as eight have been recorded (Pietsch 1976, 2005). In most species, males become fused to the females by outgrowths from their lower jaw and snout, and female tissue may invade their buccal cavity; within the fused tissue there are thin-walled intercommunicating blood sinuses (Pietsch 1976). In some species the male is carried at the end of a papilla protruding from the female. In *Neoceratias spinifer* the dermis degenerates at the point of fusion; blood vascular plexuses are present in the adjacent connective tissues of the head of the male and the skin of the female (Munk 2000).

As attached 'parasitic' males are larger than conspecific unattached males, as much as 10 cm long in some species, they must obtain nourishment from the females to which they are attached (Pietsch 1976, 2005). But how this is achieved is not known (Munk 2000). It seems likely that there is an anastomosis of both male and female capillaries within the fused tissues (i.e. the union is 'placental') but the definitive histological study remains to be done (TW Pietsch, pers. comm.). In any event it seems chemical signals are passed between the male and female that synchronise their gonadal development. This is supported by the fact that in the Ceratiidae and Linophrynidae non-parasitised females and free-living males never have developed gonads. However, in the Caulophrynidae and in *Leptacanthichthys gracilispinis* developed ovaries have been found in both parasitised and non-parasitised females. In these taxa perhaps union is temporary if spawning occurs soon after the male attaches, with fusion occurring only if spawning is delayed (Pietsch 1976). Tissue fusion has been recorded in 22 species (Pietsch 2005). There are around 139 'non-parasitic' species (i.e. the males attach to females but no fusion of tissue occurs) (Pietsch 2005).

Norman and Greenwood (1975) asserted that 'true parasitism' in fishes occurs only among the Ceratioidei. In the opinion of the present author this intraspecific relationship between the sexes is not truly a parasitic one but a mutually beneficial symbiotic one; the female diverts a small amount of her resources to the attached male in exchange for his services as a sperm donor.

Fusion of the male's mouth to the female and the invasion of its buccal cavity by tissues obviously restrict the flow of water through the mouth to the gill chamber. In some cases lateral openings into the pharynx are retained which might permit at least some limited flow of oxygenated water through to the gills. However, in *Neoceratias* and some linophrynids the pharynx becomes completely obstructed (Pietsch 1976, pers. comm.). Nevertheless, the gills in attached males are well developed, which suggests 'water might be pumped in and out through the opercular openings' (Pietsch 1976). If so, both the lampreys and parasitic anglerfishes have solved the problem of supplying oxygen-rich water to the gills by resorting to tidal ventilation. An alternative solution to this problem is to resort to 'hit and run' tactics. Several species have adopted this approach.

Cleaner fish mimics

A remarkable example of the 'hit and run' approach involves the blenny, Aspidontus taeniatus, which mimicks the cleaner wrasses, Labroides spp., that remove ectoparasites from various species of tropical reef fishes. Labroides dimidiatus has the broadest distribution; both it and the mimic are widespread throughout the Indo-Pacific Region. Both species grow to the same size (11.5 cm), and are strikingly similar in superficial body form, in colours and colour pattern. The colouration of L. dimidiatus alters with age and there are also regional differences; individual mimics match these variations both by size and locality (Randall and Randall 1960, Springer and Smith-Vaniz 1972)! In some areas additional species of cleaner wrasse coexist with L. dimidiatus; their colouration may also be mimicked by some of the local populations of A. taeniatus (Russell et al. 1976).

There is also behavioural mimicry. Members of the wrasse family have a distinctive mode of swimming – stroking with their pectoral fins; blennies depend on their caudal fin for locomotion. When stalking a host, A. taeniatus paddles with its pectoral fins like a labrid, but reverts to the typical blenny manner of progression when retreating (Randall and Randall 1960). One obvious difference between cleaner and mimic is their dentition: the mimic has a pair of large fangs in the lower jaw; cleaner fishes are fangless. Individuals of Labroides operate at fixed locations on coral reefs, providing a parasite removal service to fishes that visit the sites. Aspidontus taeniatus, by deception, stalks its target slowly and then strikes rapidly, biting off a piece of fin or skin. Young, naïve fish are more vulnerable to attack than older fish which learn to distinguish mimics from cleaner fishes and drive them away (Randall and Randall 1960).

The relationship between cleaner fishes and the fishes they clean is not entirely benign; gut contents often include scales, mucus and epidermal and dermal tissues (Losey in Randall and Helfman 1972, Gorlick 1980). Fish will sometimes twitch while being cleaned and chase the wrasse away; such 'displeasure' is probably expressed when the cleaner has exhausted the supply of ectoparasites and turned to parasitic feeding. Besides the five species of Indo-Pacific cleaner wrasses in the genus Labroides, there are many species of fish that clean ectoparasites from other fishes, and some of them also engage in eating mucus, scales and skin from their hosts (e.g. the Caribbean broadstripe goby, Gobiosoma prochilos, and the Mediterranean cleaner wrasse, Symphodus melanocercus) (Arnal and Coté 2000, Arnal and Morand 2001) (see also pp. 264–278).

Fangblennies (*Plagiotremus* spp.)

Another group of 'hit and run' parasitic fishes are the fangblennies in the genus Plagiotremus, small tropical species from the Indian and Pacific Oceans. Like their relative, the cleaner fish mimic Aspidontus taeniatus, these solitary hunters employ various forms of deception that allow them to approach a potential fish host sufficiently closely to be able to dart in and detach a piece of its skin. Three species are mimics in both colouration and behaviour of three species of Meiacanthus, fangblennies that feed on zooplankton and small benthic organisms and thus are no threat to the fishes that are deceived by the resemblance of Plagiotremus to them (Springer and Smith-Vaniz 1972, Russell et al. 1976, Randall et al. 1997). Moreover, the fangs of Meiacanthus, unlike those of Plagiotremus, have associated venom glands which are used defensively (Springer and Smith-Vaniz 1972). Not only then are potential hosts duped into a false sense of security by the disguise, but the predatory species among them will be less likely to attempt to capture an individual which may prove to be equipped with venomous fangs. The resemblance of mimics to models is highly coevolved in these relationships; geographical variations in colour are closely paralleled by the mimics.

Another species of *Plagiotremus*, *P. rhinorhynchos*, uses two different strategies to enable it to get close to a potential host. It has two colour forms. The common morph is striped blue, black and silver; like *Aspidontus taeniatus* it is a mimic of the cleaner wrasse *Labroides dimidiatus*. However, the less common orange morph has been observed swimming with a school of the plankton-feeding *Pseudanthias huchtii* (=*Anthias mortoni*) from which it launched attacks against passing fish (Russell et al. 1976). A similar activity involving this morph and *Pseudanthias squamipinnis* has also been recorded (Randall et al. 1997). Apart from the orange colour, the blenny did not resemble the anthiine fishes with which it was associating; apparently blending in with a school of innocuous plankton feeders is effective in improving its chances of launching a successful attack.

Plagiotremus tapeinosoma is another species in which the mimetic association is more flexible than in those species which are mimics of *Meiacanthus*. On the Great Barrier Reef individuals associate with schools of the labrid *Thalassoma amblycephalus*; but during summer when the species extends its range south into the temperate zone, individuals associate with the plesiopid, *Trachinops taeniatus*, in southeast Australia, while in northeast New Zealand they associate with a tripterygiid, *Fosterygion* sp. (Russell et al. 1976). All the species involved in these relationships are slender and similarly coloured with alternating light and dark stripes; and all the models are midwater swimmers that feed on plankton.

The general consensus is that all 10 species of *Plagiotremus* feed exclusively on mucus, skin and scales. The host species are many and varied, almost any largish fish suffices (e.g. scarids, siganids, acanthurids, labrids, serranids and carangids).

Scale feeders

There is a group of fishes that also detach skin and scales from other fishes but they use stealth rather than disguise to approach their target. Scale feeding is particularly common among juvenile ariid catfishes. Szelistowski (1989) found scales in 60% of the stomachs of all seven species of ariid taken on the Pacific coast of Costa Rica, with the highest individual count of 596 scales. Sample size was large enough for one species, *Ariopsis seemanni*, to demonstrate that scale feeding declined with increasing size. In aquaria, this fish fed upon the fins and scales of live and dead fish. Hoese (1966) observed juvenile sea catfish, *Ariopsis felis*, chasing and biting the sides of the mullet *Mugil cephalus* in a boat harbour in Texas. Scales from nine species of fishes were found in the stomachs of 50% of a sample of *A. felis*, although if the largest fish were excluded, the frequency of occurrence rose to 75%. In aquaria the catfish approached their hosts from behind, taking scales from the posterior parts of the body and fins. The same strategy is used by *Terapon jarbua*. By contrast, however, when a sample of this fish from an estuary was divided into those larger and those smaller than 10 cm long, with increasing size the proportion of fish with scales in their stomachs increased from 52% to 76% while the mean number of scales per fish quadrupled (Whitfield and Blaber 1978). Moreover, in aquaria *T. jarbua* only removed scales from live fish. Scale feeding also occurs in two genera of carangids. The juvenile stages of three species of queenfish, *Scomberoides*, and two species of leatherjack, *Oligoplites*, have specialised dentitions used for prising scales from hosts such as mullets (Mugilidae) and anchovies (Engraulidae) (Major 1973, Smith-Vaniz and Staiger 1973, Sazima and Uieda 1980). As the fish grow from 50 mm to 150 mm in length, the outer row of spatulate teeth in the dentary is replaced by a row of conical ones and the parasitic habit is abandoned (Major 1973). Interestingly, juvenile *O. saurus* also act as cleaners for other fish.

In all the above species, although skin, scales and mucus form a significant part of their diet, various invertebrates supplement their food intake. However two triacanthodids, *Tydemania navigatoris* and *Macrorhamphosodes uradoi*, appear to feed exclusively on scales and mucus (Tyler 1968, Mok 1978). These small demersal fishes live between depths of 50 m and 700 m. Nakae and Sasaki (2002) showed that *M. uradoi* removed scales from at least 14 species; it attacks from the rear, taking scales from the caudal fin and its base. The teeth of both species are thin, wide and chisel-like, the mouth is supraterminal and the lips large. As these characters are peculiar to these two species, they are probably the only lepidophagous triacanthodids.

Cookiecutters

Deep, crater wounds found on the sides of large pelagic fishes, whales and dolphins were once thought to be the result of secondary infection at old attachment sites of invertebrate parasites. But Jones (1971) linked them to attacks by sharks of the genus *Isistius*. These small sharks (50 cm) have suctorial lips, and very large lower teeth set in powerful jaws; in *I. plutodus* these lower teeth are proportionally larger than those of any other living shark (Fig. 5.41B). The lower teeth of *Isistius* spp. are broad, flat and sharp edged, the upper ones are small and acutely pointed. These sharks feed by locking onto their hosts with their suctorial lips and sawing out a plug of flesh with their lower teeth. In *I. brasiliensis* the plug is conical, while in *I. plutodus* it is cylindrical, and up to twice the length of the diameter of the mouth (Compagno 1984). Whatever the differences in feeding actions, the official name 'cookiecutter' is especially apt.

There are two, perhaps three, species of *Isistius* (Compagno 2000). We know little about two because they have limited distributions but they probably have a similar biology to *I. brasiliensis* which is found offshore in all tropical seas. It is sometimes caught at the surface at night but more often at depths between 85 m and 3500 m. This suggests the shark undertakes an extensive daily vertical migration in search of prey. The lower surface of its trunk is luminous; such counter-illumination is common in deepsea fishes where its proposed function is to obliterate the outline of the fish to avoid detection by both its predators and its prey. However, it has been suggested that in this case the light might be used to attract large predators that are then themselves attacked (Compagno 1984).

Cookiecutter wounds have been found on the sides of large pelagic fishes such as marlin, tuna and albacore, on the rarely seen giant megamouth shark, *Megachasma*, as well as on whales and dolphins. Size of host is of no consequence to a cookiecutter, they have even attempted to gouge chunks out of the rubber sonardomes of U.S. nuclear submarines (170 m long, 19 000 tonnes) (Compagno 1984, 2001)! Cookiecutters are, however, not exclusively parasitic, their diet also includes whole large squid, small gonostomatid fishes and benthic crustaceans (Strasburg 1963).

Coral reef and deep-sea browsers

Many coral reef fishes specialise in feeding on one of the most abundant resources around them – coral polyps. For example, 23 species of butterflyfishes (Chaetodontidae) feed exclusively on the polyps of hard or soft corals (Allen 1981, Tables 1–3). Individuals or pairs of species of *Chaetodon* either defend feeding territories or have extensive home ranges (Reese 1975). Typically, their feeding does not damage the coral skeleton. By limiting their feeding on any particular colony, the damaged polyps are able to regenerate (D Woodland pers. obs., Randall 1974). Other reef species that appear to feed exclusively on coral polyps are the two species of filefish, *Oxymonacanthus* spp. (Myers 1991). Species of *Labropsis* feed primarily but not exclusively on polyps (Randall 1981). Then there are species that *do* damage the skeleton ; for example, species of trigger, file and pufferfish bite the tips off branching species; none of these feed exclusively on coral (Hiatt and Strasburg 1960).

The diet of the family Chaetodontidae as a whole is extremely varied (Allen 1981). For example, several species feed on a range of invertebrates; but in the present context *Chaetodon capistratus* is probably representative of species that feed mainly on coral polyps. Birkeland and Neudecker (1981) found these comprised 82.5% by volume of its diet. It also browsed on the tentacles of tube dwelling serpulid polychaetes (11.25%), again apparently without killing its hosts. By contrast, 68% of the diet of *C. aculaeatus* was found to consist of the tentacles of serpulids, the remainder being mainly benthic crustaceans and fish eggs; it also was seen picking at sea urchins, presumably removing pedicellariae and tube feet (Birkeland and Neudecker 1981).

Interestingly, several polyp feeders are cleaner fishes as juveniles (e.g. *Labropsis* spp. and *Diproctacanthus xanthurus*) (Randall and Helfman 1972, Randall 1981, Myers 1991) suggesting the parasitic habit may be established early in life.

There are also deep-sea dwellers that nip pieces off epibenthic and burrowing invertebrates instead of ingesting them whole. For example, the cusk eel *Barathrites parri* feeds on the tentacles of holothurians and tube-dwelling polychaetes (Gartner et al. 1997). The large spiny eel *Notacanthus chemnitzii* has knife-like dentition that it uses to clip the tentacles from anemones, corals, bryozoans and hydrozoans (McDowell 1973). The smaller *N. bonaparte* feeds on sponges, bryozoans and ophiuroids (Coggan et al. 1998). There are probably many other deep-sea fishes with this habit so far unrecorded. Though fishes that feed in the way described are usually called 'browsers', they might equally well be classified as full or part-time parasites.

Pearlfishes

Considering their respiratory requirements, it is difficult to imagine how a fish might adapt to an endoparasitic mode of life; however, one group of fishes has achieved this by exploiting an unusual habitat – the coelom and respiratory trees of sea cucumbers (holothurians). The Carapidae or pearlfishes are slender, eel-like fishes, attaining 35 cm. About half of the 35 species are free living, but the others spend at least part of their day hiding inside sea squirts (ascidians), bivalve molluscs and seastars (asteroids) – especially the giant pin-cushion star, *Culcita*, and seacucumbers (Nielsen et al. 1999). While most of these are commensals, at least three and probably all five species of the genus *Encheliophis* (as restricted by Parmentier et al. 2000, Parmentier 2004) feed on the internal tissues of their holothurian hosts.

Holothurians make ideal hosts; they are large, elongate and their respiratory trees are inviting places for any delicate, slender, weak-swimming fish seeking somewhere to hide. The trees are a pair of much branching organs that extend throughout the coelomic cavity and open into a cloaca. The flushing of the trees is extremely efficient: the anal sphincter opens wide, the cloaca contracts rhythmically forcing oxygenated seawater into the branches of the tree. In the reverse process, the trees themselves contract, forcing water out of even the terminal vesicles (Hyman 1955). Moreover, the amplitude of contractions of the cloaca is sensitive to oxygen levels (Lutz 1930), so that the host will compensate when a carapid is in residence.

Adult carapids worm their way into holothurian hosts tail first, a manoeuvre facilitated by a tapering body and an absence of spines in the fins. Furthermore, almost all species lack pelvic fins, and two of the five species of *Encheliophis* also lack pectoral fins (Parmentier 2004). In addition, in species of *Encheliophis* the dentition is reduced and the upper jaw is immovably bound by skin to the head – features that one might expect where feeding on host tissue replaces active hunting for small crustaceans and annelid worms, as practised by their commensal relatives.

Dissected *Encheliophis* usually prove to have empty alimentary canals, but pieces of internal organs presumably from the host animal have been identified: respiratory tree (Murdy and Cowan 1980), testis (Smith 1964) and 'gonads' (Strasburg 1961, Trott and Trott 1972). How a fish penetrates the wall of the respiratory tree to ingest gonads without inconveniencing the host is not known. As the occupancy level of hosts is low it is possible that fish move from one host animal to another to minimise damage; however, in this context, it is worth noting that when irritated holothurians may void their internal organs and subsequently regenerate them.

However, feeding only on gonadal tissue is a form of 'parasitic castration', leaving the host alive but with impaired reproduction. Although, Smith (1964) recorded an individual of *E. gracilis* swimming outside its host, significantly, no free-living invertebrate has ever been recorded in the gut contents of any *Encheliophis*. But there may be another explanation for Smith's observation. Usually only one specimen of *Encheliophis* is found in a host; where two have been found, the fish were sexually mature animals of the opposite sex (Trott and Trott 1972, Murdy and Cowan 1980). It seems that mature adults may go in search of a partner and be prepared to share a host at spawning time. This is just one of many unanswered questions about the biology of this fascinating group of endoparasitic fishes.

Important references

An important account of the distribution, phylogeny and taxonomy of lampreys is by Hubbs and Potter (1971). Potter and Hilliard (1987) discussed the functional and phylogenetic significance of differences in the dentition of lampreys, and Nichols and Hamilton (2004) reviewed the literature on lamprey attacks on cetaceans. The ontogeny, taxonomy, distribution and biology of ceratioid fishes are discussed by Bertelsen (1951), and dimorphism, parasitism and reproductive strategies among deepsea ceratioid anglerfishes by Pietsch (1976, 2005). A detailed discussion of mimetic relationships involving fishes of the family Blenniidae is by Springer and Smith-Vaniz (1972). Randall and Randall (1960) discussed examples of mimicry and protective resemblance in tropical marine fishes, and Springer and Smith-Vaniz (1972) mimetic relationships involving fishes of the family Blenniidae. Hoese (1966), Tyler (1968), Mok (1978), Whitfield and Blaber (1978), Sazima (1983), Szelistowski (1989) and Nakae and Sasaki (2002) studied scale and mucus feeding in fishes.

Concerning 'cookiecutters', Jones (1971) reported that a squaloid shark, *Isistius brasiliensis*, is the probable cause of crater wounds on fishes and cetaceans. Many papers deal with the browsing of coral reef fishes. Important here is the paper by Randall (1974), who discussed the effects of fishes on coral reefs. Birkeland and Neudecker (1981) described the foraging behaviour of two Caribbean chaetodontids in greater detail. Reports of deep-sea fishes that bite off pieces of various invertebrates include those by McDowell (1973) and Gartner *et al.* (1997). Useful contributions to our knowledge of the biology of pearlfishes are by Strasburg (1961), Smith (1964), Trott (1970), Trott and Trott (1972), Murdy and Cowan (1980) and Parmentier *et al.* (2000).

Chapter 6

Behavioural aspects of parasitism

The behavioural aspects discussed in this Chapter are not unique to marine parasites. Parasites do change the behaviour of their hosts in all environments, and cleaning symbiosis is widespread on land and in fresh water. The ability of some parasites to drastically change a host's behaviour, often leading to behaviour patterns favouring the parasite and ensuring its transmission to other hosts, has been known for some time. It is a fascinating area of research that deserves much greater attention in future work. As pointed out by the authors of this section, numerous examples of host manipulation by parasites remain to be discovered, but the examples already known show that manipulation has the potential to interfere with ecological and evolutionary processes. Research in this area may resolve important issues relating to ecosystem functioning and biodiversity.

With regard to cleaning symbiosis, various birds pick off parasites from terrestrial mammals, and there are many freshwater cleaning fish. However, cleaning symbiosis has been best studied in the sea. Reasons are that diversity of cleaning associations that can easily be observed is greatest in the ocean and particularly in tropical seas; and that scuba diving, the method most widely used for studying cleaning behaviour, is best suited to clear subtropical and tropical marine waters. Numerous postgraduate students have, therefore, become involved in these studies, leading to a vast literature on cleaning symbiosis, which is very competently reviewed in this Chapter. This section not only discusses current knowledge but also points the way to future studies.

Parasite induced changes in host behaviour and morphology
Pierre Sasal and Frédéric Thomas

Introduction
By definition, parasitic organisms must reach their hosts in order to live and reproduce. Two main strategies are available for optimising host encounters (Combes 1991, 1995): either increase the number of disseminated propagules and increase random encounters, or produce propagules that will reduce the randomness of host encounters. Modification of host behaviour and/or host morphology are means that facilitate transmission to definitive hosts, the main objective being an increase in the parasites' reproductive success. Facilitation of parasite transmission has been widely studied and reviewed by Dobson (1988), Poulin (1994), Lafferty (1999), Poulin and Thomas (1999) and Moore (2002). Despite these efforts our knowledge of marine host-parasite relationships is still scanty, mainly because most of the studies on fish–parasite relationships were done in fresh water systems, and also because marine biological

models are often very complex. Most of the published studies concern intermediate hosts (both invertebrates and vertebrates) that display, when infected, altered behaviour and/or morphology which make them more likely to be eaten by a predatory definitive host. In these interactions between species (the parasite modifies the intermediate host phenotype and the final host forages on parasitised prey), it is usually assumed that there should be a trade-off between the costs and benefits for both host and parasite. All the examples from the literature concerning the phenotypic changes induced by marine parasites has not been listed here (this was recently done by Moore 2002), instead several demonstrated or suspected examples of facilitation processes due to marine parasites, are presented and some ecological effects that manipulative parasites may have on marine and lagunar ecosystems are discussed.

First examples from parasites with simple, direct life cycles are discussed followed by examples from parasites with complex life cycles. Finally manipulative parasites and their effects on ecosystem functioning are discussed.

Direct life cycle parasites

There are fewer examples of direct life cycle parasites than of complex life cycle ones that modify the phenotype of their hosts to facilitate transmission. However, there are fascinating examples of strategies aimed at increasing the probability of infective stages to find a host. For instance, there are several monogenean species that are able to delay the hatching of their eggs until they are in contact with host mucus. This non-random, active host seeking has been studied in detail in *Entobdella soleae* infecting *Solea solea* (Kearn 1974, Kearn and MacDonald 1976) and reported for other species like *Diplectanum aequans* (Oliver 1987) or *Neobenedenia girellae* (Yoshinaga et al. 2000). Positive phototaxis, chemotaxis and rheotaxis in monogeneans may increase the likelihood of infecting fish which often become lethargic and swim close to the surface when infected with monogeneans (Llewellyn 1972). Although further demonstrations are necessary, the behavioural change displayed by infected fishes might be induced by the parasites already present in order to favour subsequent infection (i.e. the arrival of sexual partners).

Beside these examples of modification of fish behaviour, there are some rare examples of parasites inducing changes in host anatomy/morphology. The only known case of a parasite functionally replacing a host organ is *Cymothoa exigua*, a species that sucks so much blood from its host fish's tongue that the tongue atrophies and is destroyed, and the isopod remains attached to the remaining tongue stub and is used by the host as a replacement tongue for food manipulation (Brusca and Gilligan 1983). Although this example is not really an illustration of facilitation by the parasite, the substitution of an organ could be a way for the parasite to avoid mechanical removal by the host or by cleaning fish. Moreover, replacing the tongue enables the host to maintain its feeding ability while the parasite's gonads mature. This could be a way for the parasite to colonise different habitats, a strategy undoubtedly of selective advantage in fragmented habitats such as coral reefs.

Complex life cycle parasites

In order to complete their life cycle, many parasites use several host species and encounters with these hosts are very unlikely events. Besides having to encounter an appropriate host species at each stage of the cycle, some parasites have free-living stages (e.g. cercariae) which are in a 'find a host or die' situation in a continually changing environment (Combes et al. 1994). The alteration of the intermediate host phenotype in a way that appears to increase its predation risks by definitive hosts is a widespread phenomenon. However, there is relatively little 'true' experimental evidence that parasitised intermediate hosts are more prone to predation than unparasitised ones, and mechanisms responsible for abnormal behaviour of parasitised hosts are in many cases unknown. Among the recent studies on this topic, McCurdy et al. (1999, 2000), for instance, demonstrated that amphipods, *Corophium volutator*, infected with intermediate stages of the nematode *Skrjabinoclava morrisoni* crawled more frequently on the surface than uninfected ones, and are thus more susceptible to predation by sandpipers (McCurdy et al. 1999, 2000). These studies also highlighted that the behavioural change occurred only when parasite larvae were adequately developed to be successfully transmitted to the final host, and also that it occurred only during the day, when the final host forages.

Several studies have revealed that burrowing bivalves infected with digenean metacercariae may be closer to the surface; for example, the cockle *Cerastoderma glaucum* parasitised by *Meiogymnophallus fossarum* (Bartoli 1984, Bowers et al. 1996) or partially autotomise their siphon as does *Paratimonia gobii* (Maillard 1976). This can be interpreted as a way for the parasite to increase the predation probability by a bird final host or by a benthic gobid fish, respectively.

The life cycle of the trematode *Cainocreadium labracis* (Maillard 1976) provides a nice illustration of a host behavioural change resulting from a collective action by kin-related infective stages. The cercariae crawl on the sand surface, just like the targeted intermediate host (generally a gobiid) and this may explain why most of the encysted parasites are found on the ventral surface of the fish. However, more than 80% of the encysted metacercariae were found in muscles directly implicated in swimming. Infected *Gobius* therefore display some inability to escape when they are attacked by the definitive host, the sea bass *Dicentrarchus labrax*. Although numerous parasite species alter their host behaviour through their accumulation in specific organs of their hosts, it is unknown whether the resulting intraspecific competition is a widespread indirect cost met by manipulative parasites.

Compared with the huge effort that parasitologists have devoted to describing phenotypic changes displayed by intermediate hosts, few studies have tried to evaluate the costs for predators of foraging on infected prey. An increasing number of studies demonstrate that although predators become infected by the many parasites generally present in the prey, very few individual parasites successfully achieve their development inside the predator. Aeby (2002) has demonstrated that the coral-feeding butterfly fish, *Chaetodon multicinctus*, on Hawaiian reefs prefers foraging on polyps, *Porites* spp., that are infected by the trematode *Podocotyloides stenometra*. Infected polyps are easier to capture as they are no longer able to retract into their protective coral skeletons. Because costs of infection are low for *C. multicinctus*, the benefits of feeding on infected coral clearly outweigh the costs associated with parasitic infection.

A full understanding of the manipulation processes used by marine parasites to favour their transmission requires an exact knowledge of the selective pressures experienced by both host and parasite. Unfortunately, conditions in laboratory studies as well as in semi-natural experiments are often poor approximations of the processes that occur in the field. Most of the experiments performed do not take into account that, under natural conditions, other predators unsuitable as hosts may also take advantage of the manipulation (but see Mouritsen and Poulin 2003). Finally, an important limitation of most studies on manipulative processes is the recognition that these phenomena occur in a metapopulational context. To understand the value of host manipulation for parasites, and the optimal adaptive level of manipulation, the spatial structure of both host and parasite populations needs to be considered as well as the heterogeneity of the local environment.

Manipulative parasites and ecosystem functioning

From 1985 to 1990 considerable progress has been made in understanding the functional importance of parasites in ecosystems. Much theoretical and empirical evidence has demonstrated that parasites, in spite of their small size, are biologically and ecologically important in ecosystems (Minchella and Scott 1991, Sousa 1991, Combes 1996, Poulin 1999, Thomas et al. 2000, Thomas and Renaud 2001, Mouritsen and Poulin 2002). Little is known, however, about the more specific role(s) of manipulative parasites in these processes (Lafferty et al. 2000). In this part the different ways in which parasite manipulation has been shown, or suspected, to influence species assemblages and ecosystem dynamics is highlighted.

First, parasite manipulation can influence community structure and biodiversity in ecosystems by apparently interfering with competition between hosts. For example, this scenario has been illustrated in salt marshes of southern France for the association between the trematode *Microphallus papillorobustus* and the two congeneric and syntopic amphipods *Gammarus insensibilis* and *Gammarus aequicauda*. Cerebral metacercariae of *M. papillorobustus* induce strong behavioural alterations (i.e. positive phototaxis, negative geotaxis and an aberrant evasive behaviour) making infected gammarids (commonly called 'crazy' gammarids) more vulnerable to predation by aquatic birds. In *G. insensibilis*, metacercariae always alter the behaviour as they are always cerebral (Helluy 1981). Conversely, in *G. aequicauda*, metacercariae can also be abdominal and have in this case no particular effect on the host behaviour (Helluy 1984). In the field, two distinct infection patterns are observed in the two amphipod species (Thomas et al. 1995), indicating that the manipulation exerted by *M. papillorobustus* probably acts as an important mechanism regulating the density of *G. insensibilis* populations versus *G. aequicauda* (see Rousset et al. 1996). Because the higher reproductive success of *G. insensibilis* (Janssen et al. 1979) is offset by its lower tolerance to *M. papillorobustus*, the sympatric coexistence of the two amphipod species is likely to be mediated by this manipulative parasite (Thomas et al. 1995).

Second, manipulative parasites could influence community structure in ecosystems through their influence on the predator community. Many trophically transmitted parasites adaptively change the phenotype of their hosts in a way that increases their probability of being captured by predators (definitive hosts), making them more conspicuous or less able to escape (Combes 1991, 1995, Lafferty 1999). Predators sometimes risk infection when feeding on manipulated prey but, as mentioned before, they also often benefit from enhanced prey capture (Lafferty 1992, Norris 1999, Hutchings et al. 2000). In addition, most manipulative parasites apparently cause little harm to definitive hosts (Lafferty 1992, 1999), so that it can be safely assumed that predators would not only have no *a priori* reason to avoid manipulated prey, but that they should also even prefer foraging on that prey (see Lafferty 1992, Aeby 2002). By increasing the accessibility to prey normally difficult to capture, the net effect of manipulative parasites in ecosystems may be the enhancement of the trophic potential of these habitats. Unfortunately, this idea at the moment is only a hypothesis as, to our knowledge, no study has been published on whether there is a positive relationship between the local abundance of manipulative parasites, the food accessibility for predators and their local richness/diversity.

The third important mechanism through which manipulative parasites may influence processes of community ecology is through their interference with engineering processes. Ecosystem engineers are organisms, plants or animals that directly or indirectly modulate the availability of resources to other species, by causing physical state changes in biotic or abiotic materials (Jones et al. 1994, 1997). Manipulative parasites, by altering the phenotype of their host, can either have impacts on existing ecosystem engineers, or act as engineers themselves (Thomas et al. 1999). The idea that parasites could create new resources for other species by shifting the phenotype of their hosts from one state to another is well illustrated by the association between the cockle, *Austrovenus stutchburyi*, the trematode *Curtuteria australis* and various epibiotic invertebrates (Thomas et al. 1998). *Austrovenus stutchburyi* lives just under the surface of the mud of many sheltered shores of New Zealand. This abundant mollusc can be considered as an autogenic engineer as its shell is the only hard substrate where invertebrates like sea anemones, *Anthopleura aureoradiata*, and limpets, *Notoacmae helmsi*, can attach (see Fig. 6.1A). This cockle is also the second intermediate host of the trematode *C. australis*, a manipulative parasite which enhances its transmission to oystercatchers by altering the burrowing behaviour of cockles

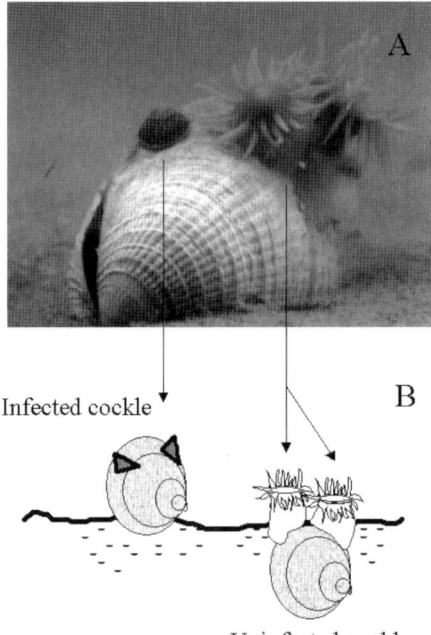

Figure 6.1 **A.** The cockle *Austrovenus stutchburyi* with the two most common invertebrate species living on its shell, the limpet *Notoacmea helmsi* and the anemone *Anthopleura aureoradiata*. **B.** Illustration of the effect of trematode infection on the fouling community. After Thomas *et al.* (1998).

(infected cockles remain at the surface of the mud, Thomas and Poulin 1998). Manipulated cockles apparently correspond to a new kind of substrate for the fouling community of invertebrates. Indeed, limpets which are normally outcompeted for space on burrowed cockles by sea anemones significantly prefer surface cockles (Thomas *et al.* 1998) (Fig. 6.1B). Conversely, manipulated cockles are less occupied by anemones because of their lower resistance to desiccation at low tide. The manipulation exerted by the trematode *C. australis* clearly turns living material (the cockle) from one physical state (buried) into a second physical state (surface) and this act of engineering modifies both the availability and the quality of habitats for invertebrates. The net effect of this manipulation on the local biodiversity is positive since, by reducing competition for space between invertebrates, the local coexistence of limpets and anemones is likely to be facilitated.

The parasite community (the trematodes *Microphallus papillorobustus*, *Maritrema subdolum* and the nematode *Gammarinema gammari*) harboured by the brackish gammarid *G. insensibilis* provides an interesting system for exploring how parasite manipulation can have both positive and negative effects on species richness. As seen above, the trematode *M. papillorobustus* is a manipulative parasite which increases the vulnerability of gammarids to predation by aquatic birds (definitive hosts of the parasite). Using the terminology of Jones *et al.* (1994), it can be considered that this manipulation turns gammarids from a phenotype A (normal behaviour) to a phenotype B (altered behaviour). Thomas *et al.* (1997) showed that the trematode *Maritrema subdolum*, which also completes its life cycle in an aquatic bird but does not alter the behaviour of the host, significantly prefers to infect phenotype B gammarids (Thomas *et al.* 1997). Shared interests seem to exist between the manipulator and the 'hitch-hiker' *M. subdolum*, but the situation is different with the nematode *G. gammari*. Indeed, because this parasite uses amphipods as a habitat and source of nutrition but not as an intermediate host, there is a clear conflict of interest between the nematode and the trematode. In accordance with theoretical expectations, Thomas et al. (2002) found in the field a negative association between the manipulator and the nematode suggesting that *G. gammari* preferred phenotype A gammarids (Thomas et al. 2002). Finally, laboratory experiments suggested that the nematode is able to 'sabotage' the manipulation exerted by *M. papillorobustus*, turning back gammarids from a phenotype B to a phenotype A (Thomas et al. 2002).

Conclusions
Interactions between manipulative parasites and free-living organisms are various and complex. Not surprisingly, parasite manipulation in marine and lagunar ecosystems seems to be as common a phenomenon as in other ecosystems. In addition, it is safe to assume that numerous examples remain to be discovered. Furthermore, parasitic manipulation has the potential to interfere with ecological and evolutionary processes that generate, maintain or reduce biological diversity in ecosystems. Concrete examples of manipulative parasites having important ecosystem functions remain scarce but probably only because of a lack of appropriate studies. Indeed, most studies on parasite manipulations have been performed without considering the ecological context in which they occur. Manipulated hosts must be considered as complete participants in ecosystem functioning, they live in the ecosystem, they feed, they interact with conspecifics as well as other species and they sometimes reproduce. Manipulated hosts keep some of the properties and attributes of uninfected conspecifics but they also display new characteristics. For this reason, it is not unrealistic to consider that manipulated hosts can be equivalent to new organisms in the ecosystem, involved in new direct and/or indirect interactions with other species (Thomas et al. 2005). Even if it appears excessively difficult to determine what ecosystem function(s) would be altered if manipulative parasites were lost, an understanding of the role of manipulative parasites in marine ecosystems (e.g. lagunar ones), would probably permit better management and protection of these habitats. Research on parasitic manipulation is not only a fascinating area of research *per se*, but has also an important potential to resolve issues related to biodiversity and ecosystem functioning.

Important references
The books by Combes (1995) and Moore (2002) contain much information on behavioural aspects of parasite transmission. Dobson (1988) reviewed the population biology of parasite-induced changes in host behaviour, and a discussion of ethological aspects of parasite transmission is by Combes (1991). Accounts of ecosystem engineering can be found in Jones et al. (1994, 1997) and Thomas et al. (1999). Foraging on prey that are modified by parasites is discussed by Lafferty (1992), and evolutionary aspects of parasite transmission by Lafferty (1999), Poulin and Thomas (1999) and Lafferty et al. (2000). Minchella and Scott (1991) showed that parasitism can be a cryptic determinant of animal community structure.

Cleaning mutualism in the sea
Alexandra S Grutter

Introduction
Cleaning behaviour involves cleaner organisms that eat ectoparasites, scales, damaged tissue, and mucus from other animals, the clients or hosts (Feder 1966, Côté 2000, Grutter 2002). These associations are considered mutualistic because it is presumed that both participants benefit from the association. There are several reviews on cleaning behaviour, with a strong empha-

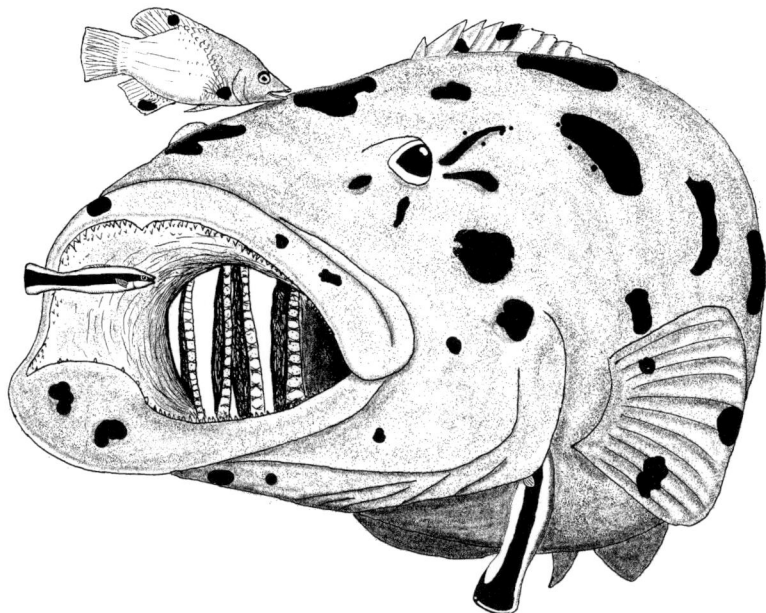

Figure 6.2 Posing with mouth and gills open in a client fish, the potato cod, *Epinephelus tukula*, while being cleaned by two cleaner fish *Labroides dimidiatus* and an adult facultative cleaner fish, *Bodianus axillaris*.

sis on cleaning in the tropics (see *Important references*, p. 278). This section reviews the literature on cleaning behaviour since 2000, but also touches on some of the main earlier topics on cleaning behaviour, while attempting not to repeat previous reviews.

Characteristics and evolution of cleaning behaviour

The interactions between cleaners and clients stand out because of the intimate association between different species. On coral reefs, cleaning behaviour is easily spotted by snorkelers or divers as client fish often posture in unusual ways. For example, clients pose with their mouths and gills wide open and their fins erect (Fig. 6.2) (Feder 1966, Nicolette 1990), or fishes pose head down or head up (Fig. 6.3). Sharks even lie on their side (Sazima and Moura 2000)!

Client fish appear to use many different types of signals in cleaning interactions (Rohde 1993) and posing is one of the most noticeable behaviours (Figs 6.2 and 6.3) (Feder 1966, Hobson 1971, Losey 1971, Côté et al. 1998). These stereotyped postures are often performed either when clients arrive at a cleaning station or after a cleaner approaches a client. Usually fish remain immobile while posing.

Posing may provide cleaners with better access to client parasites and also signal to the cleaner the client's readiness to be cleaned (Hobson 1971). Some species spend a large amount of time posing (Losey 1971, 1972a). Côté et al. (1998) found that fish that posed were more likely to be cleaned. That posing serves as appeasement behaviour by potentially dangerous fish towards cleaners, however, was not supported by higher rates of posing by piscivores (Côté et al. 1998). Hungry piscivorous clients also did not pose more than satiated ones (Grutter 2004). Clients with many parasites, however, posed more than those with few parasites (Grutter 2004). Not all clients pose, however, and not all cleaning is initiated by posing clients (Losey 1971). Côté et al. (1998) used a cost-benefit model to investigate under what conditions clients are more likely to pose.

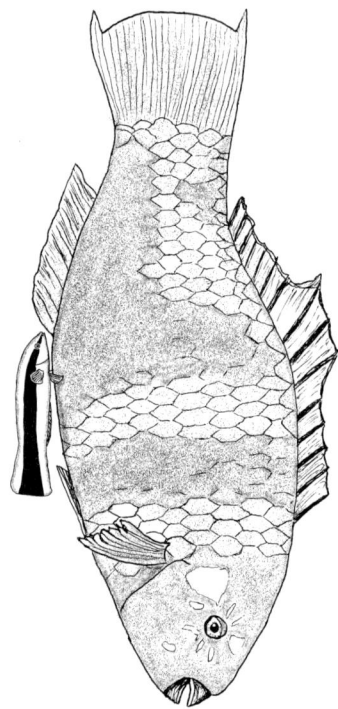

Figure 6.3 Posing head down in the client fish, the minifin parrotfish, *Scarus altipinnis*, while receiving tactile stimulation from the pelvic fins of the cleaner fish *Labroides dimidiatus*.

Even more fascinating than posturing is the apparently uninhibited behaviour of the cleaner. Cleaner shrimp, for example, descend from caves and walk into the mouth of immobile fishes (Fig. 6.4), while moving their pinchers delicately back and forth over the client's lips. Cleaner fish swim into the gaping mouth of piscivorous fish (Fig. 6.2), disappearing completely from view, then popping back out the mouth or via a gill opening.

The stimulus for cleaning in clients

The stimulus for cleaning in clients has generally been assumed to be irritation caused by ectoparasite infection (Limbaugh 1961, Feder 1966). Male garibaldi do not tolerate any other fish, including cleaners, when nest guarding and those collected during the reproductive season had more parasites than males collected out of the reproductive season (Hobson 1971). Cheney and Côté (2003b), did not find this pattern in male damsel fish, however.

The more ectoparasites that clients had, the more they visited cleaning gobies (Arnal *et al.* 2001). Sikkel *et al.* (2000) found a similar pattern, but ectoparasite load and visits to cleaners were also linked to habitat and sex differences of the clients. In contrast, potential correlates of ectoparasite load of clients, such as body size and social behaviour, were not correlated with the rate at which clients visited cleaning gobies (Arnal *et al.* 2000). Côté and Molloy (2003) found that cleaner–client interactions throughout the day did not correlate with the availability of parasitic gnathiid isopods on client fish, but Sikkel *et al.* (2004) did.

Losey (1987) proposed an alternative hypothesis: that it was the rewarding tactile stimulation provided by cleaners (Fig. 6.3) that motivated clients to seek cleaners, rather than parasite irritation. Cleaners, thus, exploited clients in order to obtain food (see Côté 2000 for a review). Losey's (1987) hypothesis predicts that clients should seek cleaners regardless of ectoparasite

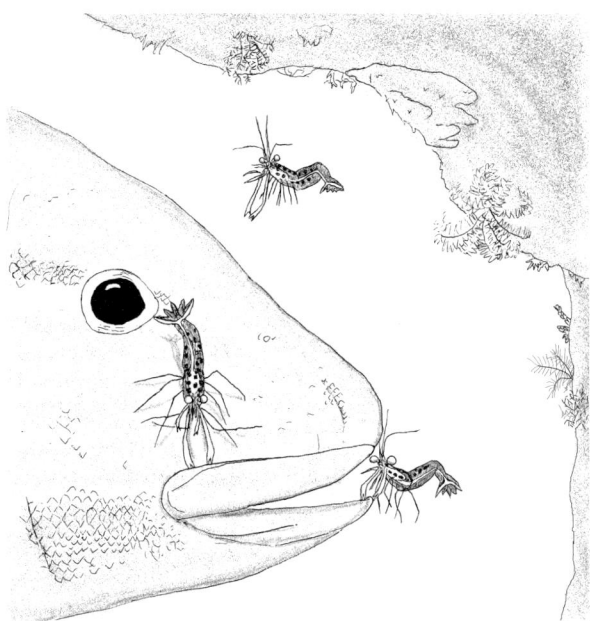

Figure 6.4 Cleaner shrimp, *Urocaridella* sp. c, cleaning a client fish, the painted sweetlips, *Diagramma labiosum*.

load, while the former predicts that if parasite irritation is the stimulus for cleaners to seek clients, then clients with more parasites should seek cleaners more often. Grutter (2001) showed experimentally that parasite infection, not tactile stimulation (Fig. 6.3), was a proximate cause of cleaning behaviour in a client fish.

The evolution of cleaning behaviour

Cleaning behaviour in labrid fishes, the family with the most cleaner fishes, is generally assumed to have evolved from the general labrid behaviour of feeding on small invertebrates (Poulin and Grutter 1996). The preference of the cleaner fish *Labroides dimidiatus* for mucus (Grutter and Bshary 2003), however, suggests a different path, that cleaning originated in opportunistic feeding on mucus off clients, a behaviour found in other fishes (Gerking 1994). Losey (1979) proposed that the evolution of cleaning in *Labroides* spp. did not originate as a symbiotic function. Rather, cleaners gained access to parasites by exploiting a pre-exiting tendency in clients to respond to a tactile reward. This required that the selective pressure exerted by parasite removal began after the relationship had formed. Poulin and Vickery (1995), however, emphasise that fitness benefits for both participants were needed at the start so the association could become established. The finding that some clients seek cleaners more when parasitised than when unparasitised, even when they had no access to tactile stimulation, agrees more with the view that the evolution of cleaning behaviour was driven by fitness benefits to both partners (Grutter 2001).

The parasites involved in cleaning interactions

Gnathiid isopods are the most common parasite found in the diet of cleaner fish (Grutter 2002) and are also eaten by cleaner shrimp (Becker and Grutter 2004). Gnathiids are also among the most common ectoparasites of coral reef fishes (Grutter and Poulin 1998). Therefore, understanding their biology is important for understanding their interactions with cleaner fish and client fish (see p. 144 for a discussion of gnathiid morphology and life cycles). Indeed, only when studies on cleaning began to incorporate information on the biology of parasites, particularly gnathiid isopods (e.g. Grutter 1999a, Sikkel et al. 2000, 2004, Cheney and Côté 2003a,b) was significant progress made in understanding the role of parasites in cleaning interactions.

Only the three juvenile stages of gnathiids are parasitic (Fig. 6.5). Juveniles engorge on fish blood and other fish fluids before returning to the benthos to digest their meal and moult to the next larval stage (Monod 1926, Grutter 2003). Gnathiids leave the reef in search of fish, with some indication of a higher emergence from the reef in early morning and near-dusk (Grutter et al. 2000, Chambers and Sikkel 2002, Côté and Molloy 2003, Sikkel et al. 2004). Gnathiids remain on fish for several hours to days (reviewed in Grutter 2003). On the Great Barrier Reef, however, most *Gnathia* spp. have an engorged gut after only 30 minutes (Grutter 2003) and after 60 minutes, most have dropped off the fish. On the wrasse, *Hemigymnus melapterus*, the abundance of gnathiids declines between dawn and sunset (Grutter 1999a,b). The blood-feeding habit of some gnathiids damages fish tissues and kills captive fish (Paperna and Por 1977, Mugridge and Stallybrass 1983, Honma and Chiba 1991).

On the Great Barrier Reef, the infection rate on fish (Grutter 1996a) and the rapid feeding rate of some gnathiids (Grutter 2003) suggest that their turnover rate on fish is high. A high turnover rate would explain why some fish species seek cleaners so often, some even every five minutes (Grutter 1995). Clients, by visiting cleaner fish repeatedly with short time intervals between visits, thus have gnathiids removed from fish before gnathiids become fully engorged on fish blood.

Many studies have shown a relationship between cleaning activity and gnathiid loads on fish or the rate at which gnathiids emerge from the reef in search of fish (e.g. Grutter 1995, 1999a, b, Sikkel et al. 2000, Cheney and Côté 2001, Grutter 2001, Cheney and Côté 2003a,b, Sikkel et al. 2004). Grutter (2002) reviewed the potential counter-adaptations that parasites may have evolved against the predatory activities of cleaner organisms. Although monogenean flatworms are important ectoparasites of coral reef fishes, very little is known of their role in cleaning interactions (Grutter 2003).

The cleaners

Cleaners include ants, birds, lizards, polychaetes and turtles. The main cleaners in the sea are bony fishes and crustaceans (Jennings and Gelder 1976, Nicolette 1990, Krawchuk et al. 1997). Côté (2000) listed 131 marine species, but additional ones have been described since (e.g. Dewet-Oleson and Love 2001, Mariani 2001, Sazima and Sazima 2001, Feitoza et al. 2002).

Cleaner fish

Obligate cleaners, such as tropical cleaner fishes *Labroides dimidiatus*, *L. phthirophagus* and *Elacatinus* spp. depend on client-derived material such as ectoparasites, while facultative cleaners (Fig. 6.2) also eat other food (Côté 2000). Most cleaner fish are facultative cleaners, often as juveniles (Côté 2000).

Species of the Indo-Pacific genus *Labroides* are highly specialised cleaners with behavioural and morphological specialisations for cleaning (Feder 1966, Côté 2000). The Indo-Pacific *L. dimidiatus* cleans over 100 different fish species, whereas the Caribbean gobies, *Elacatinus* spp., and the Mediterranean cleaner fish *Symphodus melanocercus* clean 25 and 18 species, respectively (Arnal et al. 2000, Arnal and Morand 2001b, Bansemer et al. 2002). Some cleaner fish clean great numbers of fish. For example, individual *Labroides dimidiatus* have been observed to clean an average of 2297 fish per day (Grutter 1996a).

A question that often arises is who cleans cleaner fish? *Labroides dimidiatus* clean each other (Robertson and Choat 1974) and will clean the cleaner fish *L. bicolor* (Feder 1966). Cleaner fish *Oxyjulis californica* also clean each other as well as the cleaner fish *Rachochilus vacca* (Hobson 1971).

Parasites recorded in the diet of cleaner fish include isopods, copepods, other parasitic arthropods, flatworms, leeches and ciliates (Grutter 2002). On Lizard Island, Great Barrier Reef, each *Labroides dimidiatus* eats about 1218 parasites each day, mostly gnathiid isopods (Grutter 1996a).

Parasite size may influence the likelihood that a parasite is eaten by a cleaner. *Labroides dimidiatus* selectively feed on larger gnathiid isopods (Grutter 1997a), however the maximum size of gnathiids eaten may be limited by the throat width of *L. dimidiatus* (Grutter 2000). Interestingly, cleaner fish affect *Argathona macronema* corallanid isopods (found on caged fish) that are even larger than gnathiids, but only those up to a particular size (Grutter and Lester 2002).

Sex differences in the cleaner fish diet occur in the cleaner fish *Symphodus melanocercus* with females cleaning more and eating more client-derived material than males (Arnal and Morand 2001a). Similar patterns occur in the goby, *Elacatinus evelynae*, probably because males spend much time guarding eggs (Whiteman and Côté 2002). The sexes of *L. dimidiatus* cannot be differentiated externally; however, their diet varies with age and size (Grutter 2000).

Cleaner shrimp

What little is known about cleaner shrimps is based mainly on descriptive or anecdotal reports of their behaviour towards clients, mainly fishes (Jonasson 1987, Spotte 1998). Some cleaner shrimp clean invertebrates, such as nudibranchs and crabs (Schuhmacher 1973, Lindberg and Stanton 1988). A survey of scientific journals and marine, SCUBA, and aquarium hobbyist guides where cleaner shrimp where identified using photographs or behavioural observations revealed 43 species of cleaner shrimp (Becker and Grutter 2004a).

Bunkley-Williams and Williams (1998) found that the cleaner shrimp, *Periclimenes pedersoni*, removed and ate juvenile *Anilocra haemuli*, cymothoid isopods, on client fish held in captivity. The only report of parasites in the diet of wild cleaner shrimp is of gnathiid isopods, and possibly copepod larvae, in *Urocaridella* sp. c and *Periclimenes holthuisi* from the Great Barrier Reef (Becker and Grutter 2004). *Urocaridella* sp. c also ate *Benedenia* sp. monogeneans offered to them in dishes. More importantly, *P. holthuisi* significantly reduced *Benedenia* sp. loads by 74.5% on captive surgeonfish *Ctenochaetus striatus* within 48 hours (Becker and Grutter 2004). When given a choice of two fish, *Urocaridella* sp. spent most of its time cleaning cods, *Cephalopholis cyanostigma*, parasitised with *Benedenia* sp. than unparasitised cods (Becker and Grutter in press). Hungry *Urocaridella* sp. cleaned parasitised cods longer than satiated cleaner shrimp (Becker and Grutter in press).

The benefits and costs for cleaner fish

Knowing the benefits and costs of an association is important for understanding whether the association is mutualistic. The main benefit that cleaners gain from their behaviour is obviously the food they obtain from clients. This benefit might vary according to cleaner species, and spatially and temporally according to the absolute or relative abundance of food items (Grutter 1997b, Côté 2000, Cheney and Côté in press). Cleaners may also benefit from an immunity against predation, but the support for this is sparse with predation on cleaner fish usually occurring in non-cleaning contexts only (Feder 1966, Côté 2000). Fish ascent height during spawning is constrained by predation risk (Thresher 1984). For *L. dimidiatus*, however, its presumed immunity to predators apparently allows it to spawn high in the water (Sakai and Kohda 2001).

The costs to cleaners appear to be few. The risk of predation is the highest cost and it does occur occasionally (Côté 2000, Grutter 2004). Cleaners often appear cautious by avoiding 'dangerous' areas when interacting with potentially dangerous clients and this does not appear to be due to parasite distribution on clients (Côté 2000, Grutter 2004). Possibly, individual cleaners not engaged in cleaning interactions are not recognised as cleaners, increasing the risk of predation (Côté 2000), although clients seem to recognise cleaners by their small body size and lateral stripes before engaging in cleaning interactions (Stummer et al. 2004).

Another potential cost for cleaners is infection with client parasites. Several cleaner fishes apparently become infected with bucephalid digeneans when they eat encysted bucephalid metacercariae on the surface of client fish (Jones et al. 2004). In *Oxyjulis californica*, a facultative cleaner fish, individuals that engage in cleaning have similar parasites to their clients (Hobson 1971). Cleaner fish used to control the parasites of farmed fish occasionally become infected with the pathogens of salmon, bacteria *Vibrio* sp. and *Aeromonas salmonicida*, and infectious pancreatic necrosis virus (Grutter 2002). Some cleaner fishes may avoid infection because those that clean each other tend to have few parasites (Costello 1996, Costello et al. 1996).

Factors affecting the behaviour of cleaners
The cleaning behaviour of cleaners, such as the choice of client and the rate at which they clean, is affected by a wide range of factors including client characteristics and geography. Within a cleaner species, many factors affect which client individuals they clean. For example, *Elacatinus* spp. and *Labroides dimidiatus* spend more time cleaning larger clients, which also have more ectoparasites (Arnal et al. 2000, Côté 2000, Arnal et al. 2001, Bansemer et al. 2002). Gorlick (1984) showed experimentally that *L. phthirophagus* prefer to clean parasitised fish over unparasitised fish. The frequency of cleaning in some cleaner fish appears to be influenced by the availability of their clients (Côté 2000). Indices for preferences by *Elacatinus* spp. for specific clients, however, are correlated neither with client fish density, body size nor social behaviour (Arnal et al. 2000). In the Mediterranean, the bite rate of cleaners *Symphodus melanocercus* increases with client mucus load (Arnal and Morand 2001b).

The activity of cleaning gobies, *Elacatinus* spp., is negatively influenced by territorial fishes (Arnal and Côté 1998). Although female *Elacatinus evelynae* clean less when accompanied by a male at a cleaning station, this varies with the size of the male with female cleaners cleaning more when paired with a large male, probably as a result of better mate-guarding by larger males (Whiteman and Côté 2002).

In the Mediterranean, cleaner fish activity varies seasonally and geographically, probably as a result of differences in client availability (Zander and Stöje 2002). Bansemer et al. (2002) found geographical differences in the cleaning behaviour of *Labroides dimidiatus*, possibly due to variation in gnathiid isopod loads on clients. Cheney and Côté (in press) also found differences in cleaning goby activity among six Caribbean islands which were strongly correlated to the availability of ectoparasites on clients. Rohde (1993) discussed latitudinal differences in cleaning interactions. Dewet-Oleson and Love (2001) proposed that shrinking habitat resulted in a reduced food supply for kelpfishes, facultative cleaners and that they compensated for this by increasing their cleaning rates.

Communication in cleaning interactions
How do cleaners and clients recognise and communicate with each other? There is a worldwide guild of colouration in cleaners that may be recognised by clients (Côté 2000). Within the *Elacatinus* genus, cleaning species have a longer lateral stripe relative to body length than non-cleaning species (Côté 2000). No universal colour pattern among facultative cleaners, however, was found when lateral stripes of cleaning species were compared with sympatric non-cleaning

congeners. Nor were cleaners more brightly coloured than non-cleaners (Côté 2000). Strummer *et al.* (2004), using wooden models of cleaning wrasses, found that the smaller models with snout-to-tail lateral stripes attracted more fish and sustained the interest of these visitors longer than larger models or those with incomplete stripes.

Some cleaners may use dancing as a form of communication. *Labroides dimidiatus* and juvenile *L. bicolor* often swim in a dancing oscillating manner that has been proposed to signal its cleaning services (Feder 1966, Wickler 1968, Youngbluth 1968). Others propose dancing is used to reduce client aggression (Youngbluth 1968, Losey 1971, Potts 1973). Grutter (2004) found that *Labroides dimidiatus* engaged in more 'tactile dancing' when exposed to hungry piscivorous fish than when exposed to satiated ones and proposed this behaviour served as a pre-conflict management strategy to avoid conflicts with 'dangerous' clients.

The clients
The clients of cleaner organisms are mainly fishes, including sharks and rays, but lobsters, iguanas, turtles, octopuses and whales can also be clients (Côté 2000).

The benefits and costs for clients
The most obvious benefit of cleaning for clients is a reduction in ectoparasite load. Demonstrating this, however, has not been simple (Côté 2000). A few studies have found correlative evidence of a reduction of parasite load in the presence of cleaner fish. Cheney and Côté (2001) found that damselfish *Stegastes diencaeus* living with a goby cleaning station within their territories had fewer gnathiid isopods compared with those without a cleaning station within their territorial boundaries. There was also a negative correlation between damselfish ectoparasite load and amount of time spent at cleaning stations.

Similarly, Limbaugh (1961) found that removing 'all known cleaning organisms' from two reefs resulted in an increase in 'fuzzy white blotches, swelling, and ulcerated sores and frayed fins' on client species, although the study was qualitative and did not use controls. The first experimental studies to examine the effect of cleaner fish on parasites were inconclusive (Youngbluth 1968, Losey 1972b, Grutter 1996b). Gorlick *et al.* (1987), however, found that, although *Labroides dimidiatus* did not affect the abundance of parasites on a damselfish, parasitic copepods were larger in the absence of cleaners.

The first study to show a quantitative effect of any cleaner on parasite abundance was done using caged *Hemigymnus melapterus* on reefs where *L. dimidiatus* were either not disturbed or where they were all removed (Grutter 1999a). Gnathiid isopod abundance increased 3.8-fold on caged *H. melapterus* on reefs without cleaners after 12 days. In a second experiment, no differences were found between reefs with and without cleaners when caged fish were sampled at dawn after only 12 hours. However, there was a 4.5-fold increase in gnathiid abundance on fish from reefs without cleaners when caged fish were sampled the following sunset after 24 hours. This change in gnathiid abundance between dawn and sunset is probably a result of cleaners being active only during the day (Grutter 1996a). It also implies that cleaner fish predation on gnathiids may explain the daily decline in gnathiid numbers on wild fish (Grutter 1999a, b). Cleaner fish also affected the parasitic corallanid isopod *Argathona macronema* on the same caged fish (Grutter and Lester 2002).

For clients to benefit from cleaning it must ultimately be demonstrated that one of the potential consequences of cleaning – a reduction in ectoparasites – results in increased client fitness via higher reproduction or survival. This is easier said than done. Cheney and Côté (2003a) manipulated the presence of cleaning gobies in the territories of male damselfish and quantified damselfish reproductive success. Mating success, rate of egg loss and parental aggression was not affected for up to three months after the removal of cleaning gobies. Ectoparasite load (gnathiids), however, was lower on clients with access to cleaners compared to those without access. This was the second experimental study to show a reduction of parasite abundance by cleaner fish. Possibly, the lack of an effect on reproduction was due to the small reduction in parasite load (Cheney and Côté 2003a).

Another potential benefit of cleaning to clients is wound healing (Limbaugh 1961, Hobson 1971). McCourt and Thomson (1984) observed that mullet, *Mugil cephalus*, being cleaned by sergeant majors, *Abudefduf troschelii*, in Mexico had many sores and scars and proposed that the cleaner fish removed necrotic tissue. Surgeonfish, *Acanthurus coeruleus*, spend much more time at cleaning stations (25.4 min/hour) when severely wounded than they do after they begin to heal (1.6 min/hour) (Foster 1985). Because of the difficulties in showing the fitness benefits of cleaning to both partners, it is generally assumed that energy obtained from feeding on ectoparasites by cleaners and a decrease in the abundance of a known harmful ectoparasite on the client are benefits that make the association a mutualistic one. For most cleaning associations, however, the costs and benefits are mostly unknown.

Several costs of cleaning to clients have been proposed. If such costs are high, they could reduce or outweigh any reduction in parasites and result in a non-mutual outcome of cleaning interactions (Cheney and Côté 2001, in press). Arnal and Côté (1998) found that client fish visited cleaning gobies more when the cleaning stations were within the clients' territory, suggesting travel time is a cost to clients. Travel costs, in the form of increased aggression received from other territorial fish when travelling to cleaners, was also found (Cheney and Côté 2001). Clients, however, appeared willing to pay only limited costs to be cleaned, as they stayed away longer and travelled farther from their territories for social and reproductive activities than they did to seek cleaning gobies (Cheney and Côté 2001).

Increased risk of predation on benthic eggs during cleaning may be another cost of cleaning to *Stegastes diencaeus* damselfish (Cheney and Côté 2003b). When guarding their nests, breeding males visit cleaners less, thus reducing their time away from their nests. Surprisingly, this did not result in nest-guarding males having more ectoparasites. Cheney and Côté (2003b) proposed this could be due to differential exposure to parasites rather than to differential use of cleaning stations because the territories of nest-guarding males tend to be on substrata from which few gnathiids emerge.

For territorial clients, other potential costs include the risk of predation when swimming in open areas while seeking cleaners, the loss of territory resources or the territory itself when leaving it unattended, and the risk of injuries from other fish as a result of aggression with territory intruders (Côté 2000). Cleaners themselves can also injure fish as they eat fish scales (Grutter 1997b, Arnal and Côté 2000).

Transmission of disease by cleaners to clients may be a cost to clients. Lu *et al.* (2000) found DNA of the green turtle herpesvirus in *Thalassoma duperrey*, suggesting that these cleaner fish serve as vectors for the transmission of the agent causing fibropapilloma in green turtles. Cleaner fish used in salmon farming often have parasites (Treasurer 2002). The likelihood of these being transmitted to farmed salmon, however, appears to be low as most parasites require that the host be passed on to an invertebrate to complete its life cycle or be eaten, or the parasites are specific to labrid fishes and so do not readily infect salmon (Costello 1996, Treasurer 2002). Often the pathogens are rare, due to vaccination of the salmon, the pathogens do not affect the condition of the cleaner fish or cleaner fish recover quickly from the pathogen (Treasurer 2002).

Effects of cleaners on client fish distribution

The effect of cleaners on client distribution has been much debated. It is generally assumed that if clients benefit from cleaners, then cleaners should affect client distribution. On the Great

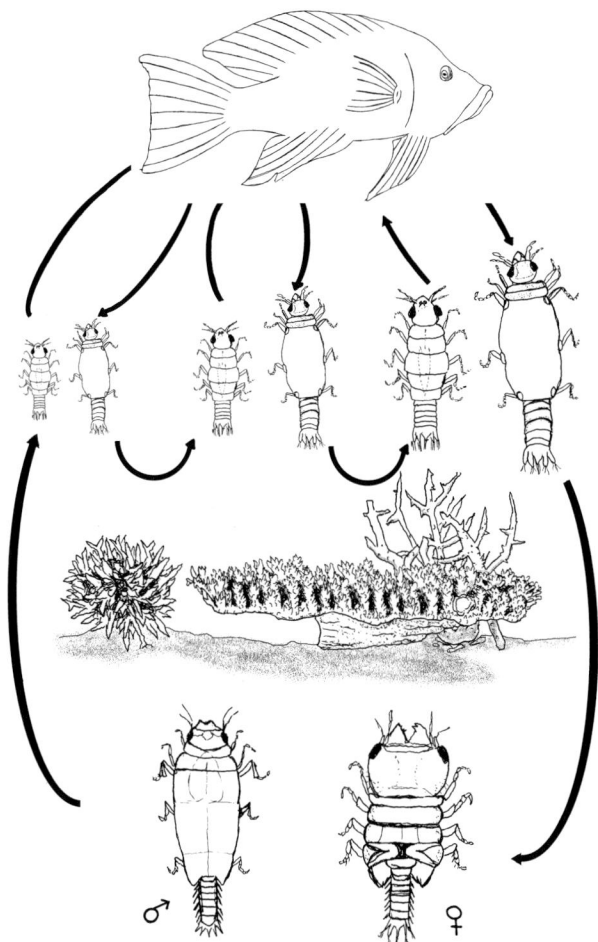

Figure 6.5 The life cycle of parasitic gnathiid isopods. Juvenile parasitic stages, both engorged (large gut) and unfed, are above the coral reef scene while the adult male and female benthic stages are below.

Barrier Reef, individuals of some client species visit *L. dimidiatus* up to a mean of 144 times per day (Grutter 1995). Such a high cleaning activity in clients raises the question of whether it affects where clients occur locally.

Increased abundance and higher local-scale species diversity of client fishes have been correlated with the presence of cleaners (Potts 1973, Slobodkin and Fishelson 1974), suggesting that clients are attracted to cleaners. Limbaugh's (1961) qualitative study suggested that removing all cleaners resulted in a reduction of non-territorial fish. In contrast, several quantitative studies found no effect of cleaners on fish abundance or diversity (Youngbluth 1968, Losey 1972b, Gorlick *et al*. 1987, Grutter 1996b, 1997c, Whiteman *et al*. 2002). This has been interpreted as cleaners responding to, rather than causing, patterns of client distribution (Côté 2000).

Recently, however, on the Great Barrier Reef, the cleaner fish *L. dimidiatus* was found to affect the distribution of 'visiting' client fish, fish that can move between small patch reefs (Grutter *et al*. 2003). Reefs without cleaners for 18 months had one-quarter the abundance and one-half the species diversity of 'visiting' fish (Fig. 6.6) compared with reefs with cleaner fish. Resident fishes (Fig. 6.6), however, were not affected. More dramatically, on coral reefs in the Red

Figure 6.6 'Choosy' or 'visiting' client fish, the great barracuda, *Sphyraena barracuda*, being cleaned by two cleaner fish *Labroides dimidiatus*, while resident client fish, the humbug, *Dascyllus aruanus*, look on.

Sea (Egypt) the species diversity of both 'visiting' and resident fish decreased when cleaners were removed for four to 20 months and increased within two to four weeks when cleaners were added (Bshary 2003).

Variation in cleaner fish densities has been investigated to better understand cleaner–client relationships. Locally, the density of *L. dimidiatus* increases with the density of reef fishes (Arnal *et al.* 1999). Cleaner fish density, however, was not explained by host fish sedentary, territorial or mobile behaviour (Arnal *et al.* 1999). Cleaner fish density increased with client fish body size, possibly because larger fish have more parasites, and with gregariousness, possibly because gregariousness is positively correlated with parasite species richness (Côté and Poulin 1995). Arnal *et al.* (2002) found that *L. dimidiatus* are more abundant where there are more fish species, more sedentary species, few fish living in large groups and few predators. The latter suggests predation risk may negatively influence *L. dimidiatus* density, which is in contrast to Carr and Hixon's (1995) findings in the Caribbean.

The positive correlation between cleaner fish density and fish species richness can be interpreted either as species richness affecting cleaner density, or that cleaner fish affect fish species richness (Slobodkin and Fishelson 1974). Experimental studies show that at the local scale, the latter does occur (e.g. Bshary 2003, Grutter *et al.* 2003).

The presence of a cleaning station in a territory did not affect the habitat choice of adult damselfish *Stegastes diencaeus* (Cheney and Côté 2003c). The benefits of proximity to a cleaning station may be outweighed by the higher rates of intrusions into territories by other fish seeking cleaners (Cheney and Côté 2001). Sluka and Sullivan (1996) found that groupers spent less time in a study area when cleaner shrimp *Periclimenes* spp. were removed from that area. Habitat utilisation in groupers was best explained by the presence or absence of a cleaning station with cleaning gobies or cleaner shrimp, *Periclimenes* spp. (Sluka *et al.* 1999).

Maintaining cooperation in cleaning interactions

Both cleaners and clients can 'cheat' in cleaning interactions. For example, instead of eating parasites, cleaners can remove a piece of client skin. Or worse, a client can eat a cleaner rather than offering it parasites! Clients regularly respond negatively to cleaner bites with a jolt, rapid shaking of the body that is sometimes followed by an aggressive response towards the cleaner (Randall 1958, Losey 1971) or by swimming away (Bshary and Grutter 2002a). Diet analyses show that in addition to ectoparasites, cleaners eat client material such as mucus, scales and skin (Gorlick 1980, Grutter 1997b, Arnal and Côté 2000, Arnal and Morand 2001b) and feed on uninfected parts of fish (Bshary and Grutter 2002b). Furthermore, client jolts are not linked to parasite removal and thus are not a by-product of removing parasites (Bshary and Grutter 2002a). Most likely, many jolts are in response to the removal of mucus or other client tissues. Estimates of amount of protein and calories in the mucus of Caribbean fishes suggest they are equivalent per unit weight to that in ectoparasites (Arnal et al. 2001). For a summary of fishes that specialise on feeding on scales, and occasionally on ectoparasites, see Losey (1987).

When trained to feed off a plastic plate and given a choice of food items, L. dimidiatus preferred parrot fish Chlorurus sordidus mucus and monogenean parasites over parasitic gnathiid isopods and the control glue (Grutter and Bshary 2003). This is surprising given that gnathiids are usually the most common item in L. dimidiatus diet analyses (Grutter 1997b); however, mucus loads of most fishes are probably higher than gnathiid loads and thus cleaners may feed against their natural preference for some mucus in natural conditions (Grutter and Bshary 2003). More importantly, as gnathiids are common on fish (Grutter and Poulin 1998) this suggests a potential for conflict between cleaners and clients over what the cleaner should eat. For the outcome to be mutualistic and also benefit the client, clients would need strategies to make cleaners feed against their natural preferences (Grutter and Bshary 2003). This is supported by studies showing the importance of partner control mechanisms in maintaining the association between the cleaner and client as a mutualistic asssociation (Bshary and Grutter 2002a) (see pp. 276–277). The role of the monogenean parasites mentioned above is unclear in cleaning interactions as, although they have been found in cleaner fish gut contents, they are difficult to detect (Grutter 1997b).

To further complicate matters , when offered two types of mucus, L. dimidiatus prefer parrotfish mucus over snapper mucus (Grutter and Bshary 2004). Mucus quality varies with fish species (Gorlick 1980, Arnal and Morand 2001b). Thus, cleaner preferences and conflicts with clients may vary according to client species. The former is supported by male cleaner wrasse Symphodus melanocercus which prefer clients based on their mucus characteristics (Arnal and Morand 2001b).

Control of cheating cleaners

Biological market theory, where traders exchange goods and services, was proposed as a theoretical framework for understanding cooperation in many systems, including cleaning symbioses (Bshary 2001, Bshary and Noë 2003). Cleaning services and client ectoparasites are the main services and goods traded in the cleaner fish 'market'. Whether a client species has access to several cleaning stations (and so are 'choosy clients') or not ('resident clients') affects cleaning interactions (Fig. 6.6). In the field, choosy clients are given priority of access by cleaners, and choosy clients change cleaners if cheated by a cleaner (Bshary and Schaffer 2002). Resident clients, in contrast, punish cheating cleaners by chasing them around (Bshary and Noë 2003).

In the field, residents are usually willing to wait for a cleaner that is busy while choosy clients usually swim away (Bshary and Grutter 2002c). Second, residents may revisit the cleaner after being ignored, while choosy clients change to another cleaner. These two situations were tested using plates on levers, with equal amounts of food, which were presented to cleaners to choose from. In the first experiment, the 'resident' plate would wait until the cleaner fish was finished with it, while the 'choosy' plate would be pulled away if it was not inspected first and would only come back in the next trial. This simulated situations where residents are more willing to queue for inspections than choosy clients. In the second experiment, the plate that was not inspected first was always pulled away: if it was the 'resident' plate it would then come back immediately after the cleaner was finished with the other plate (which was removed); the 'choosy' plate, however, did not return until the next round. This simulated the natural situation where residents will try again shortly after being ignored, while clients with choice options switch cleaning stations.

In both experiments, cleaners soon fed on the 'choosy plate' first (Bshary and Grutter 2002c). This, combined with the field observations (Bshary and Schaffer 2002), shows that choices made by clients are important in determining the outcome of cleaning interactions and provides strong support for a cleaner fish 'market'.

The cleaner fish *L. dimidiatus* appears to engage in indirect reciprocal altruism (Bshary and Noë 2003). Such altruists gain 'social prestige' (Alexander 1987, Zahavi 1995). Field observations indicate clients invite cleaners more when they see a cleaner behaving positively rather than negatively (Bshary 2002a).

Punishment (Clutton-Brock and Parker 1995), in the form of aggressive chasing of cleaners, also appears to control cheating cleaners as shown by laboratory studies where the client's ability to control the cleaner *L. dimidiatus* was manipulated by lightly anaesthetising it (Bshary and Grutter 2002a). Such chasing in the wild, after cleaners cheat, results in fewer client jolts in future interactions between the same individuals. Thus punishment, a form of partner control, also serves to stabilise the cooperation by reducing cheating in cleaners (Bshary and Grutter 2002a). Punishment can only function if there is individual recognition (Ostrum 1990). It has been shown experimentally that *L. dimidiatus* can recognise familiar clients (Tebbich et al. 2002).

Control of cheating clients
In cleaning, a question that often arises is: how do cleaners avoid being eaten by piscivorous clients? Trivers (1971) used this situation to explain his theory of reciprocal altruism with predators refraining from eating cleaners when the repeated removal of parasites by a specific cleaner led to a greater benefit than did eating that cleaner.

The iterated prisoner's dilemma has long been used to explain the evolution of cooperation between unrelated individuals. This is a game theory model that is iterated (repeated) many times, where two partners (e.g. prisoners) have three choices: either both cooperate (neither prisoner 'informs' on the other), or one cooperates and the other cheats (the prisoner 'informs' on the other), or both cheat. Some of the limitations of this model have been illustrated with the cleaner fish mutualism (Hammerstein and Hoekstra 1995). For example, if clients 'cheat' by eating the cleaner fish, this obviously ends the game and so it cannot be an iterated (i.e. a repeated) game.

Although no empirical tests of cheating behaviour in clients have been done (but see Darcy et al. 1974), there are several game theory models. Poulin and Vickery (1995) suggested that mutualism could evolve without reciprocal altruism with partners benefiting directly. Honesty is the best strategy for clients if cleaning is more beneficial than eating a cleaner, and if costs from cheating cleaners and the frequency of cheating in cleaners are not too high. Johnstone and Bshary (2002) suggested that when clients can terminate an encounter, it reduces exploitation, resulting in a change from parasitism to mutualism. Freckleton and Côté's (2003) model proposed that, in contrast to Poulin and Vickery's (1995) scenario, mixed strategies of cheating and honesty evolve within the cleaner population. This is possible when clients retaliate against cheating cleaners.

Cleaners newly introduced to reefs build up relationships with resident and predatory clients before normal cleaning interactions take place, but not with choosy clients (Bshary 2002b). Cleaners do so by providing tactile stimulation (Fig. 6.3), a behaviour that is incompatible with foraging and thus is an altruistic act on their part (Potts 1973, Bshary and Würth 2001). Such cleaners are rarely cheated. Resident clients initially invest in the relationship by chasing and avoiding new cleaners, possibly to show cleaners that residents will not tolerate any exploitation and to make potential cheater cleaners more cooperative (Bshary 2002b).

Cleaners also use tactile stimulation (Fig. 6.3) to socially manipulate client decisions on how long to stay and be cleaned, and to reconcile with clients after a conflict; predatory clients receive more tactile stimulation than non-predators, possibly as a form of pre-conflict management (Bshary and Würth 2001).

Cheating is not an option for non-predatory clients, as they cannot cheat by eating the cleaner, and so the game is asymmetrical because only the cleaner can cheat (Bshary and Grutter 2002a). Control mechanisms, such as the option to change partners, appears to keep such asymmetric cooperations stable (Bshary and Noë 2003).

Mimics of cleaner fish – effects on cleaning interactions

The fang-blenny, *Aspidontus taeniatus*, mimics *Labroides dimidiatus* so well that it appears to fool other fishes into thinking it is a cleaner fish (reviewed in Côté 2000). This apparently attracts a client close enough for the mimic to dart in and tear client scales and tissue. The mimicry might also give them some immunity from predation, such as has been proposed for cleaners (Côté 2000). Côté and Cheney (2004) showed that the presence of bluestriped fangblennies, *Plagiotremus rhinorhynchus* – mimics of juvenile *L. dimidiatus* – near cleaning stations reduced substantially the rate of visits by clients and the feeding time of juvenile cleaners. A few other species are potential mimics of cleaner fish but the resemblance to the cleaner is less (Côté 2000). Poulin and Vickery (1995) proposed that unless the client can recognise and avoid mimics, their presence will have the same effect on client fitness as that of cheating cleaners.

The fishes *Gobius vittatus* and *Parablennius rouxi* appear to mimic the colour patterns of young *Coris julis*, which are facultative cleaners, possibly to avoid being eaten by predators (Zander and Nieder 1997). Juvenile *Serranus cabrilla*, which are fish predators, however, appear to signal harmlessness to potential prey by resembling young *Coris julis* in order to obtain a meal (Zander and Nieder 1997).

Cleaner fish as a biological control of parasites

Some cleaner fish control the parasites of captive fish. Cowell *et al.* (1993) found that cleaning gobies were more effective than juvenile *Thalassoma bifasciatum* at removing *Neobenedenia melleni* monogeneans from seawater cultured Florida red tilapia. *Labroides dimidiatus* reduced the abundance and affected the size of the monogenean, *Benedenia lolo*, on *Hemigymnus melapterus* held in aquaria (Grutter *et al.* 2002).

Cleaner fish are used for the biological control of sea lice *Lepeophtheirus salmonis* on farmed salmon (Sayer *et al.* 1996, Treasurer 2002). The cleaner fish *Crenilabrus melops*, *Centrolabrus exoletus*, *Ctenolabrus rupestris*, *Labrus bergylta* and *Labrus mixta* are sourced from the wild (Treasurer 2002). Over five million wrasse are stocked annually in Norway, and successful trials have been done in Shetland and Ireland (Treasurer 2002).

Recent work has focused on improving the efficacy of cleaner fish in salmon farms. Recommendations include providing shelters for the cleaner fish to survive the winter and reduce their escapement rates, reduce salmon net fouling which cleaner fish feed on rather than sea lice, stocking *L. bergylta* which apparently remove all adult sea lice from salmon, sourcing cleaner fish locally to avoid introducing new pathogens, testing for pathogens, vaccination of wrasse against furunculosis (caused by the bacterium *Aeromonas salmonicida*), and rearing cleaner fish to guarantee a supply of disease-free fish and to avoid overfishing the wild stock (Treasurer 2002).

Protection and management of cleaners

Cleaner fish are commonly available throughout the ornamental fish trade (Edwards and Shepherd 1992, Wood 2001, Monteiro-Neto *et al.* 2003). In Sri Lanka, *Labroides dimidiatus* are the most commonly collected aquarium fish with an estimated 20 000 exported annually (Wood and Rajasuriya 1998). *Labroides dimidiatus* is also one of the species for which there is a constant demand for export (Wood 2001). *Labroides bicolor* in the Maldives is potentially at risk if the industry expands (Edwards and Shepherd 1992). In Fortaleza, Brazil, the second most commonly collected aquarium fish is *Pomacanthus paru* which makes up 12% of the trade, with five other cleaner fishes making up 7% of the trade (Monteiro-Neto *et al.* 2003). Many species of cleaner shrimp are also collected (Calado *et al.* 2003). Because the presence of cleaner fish increases the abundance and diversity of fishes at the local scale, removing cleaners from reefs on a commercial scale should be done cautiously (Grutter *et al.* 2003). Alternatively, introducing cleaner fish to reefs may be useful for increasing fish diversity on artificial or damaged reefs (Grutter *et al.* 2003).

Conclusions

Much of the research on cleaning symbiosis since 1995 has focused on determining whether cleaning interactions are mutualistic. Some support for mutualism has been obtained; however, the evidence is patchy and the studies are generally restricted to a few cleaner and client species from a few geographical areas. Much of the progress has been a result of increasing information on the parasites involved in cleaning interactions. A more recent focus has been on how cooperation is maintained in cleaning interactions. Laboratory experiments using models of client fish provide a new tool for testing ideas about how cleaners and clients interact. More information is needed on the biology of parasites and their effects on client health; the ecological role of other less-studied cleaner species, particularly cleaner shrimp, facultative cleaner fishes, and cleaner species outside the tropics; and the management and protection of cleaner species involved in the marine ornamental trade.

Important references

The role of parasites in cleaning behaviour in the marine environment was summarised by Grutter (2002). There are several reviews on cleaning behaviour, with a strong emphasis on cleaning in the tropics (e.g. Randall 1958, Feder 1966, Losey 1987, Nicolette 1990, Rohde 1993, Van Tassell *et al.* 1994, Poulin and Grutter 1996, Losey *et al.* 1999, Zander *et al.* 1999, Côté 2000). Dewet-Oleson and Love (2001) summarised the cleaner fishes of California. Cleaner fish in the South Atlantic were examined only recently (e.g. Sazima *et al.* 1997, 1998a,b, 1999a,b, Francini-Filho *et al.* 2000, Sazima and Moura 2000, Sazima *et al.* 2000).

Chapter 7

Ecology

This Chapter includes discussions of ecological aspects of marine parasites that have been studied in some detail, and which are attracting much attention and can therefore be called 'hot topics'. These aspects include transmission of parasites to hosts; specificity of parasites to hosts and microhabitats; parasites as hosts for other parasites; adaptation of parasites to an extreme habitat, brackish water; metapopulation biology; the structure of marine parasite communities; and parasite communities as non-equilibrium systems.

All parasites depend for survival on hosts, at least during part of their life cycles. It is therefore essential that transmission to hosts is assured, as discussed on pp. 280–286. There is no 'universal' parasite that infects all available microhabitats on all available host species. In other words, each parasite species occupies a particular niche – it is microhabitat and host specific. But how is specificity measured? The second section gives an account of such measures. It also discusses proximate and ultimate causes of niche restriction (i.e. the immediate chemical and physical causes that direct a parasite to its niche and are necessary for its survival there), and the biological function of niche restriction.

Parasites not only parasitise their hosts, but also may be hosts to parasites (hyperparasites) themselves. On land, such parasite 'chains' may be of remarkable length. For example, an insect may harbour an insect parasitoid, which in turn is parasitised by a hyperparasitoid of the first degree, to the third degree; a hyperparasitoid of the third degree may (at least theoretically) harbour a nematode parasite infected by a protistan, and the protistan may harbour a bacterium infected by a bacteriophage. Such long chains are not known from the marine environment; however, new hyperparasites are discovered frequently. Marine hyperparasites include crustaceans, monogeneans, nematodes, myxozoans and many protistans.

Brackish water systems are intermediate between freshwater and genuine marine ones. The largest brackish water system is the Baltic Sea, which has been studied over many decades. Its salinity ranges from approximately marine in the western parts to more or less freshwater in the eastern parts. Parasites have various degrees of adaptations to these brackish water habitats, which are discussed later in the relevant section (see pp. 298–301).

The next three sections deal with aspects of population and community ecology including metapopulation biology. The concept of metapopulation is not old. It was introduced to emphasise that populations are not homogeneous collections of individuals but are often composed of subpopulations that are, to a large degree, spatially separated, with limited exchange of individuals between them. This has important consequences for population biology, affecting the survival of species, as discussed in the metapopulation biology section (see pp. 302–309).

Much effort has gone into the study of marine parasite communities. A concise and up-to-date discussion of such communities is given in the relevant section, which includes not only a discussion of community patterns but also of processes leading to the patterns.

How stable are marine parasite systems – are such systems in equilibrium or not? Ecological systems have long been interpreted as ones in equilibrium, in which habitats are saturated with species and individuals and in which interspecific competition is of overwhelming importance. However, it has become increasingly evident that this interpretation is not correct or is of limited value only. Ecological systems, rather than being in equilibrium, are almost always in non-equilibrium, and much evidence for this view comes from the study of marine parasites, reviewed here.

The Chapter concludes with a detailed discussion of an ecological system of marine parasites that has been studied over many years in different localities, that of larval trematodes in mollusc hosts. Because of the thoroughness of studies, it can be referred to as a model system and the section discusses aspects of both population and community biology.

Transmission of marine parasites
David J Marcogliese

General patterns of transmission

Parasites have evolved numerous elaborate means of ensuring that they perpetuate themselves in their assorted environments. In aquatic systems, this often involves the development of free-living stages that have capacity to swim in an aqueous medium. In addition, like parasites in terrestrial habitats, they often possess stages that reside in intermediate hosts that must be ingested for transmission to occur. In the marine environment, one of the greatest challenges faced by parasites involves scale of the habitat. Marine populations are often patchy and scattered over vast distances in three dimensions, that is, across surface area and depth. Because of the dilute nature of the seas, parasites must adapt to maintain themselves in an environment where contact between successive hosts in a life cycle may be periodic or even rare.

Life cycles of parasites may be direct, involving a single host, or indirect, whereby at least one intermediate host and a definitive host are required for complete development to maturity. Virtually all their life cycles consist of multiple developmental transmission stages that may be free living or parasitic. Free-living stages may be transmitted to the next host in the life cycle actively or passively. Free-living stages that are actively transmitted usually swim and search out a susceptible host and penetrate it. Examples include monogenean oncomiracidia, trematode miracidia and cercariae and copepod copepodids. Those forms that are passively transmitted are usually ingested by the appropriate host in the life cycle for transmission to occur. Some representatives include cestode oncospheres, larval nematodes and acanthocephalan eggs. Parasitic stages reside in an intermediate host, which must be ingested by the next host in the life cycle for transmission to occur. Some well-known examples include trematode metacercariae, cestode procercoids and plerocercoids, nematode larvae and acanthocephalan cystacanths. Table 7.1 provides a list of higher taxa of parasites that occur in salt water, the general type of life cycle found in that taxon, the types of infective stages that are involved in transmission and a brief description of the mode of transmission. Both active and passive transmission are widespread. In addition, for parasites with complex or indirect life cycles, a parasitic stage occurring in an intermediate host is a common means of transmission, particularly among the helminths or worms.

The marine realm, being extremely diffuse, poses unique barriers that theoretically could impede parasite transmission. Free-living stages of parasites are typically short lived and must find a host within 24 to 48 hours. Many parasites use short-lived organisms, such as copepods, as their first intermediate host. These animals occur in vast numbers in the oceans, but their rates of parasitism are extremely low, often less than one infected animal in 1000 (Marcogliese

Table 7.1 Various higher taxa of metazoan parasites found in marine organisms, their type of life cycle, the infective stage(s) and the mode of transmission to the next host

Taxon	Life cycle	Infective stage(s)	Mode of transmission
Myxozoa	Indirect	Myxospore	Passive – ingestion
		Actinospore	Active – penetration
Turbellaria	Direct	Egg	Passive – attachment
		Ciliated larva	Active – attachment
Monogenea	Direct	Oncomiracidium	Active – attachment
Aspidogastrea	Direct & indirect	Cotylocidium	Passive – ingestion
		Cotylocidium	Passive – ingestion of intermediate host
Digenea	Indirect	Miracidium	Active – penetration
		Cercaria	Active – penetration or ingestion
		Metacercaria	Passive – ingestion of intermediate host
Amphilinidea	Indirect	Oncosphere (decacanth)	Passive – ingestion
		Procercoid	Passive – ingestion of intermediate host
Gyrocotylidea	Direct (?)	Decacanth	Passive – ingestion
Eucestoda	Indirect	Oncosphere (hexacanth)	Passive – ingestion
		Coracidium	Active – ingestion
		Procercoid	Passive – ingestion of intermediate host
		Plerocercoid	Passive – ingestion of intermediate host
Nematoda	Indirect	Larvae – free living	Passive – ingestion
		Larvae – parasitic	Passive – ingestion of intermediate host
Acanthocephala	Indirect	Egg (acanthor)	Passive – ingestion
		Acanthella, juvenile, cystacanth	Passive – ingestion of intermediate host
Copepoda	Direct & indirect	Nauplius, copepodid, chalimus, adult	Active – attachment
Isopoda	Direct	Juvenile, cryptoniscus, praniza	Active – attachment
Branchiura	Direct	Juvenile	Active – attachment
Tantulocarida	Direct	Tantalus	Active – attachment
Cirripedia	Direct	Cyprid, cypris	Active – attachment
Amphipoda	Direct	Juvenile	Active – attachment

1995). Thus, the probability of transmission to the next host, perhaps a fish, is slim. Parasites must evolve the means to maintain themselves in the environment or risk being unable to complete their development. For fragile organisms with short life spans that depend for transmission on other host organisms also with short life spans, this is indeed a daunting task. Marine food webs are unique, however, in that they have long food chains (Marcogliese 1995, 2002) that include large invertebrate predators such as chaetoganths, coelenterates and a variety of decapods. These invertebrates are typically long lived compared to the smaller copepods, isopods and amphipods that may serve as first intermediate hosts for many parasites. As a result, in marine systems parasites often use paratenic or transport hosts. These hosts are not required for development but they bridge trophic and temporal ecological gaps between intermediate and definitive hosts. The use of paratenic hosts is extremely common in marine systems (Williams and Jones 1994, Marcogliese 1995, 2002). These hosts permit the infective stages of parasites to persist for longer periods in the dilute marine environment and increase the probability of ingestion and, thus, transmission to the next host in the life cycle. There are numerous examples of extremely common parasites that infect paratenic hosts, including hemiurid and didymozoid trematodes, tetraphyllidean cestodes and anisakid nematodes (Williams and Jones 1994, Marcogliese 1995).

Many marine parasites can be transferred from one intermediate or paratenic host to another via predation. For example, the sealworm *Pseudoterranova decipiens* can be passed from one intermediate fish host to another at least twice (Burt et al. 1990). Similarly, because parasites can accumulate in large relatively long-lived invertebrate predators as well as fish, these paratenic and intermediate hosts can accumulate large numbers of parasites that may be transferred all at once to a susceptible definitive host. Furthermore, these invertebrate and fish predators can not only accumulate high numbers of single parasite species, but also a variety of species that may possess similar life cycles and use the same definitive hosts. Thus, when these hosts are eaten by the appropriate definitive host, they may transmit an entire assemblage of species as a single packet (Bush et al. 1993, Lotz et al. 1995, Marcogliese 2002).

Another adaptation of parasites to the marine realm is low specificity. Many of the parasites that use zooplankton intermediate hosts display broad host specificity at that level (Marcogliese 1995). Similarly, many parasites that have benthic life cycles demonstrate low specificity for their invertebrate intermediate hosts. The sealworm can infect copepods, mysids, cumaceans, isopods, amphipods, decapods, annelids and molluscs (Marcogliese 2001). Another anisakid nematode, *Hysterothylacium aduncum*, has been found in not only a variety of arthropods but also in cnidarians, chaetognaths and echinoderms (Marcogliese 1996). This ability to infect a wide range of hosts further serves to allow parasites to persist in an otherwise unforgiving habitat. Low specificity is also a common trait among fish intermediate hosts and even definitive hosts for numerous parasites. *Pseudoterranova* spp. have been found in >75 species of fish from diverse taxa (McClelland 2002). The related whaleworm, *Anisakis* spp., has been recorded from >75 fish species in Canadian waters alone (Margolis and Arthur 1979, McDonald and Margolis 1995).

Free-living infective stages possess further morphological and behavioural adaptations that increase the probability of transmission. Those that are passively transmitted may resemble typical prey of the target host. Actively transmitted free-living stages possess behaviours such as phototropism, geotropism or timing of hatch or emergence that serve to place them in the target host habitat (Williams and Jones 1994). These stages may be chemically attracted to the host over small spatial scales. Parasitic infective stages residing in intermediate or paratenic hosts that are trophically transmitted may modify the behaviour or appearance of their hosts such that they are more vulnerable to predation by the next host in the life cycle (Moore 2002, see also pp. 259–264).

Ecological patterns of transmission
Similar to free-living species, parasites will vary in their distribution within the oceans according to substrate texture and depth. This is because parasite transmission within habitats is based on

the spatial and temporal distribution of their invertebrate and vertebrate hosts (Marcogliese 2002). Most parasites that are trophically transmitted are associated with a particular niche and host diet. The factors controlling host specificity of most parasites of fish in the marine realm are primarily ecological, being host diet and habitat (Marcogliese 2002). Parasites are often generalists but are linked with particular types of host functional groups or feeding guilds. Members of these guilds will share life styles, diet preferences, depth ranges and a predilection for certain sediment types. As a consequence, members of these guilds often share a similar parasite fauna whose constituent species follow common transmission pathways (Marcogliese 2002).

There are certain general ecological patterns that emerge from the distribution of parasites in marine waters. Horizontal gradients in parasite species diversity exist, with more species in shallow waters as a result of the wide range of intermediate hosts available (Marcogliese 2002). For parasites of benthic organisms, species richness declines with depth. In pelagic waters, parasite species richness also declines with depth, and then increases in the bottom waters and along the sediment surface, a zone called the benthic boundary layer (Campbell *et al.* 1980, Campbell 1983).

In the benthos, crustaceans are the most common intermediate hosts, with digeneans, nematodes and acanthocephalans being the most common parasite taxa (Marcogliese 2002). Almost all digeneans require a molluscan first intermediate host, usually placing them in association with the benthos for at least part of their life cycles. Polychaetes and echinoderms may also serve as intermediate hosts for parasites, though to a lesser extent than crustaceans (Marcogliese 2002). Parasite diversity is higher in benthic waters compared to pelagic waters, as a result of the greater diversity and longevity of benthic invertebrates, thus promoting transmission to fish (Campbell *et al.* 1980, Campbell 1983). For example, in a massive and unprecedented survey of invertebrates, 24 species of parasites including 10 digeneans, six cestodes, six nematodes and two acanthocephalans were found infecting 31 species of crustacean (Uspenskaya 1960). Epifaunal organisms, those living on the sediment surface, are extremely important in the diets of many types of demersal fishes and are particularly important as intermediate hosts for parasites.

There is a relatively high diversity of parasites that occur in pelagic waters and use zooplankton as intermediate hosts, at least compared to fresh waters (Marcogliese 1995). In particular, certain groups of digeneans, cestodes and nematodes are common (Marcogliese 1995). Given that the first intermediate hosts in these waters are usually copepods, microcrustaceans that are short-lived, patchily distributed and infected at extremely low levels, the problems associated with transmission in a dilute environment, discussed earlier, become notably acute. Predatory zooplankters such as chaetognaths, coelenterates and ctenophores become especially important as paratenic hosts to ensure the transmission of numerous parasites. Although soft-bodied zooplankters such as coelenterates were traditionally considered unimportant as prey for fishes, recent evidence suggests that they are much more common in the diets of marine fishes than was believed (Arai *et al.* 2003), thus enhancing their role in parasite transmission.

The interface between pelagic waters and the benthos should not be neglected when examining transmission patterns of parasites in the oceans. The hyperbenthos, or suprabenthos, those bottom waters within 1 m of the sediment surface, contains a highly diverse and productive invertebrate fauna. This zone is important for the transmission of some of the most common parasites found in marine waters. For example, sealworm is more common in fish that feed in the hyperbenthos than those that feed on epifauna or benthic infauna (Martell and McClelland 1995). Although the sealworm can infect a plethora of invertebrate taxa, mysids, abundant in the hyperbenthos, appear to be the most important intermediate host (Marcogliese 2001).

The hyperbenthos, being relatively a spatially restricted habitat between the benthic and pelagic realms, also serves as a transition zone for parasites that can be transmitted between habitats. Vertical flow of parasites occurs across zones or habitats as a result of vertical migration of organisms up and down in the water column between the various compartments (Marcogliese 2002). A good example of a parasite that traverses ecological boundaries is the digenean *Derogenes varicus*. This helminth infects over 100 species of fish worldwide, including both pelagic and demersal species. As a digenean, it requires a molluscan first intermediate host, snails of the genus *Natica*. From there, it infects either planktonic or benthic invertebrates as its second intermediate host (Marcogliese 1995) so that it may enter different components of the food chain. Another example of a parasite found in both pelagic and benthic hosts is the anisakid nematode *Hysterothylacium aduncum* (Marcogliese 1996). Both parasites are examples of extreme generalists that also make use of paratenic hosts, helping to explain their widespread distribution.

Transmission and food webs

Because the helminth fauna depends on the habitat and diet of its fish host, the helminth community in a particular fish species demonstrates the role of the host in the local food web (Marcogliese and Cone 1997a, Marcogliese 2002, 2003). The potential pathways that trophically transmitted parasites may follow are illustrated in Figure 7.1. The highly diverse nature of the marine fauna is effectively exploited for transmission in some way by a variety of parasites. For example, the potential role of large invertebrate predators as paratenic hosts is evident (Fig. 7.1).. Parasites may be useful indicators of food web structure and function because they depend on the presence of other hosts in their life cycles for transmission. Moreover, many are transmitted through trophic interactions. In addition, Figure 7.1 suggests that small fishes that serve as intermediate or paratenic hosts to parasites of larger predators and as definitive hosts to other parasites may be the best indicators of food web structure, as their parasite fauna reflect both predators and prey of their hosts. There are numerous advantages in using parasites as indicators of trophic processes and food web structure (Marcogliese and Cone 1997a, Marcogliese 2003). Parasites can be used to indicate, clarify or resolve:

1. direct trophic links between the host and other organisms;
2. long-term data on the host diet;
3. discrepancies in the host diet;
4. ontogenetic changes in the host diet;
5. the trophic level of the host;
6. niche shifts in the host population;
7. individual feeding specialisations in the host population;
8. predators of the host;
9. temporary or seasonal visitors into the ecosystem;
10. inconsistencies in food web theory; and
11. theoretical food web models.

An examination of parasite transmission processes may yield important information on water quality and environmental status as numerous parasitic stages, either free-living or parasitic, may be vulnerable to environmental perturbation, leading to transmission failure. These sensitive parasites may serve as an early warning of environmental change due to pollution or other stressors (MacKenzie et al. 1995, MacKenzie 1999, see also pp. 421–425). Presumably, any disruption in the food web will quickly affect the transmission of certain parasites through effects on hosts that participate in their life cycles. Thus, the community of parasites in a particular host population may also be a useful indicator of pollution or other forms of ecosystem stress (Marcogliese and Cone 1997b). Given that the parasites in a host reflect not only the presence of its other intermediate or definitive hosts, but actual trophic interactions between that host along with its predators and its prey, parasite communities may be used as indicators of biodiversity both on the species and at the ecosystem level (Marcogliese 2003). There are

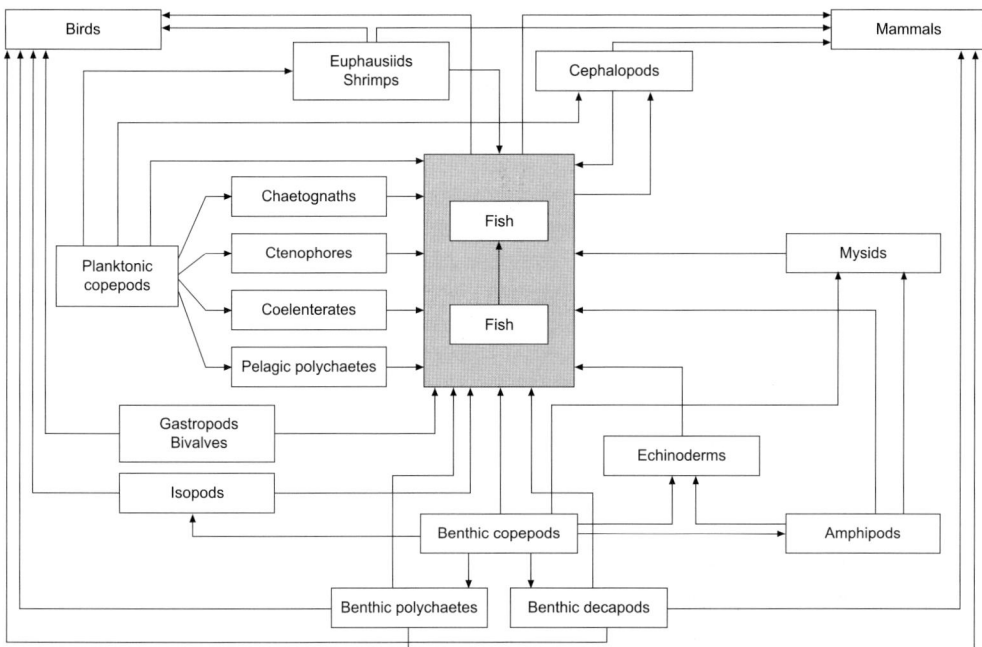

Figure 7.1 Potential transmission pathways involving predator–prey interactions for helminth parasites in marine environments. Other types of parasites are not shown for simplicity's sake, nor are interactions involving free-living infective stages depicted (e.g. cestode coracidia and digenean miracidia and cercariae). Specificity for the intermediate and definitive hosts within any one life cycle (and any one compartment in the diagram) will vary with individual parasite species. Routes of trophic pathways will also vary with parasite species, with some being obligate and others facultative, depending on the nature of the host-parasite interaction. In addition, within each life cycle, parasites may follow more than one path to reach the definitive host, again depending on the specificity of the host-parasite interaction. Reprinted (with slight modifications) from 'Trends in Ecology & Evolution', Vol. 12, Marcogliese DJ and Cone DK, Food webs: a plea for parasites, p. 320–325, 1997, with permission from Elsevier.

numerous examples of alterations in species composition and richness of parasite communities of marine organisms in response to various forms of environmental insult, including industrial waste, domestic sewage, pulp and paper mill effluents and eutrophication (MacKenzie et al. 1995, Williams and Mackenzie 2003, Marcogliese 2004).

Pollution in its variety of forms is not the only anthropogenic stress on marine ecosystems. Globally, commercial fishing has affected the structure of entire food webs. Indeed, top carnivores have been fished to the brink of extinction (Myers and Worm 2003), leading to fisheries aimed at smaller and smaller species, a phenomenon termed 'fishing down the food web' (Pauly et al. 2000). Presumably, overexploitation of marine resources causing these wholesale changes in the food web will alter or disrupt parasite transmission processes and lead to drastic changes in parasite communities in the world's oceans.

Important references

Important contributions on aspects of transmission of parasites are by Marcogliese (1995), who reviewed the role of zooplankton in the transmission of helminth parasites to fish, and Marcogliese (2002), who reviewed food webs and the transmission of parasites to marine fish. Marcogliese (2003) and Marcogliese and Cone (1997a) discussed food webs and the role of parasites in them, and Bush et al. (1993) reviewed intermediate hosts as source communities.

With regard to deep-sea communities, Campbell et al. (1980) gave an account of parasitism and ecological relationships among deep-sea benthic fishes. A comprehensive review on the use of marine parasites as indicators of pollution can be found in MacKenzie et al. (1995).

The ecological niches of parasites
Klaus Rohde and Peter P Rohde

Introduction
The ecological niche as defined by Hutchinson (1957) is a multi-dimensional hypervolume determined by environmental (biotic and abiotic) variables within which a species can exist. Rohde (1979), discussing parasites and particularly ectoparasites of fishes, emphasised that the number of niche dimensions is almost infinite, but that a few dimensions are sufficient to characterise the niche volume of a parasite with a high degree of accuracy. These dimensions are host range (host specificity), microhabitats, macrohabitats of the host, geographical range, sex and age of the host, season, food and hyperparasites. The niches of marine parasites including many examples were discussed by Rohde (1993, 1994).

Any animal or plant species has a niche that is restricted to varying degrees along all dimensions. Some parasites infect a wide range of host species (e.g. many trematodes), others are restricted to a single or a few host species (e.g. many monogeneans). Some parasites are always found in a narrowly defined microhabitat (e.g. certain didymozoid trematodes are always found in a small part of the mouth cavity of fish) whereas some larval trematodes, in contrast, infect the tissues in many body parts. Similarly, the degree of niche restriction for the other niche dimensions varies greatly. Niches are not static but may vary over time, depending on environmental conditions. For example, the size of the microhabitat of gill parasites of marine fishes may change with the oxygen content of the water. Nevertheless, niche preferences are, to a large degree, determined genetically. This is clearly shown, for example, by gill parasites. Even in the absence of potentially competing parasite species and of individuals of the same species, certain monogeneans always attach to the same parts of the gill filaments of a particular fish species.

Methods to measure niche width
Measures of niche width were discussed in Krebs (1989) and include:

1. Levin's measure of *ni*che width $B = \dfrac{1}{\sum p_j^2}$ (**Equation 7.1**);

2. Shannon-Wiener measure $H' = -\sum p_j \log p_j$ (**Equation 7.2**); and

3. Smith's measure $FT = \sum \sqrt{p_j a_j}$ (**Equation 7.3**).

In all these indices, p_j represents the proportion of individuals found in or using resource state, j. In **Equation** 7.3 a_j is the proportion of the total resources represented by resource, j.

Of particular importance for parasites is the number of hosts utilised. Poulin (1998) pointed out that the degree of host specificity found may be an artifact of sampling effort, as indicated by the strong positive correlation between the number of host species and the number of times a parasite has been recorded. Therefore, corrections for sample size should be made. Also, wrong identification may falsify the results (e.g. Combes 1995). For measuring host specificity, various host specificity indices have been proposed.

The first indices explicitly developed to measure host specificity of parasites are those of Rohde (1980a), who distinguished host range and host specificity (see also Lymbery 1989). Host range is the total number of host species found to harbour a certain parasite species, irrespective of prevalence and intensity of infection. In contrast, host specificity considers prevalence (percentage infected) and/or intensity of infection (number of parasites per host individual). The most commonly used specificity index of Rohde (1980a) (S_i) is defined as follows (**Equation 7.4**):

$$S_i = \frac{\sum_j \frac{x_{ij}}{n_{ij}h_{ij}}}{\sum_j \frac{x_{ij}}{n_{ij}}}$$

If intensity is considered, the various parameters are: x_{ij} = number of parasite individuals of ith species in jth host species; n_{ij} = number of host individuals of jth species examined; h_{ij} = rank of host species j (species with greatest intensity has rank 1); and

$$\frac{x_{ij}}{n_{ij}} = \text{intensity of infection.}$$

If prevalence (frequency) of infection is considered, the various parameters are: x_{ij} = number of host individuals of jth species infected with parasite species i; n_{ij} = number of host individuals of jth species examined; h_{ij} = rank of host species based on frequency of infection (species with highest frequency has rank 1); and

$$\frac{x_{ij}}{n_{ij}} = \text{mean frequency of infection.}$$

Numerical values for the indices vary between close to 0 and 1; the closer to 1, the higher the degree of host specificity.

The index can also be adapted to measure preferences for microhabitats. The parameters now are: x_{ij} = number of parasite individuals of ith species in jth microhabitat; n_{ij} = number of microhabitats of type j examined, h_{ij} = rank of microhabitat j.

Poulin (1998) has pointed out that the minimum value of S_i depends on the number of host species used for computing the index. Since most parasites infect fewer than 10 hosts (for which the minimum S_i is about 0.2), a high value of S_i does not necessarily mean a strong preference for a single (or a few) species, which makes a comparison of parasite species utilising different host numbers unreliable.

Many parasite species are found in several host species, but only in those that are closely related (e.g. all belong to one genus or one family). In many studies (depending on the question asked), such parasites should be considered more specific than species that infect the same or even a smaller number of host species which belong to different higher taxa (genera, families and classes). In other words, the phylogenetic position of hosts should be taken into consideration. This was done by Caira et al. (2003) and Poulin and Mouillot (2003). The former authors proposed a decimalised index S.G.F.O.C. (species, genus, family, order, class) with the following ranks: a species restricted to a single host species has rank 1.1.1.1.1 = 1 (i.e. 1 host species, 1 genus, 1 family, 1 order, 1 class), a species infecting 1000 species, 500 genera, 150 families, 75 orders and 5 classes (the maximum number in each taxon permitted by this index) has rank 11 795 988 501. Rank values are calculated by enumerating all combinations of S, G, F, O and C. Using the log(base 10) values of the ranks, index (HS) values of between 0 and about 10 are reached (a program for calculating index values can be accessed (Caira et al. 2003).

The advantage of this index is that it considers both number of higher taxa involved and number of species infected. For example, a parasite found in a single host species has an index (HS) of 0, one found in 1000 species all in one genus has an HS of 3. A disadvantage of the index is that it is not sensitive to an uneven distribution of host species among higher taxa, as pointed out by Poulin and Mouillot (2003). For example, the HS index is the same for 1 parasite infecting 3 host species in each of 10 genera, and for 1 also infecting 10 genera but 21 species in 1 genus and 1 each in the other 9. Poulin and Mouillot (2003), based on Clarke and Warwick (1998), therefore proposed the index S_{TD} (**Equation 7.5**):

$$S_{TD} = 2 \frac{\sum \sum_{i<j} \omega_{ij}}{s(s-1)}$$

where s is the number of host species used by a parasite, the double summation is over the set ($i = 1, \ldots s; j = 1, \ldots s$, such that $i < j$), and ω_{ij} is the taxonomic distinctness between host species i and j, or the number of taxonomic steps required to reach a node common to both. The index measures the average taxonomic distinctness of all host species used by a parasite species (i.e. the mean number of steps necessary to reach a taxon common to two host species, computed across all possible pairs of host species). For example, if two host species are congeners, one step (species-to-genus) is necessary; if the two species belong to different genera but the same family, two steps are required (species-to-genus, and genus-to-family). The greater the number of steps needed, the larger the value of S_{TD}. Because there are five taxonomic levels above species (genus, family, order, class, phylum) the range of values of S_{TD} is 1 to 5 (1: all host species belong to one genus, 5: all host species belong to different classes). The index measures the average taxonomic distinctness between host species, but not all the asymmetries in the taxonomic distribution of host species across higher taxa. However, this can be taken into consideration by measuring the variance of the distribution ($VarS_{TD}$) based on Clarke and Warwick (2001) and Barker (2002) as follows (**Equation 7.6**):

$$VarS_{TD} = \frac{\sum \sum_{i \neq j} (\omega_{ij} - \bar{\omega})^2}{s(s-1)}$$

where $\bar{\omega}$ is the average taxonomic distinctness, or S_{TD}. But the variance in S_{TD} can only be computed when a parasite uses not fewer than three host species.

The S_{TD} cannot be applied to parasite species infecting a single host species. Also, a disadvantage of the index is that it does not consider number of species in a genus infected. For example, the S_{TD} values for a parasite species found in 20 congeneric host species is the same ($S_{TD} = 1$) as one found in five. Neither the index of Poulin and Mouillot (Equation 7.5) nor that of Caira et al. (2003) (HS) considers intensities and prevalences of infection, although the S_{TD} index of the former authors can be modified to do this (see Clarke and Warwick 2001, Barker 2002). Also, the S_{TD} is not independent of the number of known host species.

Application of Equation 7.5 (the index of Poulin and Mouillot) to helminths of Canadian freshwater fish shows that, although acanthocephalans infect more host species than trematodes, cestodes and nematodes, there is no difference in the average taxonomic distances between host species of all these parasite groups. However, this conclusion is, at least to a degree, the consequence of an inherent weakness of the index. As pointed out above, the index does not distinguish between parasites infecting different numbers of host species within a genus; it measures taxonomic distances between hosts above the level of species and nothing else. Even in an index concentrating on the effects of phylogeny, it does not seem justified to ignore the number of congeneric host species. A species, after all, is a phylogenetic entity, although at a level lower than that of genera etc.

Also, too much emphasis on phylogenetic relationships of hosts may not only be irrelevant, but even wrong, if the question asked is ecological. A hypothetical example may make this clear: two species infect 100 host species each, but in one all hosts are congeneric, in the second all except one are congeneric. Phylogenetically based indices will yield a much greater specificity for the former than the latter species. However, in an ecological context, the difference may be meaningless, if the phylogenetically unrelated parasite species have similar roles in the ecosystem.

Similar considerations can be made for the distinction of host specificity and host preference (Lymbery 1989, Poulin 1998). The authors argue that host species are not equally available to parasites, because they may, for example, occur at different abundances, not revealed by applying a specificity index which is based on ranking abundance of parasites. Again, if emphasis is on the ecological role of a parasite, actual abundance in different host species and the range of hosts are more important than the differential 'preferences' for different host species, which may be more interesting to physiologists or evolutionists.

In summary, there are several different specificity indices with different aims. None of them is ideal nor measures everything. The kind of index suitable for a study depends on the type of question to be answered. If the major goal is to evaluate ecological aspects of host specificity, Rohde's index (Equation 7.4) still appears to be best, since it is simple and accounts for intensities and prevalences of infection, which is not done by the other indices. Rohde's index, unlike the other two, can also be used to measure microhabitat width. However, the indices of Caira *et al.* and Poulin and Mouillot (Equation 7.5) consider phylogenetic information. The former index has the advantage that it considers both number of species per higher taxon and numbers of higher taxa infected, but it has the disadvantage of being insensitive to unevenness in the distribution of host species among higher taxa, which, in the index proposed by Poulin and Mouillot, is also the case but can be compensated for by using an additional index to measure the variance. The shortcoming of the S_{TD} index, that the number of host species in a genus, and prevalences and intensities in each species are not considered, can be overcome by using it jointly with the S_I index (Equation 7.4). For example, $S_{TD} = 1$ and $S_I = 0.1$ implies that many more or less equally infected host species, all belonging to one genus, are known. This, however, would not remove the problem of sensitivity of Rohde's index to the number of host species infected.

Rohde's original index (**Equation** 7.4) can be modified to remove this problem as follows (**Equation 7.7**):

$$S_i = \frac{\sum_j \frac{x_{ij}}{n_{ij}h_{ij}}}{\sum_j \frac{x_{ij}}{n_{ij}}} - \frac{1}{j}\sum_j \frac{1}{j}}{1 - \frac{1}{j}\sum_j \frac{1}{j}}$$

where $\frac{1}{j}\sum_j \frac{1}{j}$

is the minimum possible S_i and is used to normalise S_i. This equation is based on the assumption that for minimum possible S_i all species have *approximately* the same value of

$$\frac{x_{ij}}{n_{ij}}$$

and maintain unique ranks. The possible ranges of all values are now equal (0–1), irrespective of the number of host species, with the exception of species with $j = 1$, in which case $S_i = 1$. The symbol j represents the number of host species, n_{ij} is the number of host individuals of host species j infected with parasite species i. All other symbols are as in the original index (Equation 7.4). Examples showing the differences between Rohde's original and the modified index are given in Table 7.2. In this table Rohde's original index is Equation 7.4, and the modified index, Equation 7.7. Total number of fish species examined 31; data from Shulman and Shulman-Albova (1953).

As pointed out above, for ecological studies the phylogenetic status of host species is often (probably in most cases) irrelevant. However, phylogeny can be taken into consideration by adding a phylogeny factor to above equation as follows: 1.1.1.1.1 (= 1 phylum, 1 class, 1 order, 1 family, 1 genus infected), or 1.2.5.6.7. (=1 phylum, 2 classes, 5 orders, 6 families, 7 genera infected). For example, $S_i = 1.1.1.2.3.099$ implies that 3 genera in 2 families in 1 order are infected but that almost all parasites are concentrated in a single host species, $S_i = 2.2.2.2.3.02$ implies that 2 phyla, 2 classes, 2 orders, 2 families, 3 genera are infected, and that the infection is spread more or less evenly over several host species. Although unevenness in infection between different taxa above the species level is not considered, this does not appear to be important for most studies and is more than compensated for by the possibility of seeing at one glance how many higher taxa are involved. However, if unevenness is considered to be important for a particular study, we suggest joint application of our index (Equation 7.7) and that of Poulin and Mouillot (Equation 7.5). Usually the degree of host specificity refers to the set of host species actually infected. However, in certain cases, it may be appropriate to include in the evaluation of host specificity all the potential host species occurring in a particular environment. For example, it could be useful to include in the calculation of host specificity all the hosts in Table 7.2, whether infected or not. This can be done easily, using the proposed index, by arbitrarily allocating very low infection rates to the species that are not infected.

A program (in MATLAB) for calculating

$$S_i = \frac{\sum_j \frac{x_{ij}}{n_{ij} h_{ij}} - \frac{1}{j}\sum_j \frac{1}{j}}{1 - \frac{1}{j}\sum_j \frac{1}{j}}$$

is available (Rohde 2003).

Ricotta (2004) has proposed a parametric diversity index combining relative abundances and taxonomic distinctiveness of species. Applicability of this index to parasite communities has not yet been investigated.

Proximate causes of niche selection

Proximate causes of niche selection are the causal factors that determine the niche of a species. They have been studied (incompletely) in very few species of marine parasites.

Hosts

Host specificity is often the result of ecological factors. For example, certain trematodes infect many hosts, but only those which have similar feeding habits. An important factor that limits

Table 7.2 Host specificity (as measured by Rohde's original index and the modified index) of some digenean trematodes of fishes in the White Sea, Russia (Shulman and Shulman-Albova 1953).

Fish species	1	2	3	4	5	6	7	8	9	10	11	12	S_i	$S_{modified}$
Lecithaster gibbosus													0.99	0.99
No. of parasites found	14	12	4379	3	1	2	16	32	6	1	1	13		
No. of fish examined	21	95	15	7	83	32	84	21	82	64	3	117		
Hemiurus levinseni													0.41	0.07
No. of parasites found	1	18	4	1	651	13	93	–			–	–		
No. of fish examined	15	1	83	38	83	143	21	–			–	–		
Prosorhynchus squamatus													0.98	0.97
No. of parasites found	48	4000	11	2	1	–	–	–	–	–	–	–		
No. of fish examined	41	83	38	68	143	–	–	–	–	–	–	–		
Crepidostomum farionis													0.99	0.97
No. of parasites found	6	1	–	–	–	–	–	–	–	–	–	–		
No. of fish examined	12	117	–	–	–	–	–	–	–	–	–	–		
Anisorchis opisthorchis													1.00	1.00
No. of parasites found	2	–	–	–	–	–	–	–	–	–	–	–		
No. of fish examined	3	–	–	–	–	–	–	–	–	–	–	–		

the number of hosts may be the availability of suitable host species. Poulin (1992) has shown that there is a positive correlation between the number of potential freshwater fish hosts and the number of known host species. Marine parasites have not been examined in this respect.

Factors determining host specificity have been particularly well studied in the Monogenea, in particular by Kearn and collaborators (e.g. Kearn 1967, 1974, further references see Rohde 1993, 1994). Larvae of particular parasite species react only to specific hatching factors released by particular host species, endogenous hatching rhythms are adapted to the host's behaviour, and host specific factors attract larvae. Mortality after infection also may be responsible for host restriction. The monogenean *Entobdella soleae* experimentally transferred to wrong hosts were found to become detached after 24 to 30 hours, although they survived on glass for two to six days. Experiments by McVicar and Fletcher (1970) showed that the cestode *Acanthobothrium quadripartitum*, an intestinal parasite found in the ray *Raja naevus* but not in its close relative *R. radiata*, survived for more than 24 hours in fresh serum from *R. naevus*, whereas 80% of the worms died within two hours in serum from *R. radiata*.

Microhabitats

Sometimes larvae of parasites settle in a particular microhabitat and grow up there. Often, however, microhabitats of larvae and juveniles differ from those of adults. For example, larvae of the copepod *Caligus diaphanus* infect the gill filaments, but adults prefer the wall of the mouth cavity (references in Rohde 1993).

Few studies deal with the factors involved in microhabitat selection by marine parasites. Physiological factors may partially explain site specificity of some species. For example, two gyrocotylid species are found in different parts of the digestive tract of the ratfish, *Hydrolagus colliei*. They absorb certain carbohydrates at different rates, which could be an adaptation to different carbohydrate concentrations in different parts of the intestine (Laurie 1971). Morphological factors are at least partly responsible for site specificity of some species. For example, the three gill parasites of the teleost *Seriolella brama* in New Zealand coastal waters occupy different microhabitats: the trematode *Syncoelium* is attached by means of a large ventral sucker to the spines of the gill arches, the monogenean *Eurysorchis* adheres by means of large sucker-like clamps to the smooth surface of the gill arches, and the monogenean *Neogrubea* has clamps consisting of two valves that are closed for attachment around the gill filaments. The morphological adaptations of these three species do not permit infection of other sites. Several cestode species possess attachment organs that fit into the intestinal villi or folds of their hosts like a key into a lock (for details and references see Rohde 1993, 1994).

Macrohabitats/geographical ranges

The macrohabitat of a parasite is that part of the host habitat in which the parasite is also found. Macrohabitats and geographical range cannot always be clearly distinguished. Russian authors have intensively studied factors that affect macrohabitats of marine parasites (review by Polyanski 1961). Chemical factors and especially salinity, depth resulting in different light, temperature and sometimes pressure conditions, life span and diet of hosts are most important. Temperature appears to be the most important factor affecting the geographical range of marine parasites. Hence, most parasites of marine fishes are known from tropical and subtropical surface waters (see pp. 348–349). The parasite faunas of deep-sea and cold-water surface fish are much poorer (see pp. 351 and 366–369).

Age and sex of host, seasons, food of parasites and hyperparasites

Examples and references for the niche dimensions of age and sex of host, seasons, food of parasites and hyperparasites can be found in Rohde (1993). Few species of hyperparasites that infect marine parasites have been described. They include the monogeneans *Udonella* and *Cyclobothrium* that infect copepod and isopod parasites of fish, respectively, some protozoans that infect trematode larvae in marine snails, the sporozoan *Unikaryon legeri* that infects metacercariae *of a trematode* in marine bivalves and microsporidians infecting myxozoans of fish (see pp. 293–298). Nothing is known about host finding of marine hyperparasites.

Ultimate causes of niche restriction
Ultimate causes of niche restriction are factors that relate to the biological function of niches. Factors suggested by various authors are avoidance of interspecific competition (and predation), enhancement of mating, and the selective advantage of specialisation to narrow niches. All these factors are discussed on pp. 315–321. Overall, interspecific competition is probably of less importance than often believed, and restriction of niches to facilitate mating, and segregation of niches to avoid interspecific hybridisation, are likely to be more important (e.g. Rohde 1979, 1980b).

Important references
The book by Krebs (1989) discusses measures of niche restriction. Important references on specificity indices of parasites are by Rohde (1980a), Poulin (1998), Poulin and Mouillot (2003), Caira *et al.* (2003) and Rohde (2003). Lymbery (1989) discusses the distinctions between host specificity, host range and host preference. Rohde (1979) gives a critical evaluation of intrinsic and extrinsic factors responsible for niche restriction in parasites (see also Rohde 1980b). Detailed experiments on host finding and host specificity in the monogenean skin parasite *Entobdella soleae* are described by Kearn (1967). The book by Combes (1995) contains much information on host and microhabitat specificity of many parasites, and the book chapter by Polyanski (1961) is particularly useful for macrohabitat preferences of parasites of marine fish.

Marine hyperparasites
Mark Freeman

Introduction
Probably all marine organisms are potential hosts for either ectoparasites or endoparasites, and these parasites themselves are no exception to this rule. Hyperparasitism or secondary parasitism is the parasitism of an organism which is already itself parasitic upon another. In this section, the term *hyperparasitism* is used to refer to obligate parasites only and not to organisms displaying a facultative epibiosis or acts of true ectocommensalism, such as the infestation of parasitic copepods by stalked ciliates. Such organisms, although possibly gaining a positional advantage from living on the outer surfaces of such parasitic organisms, do not derive a direct nutritional benefit from either the copepod or the copepod's fish host and are not considered to be obligate parasites as they may also inhabit other marine substrates.

Marine hyperparasites
> So, the naturalists observe, the flea,
> Hath smaller fleas that on him prey;
> And these have smaller still to bite 'em;
> And so proceed, *ad infinitum* …

Table 7.3 List of marine hyperparasites, their hosts and primary hosts

Hyperparasite	Parasite	Primary host
Microsporidian[endo]	Gregarine and eugregarine	Polychaete
	Myxozoan	Fish
	Dicyemid (mesozoan)	Cephalopod
	Trematode	Bivalve and crustacean
	Acanthocephalan	Fish
	Copepod	Fish and bivalve
Haplosporidian[endo]	Nematode, trematode and turbellarian	Bivalve
Flagellate[endo]	Trematode	Fish
Dinoflagellate[endo]	Monogenean	Fish
Myxozoan[endo]	Trematode	Fish
Monogenean[ecto]	Crustacean	Fish
Nematode[endo]	Nematode	Fish
Crustacean[ecto]	Crustacean	Crustacean

[endo] = endohyperparasite; [ecto] = ectohyperparasite.

As the above quotation from *On Poetry: A Rhapsody* (Jonathan Swift, 1667–1745; pp. 337–340) suggests, the intriguing subject of hyperparasitism has long been recognised – the first marine hyperparasites were reported as early as the 1840s. So far, there is no record of marine hyperparasites themselves being subject to further parasitism, but as many aspects of their biology remain unknown, tertiary parasites (or ultra-hyperparasites 'ultraparasites') in all likelihood do exist.

The diversity and distribution of marine hyperparasites and their hosts is considerable and while hyperparasitism may often involve members of three different phyla, hyperparasitism can also occur within the same class of organisms. Such host-parasite relationships are often extremely host specific and some hyperparasites have evolved special adaptations in order to colonise other parasites. Increasingly, newly reported cases of hyperparasitism from the marine environment emerge through our interest in fisheries and aquaculture species and our studies into their primary parasites. Table 7.3 presents a list of marine hyperparasites, their hosts and primary hosts; the list is by no means comprehensive and is intended to highlight the range of organisms involved.

Udonella

Hyperparasites found on the external surfaces of their hosts can be defined as ectohyperparasites. There are many genera of copepods, and branchiurans from the genus *Argulus*, that are parasitic upon fish and these parasitic crustaceans are sometimes found to be infected by the monogenean helminth *Udonella*. Figure 7.2 shows the parasitic copepods *Lepeophtheirus salmonis* taken from cage-cultured Atlantic salmon, *Salmo salar*, from the west coast of Scotland, and *Pseudocaligus fugu* taken from a wild-caught grass puffer, *Takifugu niphobles*, from the Pacific coast of Japan. Both copepods have numerous udonellid parasites attached to their dorsal surfaces and margins of the cephalothoracic shield.

Monogeneans are predominantly parasites of fishes, using various hooks on their posterior attachment apparatus, the opisthaptor, for attachment to fish tissues. However, *Udonella* is one of the few described monogeneans that has no hooks on its opisthaptor. The loss of such hooks in udonellids is considered to be an evolutionary adaptation to living on the hard cuticle of

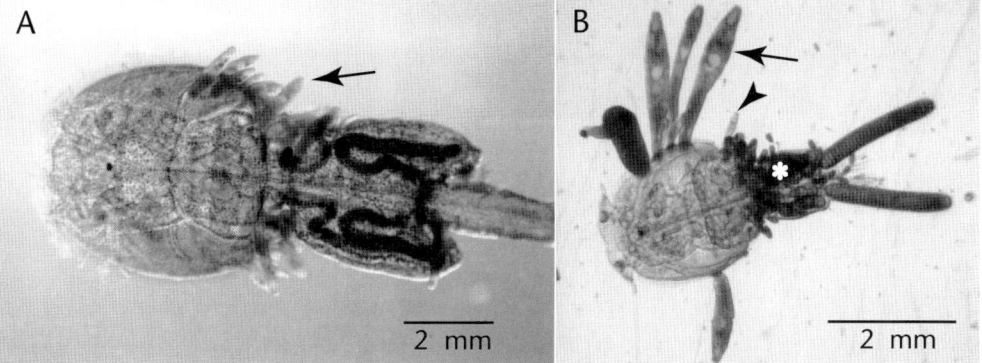

Figure 7.2 The monogenean hyperparasites *Udonella* spp. infecting the dorsal surfaces of two parasitic copepods. **A.** *Lepeophtheirus salmonis*. **B.** *Pseudocaligus fugu*. Juvenile worms (black arrowhead) hatch from eggs that are deposited on the genital segment of *P. fugu* (white asterisk). Note how the adult worms (black arrows) are positioned on the margins of the cephalothoracic shield to facilitate access to the fish's skin.

crustaceans where attachment by hooks is no longer possible (Littlewood *et al.* 1998). *Udonella* deposits eggs directly on the copepod hosts which are attached by a thin filament to the dorsal and ventral surfaces of the cephalothoracic shield, the genital segment, abdomen and egg strings; unlike most other oviparous monogeneans the eggs hatch directly into juvenile worms and no ciliated oncomiracidial stage for larval dispersal is present. Adult worms are rarely observed on the fish's skin but have been seen changing copepod hosts while the male is guarding females during copulation (Aken'Ova and Lester 1996). The udonellids move between fish while attached to their copepod hosts; this is offered as an explanation for the evolutionary loss of the ciliated larval dispersal stage usually required to find a new host, and present in most other oviparous monogeneans. Interestingly, *Udonella* has only been reported from caligid copepods and argulids, which are known to swim freely in the water column as adults in order to find new fish hosts, and free-swimming caligid copepods taken in plankton trawls have been found transporting *Udonella* (Minchin 1991). *Udonella* has not been found on more cryptic parasitic copepods, such as those with sessile non-swimming adult stages.

Udonellids, like most of their monopisthocotylean monogenean relatives, have a close association with fish and have only been reported from copepods and argulids that are parasitic upon fish and not from free-living copepods or other crustaceans. Indeed, *Udonella* are more host specific to the fish than the copepod on which they live (Byrnes and Rohde 1992). *Udonella* is frequently found on the parasitic copepod, *Caligus rogercresseyi*, that infests wild Chilean rock cod, *Eleginops maclovinus*, but is absent from the same species of copepod infesting cultured salmonids, in spite of the rock cod feeding close to the caged salmonids (Marin *et al.* 2002). In contrast to early suggestions that *Udonella* feeds directly upon their copepod hosts, it is now known that *Udonella* feeds off the fish mucus and not from the copepod (Kabata 1973). *Udonella* feeds directly on fish mucus and epithelium by locating on the margins of the copepod's cephalothoracic shield and stretching out to the fish's surface with their anteriorly positioned mouth to feed. The feeding behaviour of certain udonellids may be more pathogenic to the fish than the copepods themselves (MA Freeman and K Ogawa unpublished data 2004). Survival of adult *Udonella* on copepods removed from their fish hosts has been shown to be as little as two days, during which time no eggs were deposited and no juvenile worms matured to adults before becoming moribund and detached. However, it is possible to maintain a healthy population of *Udonella* on copepods that were maintained on experimental fish (Freeman 2002). This also suggests that *Udonella* derives some if not all of its nutrition from the fish and not directly from the copepod.

It is unclear how diverse the udonellids are, they have been reported from Japan, Australia, West and South Africa, Northern Europe, West Canada and South America; six species are recognised. The reason that *Udonella* use parasitic copepods as hosts and do not live directly on fish as other monogeneans is unexplained, but by living on copepods, they may avoid the fish's immune responses, or they may have developed this strategy in order to facilitate finding new fish hosts as 'hitchhikers'. *Udonella* spp. are not considered to be detrimental to their crustacean hosts and hence could be defined as ectocommensal, solely utilising the crustaceans as vehicles and feeding platforms (Carvajal *et al.* 2001). *Udonella* spp. are, however, obligate parasites of fish and undoubtedly totally dependent upon their crustacean hosts for continued survival.

Epicaridea

Isopods from the suborder Epicaridea are parasites of other crustaceans. In many cases they are parasites of free-living crustaceans, but they also are hyperparasitic upon parasitic crustaceans such as rhizocephalans (parasitic barnacles) and bopyrids (parasitic isopods), which belong to the same suborder, the Epicaridea. Hyperparasitism such as this, occurring within the same class of organisms, is unique among the Crustacea and highlights the remarkable variety of host-parasite relationships that can occur within parasitic crustaceans.

The rhizocephalans are parasites of benthic crustaceans, mainly decapods. After the planktonic rhizocephalan larval stage has settled on a new host, it invades the body and spreads throughout the tissues and organs forming an 'interna'. After this endoparasitic interna has matured, a posterior enlargement of the interna results in the formation of a sac that overlies the ventral abdominal wall of the host. The host's integument in this region becomes necrotic and disintegrates, forming an opening through which a sac or the externa emerges. The externa contains the reproductive structures which remain outside the host's body, attached to the interna by an isthmus of tissue (Spivey 1982). It is to this externa that hyperparasitic epicaridean isopods from the family Cryptoniscidae (Liriopsidae) attach. Isopods within the Cryptoniscidae are almost exclusively hyperparasites of rhizocephalan barnacles (Grygier and Bowman 1990), but also infect some free-living crustaceans. Figure 7.3 depicts a male decapod crab with the externa of the parasitic rhizocephalan, *Briarosaccus callosus*, protruding from its ventral abdominal surface which is hyperparasitised by a cryptoniscinid isopod.

Other epicaridean isopods from the family Cabiropsidae are typically hyperparasites within the marsupia of bopyrid isopods (Sassaman 1985), which are, in turn, branchial or dorsoabdominal ectoparasites of decapods. The Cabiropsidae cause sterilisation in their bopyrid isopod hosts. The epicaridean isopods have been the subject of much taxonomic deliberation and some ambiguity remains (see Grygier and Bowman 1990 and Grygier 1993).

Microsporidia

Microsporidia are obligate intracellular parasites and many are hyperparasites of other groups of marine parasites, they can be defined as endohyperparasites. They are hyperparasitic in unicellular eukaryotic parasites such as gregarines and eugregarines that are parasitic in polychaetes, in myxozoan parasites of fish, digenean trematodes of oysters and crabs, acanthocephalans of fish and parasitic copepods of bivalves and fish. Figure 7.4 illustrates microsporidian hyperparasites in coelozoic and histozoic myxozoans from the intestine of a cultured tiger puffer, *Takifugu rubripes*, from Nagasaki, Japan, and a microsporidian hyperparasite from the salmon louse, *Lepeophtheirus salmonis*, from farmed Atlantic salmon from the west coast of Scotland. Cultured tiger puffer in Japan are susceptible to myxozoan infections of the intestinal epithelium; three species of myxozoan infect these fish and serious emaciation and death can result

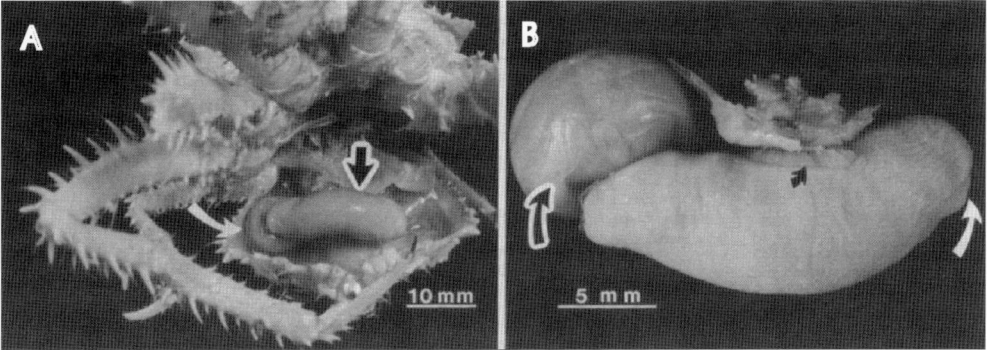

Figure 7.3 **A**. Oblique frontal view of a decapod crab with exposed abdomen, showing the externa of the rhizocephalan parasite, *Briarosaccus callosus* (black arrow) and hyperparasitic cryptoniscinid isopod (white arrow). **B**. Enlarged dorsal view of isolated *B. callosus* externa with female hyperparasitic cryptoniscinid isopod (large black arrow). The white arrow indicates the mantel opening of the rhizocephalan and the small black arrow the cuticular shield of *B. callosus* and the transition point to the interna. After Pohle GW (1992). Reprinted from Pohle, *Canadian Journal of Zoology* **70** (1992) (With permission from NRC Research Press and the author).

(Tin Tun *et al.* 2000). Two of these myxozoans are frequently hyperparatised by microsporidians, which are host specific and represent two distinct species. The infection cycle for these hyperparasites is unknown and further complicated as one of the myxozoans has a hitherto unknown obligate alternate host. However, hyperparasitism may occur after the myxozoan infections are established in the fish (MA Freeman and K Ogawa unpublished data 2004), but it remains unexplained how the histozoic trophozoites of one myxozoan, *Leptotheca fugu*, come into contact with microsporidian spores. The other species, *Enteromyxum fugu*, that is regularly found hyperparasitised by a microsporidian, is attached to the epithelial surface of the gut lumen (Fig. 7.4A), and hence could easily come into contact with microsporidian spores in the digestive system of the fish. Infected trophozoites from both species of myxozoans have not been observed to mature and form myxospores and this prolonging of sporogenesis is thought to increase the pathogenicity of these myxozoans to the fish host.

The salmon louse is sometimes found to be infected with a hyperparasitic microsporidian (Figs 7.4C and 7.4D). Experimental transmission of this hyperparasite to naive sea lice has been unsuccessful and a recent phylogenetic analysis of the microsporidian has indicated that it is more closely related to others that infect fish rather than crustaceans (Freeman *et al.* 2003). Furthermore, the same microsporidian has been isolated from the salmonid fish host, and Freeman (2002) thought that the infection in sea lice is an accidental opportunistic infection caused by the blood feeding of the copepods, and that the sea lice may represent a 'dead-end' host for the microsporidian. It is unknown whether the microsporidian infection is pathogenic to sea lice; however, the fecundity of infected sea lice may be reduced (Fig. 7.4C). Research is ongoing into this hyperparasite's potential use as a biological control agent for sea lice populations (Freeman 2002).

Important references

The feasibility of using hyperparasites as biological control agents of salmon lice is discussed by Freeman (2002, see also Freeman *et al.* 2003), and Marín *et al.* (2002). The former suggested a microsporidian as such an agent, the latter the monogenean *Udonella*. Kabata (1973) gives a detailed description of the distribution of *Udonella caligorum* on *Caligus elongatus*. Tin Tun *et al.* (2000) discuss Myxozoa and their hyperparasitic microsporeans in the intestine of emaciated

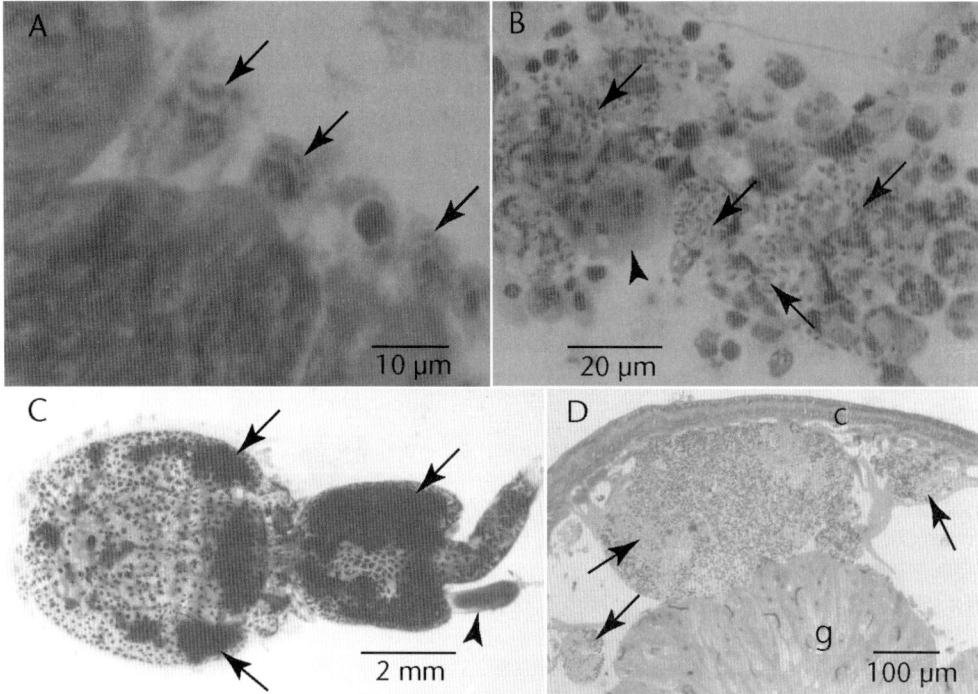

Figure 7.4 **A.** Haematoxylin and eosin stained histological section of trophozoites of the coelozoic myxozoan, *Enteromyxum fugu*, on the epithelial surface of the intestine of a cultured tiger puffer, *Takifugu rubripes*. Microspordian hyperparasites (black arrows) are present within the cytoplasm. **B.** Tissue impression from an emaciated tiger puffer's intestine, stained with Diff Quik. *Leptotheca fugu* trophozoites contain many microsporidian hyperparasites (black arrows). Note that the large trophozoite of *Enteromyxum fugu* (black arrowhead) remains uninfected. **C.** Adult female *Lepeophtheirus salmonis* infected with a microsporidian hyperparasite; the microsporidian forms large xenomas directly beneath the cuticle, seen as opaque patches (black arrows). Note the aborted egg string (black arrowhead). **D.** Transverse semi-thin section through the abdomen of a microsporidian-infected salmon louse, note the formation of large xenomas (black arrows) between the cuticle (c) and the centrally located gut (g).

tiger puffer, and Pohle (1992) recorded the decapod *Paralomis bouvieri* infected by the rhizocephalan *Briarosaccus callosus* and carrying a hyperparasitic isopod (Epicaridea).

Parasites in brackish waters
C Dieter Zander

Introduction
Brackish waters are found worldwide. They comprise small lakes and lagoons, estuaries and other river-mouth areas, pools in the supralittoral and coastal ground waters as well as larger water bodies like Hudson Bay, and the Black and Baltic Seas which comprise an area of more than 400 000 square kilometres. Whereas Hudson Bay and the Black Sea exhibit constant surface salinities of 23 parts per thousand (ppt, g of solute per kg of sea water) and 18 ppt, respectively, the Baltic Sea has a salinity gradient from almost marine conditions in the Kattegat, to limnetic conditions in the Bothnian Bay or Gulf of Finland, including the biggest region, the Baltic Proper, with 6ppt surface salinity. This may be the reason that the Baltic is best investigated with

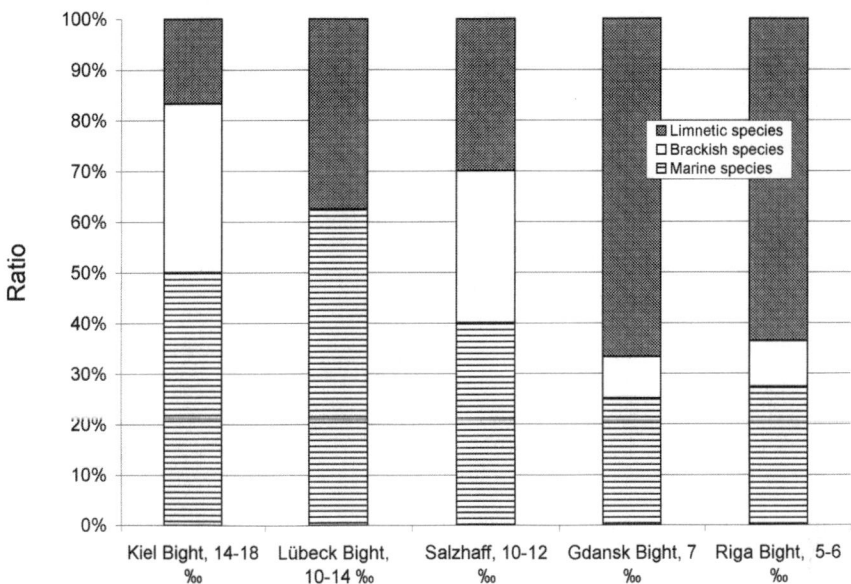

Figure 7.5 Dispersion of three categories of parasites in eelpouts, *Zoarces viviparous*, under the influence of different salinities from five localities in the Baltic Sea.

regard not only to fauna and flora, but also to ecological patterns of brackish waters (Remane 1958). Remane's investigations resulted in five rules that apply to organism inhabiting brackish water: a minimum of species, reduction in size, reduction of structures, submergence (or emergence, respectively), and an extended choice of habitat. Some of these rules are applicable to parasites; but additionally, parasites have their own patterns when in an extreme environment of decreased and highly variable salinity (Zander 1998, Zander and Reimer 2002): low host specificity, alteration of life cycles, infection of new hosts, and adaptation to genuinely brackish-water hosts.

Adaptations to decreased salinity

Pre-adaptations to life in brackish water are best evidenced by the distribution of the organisms along a salinity gradient existent in the Baltic Sea and its fjords. Remane (1958) distinguished five categories of brackish-water inhabitants, which may also be valid for parasites: species of marine origins and species of limnetic origin, genuine brackish-water species, euryhaline species which can tolerate marine and freshwater conditions, and arctic relics. Remane (1958) found a 'species minimum' at salinities of 5 ppt to 8 ppt because marine and limnetic species reach their limits of tolerance in these salinities. Parasites do not show such clear limits because, with the exception of ectoparasites, they live in hosts which often have a constant inner milieu and can tolerate migrations. But free-living larvae, or a restricted distribution of at least one of the hosts in their life cycles, make parasites also sensitive to changing salinity. The differences can best be demonstrated by parasites of stationary benthic invertebrates or fish like the eelpout, *Zoarces viviparus* (Zander and Reimer 2002) (Fig. 7.5). Here a clear trend can be observed in the Baltic, where freshwater species increase from west to east, corresponding to the decrease in salinity.

Möller (1978) tested the salinity tolerance of free cercariae of *Cryptocotyle lingua*. Under the influence of different temperatures the larvae live longest (six days) when temperatures are low

(5°C) within a wide range of salinity (8–32ppt). Cercariae survive only five days at 10°C, four days at 20°C, and only two days at 25°C. Optimal salinity was 12 ppt to 20 ppt. Therefore, conditions for dispersion of this parasite in brackish waters are most advantageous in spring and autumn, not in summer.

The phenomenon of submergence (i.e. the colonisation of deeper habitats where salinity is increased) is demonstrated by the first host of *Podocotyle atomon*, the periwinkle *Littorina saxatilis*. This species lives in the North Sea in the supralittoral and releases the short-tailed cercariae of the parasite which creep to the second intermediate host, benthic crustaceans. In the Baltic, *L. saxatilis* lives in the sublittoral where the parasite continues to exhibit the particular behaviour described above, though swimming would be more advantageous.

Specificity of brackish-water parasites

The most conspicuous pattern of brackish-water parasites is the suspension of specificity. Parasites which are specialists in marine or fresh waters infect several hosts when in brackish waters. This is apparent in snails of the Baltic, which are hosts of digenean sporocysts and rediae; it is also the case in fish parasites. *Asymphylodora demeli* (Digenea) which is distributed in fresh water as well as the Baltic and Black Seas can infect as many as eight snail species which are of marine (*Littorina* spp.) or limnetic origin (*Limnaea ovata* and *Theodoxus fluviatilis*) or are genuine brackish-water species (*Hydrobia* spp.) (Zander and Reimer 2002). *Psilochasmus oxyuris* was found in six, *Maritrema subdolum* and *Microphallus claviformis* in three, and eight other digenean in two hosts, of which mud snails, *Hydrobia stagnalis* (nine) and *H. ulvae* (11 species) were the ones most frequently used. A special example may be the bird parasite *Cryptocotyle lingua* which is usually found in the periwinkle, *Littorina littorea*; however, in the Baltic it is found not only in the related *L. saxatilis*, but also in *Hydrobia ulvae* (Zander and Reimer 2002). Hosts are the habitat of parasites; therefore, suspension of specificity can be equated with the expansion of free-living Baltic species into different habitats (Remane 1958).

Several genuine brackish-water digeneans show special affinities for the same hosts. For example, the bird parasites *Maritrema subdolum* and *Microphallus claviformis* prefer *Hydrobia stagnalis* over other *Hydrobia* spp. (Zander et al. 2000). The adults of *Aphalloides timmi* and the metacercariae of *Apatemon gracilis* occur predominantly in the genuine brackish-water gobiid, *Pomatoschistus microps*. *Cryptocotyle concavum* predominantly infects *Hydrobia stagnalis*, which release cercariae invading the skin and fins of several fish. But the same parasite penetrates the kidney of almost all specimens of *P. microps*. This example appears to be due to coevolution within genuine brackish-water species which has altered the life cycle of a parasite, by 'capturing' a new microhabitat. As many as 3000 metacercariae can be accumulated without harming the host (Zander 1998).

Other genuine brackish-water parasites reach different degrees of specialisation (Zander 2001). *Acanthostomum balthicum* (Digenea) infects several fish species as intermediate hosts but only the pipefish, *Syngnathus typhle*, as a final host. The copepod *Thersitina gasterostei* lives in great abundance in the gill chamber of the stickleback, *Gasterosteus aculeatus*; but also to a lesser degree in other sticklebacks and pipefish. The digenean *Magnibursatus caudofilomentosa*, which also colonises the gill chamber of *Gasterosteus aculeatus*, can invade not only other hosts but also another microhabitat, like the brood pouch of male pipefish (Zander and Reimer 2002).

Non-native species (neozoa) display diverse patterns in the Baltic. Parasites like *Anguillicola crassus* (Nematoda) from east Asia easily found several native hosts among invertebrates and fish which are prey of the final host, eels, *Anguilla anguilla*. In contrast, non-native invertebrates like *Marenziella arenaria* (Polychaeta), *Mya arenaria* (Bivalvia) or *Pomatopyrgus jenkinsi* (Gastropoda) were hardly infected by Baltic parasites.

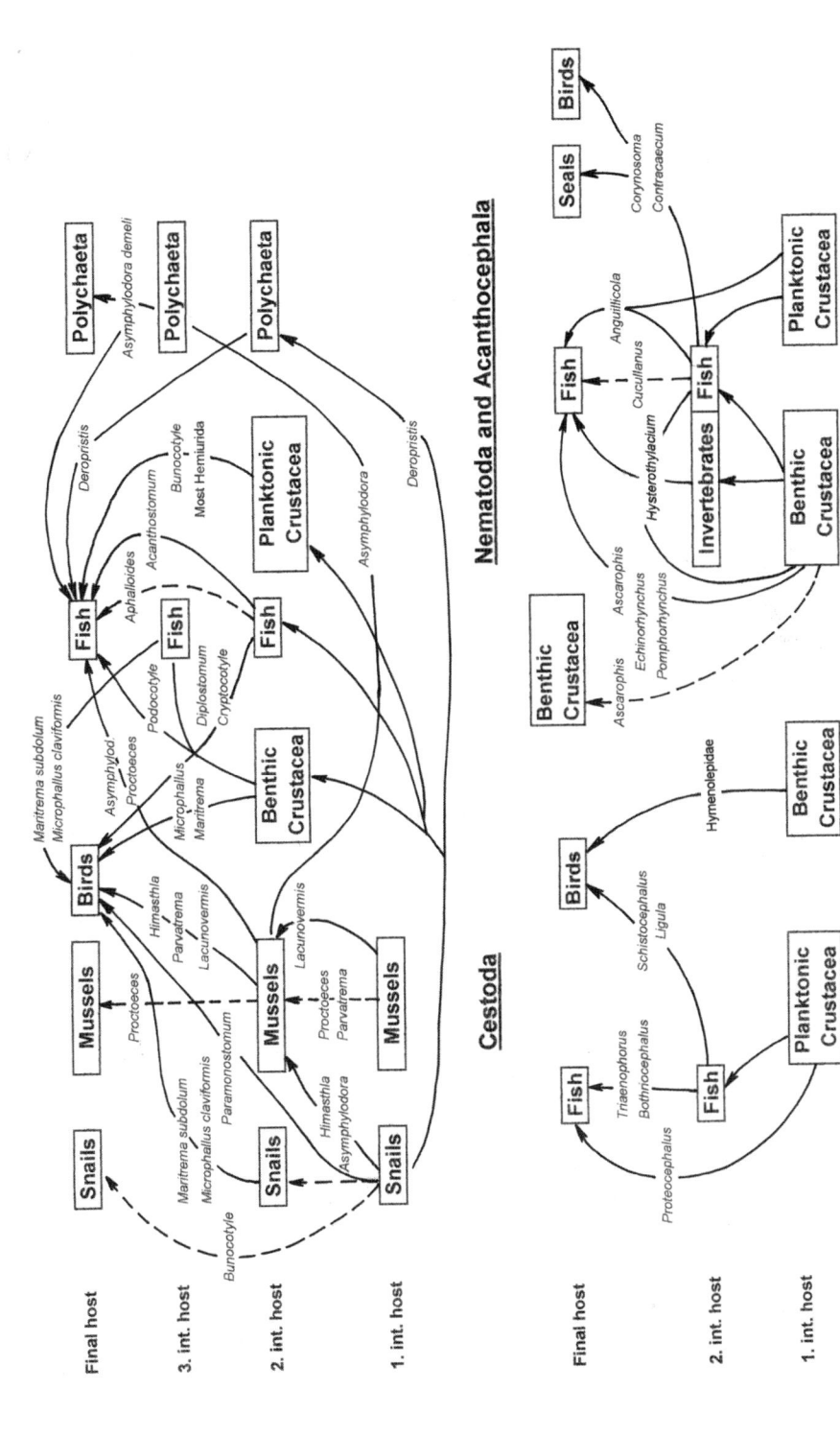

Figure 7.6 Life cycles and hosts of helminth parasites in the Baltic Sea. Dashed lines mark shortened cycles with paratenic hosts. From Zander and Reimer (2002), with permission from Cambridge University Press.

Alteration of life cycles

Alterations of life cycles especially concern the reduction and the expansion of intermediate host numbers (Zander and Reimer 2002). Most cercariae of the bird digenean *Maritrema subdolum* and some of *Microphallus claviformis* remain in mud snails, *Hydrobia* spp., which are first host and directly preyed on by birds (Fig. 7.6). The cercariae of *Parvitrema parvus* remain in mussels which are eaten by birds. In contrast, *M. claviformis* can infect fish as the third intermediate host (which prey on benthic crustaceans containing its metacercariae). Such alteration of life cycles can enlarge the spectrum of final hosts, because they now reach not only crustacean but also snail and fish-feeding birds. *Asymphylodora demeli* can mature in its second intermediate host, the polychaete *Hediste diversicolor*. The common goby, *Pomatoschistus microps*, a genuine brackish-water fish, is directly infected by cercariae of *Aphalloides timmi* which mature and live in the body cavity instead of the gut. *Bunocotyle cingulata* (Digenea) predominantly infects nine-spined sticklebacks, *Pungitius pungitius*, but adults were also found in its first host, the mud snail *Hydrobia ulvae*. The digenean *Proctoeces maculatus* reaches maturity in *Pleuronectes limanda* and also in mussels (*Mytilus edulis*), which are already themselves first hosts. In addition, the nematode *Ascarophis arctica* not only matures in fish but also in its first hosts, benthic crustaceans (Fig. 7.6). This phenomenon guarantees that parasites can exist in brackish waters without being endangered by a change in hosts.

Another strategy of parasites which ensures the completion of life cycles in brackish water is the infection of new hosts. Whereas shrimps, *Crangon crangon*, are not infected by any metazoan in the marine milieu, metacercariae of at least two Digenea were found in Baltic specimens (i.e. *Podocotyle atomon* and Microphallidae spp.) (Zander 1998). The final hosts are fish (*Podocotyle atomon*) or birds. Because shrimps are larger than most other benthic crustaceans and also widely distributed, they are the preferred prey of larger fish (flatfish and cod) or birds, respectively. This development is a kind of 'host capture' (Holmes 1990) which is promoted by a great density of parasites (Zander and Reimer 2002).

Conclusion

Most parasites in brackish waters are marine and limnetic species but there also are, to a lesser extent, genuine brackish-water species. In order to exist successfully in this extreme environment, diverse strategies were developed to reach the final hosts (usually vertebrates): shortening and expanding life cycles, maturing in invertebrates as paratenic hosts, capturing new hosts or suspension of host specificity. Though extreme environments prevent stable interactions between organisms, some partnerships of host and parasite involving only genuine brackish-water species have evolved in the Baltic Sea.

Important references

A general account of the ecology in brackish waters is by Remane (1958). Zander (1998), Zander *et al.* (2000) and Zander and Reimer (2002) give thorough discussions of host-parasite systems in the Baltic Sea. Möller (1978) studied the effect of salinity and temperature on the development and survival of fish parasites.

Metapopulation biology of marine parasites
Serge Morand and Andrea Šimková

Introduction

The metapopulation concept was developed in order to consider the spatial patterns in persistence of species in a patchy habitat. The central point of the metapopulation theory is that groups or local populations are linked to some degree among each other by exchanges of individuals. Local populations can become extinct and can re-establish through migration by individuals originating from extant local populations. Levins (1969, 1970) was the pioneer of this theory by considering the dynamics of a population of populations (i.e. a *metapopulation*). His very first simple model analyses the dynamics of spatially interconnected populations (of same size) all having the same probability of colonisation and the same probability of extinction.

In an improvement of the metapopulation theory, the term metapopulation is used for heterogeneous populations and the term *population* for local populations. Model population dynamics hypothesise that populations live in discrete habitat patches, with largely independent dynamics (i.e. local subpopulations may differ in their growth rates) (Hanski 1999).

Parasite metapopulations

Parasite ecology recognises different levels of organisation (Esch and Fernández 1993, Bush et al. 1997, Poulin 1998). The component community consists of all parasite species exploiting a host population at a given time. The component community is the local pool of parasite species from which infracommunities, comprising all parasites of different species within the same host individual, are formed. Infracommunities are assemblages of parasites generally short lived compared to the life expectancy of the host. Each individual host is parasitised by one or several infrapopulations (all individuals of the same parasite species in the host individual), which constitute the parasite infracommunity (Esch and Fernández 1993). These infrapopulations are subsets of several metapopulations and the sum of metapopulations constitutes the component community.

The analogy with metapopulation theory is evident (Hanski and Simberloff 1997, Hanski 1999). Each host individual is equivalent to a habitat patch containing infrapopulations of several metapopulations of different parasite species infecting the population of hosts. The dynamics of infection of a host population by one parasite species is equivalent to patch colonisation. However, the local extinction of each patch can be the result of infection recovering, parasite-induced mortality, or it may be due to natural birth and deathof the host.

Models of parasite population dynamics

Epidemiological models were developed to: predict the spread and maintenance of parasitic diseases in host populations; and to estimate the potential control of a host population by a given parasite (Anderson and May 1978, May and Anderson 1978). For the case of macroparasites, and in particular for the case of fish macroparasites (Morand et al. 1995), the dynamics of two populations are described by differential equations and incorporate the central point of macroparasite infection (i.e. its aggregated distribution) (well described by the parameter k of the negative binomial distribution).

Epidemiological models allow us to derive a measure of parasite invasiveness, the basic reproductive number, R_0, which can be defined as the number of female offspring produced throughout the lifetime of a female parasite that would themselves achieve reproductive maturity. As this mathematical entity is calculated at the very beginning of the infection with only one female parasite, there are no density-dependent constraints (Anderson and May 1991). R_0, then is the product of all instances of infection (from free-living stage to adult parasite including intermediate or paratenic host, if present) divided by the product of all instances of mortality (free-living mortality, mortality within the intermediate host if present and mortality in the definitive host).

However, epidemiological models applied to marine parasites are scarce (Des Clers 1990, Morand et al. 1995), mostly because information is lacking on many parameters of the parasite life cycle (e.g. longevity and adult longevity) and sometimes on the host itself. Long-term studies are also very rare and few studies have investigated the annual variation in abundance and prevalence of fish parasites (Lo et al. 1998). Most studies have, therefore, focused on parasite surveys, which may be of great help to test the usefulness of epidemiological models and metapopulation theory applied to marine parasites. These models have provided a framework for empirical epidemiology: the generalisation for describing patterns of parasite abundance and distributions has led to some success in parasite ecology.

When epidemiology meets metapopulation theory in marine fish parasites

The parasite distribution within a host population is generally measured by parasite abundance (number of parasite individuals per host) or prevalence (the percentage of infected hosts). Morand and Guégan (2000) used an epidemiology framework to explain the pattern of distribution and abundance of nematodes parasitic in mammals. Morand et al. (2002) applied the same framework to the case of ectoparasites of marine fish. The mean abundance M (mean number of parasite individuals per host infected) and its variance V were related following Taylor's power law (Taylor 1961) (**Equation 7.8**):

$$\mathrm{Log}(V) = b\mathrm{Log}(M) + \mathrm{Log}(a)$$

where M is mean abundance and V its variance, a is a constant, and b is an index of aggregation. Comparing large numbers of parasite species, this relationship was found to apply not only to nematode parasites of mammals (Morand and Guégan 2000), but also to all types of adult parasites (Shaw and Dobson 1995). From basic epidemiological models (Anderson and May 1985), the prevalence of infection can be linked to the mean abundance of parasites at any time during an infection dynamics according to the following relationship (**Equation 7.9**):

$$P = 1 - \left(1 + \frac{M}{k}\right)^{-k}$$

where P is the prevalence and k is the aggregation parameter of the negative binomial distribution (Anderson and May 1978). The parameter k is related to the parameters a and b of Taylor's power law (Equation 7.8) as follows (**Equation 7.9**):

$$\frac{1}{k} = aM^{(b-2)} - \left(\frac{1}{M}\right)$$

Morand et al. (2002) investigated this pattern for the case of marine fish ectoparasites. They used the data of Rohde et al. (1995) on 36 communities of gill and head ectoparasites of marine fish.

Treating each ectoparasite population as an independent observation, they found the positive relationship between the mean abundance (in log) and the variance of abundance (in log) (Fig. 7.7A). The estimated values of parameters $b = 1.71$ (± 0.04) and $a = 0.57$ (± 0.04) are within the ranges typically observed in other assemblages of parasites (Morand and Guégan 2000).

A positive relationship between abundance and prevalence of ectoparasites was observed (Fig. 7.7B). The frequency distribution of ectoparasite prevalence showed a bimodal pattern (Fig. 7.7C).

These theoretical relationships derived from epidemiological models allow the comparison of the prevalences of congeneric parasite species observed in nature with those predicted by the models. Across several species of monogeneans ectoparasitic on fish (Simková et al. 2002a), the congruence between observed and predicted values of the relationship linking abundance and

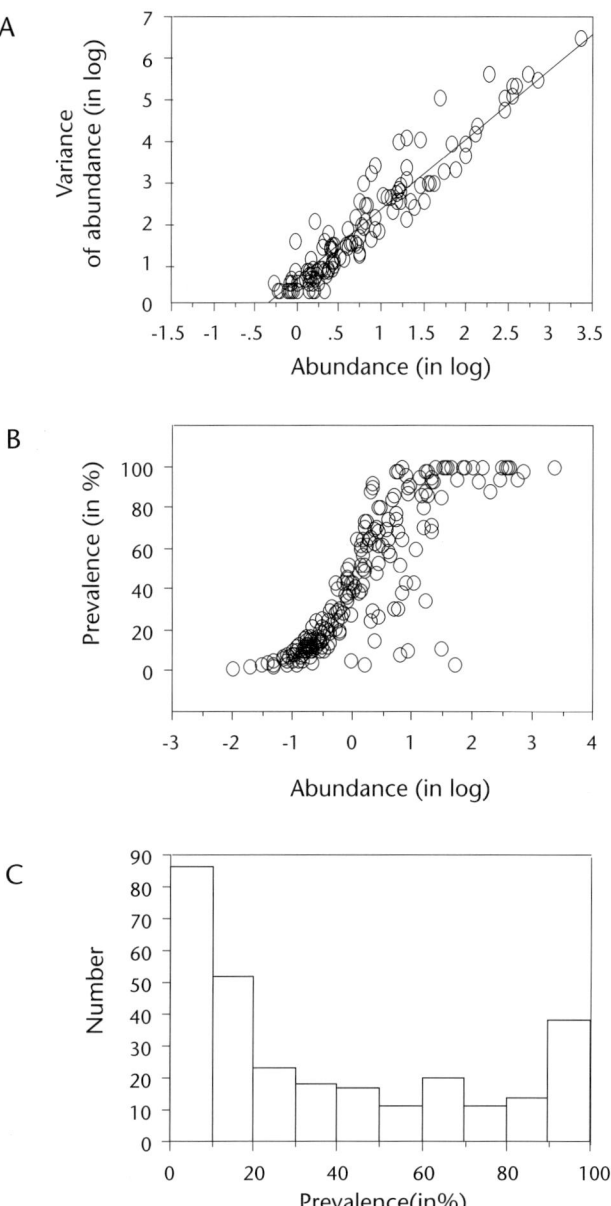

Figure 7.7 A. Relationship between the logarithms of the variance, s^2, and abundance, M, across populations of fish ectoparasites, fitted to a power function with the intercept $a = 0.57 \pm 0.04$ and the slope, $b = 1.71 \pm 0.04$ ($r^2 = 0.92$; $p < 0.0001$; $n = 171$ populations). B. Relationship between abundance (average parasite burden) and prevalence of fish ectoparasites (171 populations). C. Bimodal distribution of prevalence of fish ectoparasites. From Morand et al. (2002).

prevalence is generally good. According to Morand and Guégan (2000), a positive relationship between abundance and prevalence is purely the result of epidemiological processes. Demographic explanations, therefore, may explain the observed patterns of bimodality of prevalence when making Monte-Carlo simulations using epidemiological modeling frameworks (Anderson and May 1978, Morand and Guégan 2000).

Core-satellite hypothesis applies to metapopulation of marine fish parasites

Metapopulation dynamics models incorporating the core-satellite hypothesis (Hanski 1982, Hanski and Gyllenberg 1993) were applied to explaining the positive relationship between local abundance and prevalence of infection in parasites (Morand and Guégan 2000, Morand et al. 2002). The core-satellite hypothesis predicting bimodal distribution of organisms in their environment (i.e. most species are present in either most patches or only a small fraction of patches) was supported both for mammalian nematodes (Morand and Guégan 2000) and for marine fish ectoparasites (Morand et al. 2002). This hypothesis is not based on competition but on the ability of species to recolonise empty patches after extinction (i.e. the rescue effect). However, bimodality was not often investigated in parasites, but parasites very often show an overdispersed distribution described by a negative binomial distribution. In such cases, most hosts are infected by a few parasites or are not parasitised, or only a few hosts are strongly infected as found in the study of congeneric parasite populations in freshwater fish (Simková et al. 2002b).

Interspecific and intraspecific aggregation and the link between metapopulations and component community

One important point in metapopulation studies is the nature of the interactions between different species exploiting the same patchy resource, here adult parasites of fish. Communities are classified as interactive or non-interactive depending on whether interactions are important or not among the residents of a local habitat. Interactive communities are of two types: niche heterogeneity types, where coexistence is favoured when species differ in one or more important dimensions of their niche, or spatiotemporal heterogeneity types, where species differ in spatial or temporal occurrence. Because interactions may lead to local species extinction, there are several ways in which species can coexist. One is by reducing the overall intensity of competition via aggregated utilisation of fragmented resources (Jaenike and James 1991), formalised as the 'aggregation model of coexistence' (Shorrocks 1996). The aggregation model of coexistence postulates that species coexistence is facilitated when the distribution of species leads to the reduction of interspecific aggregation relative to intraspecific aggregation, which was confirmed in the comparative study of gill and head ectoparasites of marine fishes (Morand et al. 1999). The 'aggregation model of coexistence' implies saturation of ecological communities with species, which means that there is saturation of local species richness (here, on an individual host) independent of the size of the regional pool of species (component parasite species). This model has been applied to 36 ectoparasite communities of marine fish (Morand et al. 1999) from the original data of Rohde et al. (1995).

A measure of intraspecific aggregation J was proposed by Ives (1991). the symbol J represents the proportionate increase in the number of the same parasite species experienced by a random host relative to a random distribution (**Equation** 7.10):

$$J_1 = \frac{\sum_{i=1}^{p} \frac{n_{1i}(n_{1i}-1)}{m_1} - m_1}{m_1} = \frac{\frac{V_1}{m_1} - 1}{m_1}$$

where n_{1i} is the number in host i of parasite species 1 (and p the number of hosts), m_1 the mean numbers and V_1 the variance.

A value of $J = 0$ indicates that individuals are randomly distributed, while a value of $J = 0.5$ indicates a 50% increase in the number of parasite individuals expected in a given host compared to the random distribution.

Interspecific aggregation is measured by an index C (Ives 1991), which quantifies the proportionate increase in the number of different parasite species relative to a random association (**Equation** 7.11):

$$C_{12} = \frac{\sum_{i=1}^{P} \frac{n_{1i}n_{2i}}{m_1 P} - m_2}{m_2} = \frac{Cov_{12}}{m_1 m_2}$$

where n_{1i} and n_{2i} are the numbers of species 1 and species 2 in host i, m_1 and m_2 are mean numbers of species 1 and species 2 per host, P is the number of hosts and Cov is the covariance between a pair of parasite species.

When $C > 0$ the two parasite species are positively associated; when $C < 0$, they are negatively associated.

The last index A (Ives 1991) quantifies the reduction in competition caused by intraspecific aggregation (i.e. the relative strength of intraspecific aggregation versus interspecific aggregation) (**Equation** 7.12):

$$A_{12} = \frac{(J_1 + 1)(J_2 + 1)}{(C_{12} + 1)^2}$$

When $A_{ij} > 1$, intraspecific aggregation is stronger than interspecific aggregation.

Morand et al. (1999) found that 35 of the 36 communities showed a mean $J > 0$ indicating that parasite individuals were aggregated among their hosts (Fig. 7.8A). Fifteen communities showed negative association between parasite species with a mean $C < 0$, and 21 communities showed positive association between parasite species with a mean $C > 0$ (Fig. 7.8B).

Seventeen communities showed a mean $A > 1$ (i.e. intraspecific aggregation was stronger than interspecific aggregation), whereas 19 communities showed the opposite trend with interspecific aggregation stronger than intraspecific aggregation (Fig. 7.8C).

Finally, Morand et al. (1999) found that ectoparasite species richness of infracommunities was positively correlated with the level of intraspecific aggregation versus interspecific aggregation (i.e. A). All of these suggest that intraspecific aggregation increases compared to interspecific aggregation when component parasite species richness increases. Interspecific interactions are then reduced relative to intraspecific interactions which facilitates species coexistence.

Non-interactive fish parasite metapopulations

Competition was generally considered as one of the main factors structuring endoparasite communities (Holmes 1990). However, numerous studies on ectoparasite species of fish have shown that parasite communities are far from being saturated, with positive associations exceeding negative ones (Rohde et al. 1995, Lo and Morand 2000). Ectoparasite and endoparasite component communities of marine and freshwater fish often form random, unstructured and clumped assemblages (Worthen and Rohde 1996, Rohde et al. 1998, Simková et al. 2001, 2003, Morand et al. 2002). Some studies have described nested ectoparasite assemblages of fish (Hugueny and Guégan 1997, Matejusová et al. 2000). However, non-random structure in parasite assemblages appears not to be a result of interspecific competition but rather the consequences of demographic characteristics and biology of each parasite species (Morand et al. 2002).

Conclusion and the future of metapopulation biology of marine fish parasites

Numerous studies on marine parasites are devoted to surveys, used of parasites as biological tags (MacKenzie 2002) and parasite community ecology (Holmes 1990, Rohde et al. 1995, 1998). Very few papers are directly concerned with population dynamics (Lo et al. 1998) and hence metapopulation theory. In this section theoretical arguments are summarised that emphasise that parasites of marine fish do show dynamic behaviour that should be analysed in the framework of metapopulation theory. The patterns of distribution and abundance of fish parasites (at

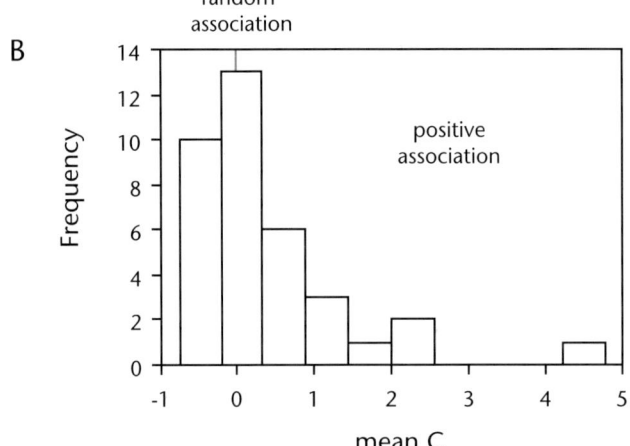
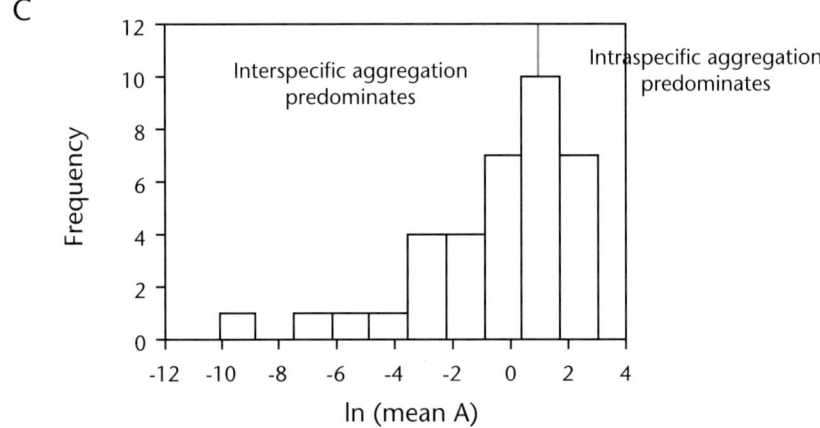

Figure 7.8 **A.** Frequency distributions of mean intraspecific aggregations, J. **B.** Frequency distributions of mean interspecific aggregation, C. **C.** Frequency distribution of the mean relative strength of intraspecific aggregation versus interspecific aggregation, A (in ln). All mean values are calculated for each of 36 communities of gills ectoparasites. From Morand et al. (1999).

least ectoparasites) are explained by epidemiological and statistical models in accordance to the theory. The most important point is the lack of interaction between ectoparasite species exploiting the same patchy resource (i.e. the fish host). Patterns of epidemiological and community structure suggest that each metapopulation of parasite behaves independently. This point is also in accordance with the theoretical prediction of Dobson and Roberts (1994) who concluded that interactions between parasites and their hosts are more important in determining community structure than direct interactions between parasite species. Hence, spatial aggregation, and demographic characteristics of both host and parasite are the crucial mechanisms permitting increases in diversity in parasite communities.

However, several questions remain unresolved, mostly because of the lack of long-term studies of fish parasite population dynamics. The most crucial question is linked to the habitat specificity both at the level of the whole resource (i.e. the host species) and at the level of the microhabitat within a local patch (i.e. individual host). For example, species coexistence of congeneric monogeneans is generally explained by pre-reproductive isolation in order to prevent hybridisation (Rohde and Hobbs 1986, Simková et al. 2002b). This necessitates the use of genetics for the analyses of metapopulation biology among and within host species, which would help to demonstrate the very necessary point that extinction and recolonisation of populations may occur, leading to a better understanding of how parasite communities are established.

Important references
A book on metapopulation ecology is by Hanski (1999). Hanski and Simberloff (1997) discussed the metapopulation approach, its history, conceptual domain, and application to conservation, and Hanski and Gyllenberg (1993) two general metapopulation models and the core-satellite species hypothesis. Among the few studies dealing with population dynamics of marine parsdites is Lo et al. (1998), but relevant to a discussion of metapopulation biology are the papers dealing with community structure of marine or freshwater parasites, such as Rohde and Hobbs (1986), Rohde et al. (1995), Morand et al. (1999, 2002,), Lo and Morand (2000), Simková et al. (2000, 2002a, 2002b, 2003) (see also the books by Poulin 1998, and Esch and Fernández 1993). Worthen and Rohde (1996), Hugueny and Guégan (1997), Rohde et al. (1998) and Matejusová et al. (2000) deal with nestedness in marine parasite communties.

Structure of parasite communities
Robert Poulin

Introduction
The search for order in nature is a central theme in ecological research. In community ecology, this often amounts to a search for non-random patterns in the species composition of naturally occurring assemblages, and for the ecological processes responsible for those patterns. Parasite communities have proven to be excellent model systems for such investigations (Holmes and Price 1986, Esch et al. 1990, Sousa 1994, Poulin 1998). They are arranged hierarchically, such that each infracommunity (all parasite individuals of all species within a host individual; Bush et al. 1997) is a subset of the component community (all parasite individuals of all species within the host population; Bush et al. 1997). Defining the pool of locally available parasite species from which infracommunities can acquire species is therefore straightforward, as is sampling several replicate infracommunities. A key question has been whether the parasite species forming each infracommunity are random subsets of the ones found in the component community. If the species composition of infracommunities departs from that expected by chance alone, then the infracommunities are most likely structured by one or more ecological processes.

Here, recent advances in the study of parasite community structure will be briefly reviewed, using communities of metazoan ectoparasites and endoparasites of marine fish as examples. Communities of fish parasites are not only the best studied assemblages of parasites in marine hosts, but also they range from species-poor to very species-rich communities, and thus include the whole gamut of parasite community types. Common patterns in the structure of these communities will first be reviewed, and this will be followed by a discussion of the likely processes acting to generate departures from random assemblages.

Detecting patterns of community structure

The distribution of parasite species among infracommunities (i.e. among individual hosts) can range from completely random to highly structured (Fig. 7.9). In the simplest scenario, the probability of occurrence of any parasite species in a host individual is equal to its prevalence in the host population, and is totally independent of the presence of other parasite species (Fig. 7.9A). The most basic departure from randomness, and the easiest one to quantify, consists in positive or negative associations between pairs of parasite species (i.e. pairs of species that co-occur in the same infracommunities more or less often than expected by chance) (Fig. 7.9B). The associations can be evaluated using presence or absence data or actual numbers of parasites per host; the two methods usually give congruent results. Any significant association between two parasite species should in itself be a sign that the composition of infracommunities is not random. When considering all possible pairwise associations between the species in a component community, the logical expectation would be that the number of positive covariances should equal the number of negative ones, regardless of statistical significance, if infracommunities are assembled randomly. In many systems, however, the number of positive associations exceeds the number of negative ones (e.g. Lotz and Font 1991). For instance, among the metazoan parasites of the red grouper, *Epinephelus morio*, sampled along the coast of the Yucatan Peninsula, Mexico, positive covariances outnumber negative ones (Vidal-Martínez and Poulin 2003). Several factors can generate spurious covariances and/or bias the sign of associations. For instance, the frequency of parasite species with low or high prevalence within a component community, and the number of host individuals (infracommunities) examined, influence the likelihood of obtaining spurious associations (Lotz and Font 1994). One needs sophisticated randomisation procedures, based on actual data on parasite abundance, to generate correct predictions about the relative number of positive and negative associations expected in a component community under random assortment (Lotz and Font 1994). Therefore, although pairwise associations of parasite species among infracommunities are commonly reported, they do not necessarily reflect any underlying community structure.

A more powerful approach is to look at the entire set of parasite species rather than tackling them one pair at a time. Using a presence or absence matrix, with the different infracommunities sampled as columns and the different parasite species found as rows, one can compare the observed pattern of parasite occurrences with that expected under random assembly of infracommunities. Various null models have been proposed to generate expected random patterns, and these can provide a solid baseline for comparisons with observed patterns. Null models must be chosen with great care, however, to make sure that they generate realistic and appropriate null distributions of species (Gotelli and Graves 1996).

The most common application of this approach to parasite communities has involved testing for the presence of nested species subsets, a pattern often seen in communities of free-living animals (Patterson and Atmar 1986, Worthen 1996, Wright *et al.* 1998). In a parasite component community, a nested pattern would imply that the species forming species-poor infracommunities are distinct subsets of progressively richer infracommunities (Fig. 7.9C). In other words, in a nested pattern, parasite species with high prevalences are found in all sorts of infracommunities, whereas rare parasite species only occur in species-rich infracommunities. To determine if

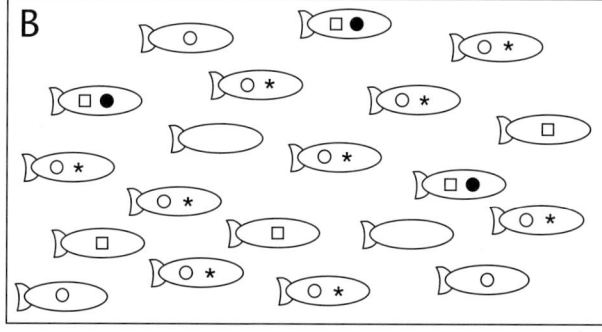

Figure 7.9 Three hypothetical distributions of parasite species among infracommunities (i.e. among host individuals) in a component community. Each example consists of 20 fish hosts and four species of parasites (represented by different symbols), each having the same prevalence in the three examples. **A.** Random distribution (i.e. no detectable structure: the presence or absence of any parasite species in a fish is statistically independent of that of other parasite species). **B.** Associative distribution, in which two pairs of species show positive associations, and the species in each pair show negative associations with the species in the other pair. **C.** Nested distribution, where a parasite species occurring in a host individual with n parasite species also tends to be found in host individuals with at least $n + 1$ species.

there is a significant nested subset pattern in a parasite component community, the observed pattern is tested against the expectations derived from a null model based on each species' prevalence, using Monte Carlo simulations (Patterson and Atmar 1986). Nestedness has been investigated extensively in communities of ectoparasites (mainly monogeneans and copepods) on marine fish hosts. Although significant nested patterns have been observed in the ectoparasite communities of some fish species, overall nestedness is not a common pattern in these communities, occurring in approximately 30% of them (Worthen and Rohde 1996, Rohde et al. 1998). Another departure from random assembly, qualitatively opposite from nestedness and dubbed anti-nestedness, is seen in another 30% or so of ectoparasite communities of marine fish (Poulin and Guégan 2000). Still, the absence of any detectable structure remains the most common pattern observed in these communities. In similar investigations of communities of endoparasites in marine fish, nested (and anti-nested) patterns were observed in some fish species but proved even less common than for ectoparasites (Rohde et al. 1998, Poulin and Valtonen 2001a). Thus, nested species subsets exist in some, but not many, communities of fish parasites, suggesting that structuring forces are occasionally responsible for the distribution of parasite species among infracommunities. Things may not be as they seem, however. Poulin and Valtonen (2001a) have found that for fish species with nested communities of endoparasites, there were strong correlations between fish size and the number of parasite species harboured by individual fish, but not for other fish species. This suggests that finding a significant pattern of nestedness in a parasite community may reflect the heterogeneity among fish in a sample: rather than indicating a non-random structure, a nested pattern may instead reflect the inclusion of different successional stages (i.e. parasite infracommunities of different ages and at different stages in their accumulation of parasite species). Some significant nested patterns may thus be sampling artifacts.

Null models have been used on data from parasite communities in marine fish to search for other departures from randomness (e.g. Gotelli and Rohde 2002). The results are the same as with analyses of nested species subsets: structured parasite communities (i.e. communities in which parasite species are non-randomly distributed among infracommunities) are found only in few fish species. Non-random structure appears to be the exception rather than the norm.

Repeatability of community structure in space and time

As stated above, departures from randomness are not common in the species composition of parasite communities of marine fish, only occurring in a few host species. If all populations of these few host species had parasite communities showing the same non-random structure, then it might be inferred that they are subject to the action of strong, consistent structuring processes. In contrast, if parasite communities are clearly structured in some host populations but randomly organised in other populations of the same host species, or if patterns of community structure come and go over time in the same host population, then a different interpretation would be required. Spatial and temporal variability in community structure would indicate that parasite communities are too dynamic for the effect of any structuring force to have any long-lasting effect. The few studies that have investigated the repeatability of parasite community structure in space and time generally suggest that non-random patterns are both ephemeral and unpredictable.

For instance, among samples of red grouper, *E. morio*, collected along the south-eastern coast of Mexico, highly significant nested patterns were observed in some localities but not in others, despite the species composition of the various parasite component communities being roughly the same among samples (Vidal-Martínez and Poulin 2003). Similarly, significant nested patterns were only found in some of the distinct populations of a pelagic fish, the anchovy *Engraulis anchoita*, sampled in the south-west Atlantic (Timi and Poulin 2003). Non-random patterns of parasite community structure are, therefore, found in some host fish species only, and only in some populations of these host species.

The processes behind the patterns

The previous synthesis has emphasised that departures from a random distribution of parasite species among infracommunities are not observed frequently, at least among communities of parasites in or on marine fish. Still, in those cases where non-random patterns are reported, what may be the processes shaping the community?

The importance of interspecific competition has received much attention in the ecological parasitology literature (Holmes 1973, Rohde 1979, 1991, Holmes and Price 1986, Sousa 1994, Poulin 1998, 2001). The existence of competitive interactions can be inferred from data on natural infections, but only experimental infections can truly assess their magnitude and directions (Poulin 2001). Classical experimental studies of single-species and mixed-species infections have demonstrated how one parasite species can have either or both numerical and functional impacts on other species; that is, the presence of a parasite species in a host can either decrease the abundance of other parasite species (e.g. Dash 1981, Holland 1984) or cause them to shift their niche (Holmes 1961, Patrick 1991). No such experiments have been performed on parasites in marine fish hosts. Several lines of evidence, including the presence of many empty niches and the low abundance of parasite populations, suggest strongly that neither competition nor any other interspecific interaction are significant in structuring communities of ectoparasites on marine fish (e.g. Rohde 1979, 1991, Morand et al. 1999). When non-random assembly patterns are found in ectoparasite communities, they must have other explanations, possibly related to individual differences among fish in their susceptibility to parasite colonisation.

Communities of endoparasites in marine fish can sometimes have all the hallmarks of interactive communities: for example, high species richness and many species with high prevalence and intensity of infection (Holmes and Price 1986, Holmes 1990, Sousa 1994). Using an index of parasite species interactivity that captures these features and collapses them into a single number, and applying it to the endoparasite communities of 37 species of marine fish from coastal Brazil, Poulin and Luque (2003) have shown that potential interactivity in these communities ranges from nil to very high. Competition and other interspecific interactions are thus potentially important structuring forces shaping the endoparasite communities of some fish species. Still, no one has yet directly linked a non-random pattern of species co-occurrences observed in a natural endoparasite community with specific data on interspecific competition obtained from the same system. Thus, the jury is still out regarding the overall importance of competition in endoparasite community structure.

Another potential structuring process can act on endoparasite communities in fish, in particular by generating positive pairwise associations between certain parasite species. With few exceptions, endoparasitic helminths join an infracommunity when their larvae, usually residing inside an intermediate host, are ingested by the host fish. Any intermediate host may contain larvae of several different helminth species, and therefore helminth larvae arrive in fish as packets, not singly (Bush et al. 1993). If the larvae of two species are positively (or negatively) associated among intermediate hosts, this association can be transferred to the adult helminth community inside fish hosts (Fig. 7.10). Each time a fish acquires larvae of parasite species A, it would also generally acquire larvae of species B. Using computer simulations, associations of parasite species in intermediate hosts, especially positive ones, can readily be transferred to parasite communities in definitive hosts (Lotz et al. 1995); this is even more likely when one of the two parasite species in a pair is capable of manipulating the behaviour of the intermediate host in ways that enhance its susceptibility to predation by the definitive host (Vickery and Poulin 2002). Associations between parasite species may persist through more than one host for parasite species with the same life cycles and sharing the same host species (Poulin and Valtonen 2001b). Within a fish population, variation among individual fish in either food preference or foraging habitat could result in different subsets of the fish population acquiring different packets of larval helminths, thus leading to a structured endoparasite community characterised by clear-cut positive associations between pairs of parasite species. This represents a likely alternative to interspecific competition among parasites within fish as a process that structures parasite communities.

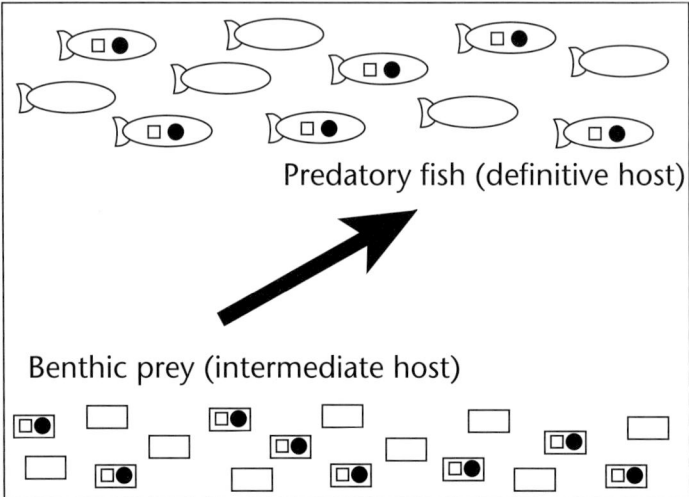

Figure 7.10 Transfer of a positive association between two parasite species (represented by different symbols) from intermediate to definitive hosts. If the two parasite species are positively associated in their intermediate hosts, then the association will be passed on to the definitive hosts. In the example, any fish feeding on infected intermediate hosts and acquiring larvae of one parasite species will also tend to automatically acquire larvae of the second species.

Conclusion

There may be no universal assembly rule governing parasite communities, no general underlying process applying to all communities. The non-random patterns observed in the parasite community of one fish species do not allow us to say much about the parasite community of another fish species, at least not with great confidence. Patterns of parasite community structure vary among the different populations of a single host species, suggesting that they are transient properties and that parasite communities are at the mercy of numerous forces simultaneously pulling them in different directions.

Perhaps our inability to find general, predictable trends stems from the focus of most parasite ecologists on either the presence or absence of parasite species in a host individual, or the numbers of individual parasites of different species in a host. This may be appropriate for communities of organisms in which the different species have body sizes that are comparable (i.e. within the same order of magnitude). Given the substantial intraspecific and interspecific variability in parasite body sizes, it is not uncommon for different members of the same infracommunity to have sizes that range over two to three orders of magnitudes (e.g. some cestodes sharing a fish gut with small trematodes). In such cases, the true patterns in community structure will be missed by an analysis based on numbers of individuals. Biomass of the different parasite species would appear to be a much better measure of their use of host resources and of their functional role in the community. Recently, Mouillot et al. (2003) looked at resource apportionment among species in the communities of metazoan parasites from six species of marine fish, using biomass as a measure of a parasite species' abundance. They found that for three of the six host species, patterns of relative biomass partitioning among parasite species followed very closely the predictions of Tokeshi's (1999) random assortment model: no theoretical model could be fitted to data from the other three fish species. The random assortment model corresponds to situations where the abundances of different parasite species vary independently of each other. Thus, even when focusing on biomass rather than just presence or absence or numerical abundance of parasite species in a community, the main outcome is that randomness is the most common pattern observed. Although a case-by-case evaluation is still necessary, the only general conclusion that one can reach at the moment is that non-random patterns of parasite community structure (in marine fish hosts at least) are not uncommon but are far from ubiquitous.

Important references
Gotelli and Graves (1996) provided a book on null models in ecology. Worthen (1996) gave a discussion of community composition and nested-subset analyses as basic descriptors for community ecology. Important general contributions on parasite communities not restricted to marine parasites are by Rohde (1979), Holmes and Price (1986), Esch et al. (1990), Sousa (1994) and Poulin (1998). Book chapters and papers dealing with more specific topics are Holmes (1973), who discussed site segregation by parasitic helminths, and Mouillot et al. (2003), who discussed how parasites divide resources: testing the niche apportionment hypothesis. Poulin (2001) reviewed interactions between species and the structure of helminth communities. Book chapters and papers dealing specifically with various aspects of the ecology of marine parasite communities are by Holmes (1990), Rohde (1991), Worthen and Rohde (1996), Rohde et al. (1998), Morand et al. (1999), Gotelli and Rohde (2002), Poulin and Guégan (2000), Poulin and Valtonen (2001a,b), Poulin and Luque (2003), Timi and Poulin (2003), Vidal-Martínez and Poulin (2003).

Parasite populations and communities as non-equilibrium systems
Klaus Rohde

Introduction
Crucial for understanding ecological systems is whether species live under equilibrium or non-equilibrium conditions. Rohde (2005) has discussed in detail the concepts of balance and non-equilibrium in nature, history of the concepts and evidence for non-equilibrium conditions in many ecological systems, including parasites. The main arguments in favour of equilibrium conditions in nature are that many studies have indeed shown that ecological systems and their components are in dynamic equilibrium; that production and respiration in communities 'must ultimately balance'; that even non-climax communities which have not reached a steady state, probably have some sort of equilibrium, determined by the frequency and pattern of disturbances; that in most communities rate of energy influx exactly balances outflow in each trophic level; that on islands, immigration and extinction of species is balanced; that birth and death rates in populations are equal over long periods; and that prey–predator and similar pairs must be 'in some sort of ecological and evolutionary balance to coexist with one another over any period of time' (Pianka 1974). Also, interspecific competition is believed to play a major role in shaping ecological communities and ecosystems.

The contrary paradigm, that non-equilibrium prevails, is based on much evidence which shows that many of the points just listed are ill supported by facts. In particular, few populations ever have a chance to reach equilibrium because of repeated and strong disturbances, that many habitats are not saturated with species, and that interspecific competition occurs but has little evolutionary significance. This has contributed to a shift in how ecological systems are interpreted. Nevertheless, equilibrium assumptions are still widely accepted, for example in Hubbell's (2001) neutral theory of biodiversity, which is based on the premise that communities are saturated with individuals leading to a zero-sum game, and that there is saturation with species. On the other hand, the neutral theory assumes that communities are open, continuously changing 'non-equilibrial' systems. Rosenzweig (1995) also stressed that ecosystems are largely balanced and in steady state.

Evidence for non-equilibrium conditions

Extensive discussions of the non-equilibrium aspects of ecology are given by Price (1980) and Price et al. (1984). An important reason for the one-sided view that competition and equilibrium prevail in nature, in the opinion of the latter authors, is that negative data (e.g. those on the absence of competition) are not found worthy of publication, and most authors try to fit data into a well-established framework without considering alternatives. Also, authors tend to select systems for hypothesis testing not at random, but only those for which expected effects are likely. Thus, for the demonstration of interspecific competition between trematodes of molluscs, species infected with many parasite species occurring at high prevalences of infection are selected, rather than much poorer communities where competitive effects are unlikely.

An important indicator of non-equilibrium is non-saturation with individuals and/or species. However, non-saturation in itself does not necessarily imply non-equilibrium, because in any system, an equilibrium level will be established below saturation. For example, if population size as a proportion of carrying capacity is plotted against reproductive rates, the resulting bifurcation diagram never reaches the carrying capacity (i.e. equilibrium is established well below saturation) (e.g. figure 12 in Rohde 2002a), because the rate of population growth decreases with increasing population size. The diversification equation (Equation 7.12) (p. 320) shows that the same holds for communities and ecosystems in evolutionary time: the number of species will always be below the maximum number possible (the 'carrying capacity for species'), even if equilibrium conditions prevail. For the demonstration of non-equilibrium, it is more important to show that an increase in population size and/or species numbers in communities over time is likely. For example, if a drought reduces population size, end of the drought is likely to lead to population increase. If two host species harbour vastly different species numbers of parasites, there is no reason to assume that species numbers at least in the poorer host cannot increase further over evolutionary time. Therefore, in both cases, it is not justified to assume equilibrium.

Many studies have shown that disturbances, such as storms, fires, droughts or predation, may lead to non-equilibrium conditions and permit coexistence of species which otherwise would be eliminated by competition. Re-establishment of an equilibrium may take very long, and combined with the frequency of disturbances, this may often be so long that populations are always kept in non-equilibrium. Examples are mainly from non-marine systems, because disturbances in the oceans are usually less pronounced: Kormondy (1969) gives some examples of populations in which equilibria are never reached. A plant community at Rothamsted, England, had not re-established equilibrium conditions 100 years after resources had been disturbed, indicative of a large time lag in tracking the resource changes (Tilman 1982).

Concerning the effects of disturbances on metapopulations, Hanski (1999, references therein) showed that habitat reduction for a butterfly species led to non-equilibrium conditions and that a new equilibrium would be established only after many (up to hundreds of) years. For parasites, the effects of disturbances are aggravated because, when population densities of hosts are suppressed below a critical level, survival of a parasite population is impossible. Re-establishment of the host population will be followed with some delay by recovery of the parasite population. If disturbances are frequent, parasites never have the chance to become established. This was shown by Koch (Koch 2004) for snail and parasite populations in a habitat at Armidale, northern NSW, Australia. The critical factor was rainfall: inset of a drought led to a collapse of the snail populations and an even more drastic collapse of their trematode populations. Many habitats with diverse faunas of snails and potential vertebrate hosts had no trematodes at all. In the marine environment, conditions in most habitats are more stable, and studies of parasite populations under such extreme environmental conditions have not been made. For hyperparasites, frequent disturbances must have even more drastic consequences, and this may explain the small number of hyperparasites described even from the relatively benign marine environment (Rohde 1989).

Non-equilibrium in marine parasite communities

Many different approaches have been used to examine whether marine parasite communities live under equilibrium or non-equilibrium conditions. Ectoparasites and endoparasites of teleost fishes have been extensively studied, and all results strongly support the conclusion that these parasites live in non-saturated, non-equilibrial assemblages rather than in saturated communities structured to a large degree by competition. This conclusion does not depend on one particular statistical test, but is based on a holistic approach using a variety of methods (see Rohde 2002a). The main evidence based on several approaches is as follows:

1. abundance and prevalence of infection are often very low (Rohde et al. 1995);
2. empty niches are common (Rohde 1979, 1998b);
3. circumstantial evidence suggests that niche restriction may be due to selection for increasing intraspecific contact and thus mating ('mating hypothesis of niche restriction') (Rohde 1979, 2002b);
4. nestedness is uncommon and, where it occurs, may be the result of epidemiological processes and differential colonisation rates (Worthen and Rohde 1996, Morand et al. 2002);
5. hyperparasites are rare (i.e. there are many empty niches for hyperparasites) (Rohde 1989);
6. spatial scaling laws do not apply to many (if not all) parasites (Rohde 2001);
7. null model analysis found little evidence for non-random co-occurrence patterns (Gotelli and Rohde 2002);
8. Tokeshi's Random Assortment model could be fitted best to parasite abundance patterns, suggesting unsaturated non-interactive communities (Mouillot et al. 2003);
9. many assemblages can best be described by log series or lognormal distributions (e.g. Rosenzweig 1995) (i.e. most species are rare with little potential for interspecific interactions) – new (rare) species could be added without much effect on species already present, an aspect not sufficiently explored for parasites; and
10. there is a linear relationship between parasite biovolume and parasite diversity in parasite assemblages from 131 vertebrate species, suggesting non-saturation and possibility of further increase in parasite diversity (Poulin et al. 2003).

Further evidence, related to the role of interspecific competition in communities, is as follows:

1. microhabitat size and infection intensities of many species are often not affected by other species (i.e. by interspecific competition) (Rohde 1991);
2. niches may be restricted even in the absence of competing species (Rohde 1977, 1980a, 1991, 1994, 2005);
3. although interspecific competition in parasites occurs and may lead to competitive exclusion or changes in microhabitat width in some or all co-occurring species (interactive site segregation), there is no evidence that such effects lead to evolutionary changes and avoidance of competition (i.e. to selective site segregation) (Rohde 1979);
4. an asymptotic relationship between infracommunity and component community richness is not necessarily due to competition, but may be the result of different colonisation rates and life spans (Rohde 1998a);

5 differences in the size of feeding organs, often given as evidence for interspecific competition, may be fortuitous (Rohde 1991);
6 differences in the size and shape of copulatory organs are not the result of competition but of reinforcement of reproductive barriers (Rohde and Hobbs 1986, 1999);
7 positive associations between parasite species are much more common than negative ones, and even negative associations are not necessarily due to competition (Rohde et al. 1994);
8 there is reduced interspecific relative to intraspecific aggregation (Morand et al. 1999);
9 there is no evidence that numbers of sympatric congeners are reduced due to interspecific competition (Rohde 1989, 1991);
10 dominance patterns are often not determined by competition but they may be the result of environmental contingencies (Rohde 1999); and
11 the probability that two species have completely coinciding niches is infinitesimally small and niche differences should not be used as evidence for competition (Rohde 1979).

In the following, results are discussed from studies based on extensive data sets in favour of some of the points listed above. Rohde et al. (1995) found 15 positive and only one negative association among metazoan ectoparasite species of 102 marine fish species from many regions. Low abundance (mean 6.7 parasites per fish, mean median abundance 4.31) and prevalences (mean 30.1%) of infection suggested scarcity of interactions between species. The most dominant species represented, on average, 90% of all parasite individuals on a fish individual, and in many cases different parasite species were dominant on different fish of the same species. Rohde (1998b) used a larger data set on fish parasites: on 5666 fish of 112 species, he found a mean number of 4.3 species of metazoan ectoparasites on the heads and gills per fish species. The maximum number was 27 on a small fish from warm-temperate waters in south-eastern Australia. Most fish species had fewer than seven, and 16 had none. Assuming that 27 species is the maximum a host species can support and that other fish species could support the same number, only 15.9% of all niches are filled.

Similarly, considering abundances (total number of all parasites of all species/number of fish of a particular species examined), maximum abundance was more than 3000 on a medium-sized fish from the tropical Great Barrier Reef, but most fish had an abundance of fewer than five. The mean abundance of 54.68 represented 1.82% of the maximum, again indicating that many more parasites could be accommodated. These data do suggest that niches not only are not saturated, but also that they are very far from saturation. Morand et al. (1999) found a positive relationship between infracommunity species richness and total parasite species richness and no evidence of ceilings (i.e. saturation). Rohde (e.g. reviews 1989, 1991, 1994, 2002b), has shown that monogenean gill parasites of marine fish from all latitudes use microhabitats that are sometimes very restricted, even when competing species do not exist or, when they exist, are not present on individual fish. Rohde suggested that one ultimate reason (i.e. one biological function) of narrow niches may be the enhancement of mating encounters.

Parasite species, like free-living species, must be specialised to varying degrees in order to optimise survival, and this alone will lead to niche restriction even in largely empty niche space, an aspect stressed by Price (1980). Packing rules derived from fractal geometry were shown to apply to plant and animal groups that either disperse well or are vagile, and whose numbers are restricted by limiting resources (Ritchie and Olff 1999). However, the rules do not apply to metazoan ectoparasites and endoparasites of marine fishes, strongly suggesting that these species do not live closely packed and do not exhaust resources (Rohde 2001).

Differences in the size of feeding organs have been used as evidence for niche segregation resulting from interspecific competition that has led to the use of differently sized food particles.

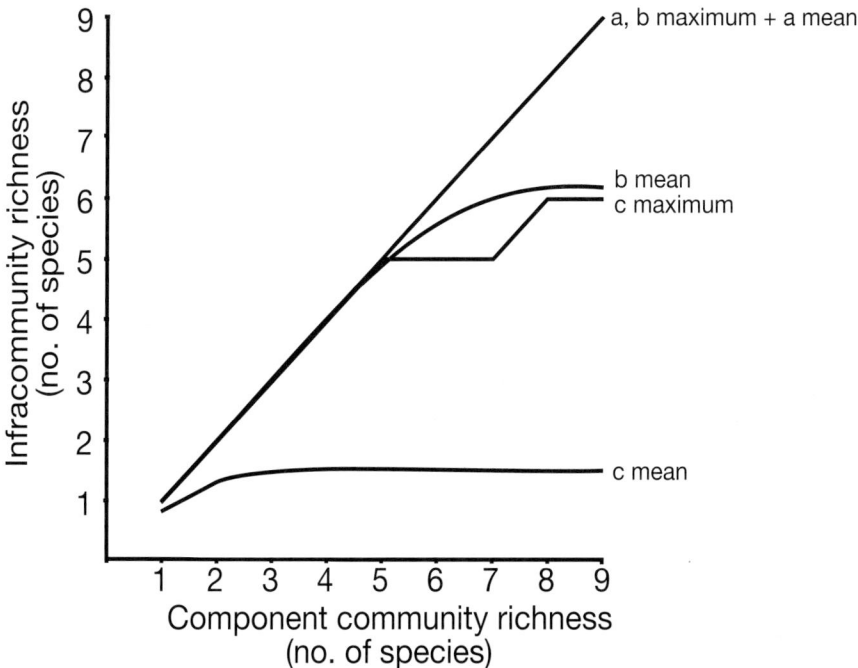

Figure 7.11 Computer simulations of the relationship between infracommunity and component community richness. The assumption is that each infracommunity can be recruited from any species in the component community, in any order, but species have different likelihoods to appear in an infracommunity due to different transmission rates and intrinsic life spans. Richnesses of communites vary between one and nine species, 1000 iterations. Legend: a, all species have a 100% likelihood to appear in the infracommunity; b, seven species have a 80% likelihood, and two have a 30% likelihood to appear in the infracommunity; c, two species have a 40% likelihood, and seven have a 10% likelihood to appear. Note: asymptotic relationships for the means and maxima of all infracommunities except maximum of a and b, and mean of a. The infra-vs component community richness program (Macintosh only) can be accessed (published Rohde 1998a) at the author's website http://www-personal.une.edu.au/~krohde.

However, Rohde (1979, 1991) has demonstrated that the feeding organs (pharynx and oral suckers) of monogeneans infecting the gills of the same host species and using the same food (i.e. blood), differ in size and shape which suggests that the differences may be fortuitous. Some authors have claimed that nestedness is a consequence of competition, but it is uncommon in marine parasites (Worthen and Rohde 1996, Rohde et al. 1998), and, where it occurs, may be the result of epidemiological processes (Morand et al. 2002). Furthermore, there is reduced interspecific relative to intraspecific aggregation in parasites of marine fishes (Morand et al. 1999), and a null model analysis found little evidence for non-random co-occurrence patterns in these parasites (Gotelli and Rohde 2002).

An asymptotic relationship between local (infracommunity) and regional (component community) species richness was explained by a ceiling to diversity in infracommunities due to increasing interspecific competition with increasing richness. However, Rohde (1998a), using computer simulations, has shown that an asymptotic relationship between local and regional species richness may simply arise from different likelihoods of species occurring in a community because of different life spans and colonisation probabilities (Fig. 7.11).

The mating hypothesis of niche restriction applies not only to bisexual, but also to hermaphroditic groups. Thus, cross-fertilisation has been demonstrated for various hermaphroditic platyhelminths. Rohde (2002b, further references and examples therein) has reviewed evidence for the hypothesis as follows: narrow host ranges and microhabitats lead to increased intraspecific contact; adult stages often have fewer hosts and narrower microhabitats than sexually immature and larval stages; microhabitats of sessile and rare species often are narrower than those of more motile and common species; and microhabitats of some species were shown to become more restricted at the time of mating.

Concerning niche segregation as a consequence of reinforcement of reproductive barriers, Rohde and Hobbs (1986) compared niche overlap in 35 parasite species on the gills and in the mouth cavity of six species of marine fish. Nineteen species occurred in congeneric pairs or triplets. They found, using an asymmetric percentage similarity index (Rohde and Hobbs 1986, 1999) that congeners overlap less than non-congeners, and that those congeners that showed considerable overlap had markedly different copulatory organs, in contrast to those that were spatially segregated. As phrased by Rohde (1991), 'if competition to reduce resource exploitation was responsible for segregation, it should affect species with similar or dissimilar copulatory organs in the same way'. Simkova et al. (2000, 2001) gave detailed morphological evidence for the effect of the morphology of copulatory organes on spatial segregation in freshwater monogeneans, supporting the above conclusions.

Effects of non-equilibrium in evolutionary history

In an evolutionary context, non-saturation of habitats with species and non-equilibrium conditions imply that ecological systems have not reached a ceiling of diversity, and that this is even clearer for 'dependent' species, such as symbionts, commensals and parasites. This was taken into account in the diversification equation developed by Rohde (1980b) and based on the logistic equation for population growth. The diversification equation is (**Equation** 7.12):

$$\frac{dS(t)}{dt} = (b(t) - e(t))S_f(t) \frac{K_f(t) - S_f + \sum_{j}^{n} a_{ij}(t)S_j(t-T)}{K_f(t)}$$

where $\frac{dS}{dt}$ = rate of species diversification, b = rate of species formation, e = rate of species extinction, S_f = number of free-living species, K_f = carrying capacity for free-living species, $\sum a_{ij}S_j$ = the sum of all opportunities provided by species, j, for the formation of new species, i, $(t-T)$ = time lag in species diversification, t = factor indicating variability due to environmental effects.

The section on latitudinal, longitudinal and depth gradients (pp. 348–351) shows that diversity at low latitudes is much greater than at high latitudes, and that this can best be explained by direct temperature effects (Rohde 1992) (i.e. the rate of species diversification in the above equation is temperature dependent).

Some parasite communities may be closer to equilibrium than others

In spite of the strong evidence for non-equilibrium conditions in many parasite communities, evidence that some rich parasite communities may be closer to equilibrium than poorer ones must not be ignored. Density-dependent processes will be important in the former but not the latter. Esch and Fernandez (1993) gave examples of density-dependent and density-independent processes in parasites, and Combes (2001) distinguished three types of density dependent mechanisms for parasites: 1, decision-dependent regulation (infective stages avoid hosts that are already infected); 2, competition-dependent regulation (population size limited by limiting resources or some active elimination processes); and 3, and host death-dependent regulation (most heavily infected host individuals die). Combes (2001, references therein) gave many examples for all three mechanisms.

Important references
Rohde (2005) published a monograph on non-equilibrium ecology, in which many examples from plant and animal populations and communities, including parasites, are discussed. The book contains large sections on interspecific competition. A general text on parasites which tends to put more emphasis on equilibrium conditions is by Combes (2001), one which emphasises non-equilibrium is by Price (1980). Important reviews and papers providing evidence for non-equilibrium in marine parasite communties are by Rohde (1977, 1979, 1980a, 1989, 1991,1994, 1998a, 2001, 2002a,b), Rohde and Hobbs (1986). Rohde et al. (1994, 1995, 1998), Worthen and Rohde (1996), Morand et al. (1999, 2002), Simková et al. (2000, 2001) and Gotelli and Rohde (2002).

Population and community ecology of larval trematodes in molluscan first intermediate hosts
Armand M Kuris and Kevin D Lafferty

Introduction
Digenea have complex multiple host life cycles, and virtually all include a mollusc (gastropods or bivalves) as a first intermediate host. A full analysis of the key factors governing the ecology of any species must include all stages of its life cycle. However, for most of the digenean trematodes, the biology of the larval stages in the mollusc includes most of the growth and reproduction in the life cycle. This merits focused attention on the factors that control the distribution and abundance of these parasites in the Molluscan First Intermediate Host (MFIH).

The most striking adaptation of larval trematodes in their MFIH is that they are parasitic castrators (Kuris 1974, Baudoin 1975, Lafferty and Kuris 2002). Infections initiated by ingestion of an egg, or penetration by a miracidium, build up through the asexual reproduction of parthenitae (sporocysts or rediae) until the trematode replaces the gonadal tissue and much of the digestive gland. This intensity-independent effect on the host has several ramifications for the population biology of larval trematodes. Because the trematode can fully exploit the host, intraspecific (little studied) and interspecific competition can be intense if two individuals infect the same individual host.

Many marine molluscs are long lived and trematode infections generally persist for the remainder of the lifetime of the parasitised host. This can often be many years. For example, a marked individual of *Cerithidea californica* infected with *Himasthla rhigedana* was recovered 12 years later (AM Kuris and KD Olson unpublished observation 1981–1993). Reproductive output is size dependent. Large infected snails can account for a disproportionately greater fraction of the reproductive output of parasites in a population.

The distribution and abundance of larval trematode species, and of the community dynamics of their species assemblages, are best understood as a sequence of events; transmission to the MFIHs preceding the impact of post-recruitment factors (growth, survivorship, fecundity and-competition).

Population ecology
The most challenging aspect of the complex life cycle of a digenean trematode appears to be infection of the MFIH. For well-studied snails such as littorines, hydrobiids, *C. californica* and

Ilyanassa obsoleta, it may take many years for prevalence to approach 100% in the largest (oldest) size classes (Robson and Williams 1970, Kuris 1990, Curtis 1997, Skirnisson *et al* 2004) of these slow-growing hosts (Sousa and Gleason 1989, Curtis 1995). Susceptibility may be size-dependent. In *C. californica* infections are virtually never seen in snails under 15 mm in length. Is this due to difficulties associated with host location, internal defences, or other factors? Answers probably await experimental investigation of this pervasive pattern. For well-studied parasites of freshwater snails, such as the human schistosomes, susceptibility *decreases* with snail size perhaps due to increased abundance of haemocytes with increasing blood volume (Loker *et al*. 1987).

Recruitment to MFIHs is clearly driven by patterns of habitat use and by behaviour of the definitive hosts. This point has been made for several studies of the distribution of avian trematodes in MFIH populations (e.g. Robson and Williams 1970, Kuris 1990, Kuris and Lafferty 1994, Skirnisson *et al*. 2004). Smith (2001) has demonstrated that this operates at a fine scale. She compared prevalence of trematodes in snails below zero to two perch sites for avian final hosts in a mangrove habitat. Bird abundance, bird droppings and prevalence of trematodes in snails near the perch sites were intercorrelated in a presumably causal fashion. Kube *et al*. (2002), however, in an extensive study, could not detect a relationship between bird abundance and trematode prevalences in snails when bird distributions were measured at large spatial scales, suggesting that transmission to MFIHs occurs at fine spatial scales. Hence, microhabitats at which final hosts congregate will tend to have high prevalences in MFIHs. Further, if these sites attract final hosts with several species of trematodes, this effect will be seen across the several species of MFIHs that may occur at such transmission foci.

Abundance of snails may affect the prevalence of trematodes. A short-term, small-scale view of transmission suggests that increasing snail density reduces prevalence. That is, prevalence could be limited by miracidial (or trematode egg) abundance. For example, if there are 100 snails at a small site, and 50 miracidia enter at that site from a final host, a maximum of 50 snails will become infected, for a prevalence of 50%. If there are 1000 snails at an identical site, the maximum number of snails that will be infected remains 50; yielding a prevalence of but 5%. Ewers (1964) ascribed to the miracidia limitation hypothesis a negative association between trematode prevalence and the population density of its snail host, *Velacumantus australis*.

However, maximum prevalence is rarely attained because some snails will be infected by more than one miracidium, some miracidia will die before infecting any snails, and snails may vary in their accessibility, susceptibility or attractiveness (Anderson 1978). Because finding and successfully penetrating a snail host is a challenge, the proportion of successful miracidia must depend on the density of snails. So, in our example above, miracidia are 10 times more likely to contact snails in the dense snail population than in the sparse snail population, suggesting that the proportion of successful miracidia should be higher where snails are dense (even though prevalence will be lower). This leads to the snail limitation hypothesis, with the prediction that the density of infected snails should be higher in areas with high snail density.

Consideration of the snail limitation hypothesis leads to a longer-term view of host-parasite population dynamics. According to simple epidemiological theory, the ability of a parasite to invade a host population increases with initial host density (initial means the density of snails if trematodes were absent, or the snail-carrying capacity). For snails, carrying capacity may be linked with productivity of the habitat and the abundance of natural enemies such as predators. In areas with high productivity and low numbers of predators, snail densities should be sufficiently high that transmission is efficient. Such efficient transmission should result in local completion of trematode life cycles. This host-density threshold hypothesis generates the prediction that a high snail carrying capacity should lead to increased success of trematode transmission which (following a buildup of trematodes in the snail population) would most likely be observed as increased prevalence. This conceptualisation leads to a focus on infected MFIH density rather than on trematode prevalence in that MFIH population. This is a critical point because it is the density of infected MFIHs that drives the risk of infection to second intermediate or final hosts.

By eliminating the reproductive output of a fraction of the population, larval trematodes might reduce snail density (Kuris 1973, Combes 1982, Lauckner 1987, Lafferty 1993, Lafferty and Kuris 1996, Fredensborg et al. 2005). Parasitic castration becomes an increasingly important regulatory factor where castrators reach high prevalence. For example, larval trematodes castrated an average of 24% of snails from 62 studies (N = 296 180 snails) (Kuris and Lafferty 1994). Theoretically, the density of snails at equilibrium is directly related to the number of reproductive individuals in the population. A reduction in snail egg output through castration should lead to a subsequent reduction in snail density. This effect may be magnified if prevalence accumulates over time such that larger snails, which would normally contribute disproportionately to egg production, were more likely to be infected. This also creates a negative association between prevalence and density as predicted under the miracidial limitation hypothesis. Unlike the miracidial limitation hypothesis, the parasitic castration hypothesis predicts that the density of infected snails has a hump-shaped relationship with prevalence; infected snails will be sparse when prevalence is low, dense when prevalence is intermediate, and less abundant when prevalence is high. This is an important prediction because density of infected snails is the best predictor of the risk of infection to trematodes of medical and veterinary importance.

There are three plausible arguments against the parasitic castration hypothesis: 1, for highly fecund organisms, rates of castration would have to be very high to affect host density (Huffaker 1964); 2, for space-limited populations, the host may have substantial surplus reproduction at or near carrying capacity and, consequently, reductions in reproductive output may be inconsequential; and 3, final hosts may disperse widely, decoupling transmission in space for different phases of the life cycle (Gaines and Lafferty 1995). Kuris and Lafferty (1992) revealed that the extent to which a parasitic castrator in a local population reduces host density depends on the amount of recruitment to that host population from other populations. Evaluating the concept that surplus larval production could insulate a marine host from the effect of a parasitic castrator, Basson (1994) used numerical simulations to address the effects of a parasitic castrator. She found that castrators decreased host density even when reproduction was in surplus.

Trematodes can also increase the mortality of infected snails (McDaniel 1969, Tallmark and Norrgren 1976, Lauckner 1987, Sousa and Gleason 1989, Jensen and Mouritsen 1992, Huxham et al. 1993, Lafferty 1993). This can have further impacts on host population density and this mortality source will be independent of the spatial scale of recruitment. For the parasite induced mortality hypothesis, the parasites should reduce host density, but that also makes it more difficult for a parasite to invade a snail population (and, therefore, generally results in lower prevalence).

Field data indicate that there are negative associations between the prevalence of trematodes and snail density (Ewers 1964, Lafferty 1993, Fredensborg et al. in review). This observation is consistent with the miracidial limitation hypothesis, the parasitic castration hypothesis and the parasite induced mortality hypothesis. One field experiment sheds some light on the effect of trematodes on snail populations. Lie and Ow-Yang (1973) added trematode eggs to a site containing uninfected snails. After several weeks, the snail population was extirpated through the effects of parasitic castration and increased mortality resulting either from multiple penetration by miracidia, cercariae or a build up of metacercariae (Lie and Ow-Yang 1973, Kuris and Warren 1980). Therefore, these results are most consistent with the parasitic castration and parasite induced mortality hypotheses.

In summary, at small spatial scales and short time scales, the density of infected snails is likely to increase with the input of trematode infective stages (eggs and miracidia). The resulting prevalence of infection will depend on snail density; higher density will result in lower prevalence. At larger spatial scales and longer time scales, areas that are productive for snails are more likely to support trematode life cycles. High trematode prevalence will impair reproductive output of the snail population, leading to a decline in snail density with trematode prevalence and a hump-shaped association between trematode prevalence and the density of infected snails. Mortality induced by trematodes will make it harder for trematodes to persist in the snail population and will reduce prevalence and snail density. Most of these predictions await empirical tests.

Community ecology

Since most, if not all, larval trematodes use the molluscan host as a resource in a comparable way, these assemblages meet the criterion of a guild. Within an individual MFIH there is general agreement that these guilds are depauperate with very few snails having multiple (2 or 3 species) infections. The lack of co-occurrences results from antagonistic interactions that usually lead to competitive exclusion of the subordinate species. (For some of these interactions intraguild predation may be a more appropriate term – rediae of dominant species ingest parthenitae of subordinate species.) However, the assembly rules structuring these guilds in MFIH populations have been subject to debate (Sousa 1990, Kuris 1990, Esch et al. 2001). The argument generally revolves around the relative roles of spatial (and temporal) heterogeneity as isolating processes versus intramolluscan competition as structuring forces. Combes (2001) provides a thoughtful review of these issues in a more general ecological context than our space permits. Lafferty et al. (1994) analysed these forces in sequence, stressing that they were not mutually exclusive hypotheses, although sometimes explicitly treated as such (Sousa 1990, 1993, Esch and Fernandez 1994). Heterogeneity in recruitment may potentially isolate some species. Trematodes using very different hosts will likely recruit to different MFIH populations (e.g. trematodes of fishes versus avian trematodes, or trematodes of piscivorous birds versus shore birds that feed on crustaceans). However, there are many factors that bring a variety of hosts together, thereby producing foci of transmission for multiple trematode species. These would include abundant local food sources andexcellent breeding sites. In a meta-analysis of 62 quantitative studies of larval trematodes in MFIHs, differential space use generally significantly *intensified* potential for competitive interactions (Kuris and Lafferty 1994). Interaction probabilities increased with size (age) of the MFIHs. Temporal variation also concentrated species rather than isolating them. Here too, consideration of natural history supports the meta-analysis. Periods of, for example, definitive host migration and breeding, will generally intensify release of the transmissive stages of larval trematodes resulting in intensified recruitment to MFIHs.

The most effective isolating mechanism was MFIH host specificity. Trematodes using different host species were usually strongly isolated. In short, patchy recruitment is quite important. However, the direction of this effect as a structuring force is opposite that originally predicted (Sousa, 1990, 1993, Esch and Fernandez 1994). At foci of transmission a concentration of infective stages of more than one species enhances the potential for interspecific interactions. Although the limiting case of competitive exclusion may be infrequent in older *C. californica* cohorts (Sousa 1990), the sequential analysis shows that a significant and considerable proportion of the subordinate species are lost to dominant competitors (Kuris and Lafferty 1994). This was conservatively estimated to range from 30% to 44% for each of the four most subordinate species. These were still relatively common, having a total prevalence of 44% (Kuris 1990).

After transmission to a MFIH, competition is generally intense; certain species are dominant over others. Dominance hierarchies are largely transitive. Some species can coexist with others.

For example, in *C. californica*, *Austrobilharzia* sp. was dominant over some very subordinate cyathocotylids and microphallids, but co-occurred in double infections with the more dominant echinostomes and heterophyids. In contrast, the renicolids were eliminated by the echinostomes, but co-occurred with the subordinate species (Kuris 1990). Priority effects may also occur, but these have been little investigated for marine trematodes.

Comparison with a null model (random assemblage), in the meta-analysis of Kuris and Lafferty (1994), estimated that each of the factors intensifying recruitment (space, time and size) contributed about 20% to community structure through intensification. Competition added an average of 23% to the structure of these communities.

Recruitment will be the predominant driver of community structure under two conditions: low overall prevalence (hence, post recruitment interactions are unlikely), or when the community includes a high proportion of subordinate or non-interactive species such as microphallids, cyathocotylids, renicolids or notocotylids. The avian/hydrobiid trematode system studied by Skirnisson and Galaktionov (2002) meets both these conditions.

Important references

Comprehensive accounts of community structure of and interactions between trematodes in snail hosts can be found in the book by Combes (2001), and in the book chapters and reviews by Kuris (1990), Sousa (1990), Esch and Fernandez (1994) and Kuris and Lafferty (1994). Further important accounts are by Combes (1982), who discussed antagonism between species of larval trematodes and sterilising effects on snails in biological control, and Lafferty *et al.* (1994) who provided a detailed analysis of larval trematode communities. Kuris (1974) drew attention to the similarity of parasitic castrators to parasitoids, and Sousa (1993) gave a detailed discussion of interspecific antagonism and species coexistence in a diverse guild of larval trematodes.

Chapter 8

Coevolution and speciation

Marine parasites have not ranked highly as evolutionary models. The only evolutionary aspect of marine parasites that has been studied in greater detail is coevolution. Much effort has gone into comparing evolution of some marine host groups, in particular various 'fishes', seabirds, and mammals, and of their parasites. The very detailed discussion of coevolution discusses these examples, but also points out that much work remains to be done: for most host and parasite taxa, information necessary to make inferences about coevolutionary patterns is insufficient. Much less is known about mechanisms of speciation. The author of the section on speciation and species delimitation stresses the importance of understanding 'what species are'. Such understanding is a prerequisite for addressing fundamental questions about biological processes and biodiversity. Various ways to delimit species are discussed, as are the few examples of studies of species delimitation and speciation in marine parasites.

Coevolution in marine systems
Eric P Hoberg

Introduction
Parasite faunas characteristic of marine invertebrates and vertebrates have been assembled through an intricate interaction of history, ecology and geography, as the determinants of organismal evolution and distribution. Elucidation of pattern and process in the origin and maintenance of biodiversity in marine systems follows from studies that integrate phylogenetic approaches and an historical context for biogeography and ecology (e.g. Brooks 1985, Brooks and McLennan 1991, 1993, 2002, Hoberg 1996, 1997, Page and Charleston 1998, Brooks and Hoberg 2000, Hoberg and Klassen 2002, Page 2003, Brooks et al. 2004). Phylogeny-based approaches are powerful because hierarchical order constrains the range of explanations for faunal structure and history in a comparative context linking host and parasite taxa. Parasites constitute exquisite phylogenetic and historical ecological indicators that reveal substantial insights into the history of the marine biosphere. Phylogenetic hypotheses for hosts and parasites are the tapestry for revealing the interaction of coevolutionary processes in shaping patterns of biodiversity, faunal structure, ecological continuity and persistence across deep temporal and geographical scales in the marine environment (Brooks and McLennan 1993, Hoberg 1997, Brooks and Hoberg 2000, Hoberg and Adams 2000, Paterson and Poulin 1999, Hoberg and Klassen 2002).

Under assumptions of coevolutionary history, or association by descent (e.g. Mitter and Brooks 1983), studies have often focused on attempts to use parasites to reveal host evolutionary relationships, or centres of origin, with research programs that were based on 'parasitological rules' and assumptions about host specificity (see Rohde 1993). As the mechanistically complex drivers for origins, diversification and persistence within assemblages of parasites and hosts are being recognised, formalised rules, such as those developed by VL Kellogg, H Fahrenholz W Eichler, L Szidat and later HW Manter in the 19th and 20th centuries, which postulated a pervasive role for coevolution and cospeciation, have seen significant transformation (e.g. Klassen 1992a, Brooks and McLennan 1993, Hoberg et al. 1997). Explicit, however, remains the view, articulated by Manter (1966), that parasites serve as keystones for understanding the history of biotas (e.g. Brooks and McLennan 1993, Hoberg 1997, Brooks and Hoberg 2000).

Marine parasite faunas have been assembled across varying temporal and geographical scales. Further, associations are historically constrained by genealogical and ecological associations (e.g. Brooks and McLennan 1993). Origins, temporal continuity and structure of marine parasite assemblages can be examined within the framework of hypotheses for coevolution or colonisation (see Boxes, pp. 328–331) that are derived from the comparative study of phylogenies for hosts and parasites generated from analyses based on morphological or molecular data (e.g. Brooks 1979, 1981, Brooks and McLennan 1993, Page 1994a, b, 2003, Hoberg and Adams 2000, Paterson and Banks 2001, Hoberg and Klassen 2002). Brooks (1979) formally recognised cospeciation in the context of patterns that resulted as a byproduct of vicariance, rather than as associations resulting from coevolutionary dynamics. Central and recurring questions arise. To what extent are coevolutionary associations based on interactive processes (e.g. structural or physiological reciprocal adaptation and microevolutionary determinants of coadaptation) or independent responses (e.g. common isolating events and macroevolutionary drivers of cospeciation) resulting in largely congruent temporal and geographical patterns for divergence among hosts and parasites (Ehrlich and Raven 1964, Brooks 1979, Mitter and Brooks 1983)? Alternatively, what has been the role of ecologically driven processes such as colonisation as a determinant of diversification or distribution for parasites, and as a core mechanism shaping historical and contemporary structure in the biosphere (Brooks 1979, Hoberg and Klassen 2002)? Finally in a synergistic sense, how can coevolutionary patterns within a biogeographic context be effectively integrated (Hoberg and Klassen 2002)?

A brief lexicon for evolution, speciation and biogeography

Biodiversity: (Wilson 1988, Brooks and McLennan 2002) biota resulting from the intricate interaction of temporal and geographical scale (local, regional and global), relationships among populations, species and higher taxa within ecosystems, and historical linkages among species, faunas and geographical regions that act as determinants of biotic structure and distribution.

Numerical diversity: numbers of species and individuals within an ecosystem.

Ecological diversity: elucidation of functional interactions among species assemblages within an ecosystem.

Genealogical diversity: numbers and distribution of monophyletic groups (or clades) within an ecosystem or community.

Historical biogeography: (Brooks and McLennan 1991, 2002, Hoberg 1997) linkage of knowledge of geographical distributions of related species to general patterns of distribution on local, regional and global scales; the study of pattern and process in the historical distribution of organismal diversity.

Historical ecology: (Brooks 1985, Hoberg 1997, Brooks and McLennan 2002) macroevolutionary effects of speciation and adaptation in the diversification of clades and multi-species associations such as those involved in complex host-parasite assemblages; study of macroevolutionary processes in community development.

Macroevolution: (Brooks and McLennan 1991, 2002, Coyne and Orr 2004) patterns of variation manifested among species; including:
1 large-scale rules that determine the origin of form, and adaptive changes; and
2 changes in species-richness within and among evolutionary groups and a focus on rates of speciation and extinction over long time frames that determine the history and structure of ecosystems and communities.

Microevolution: (Brooks and McLennan 1991, 2002; Coyne and Orr 2004) patterns of variation manifested within species; transformation or change at rates that influence populations within a single-species lineage.

Relictual faunas: (Brooks and Bandoni 1988)
Numerical relicts: the relatively few surviving members of a once-speciose group (e.g. crocodilians).
Phylogenetic relicts: 'living fossils', or members of old groups that have persisted relatively unchanged since their origins (e.g. horse-shoe crabs).
Ecological relicts: surviving lineages in complex host-parasite systems where persistence for parasites is linked to host switching and secondary radiation in ecologically equivalent host taxa (Hoberg and Klassen 2002).

Speciation processes: (Frey 1993, Brooks and McLennan 2002, Coyne and Orr 2004)
Allopatric speciation (allopatry): speciation initiated by complete geographical separation of two or more populations of an ancestral species; often considered as the general mode for species formation.
Peripheral isolates speciation (peripatric speciation): a process where the origin of new species may arise from small, isolated populations that are usually on the periphery of a larger ancestral population.
Sympatric and parapatric speciation (sympatry): origin of one or more new species in the absence of geographical segregation for populations.
Vicariant speciation (vicariance): a process when a species is geographically separated into two or more isolated populations by formation of a barrier to dispersal or gene flow; associated with large-scale perturbations such as mountain formation or changing sea levels.

A brief lexicon for coevolution and coevolutionary processes

Coevolution: (Ehrlich and Raven 1964) in the strict sense, concerted evolution in associated lineages, expressed as reciprocal adaptation; microevolutionary processes. Viewed as association by descent with macroevolutionary (cospeciation) and microevolutionary (coadaptation) components (Brooks and McLennan 1991, 2002). Mutual association without mutual modification constitutes the weak null hypothesis for coevolution.

Cospeciation: (Brooks 1979, Brooks and McLennan 1991, 2002, Page 2003) restricted here to macroevolutionary processes, stressing the degree of congruence or incongruence between phylogenies for hosts and parasites. Association by descent is signified by phylogenetic congruence and consistency. 'Cospeciation' in a strict sense is also applied to the special case of topological congruence *and* equal rates of molecular divergence (branch lengths) in associated lineages (e.g. Hafner and Nadler 1988). Synonyms are phyletic coevolution, cophylogeny and parallel cladogenesis often applied to associations among clades not involving species.

Coadaptation: (Brooks 1979, Brooks and McLennan 1991) originally as coaccommodation, these processes are microevolutionary, including anagenesis and reciprocal adaptation through arms race scenarios or where parasite lineages track host resources and fail to speciate. Host specificity is linked to coadaptation and is decoupled from cospeciation. Coadaptation is implicit in such phenomena as 'failure to speciate' and 'parasite inertia' where a parasite does not speciate in association with its host lineage (Paterson and Banks 2001).

Colonisation: (Brooks 1979, Mitter and Brooks 1983) ecological processes involved in host switching by parasites among related or phylogenetically disparate host lineages. Colonisation is the major mode of establishment of symbiotic associations, but coevolution and host switching are not mutually exclusive (Mitter and Brooks 1983, Hoberg 1986, Hoberg and Klassen 2002). Synonyms are hostal radiation (Kontrimavichus 1969), host capture (Chabaud 1965) and horizontal transfer (Page 1993).

Retro-colonisation: originally noted in studies of associations among phytophagous insects and plants (Ehrlich and Raven 1964, Mitter and Brooks 1983). Exemplified in marine systems as a specific pattern of incongruence in associations for basal parasites in crown hosts and crown parasites in basal hosts driven by sequential colonisation (Hoberg 1986, 1992, 1995, Hoberg and Adams 1992). 'Young parasites in Archaic hosts'. Synonym is 'back colonisation,' discussed as a general problem (Brooks and McLennan 1991, 1993, 2002).

Other concepts for cospeciation and colonisation

Persistence of ancestral parasites: (Brooks 1981) also 'lineage duplication' (Page 1994a) where parasite speciation within a host lineage occurs in the absence of host speciation. Such retention is a scale-dependent process for associations on a continuum from populations to species (Hoberg and Klassen 2002).

Cryptic isolation and speciation: (Hoberg 1995) a form of lineage duplication where parasite speciation is driven by allopatry for host populations. Isolation is sufficient to drive divergence and speciation for parasites, but not for hosts, leading to multiple congeneric sister-species in a single host species across its geographical range. Synonym is 'intrahost speciation' discussed as a possible example of sympatric speciation (Brooks and McLennan 1991) and as a process resulting in incongruence (Paterson and Poulin 1999, Paterson and Banks 2001, Demastes et al. 2003).

Failure to speciate: (Brooks and McLennan 1991) has been recognised as one possible consequence of host diversification (persistence and coadaptation), or may be linked to instances of parasite colonisation. Discussed as a case where gene flow for parasites exceeds that for hosts (Johnson et al. 2003).

Lineage persistence in secondary hosts: (Hoberg et al. 1999a, b, Hoberg and Klassen 2002). Persistence for parasite lineages in time and space through colonisation and radiation in a secondary host group preceding extinction of the ancestral host clade. 'Archaic parasites in young hosts' (EP Hoberg).

Sorting events: A broad class of unrelated stochastic phenomena resulting in loss of a parasite lineage from a host lineage (e.g. Brooks and Mclennan 1991), including extinction through 'missing the boat,' or 'drowning on arrival' (Johnson et al. 2003, Page 2003). Such may also reflect that a parasite was not detected (Brooks 1979, Brooks and McLennan 1991), also known as 'X events' (Paterson and Banks 2001).

Peripheral isolates speciation and non-parasitised members of a host clade: (Brooks 1979, 1981, Brooks et al. 2004) absence of associated parasite lineage(s) from a host founder population at a speciation event due to extirpation or extinction, and low population density or patchy distributions; also known as 'missing the boat' (Paterson and Gray 1997, Paterson et al. 1999, Page 2003). Absence may also reflect initial colonisation and radiation of a parasite lineage following origin and diversification of a host group (e.g. Hoberg 1986, 1992, 1995).

Drowning on arrival: (Paterson *et al.* 1999; Paterson *et al.* 2003) extinction of a parasite lineage from its host subsequent to host speciation or dispersal, possibly resulting from population bottlenecks, or from an incompatible ecological transition (Dabert and Mironov 1999).

A coevolutionary history – assumptions and expectations

Primary criteria for defining associations that have developed through coevolution versus colonisation in a phylogenetic and comparative context have been reviewed and outlined (Hoberg and Klassen 2002) (Table 8.2; Fig. 8.1). Coevolution, or association by descent, is corroborated through examination and interpretation of host-parasite associations that demonstrate:

1. consistency or congruence in host-parasite phylogenies or area relationships;
2. a high degree of cospeciation or coadaptation (Brooks 1979);
3. recognition of phylogenetic or numerical relicts (Brooks and Bandoni 1988); and
4. often widespread geographical distributions, that in marine systems may be global or antitropical in extent (Hoberg and Adams 2000).

General congruence in phylogenetic and biogeographic patterns among complex host-parasite assemblages indicates coincidental physical and biotic processes as determinants of distribution (e.g. Hoberg, 1986, 1992, 1997). In these instances, geographical scale may be linked to the relative age for the initial association of parasite and host taxa, vagility of the assemblage and persistence or duration of their history for coevolution (Hoberg and Klassen 2002). Additionally, Hafner and Nadler (1988) and Hafner *et al.* (1994) introduced the concept of temporal congruence for molecular evolution between hosts and parasites, revealing an important facet to be considered in studies of cospeciation.

Consequently, broad topological congruence is apparent in events of:

1. cospeciation (cophylogeny) (Brooks 1979, Brooks and McLennan 1993, 2002);
2. instances where there is not a direct correspondence in diversification for hosts and parasites (e.g. parasite and host phylogenies differ but are not contradictory) (Paterson *et al.* 1993);
3. coadaptation, inertia, or failure to speciate where host speciation is not accompanied by parasite divergence (Brooks 1979, Paterson and Banks 2001);
4. the persistence of ancestral parasites (lineage duplication) and intrahost speciation, where divergence for parasites occurs in the absence of cladogenesis for hosts; and
5. temporal decoupling for rates of molecular divergence (unequal branch lengths) in host and parasite lineages (e.g. Hafner and Nadler 1988, Hafner *et al.* 1994).

In contrast, incongruence arises from processes linked primarily to colonisation.

Faunas derived from a history of colonisation (Fig. 8.1) depart from coevolutionary systems in the following ways:

1. incongruent and inconsistent phylogenies for parasites and hosts (Brooks 1979, Brooks and McLennan 1991, Page 2003);
2. similarities in host trophic ecology or other ecological attributes on which transmission for parasites is dependant;
3. faunas that are geographically or regionally delimited;
4. parasite faunas in which diversification is temporally circumscribed in the context of the origin and duration of the host group (Hoberg 1992);
5. faunas of low diversity that are depauperate as opposed to relictual;

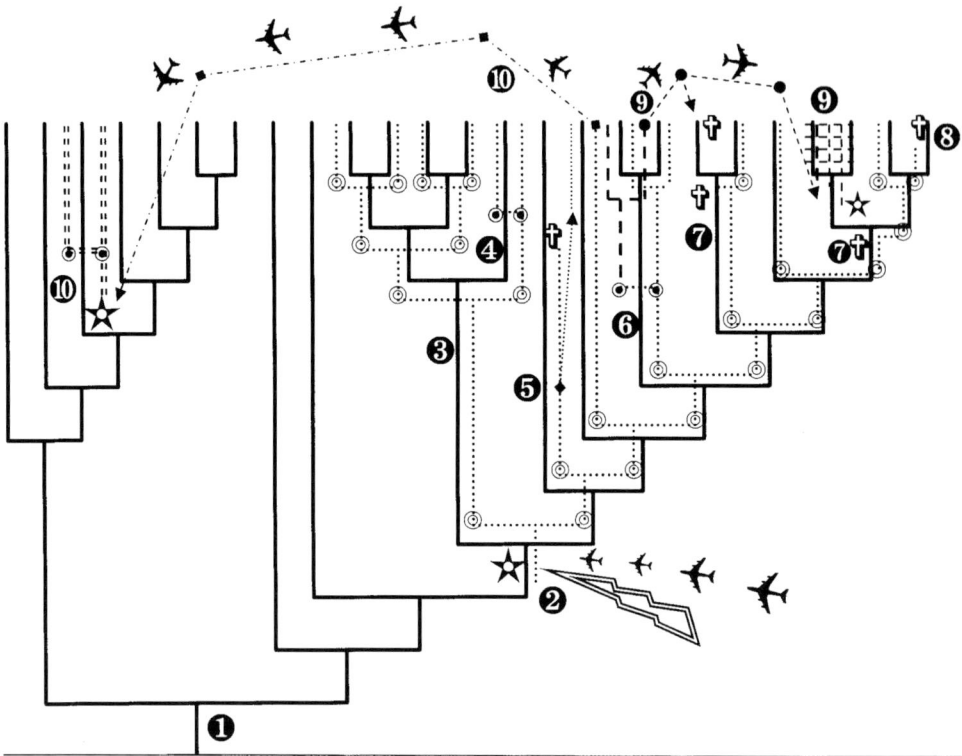

Figure 8.1 Coevolutionary processes as determinants for the origin, persistence and diversification within a host-parasite association. Principal mechanisms and patterns of association are demonstrated with parasite phylogeny and distributions mapped (patterned branches) onto a hypothetical host tree (solid branches). Processes and mechanisms are identified by numbers, symbols and varying forms of broken lines. **1.** Primitive absence of the parasite in the host clade. **2.** Origin of association through colonisation, and a subsequent complex history for diversification of parasites. **3.** Cospeciation. **4.** Intrahost speciation. **5.** Extinction in ancestral lineage subsequent to colonisation. **6.** Lineage duplication. **7.** Missing the boat. **8.** Drowning on arrival. **9.** Colonisation accompanied by extinction, *and* host switching with Coadaptation or Failure to speciate in a diverging host lineage. **10.** Retro-colonisation with subsequent diversification by Intrahost speciation. Symbols at nodes for the parasite phylogeny: ⊙, cospeciation; ★, origin by colonisation; ✝, extinction event; ⊙, lineage duplication. Other symbols: ✈, dispersal or colonisation; crosshatching at **9.** shows coadaptation. See Box pp. 329–331 and appropriate text. Air transportation, courtesy of DA McLennan and DR Brooks.

6 associations of variable temporal duration and varying degrees of cospeciation or coadaptation linked to the time frame for colonisation of the host clade(s) (Hoberg and Adams 2000).

Alternatively, fine-scale incongruence can arise from events of:

1 lineage persistence, and failure to speciate;
2 absence of a parasite lineage due to colonisation subsequent to the origin of the host group (e.g. Hoberg and Adams 1992, 2000); or
3 different patterns of stochastic 'sorting events' (Page 1994a, 1994b, Paterson *et al.* 1999, 2000, 2003, Paterson and Banks 2001, Johnson *et al.* 2003).

There should be no general expectations for any specific patterns of diversification within complex host-parasite assemblages and among host and parasite clades (Fig. 8.1) (e.g. Hoberg and Klassen 2002, Dowling *et al.* 2003), although most historical rules (i.e. those by Fahrenholz, Szidat, Eichler and Manter) are based explicitly on assumptions of cospeciation (Klassen 1992a, Brooks and McLennan 1993, Page 2003). Thus, in a mechanistically complex universe parasites may:

1. speciate in response to host speciation (allopatric cospeciation);
2. fail to speciate (non-response to vicariance) or become extinct following isolation;
3. speciate in the absence of host speciation (cryptic events, e.g. Hoberg 1995, Paterson and Banks 2001);
4. undergo 'sympatric speciation' through specialisation of ancestral parasites to different microhabitats of the host (Brooks and McLennan 1991, Dabert and Mironov 1999);
5. speciate or undergo radiation following host switching (or not) (peripheral isolates);
6. undergo post-speciation colonisation;
7. become extinct following colonisation, lineage duplication, or geographical dispersal (drowning on arrival cannot be assessed directly) (Paterson *et al.* 2003);
8. diversify in a new host group following colonisation and extinction of ancestral hosts (e.g. Hoberg *et al.* 1999a,b); or
9. undergo local extirpation (missing the boat), or regional and global extinction in an historically associated host lineage.

Rates for molecular evolution of parasites may be delayed, equal to or accelerated relative to their associated hosts (Hafner and Nadler 1988). Additionally theoretical and empirical evidence indicates that host specificity must be decoupled from a history of extended cospeciation for hosts and parasites (e.g. Brooks 1979, 1981, Hoberg 1986).

Revealing the often protracted histories for associated lineages is a process of discovery (Brooks and McLennan 2002). Criteria for coevolutionary or colonising faunas set a hypothesis-driven framework to evaluate biotic structure in marine systems. Recent studies on marine helminth systems have applied parsimony mapping (including reciprocal mapping on both host and parasite phylogenies), or Brooks Parsimony Analysis (BPA) (see Brooks 1981, Brooks and McLennan 1993, Brooks *et al.* 2001, Hoberg and Klassen 2002, Brooks *et al.* 2004). Alternative analytical methods, particularly TreeMap that define models for 'maximum cospeciation' have also been articulated (summarised in Paterson and Banks 2001, Page 2003). Such Component, Reconciliation and Jungles-based approaches have been used in evaluations of lice (Phithiraptera) and feather mites (Astigmata) on marine birds (e.g. Paterson and Gray 1997, Paterson et al. 2000, Dabert and Mironov 1999). In contrast to BPA, these latter methods are derived from a formal assumption that cospeciation is the primary driver of faunal diversification (e.g. Paterson and Banks 2001, Page 2003). Differences between a priori and a posteriori methods represented by TreeMap and BPA, respectively, are thus indicative of fundamentally disparate philosophical foundations where cospeciation is either dominant, or simply one among a diversity of processes (Dowling *et al.* 2003, Page 2003, Brooks *et al.* 2004). Beyond this point, equivalent analytical protocols are applied in both coevolutionary and historical biogeographic analyses (Dowling *et al.* 2003).

Coevolution and deep history

A growing consensus based on phylogenetic studies among neodermatans including Digenea, Monogenea, Gyrocotylidea, Amphilinidea and the Eucestoda across a diversity of host groups encompassing Chondrichthyes, Osteichthyes and the tetrapods clearly indicates a deep age (hundreds of millions of years) for the origins of parasitic groups among marine vertebrates (Brooks 1989, Kearn 1994, Rohde 1994, Boeger and Kritsky 1997, Hoberg *et al.* 1999a,b, Littlewood *et al.* 1999, Cribb *et al.* 2001). For example, tapeworms appear to have initially diversified among actinopterygian and neopterygian fishes 350 to 400 million years before present, and chondrichthyans were apparently colonised secondarily (Hoberg et al. 1999a). Patterns of association for eucestodes appear to parallel those for both Monogenea and Digenea, suggesting that basal diversification for parasitic flatworms coincided with the origins and divergence of lineages for the Chondrichthyes and Osteichthyes prior to the Mesozoic (Brooks 1989, Boeger and Kritsky 1997, Cribb et al. 2001). This is compatible with a long period of diversification of such eucestode groups as the 'Tetraphyllideans', Lecanicephalidea, Diphyllidea and Litobothriidae (Hoberg et al. 2001, Olson et al. 1999, 2001) among chondrichthyans in marine and secondarily freshwater environments and more generally is indicative of the archaic nature of the faunas in sharks and rays (Euzet 1959, Brooks et al. 1981, Bandoni and Brooks 1987a,b, Brooks and Deardorf 1988, Brooks and McLennan 1993, Nasin et al. 1997). Concepts linked to recognition of a protracted history for tapeworms and various marine host taxa have been articulated by Hoberg et al. (1999a, b), and emphasise the relictual nature of many groups (see Brooks and Bandoni 1988). Diversity may have been influenced by radiation subsequent to colonisation, or by secondary radiations in contemporary host taxa, a phenomenon clearly demonstrated by the tetrabothriideans among seabirds and marine mammals (Hoberg et al. 1999a,b). A deep history of colonisation is apparent, a further indication of the linkage between phylogeny and ecology as factors determining the historical and contemporary structure of parasite faunas in marine environments.

Recognition of deep histories for major parasite taxa has substantial implications with respect to the role of global-level extinctions through Earth history (summarised for the Phanerozoic in Briggs 1995) as determinants of faunal structure and geographical distribution (Hoberg et al. 1999a,b). Pertinent here is the idea that patterns of differential extinction for free-living taxa, across an array of potential intermediate or definitive hosts, have influenced genealogical or ecological diversity for parasites with complex indirect life cycles (Hoberg and Klassen 2002). Parasite lineages have persisted in time as a function of guild dynamics, and across a mosaic of ecological stability and perturbation, where global-scale extinctions must be viewed as a series of episodic ecological transitions for host-parasite assemblages (e.g. Hoberg 1999a).

Discovering coevolutionary associations in marine systems

Although not synoptic for any one host-parasite assemblage or taxon, there are sufficient empirical studies to derive a preliminary interpretation of the degree of contribution for cospeciation and dispersal to coevolutionary scenarios (Hoberg and Klassen 2002). Most studies have been conducted primarily by inspection and mapping and may benefit from reanalysis according to current comparative protocols, particularly with the potential insights based on inclusion of molecular-based data (see Brooks et al. 2001, Paterson and Banks 2001). Most extensive of these are Brooks et al. (1981) work on stingrays, Collette and Russo (1985) on mackerel, Klassen (1992b) on boxfishes, and Hoberg (1986, 1992, 1995) and Hoberg and Adams (2000) on the Beringian/ North Pacific fauna. Where this has been accomplished (e.g. the studies of *Alcataenia* cestodes among Alcidae, Charadriiformes), the original conclusions have been strongly upheld (Hoberg et al. 1997). Additionally, Caira and Jensen (2001) reiterated the necessity in coevolutionary studies to focus on monophyletic taxa and systems with a high level of specificity, accompanied by a robust understanding of host and parasite diversity (accurate, taxonomy, identity and comprehensive sampling), and accurate estimates of both host and parasite phylogenies (see also Page et al. 1996). The search for pattern and interpretation of process, however, is an exploratory activity rather than an attempt to identify strictly coevolving systems (Brooks and McLennan 1991, 2002). Indeed it is discovery of the departures from strict cospeciation (and support for Fahrenholz's Rule) that reveal significant insights into the complex ecological history of faunal associations as indicated, for example, in the detailed studies for the monogeneans, *Haliotrema*, and their boxfish hosts (Tetraodontiformes) (Klassen 1992b, 1994) and novel hypotheses for radiation of the Trypanorhyncha (Palm 2004).

The dominant recurring theme evident in diversification of helminth faunas among marine vertebrates including fishes, mammals, chelonians and birds has been colonisation. For example, radiation of Trypanorhyncha and the tetraphyllidean assemblage in sharks and rays appears attributable to initial colonisation, although a deep history of secondary cospeciation may be indicated by high levels of host specificity for many species and higher taxa (e.g. Euzet 1959, Hoberg et al. 1999a, Beveridge et al. 1999, Caira and Jensen 2001, Palm 2004). In general, chondrichthyan faunas have yet to be examined in great detail based on phylogenetic methods other than through the development of hypotheses for the origins of the freshwater rays, Potamotrygonidae (Brooks et al. 1981, Brooks 1992, 1995), or otherwise in groups of limited scope (Nasin et al. 1997, Caira and Jensen 2001). An exception are the detailed studies by Palm (2004) who ambitiously has attempted to develop a framework for understanding the radiation of the Trypanorhyncha and their chondrichthyan hosts. Host switching by digeneans and monogeneans has been identified among different groups of teleosts (e.g. Klassen 1992b, Gibson and Bray 1994, Barker et al. 1994, Bray and Cribb 2000, Brooks et al. 2000). Considerable details, however, remain to be revealed with respect to the coevolutionary histories of helminth faunas among osteichthyan and chondrichthyan fishes.

Interestingly, across an array of species that may be assumed to have deep temporal origins, current evidence suggests that the *Pedibothrium* fauna characteristic of nurse sharks, *Ginglymostoma*, in the Atlantic basin and Baja, Mexico, may extend to two million years ago coinciding with the closure of the Panamanian Isthmus (Caira and Euzet 2001). The intriguing disjunct distributions in the eastern and western Atlantic for *Pediobothrium* in nurse sharks and the degree of site fidelity for hosts may suggest even deeper ages extending to the Cretaceous for this assemblage. At a minimum these patterns may be consistent for several tapeworm and nematode taxa that indicate the importance of vicariance associated with closure of the Panamanian seaway in the formation and diversity of faunas in sharks and rays (Marques et al. 1995, Goshroy and Caira 2001).

The pronocephalid digeneans characteristic in marine chelonians have a complex history involving extensive colonisation, and multiple marine-freshwater transitions (Pérez-Ponce de León and Brooks 1995a). Colonisation not only involved habitat shifts for turtles, but also shifts by parasites from turtles to such phylogenetically disparate taxa as marine iguanas (Pérez-Ponce de León and Brooks 1995b). The patterns indicated a deep and complex history including vicariance and dispersal.

Among marine homeotherms including cetaceans, pinnipeds and seabirds, few taxa are indicators of historical coevolutionary linkages, or association by descent, between marine and terrestrial faunas (Deliamure 1955). Among diphyllobothriids, there is broad evidence for diversification by what has been termed 'hostal radiation' where ecologically driven host switching occurs among phylogenetically unrelated pinniped or cetacean taxa (Kontrimavichus 1969, Iurakhno 1991). Phylogenetic studies of the eucestodes have supported an hypothesis for the origin of Tetrabothriidea through host switching by ancestral tapeworms, first to basal marine birds from marine archosaurs (non-avian, consistent with tetrabothriids as ecological relicts) and second to cetaceans and pinnipeds (Hoberg and Adams 1992, 2000, Hoberg 1996); cospeciation may have been critical in later diversification of *Tetrabothrius* among avian hosts, but phylogenetic studies have yet to be completed (Hoberg 1996). Colonisation has also been recognised as a significant driver of diversification among the Tetrabothriidea in marine mammals (Hoberg and Adams 1992) and particularly for species of *Anophryocephalus* among Phocidae (Hoberg, 1992, 1995). Fernández et al. (1998a,b) and Hoberg and Adams (2000) demonstrated a complex history involving colonisation and cospeciation among odontocetes and pinnipeds for some campulid digeneans. Nadler *et al.* (2000) found that *Contracaecum* spp. associated with pinnipeds are not monophyletic, and that host switching among seabirds and pinnipeds has occurred among the ascaridoids.

Additionally, Hoberg (1986, 1992) and Hoberg *et al.* (1997) documented the pervasive nature of colonisation in the evolution of *Alcataenia* tapeworms among seabirds of the family Alcidae. Significantly, the development of marked host specificity was evident among species that had originated subsequent to relatively recent colonisation of host taxa. These studies supported the concept that strict (or 'phylogenetic') specificity should be decoupled from the process of cospeciation, and that the former was not necessarily an unequivocal indicator of the temporal duration of an association (Brooks 1979, 1985, Hoberg 1986).

Building on studies of *Alcataenia*, the Arctic Refugium Hypothesis postulated general level or congruent and synchronic patterns for phylogenetically disparate groups of tapeworms that infect Phocidae and Otariidae (Pinnipedia) and Alcidae in the North Pacific Basin and across Holarctic seas (Hoberg 1986, 1992, 1995, Hoberg and Adams 1992). Similar patterns have also been recognised for ascaridoid nematodes (species of *Contracaecum* and *Pseudoterranova*) in phocids and otariids (e.g. Paggi *et al.* 1991, Nascetti *et al.* 1993, Bullini *et al.* 1997). The underlying processes are linked to radiation of hosts and parasites in Subarctic and Arctic refugia during the late Pliocene and Quaternary where refugial effects, habitat fragmentation and isolation were significant determinants of faunal diversification (Hoberg 1992, 1995, Hoberg and Adams 2000). Although molecular clock hypotheses have been applied to studies of ascaridoid evolution and biogeography, the temporal setting for diversification among tapeworms, seabirds and pinnipeds has been estimated based on the physical history of the Holarctic region, and episodic variation in sea level as a driver for isolation at the Beringian nexus.

Arthropod ectoparasites on both fishes and seabirds may represent a contrast to the histories of colonisation being postulated for a variety of helminths and their hosts. The limited number of studies of copepods among teleosts have indicated substantial patterns of coevolution and cospeciation (summarised in Paterson and Poulin 1999). Such patterns have been demonstrated among parasite taxa that also exhibit relatively low levels of host specificity (Poulin 1992). Further detailed analyses of a wider diversity of copepod taxa and their hosts may establish the generality of these observations, and such would provide an interesting comparison to the monogeneans on the same spectrum of piscine hosts (see Rohde and Hayward 2000).

Phithiraptera among seabirds also appear to have deep coevolutionary histories with their avian hosts (Paterson and Gray 1997, Paterson *et al.* 1993, 2000). Such may reflect the constraints on the potential for transmission among conspecifics, or for host switching between phylogenetically unrelated seabirds in relative sympatry at large colony sites (e.g. Paterson *et al.* 2000). Degrees of coloniality, the physical attributes of nest sites and limited interactions during foraging in pelagic situations may serve as substantial controls on distribution. Among the assemblage of lice on both Procellariiformes and Sphenisciformes, cospeciation was postulated as a dominant driver for diversification with contributions from intrahost speciation, and patterns of host association were further influenced by sorting events (Paterson *et al.* 1999, 2000).

In a parallel manner, the primary mode of diversification among feather mites and their avian hosts appears to be coevolution and cospeciation, demonstrated empirically by the occurrence of highly characteristic faunas being associated with each of the primary avian lineages (Dabert and Mironov 1999). Although these systems have yet to be examined in great detail, phylogenetic studies of mites on Procellariiformes (species of *Promegninia*, *Rhinozachvatinia* and *Zachvatkinia*) and Charadriiformes (only species of *Zachvatkinia*) provide some insights. The presence of two genera with species inhabiting two distinct sites on procellariiforms is considered indicative of sympatric speciation, and their absence among diving petrels, *Pelecanoides*, represents secondary extinction. In contrast, the distribution of distinct faunas for *Zachvatkinia* on procellariiforms and the unrelated Lari is compatible with colonisation and secondary radiation by cospeciation among jaegers, gulls and terns (Dabert and Mironov 1999). These preliminary studies suggest a long coevolutionary association for marine birds and their highly specialised feather mites.

Among those systems that have been examined based on phylogenetic methods, recognition of widespread cospeciation has not been documented except among arthropod ectoparasites (Hoberg and Klassen 2002; Palm 2004). Among helminth faunas, and groups with complex life cycles, instances of colonisation appear more pervasive. Hoberg and Adams (1992, 2000) discussed issues related to host switching, particularly among marine homeotherms. In marine and other systems, host switching for parasites with complex life cycles is a stochastic process that may be linked to the predictably of guild associations or food web structure over extended evolutionary time frames. Life history and transmission appear to be evolutionarily conservative and structured by guild dynamics, such that selection may be for continuity in transmission rather than for associations with a specific taxonomic spectrum of definitive hosts (Hoberg and Adams 2000). It is not clear that constraints to host switching will be the same for parasites with direct versus indirect cycles, or whether ectoand endoparasites may be influenced differentially by variation in life history for their respective piscine, avian or mammalian hosts.

Allopatric speciation as a model
In those systems that have been examined, and particularly among helminth faunas in marine birds and mammals, speciation has been largely allopatric. In these systems, speciation of cestodes and ascaridoid and ancylostomatoid nematodes appears to be driven by the geographical ranges and a history for isolation of definitive hosts (Hoberg 1995, Bullini et al. 1997, Nadler 2002). Thus, isolation and speciation among diverse assemblages of marine parasites may often proceed independently from that of populations of intermediate hosts. Although different mechanisms for allopatric speciation have been identified (e.g. microallopatry, peripheral isolates) in the speciation of cestodes in pinnipeds and seabirds, all appear to be driven by the particular history of the vertebrate hosts (Hoberg and Adams 2000). Further for some parasites with direct cycles such as the monogeneans represented by *Haliotrema*, many so-called scenarios for sympatric speciation may represent examples of allopatric speciation for parasites on allopatric host populations (Klassen 1992b, Hoberg and Klassen 2002). The degree to which allopatry and geographical isolation represent a general model for marine parasites and their hosts remains to be examined in greater detail. Although speciation by geographical isolation and allopatry appears to be a generality, Brooks and Mclennan (1993) and Rohde (1993) have suggested some form of sympatric diversification may account for the numerous congeneric species that are encountered in some host individuals. The latter could be a reflection of our limited understanding of scale in marine systems (Hoberg and Klassen 2002).

Intrahost speciation
Processes for intrahost speciation represent another form or facet of cospeciation (e.g. Brooks and McLennan 1993, Paterson and Banks 2001, Demastes et al. 2003). They may be invoked based on the observation of the co-occurrence of multiple congeners in single host species, but there are few examples where such systems have been examined phylogenetically. A phylogenetic context is necessary to first demonstrate sister-species relationships and second to discriminate between hypotheses for cospeciation versus forms of colonisation. Examples of this phenomenon may be particularly common among genera and species of the Onchobothriidae and Phyllobthriidae in chondrichthyans (Caira et al. 2001) and appear to be commonly reported for species of *Acanthobothrium* and *Pedibothrium* (Caira 1992, Marques et al. 1995, 1997, Caira and Burge 2001, Caira and Zahner 2001, Caira and Euzet 2001) and among *Rhinebothroides* spp.

(Brooks and Amato 1992). Paterson and Poulin (1999) identified intrahost speciation as an important process for diversification of copepods in the genus *Chondracanthus* on a variety of marine teleosts.

Assuming that allopatric speciation is a primary determinant for parasite diversification, it may be useful to consider if such intrahost patterns indicate punctuated or cyclical/periodic pulses or bouts of geographical isolation for hosts that drive divergence and speciation among parasite lineages (Hoberg 1995). Duration of isolation may be insufficient to result in divergence for hosts, but may lead to speciation for parasites. Is this a phenomenon linked to the age or geographical extent of an assemblage, in that the influence may be most pronounced among geographically widespread taxa? The issue of geographical and temporal scale is important as it is clear that considerable discrete variation, or species-level partitions that can be demonstrated through comparative molecular analyses, are often masked by a similarity or uniformity in morphological characters. Paterson and Poulin (1999) considered that the relatively extensive level of intrahost speciation evident for species of copepods in the genus *Chondracanthus* could reflect allopatric speciation across a broad geographical range occupied by hosts. Hoberg (1995) suggested that such intrahost patterns were important indicators of cryptic isolation or speciation events for components of a host-parasite assemblage. Parasites become cryptic indicators of a complex history of episodic isolation for hosts, and this may be either reflected in the speciose and host-specific nature of some parasite taxa in respective hosts, or may also reflect the facets of biogeographic history that can no longer be recognised for the host group.

A coevolutionary future

Future studies must extend beyond descriptive coevolution and biogeography which focuses on documentation of distribution, ecological diversity and host association and include integrated approaches to phylogenetic and historical reconstruction (Brooks 1985; Hoberg 1997; Brooks and Hoberg 2000). In this manner such questions as how species are related within and between zones and regions or how higher taxonomic groups are distributed in time and space may be addressed. Evaluation of historical structure then becomes the context for identification of common mechanisms involved in distributional history for biotas including the relative roles of cospeciation or host-switching and vicariance or dispersal (Hoberg and Klassen 2002). Various facets of history are being increasingly addressed in current assessments of biodiversity and biogeography.

Combined survey and phylogenetic reconstruction represents an important foundation as exemplified by studies of Australian reef fishes (e.g. Cribb *et al.* 1992, Barker *et al.* 1994, Bray *et al.* 1998, Bray and Cribb 2000). Detailed biodiversity inventories for coastal waters of Mexico (e.g. Pérez-Ponce de León *et al.* 1999) also serve as important resources for coevolutionary analyses. These studies are critical in establishing accurate concepts for host and geographical distribution, and particularly ideas about host specificity within and among assemblages (e.g. Gibson and Bray 1994), but need to be considered in an explanatory framework derived from comparative phylogenetics. Although several faunal provinces and biotas have received focused attention, there has yet to be a synoptic and integrated approach linking survey, inventory and phylogenetic reconstruction. Such will continue as a challenge for any comprehensive work on chondrichthyans, given the exceptional diversity that remains to be discovered and described among the tetraphyllideans and other eucestodes (e.g. Caira and Jensen 2001, Caira and Euzet 2001).

The great potential for a coevolutionary program has been amply demonstrated by an array of studies across a phylogenetically diverse landscape of hosts and parasites (see also Brooks and McLennan 1993, Brooks and Hoberg 2000, Hoberg and Klassen 2002, Page 2003). Despite nearly 25 years of explicit coevolutionary studies based on phylogenetic approaches, however, there is still a lack of critical information for most host and parasite taxa and in many respects the literature is diverse but fragmented. For example, there remains a single detailed historical study of helminths among seabirds (Hoberg 1986, 1992), and our understanding of species diversity and phylogeny among the speciose tetraphyllidean taxa of chondrichthyan hosts remains to be dramatically expanded (Caira and Jensen 2001).

There are relatively few robust species-level phylogenies for parasites within the context of a detailed understanding of relationships for higher inclusive taxa. Likewise, our knowledge of host phylogeny often is inadequate as the basis for modern comparative studies in coevolution although our basic understanding for relationships among such groups as teleosts (e.g. Stiassny et al. 1996), chondrichthyans (reviewed in Caira and Jensen 2001) and marine mammals (Berta and Sumich 1999) has substantially improved since about 1995. Phylogenetic and coevolutionary studies have altered our view of global biodiversity in revealing insights about:

1. the great age of marine parasite faunas;
2. a significant role for colonisation in diversification across a taxonomic continuum at deep and relatively recent temporal scales; and
3. a primary role for allopatric speciation.

Comparative research programs now encompass a continuum from such basic but complex concerns as the delimitation of species (Nadler 2002) to the integration of coevolutionary patterns with historical biogeography (Hoberg and Klassen 2002). In this environment, continued expansion of a phylogenetic framework clearly represents a challenge and an opportunity.

Important references
A framework for concepts in coevolution and comparative historical studies of host-parasite systems is developed and reviewed in Manter (1966), Brooks and McLennan (1991, 1993, 2002) and Page (2003). A rationale for biodiversity assessment and phylogenetic studies of parasites, including the component parts of historical ecology and historical biogeography, are addressed in Hoberg (1997) and Brooks and Hoberg (2000). Seminal research exploring coevolution in the marine environment has focused on relationships of copepods on scombrid fishes (Collette and Russo 1985), helminth faunas of potamotrygonid rays (Brooks et al. 1981), cestode faunas in marine mammals and seabirds (Hoberg 1986, 1992, 1995), and lice and mites on seabirds (Paterson et al. 1993, Dabert and Mironov 1999). Finally, Hoberg and Klassen (2002) developed a synoptic review of empirical studies, hypotheses and conclusions regarding the intricate coevolutionary history for parasite faunas among marine vertebrates, and attempted a synthesis linking geographical and temporal scale with phylogeny and ecology.

Speciation and species delimitation
Steven A Nadler

Introduction
The assessment of marine parasite biodiversity, including testing hypotheses of species, remains one of the largest unexplored challenges facing parasitologists working with hosts in the marine environment. The problems of deciding what species are and actually finding them in nature (species delimitation) are prerequisite to addressing fundamental questions concerning biological processes; within parasitology this often involves determination of the number of species in a particular host or among members of an ecological community. For example, delimitation of parasite species is critical to understanding if variation in parasite-induced pathology between different host species is caused by different parasite species, or variable host responses to a single

parasite species (Nadler *et al.* 2000). Likewise, thorough species discovery operations are essential to be able to meaningfully compare host and parasite phylogenies during investigation of cospeciation (Page 1993). Studies of the speciation mechanisms responsible for the geographical distribution and host associations of parasites are of widespread scientific interest, but such investigations require accurate species delimitation. Multilocus molecular phylogenies are proving instrumental to testing hypotheses of parasite species (Littlewood *et al.* 1997, Adams 1998, Nadler *et al.* 2000), and are invaluable for revealing cryptic parasite species in many taxonomic groups. Sequence data from rapidly evolving loci are expected to be of greatest value for delimiting sister species, or testing hypotheses of species in clades with low rates of morphological change. Recent technical advances in acquisition of multilocus molecular data from small organisms have made it possible to test hypotheses of species; future advances will likely make molecular sequences the characters of choice for describing parasite biodiversity.

Hypothesis testing and species delimitation

Despite publication of an enormous volume of literature comparing definitions of species (Mayden 1997, Wheeler and Meier 2000), debates concerning species definitions have had minimal impact on species delimitation. Likewise, rigorous debate about the theory of what species are (species concepts) has not changed how most parasite systematists go about deciding that a parasite is a new species. Traditionally, species delimitation has involved a systematist deciding that a particular taxon is 'different enough' from other described species, based on the available data, to merit species status. Like species descriptions of most organisms, descriptions of new parasite species do not usually make any reference to species concepts or employ formal hypothesis testing as part of the delimitation process (Adams 1998). Instead, traditional approaches reflect the confidence of systematists in the experience, knowledge, and 'artistry' of the taxonomist in making decisions about which taxa are different enough to be recognised as species. Although most species delimited in this manner may be valid, this subjective approach lacks the framework of formal hypotheses testing common to most scientific endeavours. Admittedly, when two putative species are markedly different at the morphological or molecular level, subjective and formal hypothesis-testing methods are highly likely to arrive at the same result, namely, recognition of two species. However, if the goal is to fully characterise parasite biodiversity by finding all species, including species that may be very similar (e.g. sister species and cryptic species), subjective approaches must be replaced by species discovery operations that evaluate data by hypothesis testing. Hypothesis-testing approaches are receiving increased attention and application, and their adoption is driven by the increased use of nucleotide sequences in systematics, because defining what constitutes 'different enough' for nucleotide sequence data appears much more subjective than comparisons based on morphology, even though the underlying genetic basis for differences in morphological features are seldom understood.

Criticisms of 'genetic yardsticks'

Many recent publications employing molecular tools for parasite systematics or diagnostics have used pairwise sequence distances to evaluate if taxa merit recognition as separate species. The distance between two taxa (potential species) is compared to values obtained for comparison of 'benchmark' species, and if the taxa in question meet or exceed the benchmark threshold, they are considered separate species. Advocates of this method have argued that for a given gene, there is a minimum genetic distance that is characteristic of species. If the taxa in question meet or exceed the threshold established by the 'genetic yardstick', then they are argued to represent separate species. Analyses employing the 'genetic yardstick' approach are problematic because assumptions implicit in these comparisons are often violated, and furthermore, this use of sequence data is not fundamentally different from traditional approaches to species delimitation criticised previously. One of the more problematic evolutionary assumptions of the 'genetic yardstick' approach is that sequences are accumulating substitutions at roughly equal rates among different lineages. Heterogeneity in substitution rates among lineages will confound interpretations of genetic distance, and potentially compromise attempts to establish a distance threshold characteristic of species. Another problem involves attempting to calibrate the threshold of genetic distance characteristic of 'the species level'. Ideally, sister species would be used to determine the minimum level of genetic distance characteristic of species. Unfortunately, genetic yardsticks are usually applied without benefit of a phylogeny, and the benchmark comparison involves readily available, but less closely related, congeners. In this case, even if the sequences have evolved without significant rate heterogeneity, the benchmark distance may have little relationship to the true minimum level between species of that clade.

Evolutionary species delimitation

Evolutionary approaches to delimiting species are relatively new, but have been applied to several groups of parasitic organisms (e.g. Littlewood *et al*. 1997, Nadler *et al*. 2000). A detailed comparison of the types of errors involved in application of different species concepts demonstrated that methods of delimitation grounded in evolutionary species concepts are advantageous for reducing bias in assessments of biodiversity (Adams 1998). The main assumptions of evolutionary approaches are that speciation results from cladogenesis and that the resulting species are monophyletic. Such species are represented as independent lineages with non-reticulate (phylogenetic) relationships (Adams 1998, 2002). Testing hypotheses of species can be performed using methods that involve evolutionary analysis of character-state data, including maximum parsimony and likelihood. For cladistic (parsimony) analysis (Fig. 8.2), each species (independent evolutionary lineage) must be delimited by one or more fixed autapomorphies. Evolutionary delimitation methods are not free from potential caveats; for example, some historical patterns of gene transmission and nucleotide substitution may compromise the interpretation of phylogenetic trees as tests of lineage independence and species (Nadler 2002). Several of these potential pitfalls depend on the frequency of lineage-sorting artifacts for genes (Nichols 2001), and argue against delimiting species based on evidence from a single genetic locus. In contrast, when multiple loci support concordant patterns of lineage exclusivity, this is strong evidence that the pattern reflects the underlying common cause of speciation. From a practical standpoint, selection of loci with appropriate substitution rates is important for recovering evidence of lineage exclusivity, because the most rapidly evolving loci provide the most useful data for ensuring that hypotheses of species (including sister species) are accurately tested. Although advances in molecular methods make sequencing multiple loci more feasible for broad-based studies of parasite biodiversity, such methods remain more practical for helminths than most protozoa.

Example studies of marine parasites

Evolutionary studies of speciation in parasites have been few, but published studies have illustrated how phylogenetic approaches and sequence data can help resolve questions regarding mode of speciation (Littlewood *et al*. 1997), or the number of distinct species infecting marine definitive hosts. For example, a multilocus phylogenetic analysis of polystome monogeneans (Littlewood *et al*. 1997) revealed that congeneric *Polystomoides* infecting the same site (e.g. urinary bladder, oral cavity) of different host species were more closely related than congeners infecting different sites of the same host species. Thus, sympatric speciation is not a viable explanation for the distribution of *Polystomoides* among these turtle hosts. Parasites maturing in fish are good candidates for investigating the relative contributions of sympatric versus allopatric speciation, since about half of fish species host two or more congeneric parasite species (Rohde

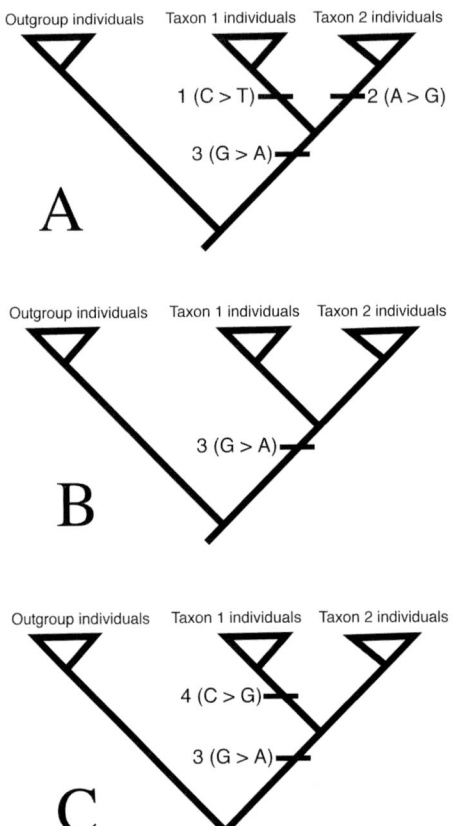

Figure 8.2 Testing hypotheses of species using nucleotide sequence data and cladistic analysis. The greater-than symbol (>) indicates direction of substitution (e.g. character 1 from cytosine to thymine). **A.** Evidence of lineage exclusivity for individuals of taxon 1 and taxon 2. Individuals of taxa 1 and 2 are defined as exclusive groups by autapomorphies (characters 1 and 2, respectively), whereas character 3 provides evidence of phylogenetic relationship among species. Based on cladistic analysis of characters 1 and 2, individuals of taxon 1 and taxon 2 are delimited as independently evolving lineages (separate species). **B.** No evidence of lineage exclusivity. In this hypothetical example, the sequenced regions are identical for individuals of taxon 1 and 2, and therefore these data provide no evidence of exclusivity for individuals of these taxa. Cladistic analysis shows a derived state (character 3) for all ingroup individuals. This character distribution is consistent with the hypothesis that individuals of taxon 1 and taxon 2 belong to a single species. **C.** No evidence of reciprocal monophyly. Individuals of taxon 1 and taxon 2 have different nucleotide states for character 4 (taxon 2 individuals have the ancestral state, cytosine), but only individuals of taxon 1 possess an autapomorphy. Although taxon 2 individuals can be distinguished from individuals of taxon 1 and the outgroup, these sequences provide no evidence that taxon 2 individuals have been evolving independently. A conservative cladistic interpretation is that individuals of taxon 1 and 2 are the same species (although such a character distribution might warrant collection of additional sequence data or examination of other loci). Reproduced with permission from SA Nadler 'Species delimitation and nematode biodiversity: phylogenies rule'; *Nematology* **4**(5), 615–625, published by Brill Academic Publishers, 2002.

1989). Similarly, certain species of marine mammals host multiple groups of parasites that differ markedly in apparent species diversity. Such cases offer the opportunity to investigate how differences in the biology of the parasites (e.g. life cycle) may contribute to differences in speciation rate.

A useful and illustrative example of evolutionary species delimitation in marine parasites is provided by studies of hookworms, *Uncinaria*, parasitising northern fur seals, *Callorhinus ursinus*, and California sea lions, *Zalophus californianus*. Morphological differences between hookworms from these two hosts are mainly morphometric (e.g. the size of the buccal capsule, oesophagus, bursa and L3). Other differences include the distribution of adults in the intestine of pups (Lyons *et al.* 1997), longer lifespan of hookworms from California sea lions (Lyons *et al.* 2000) and variation in intestinal pathology between these two host species (Lyons *et al.* 1997). Although nematodes from these pinnipeds were originally described as unique species (*U. lucasi* from *C. ursinus*, and *U. hamiltoni* from *Z. californianus*) it has been debated whether these hookworms are conspecific (George-Nascimento *et al.* 1992, Lyons *et al.* 1997, 2000, 2001, Nadler *et al.* 2000). Resolving this species-level question is interesting because the differential pathogenicity of hookworms in these hosts could result from one hookworm species infecting both host species, or parasitism by two different species of host-specific hookworms. Given that host-induced changes in morphology have been invoked to explain size differences between *Uncinaria*, this is an ideal case for using sequence data to test hypotheses of species.

To test hypotheses of species of the hookworms described in the above paragraph, sequences were obtained from three nuclear ribosomal DNA regions and one mitochondrial locus for host-associated *Uncinaria* representing different geographical population samples (Nadler *et al.* 2000, Nadler 2002). From an evolutionary species framework, testable predictions can be made about distributions of sequence character-states given different hypotheses of species. For example, if one *Uncinaria* species parasitises both California sea lions and northern fur seals, it is predicted there would either be no fixed differences between host-associated hookworms (Fig. 8.2B; consider taxon 1 to represent California sea lion hookworms, and taxon 2 northern fur seal hookworms), or diagnostic sequence differences without evidence of lineage exclusivity (Fig. 8.2C). Without fixed sequence differences, there is no possibility of phylogenetic characters demarcating separate lineages for host-associated hookworms. In the second case consistent with one species (Fig. 8.2C), hookworms from each host can be distinguished (diagnosed) based on sequence differences, but both host-associated groups are not delimited as exclusive lineages by cladistic evidence in the form of derived character states (autapomorphies). This strict requirement for evidence of reciprocal monophyly (Fig. 8.2A) to delimit species is more stringent than other phylogenetic concepts (Nixon and Wheeler 1990, Wheeler 1999). Furthermore, alternative methods of inferring character-state distributions (e.g. likelihood) that employ explicit models of character-state change have the potential to yield evidence of lineage exclusivity when parsimony does not. Although these technical considerations merit careful consideration when analysing data, they do not detract from the advantages afforded by hypothesis-testing approaches based on evolutionary concepts.

In this study, cladistic analysis of nuclear and mitochondrial sequence data revealed that hookworms from northern fur seals and California sea lions represent exclusive evolutionary species (Fig. 8.3). Nuclear rDNA sequences (large subunit, LSU, and internal transcribed spacer, ITS regions) yielded evidence of lineage exclusivity for both host-associated groups of hookworms. Comparison of these nuclear rDNA sequences (>1.5 Kb of sequence per taxon) showed that internally transcribed spacer regions provided the largest number of parsimony informative characters. This is consistent with expected differences in molecular functional constraints, and reinforces that choosing sequences with relatively high substitution rates is important to capture evidence of evolutionary history for closely related species. The other locus, mitochondrial 12S rDNA, yielded four autapomorphies from 510 nucleotides (Fig. 8.3), with two autapomorphies for the California sea lion hookworms, and two for the northern fur seal hookworms. This result suggests that variable regions of mitochondrial DNA (mtDNA) may also be informative for delimitation of closely related parasite species.

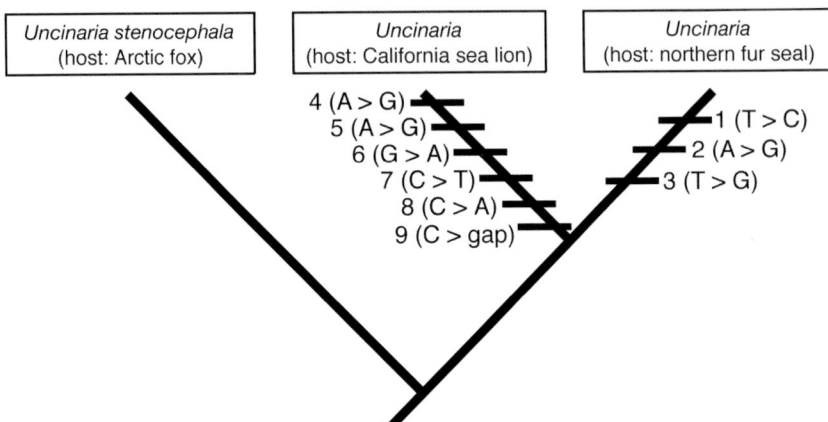

Figure 8.3 Evidence of nucleotide evolution and lineage exclusivity for *Uncinaria* hookworms from California sea lions and northern fur seals. The greater-than symbol (>) indicates direction of character-state change (e.g. character 1 from thymine to cytosine). Based on outgroup (*U. stenocephala*) comparison and cladistic analysis, 9 autapomorphies are fixed among individual hookworms from these pinnipeds. Character 1 (LSU nuclear rDNA) and characters 2–3 (12S mitochondrial rDNA) are fixed and autapomorphic for individual hookworms from northern fur seals. Characters 4–5 (12S mitochondrial rDNA) and 6–9 (ITS-1 nuclear rDNA) are fixed and autapomorphic for individual hookworms from California sea lions. These derived nucleotide states, representing two independent loci, provide evidence of historical lineage independence and species status for *Uncinaria* from northern fur seals and California sea lions. Reproduced with permission from SA Nadler 'Species delimitation and nematode biodiversity: phylogenies rule'; *Nematology* 4(5), 615–625, published by Brill Academic Publishers, 2002.

These analyses of separate genetic loci provided concordant evidence of lineage exclusivity (separate species) for host-associated hookworms from California sea lions and northern fur seals. The number of autapomorphies characterising each lineage was comparatively small, suggesting these species have recently shared a common ancestor. However, it should be emphasised that it is the pattern of lineage exclusivity, and not the amount of character-state changes (or alternatively, genetic distance) that provides strong evidence that these host-associated hookworms have been evolving independently, and represent separate species. Although these molecular data support the hypothesis that these two *Uncinaria* species are each associated with a specific pinniped host species, host fidelity is not a requirement to delimit species using genetic data. Characterisation of individual parasites, based on multiple loci, can reveal evidence of exclusive evolutionary lineages regardless of their host distributions.

Given the theoretical and practical considerations discussed previously and presented in detail elsewhere (Adams 1998, Nadler 2002), robustly testing hypotheses of species for the most closely related parasites may require datasets of moderate-to-large size representing several genetic loci. Characterising parasite biodiversity with sequences will also require population-level sampling, and this requires considerable effort and cost even for relatively small numbers of individuals. This situation is likely to change rapidly with technical advances, leading to the accumulation of a wealth of genetic data on parasites, which will facilitate comprehensive investigations of parasite biodiversity. Taking full advantage of these data will require careful analysis, including application of evolutionary principles and delimiting species through hypothesis testing.

Important references

Important papers on species concepts and species delimitation are by Mayden (1997), Adams (1998, 2002), Wheeler (1999), Wheeler and Meier (2000), Nichols (2001) and Nadler (2002). Littlewood *et al.* (1997) used molecular and morphological evidence to evaluate whether speciation in turtle polystomes has occurred within or between host species, and Nadler *et al.* (2000) used molecular and morphometric evidence to distinguish separate species of *Uncinaria* in California sea lions and northern fur seals.

Chapter 9

Zoogeography

The geographical distribution of marine parasites has not been well studied and is largely, although not entirely, limited to some aspects of the zoogeography of parasites of marine fishes. Some effort has gone into studying latitudinal, longitudinal and depth gradients of fish parasites. Interestingly, ectoparasites and endoparasites of marine fishes show different latitudinal patterns. Whereas the relative species richness (number of parasite species per host species) increases markedly from high to low latitudes for ectoparasites, this is not the case for endoparasites. The latter also show greatest diversity in the tropics, but this is entirely due to an increased diversity of host species at low latitudes. Thorson's rule, which states that benthic marine invertebrates tend to produce large numbers of small pelagic larvae in warm waters, but small numbers of large offspring by various mechanisms at high latitudes, applies to monogenean gill parasites of marine fishes. Furthermore, host ranges (but not host specificity) are greater for digeneans at high than at low latitudes, whereas monogeneans do not show differences between latitudes.

Concerning longitudinal gradients, a study of scombrid ectoparasites has shown that there is a primary centre of diversity in South-East Asian waters, and a secondary one in the Caribbean, with diversity decreasing with distance from these centres.

Concerning gradients with depth, relative species diversity of monogeneans is several times greater in surface than in deep waters off eastern Australia. Parasites can be used to study host populations and their migration, not only of marine fish but also of various invertebrates. Such parasite tags are much cheaper than other methods, such as comparative genetic studies, as discussed (on pages 351–355). The section discusses general methodology, selection criteria for tag parasites and application of the method to different host groups. Parasites also are useful for making inferences about long-term historical dispersal. The relevant section describes the only two examples studied (i.e. that of scombrid dispersal with emphasis on the role of oceanic barriers, and that on the historical migrations of Indo-Pacific whiting, Sillaginidae). The section on introduced marine parasites addresses a very important problem (i.e. that of the many parasites introduced into new regions, where they have become important pests). It also discusses the use of deliberately introduced parasites to control introduced free-living pest species, such as crabs or echinoderms. The Chapter concludes with a concise discussion of deep-sea parasites. Very little is known about the diversity and distribution of free-living deep-sea organisms (probably more than 99% of invertebrates have not yet been described), and deep-sea parasites have been studied even less. They are of great importance, considering the huge spaces of the deep-seas, and the many species of fish and invertebrates found there.

Latitudinal, longitudinal and depth gradients
Klaus Rohde

Introduction
Most plant and animal species are most diverse at low latitudes. Differences in species richness between latitudes are often truly amazing. Thus, there are about 150 species of marine fish in the North Sea, but there are thousands in tropical seas of much smaller areas. But even at a particular latitude, differences exist along longitude. For example, the south-east Asian waters are particularly rich in faunal diversity, whereas the eastern Pacific is much poorer. Little is known about species numbers in the deep sea, but the few studies that have been conducted suggest that diversity at least of some groups (e.g. nematodes) is very great in the deep-sea benthos. This section evaluates evidence for gradients not only in marine parasite diversity, but also for gradients in reproductive strategies, host specificity and latitudinal ranges.

Latitudinal gradients in species richness
Latitudinal gradients in species richness of marine parasites were last reviewed by Rohde (2002). They have not been studied for most marine groups, but are well documented for metazoan ectoparasites and endoparasites of marine fishes. In both groups of parasites, species richness is much greater in the tropics than at high latitudes. However, ectoparasites and endoparasites differ significantly in one important aspect, that is, species richness of ectoparasites increases at a greater rate towards the equator than that of host species, whereas relative species richness (number of parasite species per host species) of endoparasites is more or less the same at all latitudes. The great number of endoparasite species in tropical seas is entirely (or almost entirely) due to greater species numbers of hosts. The differences between ectoparasites and endoparasites remain after correction for phylogeny, using the method of phylogenetically independent contrasts (Poulin and Rohde 1997). The same difference is shown by two major groups of ectoparasites and endoparasites, the monogeneans and digeneans (Fig. 9.1). Consequently, infracommunity richness of ectoparasites but not of endoparasites is greater in the tropics (Rohde and Heap 1998).

Rohde (1992, 1999) and Willig (2001) have reviewed the various hypotheses proposed as explanations for the gradients. Most of them are either circular or insufficiently supported by evidence. Importantly, most of the explanations are based on the assumption that habitats are saturated with species and that species richness is determined by limiting factors (i.e. that equilibrium conditions prevail). Rohde (1992, 1999) proposed an alternative hypothesis, that the gradients can best be explained by a gradient in 'effective evolutionary time' modulated by several factors. He defined effective evolutionary time as the composite of two factors (i.e. evolutionary speed and evolutionary time under which communities have existed under relatively constant conditions). Evolutionary speed, according to Rohde (1992, 1999, references therein) is determined by mutation rates, generation times and the speed of physiological processes. These parameters are correlated with temperature: mutation rates and the speed of physiological processes are increased, and generation times are reduced in the tropics. Because niche space is largely empty, the net effect will be a faster accumulation of species at low latitudes. The possibility cannot be ruled out that different limits to species richness in different habitats do exist, but they have not been reached at this point in evolutionary time.

Several recent studies deal with the effects of temperature on generation times and the effect of generation time on speed of evolution (for a full discussion and references see Rohde 2005). Gillooly et al. (2002) described a general model, based on first principles of allometry and biochemical kinetics, that predicts generation time ('time of ontogenetic development') as a function of body mass and temperature. Development time of species in all groups tested is

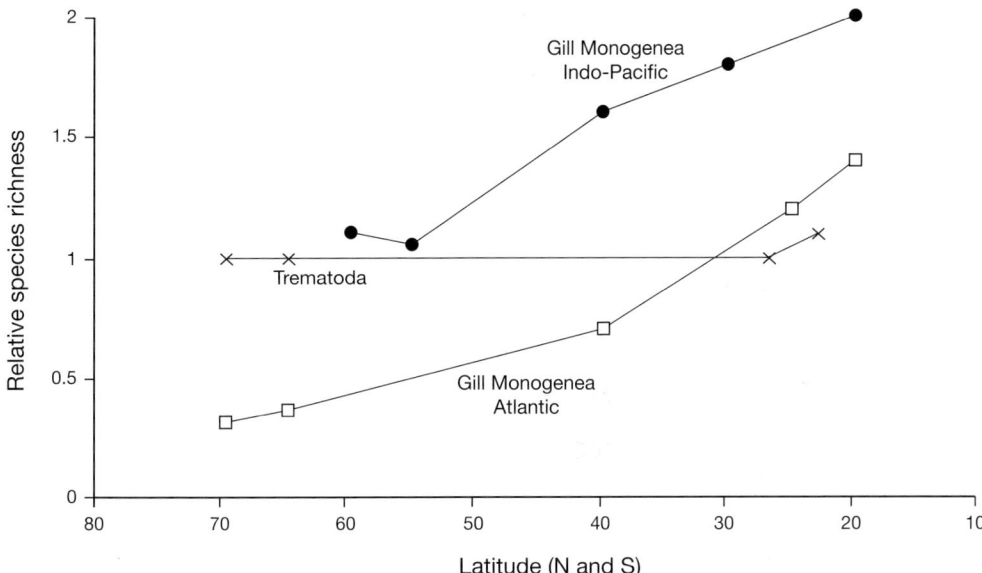

Figure 9.1 Relative species diversity (number of parasite species per host species) of digeneans and monogeneans of marine teleosts at different latitudes. Data from various sources (e.g. Rohde and Heap 1998).

negatively correlated with temperature. Mutation rates are greater at higher temperatures (references in Rohde 1992), and an acceleration of physiological processes with temperature is well documented. However, experimental evidence for direct temperature effects on speed of selection is not available. Such evidence is urgently needed.

Fossil evidence supports the view that speciation rates are higher in the tropics, but molecular evidence that relative rates of diversification per unit time increase towards the tropics is controversial (references and discussion in Rohde 2005). Of very great importance, the study of Allen et al. (2002) has shown that species diversity, including those of marine fish parasites, can be predicted from the biochemical kinetics of metabolism. Their model predicts quantitatively how species richness increases with environmental temperature. Allen et al. (2002) conclude that evolutionary rates are constrained by generation times of individuals and mutation rates, both of which are correlated with metabolic rates and show the same Boltzmann relation to temperature. The authors concluded that their results support the hypothesis that increased temperatures accelerate the biochemical reactions that control speciation rates and thereby increase the standing stock of species.

Latitudinal gradients in reproductive strategies

Thorson (1957) demonstrated that many marine benthic invertebrates in warm waters produce large numbers of small pelagic planktotrophic larvae, whereas high latitude species tend to produce fewer and larger offspring, often by viviparity or ovoviviparity, and larvae develop in egg capsules or by brooding (Thorson's rule). Many subsequent studies have confirmed this generalisation. Rohde (1985) provided evidence for a similar phenomenon in the gill Monogenea. Most species in warm or temperate seas produce numerous eggs from which free-swimming larvae hatch. In contrast, most species infecting coldwater fish (particularly in the northern hemisphere) belong to the Gyrodactylidae, all of which reproduce by viviparity (Fig. 9.2). Juveniles are transferred to other fish by contact transfer.

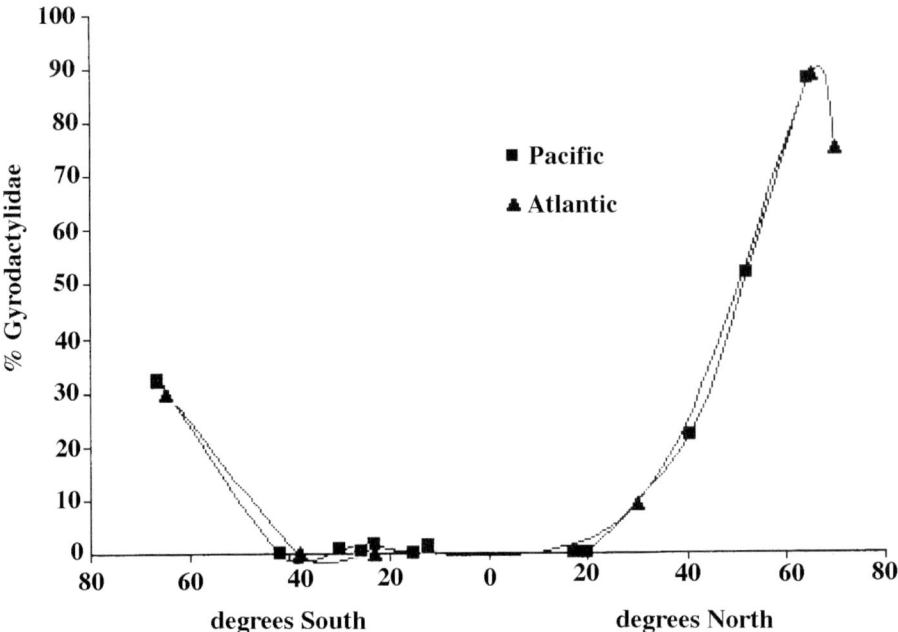

Figure 9.2 Percentage of marine Gyrodactylidae on the gills of marine teleosts at different latitudes. Data from various sources including Ernst *et al.* (2001).

Several explanations for Thorson's rule have been given; the most widely accepted one is that phytoplankton blooms at high latitudes do not last long enough to permit development of small planktotrophic larvae. This explanation does not hold for the Monogenea, because monogenean larvae do not feed on phytoplankton. Rohde (1985, 1999) suggested that the most likely explanation is direct temperature effects. The number of pelagic larvae that can be produced at low temperatures is too small to guarantee infection by host searching, and small larvae cannot locate hosts in the huge spaces of the ocean at temperatures where physiological processes are greatly slowed down. Hence, selection has led to a strategy that guarantees a suitable habitat by remaining on the same host individual. Since the trend in the Monogenea and free-living benthic invertebrates is so similar, Rohde (1985) suggested a similar explanation for the latter.

Latitudinal gradients in host ranges and host specificity

The host range of a parasite species is the number of its recorded host species, whereas host specificity takes intensity and/or prevalence of infection in each host species, and the phylogenetic relationship of hosts into consideration (for details see pp. 286–293). There is little information about latitudinal differences in host ranges and specificity for most parasite groups, but studies of monogeneans and digeneans of marine teleosts along latitudinal gradients in the Indo-Pacific and Atlantic oceans have shown that host ranges are very narrow for monogeneans at all latitudes, whereas host ranges for digeneans are much greater at high than at low latitudes (Rohde, 1978, 1980).However, host specificity is similarly great for both groups at all latitudes, if specificity is measured using prevalences and intensities of infection (Rohde 1980). Apparently, digeneans in cold waters, in spite of the greater number of hosts, infect only some host species heavily. However, only one large survey at high latitudes permits calculation both of host ranges and host specificity for Digenea. More studies are necessary, and latitudinal gradients in host ranges and specificity in other parasite groups should be assessed.

Longitudinal gradients and geographical ranges

Longitudinal differences in species diversity of marine parasites have been little studied. Rohde and Hayward (2000) studied monogenean and copepod ectoparasites of the scombrid genera *Scomberomorus* and *Grammatorcynus*, and found that the tropical/subtropical western Pacific is the primary centre of diversity, and that the tropical/subtropical western Atlantic is a secondary centre. Coastal scombrid fishes and their parasites spread between the Indo-West Pacific and Atlantic oceans via the Tethys Sea, and between the western Atlantic and the eastern Pacific, before the Central American landbridge was established. As the result of this dispersal, fish species in low diversity seas have relatively more parasite species than those found in high diversity seas. When considering only endemic fish and parasite species, however, the Indo-West Pacific and the West Atlantic (i.e. the primary and secondary centres of diversity) have parasite-host ratios of 1.75 and 1.25, respectively, versus a ratio of 1.00 for all other seas.

Studies of parasites of rabbit fish, *Siganus* spp., from the coasts of eastern Australia and East Africa have shown a much greater species richness of ectoparasites, but not of endoparasites, in Australian waters. Kleeman (2001) found 19 species of metazoan ectoparasites on *Siganus doliatus* on the Great Barrier Reef, whereas Martens and Moens (1995) and Geets *et al.* (1997) found only seven or eight species on *S. sutor* from the western Indian Ocean. Endoparasite richness did not show significant differences: seven versus five species.

The effects of geographical ranges of hosts on species richness of parasites have been little studied. There is some, but ambiguous, information that fish species with the widest geographical distribution have the greatest number of parasites species (Rohde 1989).

Depth gradient

Deep-sea parasites are discussed in detail in the section on Deep sea parasites (pp. 366–369). Relative species diversity of Monogenea is about five times greater in surface than in deepwater fish off the coast of south-eastern Australia (see Rohde 1993).

Important references

Reviews on latitudinal diversity gradients and their explanations are by Rohde (1992, 1999) and Willig (2001). Rohde (1993, 2002), Poulin and Rohde (1997) and Rohde and Heap (1998) discussed latitudinal gradients in species richness of marine parasites, in particular of gill parasites of teleost fish, Rohde and Hayward (2000) evaluated longitudinal gradients of gill monogeneans of scombrid fishes. Discussions of latitudinal gradients in host specificity and reproductive strategies can be found in Rohde (1978) and Rohde (1985), respectively. Recent evidence on a non-equilibrium explanation of latitudinal diversity gradients was presented by Allen *et al.* (2002), Gillooly *et al.* (2002) and Rohde (2005). Our knowledge of depth gradients was reviewed by Rohde (2002).

Parasites as biological tags
Ken MacKenzie

Introduction

An important applied aspect of marine parasitology is the use of parasites as biological tags to investigate the population structure of marine organisms, particularly commercially important species of fish. Different components, distinguished by varying degrees of separation, can be recognised within populations of most marine organisms. For example, two sympatric components may have a common feeding area but separate spawning areas, while others may be separated to a greater extent. The efficient and sustainable management of a commercially exploited marine species depends on a sound knowledge of its population structure. Various methods can be used to investigate this structure, including the use of naturally occurring parasites as biological tags.

The basic principle underlying this method is that the host can become infected with a particular parasite species only within the endemic area of that parasite. The endemic area is that geographical region in which conditions are suitable for transmission of the parasite and the completion of its life cycle. It is determined by the existence of suitable environmental conditions, primarily temperature and salinity for marine organisms, and the presence of all required host species for different stages of the parasite's life cycle. If an infected host is found outside the endemic area of the parasite, it can be inferred that that individual had been within the endemic area at some time.

Methods

A biological tag study should preferably begin with a survey of the entire parasite fauna of the target host in the area under study. Two different approaches can then be applied to the resultant data. First, a small number of parasite species are selected according to established criteria (see below). A large number of host individuals are then examined specifically for these species and the data are compared using fairly simple tests of statistical significance such as Fisher's Exact Test and ANOVA. Second, entire parasite assemblages are analysed using more advanced statistical methods such as a multivariate maximum likelihood model, non-parametric discriminant analysis and canonical multivariate analysis. This approach is of particular value in studies of large valuable host species that are not available for examination in large numbers. It is less applicable to studies of smaller species of fish, particularly pelagic species, because many of these do not have the diverse parasite faunas and high intensities of infection necessary for such analyses. A useful recent development (Lester et al. 2001, Moore et al. 2003) has been to divide the parasite fauna of the target host into temporary and permanent groups, depending on their life spans in that host, and to analyse the two groups separately. With large fish it may be prohibitively time consuming to count all the parasites present in individual hosts, but this problem can be minimised by subsampling the more numerous parasite taxa.

Selection criteria for tag parasites

The ideal tag parasite should satisfy the following criteria. However, parasites fulfilling all of these are rarely encountered, so compromises usually have to be made. Also, new variations in the way parasite data are analysed are continually appearing in the literature, so these criteria should be regarded as guidelines rather than hard-and-fast rules.

- The parasite should have significantly different levels of infection in different parts of the study area. Exceptions are studies using geographical variations in parasite genetics, which up to now have been applied mostly to larval anisakid nematodes (Pascual *et al.* 1996, Mattiucci *et al.* 2002).
- The parasite should have a long life span in the target host. For studies of stock identification and recruitment, parasites with life spans similar to those of the host itself are usually necessary, but for studies of seasonal migrations parasites with shorter life spans, such as adult digeneans, can be used. Encysted larval or juvenile stages of helminths, such as digenean metacercariae (Fig. 9.3) and cestode plerocercoids (Fig. 9.4) often have life spans measured in years in their fish hosts.
- Parasites with direct single-host life cycles are the simplest to use. Those with complex life cycles involving two or more stages in different hosts are more problematic because more information is required on the various biotic and abiotic factors that influence

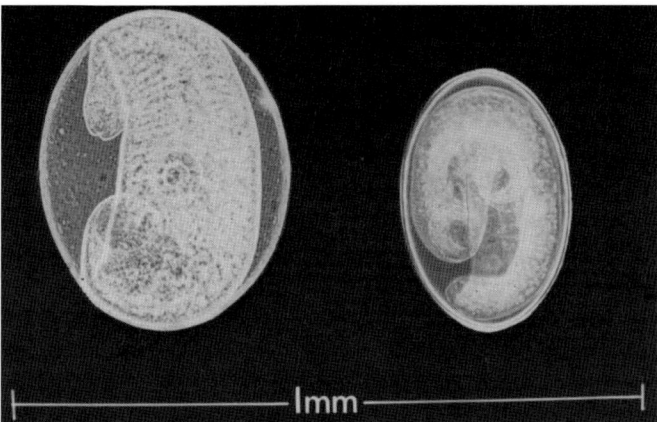

Figure 9.3 Metacercariae of *Renicola* spp. from the visceral cavity of herring, *Clupea harengus*. These infect herring in the first year of life and have life spans of many years in the fish host.

parasite transmission. Given adequate information they can, however, be used just as effectively, and continuing research on the biology and ecology of marine parasites is improving the efficiency of biological tagging.
- The level of infection should preferably show no significant variation from year to year, but the effects of annual variations can be nullified by following infection levels in single year classes of the target host over several years.
- The parasite should be easily detected and identified. Examination of the host should involve the minimum of dissection, so highly visible and site-specific parasites are particularly useful (Fig. 9.5).
- Seriously pathogenic parasites, and those that affect host behaviour, should not be considered for use as tags.

Advantages of biological tagging

The most efficient approach to a population study is the multidisciplinary one, in which several different tagging methods are used to complement one another. Each method has its own strengths and weaknesses and the use of parasites as tags is recognised as having the following advantages over other widely used methods such as artificial tagging and genetic studies.

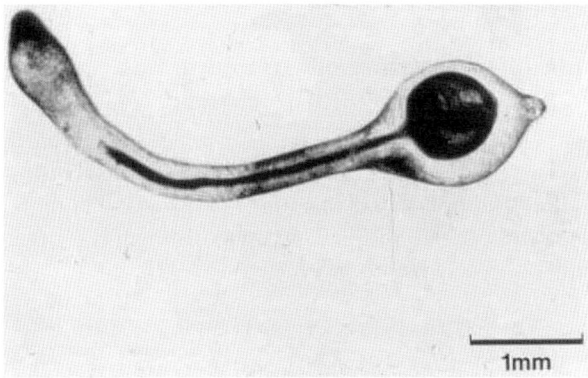

Figure 9.4 Plerocercoid of the trypanorhynch cestode *Lacistorhynchus tenuis* from the visceral cavity of teleost fish. These plerocercoids have life spans of many years in the fish host.

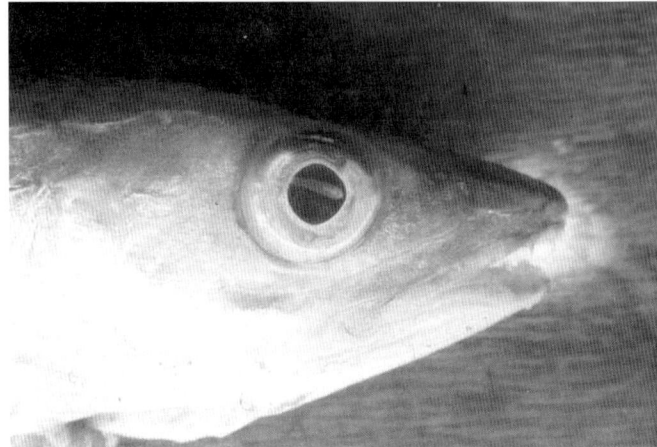

Figure 9.5 A good example of a highly visible and site-specific parasite: the plerocercoid of the trypanorhynch cestode *Gilquinia squali in situ* in the humour of the eye of whiting, *Merlangius merlangus*.

- Parasite tags are more appropriate for small, delicate and deepwater species of fish that suffer high mortality rates following artificial tagging. They are also more appropriate for crustaceans that lose artificial tags when they shed their old carapace.
- Each host specimen sampled represents a valid observation, whereas with artificial tags each individual must be sampled, tagged and recaptured to obtain a single valid observation.
- Parasites are less expensive to use because samples can be obtained from routine sampling programs or from commercial catches, rather than from dedicated tagging cruises.
- The use of parasites eliminates doubts concerning possible abnormal behaviour of artificially tagged hosts.
- Biological tagging and genetic studies operate over different time scales. Genetic studies operate over an evolutionary time scale and thus are less likely to identify population components that are at an early stage in the process of separating into subspecific groups variously known, for example as 'strains' or 'races'. Parasites, however, can often be used to identify host population components distinguished by behavioural differences, but between which there is still a considerable amount of gene flow ('ecological stocks').

Parasites as tags for different taxonomic groups of host

Crustaceans
Crustaceans are important intermediate hosts for many species of marine helminths, which can be used effectively as biological tags. Good examples are the stock identification studies of Owens (1983, 1985) and Thompson and Margolis (1987). These authors used metacestodes, digenean metacercariae and a bopyrid isopod as tags to distinguish between stocks of commercially important shrimps and prawns.

Cephalopods
The use of parasites in population studies of cephalopod hosts was reviewed by Pascual and Hochberg (1996), who discussed the problems peculiar to this host group. These included short life spans, variable growth rates and protracted spawning seasons, the latter giving rise to multi-

ple cohorts in a single population at any given time. These factors, together with the importance of squid as intermediate and paratenic hosts for marine helminths, led these authors to recommend the use of genetic studies of selected helminth larvae, particularly anisakid nematodes.

Pelagic fish
Most studies on pelagic fish have been carried out on herring, *Clupea harengus* and *C. pallasi*, and on carangid fish of the genus *Trachurus* (see MacKenzie 1987b, Williams *et al.* 1992, MacKenzie 2002). The most frequently used parasites have been larval nematodes of the genus *Anisakis*, but trypanorhynch metacestodes, digenean metacercariae and parasitic isopods have also featured prominently.

Demersal fish
The commercial importance of Atlantic cod, *Gadus morhua*, hake, *Merluccius* spp., and rockfish, *Sebastes* spp., has led to many population studies of these species using parasites as tags. Demersal fish tend to have richer and more varied parasite faunas than pelagic species and this is reflected in the wide variety of parasites that have been used as tags. Although considerable potential exists for the use of parasites as biological tags for elasmobranch fish, the only studies that have attempted to realise this potential so far have been those of Moore (2001) and Yamaguchi *et al.* (2003).

Marine mammals
The use of parasites as biological tags for marine mammals was reviewed by Balbuena *et al.* (1995), who gave examples to illustrate both the value and limitations of the method. In the few studies carried out so far, acanthocephalans have proved particularly useful for identifying stocks of cetaceans.

Important references
The first publication describing the use of a parasite as a biological tag for a marine host was that of Herrington *et al.* (1939). General accounts including reviews are by Sindermann (1961, 1983), Kabata (1963), MacKenzie (1983, 1987a, 2002), Lester (1990), Moser (1991), Williams *et al.* (1992), Balbuena *et al.* (1995), Pascual and Hochberg (1996) and Arthur (1997). MacKenzie and Abaunza (1998, 2004) provided a guide to procedures and methods for stock discrimination of marine fish by parasite tags.

Parasites as indicators of historical dispersal
Craig Hayward

Introduction
Parasites have been employed often in the study of migrations of hosts, and to distinguish different host populations (Rohde 2002). The use of parasites as biological tags in the discrimination of host stocks, for example, is considered above (pp. 351–355). Such studies are concerned with relatively small scales of distance and time. In contrast, studies using parasites as indicators of larger-scale patterns in the zoogeography of host animals, and historical dispersal in evolutionary time, are less common. This approach, dubbed the von Ihering method, was first used in the late 1800s (von Ihering 1891). In biogeography, this technique can test hypotheses about the role of particular barriers in limiting the continental distribution of living species, and also higher groups. However, studies using this approach in the marine environment are much fewer than

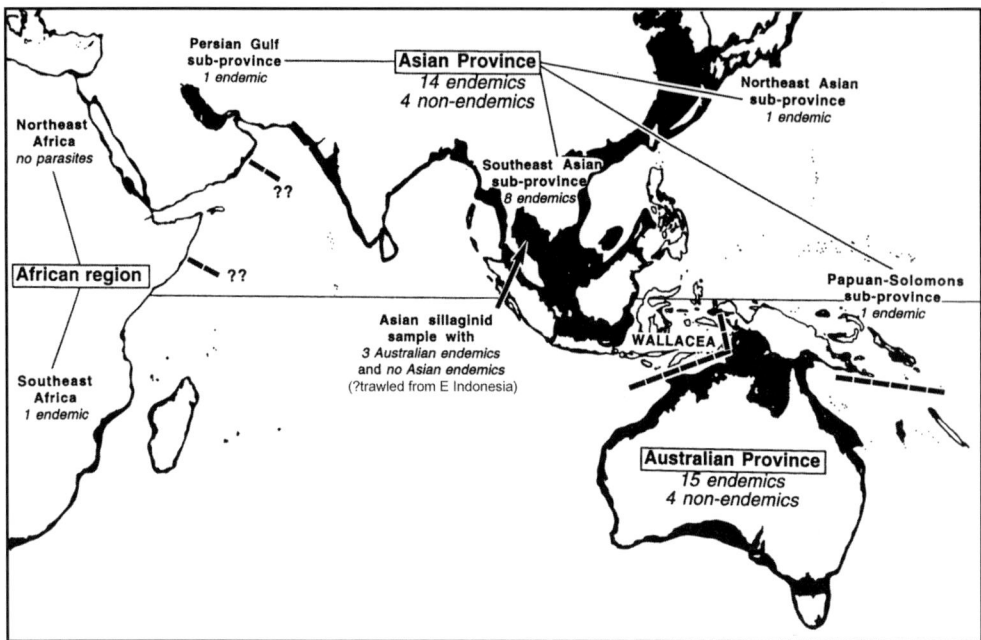

Figure 9.6 Provinces of ectoparasites of Indo-Pacific whiting (Sillaginidae), and proposed position of barriers to dispersal of these fishes. Reproduced from Hayward CJ, Distribution of external parasites indicates boundaries to the dispersal of sillaginid fishes in the Indo-West Pacific. *Marine and Freshwater Research* **48**, 391–400, with permission from CSIRO Publishing (1997).

in terrestrial and freshwater realms. This is largely because marine parasites are still relatively less well known than those in other environments.

Two recent studies have analysed the ranges of parasites of perciform fishes to investigate historical dispersal across oceanic barriers. These studies focused on external parasites having high host specificity, as the planktonic larval stages of such parasites are usually very brief, and so their long-range dispersal accurately reflects historical movements of their host fishes. The first study (Hayward 1997) dealt with the distribution of copepods and monogeneans on Indo-Pacific whiting (Sillaginidae), and the second study, by Rohde and Hayward (2000), considered the distribution of copepods and monogeneans on four genera of Scombridae.

Dispersal history of Indo-Pacific whiting

Indo-Pacific whiting are benthic schooling fishes with small, elongate bodies belonging to Sillaginidae, a family related to croakers (Sciaenidae). Most have limited regional distributions in the tropics. They live in shallow, inshore marine waters over sandy and silty substrates from Australia northwards to Asia, and westwards to eastern Africa. Hayward (1997) examined over 1500 individuals belonging to 26 species (out of 27) in three genera from throughout this region for parasites.

The distribution patterns of ectoparasites specific to these fishes lead to recognition of two provinces (regions of endemism) with high parasite diversity: one on the continental shelf of Australia, and one on the shelf of Asia (Hayward 1997; Fig. 9.6). The Australian province has 15 endemics (five monogeneans, one leech and nine copepods), and the Asian province 14 endemics (two monogeneans and 12 copepods). These provinces are separated by a region with coastlines that descend very steeply to the ocean floor (with few substrates suitable for these fishes), and by deep ocean waters that must largely inhibit sillaginid movements. This region, known as Wallacea, is already well known as a boundary region for terrestrial animals, plants, freshwater fishes and assorted inshore marine fishes.

Some sillaginids must have dispersed across this barrier in eastern Indonesia in recent geological time, leading to the occurrence of four parasites (three monogeneans and a copepod) in both provinces. At least one other less-recent invasion of Australian waters by Asian sillaginids would account for the occurrence of six pairs of copepod congeners that have one member in each province. Presumably, such dispersal occurred some time since the mid Miocene (about 15 million years ago), when the modern configuration of islands in eastern Indonesia began to form. Dispersal of sillaginids in the opposite direction seems unlikely, as a genus of polyopisthocotylean monogeneans common among Australian sillaginids, *Polylabris*, would also be expected to be present in Asia, but this is not the case.

Three species of Australian parasites also appear to be presently encroaching onto the periphery of the Asian shelf, because they were found in just a single sample of sillaginids, *Sillago aeolus*, collected from a fish market along the Gulf of Thailand (Fig. 9.6). (However, although it was assumed that the fish in this sample had been fished locally, they may have been trawled remotely in eastern Indonesia, on or near the Australian continental shelf, by long-distance trawling vessels based at Bang Saen, Thailand).

The most widespread sillaginid, *Sillago sihama*, also seems to have dispersed to African shores from the Arabian Sea as planktonic larvae only (as no Asian parasites were present in samples of 29 hosts), and relatively recently (as only one locally endemic parasite appears to have been acquired, Fig. 9.6).

Dispersal history of Spanish mackerels and mackerels

Scombrid fishes (mackerels, tunas and bonitos) are relatively large, epipelagic fishes of the tropics and subtropics. Rohde and Hayward (2000) compiled previously published and new data on the distribution of ectoparasites of four scombrid genera (*Scomberomorus*, *Grammatorcynus*, *Scomber* and *Rastrelliger*) belonging to 26 species. The distributions of 32 copepods and 25 monogeneans on these fishes were then analysed to investigate the hypothesis that the wide stretch of deep ocean between the East and West Pacific (the East Pacific Barrier) is responsible for the pronounced break in the circumtropical warm water fauna of the continental shelves. An alternative boundary to historical dispersal that is often considered important is the New World Land Barrier, which emerged to join the continents of North and South America several million years ago.

At the level of parasites species, analysis showed that there is a primary centre of diversity in the West Pacific, and a secondary centre in the West Atlantic (Rohde and Hayward 2000, Fig. 9.7). The West Pacific centre shares its species of the largely coastal *Scomberomorus* and *Grammatorcynus* and their parasites almost entirely with seas located to the west. Predominantly coastal fish species and their parasites have not dispersed across the eastern Pacific, but rather dispersed across the New World Land Barrier, before the Central American land bridge was formed. The only species of parasites shared by the western and eastern Pacific (four copepods) also have a circumtropical distribution, and may have dispersed to the Eastern Pacific through the Western Atlantic. In contrast with the West and East Pacific areas, all four other neighbouring areas (West Pacific–Indian, Indian–East Atlantic, East Atlantic–West Atlantic, West Atlantic–East Pacific) share species, indicating that the Eastern Pacific Barrier has been highly effective in limiting dispersal.

In contrast with the parasites of the relatively near-shore *Scomberomorus* and *Grammatorcynus*, the parasites of the more pelagic *Scomber* spp. in the West Pacific share species with seas both to the east and west, although at the genus level, only two circumtropical monogenean

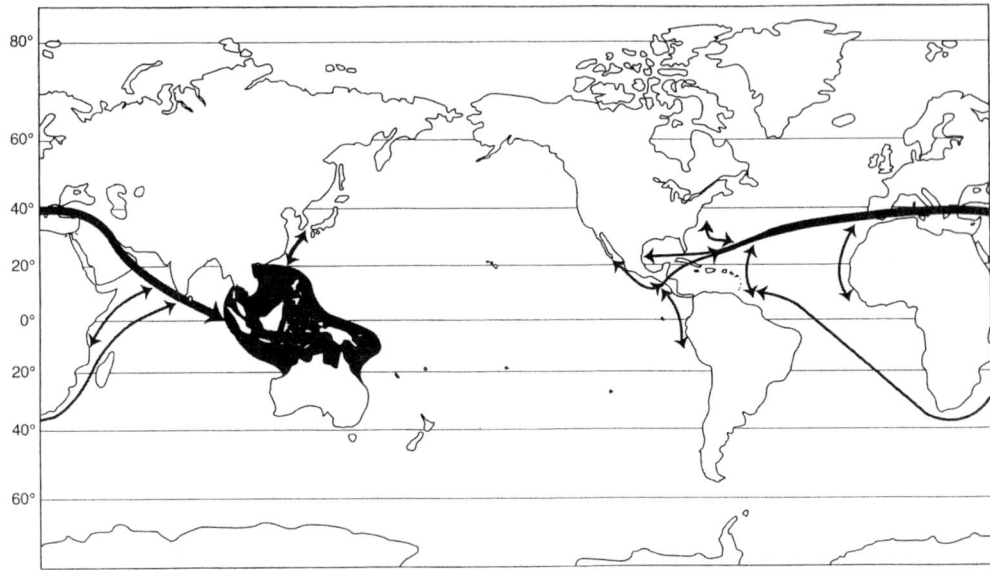

Figure 9.7 Geographical relationships of scombrids, *Scomberomorus* and *Grammatorcynus*, and their copepod and monogenean ectoparasites. Note the primary centre of species richness in the western Pacific; the East Pacific Barrier has been an effective barrier to dispersal. Reprinted from *International Journal for Parasitology* **30**, Rohde K, Hayward CJ, Oceanic barriers as indicated by scombrid fishes and their parasites, pages 579–583, Copyright (2000), with permission from Elsevier.

genera are shared by the East and West Pacific (Rohde and Hayward 2000). Extant distribution patterns are probably also affected by the fragmentation of habitats at the time the Tethys Sea closed, during the Miocene (5–25 million years ago), as a result of changes in plate tectonics. (The Tethys was a large ancient sea that existed between Europe–Northern Asia and the African and Indian continents, and once connected the Atlantic and Pacific oceans.) In conclusion, the East Pacific Barrier has been thoroughly effective as a barrier to dispersal of species of *Scomberomorus*, *Grammatorcynus* and their parasites, whereas it has been less effective for species of *Scomber* and their parasites.

Important references
Von Ihering (1891) pioneered this approach, postulating ancient relations between New Zealand and South America. The only recent work done in this area using marine parasites is by Hayward (1997) on ectoparasites of sillaginid fishes, and Rohde and Hayward (2000) on oceanic barriers as indicated by scombrid fishes and their parasites. Rohde (2002) reviewed the work.

Introduced marine parasites
Mark E Torchin and Armand M Kuris

Introduction
The deleterious effects of introduced species on biodiversity, ecosystem function and human economies are becoming increasingly apparent (Elton 1958, Vitousek 1990, Wilcove *et al.* 1998, Ruiz *et al.* 1999, Chapin *et al.* 2000). While marine and estuarine environments are among the most heavily invaded systems on earth (Ruiz *et al.* 2000, Grosholz 2002), surprisingly little is known about introduced marine parasites. Undoubtedly, more introduced marine parasites

exist than are reported. Parasites are common and important components of natural communities including marine environments (this volume). Yet, unlike the recent increase in reports of introduced free-living marine species (Cohen and Carlton 1998, Ruiz *et al.* 2000, Grosholz 2002), reports of parasites invading marine environments are relatively uncommon (Torchin *et al.* 2002). Partly, this may be because, unlike their introduced hosts, parasites are often difficult to observe and thus some exotic parasites go unreported. However, introduced 'hosts' also escape their native parasites, often resulting in significantly reduced parasite species richness and prevalence in the introduced populations compared to their native populations. Parasites from a host's native range may never reach the novel region due to stochastic processes or, if they do, stochastic and/or demographic processes may prevent their establishment (Torchin *et al.* 2002), such that, only a subset of the host's native suite of parasites actually establish introduced populations (Torchin *et al.* 2003). Escape from parasites appears to be a general phenomenon across taxa and habitats (Mitchell and Power 2003, Torchin and Mitchell 2004). Evidence from molluscs, crustaceans and fishes (Kennedy 1993, Torchin *et al.* 2003) indicate that introduced aquatic species escape most of their native parasites. Nevertheless, since parasitism is probably the most common consumer strategy, and since free-living species generally host at least one parasite species, and more commonly several (Price 1980, Toft 1986), the paucity of reported marine parasite invasions is intriguing.

Modern shipping and transport have made the global movement of ballast water a primary vector for introducing marine organisms to novel regions (Carlton and Geller 1993, Ruiz *et al.* 1997, Drake and Lodge 2004). The transport of planktonic larval life history stages in ballast water is a key invasion route (Carlton and Geller 1993). Although larval organisms can be infected by parasites and pathogens, they are often lost postrecruitment and are typically not the same types of parasites that infect adult populations (Rigby and Dufour 1996). Hence, introduced species that arrive as larvae in ballast water are presumably rarely parasitised (Lafferty and Kuris 1996, Torchin *et al.* 2002). Thus, in marine systems, invasions resulting from the release of ballast water may filter out potential introduced parasites. Although many microorganisms, including potential pathogens, have been recovered from ballast water, the extent to which this leads to successful establishment of introduced marine pathogens is uncertain (Ruiz *et al.* 2000). However, even though marine parasites may face significant obstacles to invasion compared to free-living species, some parasites do establish populations in new regions. These introduced parasites often affect native species and have deleterious consequences in their new habitat (Table 9.1).

Protozoan parasites

Protistan parasites have often caused significant and widespread effects in marine systems and aquaculture operations. Some of these pathogens are likely to be introduced species. However, for other pathogens there is limited evidence for an exotic source. Many of these are parasites of oysters, often considered to have been introduced with infected stocks. MSX disease, caused by *Haplosporidium nelsoni*, has caused massive mortalities of native oysters in Delaware and Chesapeake Bay since 1957. *Haplosporidium nelsoni* was probably introduced from Asia (Andrews 1980, Barber 1997) resulting from undocumented introductions of the Japanese oyster, *Crassostrea gigas* (Barber 1997). Recent molecular evidence supports this hypothesis and suggests that the parasite was initially introduced from Japan to California and subsequently, to the East Coast of the United States with importations of *C. gigas* from California (Friedman 1996, Burreson *et al.* 2000). Another possible introduction, *Perkinsus marinus*, agent of Dermo disease in oysters may have been introduced with infected *C. virginica* (Ruiz *et al.* 1999). The recent range extension of this oyster pathogen is a result of repeated introductions of infected native oysters to the north-eastern United States of America (USA) (Ford 1996). Both of these pathogens have

Table 9.1 Introduced parasites

Parasite taxon	Parasite	Host	Native/introduced region	Method of introduction	References
Protozoa	Haplosporidia	Oysters, bivalves	JAP/PNA	Aquaculture	Friedman (1996)
	Haplosporidium nelsoni		ASI/ANA	Aquaculture	Barber (1997), Andrews (1980)
	Bonamia ostreae		PNA/EUR	Aquaculture	Minchin (1996), Chew (1990)
	Perkinsus marinus		S. ANA/N. ANA	Aquaculture	Ford (1996)
	Paramoeba invadens	Sea urchin	Unknown/N. ANA	Unknown	Scheibling and Hennigar (1997)
Monogenea	*Gyrodactylus anguillae*	Eels	EUR?/AUS, ASI, ANA	Aquaculture	Hayward et al. (2001a), Ernst et al. (2000)
	Pseudodactylogyrus anguillae		ASI/EUR, ANA	Aquaculture	Hayward et al. (2001b)
	Pseudodactylogyrus bini		ASI/EUR, ANA	Aquaculture	Hayward et al. (2001b)
	Gyrodactylus salaris	Salmon	—/NOR	Stocking	Johnsen and Jensen (1991), Hastein and Lindstad (1991)
Trematoda	*Neobenedenia melleni*	Marine fishes	HKG, HAN/JAP	Aquaculture	Ogawa et al. (1995)
	Nitzschia sturionis	Sturgeon	CAS/ARA	Stocking	Osmanov (1971), Zholdasova (1997)
	Cercariae batillariae	Asian mud snail	ASI/PNA	Aquaculture	Torchin et al. (in press)
	Austrobilharzia variglandis	Atlantic mud snail	ANA/PNA	Aquaculture	Grodhaus and Keh (1958)
Nematoda	*Anguillicola crassus*	Eels	ASI/EUR, ANA	Aquaculture	18 Barse and Secor (1999)
	Spirocamallanus istiblenni	Marine fishes	FRP/HAW	Stocking	Font and Rigby (1999)
Polychaeta	*Terebrasabella heterouncinata*	Abalone, snails	SAF/PNA	Aquaculture	20 Kuris and Culver (1999), 21 Culver and Kuris (2000)
Copepoda	*Mytilicola orientalis*		ASI/EUR, PNA	Aquaculture	His (1977), Stock (1993), Bernard (1969), Holmes and Minchin (1995)
	Mytilicola intestinalis		MED/N. EUR	Fouling	Minchin (1996), Stock (1993)
	Myicola ostrae		ASI/EUR	Aquaculture	Stock (1993)
Rhizocephala	*Loxothylacus panopaei*	Mud crabs	GOM/ANA	Aquaculture	van Engel et al. (1965), Hines et al. (1997)
	Heterosaccus dollfusi	Swimming crab	RED/MED	Migrant	Galil and Lützen (1995)

Asterisks indicate suspected introductions. Regions are: ANA Atlantic North America, ASI Asia, AUS Australia, CAS Caspian Sea, EUR Europe, MED Mediterranean, PNA Pacific North America, ARA Aral Sea, GOM Gulf of Mexico, RED Red Sea, HKG Hong Kong, HAN Hainan, SAF South Africa, FRP French Polynesia, HAW Hawaii, NOR Norway, JAP Japan. After Torchin et al. (2002).

been a serious source of mortality for native oysters in the mid Atlantic and may limit oyster populations in the Chesapeake Bay (Ruiz et al. 1999). Oyster translocations may have also resulted in the introduction of *Bonamia ostrae* from California to Europe (Chew 1990). A shipment of the European flat oyster, *Ostrea edulis*, cultured in California from broodstock originating in the Netherlands (Barber 1997) may have initiated the spread of epizootics of this parasite throughout Europe (Barber 1997). Recent outbreaks of the marine amoeba, *Paramoeba invadens*, have caused mass mortalities in the sea urchin, *Strongylocentrotus droebachiensis*, in the North Atlantic coast of Canada. *Paramoeba invadens* may be an exotic species, but further research is needed to determine its origin (Scheibling and Hennigar 1997).

Monogeneans

Although typically restricted to the freshwater portions of their hosts' life cycles, monogeneans introduced with catadromous and anadromous fishes have also been accidentally imported along with their hosts. Stocking Norwegian rivers with infected salmon may have introduced *Gyrodactylus salaris* to wild Atlantic salmon stocks. This monogenean now causes heavy mortality in salmon parr (Johnsen and Jensen 1991). *Gyrodactylus salaris* was geographically isolated in rivers because it could not survive in brackish or marine water but the movement of infected fish among hatcheries enabled its spread (Johnsen and Jensen 1988). Introduced monogeneans are common pests in the eel trade. They often cause mortality and are capable of infecting native eel species (Hayward et al. 2001a, b). *Pseudodactylogyrus anguillae* and *P. bini* were introduced from Asia to North America and Europe with the importation of the Japanese eel, *Anguilla japonica* (Hayward et al. 2001a). Both species now infect wild populations of the native North American eel, *Anguilla rostrata* (Cone and Marcogliese 1995, Hayward et al. 2001a) and wild populations of the native European eel, *Anguilla anguilla* (Gelnar et al. 1996, Hayward et al. 2001a). Another probable invader, *Gyrodactylus anguillae*, occurs on four continents. It was likely introduced from Europe through the importation of infected European eels (Hayward et al. 2001b). Yet another monogenean introduced with the importation of fish is *Nitzschia sturionis*. Stocking stellate sturgeon from the Caspian into the Aral Sea introduced this monogenean which caused massive mortality of the native ship sturgeon in the 1930s (Osmanov 1971, Zholdasova 1997). The monogenean, *Neobenedenia melleni* (=*girellae*), is a pest in marine fisheries in Japan. Ogawa et al. (1995) suggest that this parasite was introduced to Japan with amberjack, *Seriola dumerili*, imported from Hainan and Hong Kong and spread to other fishes due to its unusually low host specificity. It causes mortality in heavily infected fish. Other parasite introductions, primarily monogeneans, resulting from fish releases have also been reported from the land-locked seas in the Ponto-Caspian region (Grigorovich et al. 2002).

Trematodes

Introduced to the West Coast of North America with the importation of oysters from Japan, the mud snail, *Batillaria cumingi* (=*attramentaria*), brought with it the Japanese trematode *Cercaria batillariae* (Torchin et al. in press). Both the snail host and parasite now occur in several estuaries from Vancouver, British Columbia to central California (Ching 1991, Torchin et al. in press). *Cercaria batillariae*, a heterophyid whose adult form has still not been described nor is its life cycle fully known (Shimura and Ito 1980), infects at least three native fish species as second intermediate hosts and native birds as final hosts in its introduced range (Torchin et al. in press). An additional consequence of this invasion is that at the southern end of its introduced range, the Japanese mud snail is outcompeting and locally extirpating populations of the California mud snail, *Cerithidea californica* (Carlton 1975, Byers 2000, Byers and Goldwasser 2001). *Cerithidea californica* serves as first intermediate host for at least 18 native trematode species throughout its range in California (Martin 1972) and, at least 10 trematode species where the

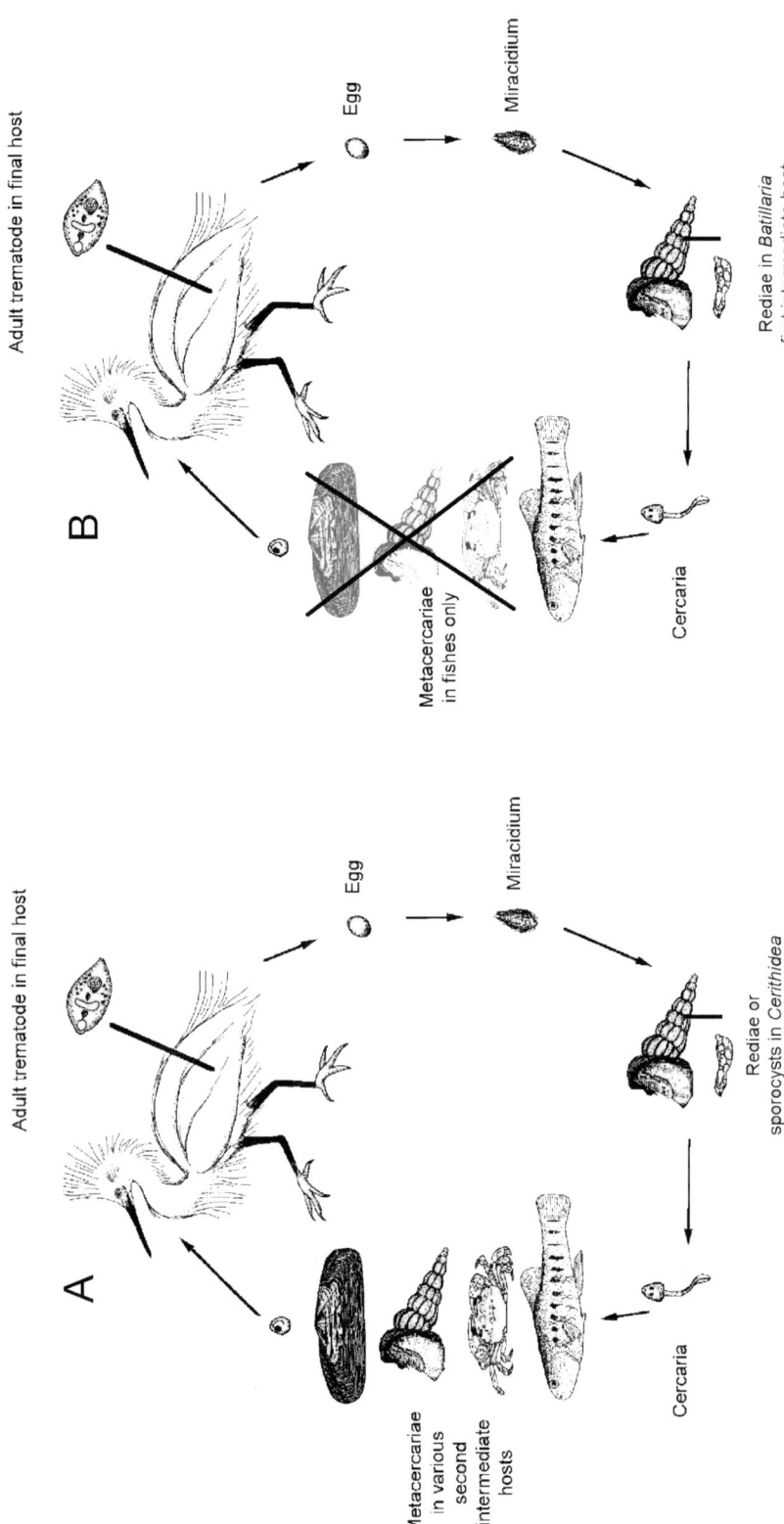

Figure 9.8 Generalised life cycle for trematode species. **A.** *Cerithidea californica* as first intermediate host. **B.** *Cercaria batillariae* using *Batillaria cumingi* as a first intermediate host. The particular type of second intermediate host used (e.g. a fish or a crustacean) depends upon trematode species, and *Cercaria batillariae* uses several native fish species as second intermediate hosts (Torchin *et al.* in press). Figure adapted from Huspeni and Lafferty (2004).

two snail species overlap (Torchin *et al.*, in press). Most of these trematodes are trophically transmitted through the marsh community, infecting multiple hosts during their life cycle. As *C. californica* is replaced by *B. attramentaria*, its parasites will also become locally extinct since none are known to infect alternative first intermediate hosts, including *B. attramentaria*. Although the precise consequenses of these local extinctions on the marsh community remain unclear, the removal of several native trematode species due to local extinction of their first intermediate host will eliminate infection by the trematodes in the second intermediate host molluscs, crustaceans and some fishes. In turn, the several native trematode species will be replaced with a single introduced trematode species (Fig. 9.8). Because some of these trophically transmitted parasites can alter foraging dynamics in these systems (Lafferty 1992, Lafferty and Morris 1996, Lafferty 1999), the replacement of this native parasite fauna may have indirect effects, potentially altering the abundances of intermediate and final hosts (Torchin *et al.* in press).

Another snail introduced with the importation of oysters is the Atlantic mud snail, *Ilyanassa obsoleta* (Demond 1952, Carlton 1999). In its introduced range, *I. obsoleta* is reported to be infected with five species of larval trematodes (Grodhaus and Keh 1958). One of these, *Austrobilharzia variglandis*, initially described from the snail's native range in coastal estuaries in eastern North America, causes swimmer's itch in humans (Miller and Northup 1926, Stunkard and Hinchliffe 1952, Grodhaus and Keh 1958). Although identifications of the four other species are not reported (Grodhaus and Keh 1958) they are possibly a subset of *I. obsoleta*'s native suite of parasites (Torchin *et al.* 2002). Studies comparing these trematodes to those in the native range are required to resolve this issue.

Nematodes

A serious pest in eel fisheries is the swim bladder nematode, *Anguillicola crassus*. Native to Asia, *A. crassus* has been introduced to Europe and North America where it now infects native eels in both in nature and in culture (Barse and Secor 1999). *Anguillicola crassus* can reach high prevalences in native eel populations and it can cause severe pathology in European and North American eels (Barse and Secor 1999).

Although the origin of the nematode, *Spirocamallanus istiblenni*, is still uncertain, Font and Rigby (1999) present evidence that this nematode may have been introduced to the Hawaiian Islands along with its host, the blue-lined snapper, *Lutjanus kasmira*. Further investigation may reveal whether this nematode, which also infects endemic Hawaiian fishes, is an introduced species (Font and Rigby 1999).

Polychaetes

In the 1980s, a polychaete worm, *Terebrasabella heterouncinata*, was accidentally imported to California from South Africa with infested abalone. It rapidly became a major pest in abalone mariculture facilities in California (Kuris and Culver 1999). Although it does not derive any nutrition from its host, it induces severe deformation of the host's shell. This reduces the market price of infested animals and may make them more susceptible to shell-cracking predators (Kuris and Culver 1999). This worm was accidentally released from a mariculture facility into the wild where it infested abalone and other susceptible native gastropods. This wild population has since been eradicated by culling potential host snails below the density threshold for transmission (Culver and Kuris 2000).

Copepods

The global transport of marine bivalves for aquaculture has lead to widespread introductions of parasitic copepods. Native to Asia, *Mytilicola orientalis* and *Myicola ostrae* are both parasitic copepods of the Pacific oyster, *Crassostrea gigas*. *Mytilicola orientalis* has been accidentally introduced to Europe and the Pacific Coast of North America, while *Myicola* has only been reported from Europe (His 1977, Stock 1993, Holmes and Minchin 1995, Minchin 1996). Both species infect native bivalves and *M. orientalis* is considered a serious pest (Holmes and Minchin 1995). *Mytilicola intestinalis*, which was presumably introduced to northern Europe from blue mussels originating in the Mediterranean Sea, may have been transported in its host, *Mytilus galloprovincialis*, on the hulls of ships (Minchin 1996). In the early 1950s an epidemic of *M. intestinalis* caused considerable damage to mussel fisheries and infections spread to other native bivalve species in the Netherlands (Stock 1993).

Rhizocephalans

Another putative aquaculture-related parasite introduction is the rhizocephalan barnacle, *Loxothylacus panopaei*. First reported in Chesapeake Bay in 1964, it was presumably introduced with infected mud crabs associated with oysters transplanted from the Gulf of Mexico (Van Engel et al. 1965). It now parasitises three crab species in its introduced range, including two which only appear to be infected within the introduced range (Hines et al. 1997). This barnacle is a parasitic castrator and anecdotal information suggests that after its introduction to Chesapeake Bay, its two primary hosts became rare. Recently this has been supported by documentation of a negative association between the prevalence of this parasite and the biomass of its host, *Rhithropanopeus harrisii* within Chesapeake Bay (ME Torchin unpublished data, summer 2002–summer 2003). Another introduced rhizocephalan barnacle, *Heterosaccus dollfusi*, recently invaded the Mediterranean Sea through the Suez Canal, even though its crab host, *Charybdis longicollis*, invaded before 1954 and is now well established (Galil and Lützen 1995, Galil and Innocenti 1999). This parasite has not been recovered from any species of native crab, nor from other introduced species of portunids, including *Charybdis hellerii*, also native to the Red Sea. Other reports of rhizocephalans introduced with their hosts are anecdotal and lack confirmation (e.g. Boschma 1972, Kinzelbach 1965).

Controlling introduced pests with parasites

The reduction or absence of parasites in introduced pest species, (Torchin et al. 2003) may explain why some introduced species proliferate in their new environment and become destructive invaders (Torchin et al. 2001). Thus, replacing some key parasites in introduced pest populations may mitigate the damage they cause. The use of biological control in marine environments, where introduced species increasingly pose ecological and economic threats (Cohen and Carlton 1998, Ruiz et al. 1997), has only been proposed recently (Lafferty and Kuris 1996, review by Secord 2003). Lafferty and Kuris (1996) developed a model for the use of natural enemies, including parasites, as biological control agents against introduced marine pests. Basing the strategy on successful examples of terrestrial biocontrol, Lafferty and Kuris (1996) emphasise the key differences between marine and terrestrial ecosystems crucial for evaluating the safety and efficacy of this approach. Kuris and Lafferty (2000) relate the concerns of safety of marine biological control to terrestrial weed pest control identifying the need for high host specificity of the proposed control agent. In contrast, the efficacy, particularly of parasitic castrators or parasitoids used as marine biological control agents, is comparable to insect pest biological control due to the regulatory similarity of these and terrestrial parasitoids (Kuris and Lafferty 2000, Kuris et al. 2002).

Serious consideration of marine pest biological control is relatively new and most research has been directed towards the control of the European green crab, *Carcinus maenas*. The green crab is a benthic marine predator and usually considered a pest where it is introduced (Lafferty and Kuris 1996, Grosholz et al. 2000). Additionally, the introduced populations of green crabs lack most of parasites common in their native range (Torchin *et al.* 2001). In particular, parasitic castrators were not present in any of the introduced populations. In Europe, however, the prevalence of parasitic castrators (which block reproduction and growth) was negatively associated with demographic success (biomass and body size) of green crabs, suggesting that these parasites may partially control green crabs in Europe. While several factors could interact with parasitism and influence this result, Torchin *et al.* (2001) found that factors such as latitude, environmental quality, limb loss (a proxy for predation) and other types of parasites had no significant effect in their analysis. Furthermore, uninfected introduced green crab populations were significantly larger and had a greater biomass compared to European populations (Torchin *et al.* 2001). Recent models also indicate that parasitic castrators, under a range of larval recruitment dynamics, have the potential to control introduced host populations (Kuris and Lafferty 1992, Lafferty and Kuris 1996). The rhizocephalan barnacle, *Sacculina carcini*, a parasitic castrator of *C. maenas* in Europe, is a candidate biological control agent of introduced green crab populations in Australia and the West Cost of North America (Lafferty and Kuris 1996). However, host specificity trials indicate that *S. carcini* is able to settle on, infect, but not complete development in non-host crabs native to Australia and the West Coast of North America (Thresher *et al.* 2000, Goddard *et al.* in press). Sands (1998) suggested that the ability to complete development and reproduce in non-target hosts is the standard for unacceptable risk of a biological control agent. However, Goddard *et al.* (in press) demonstrated that settlement of *S. carcini* on non-host native crabs often results in an infection lethal to the host but the parasite does not successfully develop. Hence, safety and efficacy of the control agent are inversely associated (Goddard *et al.* in press). Because native crabs can be infected but do not serve as hosts, they act as population sinks for the parasite. An evaluation of these impacts, plus those estimated for other species of crabs not included in the testing, could be weighed against the benefit of controlling green crabs and releasing California native crabs and other native species from predation by *C. maenas*.

The parasitic larval sea anemone, *Edwardsia liniata*, which infects the north-western Atlantic ctenophore, *Mnemiopsis leidyi*, has been proposed as a possible biological control agent for invasive populations of *M. leidyi* in the Black Sea (Bumann and Puls 1996). While the parasitic planula larvae of this anemone reduce growth rates and may control *M. leidyi* populations in their native range, they can also infect other ctenophore species. Additionally, they may cause seabather's eruption (dermatitis) in humans and the sessile benthic adults may alter benthic community structure (Bumann and Puls 1996).

Biological control has also been proposed for the Asian seastar, *Asterias amurensis* (Kuris *et al* 1996). Of four candidate species listed, the ciliate *Orchitophyra stellarum* has received the most attention (Kuris *et al.* 1996, Goggin and Bouland 1997). However, more research is needed to examine its ability to control host populations and determine its host specificity (Goggin and Bouland 1997).

Conclusion

Although introduced marine species often escape their parasites, some parasites manage to invade with their hosts and establish populations in the new region. Once established, they can spread and impact native species. Still, reports of introduced marine parasites remain limited. Undoubtedly more introduced marine parasites exist than are reported, partly because parasites are often difficult to observe and with increased attention, more introduced parasites will be inevitably be revealed. It is often difficult to determine the origin of parasites of introduced hosts (Font and Rigby 1999) and of newly discovered parasites (Scheibling and Hennigar 1997) and these should not, *a priori*, be considered exotic. Comparing parasite faunas of native and introduced hosts and native and introduced populations of the exotic species (Torchin *et al.* 2002, Torchin and Mitchell 2004) and by using historic evidence (Font and Rigby 1999) it is possible to make strong inferences as to the origins of emerging diseases and cryptogenic parasite species. Marine parasites encounter several obstacles to introduction especially if 'hosts' are introduced as larvae. Interestingly, evidence suggests that most of the exotic marine parasites in Table 9.1 resulted from aquaculture or fisheries related introductions. It is not surprising that adult hosts transferred from one aquaculture area to another (where population densities are often unnaturally high) would facilitate the introduction and transmission of parasites. It is notable, however, that most of these historical introductions often occurred without quarantine or screening for parasites and pathogens and that most current aquaculture practices account for this. Introduction of parasites can have serious impacts on native marine communities. More research is needed to discover exotic parasites and their effects as well as investigate the consequences of exotic species that lack many or all of their infectious natural enemies. In certain cases in which destructive marine invaders lack infectious agents, replacing key parasites may mitigate the damage they cause. Thus, further research is needed to fully evaluate the potential for safe and effective marine biological control.

Important references
General texts and papers on biological invasions are by Elton (1958), Dobson and May (1986), and Byers and Goldwasser (2001). Important work on marine invasions is by Chew (1990), Carlton and Geller (1993), Ruiz et al. (2000) and Grosholz (2002).

Drake and Lodge (2004) discussed global hotspots of biological invasions: evaluating options for ballast-water management. Eradication of introduced marine pests, their management and control were discussed by Lafferty and Kuris (1996), Culver and Kuris (2000) and Secord (2003). Minchin (1996) gave an account of management of the introduction and transfer of marine molluscs.

Deep-sea parasites
Rodney A Bray

Introduction
The deep sea has been 'loosely' defined as the part of the ocean below the epipelagic zone (Herring 2002), that is, all of the ocean below the level of the continental shelf margin or shelf-slope break. In most of the ocean this, in effect, means the ocean deeper than about 200 m, although the shelf can be deeper, particularly around the Antarctic continent (Fig. 9.9). The bottom drops away to the abyssal plain at 4000 m to 6000 m, with occasional trenches reaching deeper than 11 000 m. By the definition given here, the deep sea covers about 65% of the Earth's surface and over 50% is covered with water more than 3000 m deep (Gage and Tyler 1991). This vast area is not uniform and the character of the fauna of the shallower parts of the zone differs markedly from the deep areas such as the abyssal plain and, probably, the trenches. At high latitudes the typical shelf fauna may reach down the slope to more than 1000 m, and in polar regions typical deep-sea parasites may be encountered in depths of a few hundred metres. The mid-ocean ridge and hydrothermal vent fauna and cold-seep faunas probably also have a distinct character (de Buron and Morand 2004). Cold seeps occur widely, from shallow water to trenches, where methane seeping from the rocks of the sea floor nourishes bacteria and their symbiotic organisms.

Virtually everything known about deep-sea parasites is of the stages infesting fishes. The fish fauna of the deep sea is depauperate in members of the largest fish order Perciformes, and this is reflected in the parasite diversity in the deep sea. Gadiforms of the family Macrouridae constitute a major part of the deep-sea fauna in much of the world's ocean. Other gadiform families such as the Ophidiidae and Moridae and other orders such as the Scorpaeniformes, Osmeriformes, Notacanthiformes, Anguilliformes and Aulopiformes are well represented in the deep sea.

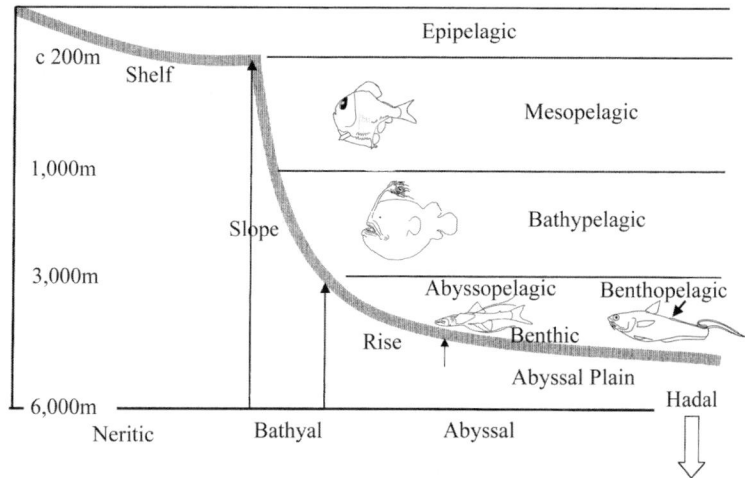

Figure 9.9 Diagram of certain oceanographic features and the terminology used.

Conditions in the deep sea

Pressure: Pressure increases by one atmosphere (0.101 megapascals) for each 10 m of increased depth (Tyler 1995). This is a problem to the fauna only when there are gas-filled cavities (e.g. swim bladder), and only then if the pressure (i.e. depth) changes rapidly. The density of water (830 times that of air) increases only by 0.5% per 1000 m. Nevertheless, this is enough to materially affect the rate of reactions involving a change in volume, such as those using enzymes. Deep-sea animal enzymes generally have reduced pressure sensitivity and reduced efficiency (relative to shallow water animals), rather than a series of sensitive enzymes relating to the pressure changes with depth (deep-sea animals normally inhabit a much greater depth range, *eurybathy*, than shallow forms, *stenobathy*).

Temperature: Temperature variation below the thermocline (800–1300 m) is low. At the thermocline the temperature is just below 4°C and decreases only slightly below that depth. Although the deep-sea fauna must be adapted to cold, the temperature is not highly variable. Cold tolerance, however, is not adequate to adapt animals to life in deep water.

Light: Photosynthesis is not well sustained below 200 m and sunlight cannot be detected below 1000 m. Bioluminescence and glow from black-smoker vents are the only sources of light in the deep sea.

Energy availability and food supply: For most of the deep sea, food descending from the photic zone supplies energy. This can be descending phytodetritus (deep-sea 'snow') or remains of animals, which may be tiny zooplankton or massive whales or fishes. The seasonal pulse of phytodetritus imposes seasonality on deep-sea communities.

Patchiness – mate finding and parasite transmission: Large food-falls attract many organisms and it is likely that these are the venues for mating and transmission. The attractiveness to animals of hydrothermal vents and cold seeps may also serve to concentrate parasite stages and their hosts.

What parasites are known

De Buron and Morand (2002) found that 57% of metazoan parasites reported in waters deeper than 1000 m were digeneans, 25% crustaceans (80% copepods), 10% cestodes, 4% acanthocephalans, 2% nematodes and 2% monogeneans. Protistans (in which they included the myxozoans) make up less than 17% of parasite records.

Protistans
This covers a wide variety of organisms, whose relationships are now being elucidated. Few have been reported from deep waters. A few blood parasites, such as haemogregarines and coccidians, are known from deep-sea fishes (Davies and Merrett 2000). One *Eimeria* species is known to damage the swim bladder of macrourids (Grabda 1983). Moreira and Lopez-Garcia (2003) surveyed protist diversity at hydrothermal vents using rDNA and identified sequences of several parasitic protist lineages, including Apicomplexa, Perkinsozoa, Syndiniales and Kinetoplastida. They concluded that many vent organisms must be infested.

Myxozoa
Identified myxosporean members of the families Myxobolidae, Auerbachiidae, Alatosporidae, Ceratomyxidae and Myxidiidae are known. *Myxidium* species are commonly found in macrourids and have been used as an indicator to separate stocks (Szuks 1980).

Platyhelminthes
Monogenea: Most records are members of the family Diclidophoridae, which are frequently found on macrourids. There are also records of Acanthocotylidae, Capsalidae, Chimaericolidae, Dactylogyridae, Discocotylidae, Hexabothriidae, Mazocraeidae, Microcotylidae, Monocotylidae, Plectanocotylidae and Tetraonchoididae. Monogeneans are successful until the lower slope where, apparently, fish densities are too low for successful transmission (Campbell *et al.* 1980).

Cestoda: The unsegmented 'cestodarian' *Gyrocotyle* is common in holocephalans, most of which occur in deep water, and numerous trypanorhynch and tetraphyllideans are reported from deep-sea elasmobranchs. All reports of adult cestodes from deep-sea teleosts are of members of the order Pseudophyllidea, from the families Bothriocephalidae, Philobythiidae, Echinophallidae and Triaenophoridae. Members of the latter family show some predilection for deep-sea fishes.

Digenea: This group is probably the largest group of metazoan parasites in deep-sea fishes, but only 17 families out of about 150 known are reported. Depth profiles are marginally better known for this group (Bray 2004). Four families, the Derogenidae, Fellodistomidae, Hemiuridae and Lepocreadiidae, have species reaching to abyssal depths (i.e. about 5000 m). Three further families, the Lecithasteridae, Opecoelidae and Zoogonidae, reach below 3000 m, one, the Gorgoderidae, deeper than 2000 m and three, the Acanthocolpidae, Accacoeliidae and Sanguinicolidae, just deeper than 1000 m. Four further families, Bucephalidae, Cryptogonimidae, Faustulidae and Monorchiidae, reach below the shelf break at 200 m. Two further families, Bivesiculidae and Hirudinellidae, include species reported from deep-sea fishes, but with no detailed bathymetric data. It is apparent that the Bucephalidae are common parasites in cold-seep regions as Powell *et al.* (1999) found that the prevalence of 'Bucephalus-like trematodes' in cold-seep mussels at a depth of 550 m to 650 m was 70% or more at three sites.

Nematoda
Relatively few nematodes are known from the deep sea. The families Capillariidae (Enoplida), Anisakidae and Cucullanidae (Ascaridida) and Philometridae, Rhabdochonidae and Cystidicolidae (Spirurida) are reported from fishes. Many anisakids are larval forms, which are likely to be found as adults in cetaceans and pinnipeds. One cystidicolid species is reported from a hydrothermal vent fish (Justine *et al.* 2002).

Acanthocephala
Adult palaeacanthocephalans of the families Echinorhynchidae and Rhadinorhynchidae are most commonly reported. Less common are members of the families Heteracanthocephalidae

and Hypoechinorhynchidae. Several polymorphid larvae are reported in deep-water fishes, presumably as adults these are cetacean or possibly pinniped parasites. One species is described from a hydrothermal vent fish (de Buron 1988).

Crustacea
A few families of copepods have exploited deep-sea demersal fishes. According to Boxshall (1998) two families of the order Poecilostomatoidea and five of the order Siphonostomatoidea are reported. Three families are commonly encountered, the Sphyriidae, Lernaeopodidae and Chondracanthidae, and members of four others, the Hatschekiidae, Pennellidae, Philichthyidae and Hyponeoidae, are occasionally found. Several copepod species are reported from hydrothermal vent fishes (see de Buron and Morand 2004).

Areas of ignorance
A huge proportion of the area covered by the deep sea has not been sampled, let alone for parasites. No data are available from trenches. The data from hydrothermal vents and cold seeps are rudimentary. The life cycles, host relationships, distribution and zoogeography of deep-sea parasites have not been studied.

Important references
Important general texts on deep-sea biology are those by Marshall (1979), Gage and Tyler (1991) and Herring (2002), and those on deep-sea fishes by Merrett and Haedrich (1997) and Randall and Farrell (1997). Klimpel *et al.* (2001) produced a checklist of the 'metazoan' parasites of deep-sea fishes, including Myxozoa. Major review articles on deep-sea parasites include those of Noble (1973), Campbell (1983) and Bray *et al.* (1999). The most important paper on deep-sea parasite ecology is Campbell *et al.* (1980). For speculations on the parasite fauna of hydrothermal vents, see de Buron and Morand (2004).

Chapter 10

Economic and environmental importance

Almost all (if not all) groups of marine animals including the various invertebrates and vertebrates are hosts to parasites. Parasites of these groups have been discussed by various authors in the four volumes edited by Kinne (1980–1985). This chapter is restricted to discussions of parasitic infections of the more important groups.

Marine parasites are of very great ecological importance, as suggested by their great diversity, and by their often high prevalences and intensities of infection. However, very little is known about how they affect host populations in the oceans. An exception is the long-term study of salmon lice on Atlantic salmon not only in aquaculture but also in wild salmon populations, conducted because of the great economic importance of salmon. Mass mortalities caused by parasites have not been well documented for marine fish and invertebrates because of the large spaces in the oceans involved and the difficulties in monitoring such effects. Effects on marine birds are little understood and need much further study: mass mortalities of seabirds have been attributed to several causes but parasite examinations which could have revealed the role of parasites in the mortalities were not conducted. Effects on mammals such as whales are well documented, and hundreds of parasite species of mammals have been described; nevertheless, much needs to be done in this area as well.

Much greater effort has gone into the study of parasite diseases in aquaculture, which is not surprising in view of the enormous and rising economic value of aquacultured fish, molluscs and crustaceans, estimated at many billions (10^9) of dollars annually. For example, the commercial value of cultured salmonid fishes alone in 2001 was around US$3.84 billion (US$7.44 billion for all finfish). Disease, and much of it resulting from parasites, is the single most important factor threatening the aquaculture industry.

Beside these important negative aspects of marine parasites, this Chapter also discusses a more positive aspect of parasites: the use of parasites in pollution monitoring. Pollution may affect the composition of parasite communities, and some parasites store certain pollutants to a higher degree than their hosts; both these phenomena can be used for monitoring pollution.

Mass mortalities in the oceans
Brian Jones

Introduction
Observations of mass mortalities of aquatic organisms in the open ocean have long been reported by sailors and naturalists but have seldom been studied in any detail, usually due to their catastrophic, transient nature and the distance from scientific laboratories. Of those events

that have been studied, most were found to result from environmental conditions including toxic algal blooms (Hornell 1918) and viral or bacterial infections (Harvell et al. 1999) and as such are outside the scope of this book. Likewise, although parasites can affect the fecundity of fish populations (Rohde 1982), this does not usually result in 'mass mortalities' and so reductions in fecundity are not considered here.

Oceanic mass mortalities attributed to protozoan and metazoan marine parasites are rare (Rohde 1982). There are several factors that contribute to this. Parasite life cycles are not usually adapted to utilise mass host mortality events but instead cause low-level background mortality or morbidity. In addition, the biology of oceanic animals mitigates against epizootics of parasites (Munro et al. 1983, Marcogliese 2002). The spawning area is often remote from the juvenile nursery areas that, in turn, are separated from areas where adults live. Larger adult fish also inhabit deeper water than small fish and most oceanic fish school by size with sick fish schooling together, having different migration patterns and different habitat requirements from healthy fish (B Jones pers. obs. 1994, Holst et al. 1997).

However, it is also now apparent that oceanic 'wanderers' do not meander in a random fashion over the ocean but are concentrated around fronts, convergences and eddies where prey species (and parasites) are concentrated (Jones 1998). The technology to reliably detect these ocean features and thus find fish aggregations in the open ocean became available to fishers in the 1980s with the advent of satellite-derived sea surface temperature data and Global Positioning Systems (Kirby et al. 2003).

Time-series datasets are now available which show that clupeid fish populations may fluctuate in prevalence and abundance by orders of magnitude (Southward et al. 1988), but the causes of such fluctuations are unclear and there is still no direct evidence that fluctuations have been caused by parasites. First, dead animals do not last long in the ocean – they are soon eaten. It is only when predators are satiated that the evidence remains long enough to be collected. Second, fisheries management is not an exact science. The errors around stock assessment surveys are usually so great that significant mortalities due to disease cannot be discriminated from 'measurement error'. Even where such 'discrepancies' in stock sizes between years are noted, fisheries biologists are seldom inclined to attribute the missing biomass to disease. If juvenile abundance in a nursery area is lower than expected, it is usually attributed to 'spawning failure'; if abundance of bigger fish is reduced, it is often ascribed to 'changed migration patterns' or 'fishing mortality'. For example, fluctuations in clupeid populations due to climate variations have been documented (Southward et al. 1988) which is why mass mortalities of pilchards in Australia and New Zealand in 1995 were attributed to climate (O'Neil 1995) despite the absence of environmental cues (Griffin et al. 1997) and the presence of herpes virus associated with the lesions which killed the fish (Jones et al. 1997). Oceanic parasite populations can also undergo large fluctuations in abundance. In 1981 the large (up to 70 mm long) external parasitic copepod *Pennella* sp., which embeds its head in the flesh of the saury, *Cololabis saira*, was first recorded from the western North Pacific. The copepod rapidly increased in prevalence to 33% in the western and central North Pacific Ocean and to 12% in the Sea of Okhotsk before disappearing in 1985 (Nagasawa et al. 1988). The impact of such a large parasite on the saury population was not studied.

Mass mortalities of adults in the ocean resulting from parasites

A parasite that does cause serious systemic disease and mortalities of fishes in the open ocean is the proctistan fungus-like *Ichthyophonus* spp., now assigned to the Class Ichthyosporea (Sinderman 1990, Rand et al. 2000). *Ichthyophonus* spp. affect more than 80 species of fish, have a worldwide distribution and have been studied for over 100 years. Irregular epizootics have been reported in: British mackerel, *Scomber scomber*, North Atlantic herring, *Clupea harengus*,

Scottish plaice, *Pleuronectes platessa*, and haddock, *Melanogrammus aegelfinus*, as well as Pacific herring, *Clupea pallasi* (Sinderman 1990, Kocan et al. 1999). Sinderman (1963) suggested that *Ichthyophonus* might be the single most important limiting factor for herring population growth in the western North Atlantic. Sporadic epizootics also occur along the eastern Atlantic and in the Skagerrak and Kattegat (Munro et al. 1983, Spanggaard et al. 1994, Rahimian and Thulin 1996). The pathogenicity of the disease agent is related to water temperature which may explain why occurrence of epizootics is erratic (Okamoto et al. 1987).

With the increased deployment of deep-sea cameras, evidence in the form of carcases on the sea floor lead to the discovery that parasitic apostome ciliates *Collinia* sp. were implicated in mass mortalities of krill *Euphausia pacifica*, *Thysanoessa spinifera* and *T. gregaria* along the continental shelf breaks in the western North Atlantic (Gómez-Gutiérrez et al. 2003). The myxosporidan *Myxobolus exiguus*, which occurs under the scales, on the gills and in the stomach and pyloric caecae of a variety of cyprinids in Eurasia and in some cases causing severe haemorrhage, has been implicated in mass mortalities of mullet, *Mugil cephalus* and *M. auratus*, in the northern Black Sea and Azov Sea (Petrushevski and Shulman 1961, Lom and Dykova 1992). Monogeneans cause serious losses in aquaculture fish and have been associated with mass mortalities in the ocean. A mass mortality of anchovy, *Engraulis japonica*, in the Sea of Iyo in 1983, which was estimated to have killed over 87 000 fish, was attributed by Yamamoto et al. (1984) to infection by the monogenean *Pseudanthocotyloides* sp. (Mazocraeidae). An estimated 99% of the fish sampled had the monogenean and the fish showed anaemia and wasted appearance. Translocation of fish for commercial purposes can also cause trouble if parasites are introduced with their hosts. The introduction of the sturgeon *Acipenser stellatus* into the Aral Sea resulted in the establishment of the monogenean *Nitzschia sturionis* in the population of the local sturgeon *A. nudiventris* and caused a major population decline (Petrushevski and Shulman 1961). Sea otters, *Enhydra lutris*, along the pacific coast of northern California exhibit nervous signs and mortalities due to a combination of *Toxoplasma gondii* and *Sarcocystis neurona*. Both of these parasites are terrestrial animal parasites that may have entered the ocean through sewerage (Rayl 2001) and now cause significant mortalities.

As already explained, parasites clearly have pathogenic effects, but evidence of mass mortalities is still absent, and the mortalities must be inferred. Examples of this category include the coccidian parasite *Goussia gadi* that infects the swim bladder of gadoid fish including haddock *Melanogrammus aegelfinus*, cod *(Gadus morhua* and *G. virens)* in the western North Atlantic and Baltic Sea. Over 30% to 58% of haddock on the Nova Scotia banks are infected. The infection fills the swim bladder with a semi-solid waxy mixture of parasite stages, cellular debris and lipid material rendering the organ non-functional. Heavily infected fish are then unable to spawn and die during winter months (Odense and Logan 1976, Lom and Dykova 1992).

In marine mammals, the trematode *Nasitrema* sp. (Nasitrematidae) as well as the nematodes *Crassicauda* sp. (Crassicaudinae) and *Stenurus* sp. (Metastrongyloidea) can infect the brain, sinuses and middle ear of cetaceans. The resultant lesions and interference with echolocation has been implicated in mass strandings of large schools involving hundreds of animals (Geraci and St Aubin 1986, Morimitsu et al. 1987, Zylber et al. 2002).

Mass mortalities of juveniles in the ocean due to parasites

While many parasites are tolerated by adult hosts, they can be lethal to juvenile animals, theoretically resulting in many deaths in a short period. However, because the background mortality due to predation is already high, it is difficult to obtain direct evidence for mass mortalities of juveniles.

Burreson and Zwerner (1984) showed that the haemoflagellate *Trypanoplasma bullocki* caused 100% mortality in inoculated 140 mm to 200 mm juvenile flounder *Paralichthys dentatus* when water temperatures were 0.5°C to 1.5°C, and they hypothesised that mortalities of up to 50% may be occurring in wild juveniles during low temperature periods. Another example is provided by the diclidophoran monogenean *Neoheterobothrium hirame* which first appeared in the western Sea of Japan in 1993 where it now infects the gills and mouth of the Japanese (olive) flounder, *Paralichthys olivaceus*. It killed most of the young-of-the-year flounder stocks by early summer (Ogawa 2002). An example in marine mammals is provided by the hookworm *Uncinaria lucasi*. The infective third-stage larvae transmit through the mothers' milk and cause mass mortalities of fur seal pups, *Callorhinus ursinus* (Geraci and St Aubin 1986).

The parasitic copepod *Lepeophtheirus salmonis* can heavily infect salmonid smolts in estuaries, causing emaciation. It is considered responsible for an absence of fish returning to freshwater to spawn, effectively reducing ova deposition rates by returning fish populations by up to 90% in some catchments. Whether this copepod-induced mortality is occurring at the smolt stage or in adult fish, or in estuaries or in the open ocean where the copepod is also found, is still unknown (Tully and Nolan 2002, see also pp. 374–378).

Summary
It has been known for some time that parasites do have an impact on host populations in the open ocean yet the words written by Rohde (1982) are still relevant. Practically nothing is known about how parasites affect host populations in the open ocean and especially how they contribute to mass mortalities. The ecology of the oceans has been perturbed by industrial fishing and other changes associated directly with human activities (such as pollution) and by climate perturbations that may be due to human activity (Daszak *et al.* 2001). How the resulting increased environmental stress has affected parasite numbers is also largely unknown. The host populations have been severely affected yet how the management and recovery of oceanic fish stocks is being affected by the parasite fauna is also still almost entirely unknown.

Important references
The extensive work of Russian parasitologists on diseases of fishes caused by parasites including mass mortalities was reviewed by Petrushevski and Shulman (1961). Rohde (1982) discussed evidence for marine mass mortalities due to parasites, and Munro *et al.* (1983), the epidemiology of infectious disease in commercially important wild marine fish. The salmon louse, *Lepeophtheirus salmonis*, is responsible for widespread disease in wild and farmed salmon (Tully and Nolan 2002). Among reports of mass moralities in particular species are those by Yamamoto *et al.* (1984) and Morimitsu *et al.* (1987). The former reported mass mortality of Japanese anchovy caused by a gill monogenean, and the latter mass stranding of Odontoceti caused by parasites.

Effects of salmon lice on Atlantic salmon
Peter Andreas Heuch

Introduction
The salmon louse, *Lepeophtheirus salmonis*, is a natural ectoparasite on salmonids in the Northern hemisphere. Its irritating effect on the salmon has been known for centuries. The bishop and naturalist Erik Pontoppidan (1753) saw lice as God's means of delivering the fish to man: by forcing the salmon from the ocean to rivers to rinse the lice off, He made the fish easier to catch. Salmon lice attracted little attention from scientists before the introduction of salmon farming in the 1970s, when they emerged as serious pathogens in fish cages (Pike and Wadsworth 1999). In the late 1980s reports of damaged, prematurely returning sea trout appeared in Ireland

(review by Tully *et al.* 1999) and Norway (Grimnes *et al.* 1996). At the same time, salmon returns to rivers in the Northern Atlantic were low, and sea lice were identified as a potential danger for running salmon smolts (Anonymous 1999). Salmon population oscillations were not a new phenomenon, and historical data indicated these were frequent through the 19th century (e.g. Shearer 1992). The question was had this new situation, with millions of new hosts for the parasite, generated an infection pressure which had increased the already high sea mortality of salmon?

The complex life history of the different hosts, and that little was known about the ecology of this host–parasite system, precluded quick and clear conclusions. This section is a brief introduction to recent research on the fascinating biology of this aquaculture pest and potentially significant mortality factor of wild salmonids.

The parasite and its host

The salmon louse is a parasitic copepod in the order Siphonostomatoida, family Caligidae, class Copepoda. 'Sea lice' are any member of this family. Note that 'fish lice' are not the same, but from the class Branchiura, and 'whale lice' are isopods (order Isopoda) from the class Malacostraca. The salmon louse has a simple life cycle comprising 10 stages separated by moults (Johnson and Albright 1991a) and only one host, fish of the genera *Salmo*, *Salvelinus* and *Onchorhyncus* (Kabata 1988). The first two stages are planktonic nauplius larvae, and the third stage is the planktonic, infective copepodid. The pelagic phase may last up to one month at 8°C, thus there is a great potential for horizontal dispersal with currents. The about 1 mm long copepodid settles and anchors itself by a chitinous thread which is injected into the fish skin, before it moults to the first parasitic stage, the chalimus 1. The fourth chalimus stage moults to a miniature, unattached adult, the first preadult stage. There are two preadult stages before the final moult to adult, and the mating takes place on the host. The fertilised eggs lie in uniseriate egg strings which trail off the female (Fig. 10.1). A female may produce at least 10 pairs of egg strings, each string with about 300 embryos, in her lifetime (Heuch *et al.* 2000).

Atlantic salmon, *Salmo salar*, brown trout, *Salmo trutta*, and Arctic charr, (*Salvelinus alpinus*, all have sea-going populations along the Norwegian coast, and these are hosts for salmon lice (Bjørn and Finstad 2002). These fish spawn in rivers, and the fry grow up there until they

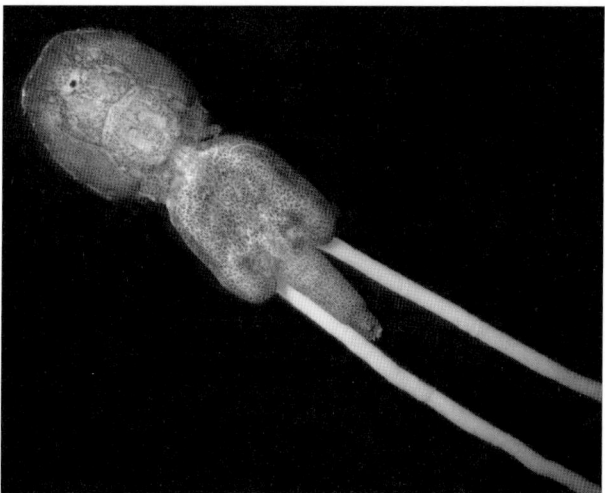

Figure 10.1 A mature female salmon louse, *Lepeophtheirus salmonis*, with newly extruded, whitish egg strings. Total length of animal is about 1.5 cm. Photo by Øivind Øines.

migrate to the sea in the spring as smolts (Crisp 2000). Whereas salmon smolts quickly pass through the coastal zone to forage on the high seas (Holst et al. 2000a, Moore et al. 2000), sea trout and sea charr (the sea-going populations of these species) forage in the littoral zone as young fish (Lyse et al. 1998). Individuals of all three species return to the rivers later in the year, but salmon may spend several years in the ocean before the return migration (Shearer 1992). Between 2% and 10% of the salmon return to spawn.

The infection biology of salmon lice in the Norwegian coastal zone

The parasite can only infect the host and reproduce in salt water (Hahnenkamp and Fyhn 1985, Johnson and Albright 1991b). Homing salmon frequently carry lice with them into the river, but these quickly fall off (McLean et al. 1990). Experiments have shown that the copepodids are positively phototactic, accumulating in surface layers of the water column at day and dispersing downwards at night (Heuch et al. 1995). They are also sensitive to water layering due to salinity (Heuch 1995), and will most likely stay below low salinity surface layers in fjords and estuaries. Salmon smolts migrate in the surface layers of the water on their way through the coastal zone to the ocean (Levings et al. 1994, Holm et al. 2003). The thickness of the fresh or brackish water layer in the fjord will strongly influence the degree to which parasite and host come into contact. Monitoring has shown that in years with high freshwater run-off the smolts are less infected (Holst et al. 2003).

The degree to which salmon smolts are infected by salmon lice will also depend on the temporal overlap between peak production of lice copepodids and the smolt runs. This is may be different from fjord to fjord due to physical factors such as different run-off from rivers and water temperature, which control smolt run timing. In areas without fish farming, such as the Norwegian Skagerrak coast, wild hosts overwinter in fresh or brackish waters where most of the lice die (Schram et al. 1998, Heuch et al. 2002), giving low prevalence and intensity of adult females in spring and summer. The natural infection pressure for running smolts is, therefore, low. This may also have been the situation on the south-west coast, where salmon farming now is ubiquitous. In Finnmark, the northernmost county in Norway, peak infections of sea trout and sea charr occur in July and August (Bjørn et al. 2001b, 2002, Bjørn and Finstad 2002). Smolt trawling (Holst and McDonald 2000) in a farmed area in Finnmark produced only uninfected salmon smolts (Bjørn et al. 2001b, 2002, 2003), probably because the salmon runs are earlier than peak abundance of lice in the water.

Heuch and Mo (2001) modelled the sources of salmon lice infection for running smolts on the Norwegian coastline from Nordland to Vest-Agder counties as a whole. About 50 000 salmon were estimated to arrive at the coast in the spring, and 700 000 sea trout may have overwintered with attached lice (Schram et al. 1998, Heuch et al. 2002). The contribution of lice from sea charr was considered to be negligible. Wild salmon arriving at the coast carry on average 10 adult female lice/fish (Johannessen 1975, Berland 1993). These fish have been infected on the high seas (Jacobsen and Gaard 1997). In areas with fish farming, they also harbour sessile stages (Grimnes et al. 1998, 1999, 2000). If low river discharge prevents the spawners from ascending and they must remain in salt water, these lice may develop to more pathogenic preadults and adults (Johnson et al. 1996), which again may contribute to the local infection pressure.

Changes in the host–parasite population dynamics due to salmon farming

The greatest change brought about by salmon farming is the addition of many new hosts. Many of these are present through the winter, when natural hosts are scarce (Heuch et al. 2002). About 160 million new smolts were stocked in net cages along the Norwegian coast in 2002, giving a standing stock of about 200 million salmon in December. The pelagic lice stages originating from these fish can pass through the net walls and disperse in the environment (Costelloe et al. 1995).

In addition, about 500 000 farmed fish escape every year (reported escapes, Norwegian Directorate of Fisheries) adding up to about 1.5 million escaped salmon and rainbow trout, *Oncorhyncus mykiss*, on the coast (Heuch and Mo 2001). The escapees have about the same infection intensity as wild homing salmon (Grimnes *et al.* 2000).

In our model of louse egg production (Heuch and Mo 2001), farmed salmon and rainbow trout, at a mean intensity of 2 lice/fish, produced 111 billion eggs in the spring of 1999, whereas wild salmonids (mean intensity of 3 lice/sea trout and 10 lice/salmon) produced 2.6 billion eggs. Escaped fish, with a mean intensity of 10 lice/fish, would have added another 15 billion lice eggs. Increasing numbers of farmed fish thus strongly affect the number of lice larvae in the area. The model predicts that if wild populations can tolerate only a doubling of the natural infection pressure in spring, the mean number of lice on the farmed fish should be no greater than 0.05 lice/fish. At present, Norwegian regulations require lice counts from all salmon and rainbow trout farms every two weeks, and if a limit of 0.5 adult female lice/fish in spring–summer is exceeded, treatment with medicine is required. The model predicts that the infection pressure is more than twice the natural level even if all farmers on average have this limit.

Possible effects of sea lice on salmon on individual and population levels

Laboratory studies of the pathological effects of salmon lice were recently reviewed by Tully and Nolan (2002). The parasite feeds on mucus, skin and blood, and pathological effects of sea lice range from minor haemorrhages in the skin to complete osmoregulatory failure and death. The extent of damage depends on the number and development stage of the parasites. The chalimii, being limited in their feeding to a small area around the anchoring thread, produce little damage to the host. However, heavy infections may erode fins, where these stages usually are found (Bjørn and Finstad 1997). When chalimus 4 larvae moult to preadult 1, major physiological disturbances may occur in the host. These include high levels of blood cortisol and glucose, and reduced levels of non-specific immunity. Field evidence indicates that sea trout and sea charr may return to fresh water prematurely to compensate for high plasma chloride levels (Birkeland 1996, Bjørn *et al.* 2001a).

The salmon, however, will carry their parasites to the ocean and may die there as they develop. Finstad *et al.* (2000) estimated from laboratory dose-response studies that wild smolts would die at 0.75 lice/gram body weight. In the fjords, the mean intensity of infection ranged from 0 to 104 copepodids and chalimus larvae per smolt (Holst *et al.* 2003). However, none of more than 3000 post-smolts caught in the North Sea in 1991–2002 carried more than 10 adult lice (Holst *et al.* 2003). An experimental study using such naturally infected, live smolts showed that only fish with 10 adult lice or fewer survived. Together, these observations indicate that if a migrating salmon smolt acquires an infection of more than 11 chalimii, it is unlikely to survive in the ocean.

Norwegian statistics show a decline in wild salmon catches from a mean of 1474 tonnes in the 1970s to 847 tonnes in the 1990s. In this period, salmon fisheries were closed to increase the spawning stock, and salmon production in farms grew from 3200 to 268 000 tonnes. It is, however, important to emphasise that the size of wild stocks is decided by both natural and human factors. In periods with a natural decline, even small losses as a result of human activity may have a large effect. Factors contributing to reduced smolt runs include reduction of river habitat from acid rain and hydroelectric dams, and the introduced parasite *Gyrodactylus salaris* (Anon. 1999). Variations in oceanic feeding areas (Friedland *et al.* 1993) and the match between river and marine temperature (Hansen *et al.* 2003) are important for survival and growth of salmon at sea. Given the physiological consequences of salmon lice infection and the high infection intensities found on running smolts, it is possible that salmon lice have produced additional mortality which may have assisted in the observed decline in wild salmon stocks.

Recent monitoring has given cause for optimism concerning the wild salmon, as catches have increased to a mean of 1150 tonnes in 2000–2004. Whether salmon lice was a decisive factor for the stock decline in the 1980s is debated, but the current increase has occurred when an intense focus on salmon lice has greatly strengthened both government and fish farmers' awareness of the possible detrimental effects of lice from aquaculture on wild salmonids. The mean intensity of adult female lice farms in farms in spring is now kept close to zero by strategic winter treatments. Both in 2002 and 2003 trawling in most regions has shown mean chalimus numbers well below the assumed threshold of 10 lice/smolt. The occurrence of prematurely returned sea trout is now less frequent, occurring only in a few fjords, again indicating a lower infection pressure. Thus, by acting on the knowledge that salmon lice from farms add to the pool of lice which harm wild salmonids, the stakeholders have managed to envisage a future where wild and farmed salmon can coexist.

Important references
The book by Shearer (1992) gives a detailed account of the Atlantic salmon, and the collection of papers (Mills 2003) gives the status of knowledge as to the many different factors influencing salmon survival in the North Atlantic. Infections of high seas Atlantic salmon are described by Jacobsen and Gaard (1997). Pike and Wadsworth (1999) review sea lice biology and control in salmon farms. Heuch and Mo (2001) presented a model of salmon louse production in Norway, with reference to effects of increasing salmon production and public management measures. Johnson *et al.* (1996) discussed disease caused by the sea louse in wild sockeye salmon in British Columbia, and Tully and Nolan (2002) reviewed the host–parasite interactions of the salmon louse. Johnson and Albright (1991a) first described the developmental stages of *Lepeophtheirus salmonis*.

Effects in finfish culture
Kazuo Ogawa

Introduction
Among organisms used in marine aquaculture, finfish rank first. A very wide range of species are cultured, and the increase in production is accelerating. Whereas finfish aquaculture is well advanced in countries like Japan, it is just beginning to become economically important in many developing countries. As a consequence of culture conditions, parasites are responsible for widespread disease and mortality.

Finfish culture of the world
According to FAO statistics, the culture production of marine finfish has been growing more than threefold every 10 years during the last 30 years, amounting to 2.46 million tonnes, worth US$7.38 billion in 2002 (FAO Fisheries Department 2002). Particularly, the production of salmonid fishes (Atlantic salmon, *Salmo salar*, rainbow trout, *Oncorhynchus mykiss*, coho salmon, *Oncorhynchus kisutch* and chinook salmon, *Oncorhynchus tshawytscha*) has greatly increased since the 1980s, to a total of 1.41 million tonnes (57.4% of total global production), worth US$3.67 billion (51.9%) in 2002. Salmonid fishes are mainly cultured in the Northern hemisphere, but in recent years salmonid culture in Chile has been rapidly expanding.

In Asia, *Seriola* spp. (yellowtail, *S. quinqueradiata*, amberjack, *S. dumerili* and golden amberjack, *S. lalandi*) as well as red seabream, *Pagrus major*, are the most intensively cultured species, both cultured almost exclusively in Japan with an annual production of 163 000 million tonnes and 73 000 million tonnes, respectively. In Europe, gilthead seabream, *Sparus aurata*, (62 000 million tonnes) and European seabass, *Dicentrarchus labrax* (39 000 million tonnes) are important aquaculture species apart from salmonids. Other fish species with an annual production of more

Figure 10.2 Culture systems of marine fish. **A**. Net cages of different shape in Japan. **B**. Land-based tank in China. Photos by Mr Zhoujin Huang Jie.

than 3000 tonnes are: olive flounder, *Paralichthys olivaceus*, tiger puffer, *Takifugu rubripes*, and black rock fish, *Sebastes schlegeli*, in Eastern Asia, groupers, *Epinephelus* spp., barramundi, *Lates calcarifer*, and milk fish, *Chanos chanos*, in south-east Asia, southern bluefin tuna, *Thunnus maccoyii*, in Australia, and turbot, *Scophthalmus maximus*, in Europe (Table 10.1).

Occurrence of parasitic diseases among cultured marine fish largely depends on the species cultured, the types of culture systems, types of seeds (artificial or wild) and feeds (fresh fish or pellets), environment and culture management.

There are several different types of culture systems. Most typically marine finfish are cultured in round or square floating net cages (Fig. 10.2A), set in coastal or offshore areas, their size ranging from less than 40 m in perimeter for small fish to 40 m in diameter, or more than 200 m in perimeter for tuna. Turbot and olive flounder, and occasionally tiger puffer and salmonids, are cultured in land-based tanks using pumped sea water (Fig. 10.2B), whereas milk fish are cultured in mud-based ponds with introduced tidal water.

With regard to seeds used for the culture of marine fish, the culture of salmonids, seabreams (gilthead seabream in Europe and red seabream in Japan), turbot, flounder and seabass is based on the use of artificially spawned seeds, whereas wild seeds are used for *Seriola* spp. in Japan and groupers in south-east Asia. Wild yellowtail and amberjack seeds are generally abundant and cheaper than artificially produced ones, and artificial seed production of these fishes is still at an experimental stage. Besides, artificial seeds often lead to abnormalities like deformed vertebrae in red seabream and pigment anomalies in olive flounder. Despite these problems, they are superior to wild ones as modern techniques have made it possible to produce seeds with practically no serious pathogens or parasites. Wild seeds, in contrast, are always potential carriers of pathogens.

Locally available fresh fish were traditionally used in finfish culture as feeds, but are used much less frequently. Feeding fresh fish was responsible for infections with some helminth parasites like the cestode *Callotetrarhynchus nipponica* in yellowtail (see Ogawa 1996). However, in many culture practices, the feeding of fresh fish has been replaced by fish that has been deep frozen, followed by moist or dry pellets. Thus, foodborne parasite infections now represent minor problems.

Parasitic diseases of salmonid fishes

Neoparamoeba pemaquidensis is the causative agent of Amoebic Gill Disease (AGD) of net-pen cultured Atlantic salmon and rainbow trout in Australia, United States of America (USA), Chile and European countries; coho salmon in net pens and land-based tanks in North America;

Table 10.1 List of important parasites of cultured marine fish[A]

Host		Parasite
Salmoniformes		
Atlantic salmon *Salmo salar*		
	F	*Neoparamoeba pemaquidensis*
	F	*Ichthyobodo necator*
	C	*Spironucleus barkhanus*
	My	*Kudoa thyrsites*
	Cr	*Caligus elongatus*
	Cr	*Caligus flexispina* (=*C. rogercresseyi*)
	Cr	*Lepeophtheirus salmonis*[B]
	Cr	*Ceratothoa gaudichaudii*
	?	Rosette agent
Rainbow trout, *Oncorhynchus mykiss*, coho salmon, *Oncorhynchus kisutch*, chinook salmon, *Oncorhynchus tshawytscha*		
	F	*Neoparamoeba pemaquidensis*
	F	*Ichthyobodo necator*
	C	*Spironucleus barkhanus*
	Mi	*Loma salmonae*
	My	*Kudoa thyrsites*
	Cr	*Caligus elongatus*
	Cr	*Caligus flexispina* (=*C. rogercresseyi*)
	Cr	*Caligus teres*
	Cr	*Lepeophtheirus salmonis*
	Cr	*Ceratothoa gaudichaudii*
Perciformes		
Barramundi, *Lates calcarifer*		
	C	*Cryptocaryon irritans*
	M	*Neobenedenia melleni*
	T	*Cruoricola lates*
	Cr	*Caligus epedimicus*
Groupers, *Epinephelus* spp.		
	C	*Cryptocaryon irritans*
	C	*Trichodina* spp.
	M	*Pseudorhabdosynochus* spp.
	M	*Benedenia epinepheli*
	M	*Neobenedenia girellae*
Cobia, *Rachycentron canadum*		
	My	*Sphaerospora*-like myxozoan
	M	*Neobenedenia* sp. (=*N. girellae*[C])
Yellowtail, *Seriola quinqueradiata*, amberjack, *Seriola dumerili*		
	Mi	*Microsporidium seriolae*
	My	*Myxobolus acanthogobii* (syn. *Myxobolus buri*)
	My	*Kudoa amamiensis*
	M	*Benedenia seriolae*
	M	*Neobenedenia girellae*
	M	*Heteraxine heterocerca*
	M	*Zeuxapta japonica*
	T	*Paradeontacylix grandispinus*
	T	*Paradeontacylix kampachi*

Table 10.1 List of important parasites of cultured marine fish^A (Continued)

Host		Parasite
Red seabream, *Pagrus major*		
	C	*Cryptocaryon irritans*
	M	*Bivagina tai*
	N	*Philometra lateolabrasis*
	A	*Longicollum pagrosomi*
Gilthead seabream, *Sparus aurata*		
	F	*Amyloodinium ocellatum*
	C	*Cryptocaryon irritans*
	My	*Enteromyxum leei* (syn. *Myxidium leei*)
European seabass, *Dicentrarchus labrax*		
	F	*Neoparamoeba pemaquidensis*
	C	*Philasterides dicentrarchi*
	C	*Cryptocaryon irritans*
Southern bluefin tuna, *Thunnus maccoyi*		
	C	*Uronema nigricans*
	T	*Cardicola forsteri*
Scorpaeniformes		
Black rock fish, *Sebastes schlegeli*		
	F	*Ichthyobodo* sp.
	M	*Microcotyle sebastis*
Tetraodontiformes		
Tiger puffer, *Takifugu rubripes*		
	F	*Ichthyobodo* sp.
	C	*Cryptocaryon irritans*
	My	*Enteromyxum leei*
	My	*Leptotheca fugu*
	M	*Neobenedenia girellae*
	M	*Heterobothrium okamotoi*
	Cr	*Pseudocaligus fugu* with *Udonella* sp.
Pleuronectiformes		
Olive flounder, *Paralichthys olivaceus*		
	F	*Ichthyobodo* sp.
	C	*Uronema marinum*
	C	*Miamiensis avidus*
	C	*Scuticociliatida* gen. sp.
	C	*Cryptocaryon irritans*
	M	*Neobenedenia girellae*
	M	*Neoheterobothrium hirame*
Turbot, *Scophthalmus maximus*		
	F	*Neoparamoeba pemaquidensis*
	C	*Philasterides dicentrarchi*
	Mi	*Tetramicra brevifilum*
	My	*Enteromyxum scophthalmi*
Gonorynchiformes		
Milk fish, *Chanos chanos*		
	Cr	*Caligus epedimicus*
	Cr	*Caligus orientalis*
A: Acanthocephala, C: Ciliata, Cr: Crustacea, F: Flagellata, M: Monogenea, Mi: Microsporidia, My: Myxozoa, T: Trematoda, ?: Taxonomic status unspecified (class Mesomycetozoea?) [A] Fish with annual production of more than 3000 tonnes in 2001 are listed. [B] See the previous section of this chapter. [C] Unpublished data (see text p. 387).		

chinook salmon in New Zealand and net-pen cultured brown trout in France (Munday et al. 2001). Atlantic salmon are apparently most susceptible to AGD among these salmonid fishes. This amoeba is a facultative pathogen that is usually free-living in seawater. The causative agent of AGD may not be a single species, but an assemblage of similar species (Kent 2000). Infected fish show excessive mucous secretion, hyperplasia of the gill epithelium and fusion and oedema of the gill lamellae (Fig. 10.3A). Outbreaks of AGD among salmonid fishes in seawater occur at water temperatures of 8.9°C or above (Douglas-Helders et al. 2001). In relation to salinity, all reported long-term infections occurred in sea water with high salinity (32 parts per thousand – ppt; g of solute per kg of sea water) (Munday et al. 2001), but the disease can occur at a minimum salinity of 7.2 ppt (Clark and Nowak 1999). Chemotherapy is not used commercially and freshwater bathing of diseased fish for two to six hours is the only effective treatment available for the farmers. Freshwater bathing may not eradicate the amoebae from the gills and provides only temporary relief for the fish (Munday et al. 2001, Gross et al. 2004).

The flagellate *Ichthyobodo necator* in freshwater can survive and multiply on anadromous salmonid in sea water, and seawater-adapted *I. necator* can survive when fish are transferred to fresh water (Urawa 1998). Mass mortality occurred among Atlantic salmon smolts in Scotland immediately after transfer to sea cages due to heavy infection on the gills by *I. necator* which had probably survived on fishes transferred from fresh water (Urawa 1998).

The flagellate *Spironucleus barkhanus*, first identified as *Hexamita* sp. (Poppe et al. 1992), infects all internal organs of cage-cultured Atlantic salmon in Norway. The parasite forms whitish nodules in the liver and kidney, causing coagulative and caseous necrosis. Arctic charr, *Salvelinus alpinus*, were found infected with this flagellate in the gall bladder and intestine. The *S. barkhanus* infection of Atlantic salmon may have resulted from transmission of flagellates from feral Arctic charr to the salmon (Sterud et al. 1998).

'Rosette agent', an unclassified protist, is an intracellular parasite of macrophages of chinook salmon cultured in net pens inthe USA. This protist may be assigned to the recently proposed class Mesomycetozoea (phylum Choanozoa), evolutionarily positioned between fungi and animals (Mendoza et al. 2002). Up to 90% mortality due to rosette agent infections were recorded in chinook salmon during their second summer in the sea. The life cycle is unknown, and it can also infect Atlantic salmon in net pens in Canada (Kent 2000).

The microsporidian *Loma salmonae* forms xenomas in the gills and some other organs of salmonids of the genus *Oncorhynchus*, leading to mortality in heavily infected fish. Infection primarily occurs in fresh water, but can persist in sea water. The parasite transmits from fish to fish in sea water, where wild salmonids may be the source of infection. In the Pacific North-West, *L. salmonae* is an important pathogen in the net-pen reared coho and chinook salmon (Kent 2000).

Kudoa thyrsites is a myxosporean with low host specificity and worldwide distribution, infecting the muscle of many marine fish. There is a report that *K. thyrsites* caused mortalities to net-pen cultured Atlantic salmon smolts. However, it is better known as the causative agent of soft flesh in Atlantic and coho salmon, but interestingly never in chinook salmon (Kent 2000). As high as 60% to 70% prevalence of infection occurred in Atlantic salmon after transfer to sea water. An infectious, possibly extrasporogonic, stage was found in the blood of coho salmon, but its whole life cycle has not yet been elucidated.

The parasitic copepods *Lepeophtheirus salmonis* (salmon lice) and *Caligus elongatus* are the major pathogens of cultured salmonids. In Chile, where no salmonids occur naturally, parasites infecting salmonids are those which are indigenous and low in host specificity. *Caligus teres* infects rainbow trout and *Caligus flexispina* (=*C. rogercresseyi*) infects rainbow trout, coho salmon and Atlantic salmon, cultured in net-pens in sea water (Carvajal et al. 1998). *Ceratothoa gaudichaudii*, an isopod parasite in the mouth cavity of a wide variety of local marine fish, also

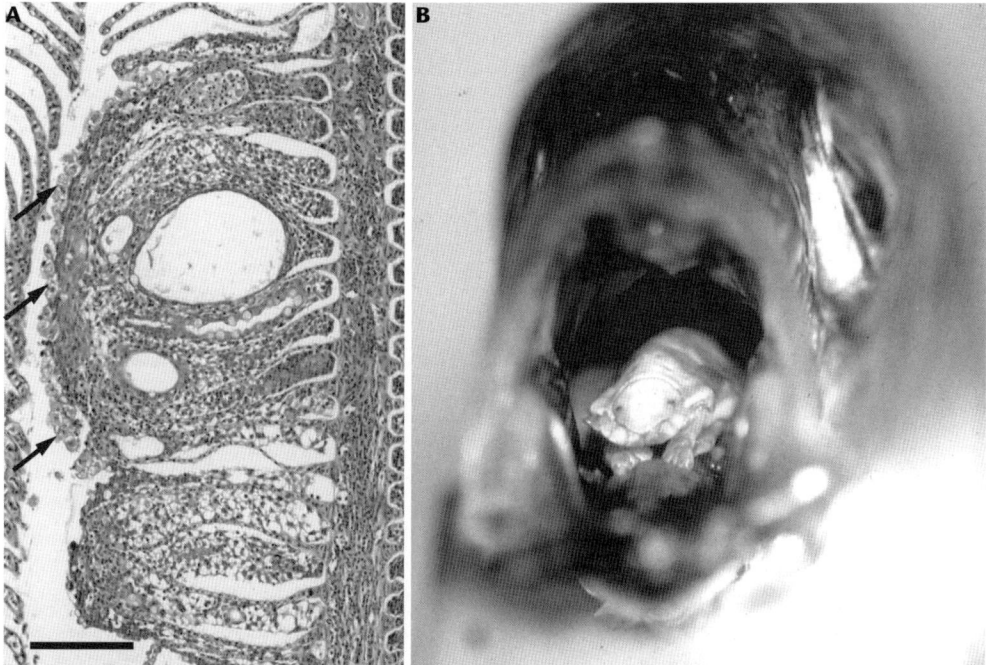

Figure 10.3 Parasitic diseases of Atlantic salmon, *Salmo salar*, and Pacific salmon, *Oncorhynchus* spp. **A**. Amoebic gill disease of Atlantic salmon caused by *Neoparamoeba pemaquidensis* (arrows), with clinical signs of hyperplasia of the gill epithelium, fusion and oedema of the gill lamellae. Scale: 100 μm. Photo by Dr Mark Adams. **B**. Isopod *Ceratothoa gaudichaudii* in the mouth of Chilean jack mackerel, *Trachurus murphyi*. Photo by Dr Juan Carvajal. This parasite also infects cultured coho salmon in Chile.

infects coho and Atlantic salmon in Chile (Fig. 10.3B), causing severe erosion of gill lamellae and ulcers on the gill arch and inside the mouth (Kent 2000). Heavy infection affects the growth of host fish (Horton and Okamura 2001). Chemotherapy, that is bathing of infected fish in organophosphate solution, was found to be effective (Kent 2000).

Parasitic diseases of perciform fishes

With a relatively long culture history, many parasitic diseases have been reported from yellowtail and amberjack cultured in Japan (Ogawa 1996, Ogawa and Yokoyama 1998).

The microsporidian *Microsporidium seriolae* infects juvenile yellowtail, forming irregular-shaped cysts in the skeletal muscle. This infection is not lethal except in heavily infected fish. However, the skeletal muscle degenerates and infected fish show a grossly irregular body surface. Its life cycle is unknown and no control measure has been developed.

Several myxosporeans infect yellowtail and amberjack. Among them, *Myxobolus acanthogobii* (syn. *Myxobolus buri*; Yokoyama *et al*. 2004) and *Kudoa amamiensis* infections are particularly important. *Myxobolus acanthogobii* is the causative agent of scoliosis of yellowtail (Fig. 10.4A). Developing cysts are observed in the brain cavity in summer and settle on the brain surface later in autumn. Cyst aggregates on the fourth ventricle possibly damage the brain mechanically, causing the curvature of vertebrae. Once deformed, fish do not recover. *Kudoa amamiensis* forms visible cysts in the skeletal muscle in yellowtail and amberjack that are cultured in the Amami and Okinawa districts of southern Japan. Infection only occurs in the limited areas of these districts, possibly reflecting the distribution of an alternate host of the

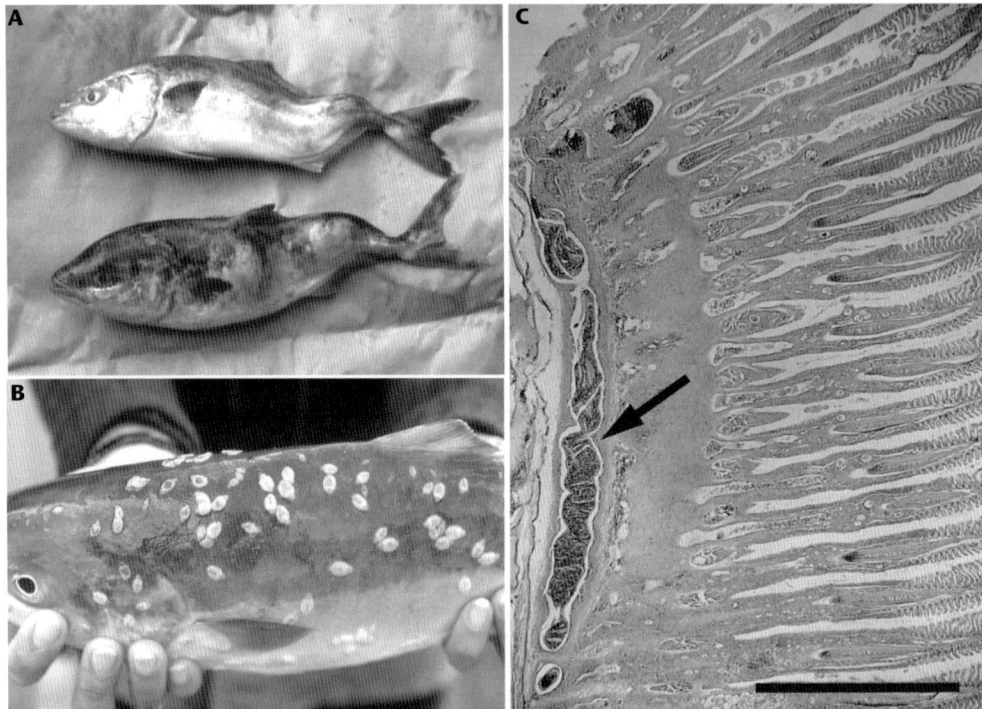

Figure 10.4 Parasitic diseases of yellowtail, *Seriola quinqueradiata*, and amberjack, *Seriola dumerili*. **A**. *Myxobolus acanthogobii* infection in the brain of yellowtail, causing scoliosis. **B**. *Benedenia seriolae* infection on the body surface of yellowtail, treated with freshwater bath. **C**. *Paradeontacylix kampachi* (arrow) in the afferent branchial artery of amberjack. Scale: 1 mm.

myxosporean, though the life cycle has not yet been elucidated (Sugiyama *et al.* 1999). No such infection is known among these fish cultured in the other areas of Japan. Fish with visible cysts in the musculature have no commercial value.

The capsalid monogenean *Benedenia seriolae* infects the body surface of yellowtail and amberjack, grazing on the epidermal tissue. Infected fish show severe erosion and haemorrhages of the skin, and the lesions deteriorate when infected fish rub their body against the culture nets in an attempt to remove the parasites. The parasite's eggs become entangled on the net meshing by a long filamentous appendage. Regular change of culture nets to remove attached eggs has limited efficacy because the eggs can hatch within one week during summer. Moreover, net change is not practical for metal or large cages. Freshwater bathing of infected fish for up to 10 minutes can eradicate the attached parasites (Fig. 10.4B), but it can be difficult to treat net cages far from a freshwater source. Instead, hydrogen peroxide bath and oral administration of praziquantel are officially approved treatments in Japan.

Amberjack seeds imported from China were infected with another capsalid monogenean *Neobenedenia girellae* (=*N. melleni*?), which until then had been unreported from Japanese fish. *N. girellae* has a comparable pathogenicity to *B. seriolae*, with a preference for higher water temperature (>15°C). Since the first confirmation of this parasite in Japanese waters in 1991, *N. girellae*, which shows no host specificity, is a constant threat not only to amberjack but also to tiger puffer and other marine fish.

The polyopisthocotylean *Heteraxine heterocerca* infects the gills of yellowtail, while *Zeuxapta japonica* infects amberjack and golden amberjack. Eggs of these parasites are deposited in a mass and entangle the culture mesh. Similar to *B. seriolae* and *N. girellae* infections, these monogeneans easily propagate within the net cages. Heavy infection causes anemia, but no approved chemotherapeutical agent has been developed.

The blood flukes *Paradeontacylix grandispinus* and *Paradeontacylix kampachi* infect the gill blood vessels of amberjack (Fig. 10.4C), though a recent report suggests their preferred site of infection is the 'girdle' (Montero *et al.* 2003). Deposited eggs accumulate in the afferent branchial arteries, occluding the blood vessels. The life cycle of these trematodes is still unknown, but field experiments indicate that the cercarial invasion into amberjack starts in September. Subsequently, mortality persists throughout the low water temperature period from December until April. The infection was first confirmed among amberjack many years before amberjack seeds started to be shipped to Japan. However, recently, heavy mortality has occurred in May to June among 0-year fish imported from China. There is evidence that some of the imported amberjack had already been infected with the blood flukes before arriving to Japanese waters.

Red seabream *Pagrus major* is intensively cultured in net pens in Japan. The most devastating parasitic infection is white spot disease caused by *Cryptocaryon irritans*. This ciliate infection is well known among fish maintained in aquaria and tanks. However, severe outbreaks have recently been recorded in net-cage cultured seabream. The optimal temperature for the parasite propagation is known to be 25°C or higher, but the outbreaks also occur in colder water temperatures (from the latter half of September to the first half of October). In this period, as the temperature of surface water decreases and exchange of water between surface and bottom of the sea occurs. This exchange of water might induce massive hatching of dormant tomites from the cysts on the bottom and that theronts are carried to the upper water column, where red seabream are cultured (Yoshinaga 2001). Frequent typhoons at this time of year may also contribute to this water mixing. However, this rarely happens in summer; the water is static and no such exchange of water occurs, even though the temperature is optimal for the ciliate. Early diagnosis and temporary evacuation of net cages to an uncontaminated area is the only method to avoid mass mortality.

Bivagina tai is a microcotylid monogenean infecting the gills. Heavy infection is more common in winter than in summer (Ogawa and Yokoyama 1998). The optimal temperature for the parasite is estimated to be around 20°C, but it can propagate in winter when the young red seabream are more stressed due to low temperature.

The nematode *Philometra lateolabrasis* in the gonad and the acanthocephalan *Longicollum pagrosomi* in the rectum sometimes occur in large numbers (Ogawa and Yokoyama 1998). The nematode and the acanthocephalan might have negative effects on the spawning and growth, respectively, of the host, but their pathogenicity remains to be studied in detail.

In Europe, gilthead seabream *Sparus aurata* has several important parasites, most of which have a low host specificity. The dinoflagellate *Amyloodinium ocellatum* and the ciliate *Cryptocaryon irritans* are probably widespread in temperate to tropical waters (Dickerson and Dawe 1995, Noga and Levy 1995). When gilthead seabream are cultured in tanks, these protozoans can propagate rapidly and mortalities are high. Attachment of *A. ocellatum* trophozoites causes necrosis of host epidermal cells. Early detection of the parasites, removal of heavily infected fish and regular cleaning of culture tanks to remove their cysts on the bottom are basic culture management techniques to keep the infection at a low level. Other control measures, such as a long-term bath in a copper solution or a short-term bath in hydrogen peroxide of *A. ocellatum* infected fish, and a long-term hyposalinity treatment for the control of *C. irritans* infection are also effective. The flagellate *Ichthyobodo* sp. also shows a low host specificity and is especially dangerous to fish during their early life stages. High mortalities of gilthead seabream larvae occurred in a hatchery in Spain due to heavy infection of *Ichthyobodo* sp. on the skin (Urawa 1998).

Recently, a heavy infection of the myxosporean *Enteromyxum leei*, originally described as *Myxidium leei* by Diamant et al. (1994) but later assigned to the newly proposed genus by Palenzuela et al. (2002), in the intestinal mucosa is a serious problem, causing emaciation to the host. *Enteromyxum leei* has been recorded from about 30 species of fish, mostly aquarium fish, but also in net-cage culture, several sparid fishes including gilthead seabream are hosts for this parasite. Interestingly, this myxosporean, most probably as presporogonic stages, can transmit from fish to fish, but it is unclear if the life cycle of this parasite also involves yet unspecified invertebrates as the alternate host (Diamant 1997). Recently, this parasite was found to infect tiger puffer and several other cultured fish in Japan. No effective control measure has been developed.

The isopod *Ceratothoa oestroides* infects the branchial and buccal cavity of cage-cultured gilthead seabream fingerlings, with 10% to 20% mortality and a 20% reduction in growth rate (Horton and Okamura 2001).

There is a report of larval European seabass infected with *Neoparamoeba pemaquidensis*, the causative agent of AGD, but no information is available on the infection of larger seabass in tank and net-cage culture (Munday et al. 2001).

The ciliate *Cryptocaryon irritans*, as with many other cultured marine fish, is an important parasite for European seabass cultured in tanks. Seabass are also susceptible to other ciliate parasites and have suffered mortalities, showing symptoms of a swollen urogenito-anal papilla, severe congestion in major visceral organs and body fluid tinged with blood. Major organs and blood contained numerous ciliates, identified as *Philasterides dicentrarchi*, the apparent causative agent of the mortality. This scuticociliatid ciliate is considered a euryhaline and opportunistic parasite. Experimental infection was unsuccessful and the route of contamination was not clear (Dragesco et al. 1995). Song and Wilbert (2000) suggested possible synonymy of *P. dicentrarchi* with *Miamiensis avidus*. Identification of this ciliate needs further study.

The isopod *Ceratothoa oestroides* has been recorded in net-pen cultured European seabass from different localities of the Mediterranean. Prevalences of infection varied dramatically between cages, possibly due to genetic differences of the broodstook in susceptibility to infection (Horton and Okamura 2001). In general the cymothoid infection negatively affects the growth of seabass, and can induce mortality, particularly in younger fish.

Culture of southern bluefin tuna has a relatively short history; its production was first recorded in the FAO statistics in 1992. In Australia, young fish, 15 kg to 35 kg in body weight, are caught in the wild and fed in cages for about three months prior to harvest. Munday et al. (1997) reported mortality of captive southern bluefin tuna as as result of the ciliate *Uronema nigricans* infection in the brain. The invasion route to the brain was probably the olfactory rosette, and possible immunosuppression of the diseased tuna due to oxidised feed might be responsible for the invasion of this facultative pathogen into the brain.

No mortality has yet been recorded, but the blood fluke *Cardicola forsteri* infection in the gills and heart is worth noting (Colquitt et al. 2001). No severe infection may indicate that the life cycle is completed outside the culture system, and this is in clear contrast with the *Paradeontacylix* infection of amberjack in Japan. Similar blood fluke infection is also noticed among cultured northern bluefin tuna *Thunnus thynnus* in Japan (K. Ogawa, unpublished data). Compared with southern bluefin tuna, northern bluefin tuna in Japan are cultured for a longer period, usually for two years or longer. This culture practice could increase the risk of the completion of the parasite life cycle within the culture sites.

When young barramundi *Lates calcarifer* are cultured in tanks, the dinoflagellate *Amyloodinium ocellatum* and the ciliate *Cryptocaryon irritans* can be serious pathogens, as described for gilthead seabream.

Mass mortality occurred in net-pen cultured barramundi due to *Neobenedenia melleni* infection in Australia (Deveney *et al.* 2001). Since the monogenean was unknown in Australian waters at the time of this outbreak, the origin of the parasite was not specified.

The blood fluke *Cruoricola lates* infects visceral blood vessels of barramundi cultured in Malaysia (Herbert *et al.* 1995). No mortality has been recorded, but haemorrhage in the gills upon the release of miracidia and encapsulation of parasite eggs in the visceral organs were observed.

Ho (2000) listed 10 species of sea lice of the genera *Caligus* and *Lepeophtheirus* responsible for mortality in aquaculture in Asia. Among them, *Caligus epidemicus* is considered the most serious, causing mortality to more than 10 marine fish species cultured in Asia, including barramundi. Grouper *Epinephelus malabaricus* experimentally infected with this sea lice died without any significant signs of disease. *Caligus epidemicus* has been found in plankton samples, indicating that it is capable of dispersal without 'hitch hiking' on wild fish.

Many parasites have been recorded from cultured groupers *Epinephelus* spp. and *Plectropomus* spp. (Leong 2000). The ciliates *Cryptocaryon irritans* and *Trichodina* spp. are highly pathogenic to young groupers during the first week after stocking in cages.

Capsalid monogeneans *Benedenia epingheli*, *Benedenia lutjani*, *Neobenedenia girellae* and *Megalocotyloides epingheli* are often responsible for mortality of groupers (Leong 2000). Among them, *Epinephelus coioides* and *E. lanceolatus* are highly susceptible to capsalid infections. Infections with *Pseudorhabdosynochus* spp. and *Diplectanum grouperi* on the gills may lead to secondary bacterial infections.

Cobia *Rachycentron canadum* is a relatively new species for aquaculture, its production being first recorded in the FAO statistics in 1995. It is cultured almost exclusively in net cages in Taiwan. Mass mortality of cobia was associated with a *Sphaerospora*-like myxosporean infection in the kidney (Chen *et al.* 2001). Probable extrasporogonic stages were found in the blood, and presporogonic and sporogonic stages in the kidney, which was extremely enlarged by the infection. Sporogony was observed in the lumen of renal tubules, which were completely occluded by sporogonic pseudoplasmodia at differing degrees of maturity.

Infection of cobia with the monogenean *Neobenedenia* sp. (=*N. girellae*; C-F Lo, G-H Kou and K Ogawa, unpublished data 2003) on the body surface is associated with dark skin, haemorrhaging and ulcers on the head (Lopez *et al.* 2002). More importantly, this capsalid tends to concentrate on the head, including the eyes, causing blindness of the host fish (C-F Lo, G-H Kou and K Ogawa, unpublished data).

Parasitic diseases of flat fishes

Olive flounder *Paralichthys olivaceus* is cultured mainly in land-based tanks, and less frequently in net cages, in Japan and more recently in intensive culture in Korea.

The flagellate *Ichthyobodo* sp., which is morphologically indistinguishable from freshwater *I. necator*, infects the dorsal body surface of olive flounder cultured in tanks (Urawa 1998). The parasite is responsible for mortalities of both brood stock and juvenile flounder reared in tanks. In the latter case, the parasite density reached 30 000 individuals per square millimetre.

In Japan, the ciliates *Cryptocaryon irritans* and an unidentified scuticociliate are both serious parasites for tank-cultured flounder (Ogawa and Yokoyama 1998). The latter ciliate was reported to be the causative agent of the systemic ciliate infection of young flounder. It was found not only on the external surface of the body and gills, but also in the skin (Fig. 10.5A) and, most typically, in the brain. This ciliate is considered a facultative pathogen, since it can be easily propagated in several media. No effective method has been developed to eradicate ciliates in the brain. Since then, a similar disease has been reported from olive flounder cultured in China and Korea, and the causative agent was identified as *Miamiensis avidus* and *Uronema marinum*, respectively (Song and Wilbert 2000, Jee et al. 2001). It remains to be studied whether these ciliates from the three countries are different from each other.

The monogeneans *Neobenedenia girellae* and *Neoheterobothrium hirame* cause problems both in tanks and in net cages (Ogawa and Yokoyama 1998). The capsalid *N. girellae* is recorded from many other cultured fish, but olive flounder appears to be very susceptible to infection. The diclidophorid *N. hirame* was first recorded from wild flounder from the Sea of Japan in the mid 1990s. Subsequently the infection spread to cultured flounder (Ogawa 2002). In most cases, pumped water contaminated with the eggs and/or oncomiracidia was suspected to be the source of infection. Immature worms attach to the gills, gill rakers and arches, whereas in adults, the posterior half of body is embedded in the host tissue, and worms are found in the buccal cavity wall. Immature worms can be eradicated by bathing infected fish in NaCl-added sea water. Effective control in tanks uses a combination of the bathing treatment and increasing water exchange rate, which helps to wash out eggs before oncomiracidia emerge to reinfect the fish.

Neoparamoeba pemaquidensis, the causative agent of AGD, primarily an important pathogen for salmonid culture in the marine environment, is also known to infect turbot *Scophthalmus maximus* cultured in tanks in Spain (Munday et al. 2001). AGD in turbot was recorded at maximum temperatures ranging from 14°C to 18.8°C. Pathological features are similar to those observed in salmonids, but an interesting difference is that the parasite from turbot has an optimal salinity of 22 ppt, whereas in salmonids it is 32 ppt or higher.

There is a record of mass mortality of turbot fry due to infection with the flagellate *Ichthyobodo* sp. (Urawa 1998).

Outbreaks of mortalities of tank-reared turbot of different sizes occurred in Spain and possibly in Norway as well, caused by the ciliate *Philasterides dicentrarchi*, originally described from European seabass, which can invade most organs and tissues (Iglesias et al. 2001). Moribund fish showed clinical signs of dark skin, abnormal swimming behaviours, and, occasionally, haemorrhagic skin ulcers. Diagnosis can be made by confirming ciliates from lesions or ascitic fluid in the body cavity.

The microsporidian *Tetramicra brevifilum*, which forms xenomas in the skeletal muscle of turbot, was responsible for 11.5% mortality and 50% growth reduction in tank-cultured turbot in Spain (Figueras et al. 1992). Diseased fish showed swelling on different parts of the body, and in heavily infected fish, 'jelly-like' muscles. The origin of infection remains to be clarified, since infection experiments through intraperitoneal injection of spores or bathing in a spore suspension were not successful.

The myxosporean *Enteromyxum scophthalmi* infects the gut epithelium, causing up to 100% mortality in tank-cultured turbot in Spain (Branson et al. 1999). Diseased fish were emaciated, associated with severe enteritis, detachment of gut epithelium and oedema in the gut subepithelial connective tissue (Fig. 10.5B). The causative agent was later described as a new species of the genus *Enteromyxum* (Palenzuela et al. 2002).

Parasitic diseases of other groups of fishes

Tiger puffer *Takifugu rubripes* is cultured in tanks and net cages in Japan. The flagellate *Ichthyobodo* sp. often parasitises net-cage cultured tiger puffer, causing severe erosion and haemorrhage on the body surface and epithelial hyperplasia and lamellar fusion on the gills (Urawa 1998).

When cultured in tanks, white spot disease caused by the ciliate *Cryptocaryon irritans* is responsible for mortalities, as is the case of tank-cultured olive flounder and gilthead seabream. Since 1996, emaciation disease has been widespread among net-cage cultured tiger puffer, with clinical signs of severe emaciation, sunken eyes, bony ridges on the head and a tapered body (Tin Tun et al. 2000). Etiological studies revealed that two species of myxosporeans, *Enteromyxum leei* (formerly described as *Myxidium* sp. or *Myxidium* sp. TP; Yanagida et al. 2004) and

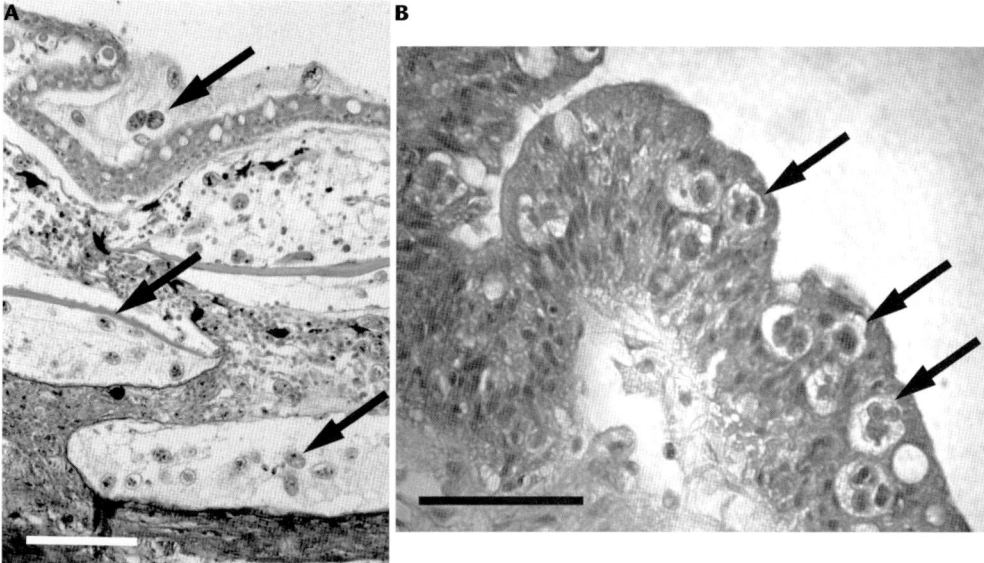

Figure 10.5 Parasitic diseases of olive flounder, *Paralichthys olivaceus*, and turbot, *Scophthalmus maximus*. **A**. Scuticociliatida gen. sp. (arrows) in the skin and muscle tissue of olive flounder. Scale: 200 μm. **B**. Turbot infected with *Enteromyxum scophthalmi* (arrows) in the intestine. Scale: 50 μm. Photo by Mr T Yanagida.

Leptotheca fugu in the intestinal mucosa are involved in the manifestation of these symptoms. Puffers are often infected with both myxosporeans, but a single infection can induce emaciation. Fish-to-fish transmission by *E. leei* was demonstrated experimentally, but it was not successful for *L. fugu*, suggesting possible transmission through an unknown actinosporean stage produced in an unspecified alternate host in the infection cycle of the latter myxosporean. The infection can be lethal but no control measures have been developed.

The diclidophorid monogenean *Heterobothrium okamotoi* is the most important parasite in tiger puffer culture (Ogawa 2002). It grows first on the gills of the host up to 5 mm in length, and then migrates onto the branchial cavity wall for maturation. Eggs in the uterus form a long string or chain through connection of their polar filaments and are deposited in a mass. Highest number of eggs in the uterus of a gravid adult exceeded 1500, forming an egg string longer than 2 m in the water. Almost all egg strings deposited from *H. okamotoi* entangle the net meshing. The parasite is resistant to freshwater bathing of infected fish. The only available chemotherapy is a hydrogen peroxide bath, but this can kill only immature worms on the gills. Frequent change of culture nets is also recommended. The capsalid monogenean *Neobenedenia girellae* is a serious parasite, occurring only in the summer months. Late detection of the parasite can lead to mortality. It is unclear whether the infection cycle of this monogenean involves other fish cultured adjacently to puffers and wild fish around the culture nets.

Pseudocaligus fugu is a common caligid on the body surface (Ogawa and Yokoyama 1998). It can be eradicated with hydrogen peroxide bathing of infected fish. Interestingly, this copepod is sometimes hyperparasitised with the monogenean *Udonella* sp., which appears to be more pathogenic than the host copepod itself (M Freeman and K Ogawa, unpublished data 2004).

Black rock fish is intensively cultured in net cages in Korea (Kim and Cho 2000). The flagellate *Ichthyobodo* sp. often heavily infects rock fish larvae in a hatchery in Hokkaido, Japan (Urawa 1998).

Microcotyle sebastis, the blood-feeding monogenean on the gills of black rock fish, causes high cumulative mortality in summer of net-cage cultured juvenile fish in Korea. Almost all parasites were eradicated after 100 ppm (µg/g) praziquantel bathing of fish for four minutes (Kim and Cho 2000).

Milk fish cultured in coastal brackish ponds in Taiwan is parasitised with the copepod *Caligus epedimicus* and *Caligus orientalis*. Both sea lice, low in host specificity, caused mortality to many fish species, including milk fish (Ho 2000).

Conclusions

Marine finfish culture has been expanding rapidly. Concurrently, several new diseases have been emerging and we have almost no idea about how to predict outbreaks of such diseases. Emerging and re-emerging diseases appear at an almost constant pace in marine aquaculture environments worldwide. It is certain that disease is the most serious problem the aquaculture industry is now facing. Parasitic infections are no exceptions. It is time to stop to think about the severe outcome of these disease outbreaks.

Why so many diseases appear in aquaculture? Fish are cultured in very different environments from those where their wild counterparts are living. Increase in the production needs higher efficiency, leading to higher culture density, which is associated with the deterioration of the culture environment after the prolonged use of the same culture site. Such extraordinary and stressful environments can provide conditions in favour of parasite propagation.

The potential risks of disease outbreaks in relation to the introduction of new and exotic species, newly exploited sites and new methods to marine finfish culture need to be examined. For example, culture of some fish species depends largely on imported seeds. International trading of finfish seeds contributes to a rapid spread of pathogens including parasites to new geographical regions. This also facilitates the pathogens' chances of host switching from imported fish to domestic ones. This type of host switching is certainly one of the causes of emerging disease occurrences among cultured fish. Anthropologic activities such as new culture practices and international fish trading are responsible for the recent increase in disease outbreaks including ones caused by parasites.

Prevention of infection through vaccine administration is becoming a central strategy against viral and bacterial diseases of fish. For example, vaccination of Atlantic salmon against furunculosis, vibriosis and infectious pancreatic necrosis (IPN) was so successful that the occurrences of these diseases have drastically decreased since the late 1980s, with a much decreased use of antibiotics and a steadily increasing salmon production. In contrast, no vaccine has yet been commercially available against any fish parasitic disease, though much effort has been done to develop vaccines against parasites like *Ichthyophthirius multifiliis* and *Lepeophtheirus salmonis*. However, many fish parasites evade the host defence system and fish acquire no or minimal immunity against some parasitic infections. In such cases, development of vaccines will not be an option for the establishment of control measures. Another problem is that it is impractical to keep fish away from pathogens or to disinfect water in net-cage culture systems, in which most marine fish are produced. It is, thus, reasonable to expect that control of parasitic diseases will be a central issue for the stable growth of the marine finfish culture industry. It is really a challenge for fish parasitologists.

Important references

Important reviews on parasite-induced mortalities in farmed fish are by Ogawa (1996) and Ogawa and Yokoyama (1998). Ho (2000) discussed the effects of sea lice in Asian aquaculture, and Noga and Levy (1995), parasitic dinoflagellates as disease agents. More specifically, Dickerson and Dawe (1995) and Urawa (1998) reviewed *Cryptocaryon irritans* and *Ichthyobodo* infection in marine fishes, respectively. Munday *et al.* (2001) gave a general account of amoebic gill disease of maricultured fish.

Effects in mollusc culture
Ryan B Carnegie

Introduction
The aquaculture of molluscan species – primarily bivalves such as oysters, clams, scallops and mussels but also gastropods like abalone – is a large and growing global industry. In 2001, five of the 30 most productive global aquaculture industries, including *the* most productive in biomass terms, produced molluscs. Aquaculture of the top three species alone – the Pacific oyster *Crassostrea gigas*, the Manila clam *Ruditapes philippinarum*, and the Japanese scallop *Patinopecten yessoensis* – was worth over $US7.4 billion or 13.3% of the total value of global aquaculture production that year (FAO 2004).

The impact of diseases, and parasitic diseases in particular, on economically important molluscan species has been profound. In North America, Europe, Asia, and Australia and New Zealand, where parasites of molluscs are most notorious and best studied, their effects have ranged from serious local disruption to complete collapse of fisheries and aquaculture industries at regional and international scales. Not only by causing heavy mortality but also by *threatening* to cause mortality and thereby discouraging economic investment in aquaculture have parasites constrained the growth of molluscan aquaculture industries around the world.

Aquaculture of molluscs, like the culture of other organisms, favours parasitism. First, animals are often maintained in large numbers at high density in sometimes suboptimal environmental conditions, a stressful situation conducive to parasite transmission and disease. Second, animal populations propagated over generations in an aquaculture hatchery or artificially selected for traits such as fast growth may be less diverse genetically than wild populations; these populations may be relatively incapable of responding to parasite selective pressure with evolution of resistance or tolerance to parasitism. Third, aquaculture-related transplantation activities can introduce parasites to naive or susceptible host populations or species in which they can flourish. Indeed, we owe the discovery of numerous parasites to disease outbreaks that followed aquaculture introductions. The haplosporidian protist *Haplosporidium nelsoni* has contributed to the decline of the eastern oyster *Crassostrea virginica* along the mid-Atlantic coast of the United States since 1957 (Haskin and Andrews 1988, Ford and Tripp 1996). This parasite, enzootic at very low prevalence in *Crassostrea gigas* populations in Asia, is believed to have been introduced to the mid-Atlantic prior to 1957 with a transfer of *C. gigas* (Burreson *et al.* 2000). Another haplosporidian, *Bonamia ostreae*, caused the collapse of the European flat oyster, *Ostrea edulis*, aquaculture industry in the years following 1979 (Grizel *et al.* 1988). It was probably introduced to France with oyster seed from California, USA (Elston *et al.* 1986), where it had emerged in a small aquacultured population of the same oyster species over a decade earlier (Katkansky *et al.* 1969).

Dozens of metazoan and protistan parasites infect molluscs, but relatively few ever cause catastrophic disease and mortality, and most are insignificant in economic terms. Comprehensive treatment of the myriad molluscan parasites (as well as other pathogens) can be found in the *Diseases of Marine Animals* volumes (e.g. Kinne 1982), and in Sindermann (1990). The relatively few parasites that have a wide and chronic effect on aquacultured molluscan species will be treated here. These primarily include endoparasitic protists in three groups: *Perkinsus* spp., members of the Phylum Haplosporidia and members of the Phylum Paramyxea. *Perkinsus* spp. notably include *Perkinsus marinus*, which infects the oyster *C. virginica* in eastern North

America, and *Perkinsus olseni* (syn. *atlanticus*; Murrell *et al.* 2002) which infects several Australian abalone species but also the clams *Ruditapes philippinarum* in Korea, Japan, Portugal, Spain, France and Italy and *Ruditapes decussatus* in Europe. Parasites in the Haplosporidia include spore-forming species such as *Haplosporidium nelsoni* and *Haplosporidium costale*, which infect *C. virginica* along the Atlantic coast of North America, and *Minchinia tapetis*, a parasite of the European clam *R. decussatus*; and microcell haplosporidians in the genus *Bonamia*, including *Bonamia roughleyi* from the Sydney rock oyster *Saccostrea glomerata* and *Bonamia ostreae* in *Ostrea edulis*. Parasites in the Paramyxea include *Marteilia* spp. such as *Marteilia refringens*, which parasitises *O. edulis* in Europe, and *Marteilia sydneyi*, which infects *S. glomerata* in Australia; and the *Marteilioides* spp. *M. chungmuensis*, which infects *Crassostrea gigas* in Japan and Korea, and *M. branchialis*, which parasitises *S. glomerata* in Australia. This list may be biased toward the parasites of oysters, but for good reason: the human history of oyster culture can be measured in thousands of years, and humans have spent much of the last 200 moving these species, and their parasites, among continents and oceans, and witnessing the effects.

Parasites of significance for molluscan aquaculture

Parasites of the oyster Saccostrea glomerata *in Australia*

The Sydney rock oyster, *Saccostrea glomerata*, is a large (to 25 cm) oyster that occurs widely from Australia and New Zealand to Thailand and west to Pakistan (Carriker and Gaffney 1996). It occurs most famously near Sydney, Australia, where it has supported an off-bottom, intertidal aquaculture industry since the 1800s. The 'winter mortality' that occurred in *S. glomerata* in New South Wales in 1924 (Roughley 1926) is perhaps the earliest epizootic in aquacultured molluscs in which a parasite can confidently be said to have been the etiological agent. The disease was, and is, characterised by gross focal lesions at the gill surface and in connective and gonadal tissues (Farley *et al.* 1988) in oysters from cooler, higher salinity waters; oyster mortality in winter months can approach 70% (Adlard and Lester 1995). Since 1924, mortality has returned annually to cultured oyster populations in high salinity waters. A small protist parasitising oyster haemocytes (*Mikrocytos roughleyi*; Farley *et al.* 1988) was eventually identified as the cause. The effect of this parasite (since renamed *Bonamia roughleyi*; Cochennec-Laureau *et al.* 2003) on aquacultured oyster populations is serious: mortality is highest in near market-sized oysters, animals in which much, over three years of growth, has been invested. Winter mortality damages in 1992 totalled $A6 million (Adlard and Lester 1995).

Bonamia roughleyi has been relatively little studied. This parasite is significant, however, in being the first member to emerge, by 40 years, of a parasite genus that perhaps is second only to *Perkinsus* spp. in its destruction to molluscan species worldwide. All known congeneric species, including *Bonamia ostreae* and New Zealand dredge oyster parasite *Bonamia exitiosa* (Hine *et al.* 2001), are directly transmissible parasites of oyster haemocytes.

The paramyxean parasite *Marteilia sydneyi* (Fig. 10.6A) (Perkins and Wolf 1976) was identified in 1972 (Wolf 1972) as the cause of summer mortalities ('QX disease') since 1968 in *Saccostrea glomerata* in southern Queensland, Australia. This parasite now occurs southward into major oyster growing regions in New South Wales, causing major disease (parasite proliferation in, and disruption of, the digestive gland, causing starvation) and mortality in warmer estuarine waters throughout its range. From 1974/75 to 2000/01, *M. sydneyi* caused a 56% drop in annual *S. glomerata* production in southern Queensland; and from 1993/94 to 2000/01, a 94% drop in production in the Georges River at Sydney. *Marteilia sydneyi* kills 90% of rock oysters in the Georges River annually (Nell and Hand 2003).

To mitigate the effect of these parasites, the state of New South Wales, Australia, has begun a *Saccostrea glomerata* breeding program to select for tolerance of *M. sydneyi* and *B. roughleyi*

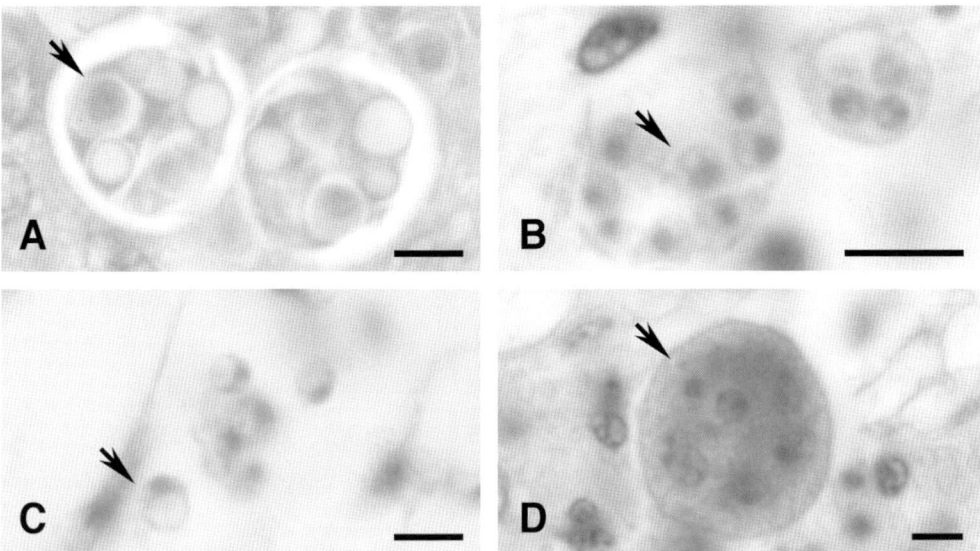

Figure 10.6 Representatives of parasite groups of primary significance in molluscan aquaculture. **A.** *Marteilia sydneyi* infecting *Saccostrea glomerata*. **B.** *Bonamia ostreae* infecting *Ostrea edulis*. **C.** *Perkinsus marinus* infecting *Crassostrea virginica*. **D.** *Haplosporidium nelsoni* infecting *Crassostrea virginica*. Arrows indicate representative parasite cells. Scale bars = 5 μm. Much information on parasites of oysters can be found in one of the four volumes on diseases of marine animals edited by Kinne (1982). Sindermann (1990) contains extensive discussion of parasites of marine fish and shellfish. A synopsis of infectious diseases and parasites of commercially exploited shellfish on a website is by Bower and McGladdery (2001). Carriker and Gaffney (1996) published a catalogue of selected species of living oysters (Ostreacea) of the world. Ford and Tripp (1996) discussed diseases and defence mechanisms of the eastern oyster, *C. virginica*.

parasitism. Initial trials with second generation *M. sydneyi-* and *B. roughleyi-*selected animals have been promising: a 22% reduction in *M. sydneyi*-caused mortality and a 5% reduction in *B. roughleyi*-caused mortality have been observed at disease-enzootic sites in the Georges River (Nell and Hand 2003).

Parasites of the flat oyster Ostrea edulis in Europe

Ostrea edulis occurs in Europe from Norway and Great Britain to Morocco, and east along the north shore of the Mediterranean Sea to Turkey and the Black Sea (Carriker and Gaffney 1996). Aquaculture of this species began during the Roman Empire. Like *Saccostrea glomerata* in Australia, *O. edulis* is seriously impacted by both a *Marteilia* species and a *Bonamia* species. In Europe, *Marteilia refringens* emerged in 1968, causing serious mortality in *O. edulis* in estuaries of Brittany and Marennes-Oléron. Between 1970 and 1977, flat oyster production in France fell by 75% as a result of *M. refringens*-associated mortality (Goulletquer and Héral 1997) as the parasite expanded its range (Fig. 10.7). *Marteilia refringens* now occurs in warmer estuarine waters from France to Greece and in Morocco (OIE 2003).

Bonamia ostreae (Fig. 10.6B) (Pichot *et al.* 1980) was discovered in Brittany, France, in 1979 following importation to Brittany of large amounts of *Ostrea edulis* seed from California, USA. This importation was very likely the source of the parasite in Europe (Elston *et al.* 1986); a similar parasite had been observed in the 1960s (Katkansky *et al.* 1969) in the same estuary at which the hatchery supplying *O. edulis* seed in the 1970s was located. Within a few years and because of unrestricted aquaculture transplantation within Europe, *B. ostreae* was widespread in cool,

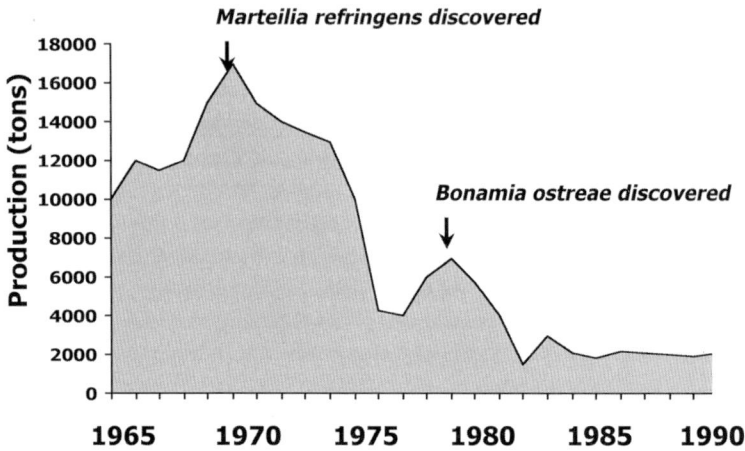

Figure 10.7 Production of *Ostrea edulis* in France, 1965 to 1990, revealing declines in production following the emergence of *Marteilia refringens* and *Bonamia ostreae*.

higher salinity waters from Ireland and England to Spain (Grizel *et al.* 1988), and the aquaculture industry for *O. edulis* had largely collapsed in Atlantic Europe (Fig. 10.7). Like *Bonamia roughleyi*, *B. ostreae* took a particularly heavy toll on older oysters approaching market size.

The Pacific oyster *Crassostrea gigas* was introduced to France in 1972, following the disappearance of the Portuguese oyster *Crassostrea angulata* due to iridoviral disease (Goulletquer and Héral 1997). Culture of *C. gigas*, which is resistant to both *Marteilia refringens* and *Bonamia ostreae*, has now supplanted *O. edulis* aquaculture throughout much of Atlantic Europe and particularly in estuaries where *M. refringens* is most active. Still, aquaculture of *O. edulis* continues on a small scale in higher salinity waters, aided by the use of oysters selected for tolerance to *B. ostreae*. In Ireland, for example, use of survivors of *B. ostreae* epizootics as broodstock has allowed recovery of the flat oyster aquaculture industry in areas such as Cork Harbour, where the parasite is enzootic (Hugh-Jones 1999, Culloty *et al.* 2001). In experimental trials in France, third-generation oysters selected for tolerance to *B. ostreae* also demonstrated higher survival than controls (Naciri-Graven *et al.* 1998).

Parasites of the eastern oyster Crassostrea virginica in North America

The eastern oyster *Crassostrea virginica* occurs in eastern North and South America from Prince Edward Island, Canada, to the coasts of Brazil and Argentina (Carriker and Gaffney 1996). It historically supported phenomenally productive fisheries, most famously in Chesapeake Bay, on the east coast of the USA. The endoparasite *Perkinsus marinus* (Fig. 10.6C) was discovered in this oyster in Louisiana, USA, in the late 1940s (Mackin *et al.* 1950). It was widespread in Chesapeake Bay, on the Atlantic coast, at the time of its discovery there in 1950 (Andrews 1996). *Perkinsus* spp. are now the most widely distributed and destructive parasites of aquacultured molluscs, causing serious disease in numerous host species in higher salinity environments where (and at times that) summer temperatures exceed 20°C for extended periods (OIE 2003). In the initial *P. marinus* epizootic, as much as 100% of oyster seed transplanted from low to high salinity waters in Louisiana died during culture to market size (Ray 1996). The parasite now occurs from Maine to Florida on the Atlantic coast of the USA and in the Gulf of Mexico from Florida to Mexico's Yucatan Peninsula (Ford and Tripp 1996). It has contributed to the decline of wild oyster populations throughout much of this range, and has seriously limited the growth of aquaculture for *C. virginica*. Severe mortality in wild and cultured oysters in Chesapeake Bay

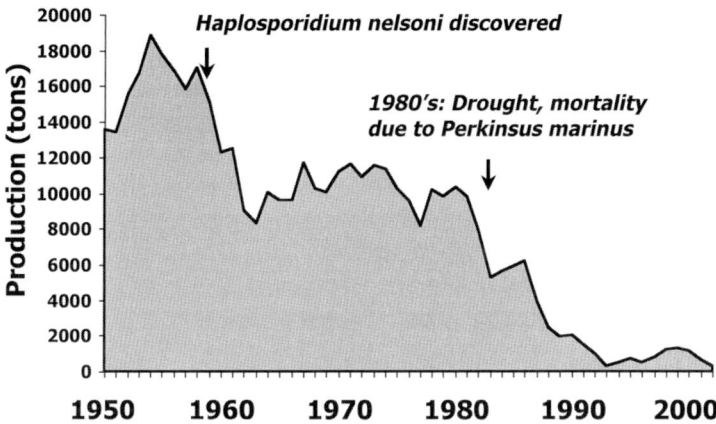

Figure 10.8 Production of *Crassostrea virginica* in Chesapeake Bay, Virginia and Maryland, USA, from 1950 to 2002, revealing the impact of *Haplosporidium nelsoni* parasitism following 1959 and intense *Perkinsus marinus* activity during drought years in the 1980s.

in the 1980s, for example, largely resulted from intense *P. marinus* activity during drought (Fig. 10.8). Oysters acquire *P. marinus* directly from nearby oysters that have succumbed and are deteriorating, and die in particularly large numbers in the second summer as overwintering infections become clinical. Where *P. marinus* has been managed at all, such as in the Gulf of Mexico, aquaculturists have reduced mortality by deploying seed oysters in fall or winter to avoid intense infection pressure in the first summer, and harvesting early in the second summer before mortality sets in (Ray 1996). Toward this end, increasingly common use of triploid *C. virginica* can stack the odds in the oysters' (and aquaculturists') favour: triploid oysters created specifically for aquaculture allocate more energy to growth during gametogenesis and less to gamete production (Longwell and Stiles 1996), and so may be more likely to reach market size before the onset of mortality due to *P. marinus*.

Haplosporidium nelsoni (the agent of Multinucleate Sphere Unknown or MSX disease; Fig. 10.6D) emerged in the eastern USA in *C. virginica* populations in Delaware Bay in 1957 and in Chesapeake Bay in 1959. Mortality caused by this parasite quickly exceeded 90% in oyster planting grounds in these bays (see Fig. 10.8) and has not abated since (Haskin and Andrews 1988, Ford and Tripp 1996). The range of the parasite has expanded such that it now occurs from Nova Scotia, Canada, to Florida, USA. *Haplosporidium nelsoni* is a spore-forming member of the phylum (Haplosporidia) that includes the non-spore-forming *Bonamia* spp. It differs also from *Bonamia* spp. in apparently not being directly transmissible; infection pressure bears no relation to oyster population density (Haskin and Andrews 1988) and the species has never been directly transmitted experimentally. In large part because the life cycle of *H. nelsoni* is unresolved and possible reservoirs in the environment are not known, *H. nelsoni* has been essentially unmanageable in an aquaculture context in higher salinity waters (>15–20 ppt) where it causes disease. In *H. nelsoni*-enzootic areas, peak times of infection acquisition are late spring to early summer (with mortality peaking late summer–early fall) and late summer to early fall (with another mortality peak in late spring to early summer; Haskin and Andrews 1988). In places along the Atlantic coast where the parasite is most active, oysters cannot be brought to market size without passing through one window of infection acquisition or mortality. In waters below 15 ppt salinity, the activity of the parasite is reduced (Haskin and Andrews 1988). As *H. nelsoni* has spread, new mortality events have followed the parasite's introduction to naive oyster populations, most recently in aquacultured *C. virginica* populations in Nova Scotia in 2002. The presence of *H. nelsoni* in *C. virginica* populations very likely has its source in aquaculture introductions of the Pacific oyster *Crassostrea gigas* from Asia. In the middle part of the 20th century, numerous introductions of the Pacific oyster were made to various locations along the Atlantic, Pacific and Gulf coasts of the USA. A haplosporidian similar microscopically and identical in small subunit ribosomal RNA gene sequence to *H. nelsoni* is enzootic at low prevalence in *C. gigas* from Korea, suggesting a long-evolved, stable host–parasite relationship and presumably revealing the source of the *C. virginica* infections (Burreson *et al.* 2000).

Selective breeding of *H. nelsoni*-tolerant oysters has met with some success (Haskin and Ford 1987). Recent efforts to develop tolerance of both *H. nelsoni* and *P. marinus* in a single *C. virginica* line were also promising. Third-generation 'DEBY' selected oysters (i.e. derived from Delaware Bay broodstocks) have shown superior performance in experimental trials in Virginia, with nearly 80% reaching market size in *H. nelsoni* and *P. marinus*-enzootic areas (Ragone Calvo *et al.* 2003). This oyster strain is now in wide use among aquaculturists in the region.

Perkinsus olseni
Perkinsus olseni was discovered in the abalone *Haliotis ruber* in South Australia in the 1970s (Lester and Davis 1981), and is now known to occur in a variety of molluscs from Australia and New Zealand north to Korea and Japan. The European *P. atlanticus* is synonymous (Murrell *et al.* 2002). This species is notable because it is a threat to one of the world's most valuable aquaculture industries: aquaculture of the Manila clam, *Ruditapes philippinarum*, was worth over $US2.4 billion worldwide in 2001. *Perkinsus olseni* causes disease in *R. philippinarum* as well as a relative, *Ruditapes decussatus*, in southern Europe (Portugal, Spain, France and Italy) and Korea and Japan (OIE 2003). Mortality caused by *P. olseni* can be severe: Korean Manila clam landings fell from 70 000 to 14 000 tonnes between 1990 and 1997 following mortality caused partly by *P. olseni* (Park and Choi 2001). The presence of *P. olseni* in Europe may be directly attributed to introductions since 1972 of *R. philippinarum* to Europe for aquaculture (Murrell *et al.* 2002).

Other parasites
Protists in the Phylum Labyrinthulomycota have emerged occasionally as parasites of cultured molluscs. *Labyrinthuloides haliotidis* produced infections and mortality in juvenile abalone, *Haliotis kamtschatkana* and *H. rufescens*, that devastated efforts to culture these species in British Columbia (Bower 1987). 'QPX' (Quahog Parasite Unknown) has occasionally caused catastrophic mortalities in cultured clams (*Mercenaria mercenaria*) in eastern North America (Whyte *et al.* 1994, Smolowitz *et al.* 1998, Ragone Calvo *et al.* 1998).

A *Perkinsus* species, *Perkinsus qugwadi*, was discovered in 1998 in Japanese scallops cultured in British Columbia, Canada (Blackbourn *et al.* 1998, Bower *et al.* 1998). This organism was never observed elsewhere and has not been reported in recent years. Still, it is a concern because of the high mortality it caused (60%) in an aquaculture species worth over US$1.5 billion worldwide (FAO 2004).

Mikrocytos mackini (Farley *et al.* 1988) is a protistan parasite of the Pacific oyster that was first discovered at Denman Island, British Columbia, Canada in 1960 (Quayle 1961). 'Denman Island disease' now occurs locally in British Columbia and adjacent waters in Washington, USA. While mortality in the initial epizootic reached 35%, the disease is of minor significance today. Typically, infection is greatest in older oysters at lower tide levels. Aquaculturists can minimise the effect of this parasite by moving older oysters to higher tide levels before onset of disease in March, and planting young oysters at lower tide levels after July (Bower 1988).

A *Marteilia* species that may well be distinct from the oyster parasite *Marteilia refringens* (Le Roux *et al.* 2001) occurs in the mussel *Mytilus galloprovincialis* in Atlantic and Mediterranean Europe. The threat that this parasite, *Marteilia maurini* (see Comps *et al.* 1982), poses to mussel aquaculture is unclear; mortality due specifically to *Marteilia* sp. activity is difficult to demonstrate, and mass mortalities of mussels are rare. Nonetheless, parasite prevalence and infection intensity can be high (Villalba *et al.* 1993), and the parasite remains a concern because of the high global value of *Mytilus* spp. aquaculture industries.

The paramyxean *Marteilioides chungmuensis* causes millions of dollars in damages annually to the *Crassostrea gigas* aquaculture industry in Japan and Korea (Itoh *et al.* 2003). While causing little or no mortality, the ovarian parasite renders high proportions of aquaculture crops unmarketable by causing large, ugly nodules in the oyster's gonad.

Managing parasitism in molluscan aquaculture

The profound (and profoundly detrimental) impacts parasites can have on molluscan aquaculture have forced aquaculturists, resource managers and governments to pursue broad strategies for managing parasitic disease. At its most basic level, management of parasites is effected by aquaculturists themselves. Whole systems of culture and husbandry have evolved under pressure from parasites. The intertidal, off-bottom growing methods used for the oyster *Saccostrea glomerata* in Australia evolved from a need to manage parasitism by the polychaete *Polydora*, which infests molluscan shells (Bower 2001). Patterns of *Crassostrea gigas* planting on beaches in British Columbia, Canada – larger oysters to the higher intertidal in March, smaller oysters planted in the lower intertidal after July – have grown out of a need to manage Denman Island disease agent *Mikrocytos mackini* (Bower 1988). Patterns of *C. virginica* planting in the Gulf of Mexico – larger seed oysters planted after temperatures begin to fall, and then harvested before the next summer – have evolved from a need to manage *Perkinsus marinus* (Andrews 1988).

Above this most basic level, management of parasites in aquaculture has consisted primarily of identifying and restricting the transplantation of infected stocks. This aspect of parasite or *aquaculture health* management has seen spectacular failures. *Bonamia ostreae* arrived in France, as noted earlier, in a shipment of seed *Ostrea edulis* that originated from the region in California, USA, where the parasite was first observed (Elston *et al.* 1986). And many parasites – including *B. ostreae* and probably *Marteilia sydneyi* (Adlard and Ernst 1995) – have rapidly expanded their local and regional distributions through imprudent aquaculture transplantations over smaller geographic scales. Sensitive detection of parasites – particularly small, endoparasitic protistans – can be very difficult with standard histopathological techniques (i.e. light microscopic analysis of stained tissue sections). The use of modern molecular techniques such as the polymerase chain reaction (PCR) and *in situ* hybridisation, however, promises sensitive detection of much lighter (or cryptic, or prepatent) infections than could previously be observed, thus more sensitive identification of infected stocks. Sensitive and specific molecular assays have been developed for many of the parasites discussed above, including *Haplosporidium nelsoni* (Stokes and Burreson 1995, Stokes *et al.* 1995), *Marteilia refringens* (Le Roux *et al.* 1999), *M. sydneyi* (Kleeman and Adlard 2000), *B. ostreae* (Carnegie *et al.* 2000, 2003, Cochennec *et al.* 2000), Quahog Parasite Unknown (QPX) (Stokes *et al.* 2002), *Perkinsus* spp. (Casas *et al.* 2002), *Mikrocytos mackini* (Carnegie *et al.* 2003) and *Marteilioides chungmuensis* (Itoh *et al.* 2003). Among many other uses, molecular tools and the PCR specifically will also help detect parasites in alternative host species, which must be identified and managed in concert with target aquaculture species if parasite management is to be truly effective; and will allow rapid identification and characterisation of newly emerging parasite species.

Short of transgenic manipulation, which is some years away for molluscan aquaculture and which may face an unreceptive public, selective breeding for parasite tolerance (or disease resistance) is perhaps the most powerful tool for mitigating the impact of parasites on aquacultured mollusc populations. As noted above, selective breeding for parasite tolerance has already been practised with success in Ireland and France (tolerance of *B. ostreae* in *O. edulis*), Australia (tolerance of *Marteilia sydneyi* and *Bonamia roghleyi* in *Saccostrea glomerata*), and the USA (tolerance of *Haplosporidium nelsoni* and *Perkinsus marinus* in *C. virginica*). There are several metrics by which parasite tolerance may be measured – reduced parasite prevalence or intensity, for example, or elevated rates of endoparasite phagocytosis by haemocytes – but in an aquaculture context the two that matter are the number of animals surviving and time to market size. (The case of *M. chungmuensis*, which does not kill but rather renders aesthetically unmarketable, is an exception.) Reducing the rate of parasite-induced mortality may be the difference between profitability and economic collapse at scales from the local to the multinational. Reducing parasite impacts on aquaculture species through selective breeding is clearly possible for many host–parasite systems, but will take time.

Important references
Much information on parasites of oysters can be found in one of the four volumes on diseases of marine animals edited by Kinne (1982). Sindermann (1990) contains extensive discussion of parasites of marine fish and shellfish. A synopsis of infectious diseases and parasites of commercially exploited shellfish on a website is by Bower and McGladdery (2001). Carriker and Gaffney (1996) published a catalogue of selected species of living oysters (Ostreacea) of the world. Ford and Tripp (1996) discussed diseases and defence mechanisms of the eastern oyster, *C. virginica*.

Effects in shrimp culture
Robin M Overstreet

Introduction
A limiting factor-if not presently *the* limiting factor-of shrimp (prawn) aquaculture is disease. Because of the rapid onset and demise of most or all shrimp in individual culture facilities resulting from a few different viral and bacterial diseases, non-microbial parasitic infections get little attention. But these parasites also can kill large numbers of shrimp or seriously influence the economic value of the product. Moreover, the corresponding infections can take a much longer time to develop than viral infections and therefore impose additional costs before loss of production. However, compared with some viral infections, many parasitic infections can be more successfully treated, controlled or managed.

This section treats parasites exclusively in cultured commercial penaeid shrimps. These shrimps form a marine warmwater group, and it includes most of the world's cultured and natural shrimp products. In contrast, most marine coldwater shrimps, pandalid and crangonid carideans, are harvested from the wild. Another caridean group, the palaemonids, are typically cultured in fresh water; most cultured palaemonids support local markets. Penaeid shrimp culture production includes about six major species and 14 minor ones in shrimp farms worldwide in warm water; most production involves *Penaeus monodon* in the eastern hemisphere and *Litopenaeus vannamei* in the western hemisphere, with a recent shift toward growing *L. vannamei* worldwide because it is easier to spawn and to produce a cost-effective supply.

Parasites and shrimp
Some parasites are highly specific, and others will infect a wide range of penaeids. In addition to the species of host shrimp, the type of farming, source of shrimp and origin of culture water have a bearing on the parasitic infections and the effect of the agents on the host. Several different groups of parasites with their different life cycles, life histories, sites of infection, effects on hosts and other biological features produce different biological effects and economic influences on their shrimp hosts. These will be treated below.

Basically three kinds of penaeid farming exist: extensive, semi-intensive and intensive. In extensive farming, wild shrimp are trapped in impoundments of up to 100 ha along bays and tidal rivers. Managers keep enclosure-gates open until wild young shrimp reach a high density and then close them. These trapped shrimp depend on tidal flow, natural food, little management and perhaps an addition of fertiliser; up to 500 kg/ha of product is harvested. Semi-intensive culture incorporates 2 ha to 30 ha ponds above high tide with stocked postlarvae, up to 25% water exchange and some commercial feeds to augment natural food. It produces 500 kg/ha to 5000 kg/ha as opposed to 5000 kg/ha to 20 000 kg/ha of product in intensive culture. The latter are grown in 0.1 ha to 1.5 ha ponds or raceways, sometimes covered or indoors. Postlarvae from hatcheries using either wild or genetically selected cultured broodstock receive commercial feeds, >30% exchange of pumped water and 24 h management with considerable risk. Super-intensive systems produce up to 100 000 kg/ha, with greater risk; experimental closed systems, like ones at our Ocean Springs campus, involve essentially no water exchange or filtration. Shrimps in each type of facility are prone to different diseases and kinds or species of parasites.

Most serious of the parasites are those that reproduce within or on the shrimp host, especially those with no intermediate or additional hosts. They are the most common kinds in hatcheries. Hatcheries remain as the critical vulnerable link in the production cycle. In hatcheries, there is a strong relationship among disease, water quality and management. Infective agents are primarily ciliates but also other protozoans and diatoms.

Protozoans

Of the ciliates there are: colonial-stalked peritrich species of *Zoothamnium*, *Carchesium* and *Epistylis*; single-stalked *Vorticella*; and loricate-covered *Legenophrys* and *Cothurnia* that commonly attach to larvae and postlarvae as do suctorian species of *Acineta* and *Thecacineta*, which have a lorica, and those of *Ephelota* and *Tokophrya*, which do not.

Parasites in hatcheries and semi-closed ponds are often those that flourish in response to changes in water quality. For example, water high in organic detritus is usually conducive to increased growth and reproduction of peritrichs. The larval and early postlarval shrimp become overwhelmed by the weight of the attached ciliates, reducing the ability to maintain position in the water column and feed. Moreover, the ciliates can attach to or attack adjacent shrimp. Careful monitoring of representative or sentinel subsamples of shrimp can reveal early infections, and thereby the amount, kind and quality of feed can be adjusted or either more or different water can be exchanged. This is because the ciliates on the gills of postlarvae or adults, and body surface of larvae or postlarvae, feed not on the host but on the bacteria and organic material. When organic material is abundant it promotes reproduction of the bacteria and ciliates and increases stress on the host (Fig. 10.9A). The older shrimp usually do not moult when stressed, and the increased number of attached ciliates compete with the shrimp for oxygen (Overstreet 1973).

Apostome ciliates also affect cultured and wild shrimp. One species, *Hyalophysa chattoni*, encysts as a phoront on the gills, and neither it nor the produced trophont, which pinocytises exuvial fluid from a moult, elicits an extensive melanistic host response (Lotz et al. 1988, Launders et al. 1999). Another species, *Gymnodinioides inkystans*, which we are presently investigating, embeds in the gills and typically produces a more conspicuous melanistic host response (Lotz et al. 1988). Even though large numbers of both agents probably influence the health of juveniles and adults, low numbers with a melanistic response produce an unmarketable product unless the cephalothorax (head of the shrimp) is removed (Fig. 10.9B). Moreover, if the common internal apostome species *Synophrya hypertrophica* ever entered a high salinity culture facility, it probably would produce mass mortalities.

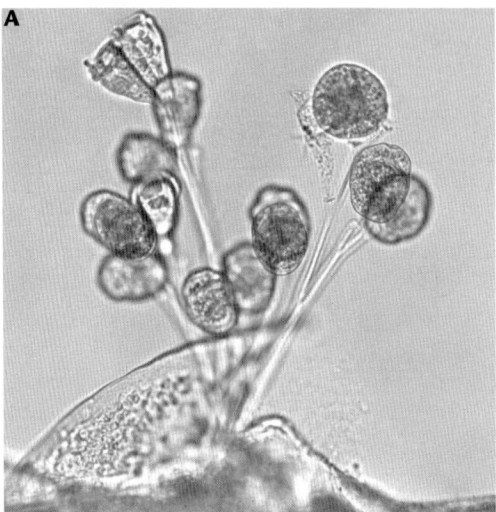 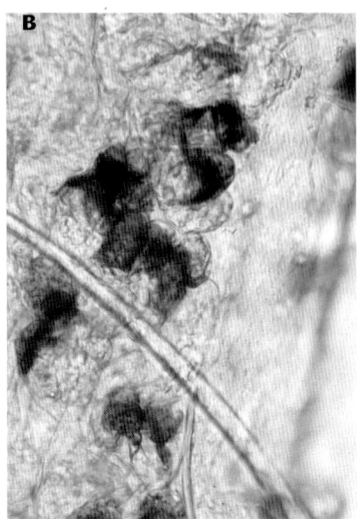

Figure 10.9 **A.** A colony of the peritrich ciliate *Zoothamnium* sp. from the gills of a cultured *Litopenaeus vannamei*. The ciliate species is one of several capable of producing mortality of larvae and postlarvae, stressing older shrimp and indicating water with high levels of organic matter.
B. A melanistic response by *Litopenaeus setiferus* to the apostome ciliate *Gymnodinioides inkystans* in the gills of a 15-day experimental infection. Heavy infections affect shrimp health and marketability of penaeid products.

Facultative ciliates such as the scuticociliates, probably a variety of species in different genera, invade the haemocoel and tissues of shrimp, especially young individuals in hatcheries. One ciliate from larval and juvenile *Farfantepenaeus aztecus* was identified as *Parauronema* sp. by Couch (1978). Overstreet (1987), Lotz and Overstreet 1990) and others (e.g. D Lightner, University of Arizona June 1996) have seen several infections with unidentified species, perhaps *Mesanophrys* sp. or members of related genera in larvae, juveniles and adults. Clearly, several facultative species can grow and reproduce in the haemolymph and feed on the blood and other tissues of penaeids that have been breached. Also, organic material adhering to or associated with shrimp larvae grown in poor quality water often attracts large numbers of scuticociliates to the shrimp.

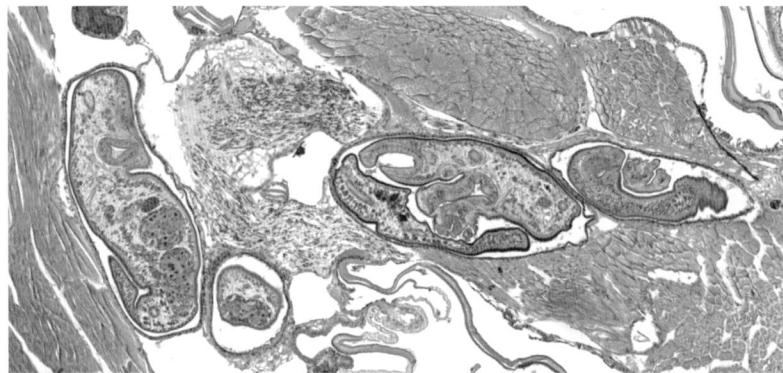

Figure 10.10 A histological section through four specimens of the metacercaria of the digenean *Opecoeloides fimbriatus* in an ectopic site in *Litopenaeus setiferus* from Mississippi. Note that the worms have disrupted nerve tissue as well as muscle tissue.

Shrimp haemolymph also serves as a culture media *in vivo* for facultative diatoms and perhaps various protists. A naturally infected *Litopenaeus setiferus* from Texas contained clusters of the diatom *Amphora* sp. within the gills associated with a melanistic response and fibrinous-like haemolymph. Experimental infections with an isolate of *A. coffaeformis* produced a similar response (Overstreet and Safford 1980). Both an amoeba involved with mass mortalities (Laramore and Barkate 1979) and a flagellate (Couch 1978) have been reported from young penaeids in American hatcheries, and these and related protozoans probably occur in shrimps worldwide.

If broodstock comes from the wild or if incoming water to a hatchery is inadequately filtered, gregarines, amoebae, microsporans and other protozoans in addition to ciliates can enter the hatchery system. In some semi-intensive grow-out ponds, the stocked postlarvae come from the wild rather than from hatcheries, and they typically introduce an abundance of protozoans. Obviously, all kinds of parasites occur in extensive grow-out farms. A few other protozoans indicated below can enter hatcheries on rare occasions, but they may occur commonly in ponds.

The aseptatorine gregarine *Paraophioidina scolecoides* heavily infected the midgut of larval and postlarval *L. vannamei* in a commercial hatchery. Probably the shrimp constituted an accidental host in which the cycle was not completed; the hosts were not noticeably harmed. Nevertheless, infected postlarvae could not be sold and shipped to farms until biological aspects and risks were determined (Jones *et al.* 1994).

The septatorine porosporid intestinal gregarines *Nematopsis* spp. have a life cycle involving gymnospores from a gametocyst in the rectum or elsewhere in the gut of the shrimp that infect a specific molluscan host. Cycles for different species differ, but in some the infective stage can be transmitted through the water, molluscan mucus or molluscan tissue. They and others pose a threat in ponds. The cephalolobids *Cephalobus* spp. also infect penaeids in the Americas, but they attach to the mouthparts, do not have an additional host and do not seem to be harmful. Perhaps that is because the optimal salinity for the ciliate is higher than that used in most farms growing *L. vannamei*.

Whereas microsporans in the muscles of cultured penaeids occur mostly in extensive grow-out facilities, some species occasionally occur in shrimp that inhabit ponds, raceways and tanks. Life cycles of those species in shrimp are not yet demonstrated, but at least infections with *Agmasoma penaei* have been shown to develop in postlarval *Farfantepenaeus duorarum* exposed to faeces containing 'conditioned' spores collected from a sciaenid fish that had been fed experimentally with one or more species of microsporans (Iversen and Kelly 1976). Infections of the same organism have also developed in tank-reared *L. vannamei* receiving water from a tank containing fish. Species in other crustaceans and arthropods have a variety of life cycles ranging from infection by a host feeding directly on spores, to transovarial infections, to indirect infections requiring a 'conditioning' or a true intermediate host. In some cases, there are as many as three different sporulation sequences (e.g. Becnel 1992, Micieli *et al.* 2000), and the spores in the intermediate and definitive hosts can appear morphologically as belonging to different microsporan genera (e.g. Becnel 1992, Sweeney *et al.* 1993).

Numerous microsporan species exist in penaeids around the world. There is uncertainty as to whether some morphologically identical individuals from geographically disparate regions in Australia and Africa as well as in the Americas, such as *Perezia nelsoni*, are conspecific or not. Numerous other species are roughly categorised as unidentified species of *Thelohania*, *Pleistophora* and *Nosema*, but some of these are now recognised in the Gulf of Mexico and elsewhere as *Agmasoma penaei*, *Thelohania duorara*, *Tuzetia weidneri*, *Perezia nelsoni* and others (e.g. Canning *et al.* 2002). All microsporans make the shrimp product unmarketable or in disfavour except possibly *A. penaei*, which entices some consumers, who confuse an infection with ripe gonads, to favour it over uninfected shrimp (Overstreet 2003). Microsporan species also can impair health, produce castration or even cause death.

Metazoans

Extensive shrimp farming is conducive to metazoan infections as well as protozoans in the product. These agents include leeches (e.g. *Myzobdella lugubris*, which deposits its egg cases, Overstreet 1973) and bopyrid isopods (*Epipenaeon ingens* in *Penaeus semisulcatus*, Overstreet 1987) as well as helminths. First, shrimp can get infected from flow-through water and small dietary prey on a daily basis. Second, shrimp can be infected when initially impounded. The less filtering of the water that enters a facility, the more likely the cultured shrimp are to become infected by parasites and other infectious agents, so even semi-intensive and intensive systems can be vulnerable. Water entering the facilities typically comes from adjacent canals, and these canals often contain wild infected shrimps or other animals that serve as hosts for various parasites that infect shrimps.

Helminths of penaeids typically incorporate the shrimp as an intermediate host. Usually, these helminths do not harm the host unless that infective stage occurs in large numbers or becomes located in an ectopic site (Fig. 10.10). For example, pathological alterations often occur when the metacercaria of the digenean *Opecoeloides fimbriatus* occurs throughout the hosts, in tissues and in the circulatory system, where it blocks haemolymph flow (RM Overstreet pers. obs. 1969–2005). The metacercaria typically lodges in its thin cyst in tissue among the hepatopancreatic tubules and adjacent haemocoel with little host response. When free or encysted in an ectopic site such as the gills, its presence often elicits considerable host inflammatory response, often including a conspicuous melanistic pigmentation and granulomata, presumably influencing shrimp health and market value. Sciaenid fishes serve as the definitive hosts.

Other digeneans also infect penaeids, and many of these are identified or unidentified microphallid digenean metacercariae. A few of these (Heard and Overstreet 1983, Overstreet 1987) as well as the ascaridoid nematode *Hysterothylacium* type MB (see Overstreet and Meyer 1981, Deardorff and Overstreet 1981) may produce a public health risk. Some shrimp helminths, more than protozoans, pose potential human health risk, with none of them being severe or even realistic unless much of the product is eaten raw or otherwise inadequately prepared. In contrast, penaeids in low-salinity ponds near appropriate rodents and terrestrial molluscs can serve as paratenic, or transfer, hosts of the metastrongyle nematodes *Angiostrongylus cantonensis*, producing eosinophilic meningitis, and *A. costaricensis*, causing abdominal angiostrongylosis, both potentially fatal (see pp. 442–446).

Additional nematodes commonly infect penaeids. Most are juveniles of several species of ascaridoids (e.g. *Hysterothylacium* spp., Deardorff and Overstreet 1981), but penaeids can also acquire spirurid species such as *Spirocamallus cricotus* (see Overstreet 1973, Fusco 1980), probably *Heliconema brooksi* (see Crites and Overstreet 1991), related species and members of free-living groups (Overstreet 1973). Most individuals of American ascaridoids consist of *H. reliquens* located in large numbers in the dorsal gland of the hepatopancreatic midgut gland, among its tubules and elsewhere. One or few individuals of species worldwide in a larval, postlarval or early juvenile penaeid can kill or make the host vulnerable to cannibalism or predation, based on observations (RM Overstreet 1970–1985) in older penaeid juveniles and experimental infections in other crustaceans. The intermediate or transport host for all these nematodes are copepods, and a variety of fishes serves as definitive hosts.

There is a variety of cestode metacestodes that infect penaeids. These include trypanorhynchans, tetraphyllideans, lecanocephalideans and cyclophyllideans. The trypanorhynchs *Prochristianella* spp. and related eutetrarhynchids typically infect the hepatopancreatic tubules, often causing an extensive inflammatory response (e.g. Sparks and Fontaine 1973, RM Overstreet pers. obs. 1969–2002) that may or may not relate to host, site and environmental conditions. In any event, a relatively low number of these typically occur in wild penaeids without a harmful effect on the shrimp, but a high mean intensity, with its associated pathological risk, could occur in restricted areas such as ponds receiving water from an area containing ray hosts. The lecanocephalans *Polypocephalus* spp. from throughout the hosts' ranges, usually with a high degree of host specificity, infect nervous tissue where many individuals can greatly enlarge nerve ganglia. Presumably once a specific threshold number is surpassed, the behaviour and ability for the host to avoid being eaten are greatly affected. For members of the first three indicated cestode groups, there needs to be a spatial relationship with the appropriate elasmobranch host nearby. Cyclophyllideans are less common in penaeids, but heavy infections of the metacestode of one encysting in the intestine of *L. setiferus* and *F. aztecus* in Texas and Mississippi (Corkern 1978, Lotz and Overstreet 1990) illustrate that one might expect other cyclophyllideans in other penaeids. Birds serve as the definitive hosts for this group.

Fouling metazoan agents are similar to the facultative protozoan agents except they occur externally. As examples, the hydroid *Obelia bicuspidata* infested caged-reared *F. aztecus*, in some cases covering the eyes and appendages (Overstreet 1973). Other hydroids remain the bane of hatchery managers. Some of these, *Moerisia lyonsi*, *Stylactis arge* and *Clytia gracilis*, competed for food with decapod larvae in South Carolina closed systems, and the medusa of *M. lyonsi* actively preyed on the larvae (Sandifer et al. 1974). Medusa from the attached stage of some of the same and probably additional species in Mississippi fed heavily on larval shrimp (Lotz and Overstreet 1990) and others certainly occur elsewhere. Acorn barnacles and other invertebrates commonly occur on both stressed cultured and wild penaeids. In ponds, diatoms are often seen attached to the gills as is the filamentous blue-green alga *Schizothrix calcicola* (e.g. Overstreet 1973). In addition to indicating or causing stress to the host, fouling agents affect marketability of the commercial product.

Management and treatment

As indicated above, management acts as the key to successful shrimp farming, and eliminating, reducing or controlling parasites and other infectious agents serve as major components of good management. Water quality in all systems is critical. Infectious agents should be removed by filtering or water treatment. Nitrogenous and other wastes should be maintained below threshold levels. There should be little accumulation of uneaten food, faeces or dead shrimp. In extensive systems where these practices are impractical, the density of the shrimp can be manipulated to a low level to keep parasites from harming their hosts. In all facilities, the higher the density of shrimp, the greater the risk of disease and mortality. Shrimp occasionally have to be harvested prematurely to reduce risk of mortality.

Monitoring both the shrimp at different stages and the water supporting those stages is critical to maintaining a healthy system. The earlier an agent is found, the earlier one can attend to the problem. As discussed above, the type of agent often indicates the problem (e.g. attached peritrichs indicate overfeeding or other reasons for excess organic material, and gregarines or microsporans can indicate poor filtration or infected broodstock). Poor water quality can serve as a predictor of what diseases to expect unless corrected.

Whereas the above methods provide useful tools to assess all systems including hatcheries and other closed systems, one needs to monitor shrimp and water in extensive and other systems. The identification of specific parasites usually indicates what other hosts in metazoan and a few protozoan cycles are transmitting agents to the facility. Consequently, such information will allow one to interrupt the cycle by excluding that host, eliminate or control the infection by modifying needs of the agent or eliminate faults in the facility.

Treatment with antibiotics or other compounds for parasites is usually not cost effective. In some cases in hatcheries, broodstock facilities, other facilities holding valuable stocks and occasionally ponds, chemical compounds can be useful for treating some bacterial and protozoan infections – but only as a stop-gap measure until the cause can be controlled. Chemicals can stop spread of the agent to adjacent ponds, tanks or raceways and save crops for an early harvest, but unless the water quality or facilities are corrected, future crops will not be similarly benefitted. In a very few cases, environmental factors such as salinity or temperature can be modified cost effectively to control a disease.

If a system is producing an abundance of healthy product, a manager or other decision maker should not try to further increase profits by reducing good monitoring and other management and feeding practices. The existence of healthy products relates to that attention, and if that attention is reduced, so will be the abundance and quality of the product within a very few years.

Important references
Comprehensive accounts of parasites of shrimp not restricted to a single parasite species or a single aspect are by Overstreet (1973), Couch (1978) and Lotz and Overstreet (1990).

Acknowledgement
Much reported research in this section results from funding from the US Department of Agriculture, CSREES Grant No. 2002-38808-01381.

Ecological aspects of parasites in the American lobster
Richard J Cawthorn

Introduction
Globally, lobster fishing is a major economic enterprise: clawed and non-clawed lobsters are very valuable seafood and also occupy highly significant positions in the food web (see Factor 1995, Phillips and Kittaka 2000). For example, non-clawed spiny (rock) lobsters inhabit tropical and temperate waters; the annual harvest is 77 000 metric tonnes with landed value exceeding US$500 million. The spiny lobster fishery is concentrated in Australia, New Zealand, South Africa, Cuba, Brazil, Mexico and the USA (Phillips and Kittaka 2000). Overall, the total annual value of global lobster fisheries (spiny and American lobsters) is US$1.25 to US$1.5 billion. Parasites represent a significant (although inadequately documented) threat to the various lobster fisheries. This discussion is restricted to parasites of the American lobster, *Homarus americanus*, one of the most important and best studied species of clawed lobsters.

General aspects of lobster fisheries in North America
Harvest of the American lobster is a highly regulated fishery in both Canada and the USA (Lawton and Lavalli 1995). In Canada in 2002, 46 000 metric tonnes of lobster were harvested, with landed value of Can.$584 million (Fisheries and Oceans Canada 2004). The fishery is concentrated in Nova Scotia (52.5% of landings) although it is very important to the other Atlantic provinces (i.e. New Brunswick, 16.5%; Prince Edward Island, 20%). In the USA in 2002, the total lobster landings were 37 093 metric tonnes valued at US$290 million (US National Marine Fisheries Service 2004). Most landings occurred in Maine (76%) and Massachusetts (16%). The definitions of inshore and offshore fisheries depend on the fisheries management regime and local community definition – presently the fishing boats are highly capable and progressive, as is the equipment and associated technology. On the northeast Atlantic coast of North America the lobster fishery is labour intensive and an important component of regional economies.

The American lobster ranges from Newfoundland Labrador south to North Carolina, inhabiting the continental shelf and upper shore and is fished to depths of 700 m. The lobster lives in various habitats dependent on the life history stage: mud, cobble, bedrock, peat reefs, eelgrass beds, sandy depressions, and rocks on sand. Its distribution is influenced by water temperature, salinity, oxygen content, light (photoperiod), and wind and oceanic currents. Different preferences are exhibited by the various life history stages. Overall the biology and ecology of the American lobster is highly complex (Lawton and Lavalli 1995).

The eastern seaboard of the USA is densely populated (by humans) and heavily industrialised – however, the short-term and long-term impacts of industrial effluents, pesticides and other human insults to the oceanic environment on the lobster fishery are not well understood. Within the context of the lobster fishery, parasites and parasitic infections can be used as indicators of climate change, food-web structures, trophic interactions and biodiversity (Marcogliese 2001).

Disease caused by parasites

The primary challenge is to recognise parasitic disease in the American lobster, whether in the wild fishery *per se* or in impoundment facilities. The initial requirement and focus is to define what constitutes a healthy lobster. Our working definition of health is based on Boyd (1970): *health* is a condition in which the organism is in complete accord with its surroundings, with that exquisite coordination of the different functions which characterises the living animal or plant. *Disease* is a change in that condition as a result of which the organism suffers from discomfort (dis-ease).

Although American lobsters are hosts to a broad range of ectoparasites and endoparasites, relatively few cause significant disease in either the wild fishery or impoundments (see review by Bower *et al.* 1994).

The three diseases of primary significance to American lobsters in captivity (i.e. impoundments) are gaffkemia (caused by the bacterium *Aerococcus viridans*) (see Stewart 1980), 'bumper car'-disease (caused by the scuticociliate *Anophryoides haemophila*) (see Cawthorn 1997) and shell disease, which is apparently caused by various pathogens depending on location (see Lavallée *et al.* 2001). Recently a novel hypothesis on shell disease involving protists, diatoms, bacteria, immunocompromised hosts and biofilms has been proposed by C O'Kelly and colleagues (March 2004). Shell disease is now manifest in at least two forms: impoundment shell disease and epidemic shell disease. The role of environmental factors in clinical manifestation of this disease is apparently highly variable, depending on locale. Various amoebae may be important to the etiology of shell disease, with bacteria-filled lesions becoming apparent months after the initiation of this disease.

In 'bumper car'-disease, relatively lobster friendly and consumer friendly methods of disease prevention by reduction of ciliate loads (Fig. 10.11) on gills and carapace surfaces in the impoundment environment are available. Ideally control of disease should be by manipulation of environmental parameters (i.e. non-chemical means) (Overstreet 1987). Synergism among pathogens can be influenced by environmental factors. Cost-benefit analysis determines management decisions on parasite control – the keys are healthy stock and excellent water quality. However, the present challenge is to demonstrate that use of formalin (50 mg/L) or low salinity (8 ppt) treatments for brief periods are cost effective means to control bumper car-disease and are also lobster friendly, user friendly and consumer friendly (Speare *et al.* 1996). Surprisingly, although infections with *A. haemophila* can be common in the wild fishery, the disease is recognised only in lobster holding facilities (Aiken *et al.* 1973, Cawthorn *et al.* 1996, Greenwood *et al.* in press).

The impoundment environment for lobsters is highly variable (cars, tank houses, tidal pounds, refrigerated lobster-holding facilities and dryland pounds) and not well evaluated and documented from either lobster health or environmental perspectives. Overstreet (1987) suggested that with the culturing of crustaceans, an increase in parasite-induced disease could

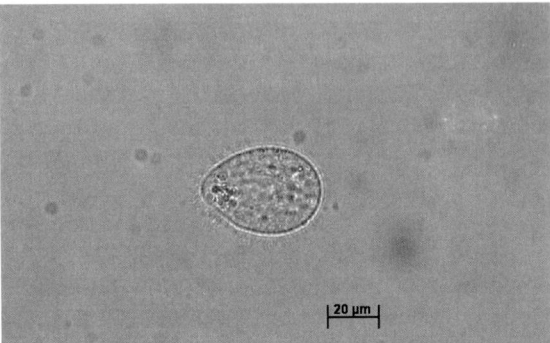

Figure 10.11 Culture form of the scuticociliate *Anophryoides haemophila*. Bright field illumination. Courtesy of Béatrice Després, Atlantic Veterinary College Lobster Science Centre.

result. Increased stock densities could adversely affect water quality and there would likely be increased transmission of pathogens and stress-influenced diseases would be exacerbated.

Control of ciliate-induced diseases may be very important in the success of farmed shrimp, crayfish or lobster operations. However, there are many challenges associated with the taxonomy, life cycles, ecology, treatment, diagnosis, host specificity, host responses and disinfection of ciliates (see Morado and Small 1995). These authors also suggested that in addition to obvious mortalities which may occur rapidly, ciliates could negatively impact host fecundity and swimming performance with significant ecological consequences, including the ability to reproduce, escape predators or capture prey.

Ciliates can act as environmental indicators. There are few sophisticated monitoring programs or inventory management of lobster impoundments, however, which makes diagnosis of parasitic disease very difficult. Additionally, the lobsters are subject to a multitude of stressors including handling and water quality issues.

A major challenge is to recognise disease outbreaks (epidemics) in the wild lobster fishery. Usually with parasitic disease there is a lack of specific clinical signs. Couch (1983) suggested that it is very difficult to assess the role of protozoan diseases as natural factors in the population dynamics of commercially and ecologically important Crustacea. Very little is known of the impact of parasites on larval stages and economic losses are undocumented. In the wild fishery, mass mortalities are not often recognised; sick or moribund crustaceans may be secretive and quickly eaten by predators or scavengers. Both Stewart (1980) and Couch (1983) suggested that to evaluate the impact of diseases on lobster populations, one must sample all life stages, including eggs, larvae, juveniles and adults. Notwithstanding the challenge of field studies in the oceanic environment, Stewart (1993) advocated a holistic approach to the study of infectious diseases of Crustacea.

Parasites can be important indicators of environmental stress (perturbations – i.e. temperature change, anoxia, toxicants and oceanic circulation) and parasites reflect the host's position in the food web and are indicative of changes in ecosystem structure and function (Marcogliese 2004). The implications of climate change (i.e. global warming) on aquatic host–parasite relationships, which are complex and context-dependent, were recently reviewed by Marcogliese (2001). There are significant adverse impacts of higher oceanic temperatures including more epidemics, thermal stress on hosts, increased reproductive rates of parasites, eutrophication of water bodies, anoxic stress on hosts, and shifts in oceanic currents (northward movement of inshore warmwater conditions). The occurrence of the Long Island Sound (LIS) lobster die-off in 1999 may reflect the above environmental factors. The amoebic parasite *Neoparamoeba*

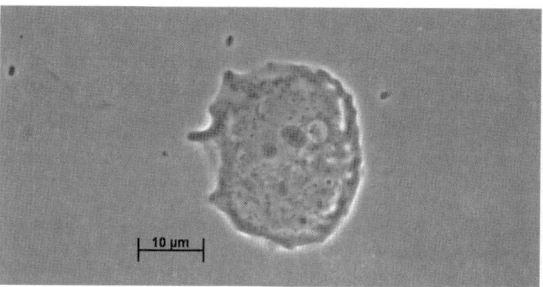

Figure 10.12 Culture form of the amoeba *Neoparamoeba pemaquidensis*. Phase contrast illumination. Courtesy of Béatrice Després, Atlantic Veterinary College Lobster Science Centre.

pemaquidensis is likely the causative agent of paramoebiasis in lobsters and other crustaceans, Atlantic salmon and sea urchins (Fig. 10.12) (see recent molecular analysis of taxonomy by Fiala and Dykova 2003, Peglar *et al*. 2003, Wong *et al*. 2004) and has been recovered from marine sediments (Crosbie *et al*. 2003).

In the LIS lobster mortality of 1999, the initial diagnosis of paramoebiasis as a systemic inflammatory disease of multiple tissues and primarily the nervous system, was made by French *et al*. (2001) and later confirmed by Mullen *et al*. (2004). Subsequently major research efforts have explored several facets of the disease in LIS, including taxonomic, diagnostic and pathologic studies (Frasca *et al*. 2003, Gast 2003, Gillevet and O'Kelly 2003, Shields 2004). The various natural and human-based disease-inducing stressors are likely to continue in LIS. For example, alkyphenols from various industrial sources and present in marine sediments are likely to function as endocrine disruptors in lobsters. In the stressful environment (i.e. high water temperatures, hypoxia) of LIS, these phenols may continue to contribute to lobster mortalities (Biggers and Laufer 2004). Apparently lobsters are very sensitive to the immunotoxic effects of malathion which was used to control mosquitoes carrying West Nile virus in the LIS region (De Guise *et al*. 2004). Although documentation of exposure of lobsters to malathion at environmental application is unavailable, these authors suggested that malathion-induced immunosuppression could increase susceptibility of lobsters to pathogens including *N. pemaquidensis* and enhance the resulting mass mortality. Recently Dohoo *et al*. (2003) suggested that epidemiological research using a naturalistic paradigm would be a better approach to multi-factorial disease scenarios, such as those occurring in the wild lobster fishery.

Important references

The various contributions in Factor (1995) give detailed accounts of many aspects of the biology of the American lobster, and those in Phillips and Kittaka (2000) of fisheries and culture of spiny lobsters. Infectious diseases including parasites of marine crustaceans are discussed by Stewart (1993), and Bower *et al*. (1994) presented a synopsis of infectious diseases and parasites of commercially exploited shellfish. Morado and Small (1995) reviewed ciliate parasites and related diseases of Crustacea, and Mullen *et al*. (2004) discussed paramoebiasis associated with mass mortality of American lobster. An overview of the impact of bumper car-disease on the North American lobster fishery was given by Cawthorn (1997).

Acknowledgement

Funding to the Atlantic Veterinary College Lobster Science Center is from a consortium of the lobster industry (fishers, processors, exporters), provincial agencies, First Nations and federal agencies (including the Atlantic Canada Opportunities Agency).

Parasites of marine mammals
Murray D Dailey

Introduction
When looking at marine parasites one certainly has to consider those that infect the marine mammals as some of the most interesting and challenging. The parasitic groups discussed in this section contain some of the smallest and most primitive animals known yet they can destroy one of the earth's largest creatures.

The study of marine mammal parasites has lagged behind those of terrestrial, freshwater and commercially important consumable species. The first attempt at putting together a comprehensive review was not published until 1955 by Delyamure. Since then a worldwide interest in marine mammals has developed through large numbers of captive animals trained to entertain the public and carry out specific missions of military importance. More recently attention to the wild populations has developed in response to the growing concern over ocean degradation.

Hundreds of species of parasitic organisms have been reported from marine mammals. This section presents an overview of the parasitic groups found in marine mammals, elaborating on those that are the most abundant and destructive as well as discussing modifications in their life cycles due to the marine environment.

Protozoa
The single-celled organisms that make up this phylum have parasitic representatives found in marine mammals throughout the world (Dailey 2001).

Apicomplexans
Sarcocystis sp., *Toxoplasma* sp. and *Eimeria phocae* have been reported from numerous marine mammals including pinnipeds, cetaceans, sea otters and the West Indian manatee. Recently a fatal case of hepatic sarcosystosis was diagnosed in a striped dolphin, *Stenella coeruleoalba*, from the northeastern Spanish Mediterranean coast (Resendes et al. 2002). *Sarcocystis* sp. generally have a two-host predator–prey life cycle with an asexual cycle in herbivores and a sexual cycle in carnivores (Dubey et al. 1989). In the last few years, however, a sea-going variety, *S. neurona*, has been found to cause a fatal case of meningoencephalitis in a naturally infected free-living southern sea otter, *Enhydra lutris nereis* (Miller et al. 2001). Through experimental infections researchers suspect that marine mammals exposed to *S. neurona* may support development of mature sarcocysts that are infectious to competent definitive hosts and could be spread by the common opossum, *Didelphis virginiana*, through runoff (Dubey et al. 2003).

Toxoplasmosis appears to be widespread in marine mammals. The first report in a dolphin was from Brazil (Bandoli and De Oliveira 1977). Since that time *T. gondii* has been isolated from striped dolphins in Spain and the Atlantic bottlenose dolphin, *Tursiops truncatus*, California sea lion *Zalophus californianus*, Pacific harbor seal, *Phoca vitulina richardsi*, and sea otters from North America (Jardine and Dubey 2002). The finding of *T. gondii* in marine mammal hosts is interesting in that only members of the family Felidae are known to serve as the definitive hosts of this parasite and are the only known species that shed oocysts in the faeces. Sea otters might become infected by oocysts in cat droppings washed into the sea by sprinklers, and by rain runoff into coastal-bound storm drains and creeks (Miller et al. 2002).

The coccidian *E. phocae* has been reported as the cause of death in captive harbor seals from North America and Scotland. This parasite may not affect the health of the seal unless the host is stressed through capture, handling or changes in diet (Munro and Synge 1991).

Ciliates

With the exception of *Haematophagus megapterae* found attached to the baleen plates of mystocete whales (humpback, fin and blue) all reported ciliates from cetaceans are found around the blow hole. The most common is the large (60–100 μm in diameter) *Kyaroikeus cetarius* which has been isolated from necrotic cutaneous lesions but never from healthy tissue in many species of cetaceans. These ciliates are thought to have direct life cycles and are spread by contact (Poynton *et al.* 2001).

Flagellates

Little is known of the flagellates from marine mammals. Flagellates reported from cetaceans have been reported primarily from the blowhole mucus. The only published description of a marine mammal flagellate is *Jarrellia altramenti* from a Pygmy sperm whale, *Kogia breviceps*, found stranded on the east coast of North America. This parasite combines the characteristics of two fish parasites of the genera *Trypanoplama* and *Cryptobia*. The principal mechanism of transmission may be via an aerosol when whales exhale (Poynton *et al.* 2001).

Giardia spp. oocysts have been identified in the faecal samples of ringed seals, *Phoca hispida*, harp seals, *P. groenlandica*, grey seals, *Halichoerus grypus*, and harbor seals in the arctic, subarctic and eastern regions of Canada. Infected animals have not demonstrated any illness. The zoonotic potential of infected seals to act as reservoirs for human infection is unknown. To date all cetaceans examined have been negative (Munro *et al.* 1999).

Helminths

Nematoda

The roundworms contained in this group make up the largest number and some of the most destructive parasites of marine mammals. The life cycles that are known experimentally for helminths that infect marine mammals are all from the nematodes (see pp. 104–115).

Lungworms

Lungworms are found worldwide in the air sacs and large airways of all groups of marine mammals. Small parasitic nematodes belonging to subgenus *Parafilaroides* (Metastongyloidea: Filaroididae), are found coiled in the respiratory parenchyma of pinnipeds (otariids and phocids) and have a cosmopolitan distribution. The pathogenesis of this worm depends on the intensity of infection and the age of the host. Healthy animals with heavy infections might be predisposed to respiratory disease during times of stress (Measures 2001). The life cycle of this lungworm demonstrates one of the interesting transitions that marine mammal parasites have had to make in order to survive in an ocean environment (Fig. 10.13). A typical terrestrial lungworm uses either a mollusk (genera *Dictyocaaulus*, *Protostrongylus*, *Muellerius*) or an earthworm (genus *Metastrongylus*) as the intermediate host. Given that invertebrates are not normal prey for pinnipeds these marine lungworms have had to switch to a vertebrate (fish) host. With *P. decorus*, which is a species commonly found in the California sea lion, that transition was made easier by the feeding habits of the tidepool fish *Girella nigricans*. This fish lives at breeding rookeries where it feeds on sea lion excrement containing the first-stage larvae of the ovoviviparous *P. decorus*. Sea lion pups learn to catch live prey during weaning in these very pools making *G. nigricans* the perfect host for this transition (Dailey 1970).

Otostrongylus circumlitus (Metastrongyloidea Crenosomatidae) is a large nematode found in the principal airways of primarily seals (phocids). Distribution is holarctic and circumpolar. Host response to this parasite varies between species and age. This worm has been reported as the cause of death in 73 stranded, juvenile northern elephant seals, *Mirounga angustirostris*,

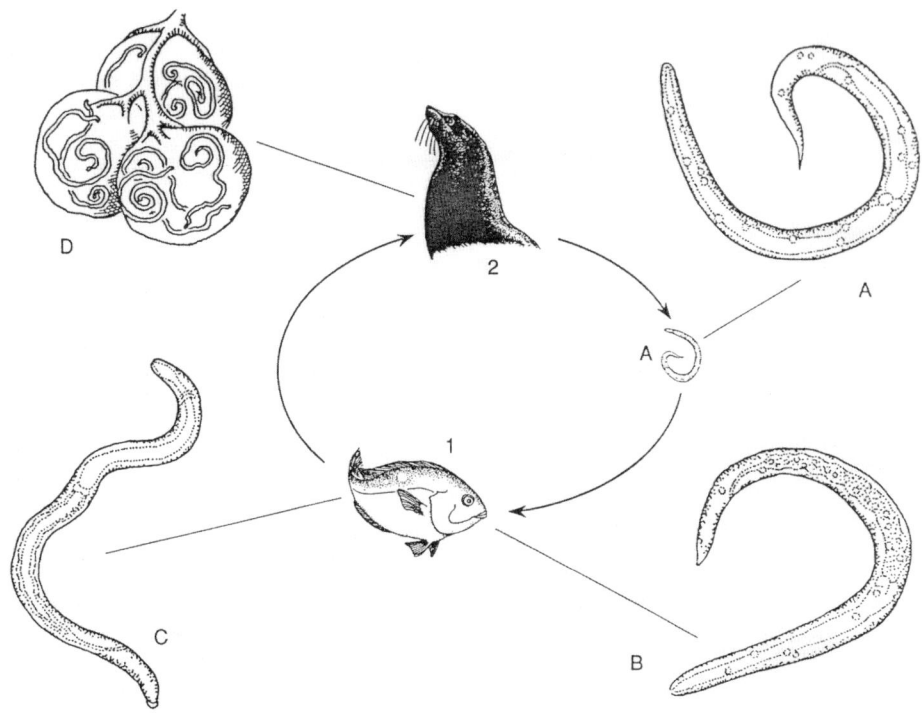

Figure 10.13 Life cycle of *Parafilaroides decorus*. **A.** First-stage juvenile (L1) into the water either via nasal mucus or faeces; L1 eaten by fish intermediate host, *Girella nigricans* (1). **B.** Second-stage juvenile (L2) in the intestinal wall of fish in 12–15 days. **C.** Third-stage juvenile (L3) found on outer surface of fish intestine in 24–36 days; fish eaten by sea lion (2) and worms migrate to lungs. **D.** Adult worms develop in alveoli of lungs, reach sexual maturity within 21 days. From Dailey, Murray D (1996) *Meyer, Olsen and Schmidt's Essentials of Parasitology*, 6th Edition, Wm. C. Brown Publishers. With permission of The McGraw-Hill Companies.

where it was found to precipitate disseminated intravascular coagulation. The life cycle of this worm is thought to echo that of *P. decorus*. In experimentally infected flat fish (American plaice) the infective third-stage larvae were recovered from the serosal lining of the intestine after 56 days post infection (Measures 2001).

The family Pseudaliidae (Metastrongyloidea) contains seven genera that infect the lungs, middle ear and cranial sinuses of toothed (odontocete) cetaceans. Pseudaliids are cosmopolitan but most of the work has been carried out in the northern hemisphere, primarily with coastal or inshore populations. Most prevalence and intensity data of pseudaliid nematodes were obtained from animals that strand or die in nets. However, studies on the harbor porpoise, *Phocoena phocoena*, from the Western Mediterranean Sea and Danish waters has provided data on five pseudaliids found in this host of which three (*Pseudalius inflexus*, *Torynurus convolutes* and *Halocercus invaginatus*) are frequently sympatric. Young-of-the-year or neonate porpoises are generally not infected with pseudaliids, and prevalence increases with age, suggesting horizontal transmission through the food chain. This trend does not appear to hold for the genus *Halocercus*, however. Prevalence in this genus in young porpoises is higher than in older animals, suggesting that transmission is transplacental or transmammary. Lungworms have not been reported from river dolphins (iniids, pontoporiids, platanistids) or sperm whales (physeterid, kogiids) (Measures 2001). To date there are no experimental data on the life cycles of any cetacean helminth parasites.

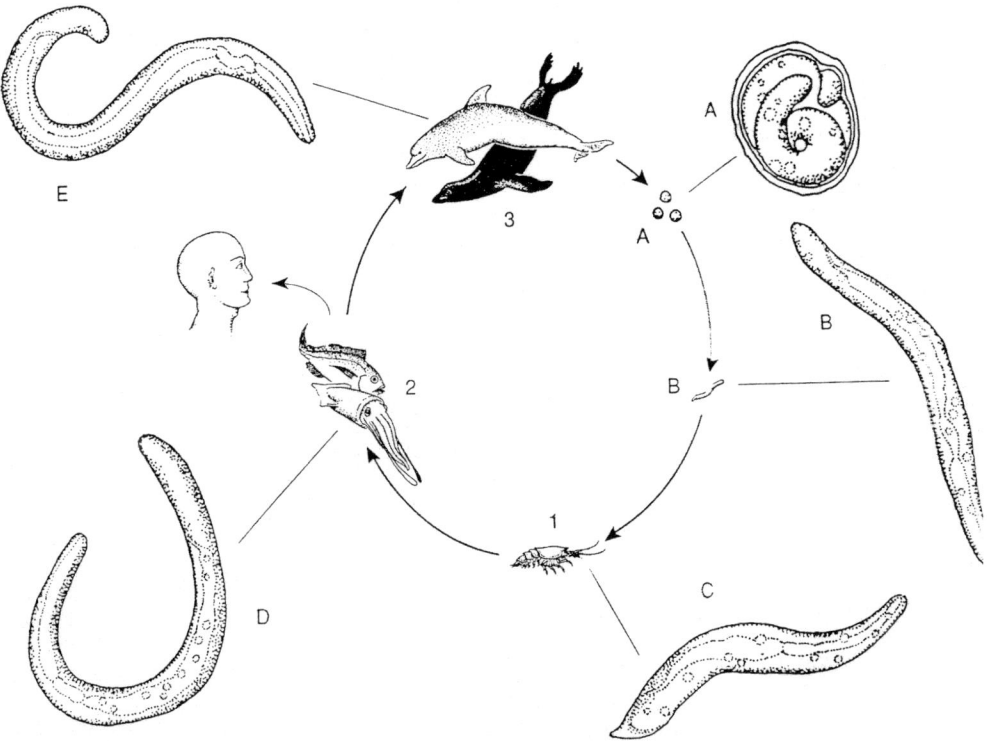

Figure 10.14 Life cycle of *Anisakis* spp. **A**. Eggs are released from the female worms and pass out with the faeces into the ocean. **B**. First-stage juveniles (L1) develop inside egg. **C**. Second-stage juvenile (L2) hatches out of egg. (1) A small crustacean (such as krill) eats the L2 and becomes the first intermediate host. (2) Second intermediate host (such as fish or squid) feeds on first intermediate and develops (**D**) infective third-stage juvenile. This stage then encysts in the body cavity, organs or muscle to await consumption by the definitive host (3) such as a pinniped or cetacean where it matures to an adult (**E**) in the stomach or intestine. Humans are accidental hosts in this life cycle. From Dailey, Murray D (1996) *Meyer, Olsen and Schmidt's Essentials of Parasitology*, 6th Edition, Wm. C. Brown Publishers. With permission of The McGraw-Hill Companies.

Gastrointestinal worms

Two families of gastrointestinal worms are found in marine mammals. Members of the family Anisakidae (large stomach roundworms) in marine mammals consist primarily of three genera (*Anisakis, Contracaecum* and *Pseudoterranova*,) with *Anisakis* reaching maturity in cetaceans and the latter two being more prevalent as adults in pinnipeds and sea otters. These worms are normally found in the stomach of the marine mammal host where, despite a high intensity, may cause no apparent ill effects. They do cause a problem, however, when humans accidentally become an incidental host in the cycle through eating infected raw or undercooked seafood (Fig. 10.14, see also pp. 430–434). Anisakiasis is the term used when referring to the disease in humans even though it can also be brought about by worms in the other two genera. This disease is found worldwide, with higher incidence in areas where raw fish is eaten (e.g. Japan, Pacific coast of South America and the Netherlands). An interesting aspect of the life cycle takes place when the second-stage larvae (L2) hatches from the egg in the ocean and becomes free swimming. The L2 is then ingested by crustaceans where they mature to an L3. Subsequently, the infected crustaceans are eaten by fish and squid where these larvae migrate to the muscle tissues. Through predation, these infective larvae are transferred from fish to fish where larval

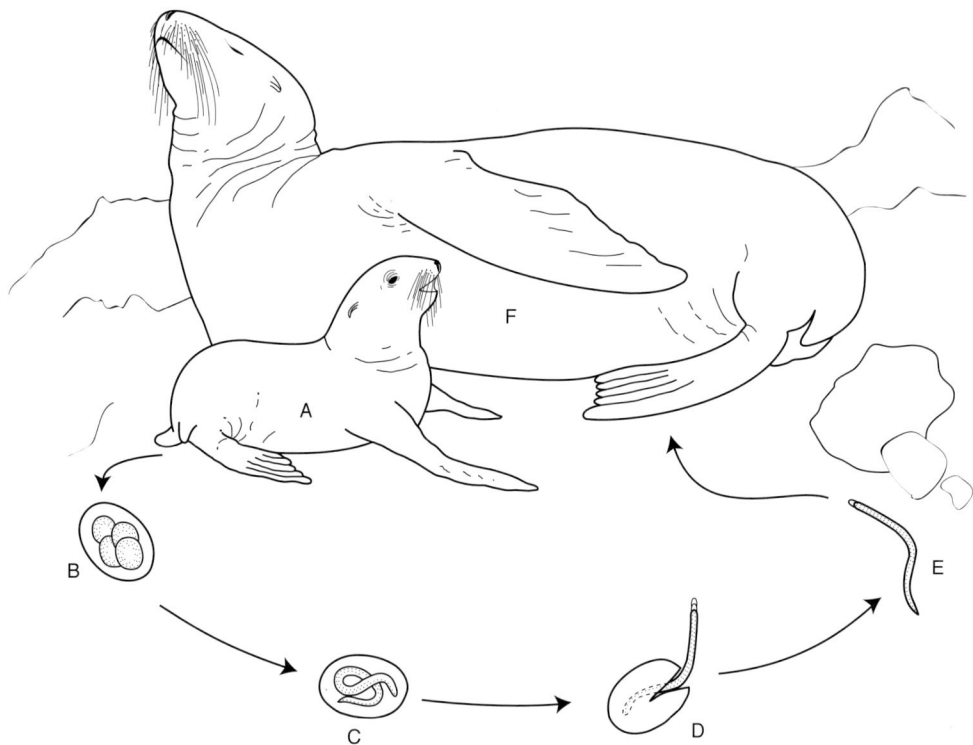

Figure 10.15 Life cycle of *Uncinaria lucasi*. **A**. Adult worms in pup intestine. **B**. Developing egg on beach. **C**. Developing juveniles (L1, L2, L3) in egg. **D**. Third-stage juvenile (L3) hatching. **E**. Free-living L3 penetrate flipper of adult female. **F**. L3 migrates via circulatory system to mammary glands of pregnant cows and pups (A) are infected while nursing. Original.

intensities are increased. These parasites are known to occur most frequently in the flesh of bottom fishes (rock fish, halibut, flounder), cod, haddock, pacific salmon, herring and monkfish. Severe cases of anisakiasis are extremely painful and require surgical intervention. Physical removal of the nematodes from the lesion is the only known method of reducing the pain and eliminating the cause (other than waiting for the worm to die). The symptoms apparently persist after the worm dies since some lesions are found after surgery to contain only nematode remnants (Lauckner 1985).

Members of the family Ancylostomatidae (hookworms) also infect the pinniped gastrointestinal tract. Hookworms in the genus *Uncinaria* have been indicated as the cause of high mortality in young pinnipeds through haemorrhagic enteritis and anaemia. The life cycle of *U. lucasi*, and probably other species in the genus, again demonstrates the adaptation this parasite had to make in the transition from terrestrial to marine habitat (Fig. 10.15). Hookworms in humans and other terrestrial animals have infective third-stage larvae that penetrate the skin of the host and grow to adults in the intestine. In fur seals and sea lions the cycle had to be modified to fit the seasonal at-sea habits of the new hosts. After breeding the females go to sea until the following spring when they return to land to give birth. Unlike cetaceans, pinniped young do not survive when born at sea. The hookworm larvae that penetrate these pregnant cows do not go to the intestine and mature to adults but migrate to the mammary glands. Here they wait until the birth of the pup the following spring at which time they are consumed with the milk during nursing. The reproducing adult worms are found only in the pups where they feed on blood causing the above symptoms and subsequently the possible death of the animal (Dailey 2001).

Trematoda
The digenetic trematodes or flukes are found worldwide in all groups of marine mammals. Some of the most destructive of these are those found in the genus *Nasitrema*. *Nasitrema* spp. may be a significant cause of single strandings in a variety of toothed whales (odontocetes). They have been cited as the cause of parasitogenic eighth cranial neuropathy in a stranding of Risso's dolphins, *Grampus griseus*, in Japan, encephalitis in a stranded striped dolphin in the Atlantic and cerebral necrosis in stranded common dolphins, *Delphinus deiphis*, in California. Symptoms include loss of equilibrium rendering the animal unable to keep its blow-hole out of the water resulting in stranding on the beach in an attempt to high center itself. Although the life cycle of this parasite is not known, attributes of the larval stage (miracidia) would indicate that it involves an intermediate host or hosts (Dailey 1985).

Acanthocephala
The spiny-headed worms are found as parasites in pinnipeds, cetaceans and otters worldwide. Generally it appears that pelagic forms are primarily infested with members of the genus *Bolbosoma* and offshore species with genus *Corynosoma*. Transmission of marine acanthocephalans appears to happen in two ways. The *Corynosoma* spp. cycle using an invertebrate (amphipod) and fish intermediate hosts while the pelagic *Bolbosoma* spp. needs only the invertebrate. These parasites penetrate the intestinal wall with a proboscis containing hooks to anchor themselves in the gut. There are reports of these parasites contributing to the death of the host(s) in the case of an eastern Pacific grey whale, *Eschrichtius robustus*, and southern sea otter (Dailey *et al.* 2000; Mayer *et al.* 2003).

Cestoda
Marine mammals harbour a rich tapeworm fauna consisting primarily of three families: Tetrabothriidae, Diphyllobothriidae and Phyllobothriidae. Very few parasites of this group are considered disease problems. The prime site of infestation of these parasites is the intestine. However, the most common are tissue-invading larval forms reported from nearly every species of Cetacea in all regions of the world's oceans. *Phyllobothrium delphini* and *Monorygma grimaldii* are found in the subcutaneous fat and body cavity, respectively, and are thought to be different species that grow to adulthood in elasmobranch (sharks, rays and skates) fishes. No tissue response is noted surrounding these cysts. Whether the cetacean represents a 'dead end' to the worm, or whether it is the natural intermediate host, is an interesting, still unanswered question. If the cetaceans are the natural hosts, this would represent the only cycle where a helminth uses a warm-blooded animal to infect a cold-blooded definitive host (Dailey 1985).

Insecta

Anaplura
Sucking lice are common ectoparasites of Pinnipedia. The significance of these parasites depends on the numbers infesting the host. Geraci and Lounsbury (2001) point out that seal lice, *Echinophsthirius horridus*, are often the indicators of a progressive debilitation and eventual death in seal pups. Lice feed on blood and are thought to transmit the seal heartworm, *Acanthochilonema spirocauda*. As the pup becomes weaker it hauls out for longer times, allowing more lice to attach, starting the downward spiral that eventually ends in death. Lice on pinnipeds tend to be host specific. Transmission is by direct spread of larvae from animal to animal (see also pp. 226–230).

Acari

Acarina
Mites of the family Halarachnidae that infest marine mammals occur in two genera. Those found in the nares, nasopharynx, airways and lungs of true seals (phocids) belong to

Halarachne, whereas eared seals (otariids) and walrus have *Orthohalarachne*. These mites cause sneezing, mucus discharge, nasal infections, emphysema and lung infection. The skin mites, *Demodex* and *Sarcoptes*, are found on all groups of pinnipeds and depending on the number are characterised by hyperkeratosis, alopecia and pruritus. Transmission is direct through animal contact (Dailey 2001) (see also pp. 216–222).

Copepoda

Parasitic copepods of the genus *Pennella* are usually found attached to large whales. There has been only one report of a parasitic copepod from a pinniped (northern elephant seal). These parasites have a complex life cycle with many stages of development. They penetrate the skin as a larval form and anchor in the muscle beneath the blubber layer. They appear to cause no major disease problems. However, in the elephant seal, the infection site resulted in multiple abscesses (Dailey 1985, Dailey *et al*. 2002).

Important references

The first comprehensive treatment of helminths of marine mammals is by Delyamure (1955). Dailey (1985, 2001) and Lauckner (1985) published detailed book chapters on diseases of marine mammals including parasites. Another comprehensive treatment is by Geraci and Lounsbury (2001).

Marine birds and their helminth parasites

Eric P Hoberg

Introduction

Seabirds are significant and conspicuous components of marine ecosystems occurring on estuarine, neritic and pelagic habitats encompassing latitudinal extremes from the Southern Ocean to the north polar seas (Ashmole 1971, Croxall 1987). More than 300 species are represented among six orders including the Podicipediformes, Gaviiformes, Sphenisciformes, Procellariiformes, Charadriiformes and the paraphyletic 'Pelecaniformes' (Harrison 1983); Anseriformes is excluded here. Obligate seabirds, those that only return to land for reproduction and forage exclusively at sea, characterise four orders consisting of penguins, the albatrosses, shearwaters and petrels, the auks and some larids, and the assemblage of shags, tropic birds, frigate birds, gannets and boobies. In contrast, facultative seabirds such as pelicans and cormorants, many larids, the grebes and the loons may breed adjacent to inland freshwater systems while dispersing into coastal marine zones over the winter. Marine birds are typically highly vagile, secondary and tertiary predators that occupy specific geographical ranges and habitats, show site fidelity for reproduction, and often follow predictable migratory paths linking breeding and wintering grounds (Ashmole 1971, Croxall 1987) (Fig. 10.16). As a consequence seabirds are excellent indicators of the state of marine ecosystems on short and long time scales (e.g. Ainley and Boekelheide 1990, Springer 1998, Croxall *et al*. 2002).

Parasitic helminths are elegant markers of contemporary ecology and the stability of trophic linkages for marine birds. Dependence on an array of intermediate, paratenic and definitive hosts indicates that each parasite species represents an assemblage of organisms within a community and tracks broadly and predictably across trophic levels (Hoberg 1996, 1997, Brooks and Hoberg 2000). Host and geographical ranges for parasites are historically constrained by genealogical and ecological associations, and these determinants interact resulting in characteristic and predictable community structure and biogeography. Understanding or predicting the range of possible effects that parasites may exert on seabirds, across the continuum linking individuals,

Figure 10.16 Short-tailed, *Puffinus tenuirostris*, and sooty shearwaters, *P. griseus*, from a rafting flock of 10–12 million birds in Unimak Pass, Alaska, during the southern migration in late August 1982. Seabirds are conspicuous and abundant top predators in marine ecosystems. Among shearwaters we may expect each bird to host four to five species of *Tetrabothrius* cestodes numbering from several hundred to thousands of specimens, along with spirurid and ascaridoid nematodes. Seabirds and their constituent parasite faunas are integral and characteristic facets of marine biodiversity and exquisite ecological indicators.

populations, species and complex communities, requires a biodiversity context (Hoberg 1996, 1997, Brooks and Hoberg 2000).

Parasite biodiversity, general patterns

Major groups of helminths are represented among seabirds, and over 700 species of Digenea, Eucestoda, Nematoda and Acanthocephala have been described or reported from 165 host species, or about 50% of the extant marine avifauna (Hoberg 1996, Storer 2000, 2002). Consequently the helminth faunas of seabirds remain incompletely known and represent a substantial facet of marine biodiversity that has yet to be evaluated in detail. Despite this apparent lacuna, ecologically discrete assemblages determined by foraging, prey selection and distribution are indicated by patterns of parasite prevalence, abundance and host associations across taxonomic, geographical and temporal scales (Hoberg 1996). Faunal diversity as briefly outlined in subsequent sections should not be considered as an exhaustive treatment, and interested readers are encouraged to consult the primary literature about these fascinating host–parasite assemblages.

Pelagic (oceanic) birds support depauperate faunas of low diversity compared to those found on neritic (over the continental shelf) and littoral (nearshore and shoreline) waters. The core component of the parasite faunas of pelagic birds are cestodes, and primarily species of *Tetrabothrius* and *Chaetophallus* (Tetrabothriidae) that may be exceptionally abundant (in excess of 20 000 to 30 000 tapeworms per host) in some procellariiforms and pygoscelid penguins (EP Hoberg unpublished data 1983). Species of *Alcataenia* in puffins, murres and auklets, and *Parorchites zederi* in penguins are the only important Cyclophyllidea in pelagic systems. These associations are apparently a consequence of broad oceanic distributions for macrozooplankton such as euphausiids, cephalopods and fishes that serve as suitable intermediate hosts in pelagic environments. In contrast, larids, grebes and loons, which exploit a broader range of invertebrates and fishes in nearshore areas, support speciose faunas including numerous cestodes (e.g. Ryzhikov et al. 1985, Hoberg 1992, Storer 2000, 2002). In addition to Tetrabothriidea, these include some Pseudophyllidea or Diphyllobothriidae (species of *Diphyllobothrium*) and very rarely Cyclophyllidea (among others, species of *Lateriporus*, *Neovalipora*, *Wardium*, *Microsomacanthus* and *Diorchis*); many dilepidids and hymenolepidids among facultative seabirds are likely to have been acquired in freshwater habitats.

Digenea are virtually absent in seabirds from pelagic zones due to the limitations imposed by life cycles, the focal nature of transmission near island systems, and the dilution effect of the marine offshore environment (Hoberg 1996). An exception is *Cardiocephaloides physalis*, a strigeid that is often abundant in some penguins, procellariiforms, shags and cormorants in the Southern Hemisphere (e.g. Randall and Bray 1983, Hoberg and Ryan 1989). Digeneans with marine life cycles, are otherwise limited in obligate seabirds, with major groups being represented by species among the Heterophyidae, Echinostomatidae, Gymnophallidae, Diplostomatidae, Strigeidae, Microphallidae, Schistosomatidae and Renicolidae (e.g. Hoberg 1992, Galaktionov 1996, Storer 2000, 2002).

Acanthocephalans also appear to be limited in host distribution to those species of seabirds feeding in nearshore environments and they are generally absent from pelagic birds. For example, species of *Corynosoma* were found only in blue-eyed shags, *Phalacrocorax atriceps*, southern black-backed gulls, *Larus dominicanus*, sheathbills, *Chionis alba*, brown skuas, *Catharacta lonnbergi*, and gentoo penguins, *Pygoscelis papua*, but not in other pygoscelid penguins, nor in a diverse assemblage of procellariiforms near the Antarctic Peninsula (Hoberg 1986, Zdzitowiecki 1986). Thus, acanthocephalans were being acquired in littoral habitats from piscine or amphipod prey, whereas seabirds representing a guild of zooplanktivores foraging on euphausiids were not infected. A general trend is evident as large diverse collections of marine birds from the North Pacific have also revealed species of *Corynosoma*, *Polymorphus*, *Bolbosoma* and *Andracantha* primarily among shags, loons and grebes, but rarely in alcids and procellariiforms (EP Hoberg and A Adams unpublished data 2002). Notably, many species of acanthocephalans that are found in seabirds only develop to maturity in cetacean or pinniped definitive hosts.

Among obligate and facultative seabirds, nematodes are often characteristic, prominent and abundant components of the overall parasite fauna. Species of *Contracaecum* and *Anisakis* are large ascaridoids commonly occurring in the proventriculus and oesophagus among piscivorous hosts such as shags, cormorants and pelicans, many large alcids, larids and loons (e.g. Fagerholm 1996, Storer 2002, EP Hoberg unpublished data 1975–1977). Species of *Contracaecum* are conspicuous parasites, and moderate to heavy infections are often typical with several hundred larval and adult nematodes occurring in a single host. Spirurid nematodes such as species of *Seuratia*, *Stegophorus*,

Paracuaria and *Tetrameres* also occur in the proventriculus or gizzard and are characteristic in penguins, procellariiforms, larids and alcids and other marine birds (Barus et al. 1978). Additional nematodes include species of *Desmidocercella*, spirurids that parasitise the airsacs of shags and cormorants, species of *Eulimdana*, members of the Splendidofilariidae that occur in subcutaneous tissues of larids, and Capillariidae from the intestine (Barus et al. 1978).

Helminths that are characteristic of respective groups and species of seabirds occupy diverse sites in the gastrointestinal, pulmonary and renal systems, and in organs, tissues and tissue spaces of their hosts. These helminths are predicted to exert varying effects on the health status of seabirds that may be manifested as morbidity and mortality among individuals, or may have a more pervasive influence on host populations across colonies and regions (Galaktionov 1996, Skorping 1996). The consequences of parasitism are determined by transmission dynamics, prevalence and intensity of infection, the interface between host and parasite populations, host population density, the reproductive and nutritional status of potential hosts, and interactive or cumulative effects of multispecies assemblages of helminths along with the co-occurrence of protozoan, fungal, bacterial or viral pathogens, in the context of abiotic environmental variables.

Interactive effects of helminth parasites

Sporadic empirical evidence for pathogenicity of helminths with morbidity and mortality linked to parasitism in marine birds has been documented (e.g. Lauckner 1985, Fagerholm 1996, Galaktionov 1996). Physiological indicators such as abdominal (and total) fat and or total mass, however, do not appear to be correlated with either prevalence or intensity of infection by gastrointestinal helminths where they have been examined in some procellariiforms and charadriiforms (e.g. Hoberg and Ryan 1989, EP Hoberg unpublished data 1976, 1977). During breeding and incubation, adult birds may be expected to have substantial levels of body fat, and the direct influence of parasitism on body condition may be negligible, or a challenge to recognise, in a regime of adequate prey availability. Factors such as age of host, recent acquisition of the helminth fauna, and intensity of infection below a threshold level could be important determinants of body condition before and during the reproductive cycle.

Tapeworms are often abundant parasites in both obligate and facultative seabirds, but direct effects are difficult to document. Characteristic lesions, where the scolex and neck of strobilate worms is deeply embedded into the submucosa, are seen with infections of *Parorchites zederi* in penguins (*Pygoscelis* spp. and *Aptenodytes* spp.) (e.g. Fuhrmann 1921). Multiple worms (often up to 20 or more per lesion) extend into the intestinal lumen from massive cysts that are visible on the serosal surface of the intestine; occasional perforation and adhesions are evident. Species of *Alcataenia* in murres, *Uria* spp., and whiskered auklets, *Aethia pygmaea*, exhibit a similar mode of deep attachment, although single cysts with large numbers of strobila are rare. Over time such lesions apparently become fibrose and necrotic, and secondary bacterial infections are likely, but the effect on either nestling or adult hosts remains undetermined.

Infections associated with ascaridoid and acuarioid nematodes also can be highly visible (Fagerholm 1996, Fagerholm et al. 1996). Larval and adult *Contracaecum* and *Anisakis* in the proventriculus can be the source of severe lesions, ulceration, perforation and peritonitis with secondary bacterial infections in procellariiforms such as northern fulmars, *Fulmarus glacialis*, and shearwaters, *Puffinus* spp., shags, *Stictocarbo* spp., and many larids and alcids including puffins and murres (Riley 1972, Fagerholm et al. 1996). Erosion of koilon lining of the gizzard is characteristic for infections of *Stegophorus* spp. Although often highly dramatic, the lesions associated with *Contracaecum*, *Seuratia* and *Stegophorus* appear to resolve over time and effects of such tissue damage may be ephemeral. Massive infections by *Contracaecum*, however, are space-occupying where a substantial volume of the lower oesophagus, proventriculus and gizzard may be supplanted by nematodes. It would be predicted that the frequency of foraging could be altered dramatically in such heavily parasitised seabirds, thus infections are unlikely to be benign.

Heavily infected breeding birds may also exhibit compensatory behaviour. Altered patterns of activity, however, including increased frequency and duration for bouts of foraging, and modifications in diurnal or nocturnal cycles may predispose heavily parasitised adult seabirds to predation or may influence growth rates and fledging success of chicks. The potential of heightened predation for adults transiting between colonies and foraging zones may be exacerbated for alcids, small procellariiforms such as storm petrels and some penguins. For example, parasitism may be linked to predation of Adélie penguins, *Pygoscelis adeliae*, by leopard seals, *Hydruga leptonyx*, and brown skuas in Antarctica, and puffins, *Fratercula* spp. and storm petrels, *Oceanodroma* spp., by large *Larus* gulls and peregrine falcons, (*Falco peregrinus*, in the North Pacific, North Atlantic and Arctic. Mortality for nestlings that are 'abandoned' for prolonged periods while adults are foraging could be also be a function of starvation, parasitism and predation.

Nestling birds acquire substantial parasite faunas (species richness and intensity) within days of hatching as actively foraging adults provide food and infectious stages of helminths (Hoberg 1992, 1996). Faunas characteristic of nestling and adult conspecifics may differ considerably as a reflection of preferential prey selection and differential effects are apparent with respect to morbidity and mortality prior to and following fledging. For example, species of *Alcataenia* are transmitted to alcids and larids by pelagic euphausiid crustaceans in the diet rather than littoral or demersal fishes. Consequently, a different spectrum of parasites in nestlings receiving piscine prey would be expected; for example, species of *Contracaecum*, possibly *Tetrabothrius* and such digeneans as *Cryptocotyle lingua* (Hoberg 1992, 1996). Although parasitism may be only tenuously linked with breeding failures across the Gulf of Alaska and Bering Sea in 1976 and 1982, the specific diversity of the helminth fauna in black-legged kittiwakes, *Rissa tridactyla*, and thick-billed and common murres, *Uria lomvia* and *U. aalge*, was indicative of a shortage of primary piscine prey required for development and fledging of chicks. Additionally, differential mortality for chicks and fledglings of jackass penguins, *Spheniscus demersus*, at colonies at St Croix Island, South Africa, was attributed to intense infections by *Cardiocephaloides physalis* (Randall and Bray 1983); the necessity to distinguish between starvation and parasitism was stressed. This large strigeid has also been regarded as a pathogen in gaunay cormorants, *Phalacrocorax bougainvillea*, Humboldt penguins, *Spheniscus humboldti*, and Magellanic penguins, *S. magellanicus*, along the coasts of South America, although most reports have emanated from adult hosts (Randall and Bray 1983).

Epizootics associated with helminths are not limited to the reproductive period for marine birds. Among facultative seabirds such as common loons, *Gavia immer*, periodic disease and mortality has been observed in wintering areas adjacent to the coast of Florida over many years (Forrester *et al.* 1997). Debilitated birds generally supported more diverse helminth faunas at higher intensities of infection during periodic epizootics over a 23-year period in Florida coastal waters (Forrester *et al.* 1997, Kinsella and Forrester 1999). Large numbers of microphallid trematodes, dominated by *Microphallus forresteri*, signified a shift in diet from fishes to benthic crustaceans and shrimp. A major epizootic during 1983 involving over 13 000 birds was linked to a shift in prey availability during an extended period of poor weather and reduced water clarity that influenced the ability of loons to effectively feed on fishes (Forrester *et al.* 1997).

Parasite-induced effects may have a synergistic interaction with prey availability and starvation as was noted at St Croix with jackass penguins and in Florida with common loons. Alterations in food habits drive shifts in parasite faunas, exposing naïve birds to a different spectrum of helminths. Essentially this is a function of a breakdown in ecological isolation or barriers and reflects fundamental ecological fluctuations involved in prey selection, foraging behaviour and habitat. Such transformations may reveal responses to ecological perturbation linked to patterns of oceanographic and water-mass structure on varying temporal and geographical scales. Among these we can identify extremes ranging from short-term weather induced variation to long term 'regime shifts' of multi-year or decadal duration on a regional or oceanic extent (e.g. El Nino Southern Oscillation or ENSO, Pacific Decadal Oscillation or PDO) that influence the distributions of macrozooplankton, fishes and seabirds (e.g. Ainley and Boekelheide 1990, Springer 1998, Croxall *et al.* 2002, Chavez *et al.* 2003). Parasite faunal diversity in the Gulf of Alaska and Bering Sea during 1976 and 1982 may reflect such alterations in marine water masses and patterns of primary and secondary productivity (Hoberg 1996). Periodic population crashes for little penguins, *Eudyptila minor*, along the south-eastern coast of Australia have been linked to intense parasitic infections by digeneans, cestodes and nematodes in conjunction with annual variation in prey availability and starvation associated with prevailing storms (Norman *et al.* 1992). Consequently, mortality events driven by parasitism could be episodic in the context of regime shifts. With increasing ecological perturbation in nearshore marine systems associated with both anthropogenic and natural events we may also anticipate or predict a heightened frequency for mortality events linked to parasitism.

The dynamics of sympatric avian assemblages at colony sites or traditional foraging zones such as oceanic fronts may further act as determinants of parasite distribution and as proximate factors in morbidity and mortality for some host species (e.g. Hoberg 1996, Galaktionov 1996). Certain seabirds such as larids may serve as sources for the local maintenance and transmission for some helminths. For example, the digenean *Cryptocotyle lingua* infects a wide range of avian hosts, but is probably most typical in *Larus* spp., and large populations of these parasites are most often associated with these littoral marine birds (Lauckner 1985). Anthropogenic effects linked to eutrophication and pollution and other potential impacts can further lead to a shift in both the distribution of parasites and birds in coastal communities (e.g. Galaktionov 1996). High density for host populations facilitates hyperparasitism. Thus, the frequency and intensity of infections and disease associated with pathogens including species of *Cryptocotyle* and *Microphallus* in kittwakes is related to large populations of gulls that are concentrated in relatively small areas as exemplified in some coastal zones of northern Norway and the Kola Peninsula, Russia (Galaktionov 1996) and the Sea of Okhotsk (Hoberg 1992). Notably, great abundance of *C. lingua* was implicated in mortality of kittwake chicks at Talan Island and further indicated extensive foraging by adult birds in nearshore habitats on demersal piscine prey in contrast to more usual pelagic environments.

Subtle effects of parasites that may have substantial and long-term consequences have also been postulated. For example, the extent and timing of annual moulting and future reproductive fitness appear to be influenced by infections of *Seuratia shipleyi* in laysan albatrosses, *Diomedea immutabilis* (Langston and Hilgarth 1995). The presence of these nematodes, even at low levels of abundance, may determine the overall condition of individual birds and their ability to moult primary flight feathers. Delayed moult where fewer flight feathers than usual are replaced in parasitised birds may be the driver for albatrosses that skip some reproductive cycles, thus acting as a constraint on overall fitness (Langston and Hilgarth 1995). As infections by *Seuratia*, other nematodes and a diversity of cestodes are common, such a phenomenon may be a general factor among large seabirds. A linkage between parasitism and fitness could be particularly pronounced among procellariiforms that are characterised by clutches with single eggs, extended breeding seasons and an incomplete annual moult. Additionally, cryptic constraints on fitness, or even mortality at sea where breeding age birds are involved, may not be immediately evident at colony sites. The impact of adult mortality can be masked by life history patterns for large marine birds that have extended life spans, defer the age of first breeding and produce relatively few chicks.

Likely the most bizarre instance of mass mortality attributable to parasites occurred among white-faced storm petrels, *Pelagodroma marina*, breeding on Chatham Island near New Zealand (Claugher 1976). An estimated 200 000 birds in a colony of one million died of starvation following accidental entanglement by filamentous metacercariae of *Distomum filiferum*. Metacercariae of these trematodes in the genus *Syncoelium* are released from euphausiids such as *Nematoscelis megalops*, and after drifting in the water column, infect the branchial cavity and mouth of certain pelagic marine fishes that serve as definitive hosts. Pattering behaviour of white-faced storm petrels while foraging for euphausiids at the surface (see Ashmole 1971) would predispose them to becoming entangled by metacercariae released by vertically migrant crustacean intermediate hosts. Mortality for such birds as petrels that feed by pattering could be considerable given atmospheric and surface-ocean conditions that may facilitate contact between birds and flukes. Although entanglement also involved substantial numbers of five additional species of procellariiforms, including common diving petrels, *Pelecanoides urinatrix*, sooty shearwaters, *Puffinus griseus*, broad-billed prions, *Pachyptila vittata*, and black-winged petrels, *Pterodroma nigripenis*, the specific behaviour and size of these birds relative to *P. marina* may explain why they were not involved in the mass mortality (Claugher 1976, Imber 1984).

Large-scale mortality events among populations of seabirds have often not been studied in detail, and parasitological examinations in this context continue to be rare (Galaktionov 1996). Incidents since the mid 1980s of considerable scope involving putative starvation and breeding failures have been documented in several regions. For example, in the Barents Sea breeding populations of murres were dramatically reduced in the winter of 1986–1987 during a population crash attributed to overfishing of capelin *Mallotus villosus* (Galaktionov 1996). Additionally, short-tailed shearwaters, *Puffinus tenuirostris*, died in large numbers, estimated as hundreds of thousands, during the Fall migration of 1997 in Unimak Pass (Aleutian Islands), Alaska, when their principal prey, euphausiids, *Thysanoessa* spp., were not available in the water column; in 1998 shearwaters shifted to piscine prey but overall body conditions remained poor. During these events, both attributed to starvation, weather and shifting abundance of primary prey, concurrent parasitological examinations were not a component of any investigations, nor have they been used to reveal the dynamics of changing oceanographic domains and trophic pathways as advocated (Hoberg 1996). Yet, all examples of a major involvement for parasites in morbidity or mortality among populations of seabirds have implicated ecological instability as a proximate cause. In these instances it is necessary to establish comparative baselines against which changing faunal diversity may be assessed (Hoberg 1996). To the extent that seabirds can tell us about fundamental shifts in marine habitats, and parasites are exquisite indicators of such variation, ecologists and parasitologists should take the opportunity for reciprocal illumination about the status of large avian populations, a diversity of parasites and the environments in which they live (Galaktionov 1996, Hoberg 1996, 1997, Brooks and Hoberg 2000).

Important references

The various chapters in Croxall (1987) contain much information on the feeding ecology of seabirds and their role in marine ecosystems. Lauckner (1985) gave a detailed and well-illustrated account of diseases including those attributed to parasites of marine birds. Barus *et al.* (1978) discussed nematodes of fish-eating birds of the Palaearctic, and Fagerholm (1996) reviewed nematodes of marine and shore birds, and their role as pathogens. Galaktionov (1996) evaluated the impact of seabird helminths on host populations and coastal ecosystems, and Hoberg (1996) discussed faunal diversity among avian parasite assemblages. Storer (2000, 2002), in extensive monographs on the parasites of grebes and loons, provides a strong rationale for cooperation and synergy among parasitologists and ecologists.

Effects of pollution on parasites, and use of parasites in pollution monitoring
Bernd Sures

Introduction

An increasing number of papers has focused on the interrelations between parasites and environmental pollution (reviewed by Khan and Thulin 1991, Poulin 1992, MacKenzie et al. 1995, Lafferty 1997, Overstreet 1997, Lafferty and Kuris 1999, MacKenzie 1999, Sures et al. 1999a, Sures 2001, Williams and MacKenzie 2003, Sures 2003, 2004). Environmental parasitology was the subject of international symposia at the VIIth European Multicolloquium of Parasitology in Parma, Italy (Paperna 1997) and at the Xth International Conference on Parasitology in Vancouver, Canada (Proceedings ICOPA 2000). Furthermore, the British Society of Parasitology dedicated their autumn meeting 2002 to the topic 'parasitism and environmental pollution' (Lewis and Hoole 2003).

Effects of pollution on parasites may be positive or negative: pollution may increase parasitism, or it may be fatal for certain parasite species, leading to a decrease in parasitism. Generally, infections with ectoparasites tend to increase, whereas infections with endoparasitic helminths tend to decrease with increasing levels of pollution. Hence, parasites may be used as bioindicators to monitor pollution. Although we are just at the beginning of understanding the complex and diverse relationships between parasites and pollution, an increasing number of papers shows that we are dealing with an extremely interesting and important field of interdisciplinary research. This section gives an account of current research on marine parasites and pollution.

Marine pollution is a complex phenomenon resulting in a variety of effects on parasites (see a comprehensive list in Williams and MacKenzie 2003). Pollution may increase parasitism if, for example, host defence mechanisms are negatively affected, thereby increasing host susceptibility, or by simply increasing the population densities of suitable intermediate or final hosts. An increase in parasitism as a result of pollution is especially likely in the case of monoxenic ectoparasites (Table 10.2). Ectoparasites are in close contact with the external environment of their hosts and should consequently have adapted to the external conditions at least as well. A good example of increased ectoparasitism due to pollution is trichodinid gill ciliates of fish. Eutrophication due to human activities often coincides with an increase in trichodinids of fish (Broeg et al. 1999). As eutrophic waters contain many bacteria on which ciliates can feed, eutrophication favours the occurrence of these protozoans (Palm and Dobberstein 1999). Other types of pollution such as crude oil and industrial effluents may also increase the numbers of trichodinids (e.g. Khan et al. 1994).

Other groups of ectoparasites such as monogeneans and crustaceans are also directly affected by contaminated water. Experimental and field studies have demonstrated that monogenean abundance usually increases in systems polluted by hydrocarbons, polycyclic aromatic hydrocarbons (PAHs), polychlorinated biphenyls (PCBs) and others (e.g. Khan and Kiceniuk 1988, Marcogliese et al. 1998, Williams and MacKenzie 2003). The most likely explanation for the increase in monogeneans in these studies is that fish were immunologically compromised by pollutants (Rice and Schlenk 1995, Regala et al. 2001, Sures and Knopf 2004).

Parasitic crustaceans, another major group of marine ectoparasites of fish, are less well studied in respect to their sensitivity to pollution. Broeg et al. (1999) found five species of copepod crustaceans on the skin and gills of flounder caught in different areas of the North Sea. Although flounders caught at the most polluted sampling site had the lowest prevalence and intensity of infection with copepod parasites, correlation analysis showed no connection between infection statistics and environmental data.

Table 10.2 Effects of different types of pollution on different parasite taxa

Type of pollution	Parasite taxa					
	Ciliophora	Monogenea	Digenea	Cestoda	Nematoda	Acanthocephala
Eutrophication	+	+	+	+	+	+
Thermal effluent	+	+/–	–	=	+	+/–
Pulp-mill effluent	+	+/–	+/–	n	+/–	n
Crude oil	+	+	–	n	+	–
Industrial effluent	+	+/–	–	–	+/–	+/–
Sewage sludge	+	=	+/–	+	=	–
Acid precipitation	n	–	–	–	n	+
Heavy metals	n	n	–	–	n	–

The symbols represent the direction of the effects on parasite abundance: +, increase; –, decrease; =, no effect, n, no data available), modified after Lafferty (1997) and MacKenzie (1999).

Under certain conditions, pollution may occasionally also increase infection with endoparasites. For example, if eutrophication favours the occurrence of invertebrates which are used as intermediate hosts (e.g. molluscs), heteroxenic trematodes might profit from such an increase in intermediate host numbers and abundance. However, more frequently the numbers of digeneans, cestodes and acanthocephalans with complex life cycles decrease with most types of pollution (Table 10.2). Pollution can negatively affect parasitism provided that:

1. infected hosts suffer more from environmental exposure than uninfected hosts;
2. parasites are more susceptible to the particular pollutant than their host; and
3. pollution drives the necessary intermediate and final hosts to extinction.

Furthermore, effects of pollution may vary between parasite species and between developmental stages, as larval and adult parasites may be affected in different ways. Susceptibility to pollution appears to be correlated with the number of developmental stages which are involved in the life cycle of a particular parasite species. Endoparasites with indirect life cycles can be affected directly or indirectly by pollutants. Effects are direct when free-living transmission stages (e.g. miracidia and cercariae) or adults in the intestine are in contact with the chemical, they are indirect when there are adverse effects of the pollutant on other hosts in the parasite's life cycle. In spite of numerous parasites that have free-living larval stages, most studies dealing with toxic effects of chemicals on parasites were conducted with miracidia and cercariae of trematodes (for review see Pietrock and Marcogliese 2003). Generally, environmental pollutants such as metals reduce survival of the test organisms. However, the reduction of the longevity of cercariae is not very important for the transmission of the respective trematodes, as infectivity of cercariae is optimal only in the first few hours after emergence from the snail intermediate host (Whyte et al. 1991). But metal exposure may also affect cercarial activity during maximum infectivity, thereby causing a significant reduction in transmission success (Morley et al. 2003).

The general conclusion that emerges from the examples mentioned above and the studies reviewed earlier (see e.g. Williams and MacKenzie 2003) is that environmental pollution does affect parasite populations and communities, both directly and via effects on intermediate, paratenic and final hosts. But as population or community changes could also be stochastic, parasitologists still need to demonstrate that pollution is actually the cause of changes in many parasite infections. In those cases where it is possible to relate parasite changes to pollution, such changes can be used as measures of environmental change, that is, parasites can be used as indicators for particular kinds of pollution.

Use of parasites in monitoring environmental changes

Biological monitoring can be defined as the use of biological responses to evaluate changes in the environment with the intent to use this information in a quality program. The use of organisms as bioindicators is advantageous compared with the analysis of chemicals in the water or sediment, as indicator organisms bioconcentrate chemicals and integrate their effects over time. Furthermore, analysis of chemical levels in the water column or in sediments is important only if there is information on the solubility and bioavailability of a particular substance.

As understood presently, bioindicators are either used as effect indicators or as accumulation indicators. Effect indicators permit the recognition of pollution by changes in their physiology and/or behaviour, whereas accumulation indicators bioconcentrate environmental pollutants above environmental levels. There is evidence from the literature that parasites might be used in both types of bioindication procedures.

Parasites as effect indicators

The best known example for effect indication is the so-called mussel monitor, where the behaviour of mussels such as *Mytilus edulis* is analysed in respect of pollution. In clean water, mussels move the two halves of their shells in a characteristic pattern. They remain open for most of the time, and only close for short periods. A mussel in contaminated water behaves differently. The frequency at which the mussel opens its shell is measured in such a mussel monitor and changes of shell movements are analysed by a microprocessor. Comparable online monitoring procedures using parasites appear to be rather unlikely.

Concerning parasites, effect indication might be possible using the direct toxicity of substances, for example on free-living parasitic stages and the subsequent alteration of parasite populations and communities as described (see above). One way to indicate the combined effects of environmental pollution on parasites might be the use of the ratio of the sums of heteroxenous and monoxenous (H/M) parasite individuals per host, as well as the ratio of heteroxenous and monoxenous parasite species richness found on the host fish (S_H/S_M) as suggested by Diamant et al. (1999). The authors found higher H/M ratios and higher S_H/S_M species richness ratios when comparing parasite communities of rabbitfish, *Siganus rivulatus*, sampled from an ecologically stable habitat of the Red Sea as compared with the community of fish sampled from anthropogenically impacted environments. In contrast, the application of the model by the same researchers in the North Sea using parasite communities of flounder, *Platichthys flesus*, revealed that it was not useful in discriminating differently polluted habitats because of the dominance of heteroxenous species at all sites (Broeg et al. 1999). Another problem in the use of parasite community changes as an indication for environmental pollution or disturbance is that these changes could also be stochastic rather than related to environmental changes (Kennedy 1990). Accordingly, there is a high degree of complexity and uncertainty in the interpretation of parasitological data in the context of pollution. In addition, analysis of the available literature reveals that there are only few parasite–pollution combinations which show predictable changes despite the considerable effort which had been put into linking levels of environmental pollution with parasitism (see Lafferty 1997, Kennedy 1997, Williams and MacKenzie 2003). However, analyses of parasite population and community structures may be advantageous compared to established effect indication procedures using free-living animals, as it integrates effects of

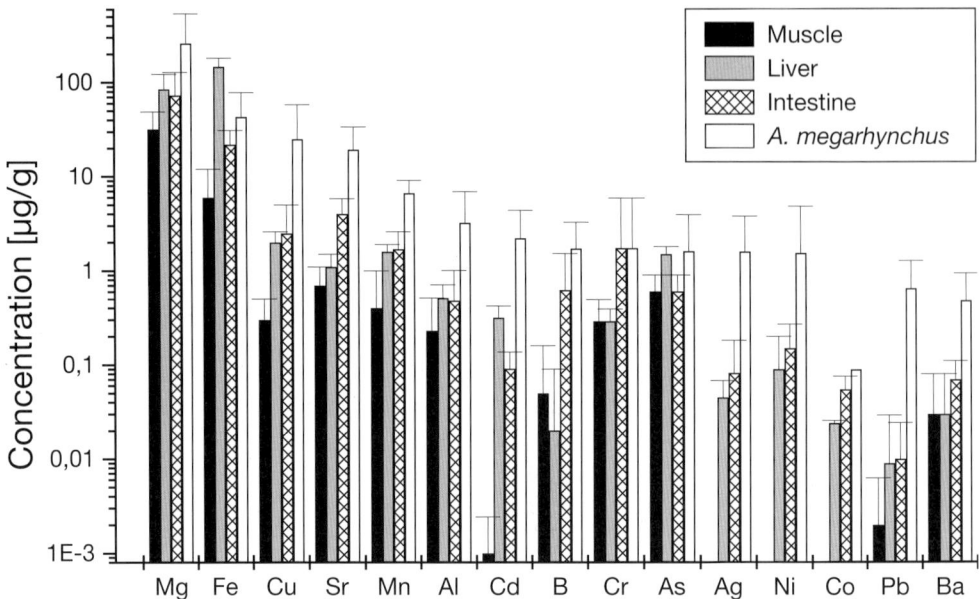

Figure 10.17 Element concentrations in different organs of *Notothenia coriiceps* and in *Aspersentis megarhynchus* caught at King George Island, Antarctica. Nearly every element was concentrated to a higher degree in the acanthocephalan parasite than in the host tissues. Data from Sures and Reimann (2003).

environmental parameters over different trophic levels. Therefore, it is still desirable to intensify research in this field as suggested by Lafferty (1997).

Parasites as accumulation indicators

Accumulation indicators (sentinels) should above all efficiently take up substances and reach an equilibrium at which the uptake of the respective substance is balanced by its excretion. This uptake and bioconcentration of substances should not be accompanied by toxic effects on the organism. Concerning parasites as accumulation indicators, studies published so far have mainly focused on metals; less frequently, investigations deal with organic pollutants (Sures 2003, 2004). This is obviously related to different accumulation patterns of hydrophilic and lipophilic substances. Lipophilic chemicals mainly accumulate in fat, whereas hydrophilic substances are distributed more evenly among tissues. Parasites, having a low percentage of fat, are not able to bioconcentrate lipophilic substances above the levels of the host tissues (Heinonen *et al.* 2000, Ruus *et al.* 2001). For example, Ruus *et al.* (2001) described lower concentrations of lindane in metacercariae of *Bucephaloides gracilescens* after experimental exposure compared with inner organs such as the gall bladder, the urinary bladder and the intestine of the intermediate host *Myoxocephalus scorpius*.

Intestinal helminths of fish such as acanthocephalans and cestodes, however, are very efficient in bioconcentrating metals (reviewed in Sures *et al.* 1999a; Sures 2001, 2003, 2004). Although most research on metal accumulation in parasites was conducted in freshwater habitats, it emerged that the same pattern of intensive metal uptake and accumulation by the parasites also occurs in marine habitats (e.g. Zimmermann *et al.* 1999, Sures and Reimann 2003). In a recent field study the marine fish *Notothenia coriiceps* infected with *Aspersentis megarhynchus* was caught at King George Island, South Shetland Islands, and analysed for 14 different elements (Sures and Reimann 2003) (Figure 10.17). Most of the elements were found in significantly higher concentrations in the acanthocephalan than in muscle, liver and intestine of its host. The value of acanthocephalans for the detection of metals becomes evident when comparing levels of lead (Pb) in the host tissues and in the parasites. The Pb levels in muscle, liver and intestine of the fish host were below the detection limit or very close to it, whereas the parasites accumulated this metal much more strongly (Fig. 10.17). Accordingly, if only the fish tissues had been analysed for lead, one could conclude that this metal was not present in concentrations high enough for biological uptake. But the enormous accumulation potential of the parasites clearly showed that lead is available in concentrations high enough for biological uptake processes.

Therefore, the use of parasites, particularly of acanthocephalans as sentinels, is advantageous for the determination of extremely low concentrations of pollutants. Metal uptake rates in the parasites surpass those of the host tissues, and the accumulation capacity of acanthocephalans is even higher than that of established bioindicator species such as the zebra mussel (Sures *et al.* 1999b). In conclusion, it appears that intestinal helminths, particularly acanthocephalans, are very useful in assessing the presence and availability of metals in aquatic (limnetic and marine) biotopes.

Important references

Khan and Thulin (1991) reviewed the influence of pollution on parasites of aquatic animals, and Broeg *et al.* (1999) and Diamant *et al.* (1999) discussed fish metabolic, pathological and parasitological indices as indicators of pollution. Further important discussions of the use of parasites in monitoring environmental pollution are by MacKenzie *et al.* (1995), Lafferty (1997), Overstreet (1997), MacKenzie (1999), Lewis and Hoole (2003), Williams and MacKenzie (2003) and Sures (2004). The use of parasites as bioindicators of heavy metals was discussed by Sures (2001, 2003).

Chapter 11

Medical importance

Well over 100 species of eukaryotic parasite species have been reported from humans. Some of them are among the most important agents of human disease. However, almost all important species have terrestrial and/or freshwater life cycles. Examples include the malaria parasites, blood flukes and various nematodes, infecting hundreds of millions of people with often fatal consequences. Nevertheless, many marine parasite species have been reported from humans as well, although all of them can complete their life cycle without involvement of humans. They include a few cestodes, many trematodes, some nematodes and many protistans. For most of these, humans are accidental hosts and the typical life cycle is either completed in marine mammals or birds, or they live primarily in fresh water and have only secondarily become marine, or infections are entirely or largely restricted to immunocompromised persons. Among the first group are many larval trematodes, some cestodes and nematodes, among the second group are also many trematode larvae, cestodes and nematodes, and among the third are an ever increasing number of protistans such as apicomplexans and microsporidians. Most important agents of disease are nematodes causing anisakiasis, acquired by eating raw or undercooked marine fish and invertebrates, and the nematode *Trichinella* acquired by eating undercooked marine mammals such as walrus. Schistosome cercariae may cause a severe dermatitis in marine and brackish waters, lungworms of the genus *Angiostrongylus* occur mainly in fresh water but may use marine invertebrates as transport hosts and cause severe symptoms and even death in infected persons, and cestodes and trematodes accidentally acquired may occasionally also cause symptoms of varying degrees of severity.

Cestode and trematode infections
David Blair

Introduction
Given that much of the human race has taken sustenance from the sea for millennia, remarkably few species of truly marine parasitic platyhelminths have taken advantage of this. All that do are probably zoonotic (pathogens that can be exchanged between vertebrate animals and humans) and their life cycles can be maintained naturally in the absence of humans. For both trematodes and cestodes, taxonomic and diagnostic uncertainties make it difficult to be dogmatic about the numbers of species that can occur in humans. Attempts to provide comprehensive lists are in Coombs and Crompton (1991) and in the appendices to Taylor *et al.* (2001).

The route of human infection is by ingestion of marine organisms in which larval parasites (metacercariae or plerocercoids) remain alive. Thus, raw, salted, marinated or lightly cooked fish are risky eating for millions of people in parts of the world where such a diet is culturally entrenched. The advent of cheap air travel has meant that people infected in one part of the world can take their souvenir parasites home to delight medical practitioners there (e.g. Chai et al. 1986). Furthermore, rapid transport of delicacies such as raw fresh fish to all parts of the planet is increasingly possible: the consequence of an exotic meal might be an exotic infection (e.g. Adams et al. 1986, Clavel et al. 1997).

Trematodes

Some of the relevant trematodes are listed in Table 11.1. All have a gastropod first intermediate host (where known), molluscan or fish second intermediate hosts and a range of birds or terrestrial mammals as final hosts (none appears to cycle naturally through marine mammals). Metacercariae, ingested along with the (inadequately cooked) second intermediate host, mature in the human intestine (usually the small intestine). Adults of almost all species are very small. Several points are noteworthy. First, of the 77 species of trematodes listed in Coombs and Crompton (1991) as likely to be transmitted to humans in this way, remarkably few are unambiguously of marine origin. Most involve mainly freshwater hosts and included among these are pathogens causing widespread and serious conditions (notably opisthorchiasis/clonorchiasis) which will not be discussed here (see Miyazaki 1991). All species listed in Table 11.1 are found in estuaries and along coastlines. Many utilise brackish-water hosts including those that can move between marine and freshwater environments. Species of mullet, especially the widespread *Mugil cephalus*, feature prominently among the reported fish hosts.

Second, almost all species listed in Table 11.1 belong to the family Heterophyidae, suggesting that members of this family are unusually broad in their host specificity. Indeed, of the approximately 70 species of trematodes known to infect the human intestine (as opposed to lungs and other sites), about 31 are heterophyids and most of the rest belong to the Echinostomatidae (Chai and Lee 2002). Apart from the Heterophyidae, the only families listed in Table 11.1 (each represented by a single species) are the Echinostomatidae and the Gymnophallidae.

Third, there is strong geographical bias in records of human infection. Japan and the Republic of Korea are sources of many recent records (e.g. Chai and Lee 2002). Elsewhere in Asia there has been relatively little published work recently. Human infection with marine trematodes is commonly recorded from the eastern Mediterranean but from few other places outside Asia. The geographical bias may simply reflect the distribution of interested researchers. However, the bias may also have a cultural component: preparation of uncooked or undercooked fish for human consumption is a culinary art in Japan and surrounding regions. Nevertheless, infection with marine trematodes is undoubtedly grossly under-reported from many parts of the world.

Finally, infection by most species listed in Table 11.1 may produce no symptoms, or symptoms that are generally vague and non-specific. This helps to explain the lack of diagnostic interest in many parts of the world and also makes accurate estimation of prevalence impossible. In Korea, anthelmintic treatment followed by purging demonstrated that many people in coastal areas unknowingly harbour fish-borne trematodes. For example, Chai et al. (1997) found 75% of people sampled in one coastal village were infected with heterophyids. Numbers of worms recovered from purged patients can be extraordinary. Chai et al. (2002) recovered 69 125 specimens of *Gymnophalloides seoi*, 328 specimens of *Heterophyes nocens* and one specimen of *Stictodora lari* from a 69-year-old male in Korea. This patient had experienced only episodes of epigastric pain and discomfort, diarrhoea and indigestion. Symptoms of this sort may be more pronounced in people with many worms but apparently rarely cause them to seek medical help. More serious consequences of infection with these small trematodes can occur. *Gymnophalloides seoi* might invade the pancreatic duct. This organism has been found in people with pancreatitis and diabetes mellitus, but whether the trematode can cause these conditions remains unclear. In

Table 11.1 Selected zoonotic trematodes with life cycles involving marine and estuarine hosts

	Notable vertebrate hosts (natural unless otherwise stated)	Localities of known human infections
Heterophyidae		
Heterophyes heterophyes	Metacercariae in *Mugil, Tilapia, Solea* and *Sciaena*; adults in various fish-eating birds, domestic dogs and cats, wild canids	Eastern Mediterranean
Heterophyes nocens	Metacercariae in *Mugil, Acanthogobius*; adults in domestic cats	Japan, Korea
Heterophyes dispar	Metacercariae in *Mugil, Tilapia* and *Epinephelus*; adults in domestic dogs and cats, wild canids and some fish-eating birds	Middle East and northern Africa
Heterophyopsis continua	Metacercariae in marine and brackish water fish – *Clupanodon, Mugil, Konosirus, Lateolabrax, Acanthogobius* and *Conger*; adults in domestic cats, ducks, seagulls and other fish-eating birds	Philippines, Japan, Korea
Cryptocotyle lingua	Metacercariae in many marine fish; adults in many birds and mammals	Alaska, Greenland
Pygidiopsis genata	Metacercariae in *Mugil* and *Tilapia*; adults in domestic dogs and cats and many fish-eating birds	Egypt
Pygidiopsis summa	Metacercariae in *Mugil, Acanthogobius*;c adults in domestic cats and dogs and fish-eating birds	Japan, Korea
Stictodora fuscata	Metacercariae in *Mugil, Acanthogobius*; adults probably in fish-eating birds	Japan, Korea, Philippines
Stictodora lari	Metcercariae in *Mugil, Acanthogobius* and other brackish-water fish, adults in seagulls	Korea
Stellantchasmus falcatus	Metacercariae in *Mugil, Liza, Acanthogobius*; adults in fish-eating birds	Thailand, Philippines, Japan, Korea, Hawaii
Echinostomatidae		
Acanthoparyphium tyosenense	Metacercariae in estuarine bivalves *Mactra veneriformis, Solen grandis* and a gastropod *Neverita bicolor*; adults in ducks	Korea
Gymnophallidae		
Gymnophalloides seoi	Metacerceriae in oysters, *Crassostrea gigas*; adults in oystercatchers, *Haematopus ostralegus*, and other wading birds	Korea
Main sources include: Williams and Jones (1976, 1994), Yamaguti (1971), Seo et al. (1984), Marty and Andersen (2000), Chai et al. (2001), Chai and Lee (2002), Lee and Chai (2001), Chai et al. (2002), Acha and Szyfres (2003).		

studies using mice, both *G. soei* and the freshwater heterophyid *Metagonimus yokogawai* were found to live among the villi of the small intestine but did not penetrate into the submucosal region in immunocompetent mice. In immunosuppressed mice, however, worms were observed to penetrate into and beyond the submucosa (Chai and Lee 2002). Movement of heterophyids and their eggs beyond the limits of the intestine has been noted in human cases in the Philippines and elsewhere. Embolisms caused by adult worms or (more usually) their eggs have been noted in several sites such as the brain, spinal cord and heart, often with fatal consequences (brief reviews in Williams and Jones 1976, Chai et al. 1986).

Echinostomes may cause more serious damage to the intestinal walls than do heterophyids (Chai et al. 2001). Some people infected with *Acanthoparyphium tyosenense*, the only echinostome included in Table 11.1, complained of gastrointestinal trouble and indigestion. However, concomitant infection with other flukes could have been the cause of pathology (Chai et al. 2001).

Cestodes

All cestodes of interest here are pseudophyllideans of the family Diphyllobothridiidae. Most are in the genus *Diphyllobothrium*, but some nominal species of *Diplogonoporus*, that normally mature in baleen whales and pinnipeds, have also been found in the human intestine (Oshima and Kliks 1986). The taxonomy of species of both genera remains poorly resolved, especially in the case of *Diphyllobothrium*, which contains some 80 nominal species (Andersen 1987, Williams and Jones 1994). Where known, first intermediate hosts are copepods: plerocercoids develop in fish that eat infected copepods. In some species, plerocercoids can concentrate up the food chain through a series of paratenic hosts. This, coupled with movements of anadromous fish hosts, can lead to infective stages being transferred between marine and freshwater habitats (Rausch and Adams 2000).

Most human infections are due to *Diphyllobothrium* species (e.g. *D. latum*) that cycle through freshwater hosts and mature naturally in a variety of fish-eating birds and terrestrial mammals. Numerous *Diphyllobothrium* species mature in marine mammals (cetaceans and pinnipeds) and some of these occur infrequently in humans. Some apparently marine species have been described only from humans (listed in Coombs and Crompton 1991), but presumably these are also zoonotic with the usual final hosts remaining unknown. It seems possible to distinguish between marine and freshwater species by the structure (thickness, degree of pitting) of the egg shell (Hilliard 1960, Maejima et al. 1983) and the ability of the coracidium to hatch and survive in sea water.

Table 11.2 lists species known to infect humans. *Diphyllobothrium pacificum* is by far the commonest and most widespread marine representative of its genus to occur in humans. Probably more common are *Diplogonoporus* spp.: representatives of this genus have been found in over 180 people in the 20th century (Kino et al. 2002), mostly in Japan.

Infections are often asymptomatic. Symptoms due to marine species are similar to those of *Diphyllobothrium* species in general (Marty and Neafie 2000): diarrhoea, abdominal discomfort, nausea, anorexia and fatigue. However, medical advice is often sought only after the patient has been alarmed by the presence of strings of proglottids in the faeces. The anaemia produced in some people by *Diphyllobothrium latum* is not seen in infections due to marine species.

Important references

Williams and Jones (1976) discussed marine helminths and human health, Acha and Szyfres (2003) gave a comprehensive account of zoonoses and communicable diseases common to humans and animals, including marine parasites, and Chai and Lee (2002) discussed foodborne intestinal trematode infections in Korea, a country where infections with marine helminths are common. The guide to human helminths by Coombs and Crompton (1991) includes marine parasites.

Anisakiasis

Kazuya Nagasawa

Introduction

Anisakiasis is a disease caused by larval worms of the ascaridoid nematode family Anisakidae (Oshima 1972, 1987, Smith and Wooten 1978). The infection is acquired by eating uncooked or

Table 11.2 Species of diphyllobothriid cestodes with marine life cycles that are known or suspected to infect humans

Species	Natural geographic distribution	Natural final hosts	Notes and references
Diphyllobothrium alascense	Alaska	Domestic dog	Dog only host (Rausch and Adams 2000). Early identification of human cases tentative (Rausch et al. 1967)
Diphyllobothrium cameroni	Midway Atoll, Pacific	Pinnipedia; Phocidae; (Monachinae)	Single human case detected in Japan
Diphyllobothrium cordatum	N. Pacific, Bering Sea; Greenland, Spitzbergen,	Pinnipedia; Otariidae, Odobenidae, Phocidae (Phocinae): domestic dog	Single human case in Greenland but identity questioned by Rausch and Adams 2000. Scolices possibly of this species from humans in Alaska (Rausch et al. 1967)
Diphyllobothrium elegans	N. Sea, Greenland	Pinnipedia; Otariidae, Phocidae (Phocinae, Monachinae)	Human record from Japan (Coombs and Crompton 1991)
Diphyllobothrium hians	N. Atlantic, Siberia; Hawaii	Pinnipedia; Phocidae (Phocinae, Monachinae)	Single human case detected in Japan but probably acquired elsewhere
Diphyllobothrium lanceolatum	N. Atlantic; N. Pacific; Bering Sea	Pinnipedia; Phocidae (Phocinae): Cetacea; Phocoenidae: domestic dog	Single plerocercoid passed by human in Alaska (Rausch and Hilliard 1970)
Diphyllobothrium nihonkaiense	Japan, Korea	unknown	Many human cases in Japan. This species formerly reported in Japan as *D. latum* – see Yamane & Shiwaku 2003
Diphyllobothrium orcini	Japan	Cetacea; Delphinidae	Single case detected in Japan
Diphyllobothrium pacificum	Peru, Chile, Alaska, Japan	Pinnipedia; Otariidae: domestic dog	Human records throughout range and in archaeological material
Diphyllobothrium scoticum	Falkland Islands; Southern Ocean	Pinnipedia; Otariidae; Phocidae (Monachinae)	Single case detected in Japan but probably acquired elsewhere
Diphyllobothrium stemmacephalum (syn. *D. yonagoense* according to Andersen, 1987)	N. Atlantic, N. Sea Japan, Korea	Cetacea; Delphinidae	Occasional human cases from Japan and one from Korea
Diplogonoporus balaenopterae and other nominal species in this genus	Circumboreal and recorded from the Antarctic	Cetacea; Balaenopteridae: also recorded from phocids and otariids: other mammals rarely	Human records from Japan, Korea and Spain (the last probably imported). Most common marine cestode in humans (Oshima and Kliks 1986, Clavel et al. 1997, Kino et al. 2002)

undercooked marine fish and invertebrates. Symptoms vary greatly (for details of life cycles and hosts see pp. 107–115).

Species responsible and geographical distribution

The nematodes that cause human anisakiasis are larval *Anisakis simplex* (frequently called *Anisakis* type I larvae) in most cases, followed by larval *Pseudoterranova decipiens*. Other larval anisakid nematodes, such as *A. physeteris* (*Anisakis* type II larvae), *Contracaecum osculatum* and *Hysterothylacium aduncum*, are very rarely found in humans. When humans eat raw or insufficiently cooked marine fish or squid harbouring anisakid larvae, they may get infected and develop anisakiasis. Based on the definitive hosts in which adult worms occur, *A. simplex* and *P. decipiens* are called 'whaleworm' and 'sealworm', respectively.

Since the first discovery of human infection by larval *Anisakis* from The Netherlands in 1955, more than 13 000 cases of the disease caused by anisakid worms have been reported from nearly 20 countries up to the early 1990s (Ishikura et al. 1992). Most (>95%) of these cases are from Japan, followed by European countries (mainly from The Netherlands, Germany, France) and the United States of America (USA). In Japan, where raw marine fish and squid are consumed as sashimi and sushi in large quantities, diagnostic methods and treatment for the disease are well developed. Anisakiasis due to larval *Anisakis* is far more common (>97%) than infections by other anisakids like *P. decipiens*. *Anisakis simplex* is responsible for most cases of human anisakiasis but *A. physeteris* is known to cause a few cases of the disease.

Disease symptoms

Based on the location of the lesion, anisakiasis is categorised into gastric, intestinal and extragastrointestinal types, all of which are further divided into fulminant (acute) and mild (chronic) forms (Ishikura and Namiki 1989, Ishikura and Kikuchi 1990). In Japan, gastric anisakiasis is exclusively predominant (95.0%), followed by intestinal (4.6%) and extragastrointestinal (0.4%) anisakiasis (Ishikura et al. 1992).

Most cases of gastric anisakiasis are diagnosed as the fulminant form based on their clinical manifestations of a sudden onset of severe abdominal pain (Ishikura and Namiki 1989, Ishikura et al. 1992). Nausea and vomiting often occur with the abdominal pain. Other symptoms, such as abdominal distention, diarrhoea, urticaria, anorexia and chest pain, are less frequently found. The abdominal pain usually occurs two to seven hours after eating raw seafood containing *Anisakis* larvae. The severity of the pain is different among patients. Larvae penetrating the stomach wall are most frequently found in the greater curvature, followed by the posterior wall, anterior wall and lesser curvature. The larva moves actively during the early stage of penetration but its movement gradually becomes slow. A non-penetrating larva is sometimes seen. One larva is usually found in the stomach lumen but there are also multiple infections (up to 10 worms). Oedema is observed at the penetration sites, where erosion and bleeding may be found. Histologically, both submucosal oedema and inflammatory cell infiltration are common at the penetration sites. Mucosal changes, such as widespread oedema, severe erosion and spotty haemorrhage, are also present in other areas. The fulminant form of gastric anisakiasis may be a result of allergic reactions induced by reinfection with *Anisakis* larvae. Various aspects of problems and disease manifestations related to allergy to larval *Anisakis* have been reviewed (Audicana et al. 2002). The fulminant form is diagnosed by gastrofiberscopy. Without surgical operation, a larva can be easily removed with a gastrofiberscopic biopsy forceps and the pain disappears after removal of the larva.

In the mild form of gastric anisakiasis, the abdominal pain is not severe and lasts for a long time (more than two years if the larva is not removed) (Ishikura and Namiki 1989, Ishikura et al. 1992). There also are cases in which patients do not develop any severe symptoms but are later diagnosed as having mild gastric anisakiasis on endoscopic examination. Nausea and anorexia may occur as other clinical symptoms. Erosion and ulceration are found in the affected area. The mild form of gastric anisakiasis is histologically divided into four stages:

1. phlegmon formation;
2. abscess formation;
3. abscess-granuloma formation; and
4. granuloma formation.

With proceeding to the next stage, degeneration and destruction of the larva may progress, and it might take more than six months after larval infection to reach the third stage, in which a degenerating larva or its debris is seen in an abscess surrounded by granulation tissue. In the final fourth stage, only larval debris is surrounded by granulomatous tissue with collagenisation, foreign body giant cells and eosinophilic infiltration. Primary infection with larval *Anisakis* may cause the mild form of gastric anisakiasis. This form is diagnosed on the basis of a result of microscopic examination of the affected gastric lesion showing an eosinophilic granuloma.

In intestinal anisakiasis, most cases are classified as the fulminant form and there are only a few cases of the mild form (Ishikura and Kikuchi 1990, Ishikura *et al.* 1992). Symptoms of the fulminant form usually appear within a few days after ingesting raw sea fish or squid infected by larval *Anisakis*. The principal clinical symptom of this form is severe pain in the lower abdomen, often accompanied by nausea, vomiting, abdominal bulging, constipation and diarrhoea. The lesion of the small intestine is thickened, causing obstruction and proximal dilation of the intestine. This thickening results from intensive oedema and cell infiltration in the submucosal through to the serosal layer. As much as 300 mL to 500 mL of straw-coloured ascites is present in the peritoneal cavity. Fulminant intestinal anisakiasis can be cured without surgical operation when the disease is correctly diagnosed based on clinical symptoms and information from roentgenographic, ultrasonographic or seroimmunological examination. In the mild form of intestinal anisakiasis, clinical symptoms are very slight and slowly progressing and this form is thus accidentally found when the patient is operated on for other diseases.

In cases of extragastrointestinal anisakiasis, larvae can be found at various sites, such as the mucous membrane of the pharynx, abdominal cavity, large omentum or mesenteries, lymph nodes and subcutaneous tissues (Ishikura *et al.* 1992). The disease usually manifests mild clinical features.

Almost all cases (>94%) of human infections with larval *Pseudoterranova decipiens* are from Japan. Clinical symptoms of the disease are similar to those of fulminant gastric anisakiasis caused by *Anisakis* larvae (Margolis 1977, Ishikura and Namiki 1989). Acute abdominal pain is the principal clinical sign. The larva of *P. decipiens* is located and removed with a gastrofiberscope fitted with biopsy forceps.

A few cases of human anisakiasis caused by larval worms of *Contracaecum* (*C. osculatum* or *Contracaecum* spp.) have been reported from Germany, Japan and Korea (Schaum and Müller 1967, Ishikura 2003). In the German case, the worms were found in the eosinophilic granulomas of the intestine. Also, there was a case of human infection in northern Japan, in which an immature female adult *Hysterothylacium aduncum* was evacuated with the faeces (Yagi *et al.* 1992).

Prevention of infection

Larval *A. simplex*, *P. decipiens* and other anisakid nematodes can be transmitted to humans when raw or inadequately cooked fish or squid are ingested. It is thus important to kill larvae in the seafood and remove them from it before ingestion. However, it is practically impossible to detect and remove all anisakid larvae from fish and squid flesh, and freezing and cooking (i.e.

low and high temperatures) are the effective treatments to kill larval worms. The time to death is temperature dependent, but it is recommended, for example, that fish fillets are held at –20°C for one or several days. Cooking at >60°C for 10 minutes is lethal to anisakid larvae.

Important references
Important reviews and accounts in books are by Oshima (1972), Margolis (1977), Smith and Wooten (1978), Ishikura and Namiki (1989), Ishikura *et al.* (1992) and Ishikura (2003).

Zoonotic potential of Protozoa
Mark Freeman

Introduction
Since about 1990 the World Health Organization reported that over 65% of emergent diseases infecting humans are believed to have an animal origin. Parasitic zoonotic diseases present a particular challenge to control as a result of the sometimes unknown animal hosts or reservoirs and often complicated transmission routes and life cycles, which may involve two or more different host species.

Microsporidia
Microsporidians are obligate, intracellular, spore-forming protist parasites that infect virtually all invertebrate phyla and five classes of vertebrates including mammals. More than 1300 species from about 150 genera of microsporidian parasites are described, with by far the greatest proportion of these infecting fish and arthropods. All microsporidian parasites produce spores, which are the resistant infectious stages that are capable of surviving environmental conditions outside of their hosts for sustained periods in order to infect a new host. At least six genera including 14 species have been reported to infect humans (Franzen and Müller 1999); these are primarily considered to be opportunistic infections in immunocompromised patients such as organ transplant recipients and those suffering from Acquired Immune Deficiency Syndrome (AIDS). However, there are an increasing number of reports of microsporidian infections from individuals with fully functional immune systems (Kotler and Orenstein 1999).

The most common microsporidian infection found in humans is caused by *Enterocytozoon bieneusi*, which infects the enterocytes of the intestine causing severe diarrhoea. *Enterocytozoon bieneusi* is the only described member in the genus because of the reassignment of the other members infecting fish to the genus *Nucleospora*, which is the only other genus in the family Enterocytozoonidae. *Enterocytozoon bieneusi*, although reported widely from immunocompromised patients, is rarely present in immunocompetent individuals, which has led to the suggestion of a zoonotic origin. *Enterocytozoon bieneusi* infections have been observed in simian immunodeficiency virus-infected macaque monkeys; *E. bieneusi* spores have been found in the faeces of farm and companion animals and more recently in the faeces of wild fur-bearing mammals such as beavers, racoons, muskrats and otters (Sulaiman *et al.* 2003). The detection of spores in the faecal material of such mammals, especially those associated with surface water and having fish or fish-meal included in their diets, may indicate only a transient relationship resulting from dietary intake of spores and not their role as definitive hosts. All other known microsporidian parasites from the Enterocytozoonidae are found infecting either freshwater fish, anadromous fish and their marine hyperparasites or marine fish (Freeman *et al.* 2003). Furthermore, Cali and Takvorian (2003) demonstrated that a microsporidian from the genus *Pleistophora*, a genus only previously known to infect poikilothermic animals, primarily marine and freshwater fish, infected the skeletal muscle of an immunocompromised human. The increasing number of reports of novel microsporidian infections in humans with an unknown origin, sharing taxonomic similarities with described microsporidia infecting fish, suggests that fish themselves cannot be excluded as potential sources of human infection with microsporidia through dietary intake. Recent scientific data suggest that meat from fish or crustaceans that is raw or only lightly cooked should be considered as a potential environmental source for human microsporidiosis (Slifko et al 2000).

Microsporidia are considered to be opportunistic but basically host-specific parasites, and there are believed to be two distinct barriers that microsporidian genera cannot traverse. First, the borders between invertebrate and vertebrate hosts, and second, the borders between poikilothermic and homoiothermic vertebrate hosts. Research on microsporidia has increased significantly since 1990, largely because they have been recognised as opportunistic pathogens in AIDS patients. The transmission of invertebrate microsporidia to mammals is normally considered to be impossible, as the temperature differences are deemed too great to overcome. Nevertheless, some experimental exceptions to this rule do exist. The microsporidian *Brachiola algerae* (syn. *Nosema algerae*) from culicine mosquitoes has been injected into the 'colder tissues' (feet, pads and tails) of immunosuppressed mice and successfully caused infection. This microsporidian was shown to cause a severe liver infection after the ocular bathing of spores in similarly immunosuppressed mice (Koudela et al. 2001). *Brachiola algerae* can also be successfully maintained *in vitro*, cultured in human fibroblasts at 38°C (Trammer et al. 1999). Temperature differences between hosts are still regarded as one of the main obstacles that microsporidia must overcome in order to infect new hosts. Therefore, species that maintain similar body temperatures are more likely to successfully share microsporidian parasites, but immunosuppression in the novel host can facilitate this transition.

Apicomplexa

Coccidian parasites from the phylum Apicomplexa are parasitic in a wide range of vertebrates and cause both anthroponotic and opportunistic zoonotic infections in humans. Infections in humans have traditionally been associated with the consumption of contaminated drinking water or with the ingestion of oocysts by faecal–oral contact. Parasites such as *Cryptosporidium parvum* are responsible for the zoonotic diarrhoeic illness known as cryptosporidiosis in humans (Fayer et al. 2000). Other species from the genus *Cryptosporidium* are now recognised as being capable of infecting humans, particularly immunocompromised individuals (Cacciò et al. 2002). Although the infective oocyst stage is normally associated with fresh water, rivers that become polluted by anthropogenic and livestock faecal discharge can lead to the contamination of estuaries and coastal environments with *Cryptosporidium* oocysts, which have been shown to remain viable in sea water for up to one year (Tamburrini and Pozio 1999). Many species of commercially important marine bivalve molluscs, including mussels, oysters, clams and cockles have been shown to contain viable *Cryptosporidium* oocysts, with prevalences as high as 18%. Furthermore, standard shellfish depuration techniques have been shown to be ineffective in totally removing *Cryptosporidium* oocysts from contaminated bivalves (Gómez-Couso et al. 2003). The infective dose of *C. parvum* for immunocompromised individuals has not been established, but may be as low as a single oocyst, and oocysts recovered from contaminated bivalves have been used to successfully infect experimental neonatal mice (Gomez-Bautista et al. 2000). These data suggest that filter-feeding marine bivalves, particularly those associated with areas of fluvial input or sewage discharge, may act as reservoirs for coccidian parasites, especially *Cryptosporidium parvum*, and the consumption of such food may lead to cryptosporidiosis in humans with immunocompromised individuals being at an increased risk.

Other protozoan parasites capable of infecting humans that are normally only associated with fresh water but have an environmentally resistant resting stage, such as *Giardia*, have also been isolated from coastal bivalves and may also represent a human health risk (Gómez-Couso et al. 2004).

Important references
Kotler and Orenstein (1999) discussed clinical syndromes associated with microsporidiosis, Tamburrini and Pozio (1999) demonstrated long-term survival of *Cryptosporidium parvum* oocysts in sea water and in experimentally infected marine mussels, and Fayer *et al.* (2000) gave an account of epidemiology of *Cryptosporidium*, including a discussion of transmission, detection and identification. Gómez-Couso *et al.* (2003) demonstrated contamination of bivalve molluscs by *Cryptosporidium* oocysts, and Slifko *et al.* (2000) gave a general account of emerging parasite zoonoses associated with water and food.

Zoonotic aspects of trichinellosis
Lorry B Forbes

Introduction
Trichinellosis is a potentially fatal disease caused by larvae of the nematode genus *Trichinella*, acquired by eating raw or undercooked meat. The genus *Trichinella* is composed of small parasitic nematodes with direct life cycles and a broad range of mammalian hosts. The adult worms are short lived in the small intestine, where they reproduce rapidly, and shed live larvae into the circulatory system. First-stage larvae of most species encyst in the cells of voluntary muscles throughout the body, where they remain until consumed by a new host. *Trichinella spiralis* was considered to be the only species in this genus until the 1960s when accumulating biological evidence indicated that there was more than one genotype. Since that time, molecular techniques have identified 11 genotypes of *Trichinella*, some of which have species status: *T. spiralis* (T1), *T. nativa* (T2), *T. britovi* (T3), *T. pseudospiralis* (T4), *T. murrelli* (T5), *T. nelsoni* (T7), *T. papuae* (T10), *T. zimbabwensis* (T11) and genotypes T6, T8, T9 (Murrell *et al.* 2000, Pozio *et al.* 2002, 2004). Mammalian carnivores and omnivores are the usual hosts, although all mammals are susceptible. The number of larvae required for infection varies according to the genotype of the parasite and the host animal. Genotype T4, which primarily infects birds, and genotypes T10 and T11, which infect mammals and crocodiles, do not form a cyst in the muscle stage (non-encapsulated). *Trichinella nativa* (T2) is the only genotype reported in marine mammals (Forbes 2000) (see also pp. 104–114).

Historically, human trichinellosis has been associated with the consumption of meat from *T. spiralis*-infected swine, and meat inspection regulations to detect and control infected pork have been in place in many countries for over 100 years. Generally accepted food safety guidelines for inactivating encysted *Trichinella* larvae in the meat of domestic animals by freezing are based on *T. spiralis*. These are inadequate for use with the meat of marine mammals as *T. nativa* is freeze tolerant (Forbes *et al.* 2003) (see pp. 438–439).

Life cycle
The life cycle of *Trichinella spiralis* is considered characteristic of the genus (Anderson 2000). Stage 1 (L1) larvae in muscle are released by digestion following consumption by a suitable host and penetrate the region between the lamina propria and the columnar epithelium of the small intestine within the first hours following infection. During the next 40 hours they develop and undergo four moults to reach adulthood. Adult males and females are small, about 1.0 mm to 1.8 mm and 1.3 mm to 3.7 mm in length, respectively, and live for about one month or less. Following copulation and a short gestation period, females release newborn larvae (pre-L1) which migrate via venules and lymphatics into the general circulation and are distributed throughout the body. They then invade striated muscle cells and show predilection for specific muscle groups such as the tongue, diaphragm and masseter muscles. In low level infections, the concen-

tration of larvae in predilection sites increases the chance of successfully transferring an infective dose to a new host. Sites of predilection may vary according to host; however, if infection is severe, most voluntary muscles contain high numbers of larvae. Infected muscle cells are referred to as nurse cells. Following infection with all genotypes except T4, T10 and T11, the nurse cell produces a double-layered protective membrane (cyst) around the invading larva. The larvae grow from 0.07–0.10 mm to 0.61–1.0 mm in about three weeks following muscle invasion. There is typically one coiled larva per cyst, although occasionally a single cyst may contain more than one larva. Intramuscular larvae can remain infective for years. There are reports of larvae remaining infective in polar bears, *Ursus maritimus*, for 11 to 20 years.

Epidemiology

All occurrences of *Trichinella* in marine mammals have been from the circumpolar arctic (Forbes 2000). The parasite is common in polar bears with a prevalence of over 60% in some populations. Trichinellosis had been described in walruses, *Odobenus rosmarus*, on numerous occasions since 1954, and more recently, outbreaks of human trichinellosis have been linked with increasing frequency to the consumption of uncooked walrus meat. Although fewer numbers of cases have been reported in walruses than polar bears, survey data indicate that prevalence in some walrus populations might exceed 60%. Additional field data are needed to provide an accurate estimate of prevalence in this species. Trichinellosis is rare in other pinnipeds and cetaceans. It has been confirmed in three bearded seals, *Erignathus barbatus*, three ringed seals, *Phoca hispida*, and a single beluga whale, *Delphinopterus leucas*. Grey seals, *Halichoerus grypus*, have been experimentally infected (Kapel et al. 2003). There have been no reports in carnivorous whales such as killer whales, *Orcinus orcus*, although these whales feed on other marine mammals and are suitable candidates for infection with *Trichinella*. A few reports of trichinellosis in Greenland seals, *Phoca groenlandica*, ribbon seals, *Phoca fasciata*, and a single narwhal, *Monodon monoceros*, have appeared in reviews and textbooks, but lack appropriate references and should not be included in the list of hosts until confirmatory data are available (Forbes 2000).

The prevalence of trichinellosis in the prey species of polar bears, primarily seals, is too low to account for the high prevalence of infection in polar bears. Cannibalism among polar bears is well documented and for many years bear-to-bear transmission was thought to be the only sylvatic cycle (i.e. a cycle not involving domestic animals) among marine mammals. The low prevalence of trichinellosis in pinnipeds and cetaceans was attributed to inefficient transmission mechanisms. These included direct scavenging of infected polar bear, arctic fox and sled dog carcasses that were deposited in the ocean as pack ice melted, and indirect transmission via ingestion of fish or copepods that scavenged these carcasses and contained undigested *Trichinella*-infected muscle tissue in their gut. However, these mechanisms do not adequately explain past occurrences in walruses or the continued identification of new cases through testing programs for harvested walrus meat. *Trichinella nativa* may have an independent sylvatic cycle in walruses similar to that seen in polar bears as walruses are long-lived carnivores (>20 years) and both seal and walrus tissue have been reported in their stomach contents. A proposed sylvatic cycle for trichinellosis in marine mammals is shown in Figure 11.1.

Trichinellosis in humans

Human trichinellosis associated with the consumption of meat from marine mammals is a significant food safety hazard in arctic areas. As early as 1914, trichinellosis was mentioned as a possible cause of illness in native people who had consumed beluga whale meat in arctic Canada. Since that time, outbreaks of trichinellosis in humans have been reported on numerous occasions and were most frequently associated with the consumption of walrus meat, followed by polar bear meat (Forbes 2000). As the source food is usually no longer available for analysis

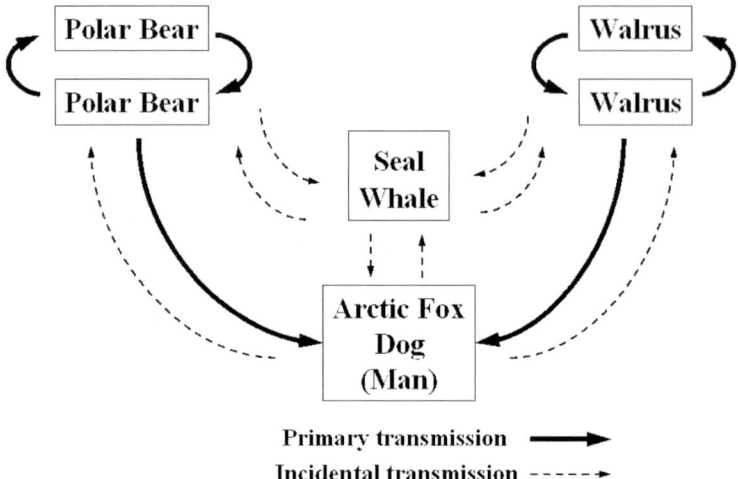

Figure 11.1 Proposed sylvatic cycle of *Trichinella nativa* in marine mammals.

by the time clinical signs appear, retrospective epidemiological studies are commonly used to link human disease to the prior consumption of infected meat. The first human outbreaks in which larvae were identified in the source meat from a marine mammal occurred in Alaska in 1975 and in Canada in 1987, and both involved consumption of infected walrus meat.

The course of clinical trichinellosis in humans varies with the extent of intestinal and muscular invasion, the genotype of *Trichinella* and the immune status of the human host (Kociecka 2000). In general, the severity of disease is closely associated with the number of infective larvae ingested. Small numbers of larvae usually result in asymptomatic infections. Clinical signs appear about five to 15 days after exposure and usually include one or more of the following symptoms: periorbital and facial oedema, myalgia, fever, conjunctivitis, photophobia, gastrointestinal upset, headache and skin rash. Myocarditis, endocarditis, encephalitis or meningitis, if they occur, are serious and may be life threatening. Clinical reports indicate that *T. nativa* from marine mammals is at least as pathogenic in humans as *T. spiralis*. In some human outbreaks associated with walrus meat, diarrhoea was the predominant clinical sign and this was attributed to reinfection of previously sensitised individuals. During the early stages of infection, trichinellosis in humans can be successfully treated with anthelminthic and glucocorticosteroid drugs with the course of treatment determined by the severity and duration of disease and the immunological and reproductive status of the patient.

Food safety

An important biological characteristic of *T. nativa* is the ability of frozen encysted larvae to remain infective for long periods. Infective larvae have been found in host musculature after four years of storage at $-20°C$. In contrast, food safety regulations for domestic animals are usually based on *T. spiralis*, and it has been recommended that meat up to 69 cm thick be frozen at $-15°C$ for four weeks to inactivate *Trichinella* (Gamble et al. 2000). In Canada, a combination of internal temperature and freezing time ranging from $-18°C$ for 106 hours to $-37°C$ for 30 minutes is recommended (Anonymous 2004). Although freeze tolerance is an important survival mechanism for the parasite under arctic conditions, freezing is also a common food preservation method in these areas. The health risk is exacerbated by traditional habits of consuming food raw or undercooked. Other methods of preserving meat from harvested marine mammals in northern areas include drying, smoking, salting and fermenting. Igunaq, a traditional food made by placing meat and blubber into an animal skin bag and allowing it to ferment, has often been associated with human outbreaks of trichinellosis when prepared using infected walrus meat. Experimentally, infective larvae have been recovered from seal sausage, dried seal meat and seal igunaq for up to five months post preparation (Forbes et al. 2003).

Although *Trichinella* larvae in meat are easily inactivated by cooking to an internal temperature of 71°C, traditional methods of preparation and consumption of food in arctic areas are important cultural activities and are likely to continue. In recognition of this, health agencies in arctic Canada have began to institute food safety testing to certify wild game meat as *Trichinella*-free (Proulx et al. 2002). While this is important in alleviating immediate food safety concerns, additional field surveillance is required to determine the true spectrum of marine mammal hosts and the prevalence of trichinellosis in host populations.

Important references
A review of the occurrence, ecology and human health aspects of marine mammal trichinellosis was published by Forbes (2000) and contains an extensive reference list. A compilation of the work of Hans Roth, by Madsen (1961) provides relevant information and a useful historical perspective. The reports by Serhir et al. (2001) and Proulx et al. (2002) provide typical information on outbreaks of human trichinellosis associated with marine mammals.

Marine schistosome dermatitis
John Walker

Introduction
Schistosome dermatitis is a maculopapular skin eruption that follows penetration of the skin by cercariae of some schistosome species. These are usually the larvae of bird parasites, but some mammalian species have also been implicated.

Geographical distribution
Schistosome dermatitis has a cosmopolitan distribution throughout which a variety of names including cercarial dermatitis, swimmer's itch, clam digger's itch, paddy itch, sedge pool itch and pelican itch have been applied. Cort (1928) first elaborated the aetiology when he proved that dermatitis afflicting staff and students at a field station at Douglas Lake, Michigan, was caused by the cercaria of *Trichobilharzia ocellata*, a schistosome parasite of ducks, developing in the freshwater snail *Lymnaea stagnalis*. In a review of early studies on freshwater schistosome dermatitis, Cort (1950) briefly mentioned reports of dermatitis produced by schistosome cercariae emerging from marine snails on both the west and east coasts of the USA.

Diagnosis of schistosome dermatitis and symptoms
The characteristic lesions of schistosome dermatitis are restricted to uncovered areas of the body immersed in water. An individual walking through shallow water might have lesions on the feet and ankles, while in deeper water the rash might occur around the knees (Fig. 11.2). The reason for this zonal restriction lies in the behaviour of the cercariae. When they emerge from the snail they swim upward and rest beneath the surface film, waiting to infect an appropriate definitive host. However, if they contact human skin they can penetrate, often producing a prickling sensation lasting for up to an hour. There may also be a macular eruption, with diffuse erythema or urticaria. Ten to fifteen hours later an intensely itchy macropapular eruption develops. This lasts for about one week with the papules progressing to fluid-filled vesicles in some cases and

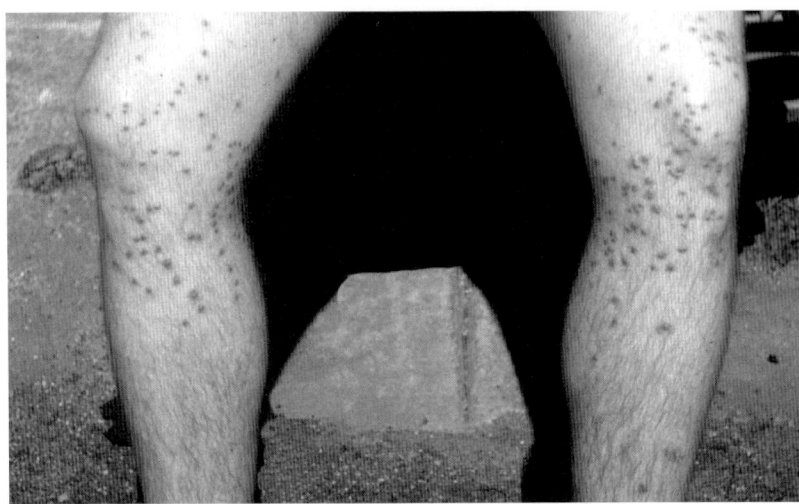

Figure 11.2 Marine schistosome dermatitis caused by the cercariae of *Austrobilharzia terrigalensis* at Narrabeen Lakes, Sydney. The individual had been fishing in knee-deep water. Photograph by the late AJ Bearup.

pustules if secondary infection occurs. The lesions usually become pigmented and persist for around 10 days but occasionally for weeks or months.

Individuals exposed to schistosome cercariae for the first time rarely develop more than a mild, transient erythematous reaction at the site of penetration. It is after subsequent infections that severe dermatitis develops, its intensity reflecting the degree of hypersensitivity induced by previous exposure. Systemic symptoms, such as fever, lymphadenopathy and oedema are uncommon but may occur in repeatedly infected individuals. Although bird schistosomes do not mature in mammals, their schistosomulae have been shown to cause pulmonary haemorrhages in laboratory rodents (Olivier 1953). Whether this happens in humans is unknown.

Marine schistosome dermatitis usually occurs after bathing or wading in calm estuarine waters. Wave swept coasts are less important because, although molluscs there can be infected, the turbulent water disrupts the typical cercarial behaviour. Acute dermatitis acquired after contact with coastal waters is commonly called seabather's eruption and is often blamed on 'sea lice'. The lesions mimic those of schistosome dermatitis except that they are usually on regions of the body covered by clothing. It has been shown that ephyrae, medusae and planulae of the thimble jellyfish, *Linuche unquiculata*, cause seabather's eruption in southern Florida and the Caribbean (Segura-Puertas et al. 2001). The cnidae are triggered by pressure from the bathing suit or by other factors, including changes in osmotic pressure as clothes dry out after leaving the water. Systemic symptoms are more common than with schistosome dermatitis, especially when the infection is caused by planulae.

The agents of marine schistosome dermatitis

The first specific account of marine schistosome dermatitis was by Penner (1950) who described *Cercaria littorinalinae* from *Littorina planaxis* on the coast of southern California. Stunkard and Hinchliffe (1952) concluded that dermatitis affecting fishers and bathers in the Narragansett Bay region of Rhode Island was caused by *Cercaria variglandis*. Experimental infections of several bird species produced adult worms indistinguishable from *Microbilharzia chapini* and, consequently, the parasite was named *M. variglandis*. Penner (1953) suppressed the genus *Microbilharzia* and transferred the species to *Austrobilharzia*.

Bearup (1955) investigated papular dermatitis acquired by people wading in Narrabeen Lagoon, Sydney, Australia, and concluded that a schistosome cercaria, *Cercaria variglandis* var. *pyrazi*, was the cause. Further study proved that it was the cercaria of *Austrobilharzia terrigalensis*, that the primary definitive host was the silver gull, *Larus novaehollandiae*, and the intermediate host *Pyrazus australis* (=*Velacumantus australis*), now called *Batillaria australis* (W. Ponder, pers. comm., 2004).

Farley (1971) and Rohde (1977) both reviewed the species of *Austrobilharzia* and concluded that *A. variglandis* is a synonym of *A. terrigalensis*. The validity of other species in the genus remained uncertain. In a major review of the phylogeny of the Schistosomatidae, Lockyer *et al.* (2003), using sequences of three genes (complete ribosomal small subunit rRNA, large ribosomal subunit RNA and mitochondrial cytochrome oxidase 1), determined that *A. terrigalensis* and *A. variglandis* are valid species. As they did not have access to specimens of the other species of *Austrobilharzia* reviewed by Farley, their status remains uncertain.

Gigantobilharzia huttoni is an avian schistosome known to cause schistosome dermatitis (Leigh 1955). Adult worms have been found in pelicans and the larval stages in the opisthobranch *Haminoea antillarum guadalupensis*. Rohde (1978) recorded the presence of an unnamed species of *Gigantobilharzia* in two silver gulls at Heron Island, Great Barrier Reef. The intermediate host of this parasite has not been determined but is possibly a local opisthobranch such as *H. cymbalum*.

The pulmonate limpet *Siphonaria denticulata* is the intermediate host of a schistosome parasite on New South Wales coasts. Ewers (1961) found schistosome infections in 35% of *S. denticulata* at Cape Banks, Sydney, and the cercariae produced dermatitis when applied to the skin of the forearm. Only immature worms, thought to belong to the Bilharziellinae, were recovered from experimentally infected budgerigars, *Melopsittacus undulatus*, so no specific identification was made. *Siphonaria denticulata* is a widely distributed and a common mollusc on rocky shores in eastern Australia. The high rate of infection found by Ewers (1961) suggests that this schistosome may cause cases of dermatitis occasionally reported from such localities.

Ornithobilharzia canaliculata is a parasite of terns with *Batillaria minima* as intermediate host. Malek and Cheng (1974) state that this parasite can cause schistosome dermatitis in humans but give no citation for the source of their claim.

Species of *Austrobilharzia* infect an unusually wide range of molluscan hosts in five different families: Batillariidae, Planaxidae, Littorinidae, Nassariidae and Potamididae. Rohde (1977) suggested that trematodes parasitic in birds with extensive geographic ranges, especially migratory species, have evolved the ability to infect distinct molluscan hosts in different localities. *Austrobilharzia terrigalensis* is particularly interesting in this respect. Walker (1979) demonstrated that this schistosome does not develop in *Batillaria australis* unless the snail is infected with another trematode species. *Austrobilharzia terrigalensis* not only requires the presence of another trematode if it is to infect *B. australis*, but also retards the development of the other species by indirect, interspecific antagonism. Species subordinate to *A. terrigalensis* include an echinostome and two philophthalmids with large rediae. In most interspecific interactions between trematode larvae in molluscs, species with rediae dominate schistosome sporocysts (Lim and Heyneman 1972). In contrast to the situation involving *B. australis*, *A. terrigalensis* infects *Planaxis sulcatus* on Heron Island, Great Barrier Reef, and is found in both single infections and in combinations with other trematode species (the same species occur in *B. australis* in Sydney). *Austrobilharzia terrigalensis* may be a relatively new introduction to southern Australia that has not yet developed mechanisms for subverting the cellular defences of *B. australis*. Unless that has already been done by another trematode, the schistosome is unable to develop in this snail.

Epidemiology of marine schistosome dermatitis

The principal factors in the epidemiology of marine schistosome dermatitis are a combination of the biology of the parasite and its hosts, human behaviour and local environmental conditions. Appleton and Lethbridge (1979) investigated a persistent problem of dermatitis associated with water contact in the Swan River estuary, Perth, Western Australia. The causative agent proved to be *Austrobilharzia terrigalensis*, with the same intermediate and definitive hosts as on the east coast of Australia. Most cases of dermatitis in the Swan estuary were contracted in shallow areas where both silver gulls and *B. australis* were common and where, between November and March, water temperatures frequently exceeded 30°C. Cercariae began emerging from snails once the temperature reached 23°C, and reached maximum densities in the water between 11 am and 12 noon. These warm, shallow waters are safe for swimming and, consequently, most of the infections occurred in children from five to 14 years of age or their accompanying parents.

Schistosome dermatitis can also be an occupational hazard. People collecting the clams *Venus mercenaria* and *Mya arenaria* in Narragansett Bay, Rhode Island, frequently developed a form of dermatitis called 'clam digger's itch'. It was caused by cercariae of *A. variglandis*, developing in the marine snail *Nassarius obsoletus* that occurs in large numbers in the same environment as the clams (Stunkard and Hinchliffe 1952).

Gulls are important definitive hosts of marine schistosomes, especially the species of *Austrobilharzia*. It is not only their susceptibility to infection that is significant, but their scavenging behaviour and close association with humans. Where there are humans, gulls often congregate and, if suitable molluscan hosts are also present, the life cycles of the dermatitis causing schistosomes become established.

Prevention and treatment of schistosome dermatitis

Prevention of freshwater schistosome dermatitis by snail control has been attempted with varying degrees of success, usually in artificial environments such as rice fields where the infection is an occupational hazard causing economic loss. In marine ecosystems mollusc control is usually not feasible and wearing protective clothing has been the preferred approach. One, yet untried, preventive measure is the use of water-resistant sunscreens containing insect repellents. The active ingredient of some repellents, N, N-diethyl-m-toluamide (DEET), has been shown to prevent infection of mice with the human schistosome, *Schistosoma mansoni* (Salafsky et al. 1999). Treatment of established infections is largely limited to symptomatic relief by application of lotions or ice to reduce itching.

Important references

Cort (1928) first elaborated the aetiology of marine cercarial dermatititis, Cort (1950) gave information on the status of knowledge of schistosome dermatitis after more than 20 years of studies, and Appleton and Lethbridge (1979) discussed schistosome dermatitis in Western Australia. Farley (1971) reviewed the family Schistosomatidae.

Infections by the rat lungworm, *Angiostrongylus cantonensis*
Robin M Overstreet

Introduction

The metastrongylid 'rat lungworm' nematode, *Angiostrongylus cantonensis*, is a nematode that typically matures in the lungs of rodents, including rats. As an adult, the female (33 mm long) can be recognised grossly by the barber-pole appearance of its haematin-laden intestine spiralling around the pale-coloured ovary and uterus. The posterior end of the relatively small male has a bursa and two long narrow spicules.

Angiostrongylus cantonensis is not a true marine parasite, even though the initial hypothesis for the source of human infections was *Katsuwonus pelamis* (known as skipjack tuna, oceanic bonito and other common names) or possibly other pelagic fishes eaten raw (Rosen et al. 1961). However, the nematode belongs in this book because marine animals can serve as a paratenic host, also known as a transport, transfer or carrier host (a host in which further development does not occur), as well as a true intermediate host in which development progresses from the first-stage to third-stage juvenile. Humans serve as an abnormal host in which the infective third-stage juvenile develops though usually not into a fully mature adult as occurs in its rodent definitive host.

Geographical distribution

Initially, knowledge about human infections was restricted to the Indo-Pacific region. The parasite now has become established in many tropical and subtropical areas by means of rats on ships and introduction of snails. The worm appeared to spread primarily by the introduction of the giant African land snail, *Achatina fulica*, which can harbour up to 39 000 third-stage juveniles, from East Africa to Madagascar before 1800 and then gradually to various areas between there and Tahiti and Hawaii, including American Samoa by about 1975 (e.g. Alicata 1988). In the western hemisphere, it first showed up in Cuba, Puerto Rico and the USA mainland (New Orleans, Louisiana) all in the early 1980s and probably from rats coming off ships (e.g. Campbell and Little 1988). It quickly became established in 21% of *Rattus norvegicus* along the Mississippi River levee in New Orleans, infected zoo primates and a boy and spread to New Iberia and Baton Rouge, Louisiana, infecting a young horse, more zoo primates and wild specimens of the wood rat, *Neotoma floridanus*, and Virginia opossum, *Didelphis virginiana* (e.g. Kim et al. 2002).

Life cycle

A more complete explanation of the life cycle of *A. cantonensis* will help the reader understand how marine hosts and humans fit into the life history of the species. Adults occur primarily in the pulmonary arteries of rodents (Fig. 11.3), but, when in large numbers or in infections with one or unbalanced sexes, they also occur in the right ventricle (Winsor 1967). The nematode infects numerous rodents, including the Norway, or brown, rat, *Rattus norvegicus*, and other common rats. The produced eggs lodge in the arteries or lung capillaries where they hatch. The first-stage juvenile enters the alveoli, migrates up the trachea and is swallowed down the alimentary tract to be combined with the faeces. Terrestrial or aquatic molluscs, primarily snails and slugs, can either eat or be penetrated externally by this first-stage juvenile. The stage takes about two to three weeks to develop into the third-stage juvenile infective to appropriate rodents. The ingested third-stage juvenile migrates in the circulatory and central nervous systems to the surface of the cerebrum as a fourth-stage juvenile. After about two weeks in the rat, the resulting young adult migrates to the subarachnoid space and then, after a couple of weeks, forces its way into the venous system to the pulmonary arteries. Once there, it matures, mates, and starts producing eggs in a little over one week (e.g. Beaver et al. 1984). The third-stage juvenile can be acquired by an abundance of paratenic hosts including crustaceans, fishes, amphibians, reptiles and turbellarians. As a terrestrial unexpected example involved with human infections, the land planarian *Endeavouria septemlineata* (as *Geoplana forsterorum*, turbellarian) feeds on snails, acquires a few juveniles and becomes dehydrated and fragmented on vegetables, which are consumed raw and probably represent the primary source of human infections in New Caledonia (Ash 1976). The juvenile *A. cantonensis* can also be emitted from molluscan hosts into the water or in the mollusc's mucus strand. Any of these many hosts or means can transmit the infection to the rodent definitive host or to the human accidental host (Fig. 11.3). Moreover, most experimentally exposed terrestrial (Campbell and Little 1988) and freshwater (Richards and Merritt 1967) molluscs can serve as intermediate hosts.

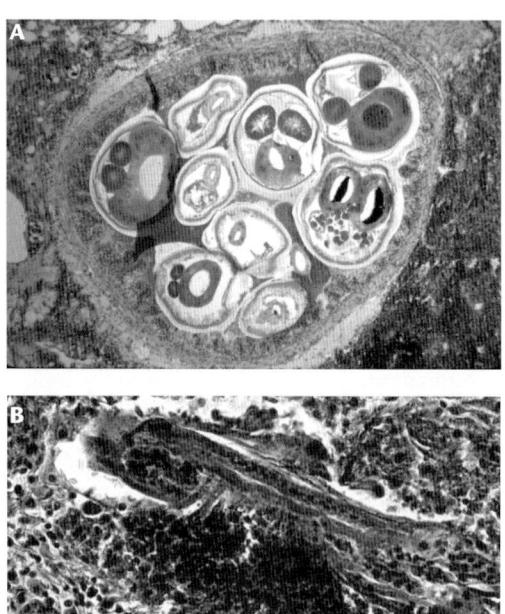

Figure 11.3 **A**. Section exhibiting a balanced ratio of five male and five female adults of *Angiostrongylus cantonensis* in the pulmonary artery of an experimentally infected white rat. Females can be recognised by the dark internal pigment, which gives an entire specimen a barber-pole appearance. This infection has the same number of pairs as an average natural rat infection, and pairing typically keeps a worm from migrating to find a mate and assures fertilisation. The produced eggs with developing larvae and associated heavy inflammation occur in the lung tissue surrounding the thick-walled artery. **B**. Section through a dead *A. cantonensis* in the brain of a person from Thailand exhibiting a relatively extensive lesion. Both photographs taken of microscope slides in the collection of the late Paul Beaver, Tulane University.

Experimental infections of *A. cantonensis* have been successfully produced in freshwater fishes and invertebrates that often live in brackish water as well as in marine fishes and invertebrates. Since people consume raw and inadequately cooked animals and animal products, a public health risk exists. Snails and slugs in either injured or healthy condition when placed in water released infective third-stage juveniles that survived at least 72 hours (Cheng and Alicata 1964) or seven days (Richards and Merritt 1967) in tap water. Death of infected freshwater snails occurred almost immediately when immersed in sea water, but the infective juveniles survived in the carcasses for five but not six days (Richards and Merritt 1967). Cheng and Burton (1965) demonstrated that the first-stage juvenile survived for at least 27 hours in 20 parts per thousand (ppt, g of solute per kg of sea water) at 24°C, and Richards and Merritt (1967) infected snails with individuals in sea water at three days and with individuals in rat faeces in sea water at 14 days. These and other findings demonstrate that juveniles in aquatic and even marine habitats can be available to infect freshwater, marine or terrestrial molluscs or other hosts as true intermediate or as paratenic hosts. When in the cultured cichlid *Oreochromis mossambicus* (as Mozambique tilapia, *Tilapia mossambica*) in fresh water and salt water and in the marine carangid *Selar crumenophthalmus* (as bigeye scad, *Trachurops crumenophthalmus*), the number of ingested or gavaged third-stage juveniles gradually diminished over time from both the viscera and muscle during the length of the four-week study. At termination of the experiment, some worms tested from the muscle could still infect experimental rats (Wallace and Rosen 1967). These fish species were tested as suspect paratenic hosts because they represented:

1. known fish species consumed as sushi or sashimi (raw fish dishes) in Tahiti and other Pacific islands where human infections occurred; and
2. fishes in general as potential hosts.

The pelagic carangid occurred close to shore at river mouths in Tahiti concurrently with infected veronicellid slugs, which get washed to sea during storms. Similar findings of reduced infections with time post infection occurred in experimental infections in the sea snake, *Laticauda colubrina* (giant banded, or yellow-lipped, sea krait) from New Caledonia (Ash 1968). An abundance of infective juveniles was still in the snake's muscles at four weeks post infection, the termination of the experiment.

The decapod phantom shrimp, *Macrobrachium lar*, from fresh water and ghost crab, *Ocypode ceratophthalma*, marine, from the high supralittoral sand zone were fed third-stage juveniles of *A. cantonensis* by Wallace and Rosen (1966), and some remained infective at day 29, the end of the study. They did not survive as long in the shore crab (ghost crab), perhaps because that crab grinds up its food more than the palaemonid shrimp. Neither paratenic host was able to serve as a true intermediate host; when first-stage juveniles were fed to the crustaceans, no development progressed to the third-stage juvenile, the stage infective to rats and humans. The third-stage juvenile acquired from a mollusc occurred in many sites throughout the host, with an abundance in the hepatopancreas ('liver'). Some specialty dishes in Tahiti include the sauce 'taioro', prepared with the uncooked hepatopancreas and surrounding tissues of *M. lar* ground up in water and mixed with grated coconut and the associated 'milk' (e.g. Deardorff and Overstreet 1990).

Perhaps the most apt marine paratenic host for *A. cantonensis* to infect humans are aquacultured river prawns (Palaemonidae) or penaeid shrimp (Deardorff and Overstreet 1990). Various palaemonid species of *Macrobrachium* occur in fresh to brackish waters throughout the tropics and subtropics. They or corresponding hybrids are most easily grown in fresh to low-salinity water. The marine commercial penaeids are usually grown in full to half-strength (e.g. 17 ppt) but sometimes grown in low salinity and even fresh water. The shrimps are typically grown in ponds separated by levees sloping into the ponds. Commercial pelleted diets are routinely or occasionally broadcast from the levees, leaving some accumulated pellets on the levees immediately adjacent to the water. Rats and snails feed on these pellets, and both can transmit the infection to each other. Third-stage juveniles can be released from the snails directly or indirectly into the ponds, where they accumulate in the shrimps. Most of the shrimp products are cooked, occasionally at the table where live infective juveniles could possibly contaminate the area, but some of the product is consumed raw, especially by specific cultural groups.

As indicated above, the intermediate host of *A. cantonensis* includes a variety of species, all molluscs, mostly land snails and slugs. However, marine molluscs also have been demonstrated to serve in that capacity. Introduced first-stage juveniles developed to the third-stage infective to rats in both the eastern oyster, *Crassostrea virginica*, and the northern quahog, *Mercenaria mercenaria*, in 15 ppt salinity (Cheng and Burton 1965). Cheng (1966) studied the infection in the oyster more carefully, and the juvenile had to be ingested. It localised throughout the body, but primarily in the Leydig tissues. The worms remained active with no encapsulation, and a cellular response did not occur until 10 to 14 days post infection, apparently in response to the juvenile's moulting fluid. The two bivalves, especially the oyster, constitute raw seafood delicacies, and either could release infective third-stage juveniles that could infect both humans directly and a variety of estuarine species that could serve as paratenic hosts.

Disease signs

In humans the nematode typically produces eosinophilic meningoencephalitis, or eosinophilic meningitis, a disease that produces various signs and symptoms, including mental impairment and even death. The nematode also harms other abnormal mammalian hosts.

Symptoms and signs of infection include primarily severe headache, but some cases include convulsions, weakness in limbs, paraesthesia (abnormal skin sensations such as tingling, tickling, itching or burning), vomiting, facial paralysis, neck stiffness and fever. In some cases, there can be painful ocular and more rarely pulmonary involvement. The presence of peripheral eosinophilia and eosinophils in the cerebrospinal fluid (CSF) is highly suggestive of infection, but these white blood cells are not always present in the CSF (e.g. Lindo et al. 2004). The presence of worms as well as large numbers of eosinophils may occur in the aspiration of CSF. Confirmation can also be achieved by serologic enzyme-linked immunosorbent assay (Garcia 2001). An indirect fluorescent antibody test can be diagnostic, even though titres decline in infections after six to eight weeks (Welch et al. 1980).

Prevention and treatment of human infections

Freezing or heating can destroy third-stage juveniles. Refrigerating juveniles embedded in host tissues at 0°C for one to seven days did not inactivate all the worms, but in −15°C for 12 hours, the worms were not infective to rats. Moreover, one has to boil hosts containing the juveniles for at least two to three minutes as boiling for one minute is not effective (Alicata 1967).

Symptomatic therapy is normally recommended for human infections. This includes surgery to remove worms from the eye, and when anthelminthic agents, usually not recommended, are used, mebendozole is the drug of choice, with thiabendazole and levamisole also used; all three have doubtful efficacy (e.g. Garcia 2001, Lindo et al. 2004). Corticosteroids, repeated lumbar punctures and analgesics may alleviate some symptoms and reduce allergic reactions. An infection, however, is usually self-limiting without mortality and with recovery taking about four weeks (e.g. Lindo et al. 2004).

Other species of Angiostrongylus

There are other species of *Angiostrongylus*, but transmission from a marine host or marine environment is less likely. One of these, *A. costaricensis*, is very similar to *A. cantonensis* in many respects. Its adults in rodents, however, occur in the mesenteric arteries of the small intestine and caecum, and in humans it causes abdominal angiostrongylosis, potentially fatal. It infects various rodents and molluscs in Latin America and probably uses paratenic hosts. The worm typically causes inflammatory lesions in the bowel wall of humans; and worms, juveniles and eggs can be found in the terminal ilius, caecum, colon, regional lymph glands and mesenteric arteries. This species and others with biology similar to either *A. cantonensis* (i.e. *A. mackerrasae*) or *A. costaricensis* (i.e. *A. siamensis*) are summarised by Beaver et al. (1984). There are others with no known human involvement that have adults in the pulmonary arteries or lung bronchioles.

Important references

Alicata (1988) gave an historical account of *Angiostrongylus* infections in humans. Medical aspects of the infections are discussed in several books: Beaver et al. (1984), Deardorff and Overstreet (1990) and Garcia (2001).

This section was partially funded by the US Department of Agriculture, CSREES Grant No. 2002-38808-01381.

References

Chapter 1

Definitions, and adaptations to a parasitic way of life
Baer JG (1952) 'Ecology of Animal Parasites'. (University of Illinois Press: Urbana).
Esch GW, Fernández JC (1993) 'A Functional Biology of Parasitism. Ecological and Evolutionary Implications'. (Chapman & Hall: London).
Rohde K (1993) 'Ecology of Marine Parasites'. 2nd edn. (CAB International: Wallingford, Oxon., UK).
Rohde K (1999) Aspidogastrea. In 'Tree of Life'. (Eds DR Maddison, WP Maddison.) http://tolweb.org/tree?group=Aspidogastrea&contgroup=Platyhelminthes.
Rohde K (2001) Parasitism. In 'Encyclopedia of Biodiversity'. Vol. I. (Ed. S Levin.) pp. 463–484. (Academic Press: New York).

Marine parasites and the tree of life
Balter M (1997) Evolutionary biology: morphologists learn to live with molecular upstarts. *Science* **276**, 1032–1103.
Blaxter ML (2003) Nematoda: genes, genomes and the evolution of parasitism. *Advances in Parasitology* **54**, 102–195.
Combes C (2001) 'Parasitism: The Ecology and Evolution of Intimate Interactions'. (University of Chicago Press: Chicago).
Cracraft J, Donoghue MJ (Eds) (2004) 'Assembling the Tree of Life'. (Oxford University Press: New York).
Cribb TH, Bray RA, Olson PD, Littlewood DTJ (2003) Life cycle evolution in the Digenea: problems solved and puzzles revealed. *Advances in Parasitology* **54**, 197–254.
Natural History Museum (2004) The Natural History Museum, Department of Zoology Host-Parasite Database http://www.nhm.ac.uk/zoology/hp-dat.htm.
Frank SA (1993) Evolution of host-parasite diversity. *Evolution* **47**, 1721–1732.
Page RDM (Ed.) (2003) 'Tangled Trees: Phylogeny, Cospeciation, and Coevolution'. (University of Chicago Press: Chicago).
Pennisi E (2003) Modernizing the tree of life. *Science* **300**, 1692–1697. (Available at: http://www.sciencemag.org/feature/data/tol/).
Poulin R (1998) 'Evolutionary Ecology of Parasites: From Individuals to Communities'. (Chapman & Hall: London).

The American Association for the Advancement of Science (2004) A Tree of Life. Science Magazine. http://www.sciencemag.org/feature/data/tol/ reproduced with permission
TreeBASE (2005) TreeBASE: A database of phylogenetic knowledge http://treebase.org.
Tree of Life Web Project (2005) The Tree of Life Web Project, hosted by The University of Arizona College of Agriculture and Life Sciences and The University of Arizona Library http://tolweb.org.
Whitfield J (2004) Origins of life: born in a watery commune. *Nature* **427**, 674–676.

Chapter 2
Protistan biodiversity
Harrison FW, Corliss JO (1991) 'Microscopic Anatomy of Invertebrates'. Vol. 1. Protozoa. (Wiley-Liss: New York).
Lee JJ, Leedale GF, Bradbury P (Eds) (2000) 'An Illustrated Guide to the Protozoa'. Vols I & II, 2nd edn. (Allen Press: Lawrence).
Lom J, Dykova I (1992) 'Protozoan Parasites of Fishes'. Developments in Aquaculture and Fisheries Science, Vol. 26. (Elsevier: Amsterdam).
Margulis L, Corliss JO, Melkonian M, Chapman DJ (Eds) (1990) 'Handbook of Protoctista'. (Jones & Bartlett Pub.: Boston).
Mehlhorn H (Ed.) (2001) 'Encyclopedic Reference of Parasitology' Vols I & II, 2nd edn. (Springer-Verlag: Berlin).

'Sarcomastigophora' (amoebae and flagellates)
Burkholder JM, Noga EJ, Hobbs Ch, Glasgow HBJr (1992) New 'phantom' dinoflagellate is the causative agent of major estuarine fish kills. *Nature* **358**, 407–410.
Douglas-Helders M, Dawson D, Carson J, Nowak BF (2002) Wild fish are not a significant reservoir of Amoebic Gill Disease. *Journal of Fish Diseases* **25**, 569–574.
Dyková I, Figueras A, Peric Z (2000) *Neoparamoeba* Page, 1987: light and electron microscopic observations of six strains of different origins. *Diseases of Aquatic Organisms* **43**, 217–223.
Dyková I, Fiala I, Lom J, Luke J (2003) *Perkinsiella amoebae*-like endosymbionts of *Neoparamoeba* spp., relatives of the kinetoplastid *Ichthyobodo*. *European Journal of Protistology* **39**, 37–52.
Dyková, I, Nowak BF, Crosbie PBB, Fiala I, Pecková H, Adams M, Macháckova B, Dvoráková (2005) *Neoparamoeba branchiphila* n. sp. and related species of genus *Neoparamoeba* Page, 1987: morphological and molecular characterisation of selected strains. *Journal of Fish Diseases* **28**, 49–64.
Hollande A (1980) Identification du parasome (Nebenkern) de *Janickina pigmentifera* á un symbionte (*Perkinsiella amoebae* nov gen-nov sp.) apparenté aux flagellés Kinetoplstidiés. *Protistologica* **16**, 613–625.
Khan RA, Murphy J, Taylor D (1980) Prevalence of a trypanosome in Atlantic cod (*Gadus morhua*) especially in relation to stocks in the Newfoundland area. *Canadian Journal of Fisheries and Aquatic Sciences* **37**, 1467–1475.
Lawler AR (1980) Studies on *Amyloodinium ocellatum* (Dinoflagellata) in Mississippi Sound: natural and experimental hosts. *Gulf Research Reports* **6**, 403–413.
Litaker RW, Tester PA, Colorni A, Levy AG, Noga EJ (1999) The phylogenetic relationship of *Pfiesteria piscicida*, Cryptoperidiniopsoid sp., *Amyloodinium ocellatum* and a *Pfiesteria*-like dinoflagellate to other dinoflagellates and apicomplexans. *Journal of Phycology* **35**, 1379–1389.

Lom J, Dyková I (1992). 'Protozoan Parasites of Fishes'. Developments in Aquaculture and Fisheries Science, Vol. 26. (Elsevier: Amsterdam).

Page FC (1983) 'Marine Gymnamoebae'. (Institute of Terrestrial Ecology: Cambridge).

Poynton SL, Sterud E (2002) Guidelines for species description of diplomonad flagellates from fish. *Journal of Fish Diseases* **25**, 15–31.

Todal JA, Karlsbakk E, Isaksen TE, Plarre H, Urawa S, Mouton A, Hoel E, Koren CWR, Nylund A (2004) *Ichthyobodo necator* (Kinetoplastida) – a complex of sibling species. *Diseases of Aquatic Organisms* **58**, 9–16.

Vogelbein WK, Lovko VJ, Shields JD, Reece KS, Mason PL, Haas LW, Walker CC (2002) *Pfiesteria shumwayae* kills fish by micropredation not exotoxin secretion. *Nature* **418**, 967–970.

Labyrinthomorpha (labyrinthomorphs)

Bower SM (1987a). *Labyrinthuloides haliotidis* n. sp. (Protozoa: Labyrinthomorpha), a pathogenic parasite of small juvenile abalone in a British Columbia mariculture facility. *Canadian Journal of Zoology* **65**, 1996–2007.

Bower SM (1987b) Pathogenicity and host specificity of *Labyrinthuloides haliotidis* (Protozoa: Labyrinthomorpha), a parasite of juvenile abalone. *Canadian Journal of Zoology* **65**, 2008–2012.

Bower SM (1987c) Artificial culture of *Labyrinthuloides haliotidis* (Protozoa: Labyrinthomorpha), a pathogenic parasite of abalone. *Canadian Journal of Zoology* **65**, 2013–2020.

Bower SM, McLean N, Whitaker DJ (1989) Mechanism of infection by *Labyrinthuloides haliotidis* (Protozoa: Labyrinthomorpha), a parasite of abalone (*Haliotis kamtschatkana*) (Mollusca: Gastropoda). *Journal of Invertebrate Pathology* **53**, 401–409.

Fisheries and Oceans Canada (2003) Synopsis of infectious diseases and parasites of commercially exploited shellfish, *Labyrinthuloides haliotidis* of Abalone, Susan Bower, http://www.pac.dfo-mpo.gc.ca/sci/shelldis/pages/labhalab_e.htm.

Fisheries and Oceans Canada (2004) Synopsis of infectious diseases and parasites of commercially exploited shellfish, QPX, a Thraustochytrid-like disease of clams, Susan Bower, http://www.pac.dfo-mpo.gc.ca/sci/shelldis/pages/chydcc_e.htm.

Ford SE, Kraeuter JN, Barber RD, Mathis G (2002) Aquaculture-associated factors in QPX disease of hard clams: density and seed source. *Aquaculture* **208**, 23–38.

Kleinschuster SJ, Smolowitz R, Parent J (1998) *In vitro* life cycle and propagation of quahog parasite unknown. *Journal of Shellfish Research* **17**, 75–78.

Levine ND, Corliss JO, Cox FEG, Deroux G, Grain J, Honigberg BM, Leedale GF, Loeblich AR, Lom J, Lynn D, Merinfeld EG, Page FC, Poljansky G, Sprague V, Vavra J, Wallace FG (1980) A newly revised classification of the Protozoa. *Journal of Protozoology* **27**, 37-58.

MacCallum GS, McGladdery SE (2000) Quahog parasite unknown (QPX) in the northern quahog *Mercenaria mercenaria* (Linnaeus, 1758) and *M. mercenaria* var. *notata* from Atlantic Canada, survey results from three maritime provinces. *Journal of Shellfish Research* **19**, 43–50.

Patterson DJ (2000) Residual heterotrophic stramenopiles. *In* 'An Illustrated Guide to the Protozoa'. (Eds JJ Lee, GF Leedale, P Bradbury.) pp. 751–754. (Society of Protozoologists; Allen Press: Lawrence, KS).

Ragone Calvo LM, Walker JG, Burreson EM (1998) Prevalence and distribution of QPX, quahog parasite unknown, in hard clams *Mercenaria mercenaria* in Virginia, USA. *Diseases of Aquatic Organisms* **33**, 209–219.

Smolowitz R, Leavitt D, Perkins F (1998) Observations of a protistan disease similar to QPX in *Mercenaria mercenaria* (hard clams) from the coast of Massachusetts. *Journal of Invertebrate Pathology* **71**, 9–25.

Whyte SK, Cawthorn RJ, McGladdery SE (1994) QPX (quahaug parasite X), a pathogen of northern quahaug *Mercenaria mercenaria* from the Gulf of St. Lawrence, Canada. *Diseases of Aquatic Organisms* **19**, 129–136.

Haplosporidia (haplosporidians)

Andrews JD (1968) Oyster mortality studies in Virginia. VII. Review of epizootiology and origin of *Minchinia nelsoni*. *Proceedings of the National Shellfisheries Association* **58**, 23–36.

Barrow, Jr JH (1961) Observations of a haplosporidian, *Haplosporidium pickfordi* sp. nov. in fresh water snails. *Transactions of the American Microscopical Society* **80**, 319–329.

Burreson EM, Ford SE (2004) A review of recent information on the Haplosporidia with special reference to *Haplosporidium nelsoni* (MSX disease). *Aquatic Living Resources* **17**(4), 499–517.

Burreson EM, Stokes NA, Friedman CS (2001) Increased virulence in an introduced pathogen: *Haplosporidium nelsoni* (MSX) in the eastern oyster *Crassostrea virginica*. *Journal of Aquatic Animal Health* **12**, 1–8.

Carnegie RG, Barber BJ, Culloty SC, Figueras AJ, Distel DL (2000) Development of a PCR assay for detection of the oyster pathogen *Bonamia ostreae* and support for its inclusion in the Haplosporidia. *Diseases of Aquatic Organisms* **42**, 199–206.

Carnegie RG, Cochennec-Laureau N (2004) Microcell parasites of oysters: recent insights and future trends. *Aquatic Living Resources* **17**(4), 519–528.

Elston RA, Farley CA, Kent ML (1986) Occurrence and significance of bonamiasis in European flat oysters *Ostrea edulis* in North America. *Diseases of Aquatic Organisms* **2**, 49–54.

Farley CA, Wolf PH, Elston R (1988) A long-term study of 'microcell' disease in oysters with a description of a new genus , *Mikrocytos* (g. n.), and two new species, *Mikrocytos mackini* (sp. n.) and *Mikrocytos roughleyi* (sp. n.). *Fishery Bulletin* **86**, 581–593.

Ford SE, Haskin HH (1982) History and epizootiology of *Haplosporidium nelsoni* (MSX), an oyster pathogen, in Delaware Bay, 1957–1980. *Journal of Invertebrate Pathology* **40**, 118–141.

Grizel H, Mialhe E, Chagot D, Boulo V, Bachere E (1988) Bonamiasis: a model study of diseases in marine molluscs, *American Fisheries Society Special Publication* **18**, 1–4.

Haskin HH, Andrews JD (1988) Uncertainties and speculations about the life cycle of the eastern oyster pathogen *Haplosporidium nelsoni* (MSX). *American Fisheries Society Special Publication* **18**, 5–22.

Hine PM (1996) The ecology of *Bonamia* and decline of bivalve molluscs. *New Zealand Journal of Ecology* **20**, 109–116.

Perkins FO (1990) Phylum Haplosporidia. *In* 'Handbook of Protoctista'. (Eds L Margulis, JO Corliss, M Melkonian, DJ Chapman.) pp. 19–29. (Jones and Bartlett Publishing: Boston).

Perkins FO (2000) Phylum Haplosporidia Caullery & Mesnil, 1899. *In* 'An Illustrated Guide to the Protozoa'. 2nd edn. Vol. 2. (Eds JJ Lee, GF Leedale, P Bradbury.) pp. 1328–1341. (Society of Protozoologists: Lawrence, KS).

Reece KS, Siddall ME, Stokes NA, Burreson EM (2004) Molecular phylogeny of the phylum Haplosporidia based on two independent gene sequences. *Journal of Parasitology* **90**, 1111–1122.

Apicomplexa (sporozoans)

Davies AJ (1995) The biology of fish haemogregarines. *Advances in Parasitology* **36**, 118–203.

Davies AJ, Ball SJ (1993) The biology of fish coccidia. *Advances in Parasitology* **32**, 293–366.

Davies AJ, Johnston MRL (2000) The biology of some intraerythrocytic parasites of fishes, amphibia and reptiles. *Advances in Parasitology* **45**, 2–89.

Duszynski DW, Upton SJ, Couch L (1999) The Coccidia of the world. Worldwide Web Electronic Publication. http://biology.unm.edu/biology/coccidia/table.html

Ferguson HW, Roberts RJ (1975) Myeloid leucosis associated with sporozoan infection in cultured turbot (*Scophthalmus maximus* L.). *Journal of Comparative Pathology* **85**, 317–326.

Fiebiger J (1913) Studien über die Schwimmblasen-Coccidien der Gadusarten (*Eimeria gadi* n.sp.). *Archiv für Parasitenkunde* **31**, 95–137.

Fournie JW, Overstreet RM (1983) True intermediate hosts for *Eimeria funduli* (Apicomplexa) from estuarine fishes. *Journal of Protozoology* **30**, 672–675.

Grabda E (1983) *Eimeria jadvigae* n.sp. (Apicomplexa: Eucoccidia), a parasite of swimming bladder of *Coryphaenoides holotrachys* (Günther, 1887) off the Falklands. *Acta Ichthyologica et Piscatoria* **13**, 131–140.

Khan RA (1980) The leech as a vector of a fish piroplasm. *Canadian Journal of Zoology* **58**, 1631–1637.

Kirmse P (1980) Observations on the pathogenicity of *Haemogregarina sachai* Kirmse 1978, in farmed turbot *Scophthalmus maximus* (L.). *Journal of Fish Diseases* **3**, 101–114.

Landau I, Marteau M, Golvan Y, Chabaud AG, Boulard Y (1975) Hétéroxenie chez les coccidies intestinales de poissons. *Comptes Réndus des de l'Académie des Sciences* **281**, 1721–1723.

Landsberg JH, Paperna I (1985) *Goussia cichlidarum* n. sp. (Barrouxiidae, Apicomplexa), a coccidian parasite in the swimbladder of cichlid fish. *Zeitschrift für Parasitenkunde* **71**, 199–201.

Levine ND (1988) 'The Protozoan Phylum Apicomplexa'. 2 vols. (CRC Press: Boca Raton, Florida).

Lom J, Dyková I (1992) 'Protozoan Parasites of Fishes'. Developments in Aquaculture and Fisheries Science, Vol. 26. (Elsevier: Amsterdam).

Mackenzie K (1978) The effect of *Eimeria* sp. infection on the condition of the blue whiting *Micromesistius poutassou* (Risso). *Journal of Fish Diseases* **4**, 473–486.

MacLean SA, Davies AJ (1990) Prevalence and development of intraleucocytic haemogregarines from Northwest and Northeast Atlantic mackerel, *Scomber scombrus* L. *Journal of Fish Diseases* **13**, 59–68.

Molnár K (1979) Studies on Coccidia of Hungarian Pond Fishes and Further Prospects of their Control. *Proceeding of the International Symposium on Coccidia*. Prague pp. 173–183.

Molnár K (1995) Phylum Apicomplexa. In 'Fish Diseases and Disorders. I. Protozoan and Metazoan Infections'. (Ed. PTK Woo.) pp. 263–287 (CABI Publishing, University Press: Cambridge).

Negm-Eldin MM (1999) Life cycle, host restriction and longevity of *Cyrilia nili* (*Haemogregarina nili* Wenyon, 1909) n. comb. *Deutsche Tierärztliche Wochenschrift* **106**, 191–199.

Odense PH, Logan VH (1976) Prevalence and morphology of *Eimeria gadi* in the haddock. *Journal of Protozoology* **23**, 564–571.

Paterson WB, Desser SS (1982) The biology of two *Eimeria* species (Protista: Apicomplexa) in their mutual fish hosts in Ontario. *Canadian Journal of Zoology* **60**, 164–175.

Pinto JS (1956) Parasitic castration in males of *Sardinia pilchardus* (Walb.) due to testicular infestation by the coccidia *Eimeria sardinae* (Thelohan). *Revista de la Faculdade de Ciencias de la Universidade de Lisboa Serie C* **5**, 209–214.

Solangi MA, Overstreet RM (1980) Biology and pathogenesis of the coccidium *Eimeria funduli* infecting killifishes. *Journal of Parasitogy* **66**, 513–526.

Steinhagen D, Körting W (1990) The role of tubificid oligochaetes in the transmission of *Goussia carpelli*. *Journal of Parasitology* **76**, 104–107.

Upton SJ, Stamper MA, Osborn AL, Mumford SL, Zwick L, Kinsel MJ, Overstreet RM (2000) A new species *of Eimeria* (Apicomplexa, Eimeriidae) from the weedy sea dragon *Phyllopterix teniolatus* (Osteichthyes: Sygnathinae). *Diseases of Aquatic Organisms* **43**, 55–59.

Microsporidia (microsporans)

Azevedo C, Matos E (2003) *Amazonspora hassar* n. gen. and n. sp. (Phylum Microsporidia, fam. Glugeidae), a parasite of the Amazonian teleost *Hassar orestis* (fam. Doradidae). *Journal of Parasitology* **89**, 336–341.

Barker DE, Khan RA, Hooper R (1994) Bioindicators of stress in winter flounder, *Pleuronectes americanus*, captured adjacent to a pulp and paper mill in St George's Bay, Newfoundland. *Canadian Journal of Fisheries and Aquatic Science* **51**, 2203–2209.

Becnel JJ, Andreadis TG (1999) Microsporidia in insects. *In* 'The Microsporidia and Microsporidiosis'. (Eds M Wittner, LM Weiss.) pp. 447–501. (ASM Press: Washington DC).

Cali A, Takvorian PM (1999) Developmental morphology and life-cycles of the Microsporidia. *In* 'The Microsporidia and Microsporidiosis'. (Eds M Wittner, LM Weiss.) pp. 85–128. (ASM Press: Washington DC).

Canning EU (1988) Nuclear division and chromosome cycle in microsporidia. *Biosystems* **21**, 333–340.

Canning EU, Vavra J (2000) Phylum Microspora. *In* 'An Illustrated Guide to the Protozoa'. 2nd edn. (Eds JJ Lee, GF Leedale) pp. 39–126. (Society of Protozoologists: Lawrence, KS).

Canning EU, Refardt D, Vossbrinck CR, Okamura B, Curry A (2002) New diplokaryotic microsporidia (Phylum Microsporidia) from freshwater bryozoans (Bryozoa, Phylactolaemata). *European Journal of Protistology* **38**, 247–265.

Capaul M (2003) Parasite-mediated selection in experimental *Daphnia magna* populations. *Evolution* **57**, 249–260.

Clausen C (2000) Light and ultrastructural observations on a microsporidium in the hydrozoan *Halammohydra intermedia* (Cnidaria). *Sarsia* **85**, 177–180.

Dunn AM, Terry RS, Smith JE (2001) Transovarial transmission in the Microsporidia. *Advances in Parasitology* **48**, 57–100.

Ebert D (1994) Virulence and local adaptation of a horizontally transmitted parasite. *Science* **265**, 1084–1086.

Ebert D, Mangin KL (1997) The influence of host demography on the evolution of virulence of a microsporidian gut parasite. *Evolution* **51**, 1828–1837.

Fisheries and Oceans Canada (2003) Synopsis of infectious diseases and parasites of commercially exploited shellfish, *Mikrocytos mackini* (Denman Island Disease) of oysters, Susan Bower, http://www.pac.dfo-mpo.gc.ca/sci/shelldis/pages/mikmacoy_e.htm.

Flegel TW, Pasharawipas T (1995) A proposal for typical eukaryotic meiosis in microsporidians. *Canadian Journal of Microbiology* **41**, 1–11.

Iversen ES, Kelly JF (1976) Microsporidiosis successfully transmitted experimentally in pink shrimp. *Journal of Invertebrate Pathology* **27**, 407–408.

Keeling PJ (2003) Congruent evidence from α tubulin and β tubulin gene phylogenies for a zygomycete origin of microsporidia. *Fungal Genetics and Biology* **38**, 298–309.

Keeling PJ, Fast NM (2002) Microsporidia: biology and evolution of highly reduced intracellular parasites. *Annual Review of Microbiology* **56**, 93–116.

Langdon JS (1991) Description of *Vavraia parastacida* sp. nov. (Microspora: Pleistophoridae) from marron, *Cherax tenuimanus* (Smith), (Decapoda: Parastacidae). *Journal of Fish Diseases* **14**, 619–629.

Larsson JIR (1999) Identification of Microsporidia. *Acta Protozoologica* **38**, 161–197.

Lom J, Dyková I (1992) Microsporidia (Phylum Microspora Sprague, 1977). Chapter 6. *In* 'Protozoan Parasites of Fishes'. pp. 125–157. (Elsevier: Amsterdam).

Lom J, Nilsen F (2003) Fish microsporidia: fine structural diversity and phylogeny. *International Journal for Parasitology* **33**, 107–127.

Mathews JL, Brown AM, Larison K, Bishop-Stewart JK, Rogers P, Kent ML (2001) *Pseudoloma neurophila* n. g., n. sp. a new microsporidium from the central nervous system of the zebrafish (*Danio rerio*). *Journal of Eukaryotic Microbiology* **48**, 227–233.

Mathis A (2000) Microsporidia: emerging advances in understanding the basic biology of these unique organisms. *International Journal for Parasitology* **30**, 795–804.

Moodie EG, Le Jambre LF, Katz M (2003) Ultrastructural characteristics and small subunit ribosomal DNA sequence of *Vairimorpha cheracis* sp. nov., (Microspora: Burenellidae), a parasite of the Australian yabby, *Cherax destructor* (Decapoda: Parastacidae). *Journal of Invertebrate Pathology* **84**, 198–213.

Nilsen F (1999) Small subunit rDNA phylogeny of *Bacillidium* sp. (Microspora, Mrazekiidae) infecting oligochaetes. *Parasitology* **118**, 553–558.

Overstreet RM (1973) Parasites of some penaeid shrimps with emphasis on reared hosts. *Aquaculture* **2**, 105–140.

Petry F (2000) 'Cryptosporidiosis and Microsporidiosis'. (Basel: Karger).

Ramsay JM, Speare DJ, Dawe SC, Kent ML (2002) Xenoma formation during microsporidial gill disease of salmonids caused by *Loma salmonae* is affected by host species (*Oncorhynchus tshawytscha, O. kisutch, O. mykiss*) but not by salinity. *Diseases of Aquatic Organisms* **48**, 125–131.

Shaw RW, Kent ML (1999) Fish microsporidia. In 'The Microsporidia and Microsporidiosis'. (Eds M Wittner, LM Weiss.) pp. 447–501. (ASM Press: Washington DC).

Sindermann CJ (1990) 'Principal Diseases of Marine Fish and Shellfish'. (Academic Press: San Diego).

Speare DJ, Beaman HJ, Jones SRM, Markham RJF, Arsenault GJ (1998) Induced resistance of rainbow trout to gill disease associated with the microsporidian gill parasite *Loma salmonae*. *Journal of Fish Diseases* **21**, 93–100.

Sprague V, Becnel JJ, Hazard EJ (1992) Taxonomy of Phylum Microspora. *Critical Reviews in Microbiology* **18**, 285–395.

Undeen A (1997) Microsporidia (Protozoa): A Handbook of Biology and Research Techniques. Southern Cooperative Series Bulletin. Center for Medical, Agricultural and Veterinary Entomology, Gainesville, Florida, USA. http://pearl.agcomm.okstate.edu:16080/scsb387/content.htm, accessed 6 December 2003.

Vavra J, Larsson JIR (1999) Structure of the Microsporidia. In 'The Microsporidia and Microsporidiosis'. (Eds M Wittner, LM Weiss.) pp. 447–501. (ASM Press: Washington DC).

Weiss LM (2000) Molecular phylogeny and diagnostic approaches to Microsporidia. In 'Cryptosporidiosis and Microsporidiosis'. (Ed. F Petry) pp. 209–235. (Karger: Basel).

Weiss LM, Vossbrinck CR (1999) Molecular biology, molecular phylogeny, and molecular diagnostic approaches to the Microsporidia. In 'The Microsporidia and Microsporidiosis'. (Eds M Wittner, LM Weiss.) pp. 129–171. (ASM Press: Washington DC).

Williams BAP, Hirt RP, Lucocq JM, Embley TM (2002) A mitochondrial remnant in the microsporidian *Trachipleistophora hominis*. *Nature* **418**, 865–869.

Wittner M, Weiss LM (1999) 'The Microsporidia and Microsporidiosis'. (ASM Press: Washington DC).

Mikrocytos mackini (microcell)

Bower SM (2001) Hazards and risk management of *Mikrocytos mackini* in oysters. In 'Proceedings of the OIE International Conference on Risk Anaysis in Aquatic Animal Health'. (Ed. CJ Rodgers.) pp. 164–166. (World Organisation for Animal Health: Paris).

Bower SM, Hervio D, Meyer GR (1997) Infectivity of *Mikrocytos mackini*, the causative agent of Denman Island disease in Pacific oysters *Crassostrea gigas*, to various species of oysters. *Diseases of Aquatic Organisms* **29**, 111–116.

Carnegie RB, Meyer GR, Blackbourn J, Cochennec-Laureau N, Berthe FCJ, Bower SM (2003). Molecular detection of the oyster parasite *Mikrocytos mackini* and a preliminary phylogenetic analysis. *Diseases of Aquatic Organisms* **54**, 219–227.

Cochennec-Laureau N, Reece KS, Berthe FCJ, Hine PM (2003) *Mikrocytos roughleyi* taxonomic affiliation leads to the genus *Bonamia* (Haplosporidia). *Diseases of Aquatic Organisms* **54**, 209–217.

Farley CA, Wolf PH, Elston RA (1988) A long-term study of 'microcell' disease in oysters with a description of a new genus, *Mikrocytos* (g.n.) and two new species *Mikrocytos mackini* (sp.n.) and *Mikrocytos roughleyi* (sp.n.). *U.S. National Marine Fish Service Bulletin* **86**, 581–593.

Fisheries and Oceans Canada (2003) Synopsis of infectious diseases and parasites of commercially exploited shellfish, *Mikrocytos mackini* (Denman Island Disease) of oysters, Susan Bower, http://www.pac.dfo-mpo.gc.ca/sci/shelldis/pages/mikmacoy_e.htm.

Hervio D, Bower SM, Meyer GR (1996) Detection, isolation and experimental transmission of *Mikrocytos mackini*, a microcell parasite of Pacific oysters *Crassostrea gigas* (Thunberg). *Journal of Invertebrate Pathology* **67**, 72–79.

Hine PM, Bower SM, Meyer GR, Cochennec-Laureau N, Berthe FCJ (2001) Ultrastructure of *Mikrocytos mackini*, the cause of Denman Island disease in oysters *Crassostrea* spp. and *Ostrea* spp. in British Columbia, Canada. *Diseases of Aquatic Organisms* **45**, 215–227.

Quayle DB (1982) Denman Island oyster disease 1960–1980. *British Columbia Shellfish Mariculture Newsletter* **2**, 1–5.

Ciliophora (ciliates)

Corliss JO (1979) 'The Ciliated Protozoa – Characterization, Classification and Guide to the Literature'. 2nd edn. (Pergamon: Oxford).

Deroux G, Campillo A, Bradbury PC (1975) *Ascophrys rodor* (Campillo et Deroux) parasite de la crevette rose *P. serratus* en élevage. *Revue de Travaux de l'Institut de Pêche Maritime* **39**, 359–379.

Grassé PP (Ed.) (1984) Traité de Zoologie, Tome II, Infusoires ciliés, Fasc. 1, Généralités. (Masson: Paris).

Hausmann K, Bradbury PC (Eds) (1996) 'Ciliates: Cells as Organisms'. (Gustav Fischer Verlag: Stuttgart).

Lom J (1995) Trichodinidae and other ciliates (Phylum Ciliophora). *In* 'Fish Diseases and Disorders. Vol. 1. Protozoan and Metazoan Infections'. (Ed. PTK Woo.) pp. 229–262. (CAB International: Wallingford).

Lom J, Dyková I (1992) Protozoan parasites of fishes. Developments in Aquaculture and Fisheries Science, 26. pp. 315. (Elsevier: Amsterdam).

Lom J, Puytorac P de (Ed.) (1994) Sous-classe des Peritrichia Stein, 1994. *In* 'Traité de Zoologie, Tome II, Infusoires Ciliés' (Ed. PP Grassé.), 'Fasc. 2, Systématique', (Ed. P de Puytorac.) pp. 681–737. (Masson: Paris).

Lynn DH (2003) Morphology or molecules: how do we identify the major lineages of ciliates (Phylum Ciliophora)? *European Journal of Protistology* **39**, 356–364. (= latest classification of ciliates)

Lynn DH, Small EB (2000) Phylum Ciliophora. *In* 'Illustrated Guide to the Protozoa'. 2nd edn. Vol. I. (Eds JJ Lee, GF Leedale, PC Bradbury.) pp. 371–656. (Allen Press: Lawrence, KS).

Myxozoa (myxozoans)

Canning EU, Okamura B (2004) Biodiversity and evolution of the Myxozoa. *Advances in Parasitology* **56**, 44–131.

El-Matbouli M, Hoffmann RW (1998) Light and electron microscopic studies on the chronological development of *Myxobolus cerebralis* to the actinosporean stage in *Tubifex tubifex*. *International Journal for Parasitology* **28**, 195–217.

El-Matbouli M, Hoffmann RW, Mandok C (1995) Light and electron microscopic observations on the route of the triactinomyxon-sporoplasm of *Myxobolus cerebralis* from epidermis into rainbow trout cartilage. *Journal of Fish Biology* **46**, 919–935.

Hedrick RP, Baxa DV, DeKinekelin P, Okamura B (2004) Malacosporean-like spores in urine of rainbow trout react with antibody and DNA probes to *Tetracapsuloides bryosalmonae*. *Parasitology Research* **92**, 81–88.

Kent ML, Moser M, Marques A, Lom J (2000) Phylum Myxozoa. *In* 'Illustrated Guide to the Protozoa'. 2nd edn. Vol. I. (Eds JJ Lee, GF Leedale, PC Bradbury.) pp. 127–148. (Allen Press: Lawrence, KS).

Kent ML, Andree KB, Bartolomew JL, El-Matbouli M, Desser SS, Devlin RH, Feist SW, Hedrick RP, Hoffman RW, Khattra J, Hallett SL, Lester RJG, Longshaw M, Palenzuela O, Siddall ME, Xiao C (2001) Recent advances in our knowledge of the Myxozoa. *Journal of Eukaryotic Microbiology* **48**, 395–413.

Køie M, Whipps CM, Kent ML (2004) *Ellipsomyxa gobii* (Myxozoa: Ceratomyxidae) in the common goby *Pomatoschistus microps* uses *Nereis* spp. (Annelida: Polychaeta) as invertebrate hosts. *Folia Parasitologica* **51**, 14–18.

Kudo RR (1919) Studies on Myxosporidia. *Illinois Biological Monographs* **5**, 241–503.

Lom J (1990) Phylum Myxozoa. *In* 'Handbook of Protoctista'. (Eds L Margulis, JO Corliss, M Melkonian, DJ Chapman.) pp. 36–52. (Jones and Bartlett Publishers: Boston).

Lom J, Dyková I (1992) Protozoan parasites of fishes. Developments in Aquaculture and Fisheries Science, 26. pp. 315. (Elsevier: Amsterdam).

Lom J, Dyková I (1995) Myxosporea (Phylum Myxozoa). *In* 'Fish Diseases and Disorders, Vol. 1. Protozoan and Metazoan Infections'. (Ed. PTK Woo.) pp. 97–148. (CAB International: Wallingford, Oxon, UK).

Okamura B, Curry A, Wood TS, Canning EU (2002) Ultrastructure of *Buddenbrockia* identifies it as a myxozoan and verifies the bilaterian origin of the Myxozoa. *Parasitology* **124**, 215–223.

Padrós F, Palenzuela O, Hispano C, Tosas O, Zarza C, Crespo S, Alvarez-Pellitero P (2001) *Myxidium leei* (Myxozoa) infections in aquarium-reared Mediterranean fish species. *Diseases of Aquatic Organisms* **47**, 57–62.

Palenzuela O, Sitjà-Bobadilla A, Alvarez-Pellitero P (1997) *Ceratomyxa sparusaurati* (Protozoa: Myxosporea) infections in cultured gilthead sea bream *Sparus aurata* (Pisces: Teleostei) from Spain: aspects of the host-parasite relationship. *Parasitology Research* **83**, 539–548.

Redondo MJ, Palenzuela O, Alvarez-Pellitero P (2004) Studies on transmission and life cycle of *Enteromyxum scophthalmi* (Myxozoa), an enteric parasite of turbot (*Scophthalmus maximus* L.). *Folia Parasitologica* **51**, (in press).

Shulman SS (1966) 'Myxosporidia of the fauna of the USSR'. pp. 504. (Nauka: Moscow). (In Russian).

Wolf K, Markiw ME (1984) Biology contravenes taxonomy in the Myxozoa: New discoveries show alternation of invertebrate and vetebrate hosts. *Science* **255**, 1449–1452.

Chapter 3
'Turbellaria' (turbellarians)

Ball IR, Khan R (1976) On *Micropharynx parasitica* Jagerskiöld, a marine planarian ectoparasitic on thorny skates, *Raja radiata* Donovan, from the North Atlantic Ocean. *Canadian Journal of Zoology* **8**, 419–426.

Blair D, Williams JB (1987) A new fecampiid of the genus *Kronborgia* (Platyhelminthes: Turbellaria: Neorhabdocoela) parasitic in the intertidal isopod *Exosphaeroma obtusum* (Dana) from New Zealand. *Journal of Natural History* **21**, 1155–1172.

Bellon-Humbert C (1983) *Fecampia erythrocephala* Giard (Turbellaria: Neorhabdocoela), a parasite of the prawn *Palaemon serratus* Pennant: The adult phase. *Aquaculture* **31**, 117–140.

Bresslau E (1933) Turbellaria. *In* 'Handbuch der Zoologie' Bd. 2. (Eds W Kükenthal, T Krumbach.) pp. 193–293. (Walter de Gruyer: Berlin).

Cannon LRG (1982) Endosymbiotic umagillids (Turbellaria) from holothurians of the Great Barrier Reef. *Zoologica Scripta* **11**, 173–188.

Cannon LRG (1987) Two new rhabdocoel turbellarians, *Umagilla pacifica* sp.n. and *U. karlingi* sp.n. (Umagillidae), endosymbiotic with holothurians (Echinodermata) from the Great Barrier Reef; and a discussion of sclerotic structures in the female system of the Umagillidae. *Zoologica Scripta* **16**, 297–303.

Cannon LRG (1990) *Apidioplana apluda* n.sp., a polyclad turbellarian symbiote of gorgonian corals from the Great Barrier Reef, with a review of the family Apidioplanidae. *Memoirs of the Queensland Museum* **28**, 435–442.

Cannon LRG (1998) 'Turbellaria of the World: a guide to families and genera.' (ETI: Amsterdam). CD-ROM.

Cannon LRG, Grygier MJ (1991) The turbellarian *Notoplana comes* n. sp. (Leptoplanidae: Acotylea: Polycladida), found with the intertidal brittle star *Ophiocoma scolopendrina* in Okinawa, Japan. *Galaxea* **10**, 23–33.

Cannon LRG, Lester RJG (1988) Two turbellarians parasitic in fish. *Diseases of Aquatic Organisms* **5**, 15–22.

Christensen AM, Kanneworff B (1965) Life history and biology of *Kronborgia amphipodicola* Christensen and Kanneworff (Turbellaria: Neorhabdocoela). *Ophelia* **2**, 237–251.

de Beauchamp P (1961) Classe des Turbellariés. *In* 'Traite de Zoologie'. Vol. 4. (Ed. PP Grasse.) pp. 35–212. (Masson et Cie: Paris)

Doignon G, Artois T, Deheyn D (2003) *Discoplana malagasensis* sp. nov., a new turbellarian (Platyhelminthes: Polycladida: Leptoplanidae) symbiotic in an ophiuroid (Echinodermata), with a cladistic analysis of the *Discoplana/Euplana* species. *Zoological Science (Tokyo)* **20**, 357–369.

Fleming LC, Burt MDB, Bacon GB (1981) On some commensal Turbellaria of the Canadian east coast. *Hydrobiologia* **84**, 131–137.

Gevaerts H, Moens JB, Martens EE, Schockaert ER (1995) Hard parts in the female system of *Syndesmis longicanalis* (Platyhelminthes, Rhabdocoela, Umagillidae) are basement membrane derivatives. *Invertebrate Biology* **114**, 279–284.

Hickman VV (1955) Two new rhabdocoel turbellarians parasitic in Tasmanian holothurioids. *Papers and Proceedings of the Royal Society of Tasmanian* **89**, 81–97.

Hickman VV (1956) Parasitic turbellarian from Tasmanian Echinoidea. *Papers and Proceedings of the Royal Society of Tasmanian* **90**, 169–181.

Hyman LH (1951) 'The Invertebrates, Vol. II. Platyhelminthes and Rhynchocoela: the acoelomate Bilateria'. (McGraw Hill: New York).

Hyra GS (1993) *Genostoma kozloffi*, new species and *Genostoma inopinatum*, new species (Turbellaria: Neorhabdocoela: Genostomatidae) from leptostracan crustaceans of the genus *Nebalia*. *Cahiers de Biologie Marine* **34**, 111–126.

Jägersten G (1940) Zur Kenntnis von *Glanduloderma myzostomatis* n. gen., n. sp., einer eigentümlichen, in Myzostomiden schmarotzenden Turbellarienform. *Arkiv för Zoologi* **33**, 1–24.

Jennings JB (1989) Epidermal uptake of nutrients in an unusual turbellarian parasitic in the starfish *Coscinasterias calamaria* in Tasmanian waters. *Biological Bulletin* **176**, 327–336.

Jennings JB (1997) Nutritional and respiratory pathways to parasitism exemplified in the Turbellaria. *International Journal for Parasitology* **27**, 679–691.

Jennings JB, Cannon LRG (1985) Observation of the occurrence nutritional physiology and respiratory pigment of three species of flatworms (Rhabdocoela: Pterastericolidae) entosymbiotic in starfish from temperate and tropical waters. *Ophelia* **24**, 99–215.

Jennings JB, Cannon LRG (1987) The occurrence, spectral properties and probable role of haemoglobins in four species of entosymbiotic turbellarians (Rhabdocoela: Umagillidae). *Ophelia* **27**, 143–154.

Jennings JB, Hick AJ (1990) Differences in the distribution, mitochondrial content and probable roles of haemoglobin-containing parenchymal cells in four species of entosymbiotic turbellarians (Rhabdocoela: Umagillidae and Pterastericolidae). *Ophelia* **31**, 163–175.

Joffe BI, Kornakova EE (1998) *Notentera ivanovi* Joffe *et al.*, 1997: a contribution to the question of phylogenetic relationships between 'turbellarians' and the parasitic Platyhelminthes (Neodermata). *Hydrobiologia* **383**, 245–250.

Jondelius U (1988) Epidermal ultrastructure of adults and juveniles of *Triloborhynchus astropectinis* (Dalyellioida, Platyhelminthes). *Progress in Zoology* **36**, 9–44.

Jondelius U (1991) Unciliated body surface in three species of the Umagillidae (Dalyellioida, Platyhelminthes). *Hydrobiologia*. **227**, 299–305.

Jondelius U (1992a) Adhesive glands in the Pterastericolidae, Platyhelminthes, Rhadbocoela. *Zoomorphology* **11**, 229–238.

Jondelius U (1992b) New *Pterastericola* species (Platyhelminthes, Rhabdocoela) and a cladistic analysis of the Pterastericolidae. *Zoologica Scripta* **21**, 93–101.

Karling TG (1981) *Typhlorhynchus nanus* Laidlaw, a kalyptorhynch turbellarian without proboscis (Platyhelminthes). *Annales Zoologici Fennici* **18**, 69–177.

Karling TG, Nilsson AM (1974) Further studies on the genus *Hypoblepharina* Böhmig (Turbellaria, Dalyellioida) with description of *H. serrifera* sp. n. *Zoologica Scripta* **3**, 59–63.

Kent ML, Olson AC (1986) Interrelationships of a parasitic turbellarian, (*Paravortex* sp.) (Graffillidae, Rhabdocoela) and its marine fish hosts *Fish Pathology* **21**, 65–72.

Køie M (1969) On the endoparasites of *Buccinum undatum* L. with special reference to trematodes. *Ophelia* **6**, 251–279.

Kozloff EN (1965) *Desmote inops* sp .n. and *Fallacohospes inchoatus* gen. et sp. n. umagillid rhabdocoels from the intestine of the crinoid *Florometra serratissima* (A.H. Clark). *Journal of Parasitology* **51**, 305–312.

Lesson RP (1843) *L'echo du monde savant.* **10**, 14.

Lockyer AE, Olson PD, Littlewood DTJ (2003) Utility of complete large and small subunit rRNA genes in resolving the phylogeny of the Neodermata (Platyhelminthes): Implications and a review of the cercomer theory. *Biological Journal of the Linnean Society* **78**, 155–171.

Naylor E (1955) The seasonal abundance on *Idotea* of the cocoons of the flatworm *Plagistomum oyense* de Beauchamp. *Reports of the Marine Biological Station, Port Erin* **67**, 25–30.

Phillips JI (1978) The occurrence and distribution of haemoglobin in the entosymbiotic rhabdocoel *Paravortex scrobiculariae* (Graff) (Platyhelminthes: Turbellaria). *Comparative Biochemistry and Physiology* **61A**, 679–683.

Palombi A (1926) *Digenobothrium inerme* nov. gen. nov. sp. (Crossocoela). Considerazioni sistematiche sull'ordine degli Alloecoele. *Archivio Zoologia Italiano* **2**, 143–177.

Prudhoe S (1985) 'A Monograph on Polyclad Turbellaria'. (British Museum – Natural History: London).

Reisinger E (1930) Zum Ductus genito-intestinalis Problem. I. über primäre Geschlechtstrakt-Darmverbindungen bei rhabdocoelen Turbellarien. (Zugleich ein Beitrag zur europäischen und grönländischen Turbellarienfauna). *Zeitschrift für Morphologie und ökologie* **16**, 49–75.

Robledo JAF, Caceres-Martinez J, Sluys R, Figueras A (1994) The parasitic turbellarian *Urastoma cyprinae* (Platyhelminthes: Urastomidae) from blue mussel *Mytilus galloprovincialis* in Spain: occurrence and pathology. *Diseases of Aquatic Organisms* **18**, 203–210.

Rohde K (2001) Protonephridia as phylogenetic characters. *In* 'Interrelationships of the Platyhelminthes'. (Eds DTJ Littlewood, RA Bray.) pp. 203–216. (Taylor and Francis: London).

Schell SC (1986) *Graffilla pugetensis* n.sp. (Order Neorhabdocoela: Graffillidae), a parasite in the pericardial cavity of the bent-nose clam, *Macoma nasuta* (Conrad, 1837). *Journal of Parasitology* **72**, 748–754.

Shinn G (1981) The diet of three species of umagillid neorhabdocoel turbellarians inhabiting the intestine of echinoids. *Hydrobiologia* **84**, 155–162.

Shinn G (1985a) Reproduction of *Anoplodium hymanae*, a turbellarian flatworm (Neorhabdocoela, Umagillidae) inhabiting the coelom of sea cucumbers: production of egg capsules, and escape of infective stages without evisceration of the host. *Biological Bulletin* **169**, 182–198.

Shinn G (1985b) Infection of new hosts by *Anoplodium hymanae*, a turbellarian flatworm (Neorhabdocoela, Umagillidae) inhabiting the coelom of the sea cucumber *Stichopus californicus*. *Biological Bulletin* **169**, 199–214.

Shinn G (1986) Life history and function of the secondary uterus of *Wahlia pulchella*, an umagillid turbellarian from the intestine of a northeastern Pacific sea cucumber (*Stichopus californicus*). *Ophelia* **25**, 59–74.

Shinn G (1987) Two new species of umagillid flatworms from the 20-rayed Antarctic crinoid *Promachocrinus kerguelensis*. *Canadian Journal of Zoology* **65**, 1001–1009.

Shinn GL, Christensen AM (1985) *Kronborgia pugettensis* sp. nov. (Neorhabdocoela: Fecampiidae), an endoparasitic turbellarian infesting the shrimp *Heptacarpus kincaidi* (Rathbun), with notes on its life-history. *Parasitology* **91**, 431–447.

Sluys R (1989) 'A Monograph of the Marine Triclads'. (AA Balkema: Rotterdam).

Sluys R, Cannon LRG (1989) A new marine triclad from the west Pacific (Platyhelminthes: Tricladida: Maricola). *Invertebrate Zoology* **3**, 149–153.

Tyler S, Tyler, MS (1997) Origin of the epidermis in parasitic platyhelminths. *International Journal for Parasitology* **27**, 715–738.

Villalba A, Mourelle SG, Carballal MJ, Lopez C (1997) Symbionts and diseases of farmed mussels *Mytilus galloprovincialis* throughout the culture process in the Rias of Galicia (NW Spain). *Diseases of Aquatic Organisms* **31**, 127–139.

Watson NA (1997) Proximo distal fusion of flagella during spermiogenesis in the 'turbellarian' platyhelminth *Urastoma cyprinae*, and phylogenetic implications. *Invertebrate Reproduction and Development* **32**, 107–117.

Westblad E (1950) On *Meara stichopi* (Bock) Westbald, a new representative of the Turbellaria archoophora. *Arkiv för Zoologi* **1**, 43–57.

Westervelt CA (1981) *Collastoma kozloffi* sp.n., a neorhabdocoel turbellarian from the intestine of the sipunculan *Themiste dyscrita*. *Journal of Parasitology* **67**, 574–577.

Winsor L (1990) Marine Turbellaria (Acoela) from north Queensland. *Memoirs of the Queensland Museum* **28**, 785–800.

Woods CMC, Hayden BJ (1998) An observation of the turbellarian *Paravortex* sp. in the New Zealand scallop *Pecten novaezelandiae* (Bivalvia: Pectinidae). *New Zealand Journal of Marine and Freshwater Research* **32**, 551–553.

Monogenea Polyopisthocotylea (ectoparasitic flukes)

Bychowsky BE (1957) 'Monogenetic Trematodes, their Classification and Phylogeny'. (Academy of Sciences, USSR, Moscow. (In Russian.). English translation by WJ Hargis and PC Oustinov (1961) (American Institute of Biological Sciences: Washington DC).

Jovelin R, Justine JL (2001) Phylogenetic relationships within the polyopisthocotylean monogeneans (Platyhelminths) inferred from partial 28S rDNA sequences. *International Journal for Parasitology* **31**, 393–401.

Kearn GC (1986) The eggs of monogeneans. *Advances in Parasitology* **25**, 175–273.

Kearn GC (1998) 'Parasitism and the Platyhelminths'. (Chapman and Hall: London).

Lebedev B I (1995) 'Biodiversity and evolution of Oligonchoinean Monogeneoidea'. (Institute of Biology and Pedology, The Russian Academy of Sciences, Far East Branch: Vladivostok).

Mollaret I, Jamieson BGM, Justine JL (2000) Phylogeny of the Monopisthocotylea and Polyopisthocotylea (Platyhelminthes) inferred from 28S rDNA sequences. *International Journal for Parasitology* **30**, 171–185.

Olsen PD, Littlewood DTJ (2002) Phylogenetics of Monogenea – evidence from a medley of molecules. *International Journal for Parasitology* **32**, 233–244.

Whittington ID, Chisholm LA, Rohde K (2000) The larvae of Monogenea (Platyhelminthes). *Advances in Parasitology* **44**, 139–232.

Yamaguti S (1963) 'Systema Helminthum, Volume IV, Monogenea and Aspidocotyla'. (Interscience Publishers: London).

Monogenea Monopisthocotylea (ectoparasitic flukes)

Boeger WA, Kritsky DC (2001) Phylogenetic relationships of the Monogenoidea. *In* 'Interrelationships of the Platyhelminthes'. (Eds DTJ Littlewood, RA Bray.) pp. 92–102. (Taylor and Francis: London).

Kearn GC (1986) The eggs of monogeneans. *Advances in Parasitology* **25**, 175–273.

Kearn GC (1994) Evolutionary expansion of the Monogenea. *International Journal for Parasitology* **24**, 1227–1271.

Kearn GC (1998) 'Parasitism and the Platyhelminths'. (Chapman and Hall: London).

Kearn GC (1999) The survival of monogenean (platyhelminth) parasites on fish skin. *Parasitology* **119**, S57–S88.

Rohde K (1994) Niche restriction in parasites: proximate and ultimate causes. *Parasitology* **109**, S69–S84.

Thoney DA, Hargis WJ Jr (1991) Monogenea (Platyhelminthes) as hazards for fish in confinement. *Annual Review of Fish Diseases* **1**, 133–153.

Whittington ID, Cribb BW (2001) Adhesive secretions in the Platyhelminthes. *Advances in Parasitology* **48**, 101–224.

Whittington ID, Ernst I (2002) Migration, site-specificity and development of *Benedenia lutjani* (Monogenea: Capsalidae) on the surface of its host, *Lutjanus carponotatus* (Pisces: Lutjanidae). *Parasitology* **124**, 423–434.

Whittington ID, Cribb BW, Hamwood TE, Halliday JA (2000a) Host-specificity of monogenean (platyhelminth) parasites: a role for anterior adhesive areas? *International Journal for Parasitology* **30**, 305–320.

Whittington ID, Chisholm LA, Rohde K (2000b) The larvae of Monogenea (Platyhelminthes). *Advances in Parasitology* **44**, 139–232.

Aspidogastrea (aspidogastreans)

Baer KE von (1927) Beiträge zur Kenntnis der niedern Thiere. *Nov. Acta Acad. Nat. Curios* **13**, 523–762, 881–882.

Bakker KE, Davids C (1973) Notes on the life history of *Aspidogaster conchicola* Baer, 1826 (Trematoda; Aspidogastridae). *Journal of Helminthology* **47**, 269–276.

Cunningham JT (1897) On *Stichocotyle nephropis*, a new trematode. *Transactions of the Royal Society Edinburgh* **32**, 273–280.

Huehner MK, Etges FJ (1977) The life cycle and development of *Aspidogaster conchicola* in the snails, *Viviparus malleatus* and *Goniobasis livescens*. *Journal of Parasitology* **63**, 669–674.

Littlewood DTJ, Rohde K, Clough KA (1999) The interrelationships of all major groups of Platyhelminthes – phylogenetic evidence from morphology and molecules. *Biological Journal of the Linnean Society* **66**, 75–114.

Rohde K (1972) The Aspidogastrea, especially *Multicotyle purvisi* Dawes, 1941. *Advances in Parasitology* **10**, 77–151.

Rohde K (1973) Structure and development of *Lobatostoma manteri* sp. nov. (Trematoda, Aspidogastrea) from the Great Barrier Reef, Australia. *Parasitology* **66**, 63–83.

Rohde K (1975) Early development and pathogenesis of *Lobatostoma manteri* Rohde (Trematoda: Aspidogastrea). *International Journal for Parasitology* **5**, 597–607.

Rohde K (1989) At least eight types of sense receptors in an endoparasitic flatworm: a countertrend to sacculinization. *Naturwissenschaften* **76**, 383–385.

Rohde K (1994) The minor groups of parasitic Platyhelminthes. *Advances in Parasitology* **33**, 145–234.

Rohde K (1999) Aspidogastrea. *In* 'Tree of Life'. (Eds DR Maddison, WP Maddison.) http://tolweb.org/tree?group=Aspidogastrea&contgroup=Platyhelminthes

Rohde K (2001) The Aspidogastrea, an archaic group of Platyhelminthes. *In* 'Interrelationships of the Platyhelminthes'. (Eds DTJ Littlewood, RA Bray.) pp. 159–167. (Taylor & Francis: London).

Rohde, K (2002) Subclass Aspidogastrea Faust and Tang, 1936. *In* 'Keys to the Trematoda'. Vol. 1. (Eds DI Gibson, A Jones, RA Bray.) pp. 5–14. (CABI Publishing: Wallingford).

Rohde K, Sandland R (1973) Host-parasite relations in *Lobatostoma manteri* Rohde (Trematoda, Aspidogastrea). *Zeitschrift für Parasitenkunde* **41**, 115–136.

Digenea (endoparasitic flukes)

Brooks DR, O'Grady RT, Glen DR (1985) Phylogenetic analysis of the Digenea (Platyhelminthes: Cercomeria) with comments on their adaptive radiation. *Canadian Journal of Zoology* **63**, 411–443.

Cribb TH, Bray RA, Olson PD, Littlewood DTJ (2003) Life cycle evolution in the Digenea: a new perspective from phylogeny. *Advances in Parasitology* **54**, 197–254.

Gibson DI, Jones A, Bray RA (Eds) (2002) 'Keys to the Trematoda'. Vol. 1. (CABI Publishing: Wallingford).

Helluy, S, Thomas, F (2003) Effects of *Microphallus papillorobustus* (Platyhelminthes: Trematoda) on serotonergic immunoreactivity and neuronal architecture in the brain of

Gammarus insensibilis (Crustacea: Amphipoda). *Proceedings of The Royal Society of London Series B-Biological Sciences* **270**, 563–568.

Jousson O, Bartoli P (2002) Species diversity among the genus *Monorchis* (Digenea: Monorchiidae) parasitic in marine teleosts: molecular, morphological and morphometrical studies with a description of *Monorchis blennii* m. sp. *Parasitology Research* **88**, 230–241.

Køie M (1979) On the morphology and life-history of *Derogenes varicus* (Müller, 1784) Looss, 1901 (Trematoda, Hemiuridae). *Zeitschrift für Parasitenkunde* **59**, 67–78.

Køie M (1991) Aspects of the morphology and life cycle of *Lecithocladium excisum* (Digenea, Hemiuridae), a parasite of *Scomber* spp. *International Journal for Parasitology* **21**, 597–602.

Lafferty KD (1993) The marine snail, *Cerithidea californica*, matures at smaller sizes where parasitism is high. *Oikos* **68**, 3–11.

Morimitsu T, Nagai T, Ide M, Kawano H, Naichuu A, Koono M, Ishii A (1987) Mass stranding of Odontoceti caused by parasitogenic eighth cranial neuropathy. *Journal of Wildlife Diseases* **23**, 586–590.Lively CM (1987) Evidence from a New Zealand snail for the maintenance of sex by parasitism. *Nature* **328**, 519–521.

Ogawa K, Fukudome M (1994) Mass mortality caused by blood fluke (*Paradeontacylix*) among amberjack (*Seriola dumerili*) imported to Japan. *Fish Pathology* **29**, 265–269.

Olson PD, Cribb TH, Tkach VV, Bray RA, Littlewood DTJ (2003) Phylogeny and classification of the Digenea (Platyhelminthes: Trematoda). *International Journal for Parasitology* **33**, 733–755.

Pearson JC (1972) A phylogeny of life-cycle patterns of the Digenea. *Advances in Parasitology* **10**, 153–189.

Poulin R, Cribb TH (2002) Trematode life cycles: short is sweet? *Trends in Parasitology* **18**, 176–183.

Wegeberg AM, Jensen KT (1999) Reduced survivorship of *Himasthla*-infected cockles (*Cerastoderma edule*) exposed to oxygen depletion. *Journal of Sea Research* **42**, 325–331.

Yamaguti S (1971) 'Synopsis of digenetic trematodes of vertebrates'. (Keigaku Publishing Co.: Tokyo).

Yamaguti S (1975) 'A synoptic review of life histories of digenetic trematodes of vertebrates'. Part I. (Keigaku Publishing: Tokyo).

Amphilinidea (unsegmented tapeworms)

Andreev VV, Markov GS (1971) [Influence of some helminths on sturgeon]. *Zoologicheskij Zhurnal* **50**, 15–24. (In Russian).

Dubinina MN (1982) '[Parasitic worms of the class Amphilinida (Platyhelminthes).]' (Nauka: Leningrad). (In Russian).

Littlewood DTJ, Rohde K, Clough KA (1999) The interrelationships of all major groups of Platyhelminthes – phylogenetic evidence from morphology and molecules. *Biological Journal of the Linnean Society* **66**, 75–114.

Popova LB, Davidov VG (1988) Studies on localization of *Amphilina foliacea* (Amphilinidae Dubinina, 1974) in definitive hosts. *Helminthologia* **25**, 129–138.

Rohde K (1994) The minor groups of parasitic Platyhelminthes. *Advances in Parasitology* **33**, 145–234.

Rohde K (1999) Amphilinidea. *In* 'Tree of Life'. (Eds DR Maddison, WP Maddison). http://tolweb.org/tree?group=Aspidogastrea&contgroup=Platyhelminthes.

Rohde K, Georgi M (1983) Structure and development of *Austramphilina elongata* Johnston, 1931 (Cestodaria, Amphilinidea). *International Journal for Parasitology* **13**, 273–287.

Gyrocotylidea (unsegmented tapeworms)

Bristow G (1992) On the distribution, ecology and evolution of *Gyroctotyle urna*, *G. confusa* and *G. nybelini* (Cercomeromorpha: Gyrocotylidea) and their host *Chimaera monstrosa* (Holocephalida: Chimaeridae) in Norwegian waters, with a review of the species question. *Sarsia* **77**, 119–124.

Halvorsen O, Williams HH (1968) Studies of the helminth fauna of Norway. IX. *Gyrocotyle* (Platyhelminthes) in *Chimaera monstrosa* from Oslo Fjord, with emphasis on its mode of attachment and a regulation in the degree of infection. *Nytt Magasin for Zoology* **15**, 130–142.

Rohde K (1994) The minor groups of parasitic Platyhelminthes. *Advances in Parasitology* **33**, 145–234.

Van der Land K, Dienske H (1968) Two new species of Gyrocotyle (Monogenea) from chimarids (Holocephali). *Zoologische Medelingen (Leiden)* **43**, 97–105.

Xylander WER (1989) Untersuchungen zur Biologie von *Gyrocotyle urna* (Cestoda) und Überlegungen zu ihrem Lebenszyklus. *Verhandlungen der Deutschen Zoologischen Gesellschaft* **82**, 251.

Xylander WER (2001) Gyrocotylidea, Amphilinidea and the early evolution of Cestoda. *In* 'Interrelationships of the Platyhelminthes'. (Eds DTJ Littlewood, RA Bray.) pp. 103–111. (Taylor & Francis: London.).

Xylander WER (2003) Neodermata. *In* 'Spezielle Zoologie, Teil 1: Einzeller und Wirbellose'. 2nd edn. (Eds W Westheide, RM Rieger.) pp. 230–258. (Gustav Fischer Verlag: Stuttgart).

Eucestoda (true tapeworms)

Arme C, Pappas PW (Eds) (1983) 'Biology of the Eucestoda Volume 1'. (Academic Press: London).

Bates RM (1990) 'A Checklist of the Trypanorhyncha (Platyhelminthes: Cestoda) of the World (1935–1985)'. (National Museum of Wales, Zoological Series No. 1: Cardiff).

Beveridge I (2001) The use of life-cycle characters in studies of the evolution of cestodes. *In* 'Interrelationships of the Platyhelminthes'. (Eds DTJ Littlewood, RA Bray) pp. 250–256. (Taylor & Francis: London).

Caira JN, Jensen, K, Healy CJ (2001) Interrelationships among tetraphyllidean and lecanicephallidean cestodes. *In* 'Interrelationships of the Platyhelminthes'. (Eds DTJ Littlewood, RA Bray.) pp. 135–156. (Taylor & Francis: London).

Cake EW (1978) Larval cestode parasites of edible mollusks of the Northeaastern Gulf of Mexico. *Gulf Research Reports* **6**, 1–8.

Campbell RA (1983) Parasitism in the deep-sea. *In* 'Deep-sea Biology, The Sea'. Vol. 8. (Ed. GT Rowe.) pp. 473–552. (Wiley: New York).

Campbell RA, Haedrick RL, Munroe TA (1980) Parasitism and ecological relationships among deep-sea benthic fishes. *Marine Biology* **57**, 301–313.

Chervy L (2002) The terminology of larval cestodes or metacestodes. *Systematic Parasitology* **52**, 1–33.

Dollfus R (1976) Énumération des Cestodes du plankton et des invertébrés marins. *Annales de Parasitologie Humaine et Comparee* **51**, 207–220.

Esch GW (1983) The population and community ecology of cestodes. *In* 'Biology of the Eucestoda'. Vol. 1. (Eds C Arme, PW Pappas.) pp. 81–137. (Academic Press: London).

Euzet L (1959) 'Recherches sur les Cestodes Tétraphyllides des Sélaciens des Cotes de France'. (Causse, Graille & Castelneau: Montpellier).

Hoberg EP (1987) Recognition of larvae of the Tetrabothriidae (Eucestoda): implications for the origin of tapeworms in marine homeotherms. *Canadian Journal of Zoology* **65**, 997–1000.

Hoberg EP (1995). Historical biogeography and modes of speciation across high-latitude seas of the Holarctic: concepts for host-parasite coevolution between the Phocini (Phocidae) and Tetrabothriidae (Eucestoda). *Canadian Journal of Zoology* **73**, 45–57.

Hoberg EP (1996). Faunal diversity among avian parasite assemblages: the interaction of history, ecology & biogeography in marine systems. *Bulletin of the Scandinavian Society of Parasiotology* **6**, 65–89.

Hoberg EP (2002) Colonization and diversification: historical and coevolutionary trajectories among cestodes, cetaceans and pinnipeds. In 'Proceedings of the Tenth International Congress of Parasitology'. pp. 65–69. (Monduzzi: Medimond : Bologna).

Hoberg EP, Jones A, Bray RA (1999) Phylogenetic analysis among the families of the Cyclophyllidea (Eucestoda) based on comparative morphology, with new hypotheses for co-evolution in vertebrates. *Systematic Parasitology* **42**, 51–73.

Jarecka L (1961) Morphological adaptations of tapeworm eggs and their importance in the life cycles. *Acta Parasitologica Polonica* **9**, 409–426.

Jensen K (2001) A monograph of the order Lecanicephalidea (Platyhelminthes: Cestoda). PhD dissertation. University of Connecticut.

Joyeux C, Baer JG (1961) Classe des Cestodaires. *In* 'Traité de Zoologie Anatomie, systématique, Biologie'. (Ed. PP Grassé.) pp. 327–560. (Masson et Cie: Paris).

Khalil LF, Jones A, Bray RA (1994) 'Keys to the Cestode Parasites of Vertebrates'. (CAB International: Wallingford).

Klimpel S, Seehagen A, Palm H-W, Rosenthal A (2001) 'Deep-water Metazoan Fish Parasites of the World'. (Logos Verlag: Berlin).

Mackiewicz JS (1988) Cestode transmission patterns. *Journal of Parasitology* **74**, 60–71.

Mattis TE (1986) Development of two tetraphyllidean cestodes from the northern Gulf of Mexico. PhD dissertation. University of Southern Mississippi.

Olson PD, Littlewood TJ, Bray TA, Mariaux J (2001) Interrelationships and evolution of the tapeworms (Platyhelminthes: Cestoda). *Molecular Phylogenetics and Evolution* **19**, 443–467.

Rees G (1967) Pathogenesis of adult cestodes. *Helminthological Abstracts* **36**, 1–23.

Schmidt GD (1986) 'Handbook of Tapeworm Identification'. (CRC Press: Boca Raton).

Tyler GA (2001) A monograph on the order Diphyllidea (Platyhelminthes: Cestoda). PhD dissertation. University of Connecticut.

Williams IC, Harris MD (1965) The infection of the gulls *Larus argentatus* Pont., *L. fuscus* L. and *L. marinus* L. with Cestoda on the coast of Wales. *Parasitology* **55**, 237–256.

Williams H, Jones A (1994) 'Parasitic Worms of Fish'. (Taylor & Francis: London).

Williams H, McVicar A, Ralph R (1970) The alimentary canal of fish as an environment for helminth parasites. *In* 'Aspects of Fish Parasitology'. (Eds AER Taylor, R Muller.) pp. 43–77. (Blackwell Scientific Publications: Oxford).

Nematoda (roundworms)

Adamson ML (1986) Modes of transmission and evolution of life histories in zooparasitic nematodes. *Canadian Journal of Zoology* **64**, 1375–1384.

Adamson ML, Deets GB, Benz GW (1987) Description of male and redescription of female *Phlyctainophora squali* Mudry and Dailey, 1969 (Nematoda: Dracunculoidea) from elasmobranchs. *Canadian Journal of Zoology* **65**, 3006–3010.

Adamson ML, Roth M (1990) Prevalence and intensity of *Pseudodelphis oligocotti* n. gen. n. sp. (Dracunculoidea: Guyanemidae) in the tidepool sculpin, *Oligocottus maculosus* (Scorpaeniformes, cotidae). *Journal of Parasitology* **76**, 509–514.

Anderson RC (1984) The origins of zooparasitic nematodes. *Canadian Journal of Zoology* **62**, 317–328.

Anderson RC (1996) Why do fish have so few roundworm (nematode) parasites? *Environmental Biology of Fishes* **46**, 1-5.

Anderson RC (2000) 'Nematode Parasites of Vertebrates. Their Development and Transmission'. 2nd edn. (CABI Publishing: Wallingford).

Anderson RC, Lim LHS (1996) *Synodontisia moraveci* n. sp. (Oxyuroidea: Pharyngodonidae) from *Osteochilus melanopleurus* (Cyprinidae) of Malaysia, with a review of pinworms in fish and a key to species. *Systematic Parasitology* **34**, 157–162.

Anderson RC, Bain O (1976) Keys to genera of the order Spirurida Part 3. Diplotriaenoidea, Aproctoidea and Filarioidea. *In* 'CIH Keys to the Nematode Parasites of Vertebrates'. No. 3. (Eds RC Anderson, AG Chabaud, S Willmott.) (Commonwealth Bureau of Agriculture: Farnham Royal, England).

Anderson RC, Wong PL, Bartlett CM (1996) The Acuaroid and Habronematoid nematodes (Acurioidea, Habronematoidea) of the upper digestive tract of waders. A review of observations on their host and geographic distributions and transmission in marine environments. *Parasite* **4**, 303–312.

Arnold PW, Gaskin DE (1975) Lungworms (Metastrongyloidea: Pseudaliidae) of harbour porpoise *Phocoena phoecoena* (L. 1758). *Canadian Journal of Zoology* **53**, 713–735.

Bashirullah KM, Aquada N (1993) Diseases and parasites of penaeid shrimps of commercial interest in the eastern region of Venezuela. Special publication European Aquaculture Society No. 19, Oostende.

Bergeron E, Measures LM, Huot J (1997) Experimental transmission of *Otostrongylus circumlitus* (Raillet, 1899) (Metastrongyloidea: Crenosomatidae), a lungworm of seals in eastern arctic Canada. *Canadian Journal of Zoology* **75**, 1364–1371.

Beveridge I (1987) *Echinocephalus overstreeti* Deardorff and Ko, 1983 (Nematoda: Gnathostomatoidea) from elasmobranchs and mollusks in South Australia. *Transactions of the Royal Society of South Australia* **111**, 79–92.

Borgsteede FHM (1997) Parasitology of Marine Birds. *In* 'Marine Mammals, Seabirds and Pollution of Marine Systems'. (Eds T Jauniaux, JM Bouquegneau, F Coignoul.) pp. 91–108. (Veterinary Medicine Presses of the University of Liege: Liege).

Borucinska JD, Frasca S Jr (2002) Pancreatic fibrosis and ductal ectasia associated with the nematode *Pancreatonema americanum* sp. nov., in spiny dogfish, *Squalus acanthias* L., from the north-western Atlantic. *Journal of Fish Diseases* **25**, 367–370.

Bosch M, Torres J, Figuerola J (2000) A helminth community in breeding Yellow-legged Gulls (*Larus cachinnans*): pattern of association and its effect on host fitness. *Canadian Journal of Zoology* **78**, 777–786.

Bourgeois CE, Threlfall W (1979) Parasites of the greater shearwater (*Puffinus gravis*) from Newfoundland, Canada. *Canadian Journal of Zoology* **57**, 1355–1357.

Bruce NL, Cannon LRG (1990) Ascaridoid nematodes from sharks from Australia and the Solomon Islands, southwestern Pacific Ocean. *Invertebrate Taxonomy* **4**, 763–783.

Chabaud AG (1974) Class Nematoda. Keys to the subclasses, orders, and superfamilies. *In* 'CIH Keys to the Nematode Parasites of Vertebrates'. No. 1. (Eds RC Anderson, AG Chabaud, S Willmott.) pp. 6–17. (Commonwealth Bureau of Agriculture: Farnham Royal, England).

Chabaud AG (1975) Keys to genera of the Order Spirurida. Part 1. Camallanoidea, Dracunculoidea, Gnathostomatoidea, Physalopteroidea, Rictularioidea and Thelazioidea. *In* 'CIH Keys to the Nematode Parasites of Vertebrates'. No. 3. (Eds RC Anderson, AG Chabaud, S Willmott.) (Commonwealth Bureau of Agriculture: Farnham Royal, England).

Chabaud AG (1978) Keys to genera of the Superfamilies Cosmocercoidea, Seuratoidea, Heterakoidea and Subuluroidea. *In* 'CIH Keys to the Nematode Parasites of Vertebrates'. No. 6.

(Eds RC Anderson, AG Chabaud, S Willmott.) (Commonwealth Bureau of Agriculture: Farnham Royal, England).

Couture C, Measures L, Gagnon J, Desbiens C (2003) Human intestinal anisakiosis due to consumption of raw salmon. *The American Journal of Surgical Pathology* **27**, 1167–1172.

Coyner DF, Schaack SR, Spalding MG, Forrester DJ (2001) Altered predation susceptibility of mosquitofish infected with *Eustrongylides ignotus*. *Journal of Wildlife Diseases* **37**, 556–560.

Dailey MD (1985) Diseases of Mammalia: Cetacea. *In* 'Diseases of Marine Animals'. Vol. IV, Part 2. (Ed. O Kinne.) pp. 805–844. (Biologische Anstalt Helgoland: Hamburg).

Dailey MD (2001) Parasitic diseases. *In* 'Marine Mammal Medicine'. (Eds LA Dierauf, MD Gulland.) pp. 357–379. (CRC Press: Boca Raton).

Dailey M, Walsh M, Odell D, Campbell T (1991) Evidence of prenatal infection in the bottlenose dolphin (*Tursiops truncates*) with the lungworm *Halocercus lagenorhynchi* (Nematoda: Pseudaliidae). *Journal of Wildlife Diseases* **27**, 164–165.

De Ley P, Blaxter ML (2002) Systematic position and phylogeny. *In* 'The Biology of Nematodes'. (Ed. D Lee.) pp. 1–30. (Taylor and Francis: London).

Dyer WG, Williams EHJr, Mignucci-Giannoni AA, Jimenez-Marrero NM, Bunkley-Wiliams L, Moore DP, Pence DB (2002) Helminth and arthropod parasites of the brown pelican, *Pelecanus occidentalis*, in Puerto Rico, with a compilation of all metazoan parasites reported from this host in the Western Hemisphere. *Avian Pathology* **31**, 441–448.

Forbes LB, Measures L, Gajadhar A, Kapel K (2003) Infectivity of *Trichinella nativa* in traditional northern (country) foods prepared with meat from experimentally infected seals. *Journal of Food Protection* **66**, 1857–1863.

George-Nascimento M, Carmona R, Riffo R (1994) Occurrence of larval nematodes *Proleptus* sp. (Spirurida: Physalopteridae) and *Anisakis* sp. (Ascaridida: Anisakidae) in the crab *Cancer plebejus* Poeppig, in Chile. *Scientia Marina* **58**, 355–358.

Hallett SL, Erséus C, O'Donoghue PJ, Lester RJG (2001) Parasite fauna of Australian Marine Oligochaetes. *Memoirs of the Queensland Museum* **46**, 555–576.

Hesp SA, Hobbs RP, Potter IC (2002) Infection of the gonads of *Glaucosoma hebraicum* by the nematode *Philometra lateolabracis*: occurrence and host response. *Journal of Fish Biology* **60**, 663–673.

Heupel MR, Bennett MB (1998) Infection of the epaulette shark, *Hemiscyllium ocellatum* (Bonnaterre), by the nematode parasite *Proleptus australis* Bayliss (Spirurida: Physalopteridae). *Journal of Fish Diseases* **21**, 407–413.

Hoa LV, Khue PN, Lien NT (1972) Study of two new species of nematodes of the genus Bulbocephalus Rasheed 1966, fish parasites from the sea of South Vietnam. Remarks on the subfamily Bulbocephaline, Rasheed 1966. *Bulletin de la Societe de Pathologie Exotique et de ses Filiales* **65**, 313–322.

Hoberg EP, Ryan PG (1989) Ecology of helminth parasitism in *Puffinus gravis* (Procellariiformes) on the breeding grounds at Gough Island. *Canadian Journal of Zoology* **67**, 220–225.

Hoberg EP, Brooks DR, Molina-Urena H, Erbe E (1998) *Echinocephalus janzeni* n. sp. (Nematoda: Gnathostomatidae) in *Himantura pacifica* (Chondrichthyes: Myliobatiformes) from the Pacific coast of Costa Rica and Mexico, with historical biogeographic analysis of the genus. *Journal of Parasitology* **84**, 571–581.

Hochberg FG (1990) Diseases caused by Protistans and Metazoans. *In* 'Diseases of Marine Animals'. Vol. III. (Ed. O Kinne.) pp. 47–205. (Biologische Anstalt Helgoland: Hamburg).

Jackson CJ, Marcogliese DJ, Burt MDB (1997) Role of hyperbenthic crustaceans in the transmission of marine helminth parasites. *Canadian journal of Fisheries and Aquatic Sciences* **54**, 815–820.

Jangoux M (1990) Diseases of Echinodermata. *In* 'Diseases of Marine Animals'. Vol. III. (Ed. O Kinne.) pp. 439–542. (Biologische Anstalt Helgoland: Hamburg).

Kapel CMO, Measures L, Moeller LN, Forbes L, Gajadhar A (2003) Experimental *Trichinella* infection in seals. *International Journal for Parasitology* **33**, 1463–1470.

Kirk RS, Morritt D, Lewis JW, Kennedy CR (2002) The osmotic relationship of the swimbladder nematode *Anguillicola crassus* with seawater eels. *Parasitology* **124**, 339–347.

Knoff M, De Sao Clemente SC, Pinto RM, Gomes DC (2001) Nematodes of elasmobranch fishes from the southern coast of Brazil. *Memorias do Instituto Oswaldo Cruz* **96**, 81–87.

Køie M (1993) Metazoan parasites in teleost fishes from 0 m to 1540 m depth off the Faroe Islands (the north Atlantic). *Ophelia* **38**, 217–242.

Køie M (1999) Metazoan parasites of flounder *Platichthys flesus* (L.) along a transect from the southwestern to the northwestern Baltic Sea. *ICES Journal of Marine Science* **56**, 157–163.

Køie M (2000a) Life cycle and seasonal dynamics of *Cucullanus cirratus* O.F. Müller, 1777 (Nematoda, Ascaridida, Seuratoidea, Cucllanidae) in Atlantic cod, *Gadus morhua* L. *Canadian Journal of Zoology* **78**, 182–190.

Køie M (2000b) Metazoan parasites of teleost fishes from Atlantic waters off the Faroe Islands. *Ophelia* **52**, 25–44.

Køie M (2000c) The life cycle of the flatfish nematode *Cucullanus heterochrous*. *Journal of Helminthology* **74**, 323–328.

Køie M (2001a) The life cycle of *Capillaria gracilis* (Capillaridae), a nematode parasite of gadoid fish. *Sarsia* **86**, 383–387.

Køie M (2001b) The life cycle of *Dichelyne* (*Cucullanellus*) *minutus* (Nematoda: Cucullanidae). *Folia Parasitologica* **48**, 304–310.

Kuiken T, Leighton FA, Wobeser G, Wagner B (1999) Causes of morbidity and mortality and their effect on reproductive success in double-crested cormorants from Saskatchewan. *Journal of Wildlife Diseases* **35**, 331–346.

Langston N, Hillgarth N (1995) Moult varies with parasites in Laysan Albatrosses. *Proceedings of the Royal Society of London* **261**, 239–243.

Latham AD, Poulin R (2002) New records of gastrointestinal helminths from the southern black-backed gull (*Larus dominicanus*) in New Zealand. *New Zealand Journal of Zoology* **29**, 252–257.

Lauckner G (1983) Diseases of Mollusca: Bivalvia. *In* 'Diseases of Marine Animals'. Vol. II. (Ed. O Kinne.) pp. 477–802. (Biologische Anstalt Helgoland: Hamburg).

Lauckner G (1985a) Diseases of Reptilia. *In* 'Diseases of Marine Animals'. Vol. IV, Part 2. (Ed. O Kinne.) pp. 553–613. (Biologische Anstalt Helgoland: Hamburg).

Lauckner G (1985b) Diseases of Aves. *In* 'Diseases of Marine Animals'. Vol. IV, Part 2. (Ed. O Kinne.) pp. 627–637. (Biologische Anstalt Helgoland: Hamburg).

Lauckner G (1985c) Diseases of Mammalia: Pinnepedia. *In* 'Diseases of Marine Animals'. Vol. IV, Part 2. (Ed. O Kinne.) pp. 683–772. (Biologische Anstalt Helgoland: Hamburg).

Lauckner G (1985d) Diseases of Mammalia: Sirenia. *In* 'Diseases of Marine Animals'. Vol. IV, Part 2. (Ed. O Kinne.) pp. 795–801. (Biologische Anstalt Helgoland: Hamburg).

Machida M, Araki J, Regoniel PA, Pontillas FA, Kurata Y. (1992) Three species of ascaridoid nematodes from crocodile in the Philippines. *Bulletin of the National Science Museum of Tokyo* **18**, 95–102.

Manzanell R (1986) *Oswaldofilaria kunbaya* n. sp., a new filaroid worm (Nematoda: Filarioidea) from the saltwater crocodile *Crocodilus porosus* from Australia. *Annales de Parasitologie Humaine et Comparee* **61**, 245–254.

Marcogliese DJ (2002) Food webs and the transmission of parasites to marine fish. *Parasitology* **124**, 83–89.

Martorelli SR, Navone GT, Ivanov V (2000) Proposed life cycle of *Ascarophis marina* (Nematoda: Cystidicolidae) in Argentine waters. *Journal of Parasitology* **86**, 1047–1050.

McClelland G (1990) Larval sealworm (*Pseudoterranova decipiens*) infections in benthic macrofauna. In 'Population Biology of Sealworm (*Pseudoterranova decipiens*) in Relation to its Intermediate and Seal Hosts'. (Ed. WD Bowen.) pp. 47–65. (Canadian Bulletin of Fisheries and Aquatic Sciences: Ottawa).

McClelland G (2002) The trouble with sealworms (*Pseudoterranova decipiens* species complex, Nematoda): a review. *Parasitology* **124**, 183–203.

McClelland G, Misra RK, Martell DJ (1990) Larval anisakine nematodes in various fish species from Sable Island bank vicinity. In 'Population Biology of Sealworm (*Pseudoterranova decipiens*) in Relation to its Intermediate and Seal Hosts'. (Ed. WD Bowen.) pp. 83–118. (Canadian Bulletin of Fisheries and Aquatic Sciences: Ottawa).

McCurdy DG, Forbes MR, Boates JS (1999) Evidence that the parasitic nematode *Skrjabinoclava* manipulates host *Corophium* behavior to increase transmission to the sandpiper, *Calidris pusilla*. *Behavioral Ecology* **10**, 351–357.

McDonald TE, Margolis L (1995) Synopsis of the parasites of fishes of Canada: Supplement (1978–1993). *Canadian Special Publications in Fisheries and Aquatic Science* No. 122, Ottawa.

Measures LN (2001) Lungworms of marine mammals. In 'Parasitic Diseases of Wild Mammals'. (Eds WM Samuel MJ Pybus AA Kocan.) (The Iowa State University Press: Ames).

Measures LN, Gosselin JF, Bergeron E (1997) Heartworm, *Acanthocheilonema spirocauda* (Leidy, 1858), infections in Canadian phocid seals. *Canadian Journal of Fisheries and Aquatic Sciences* **54**, 842–846.

Moles A (2003) Effect of parasitism by *Philonema agubernaculum* (Nematoda: Philometridae) on the ability of Dolly Varden to capture prey in fresh and salt water. *Alaska Fishery Research Bulletin* **10**, 119–123.

Moravec F (2001) 'Trichinelloid nematodes parasitic in cold-blooded vertebrates.' Academy of Sciences of the Czech Republic. (Academia: Praha).

Moravec F, Nagasawa K (2000) Two remarkable nematodes from sharks in Japan. *Journal of Natural History* **34**, 1–13.

Moravec F, Nagasawa K, Ogawa K (1998) Observations on five species of philometrid nematodes from marine fishes in Japan. *Systematic Parasitology* **40**, 67–80.

Moravec F, Kohn A, Santos LA (1999) New data on *Oncophora melanocephala* (Nematoda: Camallanidae), a little known parasite of scrombrid fishes. *Parasite* **6**, 79–84.

Moravec F, Borucinska JD, Salvatore FJr (2001) *Pancreatonema americanum* sp. nov. (Nematoda, Rhabdochonidae) from the pancreatic duct of the dogfish shark, *Squalus acanthias*, from the coast of Massachusetts, USA. *Acta Parasologica* **46**, 293–298.

Moravec F, Ogawa K, Masaaki S, Miyazaki K, Donai H (2002) On two species of *Philometra* (Nematoda, Philometridae) from the serranid fish *Epinephelus septemfasciatus* in Japan. *Acta Parasitologica* **47**, 34–40.

Moravec M (2000) Systematic status of *Laurotravassoxyuris bravoae* Osorio-Sarabia, 1984 (Nematoda: Pharyngodonidae) [= *Atractis bravoae* (Osorio-Sarabia, 1984) n. comb.: Cosmocercidae]. *Systematic Parasitology* **46**, 117–122.

Nadler SA, Hudspeth DSS (2000) Phylogeny of the Ascaridoidea (Nematoda: Ascaridida) based on three genes and morphology: hypothesis of structural and sequence evolution. *Journal of Parasitology* **86**, 380–393.

Nadler SA, D'Amelio S, Fagerholm HP, Berland B, Paggi L (2000) Phylogenetic relationships among species of *Contracaecum* Railliet and Henry, 1912 and *Phocascaris* Høst, 1932 (Nematoda: Ascaridoidea) based on nuclear rDNA sequence data. *Parasitology* **121**, 455–463.

Nagasawa K (1987) Prevalence of visceral adhesions in sockeye salmon (*Oncorhynchus nerka*) in the North Pacific Ocean Bering Sea. *Canadian Special Publications in Fisheries and Aquatic Science* No. 96, Ottawa.

Nagasawa K, Moravec F (2002) Larval Anisakid nematodes from four species of squid (Cephalopoda: Teuthoidea) from the central and western North Pacific Ocean. *Journal of Natural History* **36**, 883–891.

Petter AJ (1983) Description of a new genus of Benthimermithidae (Nematoda) with unusual uterine glands. *Annales de Parasitologie Humaine et Comparee* **58**, 177–184.

Petter AJ, Caberet J (1995) Ascaridoid nematodes of teleostean fishes from the eastern North Atlantic and the seas of the north of Europe. *Parasite* **2**, 217–230.

Petter AJ, Gourbault N (1985) Nématodes abyssaux (campagne Walda du N/O Jean Charcot) IV. Des nématodes parasites de nématodes. *Bulletin du Museum National d'Histoire Naturelle (Paris)* **7**, 125–130.

Petter AJ, Køie M (1993) Fellicola longispiculus gen. Nov., sp. Nov. (Nematoda, Rhabdochonidae) from the gall bladder of the marine fish *Cryphaenoides rupestris*. *Annales de Parasitologie Humaine et Comparee* **68**, 226–228.

Pierrot-Bults AC (1990) Diseases of Chaetognatha. *In* 'Diseases of Marine Animals' Vol. III. (Ed. O Kinne.) pp. 425–434. (Biologische Anstalt Helgoland: Hamburg).

Poinar GJr, Latham DA, Poulin R (2002) *Thaumamermis zealandica* n. sp. (Mermithidae: Nematoda) parasitizing the intertidal marine amphipod *Talorchestia quoyana* (Talitridae: Amphipoda) in New Zealand, with a summary of mermithids infecting amphipods. *Systematic Parasitology* **53**, 227–233.

Raga JA (1997) Parasitology of marine mammals. *In* 'Marine Mammals Seabirds and Pollution of Marine Systems'. (Eds T Jauniaux, JM Bouquegneau, F Coignoul.) pp. 67–90. (Veterinary Medicine Presses of the University of Liege: Liege).

Ranum A, Wharton DA (1996) Helminth and protozoan parasites of the alimentary tract of the yellow eyed penguin (*Megadyptes antipodes*). *New Zealand Journal of Zoology* **23**, 83–87.

Rigby MC, Adamson ML (1997) *Spirocamallanus* species of French Polynesian coral reef fishes. *Canadian Journal of Zoology* **75**, 1270–1279.

Roberts LS, Janovy JJr (2005) 'Gerald D. Schmidt & Larry S. Roberts' foundations of parasitology.' 7th edn. (McGraw-Hill: Boston).

Roca V, LaFuente M, Carbonell E (1999) Helminth communities in Audouins gulls, *Larus audouinii*, from the Chafarinas Islands (Western Mediterranean). *Journal of Parasitology* **85**, 986–989.

Rocka A, Stefanski W (2002) Nematodes of fishes in the Weddell Sea (Antarctica). *Acta Parasitologica* **47**, 294–299.

Rohde K (1984) Diseases caused by Metazoans: Helminths. *In* 'Diseases of Marine Animals' Vol. IV, Part 1. (Ed. O Kinne.) pp. 193–319. (Biologische Anstalt Helgoland: Hamburg).

Rubstov IA (1977) A new genus and species of parasitic nematode *Ananus asteroideus* (Nematoda: Marimermithidae), from the asteroid *Diplopteraster peregrinator*. *Bulletin du Museum National d'Histoire Naturelle (Paris)* **496**, 1113–1117.

Rubstov IA, Platonova TA (1974) A new family of marine parasitic nematodes. *Zoological Zhurnal* **53**, 1445–1458.

Sanmartín ML, Alvarez MF, Peris D, Iglesias R, Leiro J (2000) Helminth parasite communities of the conger eel in the estuaries of Arousa and Muros (Galicia, north-west Spain). *Journal of Fish Biology* **57**, 1122–1133.

Smith JW (1999) Ascaridoid Nematodes and pathology of the alimentary tract and its associated organs in vertebrates, including man: a literature review. *Helminthological Abstracts* **68**, 49–96.

Sprent JFA (1990) Some ascaridoid nematodes of fishes: *Paranisakis* and *Mawsonascaris* n. g. *Systematic Parasitology* **15**, 41–63.

Sprent JFA, McKeown EA, Cremin M (1998) *Dujardinascaris* spp. (Nematoda: Ascaridoidea) in Old World crocodilians. *Systematic Parasitology* **39**, 209–222.

Stein A (1999) Effects of the parasitic nematode *Echinomermella matsi* on growth and survival of its host, the sea urchin *Strongylocentrotus droebachiensis*. *Canadian Journal of Zoology* **1**, 139–147.

Tanzola RD, Gigola G (2002) *Johnstonmawsonia porichthydis* n. sp. (Nematoda: Rhabdochonidae) from *Porichthys porosissimus* (Pisces: Batrachoidiformes). *Helminthologia* **39**, 99–102.

Tchesunov AV (1996) Anatomical features of *Marimermis maritima* Rubstov & Platonova 1974 (Marimermithida) a parasite of sea urchin. Abstracts of papers presented at the First English Language International Symposium of the Russian Society of Nematologists *Russian Journal of Nematology* **4**, 77–104.

Tchesunov AV, Hope WD (1997) *Thalassomermis megamphis* n. sp. (Mermithidae: Nemata) from the Bathyal South Atlantic Ocean. *Journal of Nematology* **29**, 451–464.

Tchesunov AV, Spiridonov SE (1985) *Australonema eulagiscae* gen. et sp. n. (Nematoda, Marimermithida), a parasite from the Antarctic polychaete. *Vestnik Zoologii* **2**, 16–21.

Tchesunov AV, Spiridonov SE (1993) *Nematimermis enoplivora* gen. n., sp.n. (Nematoda: Mermithoidea) from marine free-living nematodes *Enoplus*. *Russian Journal of Nematology* **1**, 7–16.

Torres P, Ruiz E, Gesche W, Montefusco A (1991) Gastrointestinal helminths of fish-eating birds from Chiloe Island, Chile. *Journal of Wildlife Diseases* **27**, 178–179.

US Food and Drug Administration/Center for Food Safety and Applied Nutrition (1991) *Eustrongylides* sp. In 'Foodborne Pathogenic Microorganisms and Natural Toxins Handbook'. http://www.cfsan.fda.gov/~mow/chap28.html.

Williams H, Jones A (1994) 'Parasitic Worms of Fish'. (Taylor and Francis: London).

Wong PL, Anderson RC (1993) New and described species of nematodes from shorebirds (Charadriiformes) collected in spring in Iceland. *Systematic Parasitology* **25**, 187–202.

Zablotzky VI (1997) *Capillaria delamurei* sp. n. (Nematoda: Trichocephalata) from the Caspian Sea. *Canadian Translation of Fisheries and Aquatic Sciences*.

Acanthocephala (spiny-headed worms)

Garcia-Varela, M, Cummings MP, Perez-Ponce de Leon G, Gardner SL, Laclette JP (2002) Phylogenetic analysis based on 185 ribosomal RNA gene sequences supports the existence of the class Polyacanthocephala (Acanthocephala). *Molecular and Phylogenetic Evolution* **23**, 288–92.

Herlyn H, Piskurek O, Schmidtz J, Ehlers U, Zischler H (2003) The phylogeny of the Syndermate (Rotifera: Monogonota, Bdelloidea, Seisonidea; Acanthocephala Palaeacanthocephala, Eoacanthocephala, Archiacanthocephala). *Molecular and Phylogenetic Evolution* **26**, 155–164.

Klimpel S, Seehagen A, Palm H, Rosenthal H (2001) Deep-water metazoan fish parasites of the world. (Logos: Berlin).

Latham ADM, Poulin R (2002) Field evidence of the impact of two acanthocephalan parasites on the mortality of three species of New Zealand shoe crabs (Brachyura). *Marine Biology* **14**, 1131–1139.

Mayer AK, Dailey MD, Miller MA (2003) Helminth parasites of the southern sea otter *Enhydra lutris nereis* in central California: abundance, distribution and pathology. *Diseases of Aquatic Organisms* **53**, 77–88.

O'Neill G, Whelan J (2002) *The occurance of Corynosoma strumosum* in the grey seal, *Halichoerus grypus*, caught off the Atlantic coast of Ireland. *Journal of Helminthology* **76**, 231–234.

Taraschewski H (2000) Host-parasite interactions in Acanthocephala: a morphological approach. *Advances in Parasitology* **46**, 1–179.

Zdzitowiecki K, White MG. (1996) Acanthocephalan infection of inshore fish at the South Orkney Islands *Antarctic Science* **8**, 273–276.

Chapter 4
Copepoda (copepods)

Boxshall GA (1990a) The skeletomusculature of siphonostomatoid copepods, with an analysis of adaptive radiation in structure of the oral cone. *Philosophical Transactions of the Royal Society* (B) **328**, 167–212.

Boxshall GA (1990b) Precopulatory mate guarding in Copepods. *Bijdragen tot de Dierkunde* **60**, 209–213.

Boxshall GA (1998) Host specificity in copepod parasites of deep-sea fishes. *Journal of marine Systems* **15**, 215–223.

Boxshall GA, Defaye D (Eds) (1993) 'Pathogens of Wild and Farmed Fish: Sea Lice'. (Ellis Horwood: Chichester).

Boxshall GA, Halsey SH (2004) 'An Introduction to Copepod Diversity'. (The Ray Society: London).

Bresciani J (1986) The fine structure of the integument of free-living and parasitic copepods. *Acta Zoologica* **67**, 125–145.

Bresciani J, Lützen J (1961) *Gonophysema gullmarensis* (Copepoda Parasitica). An anatomical and biological study of an endoparasite living in the ascidian *Ascidiella aspersa*. II. Biology and development. *Cahiers de Biologie Marin* **2**, 347–371.

Hansen HJ (1897) 'The Choniostomatidae, a family of Copepoda, parasites on Crustacea Malacostraca'. (Host & Son: Copenhagen).

Hippeau-Jacquotte R (1987) Ultrastructure and presumed function of the pleural dermal glands in the atypical male of the parasitic copepod *Pachypygus gibber* (Crustacea: Notodelphyidae). *Journal of Crustacean Biology* **7**, 60–70.

Humes AG (1971) Cyclopoid copepods (Stellicomitidae) parasitic on sea stars from Madagascar and Eniwetok Atoll. *Journal of Parasitology* **57**, 1330–1343.

Humes AG (1973) Cyclopoid copepods associated with the ophiuroid *Astroboa nuda* in Madagascar. *Beaufortia* **21**, 25–35.

Humes AG (1985) Cnidarians and copepods: a success story. *Transactions of the American Microscopical Society* **104**, 313–320.

Huys R, Boxshall GA (1991) 'Copepod Evolution'. (The Ray Society: London).

Kabata Z (1979) 'Parasitic Copepoda of British Fishes'. (The Ray Society: London).

Pike AW, Wadsworth SL (1999) Sealice on salmonids: their biology and control. *Advances in Parasitology* **44**, 233–337.

Raibaut A (1985) Les cycles évolutif des Copépodes parasites et les modalités de l'infestation. *L'Année Biologique* **24**, 233–274.

Isopoda (isopods)

Anderson G (1975a) Larval metabolism of the epicaridean isopod parasite *Probopyrus pandicola* and metabolic effects of *P. pandicola* on its copepod intermediate host *Arcatia tonsa*. *Comparative Biochemistry and Physiology* **50A**, 747–751.

Anderson G (1975b) Metabolic response of the caridean shrimp *Palaemonetes pugio* to infection by the adult epibranchial isopod parasite *Probopyrus pandicola*. *Comparative Biochemistry and Physiology* **52A**, 201–207.

Anderson G (1977) The effects of parasitism on the energy flow through laboratory shrimp populations. *Marine Biology* **42**, 239–251.

Anderson G (1990) Post infection mortality of *Palaemonetes* spp. (Decapoda: Palaemonidae) following experimental exposure to the bopyrid isopod *Probopyrus pandicola* (Packard) (Isopoda: Epicaridea). *Journal of Crustacean Biology* **10**, 284–292.

Anderson G, Dale (1981) *Probopyrus pandicola* (Packard) (Isopoda: Epicaridea). Morphology and development of larvae in culture. *Crustaceana* **41**, 143–161.

Bass CS, Weis JS (1999) Behavioural changes in the grass shrimp, *Palaeomonetes pugio* (Holthuis), induced by the parasitic isopod, *Probopyrus pandicola* (Packard). *Journal of Experimental Marine Biology and Ecology* **241**, 223–233.

Bergey L, Weis JS, Weis P (2002) Mercury uptake by the estuarine species *Palaeomonetes pugio* and *Fundulus heteroclitus* compared with their parasites, *Probopyrus pandicola* and *Eustrongylides* sp. *Marine Pollution Bulletin* **44**, 1046–1050.

Blower S, Roughgarden J (1989) Population dynamics and parasitic castration: test of a model. *American Naturalist* **134**, 848–858.

Bragoni G, Romestand B, Trilles J-P (1984) Parasitoses a cymothoadien chez le loup, *Dicentrarchus labrax* (Linnaeus, 1758) en elevage. I. Ecologie parasitaire dans le cas de l'Etang de Diana (Haute Corse) (Isopoda, Cymothoidae). *Crustaceana* **47**, 44–51.

Brandt A, Hanssen HH (1994) Redescription of *Zonophyxus quinquedens* Barnard, 1913 (Crustacea, Isopoda, Dajidae) from the Weddell Sea, Antarctica, with notes on its biology and zoogeography. *Polar Biology* **14**, 343–350.

Bruce NL 1987 Australian species of *Nerocila* Leach, 1818, and *Creniola* n. gen. (Isopoda: Cymothoidae), crustacean parasites of marine fishes. *Records of the Australian Museum* **39**, 355–412.

Brusca RC (1981) A monograph on the Isopoda Cymothoidae (Crustacea) of the eastern Pacific. *Zoological Journal of the Linnean Society* **73**, 117–199.

Bunkley-Williams L, Williams EHJ (1998) Isopods associated with fishes: a synopsis and corrections. *Journal of Parasitology* **84**, 893–896.

Cash CE, Bauer RT (1993) Adaptations of the branchial parasite Probopyrus pandicola (Isopoda: Bopyridae) for survival and reproduction related to ecdysis of the host, *Palaemonetes pugio* (Caridea: Palaemonidae). *Journal of Crustacean Biology* **13**, 111–124.

Dreyer H, Wagele JW (2001) Parasites of crustaceans (Isopoda: Bopyridae) evolved from fish parasites: Molecular and morphological evidence. *Zoology-Jena* **103**, 157–178.

Giard A, Bonnier J (1887) Contribution à l'etude des Bopyriens. *Travaux de l'Institut Zoologique de Lille et du Laboratoire de Zoologie Maritime de Wimereux (Pas de Calais)* **5**, 3–272.

Jones MB, Smaldon G (1986) On the genus Holophryxus Isopoda Epicaridea with description of the male and redescription of the female of *Holophryxus-Acanthephyrae*. *Journal of the Marine Biological Association of the United Kingdom* **66**(2), 303–314.

Kuris AM (1974) Trophic interactions: similarity of parasitic castrators to parasitoids. *Quarterly Review of Biology* **49**: 129–138.

Lester RJG, Hayward CJ (in press) Phylum Arthropoda. *In* 'Diseases of Fish'. Vol 1. (Ed. PTK Woo.) (CABI Publishing: Wallingford).

Moles A, Pela JJ (1984) Effects of parasitism and temperature on salinity tolerance of the kelp shrimp *Eualus suckley*. *Transactions of the American Fisheries Society* **113**, 354–359.

Monod T (1926) Les Gnathiidae. *Memoires de la Societe des Ciences Naturelles du Maroc* **13**, 1–667.

Montalenti G (1948) Note sulla sistematica e la biologia di alcuni Cimotoidi del Golfo di Napoli. *Archivio di Oceanografia e Limnologia, Venezia* **5**, 25–81.

Owens L (1993) Prevalence of *Cabirops orbionei* (Epicaridea; Cryptoniscidae) in northern Australia: a biocontrol agent for bopyrids. *Australian Journal of Marine and Freshwater Research* **44**, 381–387.

Owens L, Glazebrook J (1985) Sex determination in the Bopyridae. *Journal of Parasitology* **71**, 134–135.

Papapanagiotou EP, Trilles J-P, Photis G (1999). First record of *Emetha audouini*, a cymothoid isopod parasite, from cultures sea bass *Dicentrarchus labrax* in Greece. *Diseases of Aquatic Organisms* **38**, 235–237.

Raibaut A, Trilles J-P (1993) The sexuality of parasitic crustaceans. *In* 'Advances in Parasitology'. (Eds JR Baker, R Muller.) pp. 367–444. (Harcourt Brace and Company: London).

Roccatagliata D, Jorda MT (2002) Infestation of the fiddler crab *Uca uruguayensis* by *Leidya distorta* (Isopoda, Bopyridae) from the Rio de la Plata estuary, Argentina. *Journal of Crustacean Biology* **22**, 69–82.

Schuldt M, Rodrigues-Capitulo A (1987) Infestation of *Palaemonetes argentinus* (Crustacea, Caridea, Palaemonidae) with *Probopyrus oviformis* (Crustacea, Epicaridea, Bopyridae): I. Observation on branchial histopathology and physiology of shrimp. *Revista del Museo de la Plata, seccion Zoologia* **14**, 65–82.

Sheader M (1977) The breeding biology of *Idotea pelagica* (Isopoda: Valvifera) with notes on the occurrence and biology of its parasite *Clypeoniscus hanseni* (Isopoda: Epicaridea). *Journal of the Marine Biological Association UK* **57**, 659–674.

Somers IF, Kirkwood GP (1991) Population ecology of the grooved tiger prawn *Penaeus semisulcatus* in the northwestern Gulf of Carpentaria, Australia. Growth, movement, age structure and infestation by the bopyrid parasite *Epipenaeon ingens*. *Australian Journal of Marine and Freshwater Research* **42**, 349–368.

Trilles J-P (1999) Ordre des isopodes, Sous ordre des Epicarides (Epicaridea Latreille, 1825). *Mémoire de l'Institut Océanographique de Monaco* **19**, 279–352.

Trilles J-P, Hipeau-Jacquotte R (1996) Association et parasitisme chez les Crustacés. *Traité de Zoologie (Grassé)* 7(2), 187–234.

Wagele J-W (1988) Aspects of the life-cycle of the Antarctic fish parasite *Gnathia calva* Vanhoffen (Crustacea: Isopoda). *Polar Biology* **8**, 287–291.

Branchiura (fish lice)

Abele LG, Kim W, Felgenhauer (1989) Molecular evidence for inclusion of the phylum Pentastomida in the Crustacea. *Molecular Biology and Evolution* **6**, 685–691.

Gresty KA, Boxshall GA, Nagasawa K (1993) The fine structure and function of the cephalic appendages of the branchiuran parasite, *Argulus japonicus* Thiele. *Philosophical Transactions of the Royal Society of London*, **B339**, 119–135.

Haase W (1975) Ultrastruktur und Funktion der Carapaxfelder von *Argulus foliaceus* L. (Crustacea: Branchiura). *Zeitschrift für Morphologie der Tiere* **81**, 161–189.

Overstreet RM, Dyková I, Hawkins WE (1992) Branchiura. *In* 'Microscopic Anatomy of Invertebrates'. Vol. IX, Crustacea. (Ed. FW Harrison.) pp. 345–413. (J.Wiley & Sons: New York).

Rushton-Mellor S, Boxshall GA (1994) The developmental sequence of *Argulus foliaceus*. *Journal of Natural History* **28**, 763–785.

Wingstrand KG (1972) Comparative spermatology of a Pentastomid, *Railletiella hemidactyli*, and a Branchiuran Crustacean, *Argulus foliaceus*, with a discussion of Pentastomid relationships. *Biologiske Skrifter Danske Videnskabernes Selskab* **19**, 1–72.

Tantulocarida (tantulocarids)

Boxshall GA, Lincoln RJ (1983) Tantulocarida, a new class of Crustacea ectoparasitic on other crustaceans. *Journal of Crustacean Biology* **3**, 1–16.

Boxshall GA, Lincoln RJ (1987) The life cycle of the Tantulocarida (Crustacea). *Philosophical Transactions of the Royal Society of London* **B315**, 267–303.

Huys R (1991) Tantulocarida (Crustacea: Maxillopoda): A new taxon from the temporary meiobenthos. *Marine Ecology: Pubblicazione della Stazione Zoologica di Napoli* **12**, 1–34.

Huys R, Boxshall GA, Lincoln RJ (1993) The tantulocaridan life cycle: the circle closed? *Journal of Crustacean Biology* **13**, 432–442.

Ascothoracida (ascothoracids)

Brattström H (1947) Undersökningar over Öresund XXXII. On the ecology of the ascothoracid *Ulophysema öresundense* Brattström. Studies on *Ulophysema öresundense* 1. *Lunds Universitets Årsskrift N.F. Avd. 2*, **43**, 7, 1–75.

Brattström H (1948) Undersökningar over Öresund XXXIII. On the larval development of the ascothoracid *Ulophysema öresundense* Brattström. Studies on *Ulophysema öresundense* 2. *Lunds Universitets Årsskrift N.F. Avd. 2*, **44**, 1–70.

Grygier MJ (1990) Five new species of bathyal Atlantic Ascothoracida (Crustacea: Maxillopoda) from the equator to 50° N latitude. *Bulletin of Marine Science* **46**, 655–676.

Grygier MJ (1991) Redescription, ontogeny and demography of *Parascothorax synagogoides* (Crustacea: Ascothoracida), parasitic on *Ophiophthalmus normani*, a dominant ophiuroid in the bathyal basins off southern California. *Proceedings of the San Diego Society of Natural History* **6**, 1–20.

Grygier MJ (1996a) Sous-classe des Facetotecta (Facetotecta Grygier, 1985). In 'Traité de Zoologie 7, Crustacés 2: Généralités (suite) et Systématique'. (Ed. J Forest.) pp. 425–432. (Masson: Paris).

Grygier MJ (1996b) Sous-classe des Ascothoracides (Ascothoracida Lacaze-Duthiers, 1880). In 'Traité de Zoologie 7, Crustacés 2: Généralités (suite) et Systématique'. (Ed. J Forest.) pp. 433–452. (Masson: Paris).

Høeg JT, Kolbasov GA (2002). Lattice organs in y-cyprids of the Facetotecta and their significance in the phylogeny of the Crustacea Thecostraca. *Acta Zoologica (Stockholm)* **83**, 67–79.

Itô T, Grygier MJ (1990) Description and complete larval development of a new species of *Baccalaureus* (Crustacea: Ascothoracida) parasitic in a zoanthid from Tanabe Bay, Honshu, Japan. *Zoological Science* **7**, 485–515.

Pérez-Losada M, Høeg JT, Kolbasov GA, Crandall KA (2002) Reanalysis of the relationships among the Cirripedia and the Ascothoracida and the phylogenetic position of the Facetotecta (Maxillopoda: Thecostraca) using 18S rDNA sequences. *Journal of Crustacean Biology* **22**, 661–669.

Wagin VL (1947, dated 1946). *Ascothorax ophioctenis* and the position of the Ascothoracida in the system of the Entomostraca. *Acta Zoologica (Stockholm)* **27**, 155–267.

Wagin VL (1976) Meshkogrudyye Raki (Ascothoracida). (Izdatel'stvo Kazanskogo Universiteta: Kazan). 140 pp. (In Russian.)

Cirripedia Thoracica and Rhizocephala (barnacles)

Alvarez F, Hines AH, Reaka-Kudla ML (1995) The effects of parasitism by the barnacle *Loxothylacus panopaei* (Gissler) (Cirripedia: Rhizocephala) on growth and survival of the host crab *Rhithropanopeus harrisii* (Gould) (Brachyura: Xanthidae). *Journal of Experimental Marine Biology and Ecology* **192**, 221–232.

Alvarez F, Campos E, Høeg JT, O'Brien J (2001) Distribution and prevalence records of two parasitic barnacles (Crustacea: Cirripedia: Rhizocephala) from the west coast of North America. *Bulletin of Marine Science* **68**, 233–241.

Andersen ML, Bohn M, Høeg JT, Jensen PG (1990) Cyprid ultrastructure and adult morphology in *Ptychascus barnwelli* new species, and *P. glaber* (Cirripedia: Rhizocephala), parasites on semiterrestrial crabs. *Journal of Crustacean Biology* **10**, 20–28.

Bishop HK, Cannon LR (1979) Morbid behavior of the commercial sand crab, *Portunus pelagicus*, parasitized by *Sacculina granifera* Boschma. *Journal of Fish Diseases* **2**, 131–144.

Blower S, Roughgarden J (1989a) Population dynamics and parasitic castration: test of a model. *American Naturalist* **134**, 848–858.

Blower S, Roughgarden J (1989b) Parasites detect host spatial pattern and density: a field experimental analysis. *Oecologia* **78**, 138–141.

Boone EJ, Boettcher AA, Sherman, TD O'Brien JJ (2003) Characterization of settlement cues used by the rhizocephalan barnacle *Loxothylacus texanus*. *Marine Ecology Progress Series* **252**, 187–197.

Boone E, Boettcher AA, Sherman TD, O'Brien JJ (2004) What constrains the geographic and host range of the rhizocephalan *Loxothylacus texanus* in the wild? *Journal of Experimental Marine Biology and Ecology* **309**, 129–139.

Bresciani J, Høeg JT (2001) Comparative ultrastructure of the root system in rhizocephalan barnacles (Crustacea: Cirripedia: Rhizocephala). *Journal of Morphology* **249**, 9–42.

Conway Morris S (1982) Parasites and the fossil record. *Trends and Perspectives in Parasitology* **2**, 24–44.

Cowen RK, Lwiza KMM, Sponaugle S, Paris CB, Olson DB (2000) Connectivity of marine populations: open or closed? *Science* **287**, 857–859.

Delage Y (1884) évolution de la Sacculine (*Sacculina carcini* Thomps.) crustacéendoparasite de l'ordre nouveau des kentrogonides. *Archives de Zoologie Expérimentale et Générale* Series 2, 417–736.

Galil BS; Lützen J (1995) Biological observations on *Heterosaccus dollfusi* Boschma (Cirripedia: Rhizocephala), a parasite of *Charybdis longicollis* Leene (Decapoda:Brachyura), a Lessepsian migrant to the Mediterranean. *Journal of Crustacean Biology* **15**, 659–670.

Galil BS, Lützen J (1998) Jeopardy: Host and parasite lessepsian migrants from the Mediterranean coast of Israel. *Journal of Natural History* **32**, 1549–1551.

Gannon AT, Wheatly MG (1992) Physiological effects of an ectocommensal gill barnacle, *Octolasmis muelleri*, on gas exchange in the blue crab, *Callinectes sapidus*. *Journal of Crustacean Biology* **12**, 11–18.

Gannon AT, Wheatly MG (1995) Physiological effects of a gill barnacle on host blue crabs during short-term exercise and recovery. *Marine Behavior and Physiology* **24**, 215–225.

Glenner H (2001) Cypris metamorphosis, injection and earliest internal development of the kentrogonid rhizocephalan *Loxothylacus panopaei* (Gissler) Crustacea: Cirripedia: Rhizocephala: Sacculinidae. *Journal of Morphology* **249**, 43–75.

Glenner H, Høeg JT (2002) A scenario for the evolution of the Rhizocephala. In 'Modern Approaches to the Study of Crustacea'. (Eds F Alvarez, E Escobar-Briones.) pp. 301–310. (Kluwer Academic/Plenum Publishers: New York).

Glenner H, Werner M (1998) Increased susceptibility of recently moulted *Carcinus maenas* (L.) to attachment by the parasitic barnacle *Sacculina carcini*. *Journal of Experimental Marine Biology and Ecology* **228**, 29–33.

Grygier MJ, Newman WA (1991) A new genus and two new species of Microlepadidae (Cirripedia: Pedunculata) found on western Pacific diadematid echinoids. *Galaxea* **10**, 1–22.

Hawkes CR, Meyers TR, Shirley TC, Koeneman TM (1986) Prevalence of the parasitic barnacle *Briarosaccus callosus* on king crabs of southeastern Alaska USA. *Transactions of the American Fisheries Society* **115**, 252–257.

Hines AH, Alvarez F, Reed SA (1997) Introduced and native populations of a marine parasitic castrator: variation in prevalence of the rhizocephalan *Loxothylacus panopaei* in xanthid crabs. *Bulletin of Marine Science* **61**, 197–214.

Hochberg RJ, Bert TM, Steele P, Brown SD (1992) Parasitization of *Loxothylacus texanus* on *Callinectes sapidus*: Aspects of population biology and effects on host morphology. *Bulletin of Marine Science* **50**, 117–132.

Høeg JT (1982) The anatomy and development of the rhizocephalan barnacle *Clistosaccus paguri* Lilljeborg and relation to its host *Pagurus bernhardus* (L.). *Journal of Experimental Marine Biology and Ecology* **58**, 87–125.

Høeg JT (1984) Size and settling behaviour in male and female cypris larvae of the parasitic barnacle *Sacculina carcini* Thompson (Crustacea: Cirripedia: Rhizocephala). *Journal of Experimental Marine Biology and Ecology* **76**, 145–156.

Høeg JT (1987) Male cypris metamorphosis and a new male larval form, the trichogon, in the parasitic barnacle *Sacculina carcini* (Crustacea: Cirripedia: Rhizocephala). *Philosophical Transactions of the Royal Society of London* **317B**, 47–63.

Høeg JT (1991) Functional and evolutionary aspects of the sexual system in the Rhizocephala (Thecostraca: Cirripedia). *In* 'Crustacean Sexual Biology'. (Eds RT Bauer, JW Martin.) pp. 208-227. (Columbia University Press: New York).

Høeg JT (1992) Rhizocephala. *In* 'Microscopic Anatomy of Invertebrates'. Vol. 9, Crustacea. (Eds FW Harrison, AG Humes.) pp. 313–345. (Wiley-Liss: New York).

Høeg JT (1995) The biology and life cycle of the Cirripedia Rhizocephala. *Journal of the Marine Biological Association of the United Kingdom* **75**, 517–550.

Høeg JT, Lützen J (1985) Crustacea Rhizocephala. *In* 'Marine Invertebrates of Scandinavia'. Vol. 6. pp 1–92. (Norwegian University Press: Oslo).

Høeg JT, Lützen J (1995) Life cycle and reproduction in the Cirripedia Rhizocephala. *Oceanography and Marine Biology Annual Reviews* **33**, 427–485.

Høeg JT, Lützen J (1996) Super-ordre des Rhizocéphales (Rhizocephala F. Müller, 1862). *In* 'Traité de Zoologie **7**, Crustacés: Géneralités (suite) et systématique'. (1st partie). (Ed. J Forest.) pp. 541–568. (Masson: Paris).

Høeg JT, Ritchie LE (1985) Male cypris settlement and its effects on juvenile development in *Lernaeodiscus porcellanae* Müller (Crustacea: Cirripedia: Rhizocephala). *Journal of Experimental Biology and Ecology* **87**, 1–11.

Innocenti G, Vannini M, Galil B (1998) Notes on the behaviour of the portunid crab *Charybdis longicollis* Leene parasitized by the rhizocephalan *Heterosaccus dollfusi* Boschma. *Journal of Natural History* **32**, 1577–1585.

Kuris AM, Lafferty KD (1992) Modelling crustacean fisheries effects of parasites on management strategies. *Canadian Journal of Fisheries and Aquatic Sciences* **49**, 327–336.

Lützen J (1984) Growth, reproduction, and life span in *Sacculina carcini* Thompson (Cirripedia: Rhizocephala) in the Isefjord, Denmark. *Sarsia* **69**, 91–106.

Meyers TR 1990. Diseases caused by protistans. *In* 'Diseases of Marine Animals'. Vol. 3, Diseases of Crustacea. (Ed. O Kinne.) pp. 350–368. (Biologische Anstalt Helgoland: Hamburg).

Murphy NE, Goggin CL (2000) Genetic discrimination of sacculinid parasites (Crustacea, Cirripedia, Rhizocephala): implication for control of introduced green crabs. *Journal of Crustacean Biology* **20**, 153–157.

Nuismer SL, Thompson JN, Gomulkiewicz (2003). Coevolution between hosts and parasites with partially overlapping geographic ranges. *Journal of evolutionary Biology* **16**, 1337–1345.

O'Brien JJ, Van Wyk P (1985) Effects of crustacean parasitic castrators (Epicaridean Isopods and Rhizocephalan barnacles) on growth of crustacean hosts. *In* 'Crustacean Growth: Factors in Adult Growth. Crustacean Issues 3'. (Ed. A Wenner.) pp. 191–218. (A.A. Balkema: Rotterdam).

Øksnebjerg B (2000) The Rhizocephala (Crustacea: Cirripedia) of the Mediterranean and Black seas: taxonomy, biogeography, and ecology. *Israel Journal of Zoology* **46**, 1–102.

Overstreet RM (1983) Metazoan symbionts of crustaceans. *In* 'The Biology of the Crustacea'. Vol. 6, Pathobiology. (Ed. AJ Provenzano Jr.) pp. 156–250. (Academic Press: New York).

Pasternak Z, Blasius B, Abelson A (2004a) Host location by larvae of a parasitic barnacle: larval chemotaxis and plume tracking in flow. *Journal of Plankton Research* **26**, 487–493.

Pasternak Z, Garm A, Høeg JT (2004b) The morphology of the chemosensory aesthetasc-like setae used during settlement of cypris larvae in the parasitic barnacle *Sacculina carcini* (Cirripedia: Rhizocephala). *Marine Biology* **146**, 1005–1013.

Reisser CE, Forward RB Jr (1991) Effect of salinity on osmoregulation and survival of a rhizocephalan parasite *Loxothylacus panopaei* and its crab host *Rhithropanopeus harrisii*. *Estuaries* **14**, 102–106.

Ritchie LE, Høeg JT (1981) The life history of *Lernaeodiscus porcellanae* (Cirripedia Rhizocephala)and co-evolution with its porcellanid host. *Journal of Crustacean Biology* **1**, 334–347.

Ross A, Newman WA (1995) A coral-eating barnacle, revisited (Cirripedia; Pyrgomatidae). *Contributions to Zoology* **65**, 129–175.

Rybakov AV, Høeg JT (2002) The ultrastructure of retinacula in the Rhizocephala (Crustacea: Cirripedia) and their systematic significance. *Zoologische Anzeiger* **241**, 95–103.

Shields JD, Wood FEI (1993) Impact of parasites on the reproduction and fecundity of the blue sand crab *Portunus pelagicus* from Moreton Bay, Australia. *Marine Ecology Progress Series* **92**, 159–170.

Sloan NA (1984) Incidence and effects of parasitism by the rhizocephalan barnacle, *Briarosaccus callosus* Boschma, in the golden king crab, *Lithodes aequispina* Benedict, from deep fjords in northern British Columbia, Canada. *Journal of Experimental Marine Biology and Ecology* **84**, 111–131.

Sloan NA (1985) Life history characteristics of fjord-dwelling golden king crabs *Lithodes aequispina*. *Marine Ecology Progress Series* **22**, 219–228.

Sparks AK, Morado JF (1986) Histopathology and host response in lithodid crabs parasitized by *Briarosaccus callosus*. *Diseases of Aquatic Organisms* **2**, 31–38.

Thresher RE, Werner M, Høeg JT, Svane I, Glenner H, Murphy NE, Wittwer C (2000) Developing the options for managing marine pests: specificity trials of the parasitic castrator, *Sacculina carcini*, against the European crab, *Carcinus maenas*, and related species. *Journal of Experimental Marine Biology and Ecology* **254**, 37–51.

Tindle A, Boone E, O'Brien J, Boettcher A (2004) Effects of salinity on larval stages of the rhizocephalan barnacle *Loxothylacus texanus*: survival and metamorphosis in response to the host, *Callinectes sapidus*. *Journal of Experimental Marine Biology and Ecology* **302**, 165–176.

Veillet A (1945) Recherches sur le parasitisme des crabes at des galathées par les rhizocéphales et les épicarides. *Annales de les Institute Océanographie Monaco* **22**, 193–341.

Walker G (1988) Observations on the larval development of *Sacculina carcini* (Crustacea: Cirripedia: Rhizocephala). *Journal of the Marine Biological Association of the United Kingdom* **68**, 377–390.

Walker G (2001) Introduction to the Rhizocephala (Crustacea: Cirripedia). *Journal of Morphology* **249**, 1–8.

Walker G, Lester RJG (2002) Effect of salinity on development of larvae of *Heterosaccus lunatusi* (Cirripedia: Rhizocephala). *Journal of Crustacean Biology* **18**, 650–655.

Walossek D, Høeg JT, Shirley TC (1996) Larval development of the rhizocephalan cirripede *Briarosaccus tenellus* (Maxillopoda: Thecostraca) reared in the laboratory: A scanning electron microscopy study. *Hydrobiologia* **328**, 9–47.

Wardle WJ, Tirpak AJ (1991) Occurrence and distribution of an outbreak of infection of *Loxothylacus texanus* (Rhizocephala) in blue crabs in Galveston Bay, Texas, with special reference to and coloration of the parasite's external reproductive structures. *Journal of Crustacean Biology* **11**, 553–560.

Werner M (2001) Prevalence of the parasite *Sacculina carcini* Thompson 1836 (Crustacea, Rhizocephala) on its host crab *Carcinus maenas* (L.) on the west coast of Sweden. *Ophelia* **55**, 101–110.

Yanagimachi R (1961) Studies on the sexual organization of the Rhizocephala. III. The mode of sex determination in *Peltogasterella*. *Biological Bulletin (Woods Hole)* **120**, 272–283.

Yusa Y, Yamato S (1999) Cropping of sea anemone tentacles by a symbiotic barnacle *Biological Bulletin* (Woods Hole) **197**, 315–318.

Yusa Y, Yamato S, Marumura M (2001) Ecology of a parasitic barnacle, *Koleolepas avis*: relationship to the hosts, distribution, left-right asymmetry and reproduction. *Journal of the Marine Biological Association of the United Kingdom* **81**, 781–788.

Amphipoda (amphipods)

Balbuena JA, Raga JA (1991) Ecology and host relationships of the whale-louse *Isocyamus delphini* (Amphipoda: Cyamidae) parasitizing long-finned pilot whales (*Globicephalus melas*) off the Faroe Islands (northeast Atlantic). *Canadian Journal of Zoology* **69**, 141–145.

Gruner AG (1975) Caprellidea II. Fam. Cyamidae. *Crustaceorum Catalogus* **5**, 79–104.

Laval P (1980) Hyperiid amphipods as crustacean parasitoids associated with gelatinous plankton. *Oceanography and Marine Biology. An annual Review* **18**, 11–56.

Leung YM (1976) Life cycle of *Cyamus scammoni* (Amphipoda: Cyamidae, ectoparasite of gray whale, with a remark on the associated species. *Scientific Reports of the Whales Research Institute, Tokyo* **28**, 153–160.

Madin LP, Harbison GR (1977) The associations of Amphipoda Hyperiidea with gelatinous zooplankton-I. Associations with Salpidae. *Deep-Sea Research* **24**, 449–463.

Martin JW, Heyning JE (1999) First record of *Isocyamus kogiae* Sedlak-Weinstein, 1992 (Crustacea, Amphipoda, Cyamidae) from the eastern Pacific, with comments on morphological characters, a key to the genera of Cyamidae, and a checklist of cyamids and their hosts. *Bulletin of Southern California Academy of Sciences* **98**, 26–38.

Thurston MH (2000) Pelagic amphipods. *In* 'M. Gabele: Alpha-taxonomy workshop on the world-wide status of Amphipoda'. *Polskie Archiwum Hyobiologi* **47**, 682–694.

Vinogradov ME, Volkov AF, Semenova TN (1996) 'Hyperiid Amphipods (Amphipoda, Hyperiidea) of the World Oceans'. (Science Publishers: Lebanon, New Hampshire).

Chapter 5

Fossil parasites

Andres D (1989) Phosphatisierte Fossilien aus dem unteren Ordoviz vin Südschweden. *Berliner geowissenschaftliche Abhandlungen, Rehe A* **106**, 9–19.

Baumiller TK, Gahn JJ (2002) Fossil record of parasitism on marine invertebrates with special emphasis on the platyceratid-crinoid association. *Paleontological Society Papers* **8**, 195–209.

Boucot AJ (1990) 'Evolutionary Paleobiology of Behaviour and Coevolution'. (Elsevier: Amsterdam).

Conway Morris S (1981) Parasites and the fossil record. *Parasitology* **82**, 489–509.

Conway Morris S (1990) Parasitism. *In* 'Paleobiology: A synthesis'. pp. 376–381. (Blackwell Scientific Publications: Oxford)

Feldmann RM (1998) Parasitic castration of the crab, *Tumidocarcinus giganteus* Glaessner, from the Miocene of New Zealand: coevolution within the crustacea. *Journal of Paleontology* **72**, 493–498.

Gahn FJ, Baumiller TK (2003) Infestation of Middle Devonian (Givetian) camerate crinoids by platyceratid gastropods and its implications for the nature of their biotic interaction. *Lethaia* **36**, 71–82.

Hess H, Ausich WI, Brett CE, Simms MJ (Eds) (1999) 'Fossil Crinoids'. (Cambridge University Press: Cambridge).

Kowalewski M, Simoes MG, Torello FF, Mello LHC, Ghilardi RP (2000) Drill holes in shells of Permian benthic invertebrates. *Journal of Paleontology* **74**, 532–543.

Littlewood DTJ, Donovan SK (2003) Fossil parasites: a case of identity. *Geology Today* **19**, 136–142.

Radwanski A (1972) Isopod-infected prosoponids from the upper Jurassic, Poland. *Acta Geologica Polonica* **22**, 499–506.

Wachsmuth C, Springer F (1897) The North American Crinoidea Camerata. *Harvard College Museum of Comparative Zoology, Memoir* **20, 21**, 1–897.

Walossek D, Müller KJ (1994) Pentastomid parasites from the lower Palaeozoic of Sweden. *Transactions of the Royal Society of Edinburgh: Earth Sciences* **85**, 1–37.

Welch JR (1976) *Phosphoannulus* on Paleozoic crinoid stems. *Journal of Paleontology* **50**, 218–225.

Porifera (sponges)

Annandale N (1915) Some sponges parasitic on Clionidae with further notes on that family. *Records of the Indian Museum* **11**, 457–478.

Calcinai B, Arillo A, Cerrano C, Bavastrello G (2003) Taxonomy-related differences in the excavating micro-patterns of boring sponges. *Journal of the Marine Biological Association of the United Kingdom* **83**, 37–39.

Carter HJ (1871) Parasites of the sponges. *Annals and Magazine of Natural History* **8**, 330–332.

Connes R, Paris J, Sube J (1971) Réactions tissulaires de quelques Démosponges vis-à-vis de leurs commensaux et parasites. *Naturaliste Canadien* **98**, 923–935.

Halley RBE, Shinn EA, Hudson JH, Lidz B (1977) Recent and relict topography of BooBee Patch reef, Belize. In 'Proceedings of the Third International Coral Reef Symposium'. Vol. 2. (Ed. DL Taylor.) pp. 29–35. (Rosenstiel School of Marine and Atmospheric Science, University of Miami: Miami, Florida).

Hartman WD (1958) Natural history of the marine sponges of Southern New England. *Bulletin of the Peabody Museum of Natural History* **12**, 1–155.

Hooper JNA, Van Soest RWM (2002). 'Systema Porifera. A Guide to the Classification of Sponges'. (Kluwer Academic/Plenum Publishers: New York).

Humes AG (1996) Siphonostomatoid copepods (Asterocheridae) associated with the sponge *Dysidea* in the Moluccas. *Systematic Parasitology* **35**, 157–177.

Lauckner G (1983) Diseases of Mollusca: Bivalvia. In 'Diseases of Marine Animals. Vol. 2, Introduction, Bivalvia to Scaphopoda'. (Ed. O Kinne.) pp. 477–961. (Biologische Anstalt Helgoland: Hamburg).

MacGeachy JK (1977) Factors controlling sponge boring in Barbados reef corals. In 'Proceedings of the Third International Coral Reef Symposium'. Vol. 2. (Ed. DL Taylor.) pp. 477–483. (Rosenstiel School of Marine and Atmospheric Science, University of Miami: Miami, Florida).

Perry TP, Bertling M (2000) Spatial and temporal patterns of macroboring within Mesozoic and Cenozoic coral reef systems. In 'Carbonate Platform Systems: Components and Interactions'. (Ed. M Insalaco et al.) **178**, 33–50 (Geological Society Special Publication: London).

Pomponi SA (1979) Ultrastructure and cytochemistry of the etching area of boring sponges. In 'Biologie des Spongiaires'. (Eds C Lévi, N Boury-Esnault.) pp. 317–323. (Colloques Internationaux du Centre National de la Recherche Scientifique: Paris).

Price IR, Fricker RL, Wilkinson CR (1984) *Ceratodictyon spongiosum* (Rhodophyta), the macroalgal partner in an alga-sponge symbiosis, grown in unialgal culture. *Journal of Phycology* **20**, 156–158.

Rosique MJ, Cano J, Rocamora J (1996) Influence of the sponge *Cliothosa hancocki* on the European flat oyster bed (*Ostrea edulis*) in the Mar Menor (Murcia, SE Spain). *Oebalia* **22**, 99–111.

Rützler K (2002a) Impact of crustose clionid sponges on Caribbean reef corals. *ACTA Geologica Hispanica* **37**, 61–72.

Rützler K (2002b) Family Clionaidae D'Orbigny, 1851. Family Spirastrellidae Ridley & Dendy, 1886. Family Alectonidae Rosell, 1996. *In* 'Systema Porifera. A Guide to the Classification of Sponges'. Vol. 1. (Eds JNA Hooper, RWM Van Soest.) pp. 173–185, 220–223, 281–290. (Kluwer Academic/Plenum Publishers: New York).

Rützler K, Rieger G (1973) Sponge burrowing: Fine structure of *Cliona lampa* penetrating calcareous substrata. *Marine Biology* **21**, 144–162.

Schönberg CHL, Wilkinson CR (2001) Induced colonization of corals by a clionid bioeroding sponge. *Coral Reefs* **20**, 69–76.

Thomas PA (1981) Boring sponges destructive to economically important molluscan beds and coral reefs in Indian seas. *Indian Journal of Fisheries* **26**, 163–200.

Turner HM (1985) Parasites of eastern oysters from subtidal reefs in a Louisiana estuary with a note on their use as indicators of water quality. *Estuaries* **8**, 323–325.

Uriz MJ, Rosell D, Maldonado M (1992) Parasitism, commensalism or mutualism? The case of Scyphozoa (Coronatae) and horny sponges. *Marine Ecology Progress Series* **81**, 247–255.

Wesche SJ, Adlard RD, Hooper JNA (1997) The first incidence of clionid sponges (Porifera) from the Sydney rock oyster *Saccostrea commercialis* (Iredale and Roughley, 1933). *Aquaculture* **157**, 173–180.

Wilkinson CR (1983) Role of sponges in coral reef structural processes. *In* 'Perspectives on Coral Reefs'. (Ed. DJ Barnes.) pp. 263–274. (Australian Institute of Marine Science: Townsville).

Wilkinson CR (1984) Sponges. *In* 'Reader's Digest Book of the Great Barrier Reef.' (Ed. F Talbot.) pp. 156–163. (Reader's Digest: Sydney).

Wilkinson CR (1992) Symbiotic interactions between marine sponges and algae. *In* 'Algae and Symbioses: Plants, Animals, Fungi, Viruses, Interactions Explored'. (Ed. W Reisser.) pp. 111–151. (Biopress: Bristol).

Cnidaria and Ctenophora (cnidarians and comb jellies)

Boero F, Bouillon J (1989) The life cycles of *Octotiara russelli* and *Stomotoca atra* (Cnidaria, Anthomedusae, Pandeidae). *Zoologica Scripta* **18**, 1–7.

Boero F, Bouillon J, Gravili C (1991) The life cycle of *Hydrichthys mirus* (Cnidaria: Hydrozoa: Anthomedusae: Pandeidae). *Zoological Journal of the Linnean Society* **101**, 189–199.

Boero F, Bouillon J, Kubota S (1997) The medusae of some species of *Hebella* Allman, 1888, and *Anthohebella* gen. nov. (Cnidaria, Hydrozoa, Lafoeidae), with a world synopsis of species. *Zoologische Verhandelingen* **310**, 1–53.

Boero F, Bouillon J, Gravili C (2000) A survey of *Zanclea*, *Halocoryne* and *Zanclella* (Cnidaria, Hydrozoa, Anthomedusae, Zancleidae) with description of new species. *Italian Journal of Zoology* **67**, 93–124.

Bouillon J (1967) Révision de la famille des Ptilocodiidae avec la description d'un nouveau genre et d'une nouvelle espèce. *Bulletin de l'Academie royale de Belgique* **53**, 1106–1131.

Bouillon J (1971) Sur quelques Hydroïdes de Roscoff. *Cahiers de Biologie Marine* **12**, 323–364.

Bouillon J (1987) Considérations sur le développement des Narcoméduses et sur leur position phylogénétique. *Indo-Malayan Zoology* **4**, 189–278.

Grassé P, Doumenc D (1987) 'Traité de Zoologie - Tome III 3 – Cnidaires Anthozoaires'. (Masson: Paris).

Grassé P, Doumenc D (1993) 'Traité de Zoologie - Tome III 2– Cnidaires Cténaires'. (Masson: Paris).

Kramp PL (1957) Notes on a living specimen of the hydroid *Kinetocodium danae* Kramp, parasitic on a pteropod. *Videnskabelige Meddelelser fra Dansk Naturhistorisk Forening* **119**, 47–54.

Millard NAH (1975) Monograph on the Hydroida of southern Africa. *Annals of the South African Museum* **68**, 1–513.

Palombi A (1935) *Eugymnanthea inquilina* nuova Leptomedusa derivante da un atecato idroide ospite interno di *Tapes decussatus* L. *Pubblicazioni della Stazione Zoologica di Napoli* **15**, 159–168.

Piraino S, Bouillon J, Boero F (1992) *Halocoryne epizoica* (Cnidaria, Hydrozoa), a hydroid that 'bites'. *Scientia Marina* **56**, 141–147.

Raikova EV (1973) Life cycle and systematic position of *Polypodium hydriforme* Ussov (Coelenterata), a cnidarian parasite of the eggs of Acipenseridae. *Publications of the Seto Marine Biological Laboratory* **20**, 165–173.

Schuchert P (1996) The marine fauna of New Zealand: Athecate hydroids and their medusae (Cnidaria: Hydrozoa). *New Zealand Oceanographic Institute Memoir* **106**, 1–159.

Uriz MJ, Rosell D, Maldonado M (1992) Parasitism, comensalism or mutualism? The case of Scyphozoa (Coronatae) and horny sponges. *Marine Ecology Progress Series*, **81**, 247–255.

Vervoort W (1966) Skeletal structure in the Solanderiidae and its bearing on hydroid classification. *In* 'The Cnidaria and Their Evolution'. (Ed. WJ Rees.) pp. 373–396. (Academic Press: London).

Wedler E, Larson R (1986) Athecate hydroids from Puerto Rico and the Virgin Islands. *Studies on Neotropical Fauna and Environment* **21**, 69–101.

Mesozoa Orthonectida (orthonectids)

Atkins D (1933) *Rhopalura granosa* sp. nov. an orthonectid parasite of a lamellibranch *Heteranomia squamula* L. with a note on its swimming behavior. *Journal of the Marine Biological Association of the United Kingdom* **19**, 233–252.

Caullery M, Lavallée A (1908) La fécondation et le développement de l'oeuf des Orthonectides. I. *Rhopalura ophiocomae*. *Archives de Zoologie expérimentale et générale* (ser. 4) **8**, 421–469.

Caullery M, Lavallée A (1912) Recherches sur le cycle évolutif des Orthonectides. Les phases initiales dans l'infection expérimentale de l'ophiure, *Amphiura squamata*, par *Rhopalura ophiocomae* Giard. *Bulletin Scientifique de la France et de la Belgique* **46**, 139–171.

Caullery M, Mensnil F (1901) Le cycle évolutif des Orthonectides. *Comptes Rendus des Seances Hebdomadaires de Académie des Sciences* **132**, 1232–1234.

Furuya H (2001) Phyla Dicyemida and Orthonectida. *In* 'Atlas of Marine Invertebrate Larvae'. (Ed. CM Young.) pp. 149–161. (Academic Press: London).

Giard A (1880) The Orthonectida, a new class of the phylum of the worms. *Quarterly Journal of Microscopical Science (new ser.)* **20**, 225–240.

Julin C (1882) Contribution à l'histoire des Mésozoaires. Recherches sur l'organisation et le développement embryonnaire des Orthonectides. *Archives de Biologie* **3**, 1–54.

Kozloff EN (1969) Morphology of the orthonectid *Rhopalura ophiocomae*. *Journal of Parasitology* **55**, 171–195.

Kozloff EN (1971) Morphology of the orthonectid *Ciliocincta sabellariae*. *Journal of Parasitology* **57**, 585–597.

Kozloff EN (1990) Phylum Orthonectida. *In* 'Invertebrates'. pp. 216–220. (Saunders College Publishing).

Kozloff EN (1992) The genera of the phylum Orthonectida. *Cahiers de Biologie Marine* **33**, 377–406.

Kozloff EN (1994) The structure and origin of the plasmodium of *Rhopalura ophiocomae* (Phylum Orthonectida). *Acta Zoologica* **75**, 191–199.

Metschnikoff E (1881) Untersuchungen über Orthonectiden. *Zeitschrift für wissenschaftliche Zoologie* **35**, 282–303.

Mesozoa Dicyemida (dicyemids)

Czaker R (1994) *Kantharella antarctica*, a new and unusual dicyemid mesozoan from the Antarctic. *Zoologischer Anzeiger* **232**, 151–158.

Furuya H (1999) Fourteen new species of dicyemid mesozoans from six Japanese cephalopods, with comments on host specificity. *Species Diversity* **4**, 257–319.

Furuya H, Tsuneki K (2003) Biology of dicyemid mesozoan. *Zoological Science* **20**, 519–532.

Furuya H, Tsuneki K, Koshida Y (1996) The cell lineages of two types of embryo and a hermaphroditic gonad in dicyemid mesozoans. *Development Growth and Differentiation* **38**, 453–463.

Furuya H, Hochberg FG, Tsuneki K (2003a) Reproductive traits of dicyemids. *Marine Biology* **142**, 693–706.

Furuya H, Hochberg FG, Tsuneki K (2003b) Calotte morphology in the phylum Dicyemida: niche separation and convergence. *Journal of Zoology* **259**, 361–373.

Hochberg FG (1982) The 'kidneys' of cephalopods: a unique habitat for parasites. *Malacologia* **23**, 121–134.

Hochberg FG (1990) Diseases caused by protistans and mesozoans. *In* 'Diseases of Marine Animals'. Vol. III. (Ed. O Kinne.) pp. 47–202. (Biologische Anstalt Helgoland: Hamburg).

Kalavati C, Narasimhamurti CC (1980) A new dicyemid mesozoan, *Dodecadicyema loligoi* n. gen., n. sp. From the renal appendages of *Loligo* sp. *Proceedings Indian Academy of Sciences (Animal Sciences)* **89**, 287–292.

Lapan EA (1975) Studies on the chemistry of the octopus renal system and an observation on the symbiotic relationship of the dicyemid Mesozoa. *Comparative Biochemistry and Physiology* **52**, 651–657.

Lapan EA, Morowitz HJ (1975) The dicyemid mesozoa as an integrated system for morphogenetic studies I. Description, isolation, and maintenance. *Journal of Experimental Zoology* **193**, 147–160.

McConnaughey BH (1951) The life cycle of the dicyemid Mesozoa. *University of California Publications in Zoology* **55**, 295–336.

Nouvel H (1947) Les Dicyémides. 1re partie: systématique, générations, vermiformes, infusorigène et sexualité. *Archives de Biologie, Paris* **58**, 59–220.

Nouvel H (1948) Les Dicyémides. 2^{é} partie: infusoriforme, tératologie, spécificité du parasitisme, affinités. *Archives de Biologie, Paris* **59**, 147–223.

Myzostomida (myzostomids)

Eeckhaut I (1998) *Mycomyzostoma calcidicola* gen. et sp. nov., the first extant parasitic myzostome infesting crinoid stalks, with a nomencatural appendix by M. J. Grygier. *Species Diversity* **3**, 89–103.

Eeckhaut I, Jangoux M (1997) Infestation, population dynamics, growth and reproductive cycle of *Myzostoma cirriferum* (Myzostomida), an obligate symbiont of the comatulid crinoid *Antedon bifida* (Crinoidea, Echinodermata). *Cahiers de Biologie Marine* **38**, 7–18.

Eeckhaut I, Grygier MJ, Deheyn D (1998) Myzostomes from Papua New Guinea, with related Indo-west Pacific distribution records and description of five new species. *Bulletin of Marine Science* **62**, 841–886.

Eeckhaut I, McHugh D, Mardulyn P, Tiedemann R, Monteyne D, Jangoux M, Milinkovitch C (2000) Myzostomida: a link between trochozoans and flatworms? *Proceedings of the Royal Society, Series B* **267**, 1383–1392.

Eeckhaut I, Fievez L, Müller M (2003) Larval development of *Myzostoma cirriferum* (Myzostomida). *Journal of Morphology* **258**, 269–283.

Graff L von (1877) 'Das Genus *Myzostoma* (F.S. Leuckart)'. (Verlag von Wilhelm Engelmann: Leipzig).

Graff L von (1884) Report on the Myzostomida collected during the voyage of H.M.S. *Challenger* during the years 1873–1876. *Reports of the Scientific Research Voyage of H.M.S. Challenger 1873–1876, Zoology* **10**, 1–82.

Graff L von (1887) Report on the Myzostomida collected during the voyage of H.M.S. *Challenger* during the years 1873–1876. Supplement. *Reports of the Scientific Research Voyage of H.M.S. Challenger 1873–1876, Zoology* **20**, 1–16.

Grygier MJ (1990) Distribution of Indo-Pacific *Myzostoma* and host specificity of comatulid-associated Myzostomida. *Bulletin of Marine Science* **47**, 182–191.

Grygier MJ (2000) Myzostomida. *In* 'Polychaeta and Allies: The Southern Synthesis. Fauna of Australia. Volume 4A, Polychaeta, Myzostomida, Pogonophora, Echiura, Sipuncula'. (Eds P Beesley, GJB Ross, CJ Glasby.) pp. 297–329. (CSIRO Publishing: Melbourne).

Pietsch A, Westheide W (1987) Protonephridial organs in *Myzostoma cirriferum* (Myzostomida). *Acta Zoologica* **68**, 195–203.

Rouse GW, Fauchald K (1997) Cladistics and polychaetes. *Zoologica Scripta* **26**, 139–204.

Rouse GW, Pleijel F (2001) 'Polychaetes'. (Oxford University Press: London).

Woodham A (1992) Distribution and population studies on *Myzostoma cirriferum* Leuckart (Myzostomida) in a Scottish sea loch. *In* 'Marine Eutrophication and Population Dynamics'. (Eds G Colombo, I Ferrari, VU Ceccherelli, R Rossi.) pp. 247–255. (Olsen & Olsen: Fredensborg).

Zrzavy J, Hypsa V, Tietz DF (2001) Myzostomida are not annelids: molecular and morphological support for a clade of animals with anterior sperm flagella. *Cladistics* **17**, 170–198.

Polychaeta (bristle worms)

Clark RB (1956) *Capitella capitata* as a commensal, with a bibliography of parasitism and commensalism in the polychaetes. *Ann. mag. nat Hist. (ser. 12)* **9**, 433–448.

Eibye-Jacobsen D, Kristensen RM (1994) A new genus and species of Dorvilleidae (Annelida, Polychaeta) from Bermuda, with a phylogenetic analysis of Dorvilleidae, Iphitimidae and Dinophilidae. *Zool. Scr.* **23**, 107–131.

Eisig H (1906) *Ichthyotomus sanguinarius*, eine auf Aalen schmarotzende Annelide. *Fauna und Flora des Golfes von Neapel* **28**, 1–300.

Fischer A (1999) Reproductive and developmental phenomena in annelids: a source of explanatory research problems. *Hydrobiologia* **402**, 1–20.

Fitzhugh K, Rouse GW (1999) A remarkable new genus and species of fan worm (Polychaeta : Sabellidae : Sabellinae) associated with marine gastropods. *Invertebrate Biology* **118**, 357–390.

Hartman O (1948) The polychaetous annelids of Alaska. *Pacific Science* **2**, 3–58.

Hartman O, Boss KJ (1965) *Antonbruunia viridis*, a new inquisite annelid with dwarf males, inhabiting a new species of pelecypod, *Lucina fosteri*, in the Mozambique Channel. *Ann. Mag. nat. Hist.* **8**, 177–186.

Jennings JB, Gelder SR (1976) Observations of the feeding mechanism, diet and physiology of *Histriobdella homari* van Beneden 1858: an aberrant polychaete symbiotic with north american and european lobsters. *Biological Bulletin* **151**, 489–517.

Jones ML (1974) On the Caobangiidae, a new family of the Polychaeta, with a redescription of *Caobangia billeti* Girard. *Smithsonian Contributions to Zoology* **175**, 1–55.

Martin D, Britayev TA (1998) Symbiotic polychaetes: Review of known species. *Oceanography and Marine Biology* **36**, 217–340.

Paris J (1955) Commensalisme et parasitisme chez les Annelides Polychetes. *Vie et milieu* **6**, 525–536.

Rouse GW, Pleijel F (2001) 'Polychaetes.' (Oxford University Press: London).

Tzetlin AB, Britayev TA (1985) A new species of the Spionidae (Polychaeta) with asexual reproduction associated with sponges. *Zoologica Scripta* **14**, 177–181.

Uebelacker JM (1978) A new parasitic polychaetous annelid (Arabellidae) from the Bahamas. *Journal of Parasitology* **64**, 151–154.

Hirudinea (leeches)

Anderson DT (1973) 'Embryology and Phylogeny in Annelids and Arthropods'. (Pergamon Press: Oxford).

Cruz-Lacierda ER, Toledo JD, Tan-Fermin JD, Burreson EU (2000) Marine leech (*Zeylanicobdella arugamensis*) infestation in cultured orange-spotted grouper, *Epinephelus coioides*. *Aquaculture* **185**, 191–196.

Davies RW, Govedich FR (2001) Annelida: Euhirudinea and Acanthobdellids. In 'Ecology and Classification of North American Freshwater Invertebrates'. 2nd edn. (Eds JH Thorp, APCovich.) pp. 465–504. (Academic Press: San Diego).

Govedich FR (2001) A reference guide to the ecology and taxonomy of freshwater and terrestrial leeches (Euhirudinea) of Australasia and Oceania. In 'Cooperative Research Centre for Freshwater Ecology Identification Guide No. 35'. (Ed. JH Hawking.) pp. 1–67. (Cooperative Research Center for Freshwater Ecology: Canberra).

Graf J (1999) Symbiosis of *Aeromonas veronii*, biovar sobria, and *Hirudo medicinalis*, the medicinal leech: a novel model for digestive tract associations. *Infection and Immunity* **67**, 1–7.

Kutschera U, Wirtz P (2001) The evolution of parental care in freshwater leeches. *Theory in Biosciences* **120**, 115–137.

Moore JP (1900) A description of *Microbdella biannulata* with especial regard to the constitution of the leech somite. *Proceedings of the Academy of Natural Sciences of Philadelphia* **1900**, 50–74.

Negm-Eldin M (1997) *Trypanosoma makasai* (Hoare, 1932) in its biological vector *Batracobdelloides tricarinata* (Blanchard, 1897) and their life cycles. *Deutsche Tierärztliche Wochenschrift* **104**, 215–219.

Reddy DC, Davies RW (1993) Metabolic adaptations by the leech *Nephelopsis obscura* during long-term anoxia and recovery. *Journal of Experimental Zoology* **265**, 224–230.

Sawyer RT (1986a) 'Anatomy, Physiology and Behaviour'. Vol. I. 'Leech Biology and Behaviour'. (Oxford University Press: Oxford).

Sawyer RT (1986b) 'Feeding Biology, Ecology and Systematics'. Vol. II. 'Leech Biology and Behaviour'. (Oxford University Press: Oxford).

Siddall ME, Burreson EM (1995). Phylogeny of the Euhirudinea: independent evolution of blood feeding by leeches? *Canadian Journal of Zoology* **73**, 1048–1064.

Cycliophora (wheel wearers)

Funch P (1996) The chordoid larva of *Symbion pandora* (Cycliophora) is a modified trochophore. *Journal of Morphology* **230**, 231–263.

Funch P, Kristensen RM (1995) Cycliophora is a new phylum with affinities to Entoprocta and Ectoprocta. *Nature* **378**, 711–714.

Funch P, Kristensen RM (1997) Cycliophora. *In* 'Microscopic Anatomy of Invertebrates'. Vol. 13. 'Lophophorates, Entoprocta and Cycliophora'. pp. 409–474. (Wiley-Liss: New York).

Funch P, Kristensen RM (1999) Cycliophora. *In* 'Encyclopedia of Reproduction'.Vol. 1. pp. 800-808 (Academic Press: New York).

Giribet G, Sørensen MV, Funch P, Kristensen RM, Sterrer W (2004) Investigations into the phylogenetic position of Micrognathozoa using four molecular loci. *Cladistics* **20**, 1–13.

Kristensen RM (2002a) Cycliophora. *In* 'Encyclopedia of Life Sciences'. Vol. 5. pp. 251–255. (www.els.net.) (Macmillan Reference: London).

Kristensen RM (2002b) An introduction to Loricifera, Cycliophora, and Micrognathozoa. *Integrative and Comparative Biology* **42**, 641–651.

Kristensen RM, Funch, F (2002) Chapter 11, Phylum Cycliophora. *In* 'Atlas of Marine Invertebrate Larvae'. (Eds CM Young, MA Sewell, ME Rice.) pp. 199–208 (Academic Press: London).

Obst M, Funch P (2003) Dwarf male of *Symbion pandora* (Cycliophora). *Journal of Morphology* **255**, 261–278.

Winnepenninckx BMH, Backeljau T, Kristensen RM (1998) Relations of the new Phylum Cycliophora. Scientific correspondence. *Nature* **393**, 636–638.

Nemertea (ribbon worms)

Berg G, Gibson R (1996) A redescription of *Nemertoscolex parasiticus* Greeff, 1879, an apparently endoparasitic heteronemertean from the coelomic fluid of the echiuroid *Echiurus echiurus* Pallas). *Journal of Natural History* **30**, 163–173.

Campbell A, Gibson R, Evans LH (1989) A new species of *Carcinonemertes* (Nemertea: Carcinonemertidae) ectohabitant on *Panulirus cygnus* (Crustacea: Palinuridae) from Western Australia. *Zoological Journal of the Linnean Society* **95**, 257–268.

Coe WR (1902) The nemertean parasites of crabs. *American Naturalist* **36**, 431–450.

Corrêa DD (1966) A new hermaphroditic nemertean. *Anais da Academia Brasileira de Ciencias* **38**, 365–369.

Fleming LC, Gibson R (1981) A new genus and species of monostiliferous hoplonemerteans, ectohabitant on lobsters. *Journal of Experimental Marine Biology and Ecology* **52**, 79–93.

Gibson R (1974) A new species of commensal hoplonemertean from Australia. *Zoological Journal of the Linnean Society* **55**, 247–266.Gibson R (1968) Studies on the biology of the entocommensal rhynchocoelan *Malacobdella grossa*. *Journal of the Marine Biological Association of the United Kingdom* **48**, 637–656.

Gibson R (1982) Nemertea. *In* 'Synopsis and Classification of Living Organisms'. (Ed. SP Parker.) pp. 823–846. (McGraw-Hill: New York).

Gibson R (1986) Redescription and taxonomic reappraisal of *Nemertopsis actinophila* Burger, 1904 (Nemertea: Hoplonemertea: Monostilifera). *Bulletin of Marine Science* **39**, 42–60.

Gibson R (1995) Nemertean genera and species of the world: an annotated checklist of original names and description citations, synonyms, current taxonomic status, habitats and recorded zoogeographic distribution. *Journal of Natural History* **29**, 271–561.

Gibson R, Jennings JB (1969) Observations on the diet, feeding mechanisms, digestion and food reserves of the entocommensal rhynchocoelan *Malacobdella grossa*. *Journal of the Marine Biological Association of the United Kingdom* **49**, 17–32.

Humes AG (1942) The morphology, taxonomy, and bionomics of the nemertean genus *Carcinonemertes*. *Illinois Biological Monographs* **18**, 1–105.

Ivanov VA, Bigatti G, Penchaszadeh PE, Norenburg JL (2002) *Malacobdella arrokeana* (Nemertea: Bdellonemertea), a new species of nemertean from the Southwestern Atlantic

Ocean entocommensal in *Panopea abbreviata* (Bivalvia, Heterodonta, Hiatellidae) in Argentina. *Proceedings of the Biological Society of Washington* **115**, 359–367.

Iwata F (1967) *Uchidana parasita* nov. gen. et nov. sp., a new parasitic nemertean from Japan with peculiar morphological characters. *Zoologischer Anzeiger* **178**, 122–136.

Kuris AM (1993) Life cycles of nemerteans that are symbiotic egg predators of decapod Crustacea: adaptations to host life histories. *Hydrobiologia* **266**, 1–14.

Kuris AM, Blau SF, Paul AJ, Shields JD, Wickham DE (1991) Infestation of brood symbionts and their impact on egg mortality in the red king crab, *Paralithodes camtschatica*, in Alaska: biogeographical and temporal variation. *Canadian Journal of Fisheries and Aquatic Sciences* **48**, 559–568.

Kuris AM, Wickham DE (1987) Effect of nemertean egg predators on crustaceans. *Bulletin of Marine Science* **41**, 151–164.

Roe P (1986) Parthenogenesis in *Carcinonemertes* spp. (Nemertea: Hoplonemertea). *Biological Bulletin* **171**, 640–646.

Roe P (1988) Ecological implications of the reproductive biology of symbiotic nemerteans. *Hydrobiologia* **156**, 13–22.

Sadeghian PS (2003) A new species of ribbon worm (Nemertea): the taxonomy and life history of a specialized egg predator on the Purple globe crab. (MA Thesis, University of California, Santa Barbara).

Sadeghian PS, Kuris AM (2001) Distribution and abundance of a nemertean egg predator (*Carcinonemertes* sp.) on a leucosiid crab, *Randallia ornata*. *Hydrobiologia* **456**, 59–63.

Shields JD, Kuris AM (1988) Temporal variation in abundance of the egg predator *Carcinonemertes epialti* (Nemertea) and its effects on egg mortality of its host, the shore crab, *Hemigrapsus oregonensis*. *Hydrobiologia* **156**, 31–38.

Shields JD, Kuris AM (1990) *Carcinonemertes wickhami* n. sp. (Nemertea), a symbiotic egg predator from the spiny lobster *Panulirus interruptus* in southern California, with remarks on symbiont-host adaptations. *Fishery Bulletin* **88**, 279–287.

Shields JD, Wickham DE, Kuris AM (1989) *Carcinonemertes regicides* n. sp. (Nemertea), symbiotic egg predator from the red king crab, *Paralithodes camtschatica* (Decapoda: Anomura), in Alaska. *Canadian Journal of Zoology* **67**, 923–930.

Sundet JH, Jobling M (1985) An investigation of the interactions between the nemertine, *Malacobdella grossa*, and its bivalve host, *Arctica islandica*. In 'Marine Biology of Polar Regions and Effects of Stress on Marine Organisms'. (Eds JS Gray, ME Christiansen.) pp. 185–197. (John Wiley and Sons: Chichester).

Thollesson M, Norenburg JL (2003) Ribbon worm relationships: A phylogeny of the phylum Nemertea. *Proceedings of the Royal Society Biological Sciences Series B* **270**, 407–415.

Uhazy LS, Aiken DE, Campbell A (1985) Morphology and systematics of the nemertean *Pseudocarcinonemertes homari* (Hoplonemertea: Monostilifera) from the American lobster, *Homarus americanus*. *Canadian Journal of Fisheries and Aquatic Sciences* **42**, 342–350.

Wickham DE (1980) Aspects of the life history of *Carcinonemertes errans* (Nemertea: Carcinonemertidae), an egg predator of the crab, *Cancer magister*. *Biological Bulletin* **159**, 247–257.

Wickham DE, Kuris AM (1985) The comparative ecology of nemertean egg predators. *American Zoologist* **25**, 127–134.

Rotifera and Seison (rotifers)

Ahlrichs WH (1995) Ultrastruktur und Phylogenie von *Seison nebaliae* (Grube 1859) und *Seison annulatus* (Claus 1876–). Hypothesen zu phylogenetischen Verwandtschaftsverhältnissen innerhalb der Bilateria, 1–310. (Cuvillier Verlag: Göottingen).

Ahlrichs WH (2003) Rotifera (Rädertiere) und *Seison* (Teleskophalstiere). S. 704–733. *In* 'Das Mittelmeer. Fauna, Flora, ökologie'. Band II/1. 'Bestimmungsführer: Prokaryota, Protista,

Fungi, Algae, Plantae, Animalia (bis Nemertea)'. (Ed. R Hofrichter.) ((Spektrum, Akademie Verlag: place of publication?).

De Smet WH (1996) The Proalidae (Monogononta) *In* 'Guides to the Identification of the Microinvertebrates of the Continental Waters of the World'. Vol. 9. (Eds HJF Dumont, WH De Smet.) Rotifera Vol. 4, (SPB Academic Publishing: New York).

Koste W (1975) Das Rädertiereportait, *Seison annulatus*, ein Ektoparasit des marinen Krebses *Nebalia*. *Mikrokosmos* **64**, 341–347.

Levander KM (1894) Materialien zur Kenntnis der Wasserfauna in der Umgebung von Helsingfors. II. Rotatorien. *Acta. Soc. Fauna Flora Fennica* **12**.

Remane A (1929a) Proales gonothyraeae n.sp., ein an Hydroidpolypen parasitierendes Rädertier. *Zoologischer Anzeiger* **80**, 289–295.

Remane A (1929b) Rotatoria. *In* 'Die Tierwelt der Nord und Ostsee'. L. XVI, T. VII. (Eds G Grimpe, E Wagler.) pp. 1–156. (Publisher: Place of publication?)

Thane-Fenchel A (1968) Distribution and ecology of non-planktonic brackish-water rotifers from Scandinavian waters. *Ophelia* **3**, 93–97.

Nematomorpha (horse-hair worms)

Arvy L (1963) Données sur le parasitisme protélien de *Nectonema* (Nématomorphe), chez les Crustacés. *Annales de Parasitologie* **38**, 887–892.

Bakke A (1975) En marin taggelmark *Nectonema* sp. (Nematomorpha) fra Nord-Norge. *Fauna (Oslo)* **28**, 163–169.

Beattie DA (1987) *Nectonema agile* (Nematomorpha) from Passamaquoddy Bay, New Brunswick. (Masters Thesis, University of New Brunswick).

Bock S (1913) Zur Kenntnis von *Nectonema* und dessen systematischer Stellung. *Zoologiska Bidrag fran Uppsala* **2**, 1–8.

Brattey J, Elner RW, Uhazy LS, Bagnall AE (1985) Metazoan parasites and commensals of five crab (Brachyura) species from eastern Canada. *Canadian Journal of Zoology* **63**, 2224–2229.

Bresciani J (1975) Ultrastructure of the integument of the horsehair worm, *Nectonema munidae* Brinkmann 1930 (Nematomorpha). *2nd European Multicolloquium on Parasitology*, Trogir, 75.

Bresciani J (1991) Nematomorpha. *In* 'Microscopic Anatomy of Invertebrates'. (Eds FW Harrisson, EE Ruppert.) pp. 197–218 (Wiley-Liss: New York).

Bürger O (1891) Zur Kenntnis von *Nectonema agile* Verr. *Zoologische Jahrbücher Abteilung für Anatomie und Ontogenie der Tiere* **4**, 631–652.

Fewkes JW (1883) On the development of certain worm larvae. *Bulletin of the Museum of Comparative Zoology at Harvard College* **11**, 167–208.

Feyel T (1936) Researches histologiques sur *Nectonema agile* Verr. étude de la forme parasitaire. *Archives d'Anatomie Microscopique* **32**, 197–234.

Huus J (1932) über die Begattung bei *Nectonema munidae* Br. und über den Fund der Larve von dieser Art. *Zoologischer Anzeiger* **97**, 33–37.

Leslie HA, Campbell A, Daborn GR (1981) *Nectonema* (Nematomorpha: Nectonematidae) a parasite of decapod Crustacea in the Bay of Fundy. *Canadian Journal of Zoology* **59**, 1193–1196.

Nielsen SO (1969) *Nectonema munidae* Brinkmann (Nematomorpha) parasitizing *Munida tenuimana* G.O. Sars (Crust. Dec.) with notes on host-parasite relations and new host species. *Sarsia* **38**, 91–110.

Nierstrasz F (1907) Die Nematomorpha der Siboga-Expedition. *Siboga-Expeditie* **20**, 1–21.

Nouvel H, Nouvel L (1938) Sur deux hotes nouveaux de *Nectonema*. *Bulletin du Muséum National d´Histoire Naturelle Paris* **10**, 507–508.

Oku Y, Fukumoto SI, Ohbayashi M, Koike M (1993) A marine horsehair worm, *Nectonema* sp., parasitizing a telecyclid crab, *Erimacrus isenbeckii*, from Hokkaido, Japan. *The Japanese Journal of Veterinary Research* **31**, 65–69.

Poinar G, Brockerhoff AM (2001) *Nectonema zealandica* n. sp. (Nematomorpha: Nectonematoidea) parasitizing the purple rock crab *Hemigrapsus edwardsi* (Brachyura: Decapoda) in New Zealand, with notes on the prevalence of infection and host defense reactions. *Systematic Parasitology* **50**, 149–157.

Schmidt-Rhaesa A (1996a) Ultrastructure of the anterior end in three ontogenetic stages of *Nectonema munidae* (Nematomorpha). *Acta Zoologica* **77**, 267–278.

Schmidt-Rhaesa A (1996b) The nervous system of *Nectonema munidae* and *Gordius aquaticus*, with implications for the ground pattern of the Nematomorpha. *Zoomorphology* **116**, 133–142.

Schmidt-Rhaesa A (1997) Ultrastructural features of the female reproductive system and female gametes of *Nectonema munidae* Brinkmann 1930 (Nematomorpha). *Parasitology Research* **83**, 77–81.

Schmidt-Rhaesa A (1998) Muscular ultrastructure in *Nectonema munidae* and *Gordius aquaticus* (Nematomorpha). *Invertebrate Biology* **117**, 37–44.

Schmidt-Rhaesa A (1999) Nematomorpha. *In* 'Encyclopedia of Reproduction'. (Eds E Knobil, JD Neill.) pp. 333–341. (Academic Press: San Diego).

Skaling B, MacKinnon BM (1988) The absorptive surface of *Nectonema* sp. (Nematomorpha: Nectonematoidea) from *Pandalus montagui*: histology, ultrastructure, and absorptive capabilities of the body wall and intestine. *Canadian Journal of Zoology* **66**, 289–295.

Ward HB (1892) On *Nectonema agile* Verrill. *Bulletin of the Museum of Comparative Zoology at Harvard College* **23**, 135–188.

Acari (mites and ticks)

Barton TR, Harris MP, Wanless S, Elston DA (1995) The activity periods and life-cycle of the tick *Ixodes uriae* (Acari: Ixodidae) in relation to host breeding strategies. *Parasitology* **112**, 571–580.

Bartsch I (1987) *Australacarus inexpectatus* gen. et spec. nov. (Halacaroidea, Acari), mit einer Übersicht über parasitisch lebende Halacariden. *Zoologische Anzeiger* **218**, 17–24.

Bartsch I (1989) Marine mites (Halacaroidea, Acari): a geographical and ecological survey. *Hydrobiologia* **178**, 21–42.

Butenko OM (1975) The systematic position of mites of the genus *Larinyssus strandtmanni* (Gamasoidea, Rhinonyssidae), parasitizing gulls. *Parazitologiia* **9**, 175–82. [In Russian.]

Butenko OM, Staniukovich MK (2001) New species of the rhinonyssid mites (Gamasina: Rhinonyssidae) from birds of Russia and neighboring countries. *Parazitologiia* **35**, 519–530. [In Russian.]

Cáceres-Martínez CJ, Vásquez-Yeomans R, Rentera YG, Curiel-Ramírez SC, Valdez JAO, Rivas G (2000) The marine mites *Hyadesia* sp. and *Copidognathus* sp. associated with the mussel *Mytilus galloprovincialis*. *Journal of Invertebrate Pathology* **76**, 216–221.

Dahme E, Popp E (1963) Todesfälle bei Seelöwen (*Zalophus californianus* Lesson) verursacht durch eine bisher unbekannte Milbe (*Orthohalarachne letalis* Popp). *Berliner und Münchener Tierärztliche Wochenschrift* **21**, 441–443.

Dailey MD, Nutting WB (1980) *Demodex zalophi* sp. nov. (Acari: Demodecidae) from *Zalophus californianus*, the California sea lion. *Acarologia* **21**, 423–428.

Danchin E (1992) The incidence of the tick parasite *Ixodes uriae* in kittiwake *Rissa tridactyla* colonies in relation to the age of the colony, and a mechanism of infecting new colonies. *Ibis* **134**, 134–141.

Desch CD, Dailey MD, Tuomi P (2003) Description of a hair follicle mite (Acari: Demodecidae) parasitic in the earless seal family Phocidae (Mammalia: Carnivora) from the harbor seal *Phoca vitulina* Linnaeus, 1758. *International Journal of Acarology* **29**, 231–235.

Dunlap J, Piper R, Keyes M (1976) Lesions associated with *Orthohalarachne attenuata* (Halarachnidae) in the northern fur seal (*Callorhinus ursinus*). *Journal of Wildlife Diseases* **12**, 42–44.

Evans GO (1992) 'Principles of Acarology'. (CAB International: Wallingford).

Fain A (1977) Observation sur les Turbinoptidae d'Afrique au Sud du Sahara. *Revue de Zoologie Africaine* **91**, 83–116.

Fain A, Clark JM (1994) Description and life cycle of *Suladectes hughesae antipodus* subspec. nov. (Acari: Hypoderatidae) associated with *Sula bassana serrator* Gray (Aves: Pelecaniformes) in New Zealand. *Acarologia* **35**, 361–371.

Fay FH, Furman DP (1982) Nasal mites (Acari: Halarachnidae) in the spotted seal, *Phoca largha* Pallas, and other pinnipeds of Alaskan waters. *Journal of Wildlife Diseases* **18**, 63–67.

Frenot Y, De Oliveira E, Gauthier-Clerc M, Deunff J, Bellido A, Vernon P (2001) Life cycle of the tick *Ixodes uriae* (White, 1852) in penguin colonies: relationship with host breeding activity. *International Journal of Parasitology* **31**, 1040–1047.

Gaud J, Atyeo WT (1996) Feather mites of the world (Acarina, Astigmata): The supraspecific taxa. Part 1 Text. *Annales du Musée royale de l'Afrique centrale, Série in–8^0, Sciences Zoologiques* **277**, 1–193.

Gauthier-Clerc M, Clerquin Y, Handrich Y (1998) Hyperinfestation by ticks *Ixodes uriae*: a possible cause of death in adult king penguins, a long-lived seabird. *Colonial Waterbirds* **21**, 229–233.

Gauthier-Clerc M, Jaulhac B, Frenot Y, Bachelard C, Monteil H, Le Maho Y, Handrich Y (1999) Prevalence of *Borrelia burgdorferi* (the lyme disease agent) antibodies in king penguin *Aptenodytes patagonicus* in Crozet Archipelago. *Polar Biology* **22**, 141–143.

Gylfe Å, Olsen B, Ras NM, Strasevicius D, Noppa L, Östberg Y, Weihe P, Bergström S (1999). Isolation of Lyme disease *Borrelia* from puffins (*Fratercula arctica*) and seabird ticks (*Ixodes uriae*) on Faeroe Islands. *Journal of Clinical Microbiology* **37**, 890–896.

Halliday RB, Collins RO (2002) *Histiostoma papillata* sp. nov. (Acari: Histiostomatidae), a mite attacking fish in Australia. *Australian Journal of Entomology* **41**, 155–158.

Hoogstraal H (1973) Ticks. *In* 'Viruses and Invertebrates'. (Ed. AJ Gibbs.) pp. 89–103. (North-Holland Publishing Company: Amsterdam).

Hyland KE, Bernier J, Merkowski D, MacLachlan A, Amr Z, Pitocchelli J, Myers J, Hu R (2000) Records of ticks (Acari: Ixodidae) parasitizing birds (Aves) in Rhode Island, USA. *International Journal of Acarology* **25**, 183–192.

Keirans JE, Clifford CM, Hoogstraal H (1980) Identity of the nymphs and adults of the Galapagos iguanid lizard parasites, *Ornithodoros* (*Alectotobius*) *darwini* and *O.* (*A.*) *galapagensis* (Ixodoidea: Argasidae). *Journal of Medical Entomology* **17**, 427–438.

Kenyon KW, Yunker CE, Newell IM (1965) Nasal mites (Halarachnidae) in the sea otter. *Journal of Parasitology* **51**, 960.

Kerry K, Riddle M, Clarke J (2000) Diseases in Antarctic wildlife. A Report for The Scientific Committee on Antarctic research (SCAR) and The Council of Managers of National Antarctic Programs COMNAP). pp. 1–104. (Australian Antarctic Division: Kingston).

Kethley J (1971) Population regulation in quill mites (Acarina: Syringophilidae). *Ecology* **51**, 1113–1118.

Kethley J (1982) Acariformes. *In* 'Synopsis and Classification of Living Organisms 2'. (Ed. SP Parker.) pp. 117–145. (McGraw-Hill: New York).

Kim KC (1985) Evolutionary relationships of parasitic arthropods and mammals. *In* 'Coevolution of Parasitic Arthropods and Mammals'. (Ed. KC Kim.) pp. 3–82. (Wiley-Interscience: New York).

Konishi K, Shimazaki K (1998) Halarachnid mites infesting the respiratory tract of Steller sea lions. *Biosphere Conservation* **1**, 45–48.

Krantz GW (1978) 'A Manual of Acarology'. 2nd edn. (Oregon State University Book Stores: Corvallis, Oregon).

Kudryashova NI (1998) 'Chigger Mites (Acariformes, Trombiculidae) of the East Palearctics'. (KMK Scientific Press: Moscow).

Mangin S, Gauthier-Clerc M, Frenot Y, Gendner J-P, Le Maho Y (2003) Ticks *Ixodes uriae* and the breeding performance of a colonial seabird, king penguin *Aptenodytes patagonicus*. *Journal of Avian Biology* **34**, 30–34.

McCoy KD, Tirard C (2002) Reproductive strategies of the seabird tick *Ixodes uriae* (Acari: Ixodidae). *Journal of Parasitology* **88**, 813–816.

Mironov SV, Malyshev LL (2002) Dynamics of infection the chaffinch nestlings *Fringilla coelebs* with feather mites (Acari: Analgoidea). *Parazitologiia* **36**, 356–374.

Moeller RB, Jr (1997) 'Diseases of Marine Mammals'. (LTC, VC, USA, United States Army Medical Research Institute of Chemical Defense: Aberdeen Proving Ground, MD).

Morgan IR, Westbury HA, Campbell J (1985) Viral infections of Little Blue penguins (*Eudyptla minor*) along the Southern Coast of Australia. *Journal of Wildlife Diseases* **21**, 193–198.

Murray MD, Vestjens WJM (1967) Studies on the ectoparasites of seals and penguins. *Australian Journal of Zoology* **15**, 715–725.

Nuttall PA (1984) Tick-borne viruses in seabird colonies. *Seabird* **7**, 31–41.

OConnor BM (1985) Hypoderatid mites (Acari) associated with cormorants (Aves: Phalacrocoracidae), with description of a new species. *Journal of Medical Entomology* **22**, 324–331.

Oliver JHJ (1989) Biology and systematics of ticks (Acari: Ixodida). *Annual Review of Ecology and Systematics* **20**, 397–430.

Otto JC (2000). *Spongilhalacarus longiscutus* n. gen., n. sp., a marine mite (Acari: Prostigmata: Halacaridae) associated with an alga-sponge symbiosis from the Great Barrier Reef lagoon in Australia. *International Journal of Acarology* **26**, 285–291.

Pence DB, Cole RA (1995) First record of an hypopus (Acari: Hypoderatidae) from a jaeger (Aves: Charadrifformer: Stercorariidae). *Journal of Medical Entomology* **32**, 395–396.

Pence DB, Hoberg EP (1991) *Thalassornectes* (*Alcidectes*) *aukletae* subgen. et sp. n. (Acari: Hypoderatidae) from the crested and parakeet auklets (Aves: Charadriiformes; Alcidae). *Journal of Medical Entomology* **28**, 205–209.

Pence DB, Scott WD (1998) *Chelonacarus elongatus* n. gen., n. sp. (Acari: Cloacaridae) from the cloaca of the green turtle *Chelonia mydas* (Cheloniidae). *Journal of Parasitology* **84**, 835–839.

Proctor H, Owens I (2000) Mites and birds: diversity, parasitism and coevolution. *TREE* **15**, 358–364.

Raga J (1992). Parasitismus bei den Pinnipedia. *In* 'Handbuch der Säugetiere Europas'. Vol. 6/II. (Eds J Niethammer, F Krapp.) pp. 42–75. (AULA-Verlag: Wiesbaden).

Roberts LS, Janovy J (1996) 'Foundations of Parasitology'. (WCB Publishers, Dubuque, IA).

Sonenshine DE (1991) 'Biology of Ticks'. Vol. 1. (Oxford University Press: New York).

van der Hammen L (1989) 'An Introduction to Comparative Arachnology'. (SPB Academic Publishing: The Hague).

Walter DE, Proctor HC (1999) 'Mites: Ecology, Evolution and Behaviour'. (University of New South Wales Press: Sydney).

Wikelski M (1999) Influences of parasites and thermoregulation on grouping tendencies in marine iguanas. *Behavioral Ecology* **10**, 22–29.

Pycnogonida (pycnogonids)

Arango CP (2001) Sea spiders (Pycnogonida) from the Great Barrier Reef, Australia, feed on fire corals and zoanthids. *Memoirs of the Queensland Museum* **46**, 656.

Arnaud F, Bamber RN (1987) The biology of Pycnogonida. *Advances in Marine Biology* **24**, 1–96.

Benson PHC, Chivers DC (1960) A pycnogonid infestation of *Mytilus californicus*. *The Veliger* **3**, 16–18.

Cadien DB (1997) The Arthropoda–The Pycnogonida. *In* 'Taxonomic Atlas of the Benthic Fauna of the Santa Maria Basin and Western Santa Barbara Channel'. Vol. 10. (Ed. JA Blake, PH Scott.) pp. 1–47. (Santa Barbara Museum of Natural History: Santa Barbara).

Child CA (1988) *Nymphon tortulum*, new species and other Pycnogonida associated with the coral *Oculina varicosa* on the East Coast of Florida. *Bulletin of Marine Science* **63**(3), 595–604.

Dohrn A (1881) Die Pantopoden des Golfes von Naepel und der angrenzenden Meeresabschnitte. *Monographie der Fauna und Flora des Golfes von Neapel* **3**, 1–252.

Loman JCC (1907) Biologische Beobachtungen an einem Pantopoden. *Tidschrift van die Nederlandse dierkundige Vereeninging* **10**, 255–284.

Losina-Losinsky LA (1933) Die Pantopoden der östlichen Meere der U.S.S.R. *Issledovaniya Morie SSSR Leningrad* **17**, 43–80.

Moseley HN (1879) On the structure of the Stylasteridea, a family of the hydroid stony corals. *Philosophical Transactions of the Royal Society London,* **169**, 425–504. Miyazaki K 2002 Occurrence of juvenile forms of a pycnogonid, *Ammothella biunguiculata* (Pycnogonida, Ammotheidae) in an Actinian, *Entacmaea actinostoloides* (Anthozoa, Stichodactylidae). *Proceedings Arthropodan Embryological Society of Japan* **37**, 43–44.

Ohshima H (1927) Notes on some pycnogons living semiparasitic on holothurians. *Proceedings of the Imperial Academy, Tokyo* **3**(9), 610–613.

Ohshima H (1933) Young pycnogonids found parasitic on nudibranchs. *Annotations Zoologicae Japanonenses* **14**, 61–66.

Okuda S (1940) Metamorphosis of a pycnogonid parasitic in a hydromedusa. *Journal of the Faculty of Science, Hokkaido Imperial University, (Zool.* 6) **7**, 73–86.

Ryland JS (1976) Physiology and ecology of marine bryozoans. *Advances in Marine Biology* **14**, 417–421.

Salazar-Vallejo SI, Stock JH (1987) Apparent parasitism of *Sabella melanostigma* (Polychaeta) by *Ammothella spinifera* (Pycnogonida) from the Gulf of California. *Revistas de Biologia Tropical* **35**, 269–275.

Sloan NA (1979) A pycnogonid-ophiuroid association. *Marine Biology* **52**, 171–176.

Staples DA, Watson JE (1987) Associations between pycnogonids and hydroids. *In* 'Modern Trends in the Systematics, Ecology and Evolution of Hydroids and Hydromedusae'. (Eds J Bouillion, J Oxford.) pp. 215–226. (Oxford University Press: Clarendon).

Stock JH (1953) Biological results of the 'Snellius' Expedition. XVII. Contribution to the knowledge of the pycnogonid fauna of the East Indian Archipelago. *Temminckia* **9**, 276–313.

Stock JH (1978) Experiments on food preferences and chemical sense in Pycnogonida. *Zoological Journal of the Linnean Society (London)* **63**, 59–74.

Wyer DW, King PE (1973) Relationships between some British littoral and sublittoral bryozoans and pycnogonids. 18. *In* 'Living and Fossil Bryozoa: Recent Advances in Research'. (Ed. GP Larwood.) pp 199–207. (Academic Press: London).

Insecta (insects)

Anderson DT, Fletcher MJ, Lawson-Kerr C (1976) A marine caddis fly, *Philanisus plebeius*, ovipositing in a starfish, *Patiriella exigua*. *Search* **7**, 483–484.

Anderson DT, Lawson-Kerr C (1977) The embryonic development of the marine caddis fly, *Philanisus plebeius* Walker (Trichoptera: Chathamiidae). *Biological Bulletin* **153**, 98–105.

Banks JC, Palma RL (2003) A new species and new host records of *Austrogoniodes* (Insecta: Phthiraptera: Philopteridae) from penguins (Aves: Sphenisciformes). *New Zealand Journal of Zoology* **30**, 69–75.

Banks JC, Paterson AM (2004) A penguin-chewing louse (Insecta: Phthiraptera) phylogeny derived from morphology. *Invertebrate Systematics* **18**, 89–100.

Barker SC, Whiting M, Johnson KP, Murrell A (2003) Phylogeny of the lice (Insecta, Phthiraptera) inferred from small subunit rDNA. *Zoological Scripta* **32**, 407–414.

Bell PJ, Burton HR, van Franeker JA (1988) Aspects of the biology of *Glaciopsyllus antarcticus* (Siphonaptera: Ceratophyllidae) during the breeding season of a host (*Fulmarus glacialoides*). *Polar Biology* **8**, 403–410.

Caira JN, Jensen K, Holsinger KE (2003) On a new index of host specificity. *In* 'Taxonomie, écologie et évolution des Métazoaires Parasites. (Livre Hommage à Louis Euzet)'. Tome I. (Eds I Combes, J Jourdane.) pp. 161–201. (PUP, Perpignan: France).

Durden LA, Musser GG (1994) The sucking lice (Insecta, Anoplura) of the world: a taxonomic checklist with records of mammalian hosts and geographical distributions. *Bulletin of the American Museum of Natural History* **218**, 1–90.

Geraci JR, Lounsbury VJ (2002) Marine mammal health: holding the balance in an ever-changing sea. *In* 'Marine Mammals: Biology and Conservation'. (Eds PGH Evans, JA Raga.) pp. 365–383. (Kluwer Academic/Plenum Publishers: New York).

Hoberg EP, Wehle DHS (1982) Host and geographic records of ectoparasites from Alaskan seabirds (Charadriiformes: Alcidae and Laridae). *Canadian Journal of Zoology* **60**, 472–475.

Holland GP (1985) The fleas of Canada, Alaska and Greenland (Siphonaptera). *Memoirs of the Entomological Society of Canada* **130**, 1–631.

Johnson KP, Clayton DH (2003) The biology, ecology, and evolution of chewing lice. *In* 'The Chewing Lice: World Checklist and Biological Overview'. (Eds RD Price, RA Hellenthal, RL Palma, KP Johnson, DH Clayton.) pp. 449–476. (Illinois Natural History Survey Special Publication 24: State of Illinois, USA).

Johnson PT (1957) A classification of the Siphonaptera of South America with descriptions of new species. *Memoirs of the Entomological Society of Washington* **5**, 1–299.

Kim KC (Ed.) (1985) 'Coevolution of Parasitic Arthropods and Mammals'. (John Wiley and Sons: New York).

Lewis RE (1998) Résumé of the Siphonaptera (Insecta) of the world. *Journal of Medical Entomology* **35**, 377–389.

Marshall AG (1981) 'The Ecology of Ectoparasitic Insects'. (Academic Press: London).

Mehlhorn B, Mehlhorn H, Plötz J (2002) Light and scanning electron microscopical study on *Antarctophthirus ogmorhini* lice from the Antarctic seal *Leptonychotes weddellii*. *Parasitology Research* **88**, 651–660.

Merino S, Mínguez E, Belliure B (1999) Ectoparasite effects on nestling European storm-petrels. *Waterbirds* **22**, 297–301.

Mey E, Chastel O, Beaucournu J-C (2002) A 'penguin' chewing louse *Nesiotinus* on a Kerguelen diving-petrel (*Pelecanoides urinatrix exsul*): an indication of a phylogenetic relationship? *Journal für Ornithologie* **143**, 472–476.

Murray MD (1976) Insect parasites of marine birds and mammals. *In* 'Marine Insects'. (Ed. L Cheng.) pp. 79–96. (North-Holland Publishing Company: Amsterdam).

Page RDM, Cruickshank RH, Dickens M, Furness RW, Kennedy M, Palma RL, Smith VC (2004) Phylogeny of '*Philoceanus* complex' seabird lice (Phthiraptera: Ischnocera) inferred from mitochondrial DNA sequences. *Molecular Phylogenetics and Evolution* **30**, 633–652.

Palma RL, Pilgrim RLC (2002) A revision of the genus *Naubates* (Insecta: Phthiraptera: Philopteridae). *Journal of the Royal Society of New Zealand* **32**, 7–60.

Paterson AM, Banks J (2001) Analytical approaches to measuring cospeciation of host and parasites: through a glass, darkly. *International Journal for Parasitology* **31**, 1012–1022.

Paterson AM, Wallis GP, Wallis LJ, Gray RD (2000) Seabird and louse coevolution: complex histories revealed by 12S rRNA sequences and reconciliation analyses. *Systematic Biology* **49**, 383–399.

Price RD, Hellenthal RA, Palma RL (2003) World checklist of chewing lice with host associations and keys to families and genera. *In* 'The Chewing Lice: World Checklist and Biological Overview'. (Eds RD Price, RA Hellenthal, RL Palma, KP Johnson, DH Clayton.) pp. 1–448. (Illinois Natural History Survey Special Publication 24: State of Illinois, USA).

Smit FGAM (1979) The fleas of New Zealand. *Journal of the Royal Society of New Zealand* **9**, 143–232.

Smit FGAM (1984) *Parapsyllus magellanicus largificus*, a new flea from the Bounty Islands. *New Zealand Journal of Zoology* **11**, 13–16.

Thompson PM, Corpe HM, Reid RJ (1998) Prevalence and intensity of the ectoparasite *Echinophthirius horridus* on harbour seals (*Phoca vitulina*): effects of host age and interannual variability in host food availability. *Parasitology* **117**, 393–403.

Traub R, Rothschild M, Haddow JF (1983) 'The Rothschild Collection of Fleas. The Ceratophyllidae: Key to the Genera and Host Relationships with Notes on their Evolution, Zoogeography and Medical Importance'. (Miriam Rothschild & Robert Traub: England).

Winterbourn MJ, Anderson NH (1980) The life history of *Philanisus plebeius* Walker (Trichoptera: Chathamiidae), a caddisfly whose eggs were found in starfish. *Ecological Entomology* **5**, 293–303.

Tardigrada (water bears)

Aguinaldo AM, Turbeville JM, Linford LS, Rivera MC, Garey JR, Raff RA, Lake JA (1997) Evidence for a clade of nematodes, arthropods and other moulting animals. *Nature* **387**, 489–493.

Bertolani R, Grimaldi D (2000) A new eutardigrade (Tardigrada: Milnesiidae) in amber from the Upper Cretaceous (Turonian) of New Jersey. *In* 'Studies on Fossils in Amber with Particular Reference to the Cretaceous of New Jersey'. (Ed. D Grimaldi.) pp. 159–167. (Backhuys Publishers: Leiden, Netherlands).

Cantacuzène A (1951) Tardigrade marine noveau, commensal de *Limnoria lignorum* (Rathke). *Compte Rendu de l'Academie des Sciences (Paris)* **232**, 1699–1700.

Cuénot L (1892) Commensaux et parasites des Echinodermes (I). *Revue de Biologie, Nord de la France* **5**, 1–22.

Cuénot L (1932) Tardigrades. *Faune de France* **24**, 1–96.

Cooper KW (1964) The first fossil Tardigrade: *Beorn leggi* Cooper, from Cretaceous amber. *Psyche* **71**, 41–48.

Garey JR, Nelson DR, Mackey LY, Li J (1999) Tardigrade phylogeny: congruency of morphological and molecular evidence. *Zoologischer Anzeiger* **238**, 205–210.

Giribet G, Carranza S, Baguña J, Riutort M, Ribera C (1996) First molecular evidence for the existence of a Tardigrada + Arthropoda clade. *Molecular Biology and Evolution* **13**, 76–84.

Green J (1950) Habitats of the marine tardigrade *Echiniscoides*. *Nature* **166**, 153–154.

Grell KG (1937) Beträge zur Kentnis von *Actinarctus doryphorus* E. Schulz nebst Bemerknungen zur Tardigraden fauna des Helgoländer Skitt Gatts. *Zoologisher Anzeiger* **117**, 143–154.

Kristensen RM (1980) Zur Biologie des marinen Heterotardigraden *Tetrakentron synaptae. Helgoländer Wissenschaftliche Meeresuntersuchungen* **34**,165–177.

Kristensen RM, Hallas TE (1980) The tidal genus *Echiniscoides* and its variability, with erection of Echiniscoididae fam.n. (Tardigrada). *Zoologica Scripta* **9**, 113–127.

Maas A, Waloszek D (2001) Cambrian derivates of the early arthropod stem lineage, pentastomids, tardigrades and lobopodians – An 'Orsten' perspective. *Zoologischer Anzeiger* **240**, 451–459.

Marcus E (1929) Tardigrada. *In* 'Klassen und Ordnungen des Tier-Reichs'. Vol. 5, Section 4, Part 3 (Ed. HG Bronn.) pp. 1–608. (Akademische Verlagsgesellschaft: Leipzig).

Nielsen C (2001) 'Animal Evolution. Interrelationships of the Living Phyla'. 2nd edn. (Oxford University Press: New York).

Ramazzotti G, Maucci W (1983) Il Phylum Tardigrada. Terza edizione riveduta e corretta. *Memorie dell'Istituto Italiano di Idrobiologia Dott. Marco de Marchi* **41**, 1–1012.

Schultz E (1935) *Actinarctus doryphorus* nov. gen .nov. spec. ein merkwürdiger Tardigrad aus der Nordsee. *Zoologisher Anzeiger* **111**, 285–288.

Van der Land J (1975) The parasitic marine tardigrade *Tetrakentron synaptae*. *In* 'International Symposium on Tardigrades, Pallanza, Italy, June 17–19, 1974'. (Ed. RP Higgins.) *Memorie dell'Istituto Italiano di Idrobiologia Dott. Marco de Marchi* **32** (Supplement), 413–423.

Pentastomida (tongue worms)

Abele LG, Kim W, Felgenhauer BE (1989) Molecular evidence for inclusion of the phylum Pentastomida in the Crustacea. *Moecular Biology and Evolution* **6**, 685–691.

Banaja AA, James JL, Riley J (1975) An experimental investigation of a direct life-cycle in *Reighardia sternae*, (Diesing, 1964), a pentastomid parasite of the herring gull (*Larus argentatus*). *Parasitology* **71**, 493–503.

Banaja AA, James JL, Riley J (1976) Some observations on egg production and autoreinfection of *Reighardia sternae* (Diesing, 1864), a pentastomid parasite of the herring gull. *Parasitology* **72**, 81–91.

Böckeler W (1984a) Embryogenese und ZNS-Differenzierung bei *Reighardia sternae*. Licht- und elektronen-mikroskopische Untersuchungen zur Tagmosis und systematischen Stellung der Pentatsomidenl̈. *Zoologische Jahrbücher, Anatomie* **111**, 297–342.

Böckeler W (1984b) Der Entwicklungszyklus von *Reighardia sternae* (Pentastomida) nach Untersuchungen an natürlich und experimentell infizierten Möwen. *Zoologischer Anzeiger* **213**, 374–394.

Böckeler W, Vauk-Hentzelt E (1979) Die Mantelmöwe (*Larus marinus*) als neuer Wirt des Luftsackparasiten *Reighardia sternae* (Pentastomida). *Zoologischer Anzeiger* **203**, 95–98.

Dyck J (1975) *Reighardia lomviae sp.nov.*, a new pentastomid from guillemot. *Norwegian Journal of Zoology* **23**, 97–109.

Haffner K von, Rack G (1965) Neues über die Entwicklung der Pentastomide *Reighardia sternae* (Diesing, 1864). *Zoologische Jahrbücher, Anatomie und Ontogenie* **72**, 419–444.

Riley J (1972) Some observations on the life-cycle of *Reighardia sternae* Diesing, 1864. *Zeitschrift für Parasitenkunde* **40**, 49–59.

Riley J (1992) Pentastomids and the immune response. *Parasitology Today* **8/4**, 133–137.

Riley J, James JL, Abanaja A (1979) The possible role of the frontal and subparietal gland systems of the pentastomid *Reighardia sternae* (Diesing, 1864) in the evasion of the immune response. *Parasitology* **78**, 53–66.

Riley J, Oaks JL, Gilbert M (2003) *Raillietiella trachea* n.sp. of an oriental white-backed vulture *Gyps ben-galensis* taken in Pakistan, with speculation about its life-cycle. *Systematic Parasitology* **56**, 155–161.

Stender-Seidel S, Thomas G, Böckeler W. (2000) Investigation of different ontogenetic stages of Raillietiella sp. (Pentastomida:Cephalobaenida):suboral and frontal gland. *Parasitology Research* **86**, 385–400.

Storch V (1993) Pentastomida. *In* 'Microscopic Anatomy of the Invertebrates'. Vol. 12, pp. 115–142. (John Wiley: New York).

Storch V, Böckeler W (1979) Electronmicroscopic observations on the sensilla of the pentastomid *Reighardia sternae* (Diesing, 1864). *Zeitschrift für Parasitenkunde* **60**, 77–86.

Storch V, Jamieson BGM (1992) Further spermatological evidence for including the Pentastomida (tongue worms) in the Crustacea. *International Journal for Parasitology* **22**, 95–108.

Vauk-Hentzelt E (1987) Das Vorkommen von Reighardia (Pentastomida) in acht Seevogelarten im Zusammenhang mit der Verbreitung der Nahrungsökologie der Wirtstiereĺ. *Dissertation, Hamburg*.

Vauk-Hentzelt E, Schumann K (1980) Zur Winterernährung durchziehender und rastender Sturmmöwen (*Larus canus*) aus dem Bereich der Insel Helgoland. *Angewandte Ornithologie* **5**, 178–184.

Walossek D, Möller KJ (1994): Pentastomid parasites from the Lower Paleozoic of Sweden . *Transactions of the Royal Society of Edinburgh-Earth Sciences* **85**, 41–37.

Walossek D, Repetski JE, Möller KJ (1994) An exceptionally preserved parasitic arthropod, *Heymonsi-cambria taylori n. sp.* (Arthropoda incertae sedis, Pentastomida), from Cambrian-Ordovician boundary beds of Newfoundland, Canada. *Canadian Journal of Earth Sciences* **31**, 1664–1671.

Wingstrand KG (1972) Comparative sprematology of a pentastomid, *Raillietiella hemidactyli*, and a branchi-uran crustacean, *Argulus foliaceus*, with a discussion of pentastomid relationship. *Kong Dansk. Videnskab Selskab. Biol. Skrifter* **19**, 1–72.

Zrzavy J (2001) The interrelationships of metazoan parasites: a review of phylum- and higher-level hypotheses from recent morphological and molecular phylogenetic analyses. *Folia Parasitologica* **48**, 81–103.

Mollusca (molluscs)

Bouchet P (1989) A marginellid gastropod parasitizes sleeping fishes. *Bulletin of Marine Sciences* **45**, 76–84.

Bouchet P, Perrine D (1996) More gastropods feeding at night on parrotfishes. *Bulletin of Marine Science* **59**, 224–228.

Egloff DA, Smouse DT, Pembroke JE (1988) Penetration of the radial Hemal and Perihemal systems of *Linckia laevigata* (Asteroidea) by the proboscis of *Thyca crystallina*, and ectoparasitic gastropod. *The Veliger* **30**(4), 342–346.

Elder HY (1979) Studies on the host parasite relationship between the parasitic prosobranch *Thyca crystallina* and the asteroid starfish *Linckia laevigata*. *Journal of Zoology* **187**, 369–391.

Fretter V (1946) The genital ducts of *Theodoxus*, *Lamellaria* and *Trivia*, and a discussion on their evolution in the prosobranchs. *Journal of Marine Biology* **26**, 312–351.

Fretter V (1951) *Turbonilla elegantissima* (Montagu), a parasitic opisthobranch. *Journal of the Marine Biological Association of the United Kingdom* **30**, 37–47.

Fretter V, Graham A (1949) The structure and mode of life of the Pyramidellidae, parasitic opisthobranchs. *Journal of the Marine Biological Association of the United Kingdom* **28**, 493–532.

Fretter V, Graham A (1997) 'British Prosobranch Molluscs'. (Ray Society: London).
Humphreys WF, Lützen J (1972) Studies on parasitic gastropods from echinoderms. *Biologiske Skrifter* **19**(1), 1–26.
Johnson S, Jazwinski S (1995) Parasitism of sleeping fish by gastropod molluscs in the Colubrariidae and Marginellidae at Kwajalein, Marshall Islands. *The Festivus* **27**, 121–125.
Koehler R, Vaney C (1903) *Entosiphon deimatis* Nouveau Mollusque parasite d'une Holothurie abyssale. *Revue Suisse de Zoologie, Genève* **11**, 23–41.
Liltved WR (1989) 'Cowries and their Relatives of Southern Africa – A Study of the Southern African Cypraeacean and Velutinacean Gastropod Fauna'. (Gordon Verhoef: Cape Town).
Liltved WR, Gosliner TM (1987) Further studies on the morphology of the Triviidae (Gastropoda, Prosobranchia) with emphasis on species from southern Africa. *Zoological Journal of the Linnean Society* **90**, 207–254.
Lorenz F (1996) The habitat of some coralliophilid species from Tanzania. *La Conchiglia* **278**, 22–24.
Lorenz F (2001) 'Monograph of the Living *Zoila* – A Fascinating Group of Australian Endemic Cowries (Mollusca, Gastropoda, Cypraeidae)'. (Conchbooks: Hackenheim).
Lorenz F, Hubert A (1993) 'A Guide to Worldwide Cowries'. (Christa Hemmen Verlag: Wiesbaden).
Lützen J (1972a) Records of parasitic gastropods from crinoids, with description of a new genus, *Goodingia* (Gastropoda, Prosobranchia). *Steenstrupia* **2**, 233–246.
Lützen J (1972b) Studies on parasitic gastropods from echinoderms II. *Biologiske Skrifter* **19**, 1–18.
Lützen J (1979) Studies on the life history of *Enteroxenos bonnevie*, a gastropod endoparasitic in aspidochirote holothurians. *Ophelia* **18**, 1–51.
Lützen J, Nielsen K (1975) Contributions to the anatomy and biology of *Echineulima* n.g. (Prosobranchia, Eulimidae) parasitic on sea urchins *Videnskabelige Meddelelse fra den Danske Naturhistoriske Forening* **138**, 171–199.
Mandahl-Barth G (1941) *Thyonicola mortenseni* n. gen, n. sp., eine neue parasitische Schnecke. *Videnskabelige Meddelelse fra den Danske Naturhistoriske Forening* **104**, 341–351.
Mandahl-Barth G (1946) *Diacolax cucumariae* n.gen., n.sp. A new parasitic snail. *Videnskabelige Meddelelse fra den Danske Naturhistoriske Forening* **109**, 55–68.
Morton B (1979) The population dynamics and expression of sexuality in *Balcis shaplandi* and *Mucronalia fulvescens* (Mollusca, Gastropoda, Aglossa) parasitic upon *Archaster typicus* (Echinodermata, Asteroidea) *Malacologia* **18**, 327–346.
O'Sullivan JB, McConnaughey RR, Huber ME (1987) A blood-sucking snail, Cooper's nutmeg, *Cancellaria cooperi* Gabb, parasitizes the California electric ray, *Torpedo californica* Ayres. *Biological Bulletin* **172**, 362–366.
Robertson R (1970) Review of the predators and parasites of stony corals, with special reference to symbiotic prosobranch gastropods. *Pacific Science* **24**, 43–54.
Robertson R (1981) *Epitonium millecostatum* and *Coralliophila clathrata*, two prosobranch gastropods symbiotic with Indo-Pacific *Palythoa* (Coelenterata, Zoanthidae). *Pacific Science* **34**, 1–17.
Robertson R, Mau-Lastrovicka T (1979) The ectoparasitism of *Boonea* and *Fargoa* (Gastropoda, Pyramidellidae). *Biological Bulletin* **157**, 320–333.
Rosenberg G (1993) 'The Encyclopedia of Seashells'. (Robert Hale: London).
Tikasingh ES. (1961) A new genus and two new species of endoparasitic gastropods from Puget Sound, Washington. *Journal of Parasitology* **47**, 268–272.
Vaney C (1913) L'Adaptation des Gastropodes au Parasitisme. *Bulletin Scientifique de la France et de la Belgique* **97**, 1–87.

Voigt W (1901) *Entocolax schiemenzii* n. sp. *Zoologischer Anzeiger* **24**, 285–292.
Warén A (1981) Eulimid gastropods parasitic on echinoderms in the New Zealand region. *New Zealand Journal of Zoology* **8**, 313–324.
Warén A (1983) A generic revision of the family Eulimidae (Gastropoda, Prosobranchia). *The Journal of Molluscan Studies 1983 - Supplement* **13**, 1–96.
Waren A, Crossland RM (1975) Revision of *Hypermastus* Pilsbry, 1899 and *Turveria* Berry, 1956 (Gastropoda, Prosobranchia, Eulimidae), two genera parasitic on sand dollars. *Records of the Australian Museum* **43**, 85–112.
Wise JB (1993) Anatomy and functional morphology of the feeding structures of the ectoparasitic gastropod *Boonea impressa* (Pyramidellidae). *Malacologia* **35**, 119–134.

Echiura (spoon worms)

Baltzer F (1931a) Echiurida. *In* 'Handbuch der Zoologie'. (Ed. T Krumbach.) pp. 1–160. (Walter de Gruyter & Co.: Berlin).
Baltzer F (1931b) Entwicklungsmechanische Untersuchungen an *Bonellia viridis*. *Revue Suisse de Zoologie* **38**, 361–371.
Brusca RC, Brusca GJ (2003) 'Invertebrates'. (Sinauer Associates: Sunderland, Massachusetts).
Dawydoff C (1959) Classe des echiuriens (Echiurida de Blainville *Gephyrea armata* de Quatrefages 1847). *In* 'Traité de Zoologie. Anatomie, Systématique, Biologie'. (Ed. P-P Grassé.) pp. 855–907. (Masson et Cie: Paris).
Edmonds SJ (2000) Phylum Echiura. *In* 'Polychaetes and Allies'. Volume 4A. 'Fauna of Australia'. (Eds PL Beesley, GJB Ross, CJ Glasby.) pp. 353–374. (CSIRO Publishing/Australian Biological Resource Study: Melbourne).
Hessling R, Westheide W (2002) Are Echiura derived from a segmented ancestor? Immunohistochemical analysis of the nervous system in development of *Bonellia viridis*. *Journal of Morphology* **252**, 100–113.
Jaccarini V, Agius L, Schembri PJ, Rizzo M (1983) Sex determination and larval sexual interaction in *Bonellia viridis* Rolando (Echiura: Bonelliidae). *Journal of Experimental Marine Biology and Ecology* **66**, 25–40.
Murina V (1998) 'Phylum Echiura Stephen, 1965 (Systematics, Ecology and Distribution).' http://www.ibss.iuf.net/people/murina/echiura.html.
Schuchert P, Rieger RM (1990) Ultrastructural observations on the dwarf male of *Bonellia viridis* (Echiura). *Acta Zoologica* **71**, 5–16.

Echinodermata (echinoderms)

Baker AN, Rowe FWE, Clark HES (1986) A new class of Echinodermata from New Zealand. *Nature* **321**, 862–864.
Hyman LH (1955) 'The Invertebrates: Echinodermata'. Vol. IV. (McGraw-Hill Book Company: New York).
Janies D (2001) Phylogenetic relationships of extant echinoderm classes. *Canadian Journal of Zoology* **79**, 1232–1250.
Marsh LM (1998) Hitch-hiking ophiuroids. *In* 'Echinoderms: San Francisco. Proceedings of the Ninth International Echinoderm Conference, San Francisco, California, USA, 5–9 August 1996'. (Eds R Mooi, M Telford.) pp. 393–396. (A.A. Balkema: Rotterdam).
Martin WE (1969) *Rynkatorpa pawsoni* n. sp. (Echinodermata: Holothuroidea) a commensal sea cucumber. *Biological Bulletin. Marine Biological Laboratory, Woods Hole, Mass.* **137**, 332–337.
McKnight DG (2000) The marine fauna of New Zealand: basket-stars and snake-stars (Echinodermata: Ophiuroidea: Euryalinida). *NIWA Biodiversity Memoir* 1–79.

Pannikkar N, Prassad R (1955) Association of ophiuroids, fish and crabs with *Rhopilema*. *Journal of the Bombay Natural History Society* **51**, 295–296.

Rowe FWE, Gates J (1995) 'Echinodermata'. Vol. 33. 'Zoological Catalogue of Australia'. (CSIRO Publishing: Melbourne).

Parasitic marine fishes

Allen GR (1981) 'Butterfly and Angelfishes of the World'. Vol. 2. English Language Edition. (Mergus: Melle).

Arnal C, Coté IM (2000) Diet of broadstripe cleaning gobies on a Barbadian reef. *Journal of Fish Biology* **57**, 1075–1082.

Arnal C, Morand S (2001) Importance of ectoparasites and mucus in cleaning interactions in the Mediterranean cleaner wrasse *Symphodus melanocercus*. *Marine Biology* **138**, 777–778.

Baxter EW (1956) Observations on the buccal glands of lampreys (Petromyzontidae). *Proceedings of the Zoological Society, London* 127, 95–118.

Bertelsen E (1951) 'The Ceratioid Fishes: Ontogeny, Taxonomy, Distribution and Biology'. Dana Report, No. 39, 1–276. (Andr. Fred. Høst & Søn: Copenhagen).

Birkeland C, Neudecker S (1981) Foraging behavior of two Caribbean chaetodontids: *Chaetodon capistratus* and *C. aculeatus*. *Copeia* **1981**, 169–178.

Coggan RA, Gordon JDM, Merrett NR (1998) Abundance, distribution, reproduction and diet of notacanthid fishes from the north-east Atlantic. *Journal of Fish Biology* **52**, 1038–1057.

Compagno LJV (1984) 'FAO Species Catalogue, Vol. 4. Sharks of the World: An Annotated and Illustrated Catalogue of Shark Species Known to Date. Part 1. Hexanchiformes to Lamniformes'. FAO Fisheries Synopsis, No. **125**, 1–249. (FAO: Rome).

Compagno LJV (2000) Squaliformes: Dalatiidae (sleeper sharks and allies). *In* 'A Checklist of Fishes of the South China Seas'. (Eds JE Randall, KKP Lim.) *Raffles Bulletin of Zoology* Suppl. No. **8**, 580.

Compagno LJV (2001) 'Sharks of the World: An Annotated and Illustrated Catalogue of Shark Species Known to Date'. Vol. 2. 'Bullhead, Mackerel and Carpetsharks (Heterodontiformes, Lamniformes and Orectolobiformes)'. FAO species catalogue for fishery purposes, No. 1, Vol. 2, 1–269. (FAO: Rome).

Gartner JV, Crabtree RE, Sulak KJ (1997) Feeding at depth. *In* 'Deep-Sea Fishes'. (Eds DJ Randall, AP Farrell.) pp. 115–193. (Academic Press: Orlando).

Gill HS, Renaud CB, Chapleau F, Mayden RL, Potter IC (2003) Phylogeny of living parasitic lampreys (Petromyzontiformes) based on morphological data. *Copeia* **2003**, 687–703.

Gorlick DL (1980) Ingestion of host fish surface mucus by the Hawaiian cleaning wrasse, *Labroides phthirophagus* (Labridae) and its effect on host species preference. *Copeia* **1980**, 863–868.

Hardisty MW, Potter IC (1971) The general biology of adult lampreys. *In* 'The Biology of Lampreys'. (Eds MW Hardisty, IC Potter.) Vol. 1, pp. 127–206. (Academic Press: London).

Hiatt RW, Strasburg DW (1960) Ecological relationships of the fish fauna on coral reefs of the Marshall Islands. *Ecological Monographs* **30**, 65–127.

Hoese HD (1966) Ectoparasitism by juvenile sea catfish, *Galeichthys felis*. *Copeia* **1966**, 880–881.

Hubbs CL, Potter IC (1971) Distribution, phylogeny and taxonomy. *In* 'The Biology of Lampreys'. Vol. 1. (Eds MW Hardisty, IC Potter.) pp. 1–65. (Academic Press: London).

Hyman LH (1955) 'The Invertebrates: Echinodermata, the Coelomate Bilateria'. Vol. IV. (McGraw-Hill: New York).

Jones EC (1971) *Isistius brasiliensis*, a squaloid shark, the probable cause of crater wounds on fishes and cetaceans. *Fishery Bulletin, NOAA/NMFS* **69**, 791–798.

Lethbridge RC, Potter IC (1979) The oral fimbriae of the lamprey, *Geotria australis*. *Journal of Zoology, London* **188**, 267–277.

Lutz B (1930) The effect of low oxygen tension on the pulsations of the isolated holothurian cloaca. *Biological Bulletin* **58**, 74–84.

Major PF (1973) Scale feeding behavior of the leatherjacket, *Scomberoides lysan*, and two species of the genus *Oligoplites* (Pisces: Carangidae). *Copeia* **1973**, 151–154.

Marshall NB (1979) 'Developments in Deep Sea Biology'. (Blandford Press: Poole).

McDowell SB (1973) Order Heteromi (Notacanthiformes). *In* 'Fishes of the Western North Atlantic'. Part 6. (Ed. DM Cohen.) pp. 1–228. (Sears Foundation for Marine Research: New Haven).

Mok HK (1978) Scale-feeding in *Tydemania navigatoris* (Pisces: Triacanthodidae). *Copeia* **1978**, 338–340.

Munk O (2000) Histology of the fusion area between the parasitic male and the female in the deep-sea anglerfish *Neoceratias spinifer* Pappenheim, 1914 (Teleostei, Ceratioidei). *Acta Zoologica (Stockholm)* **81**, 315–324.

Murdy EO, Cowan ME (1980) Observations on the behavior and symbiotic relationship of the pearlfish *Encheliophis vermicularis* (Osteichthys: Carapidae). *Kalikasan, Philippine Journal of Biology* **9**, 309–312.

Myers RF (1991) 'Micronesian Reef Fishes'. 2nd edn. (Coral Graphics: Barricada, Guam).

Nakae M, Sasaki K (2002) A scale-eating triacanthodid, *Macrorhamphosodes uradoi*: prey fishes and mouth 'handedness' (Tetrodontiformes, Triacanthoidei). *Ichthyological Research* **49**, 7–14.

Nichols OC, Hamilton PK (2004) Occurrence of the parasitic sea lamprey, *Petromyzon marinus*, on western North Atlantic right whales, *Eubalaena glacialis*. *Environmental Biology of Fishes* **71**, 413–417.

Nielsen JG, Cohen DM, Markle DF, Robins CR (1999) 'FAO Species Catalogue. Vol. 18. Ophidiiform Fishes of the World. (Order Ophidiiformes). An Annotated and Illustrated Catalogue of Pearlfishes, Cusk-eels, Brotulas and other Ophidiiform Fishes known to Date'. FAO Fisheries Synopsis No. 125, Vol. 18, 1–178. (FAO: Rome).

Nikol'skii GV (1956) [Some data on the period of marine life of the Pacific lamprey *Lampetra japonica*.] *Zoologicheskij Zhurnal* **35**, 585–591. [In Russian].

Norman JR, Greenwood PH (1975) 'A History of Fishes'. 3rd Edn. (Ernest Benn: London).

Parmentier E (2004) *Encheliophis chardewalli*: a new species of Carapidae (Ophidiiformes) from French Polynesia, with a redescription of *Encheliophis vermicularis*. *Copeia* **2004**, 62–67.

Parmentier E, Castillo G, Chardon M, Vandewalle P (2000) Phylogenetic analysis of the pearlfish tribe Carapini (Pisces: Carapidae). *Acta Zoologica (Stockholm)* **81**, 293–306.

Pietsch TW (1976) Dimorphism, parasitism and sex: reproductive strategies among deepsea ceratioid anglerfishes. *Copeia* **1976**, 781–793.

Pietsch TW (1999) Families Caulophrynidae, Neoceratiidae, Ceratiidae, Linophrynidae. *In* 'FAO Species Identification Guide for Fishery Purposes. The Living Marine Resources of the Western Central Pacific'. Vol. 3. 'Batoid Fishes, Chimaeras and Bony Fishes. Part 1 (Elopidae to Linophrynidae)'. pp. 2026–2037. (FAO: Rome).

Pietsch TW (2005) Dimorphism, parasitism, and sex revisited: modes of reproduction among deep-sea ceratioid anglerfishes (Teleostei: Lophiiformes). *Ichthyological Research* **52**, 207–236.

Pike GC (1951) Lamprey marks on whales. *Journal of the Fisheries Research Board of Canada* **8**, 275–280.

Potter IC, Gill HS (2003) Adaptive radiation of lampreys. *Journal of Great Lakes Research* **29**(Suppl. 1), 95–112.

Potter IC, Hilliard RW (1987) A proposal for the functional and phylogenetic significance of differences in the dentition of lampreys (Agnatha: Petromyzontiformes). *Journal of Zoology, London* **212**, 713–737.

Potter IC, Hilliard RW, Neira FJ (1986) The biology of Australian lampreys. *In* 'Limnology in Australia'. (Eds P de Deckker, WD Williams.). pp. 207–230. (CSIRO: Melbourne; W Junk: Dordrecht).

Randall DJ (1972) Respiration. *In* 'The Biology of Lampreys'. Vol. 2. (Eds MW Hardisty, IC Potter.) pp. 287–306. (Academic Press: London).

Randall JE (1974) The effects of fishes on coral reefs. *In* 'Proceedings of the Second International Coral Reef Symposium'. Vol. 1. pp. 159–166. (Great Barrier Reef Committee: Brisbane).

Randall JE (1981) Revision of the labrid fish genus *Labropsis* with descriptions of five new species. *Micronesica* **17**, 125–155.

Randall JE, Helfman G (1972) *Diproctacanthus xanthurus*, a cleaner wrasse from the Palau Islands, with notes on other cleaning fishes. *Tropical Fish Hobbyist* **20** (11), 87–95.

Randall JE, Randall HA (1960) Examples of mimicry and protective resemblance in tropical marine fishes. *Bulletin of Marine Science of the Gulf and Caribbean* **10**, 444–480.

Randall JE, Allen GR, Steene RC (1997) 'Fishes of the Great Barrier Reef and the Coral Sea'. Revised Edn. (Crawford House Publishing: Bathurst).

Reese ES (1975) A comparative field study of the social behavior and related ecology of reef fishes of the family Chaetodontidae. *Zeitschrift für Tierpsychologie* **37**, 37–61.

Roberts TDM (1950) The respiratory movements of the lamprey (*Lampetra fluviatilis*). *Proceedings of the Royal Society of Edinburgh*, ser. B, **64**, 235–251.

Rogers PA, Glenn AR, Potter IC (1980) The bacterial flora of the gut contents and environment of larval lampreys. *Acta Zoologica (Stockholm)* **61**, 23–27.

Russell BC, Allen GR, Lubbock HR (1976) New cases of mimicry in marine fishes. *Journal of Zoology, London* **180**, 407–423.

Sazima I (1983) Scale-eating in characoids and other fishes. *Environmental Biology of Fishes* **9**, 87–101.

Sazima I, Uieda VS (1980) Comportamento lepidofágico de *Oligoplites saurus* e registro lepidofagia em *O. palometa* e *O. saliens* (Pisces, Carangidae). *Revista Brasileira de Biologia* **40**, 701–710.

Smith CL (1964) Some pearlfishes from Guam, with notes on their ecology. *Pacific Science* **18**, 34–40.

Smith-Vaniz WF, Staiger JC (1973) Comparative revision of *Scomberoides*, *Oligoplites*, *Parona*, and *Hypacanthus* with comments on the phylogenetic position of *Campogramma* (Pisces: Carangidae). *Proceedings of the California Academy of Sciences* **39**, 185–256.

Springer VG, Smith-Vaniz WF (1972) Mimetic relationships involving fishes of the family Blenniidae. *Smithsonian Contributions to Zoology* No. **112**, 1–36.

Strasburg DW (1961) Larval carapid fishes from Hawaii, with remarks on ecology of adults. *Copeia* **1961**, 478–480.

Strasburg DW (1963) The diet and dentition of *Isistius brasiliensis*, with remarks on tooth replacement in other sharks. *Copeia* **1963**, 33–40.

Szelistowski WA (1989) Scale-feeding in juvenile marine catfishes (Pisces: Ariidae). *Copeia* **1989**, 517–519.

Trott LB (1970) Contributions to the biology of carapid fishes (Paracanthopterygii: Gadiformes). *University of California Publications in Zoology* No. **89**, 1–60.

Trott LB, Trott EB (1972) Pearlfishes (Carapidae: Gadiformes) collected from Puerto Galera, Mindoro, Philippines. *Copeia* **1972**, 839–843.

Tyler JC (1968) A monograph on plectognath fishes of the superfamily Triacanthoidea. *Academy of Natural Sciences of Philadelphia*, Monograph **16**, 1–364.

van Utrecht WL (1959) Wounds and scars in the skin of the Common Porpoise, *Phocaena phocaena* (L.). *Mammalia* **13**, 100–122.

Whitfield AK, Blaber SJM (1978) Scale-eating habits of the marine teleost *Terapon jarbua* (Forskål). *Journal of Fish Biology* **12**, 61–70.

Chapter 6
Parasite induced changes in host behaviour and morphology

Aeby GS (2002) Trade-offs for the butterflyfish, *Chaetodon multicinctus*, when feeding on coral prey infected with trematode metacercariae. *Behavioural Ecology and Sociobiology* **52**, 158–165.

Bartoli P (1984) Distomatoses des lamellibranches marins sur le littoral méditerranéen Français. *Haliotis* **14**, 98–107.

Bowers EA, Bartoli P, Russell-Pinto F, James BL (1996) The metacercariae of sibling species of *Meiogymnophallus*, including *M. rebecqui* comb. nov. Digenea: Gymnophallidae), and their effects on closely related *Cerastoderma* host species Mollusca: Bivalvia). *Parasitology Research* **82**, 505–510.

Brusca RC, Gilligan MR (1983) Tongue replacement in a marine fish (*Lutjanus guttatus*) by a parasitic isopod (Crustacea: Isopoda). *Copeia* **3**, 813–816.

Combes C (1991) Ethological aspects of parasite transmission. *The American Naturalist* **138**, 866–880.

Combes C, Fournier A, Moné H, Théron A (1994) Behaviours in trematode cercariae that enhance parasite transmission: patterns and processes. *Parasitology* **109**, S3–S13.

Combes C (1995) 'Interactions Durables. Ecologie et Evolution du Parasitisme'. (Masson: Paris).

Combes C (1996) Parasites, biodiversity and ecosystem stability. *Biodiversity and Conservation* **5**, 953–962.

Dobson AP (1988) The population biology of parasite-induced changes in host behaviour. *Quaterly Review in Biology* **63**, 139–165.

Helluy S (1981) Parasitisme et comportement. Etude de la métacercaire de *Microphallus papillorobustus* (Rankin 1940) et de son influence sur les gammares. PhD Thesis Université des Sciences et Techniques du Languedoc Montpellier.

Helluy S (1984) Relations hôtes-parasites du *Microphallus papillorobustus* (Rankin 1940). III Facteurs impliqués dans les modifications du comportement des *Gammarus* hôtes intermédiaires et tests de prédation. *Annales de Parasitologie Humaine et Comparée* **59**, 41–56.

Hutchings MR, Kyriazakis I, Papachristou TF, Gordon IJ, Jackson F (2000) The herbivores' dilemma: trade-offs between nutrition and parasitism in foraging decisions. *Oecologia* **124**, 242–251.

Janssen H, Scheepmaker M, Couwelaar MV, Pinkster S (1979) Biology and distribution of *Gammarus aequicauda* and *G. insensibilis* (Crustacea, Amphipoda) in the lagoon system of Bages-Sigean (France). *Bijdragen tot de Dierkunde* **49**, 42–70.

Jones CG, Lawton JH, Shachak M (1994) Organisms as ecosystems engineers. *Oikos* **69**, 373–386.

Jones CG, Lawton JH, Shachak M (1997) Positive and negative effects of organisms as physical ecosystem engineers. *Ecology* **78**, 1946–1957.

Kearn GC (1974) The effects of fish skin mucus on hatching in the monogenean parasite *Entobdella soleae* from the skin of the common sole, *Solea solea*. *Parasitology* **68**, 173–188.

Kearn GC, MacDonald S (1976) The chemical nature of host hatching factors in the monogenean skin parasite *Entobdella soleae* and *Acanthocotyle lobianchi*. *International Journal for Parasitology* **6**, 457–466.

Lafferty KD (1992) Foraging on prey that are modified by parasites. *The American Naturalist* **140**, 854–867.

Lafferty KD (1999) The evolution of trophic transmission. *Parasitology Today* **15**, 111–115.

Lafferty KD, Thomas F, Poulin R (2000) Evolution of host phenotype manipulation by parasites and its consequences. *In* 'Evolutionary Biology of Host-Parasite Relationships: Theory Meets Reality.' (Eds R Poulin, S Morand, A Skorping.) pp. 117–127. (Elsevier Science: Amsterdam).

Llewellyn J (1972) Behaviour of monogeneans. *Biological Journal of The Linnean Society* **51**, 19–30.

Maillard C (1976) Distomatoses de poisons en milieu lagunaire. PhD Thesis Université des Sciences et Techniques du Languedoc Montpellier.

McCurdy DG, Forbes MR, Boates JS (1999) Evidence that the parasitic nematode *Skrjabinoclava* manipulates host *Corophium* behavior to increase transmission to the sandpiper, *Calidris pusilla*. *Behavioral Ecology* **10**, 351–357.

McCurdy DG, Forbes MR, Boates JS (2000) Male amphipods increase their mating effort before behavioural manipulation by trematodes. *Canadian Journal of Zoology* **78**, 606–612.

Minchella DJ, Scott ME (1991) Parasitism: a cryptic determinant of animal community structure. *Trends in Ecology and Evolution* **6**, 250–254.

Moore J (2002) 'Parasites and the Behaviors of Animals.' (Oxford University Press: New York).

Mouritsen KN, Poulin R (2002) Parasitism, community structure and biodiversity in intertidal ecosystems. *Parasitology* **124**, S101–S117.

Mouritsen KN, Poulin R (2003) Parasite-induced trophic facilitation exploited by a non-host predator: a manipulator's nightmare. *International Journal for Parasitology* **33**, 1043–1050.

Norris K (1999) A trade-off between energy intake and exposure to parasites in oystercatchers feeding on a bivalve mollusc. *Proceedings of the Royal Society of London B* **266**, 1703–1709.

Oliver G (1987) Les Diplectanidae Bychowsky, 1957 (Monogenea, Monopisthocotylea, Dactylogyridea). Systématique. Biologie. Ontogénie. écologie. Essai de phylogénèse. PhD Thesis. Université des Sciences et Techniques du Languedoc Montpellier.

Poulin R (1994) Meta-analysis of parasite-induced behavioral changes. *Animal Behaviour* **48**, 137–146.

Poulin R (1999) The functional importance of parasites in animal communities: many roles at many levels. *International Journal for Parasitology* **29**, 903–914.

Poulin R, Thomas F (1999) Phenotypic variability induced by parasites: extent and evolutionary implications. *Parasitology Today* **15**, 28–32.

Rousset F, Thomas F, De Meeüs T, Renaud F (1996) Inference of parasite-induced host mortality from distributions of parasite loads. *Ecology* **77**, 2203–2211.

Sousa WP (1991) Can models of soft-sediment community structure be complete without parasites? *American Zoologist* **31**, 821–830.

Thomas F, Poulin R (1998) Manipulation of a mollusc by a trophically transmitted parasite: convergent evolution or phylogenetic inheritance? *Parasitology* **116**, 431–436.

Thomas F, Renaud F (2001) *Microphallus papillorobustus* (Trematoda): a review of its effects in lagoon ecosystems. *Revue D'Ecologie Terre et Vie* **56**, 147–156.

Thomas F, Renaud F, Rousset F, Cezilly F, DeMeeüs T (1995) Differential mortality of two closely related host species induced by one parasite. *Proceedings of the Royal Society of London B* **260**, 349–352.

Thomas F, Mete K, Helluy S, Santalla F, Verneau O, De Meeûs T, Cézilly F, Renaud F (1997) Hitch-hiker parasites or how to benefit from the strategy of another parasite. *Evolution* **51**, 1316–1318.

Thomas F, Renaud F, De Meeüs T, Poulin R (1998) Manipulation of host behaviour by parasites: ecosystem engineering in the intertidal zone? *Proceedings of the Royal Society of London B* **265**, 1091–1096.

Thomas F, Poulin R, de Meeüs T, Guégan JF, Renaud F (1999) Parasites and ecosystem engineering: what roles could they play? *Oikos* **84**, 167–171.

Thomas F, Guégan JF, Michalakis Y, Renaud F (2000) Parasites and host life-history traits: implications for community ecology and species co-existence. *International Journal for Parasitology* **30**, 669–674.

Thomas F, Fauchier J, Laffery K (2002) Conflict of interest between a nematode and a trematode in an amphipod host: test of the 'sabotage' hypothesis. *Behavioural Ecology and Sociobiology* **51**, 296–301.

Thomas F, Adamo S, Moore J (2005) Parasitic manipulation: where are we and where should we go? *Behavioural Processes* **68**, 185–199.

Yoshinaga T, Nagakura T, Ogawa K, Wakabayashi H (2000) Attachment-inducing capacities of fish tissue extracts on oncomiracidia of *Neobenedenia girellae* (Monogenea, Capsalidae). *Journal of Parasitology* **86**, 214–219.

Cleaning mutualism in the sea

Alexander RD (1987) 'The Biology of Moral Systems'. (Aldine Publishing: New York).

Arnal C, Côté IM (1998) Interactions between cleaning gobies and territorial damselfish on coral reefs. *Animal Behaviour* **55**, 1429–1442.

Arnal C, Côté IM (2000) Diet of broadstripe cleaning gobies on a Barbadian reef. *Journal of Fish Biology* **57**, 1075–1082.

Arnal C, Morand S (2001a) Cleaning behaviour in the teleost, *Symphodus melanocercus*: females are more specialized than males. *Journal of the Marine Biological Association of the United Kingdom* **81**, 317–323.

Arnal C, Morand S (2001b) Importance of ectoparasites and mucus in cleaning interactions in the Mediterranean cleaner wrasse *Symphodus melanocercus*. *Marine Biology* **138**, 777–784.

Arnal C, Côté IM, Sasal P, Morand S (2000) Cleaner-client interactions on a Caribbean reef: influence of correlates of parasitism. *Behavioral Ecology and Sociobiology* **47**, 353–358.

Arnal C, Côté IM, Morand S (2001) Why clean and be cleaned? The importance of clients ectoparasites and mucus in a marine cleaning symbiosis. *Behavioral Ecology and Sociobiology* **51**, 1–7.

Arnal C, Kulbicki M, Harmelin-Vivien M, Galzin R, Morand S (2002) Patterns of local distribution of *Labroides dimidiatus* in French Polynesian atolls. *Environmental Biology of Fishes* **63**, 9–15.

Arnal C, Morand S, Kulbicki M (1999) Patterns of cleaner wrasse density among three reefs of the Pacific. *Marine Ecology Progress Series* **177**, 213–220.

Bansemer C, Grutter AS, Poulin R (2002) Geographic variation in the behaviour of the cleaner fish *Labroides dimidiatus* (Labridae). *Ethology* **108**, 353–366.

Becker JHA, Grutter AS (2004) Cleaner shrimp do clean. *Coral Reefs* **23**, 515–520.

Becker JHA, Grutter AS (in press). Client-fish ectoparasite loads and cleaner-shrimp hunger levels affect the cleaning behaviour of the cleaner-shrimp *Urocaridella* sp. c. *Animal Behaviour*.

Bshary R (2001) The cleaner fish market. *In* 'Economics in Nature: Social Dilemmas, Mate Choice and Biological Markets'. (Eds R Noë, J van Hooff, P Hammerstein.) pp. 146–172. (Cambridge University Press: Cambridge).

Bshary R (2002a) Biting cleaner fish use altruism to deceive image-scoring client reef fish. *Proceedings of the Royal Society Biological Sciences Series B* **269**, 2087–2093.

Bshary R (2002b) Building up relationships in asymmetric co-operation games between the cleaner wrasse *Labroides dimidiatus* and client reef fish. *Behavioral Ecology and Sociobiology* **52**, 365–371.

Bshary R (2003) The cleaner wrasse, *Labroides dimidiatus*, is a key organism for reef fish diversity at Ras Mohammed National Park, Egypt. *Journal of Animal Ecology* **72**, 169–176.

Bshary R, Grutter AS (2002a) Asymmetric cheating opportunities and partner control in the cleaner fish mutualism. *Animal Behaviour* **63**, 547–555.

Bshary R, Grutter AS (2002b) Parasite distribution on client fish determines cleaner foraging patterns. *Marine Ecology Progress Series* **235**, 217–222.

Bshary R, Grutter AS (2002c) Experimental evidence that partner choice is a driving force in the payoff distribution among cooperators or mutualists: the cleaner fish case. *Ecology Letters* **5**, 130–136.

Bshary R, Noë R (2003) Biological markets: the ubiquitous influence of partner choice on the dynamics of cleaner-client reef fish interactions. *In* 'Genetic and Cultural Evolution of Cooperation'. (Ed. P Hammerstein.) pp. 167–184. (MIT Press: Cambridge).

Bshary R, Schaffer D (2002) Choosy reef fish select cleaner fish that provide high-quality service. *Animal Behaviour* **63**, 557–564.

Bshary R, Wurth M (2001) Cleaner fish *Labroides dimidiatus* manipulate client reef fish by providing tactile stimulation. *Proceedings of the Royal Society of London. Series B* **268**, 1495–1501.

Bshary R, Wickler W, Fricke H (2002) Fish cognition: a primate's eye view. *Animal Cognition* **5**, 1–13.

Bunkley-Williams L, Williams EH (1998) Ability of pederson cleaner shrimp to remove juveniles of the parasitic cymothoid isopod, *Anilocra haemuli*, from the host. *Crustaceana* **71**, 862–869.

Calado R, Naraciso L, Morais S, Rhyne AL, Lin J (2003) A rearing system for the culture of ornamental decapod crustacean larvae. *Aquaculture* **218**, 329–339.

Carr MH, Hixon MA (1995) Predation effects on early post-settlement survivorship of coral-reef fishes. *Marine Ecology Progress Series* **124**, 31–42.

Chambers SD, Sikkel PC (2002) Diel emergence patterns of ecologically important, fish-parasitic, gnathiid isopod larvae on Caribbean coral reefs. *Caribbean Journal of Science* **38**, 37–43.

Cheney KL, Côté I, M. (2001) Are Caribbean cleaning symbioses mutualistic? Costs and benefits of visiting cleaning stations to longfin damselfish. *Animal Behaviour* **62**, 927–933.

Cheney KL, Côté IM (2003a) The ultimate effect of being cleaned: does ectoparasite removal have reproductive consequences for damselfish clients? *Behavioral Ecology* **14**, 892–896.

Cheney KL, Côté IM (2003b) Indirect consequences of parental care: sex differences in ectoparasite burden and cleaner-seeking activity in longfin damselfish. *Marine Ecology Progress Series* **262**, 267–275.

Cheney KL, Côté IM (2003c) Habitat choice in adult longfin damselfish: territory characteristics and relocation times. *Journal of Experimental Marine Biology and Ecology* **287**, 1–12.

Cheney KL, Côté IM (in press) Mutualism or parasitism? The variable outcome of cleaning symbioses. *Biology Letters*.

Clutton-Brock TH, Parker GA (1995) Punishment in animal societies. *Nature* **373**, 209–216.

Costello MJ (1996) Development and future of cleaner-fish technology and other biological control techniques in fish farming. *In* 'Wrasse: Biology and Use in Aquaculture'. (Eds MDJ Sayer, JW Treasurer, MJ Costello.) pp. 171–184. (Fishing News Books: Oxford).

Costello MJ, Deady S, Pike A, Fives JM (1996) Parasites and diseases of wrasse being used as cleaner-fish on salmon farms in Ireland and Scotland. *In* 'Wrasse: Biology and Use in Aquaculture'. (Eds MDJ Sayer, JW Treasurer, MJ Costello.) pp. 211–227. (Fishing News Books: Oxford).

Côté I (2000) Evolution and ecology of cleaning symbioses in the sea. *Oceanography and Marine Biology: An Annual Review* **38**, 311–355.

Côté IM, Cheney KL (2004) Distance-dependent costs and benefits of aggressive mimicry in a cleaning symbiosis. *Proceedings Of The Royal Society Of London Series B-Biological Sciences* **271**, 2627–2630

Côté IM, Molloy PP (2003) Temporal variation in cleanerfish and client behaviour: does it reflect ectoparasite availability? *Ethology* **109**, 487–499.

Côté IM, Poulin R (1995) Parasitism and group size in social animals: a meta-analysis. *Behavioral Ecology* **6**, 159–165.

Côté IM, Arnal C, Reynolds JD (1998) Variation in posing behaviour among fish species visiting cleaning stations. *Journal of Fish Biology* **53**, 256–266.

Cowell LE, Watanabe WO, Head WD, Grover JJ, Shenker JM (1993) Use of tropical cleaner fish to control the ectoparasite *Neobenedenia melleni* (Monogenea: Capsalidae) on seawater-cultured Florida red tilapia. *Aquaculture* **113**, 189–200.

Darcy GH, Maisel E, Ogden JC (1974) Cleaning preferences of the gobies *Gobiosoma evelynae* and *G. prochilos* and the juvenile wrasse *Thalassoma bifasciatum*. *Copeia* **2**, 375–379.

Dewet-Oleson K, Love M (2001) Observations of cleaning behaviour by giant kelpfish, *Heterostichus rostratus*, Island kelpfish, *Alloclinus holderi*, bluebanded goby, *Lythrypnus dalli*, and kelp bass, *Paralabrax clathratus*, on giant sea bass, *Stereolepis gigas*. *California Fish and Game* **87**, 87–92.

Edwards AJ, Shepherd AL (1992) Environmental implications of aquarium-fish collection in the Maldives, with proposals for regulation. *Environmental Conservation* **19**, 61–72.

Feder HM (1966) Cleaning symbiosis in the marine environment. *In* 'Symbiosis'. (Ed. SM Henry.) pp. 327–380. (Academic Press: New York).

Feitoza BM, Dias TLP, Rocha LA, Gasparini JL (2002) First record of cleaning activity in the slippery dick, *Halichoeres bivittatus* (Perciformes: Labridae), off northeastern Brazil. *Aqua* **5**, 73–76.

Foster SA (1985) Wound healing: a possible role of cleaning stations. *Copeia* **4**, 875–880.

Francini-Filho RB, Moura RL, Sazima I (2000) Cleaning by the wrasse *Thalassoma noronhanum*, with two records of predation by its grouper client *Cephalopholis fulva*. *Journal of Fish Biology* **56**, 802–809.

Freckleton RP, Côté IM (2003) Honesty and cheating in cleaning symbioses: evolutionarily stable stragies defined by variable pay-offs. *Proceedings of the Royal Society of London Series B* **270**, 299–305.

Gerking SD (1994) Fish that eat other fish and some unusual sources of food. *In* 'Feeding Ecology of Fish'. (Ed. SD Gerking.) pp. 265–295. (Academic Press: New York).

Gorlick DI. (1980) Ingestion of host fish surface mucus by the Hawaiian cleaning wrasse, *Labroides phthirophagus* (Labridae), and its effect on host species preference. *Copeia* **4**, 863–868.

Gorlick D (1984) Preference for ectoparasite-infected host fishes by the Hawaiian cleaning wrasse, *Labroides phthirophagus* (Labridae). *Copeia* **3**, 758–762.

Gorlick DL, Atkins PD, Losey GS (1987) Effect of cleaning by *Labroides dimidiatus* (Labridae) on an ectoparasite population infecting *Pomacentrus vaiuli* (Pomacentridae) at Enewetak Atoll. *Copeia* **1**, 41–45.

Grutter AS (1995) The relationship between cleaning rates and ectoparasite loads in coral reef fishes. *Marine Ecology Progress Series* **118**, 51–58.

Grutter AS (1996a) Parasite removal rates by the cleaner wrasse *Labroides dimidiatus*. *Marine Ecology Progress Series* **130**, 61–70.

Grutter AS (1996b) Experimental demonstration of no effect by the cleaner wrasse *Labroides dimidiatus* (Cuvier and Valenciennes) on the host fish *Pomacentrus moluccensis* (Bleeker). *Journal of Experimental Marine Biology and Ecology* **196**, 285–298.

Grutter AS (1997a) Size-selective predation by the cleaner fish *Labroides dimidiatus*. *Journal of Fish Biology* **50**, 1303–1308.

Grutter AS (1997b) Spatio-temporal variation and feeding selectivity in the diet of the cleaner fish *Labroides dimidiatus*. *Copeia* **2**, 346–355.

Grutter AS (1997c) Effect of the removal of cleaner fish on the abundance and species composition of reef fish. *Oecologia* **111**, 137–143.

Grutter AS (1999a) Cleaner fish really do clean. *Nature* **398**, 672–673.

Grutter AS (1999b) Infestation dynamics of parasitic gnathiid isopod juveniles on a coral reef fish *Hemigymnus melapterus*. *Marine Biology* **135**, 545–552.

Grutter AS (2000) Ontogenetic variation in the diet of the cleaner fish *Labroides dimidiatus* and its ecological consequences. *Marine Ecology Progress Series* **197**, 241–246.

Grutter AS (2001) Parasite infection rather than tactile stimulation is the proximate cause of cleaning behaviour in reef fish. *Proceedings of the Royal Society of London. Series B* **268**, 1361–1365.

Grutter AS (2002) Cleaning behaviour: from the parasite's perspective. *Parasitology* **124**, S65–S81.

Grutter AS (2003) Feeding ecology of the fish ectoparasite, *Gnathia* sp. (Crustacea: Isopoda), from the Great Barrier Reef, Australia and its implications for fish cleaning behaviour. *Marine Ecology Progress Series* **259**, 295–302.

Grutter AS (2004) Cleaner fish use tactile dancing as pre-conflict management strategy. *Current Biology* **14**, 1080–1083.

Grutter AS, Bshary R (2003) Cleaner fish prefer client mucus: support for partner control mechanisms in cleaning interactions. *Proceedings of the Royal Society Biological Sciences Series B, Biology Letters Supp* **2**, S242–S244.

Grutter AS, Bshary R (2004) Cleaner fish *Labroides dimidiatus* diet preferences for different types of mucus and parasitic gnathiid isopods. *Animal Behaviour* **68**, 583–588.

Grutter AS, Lester RJG (2002) Cleaner fish *Labroides dimidiatus* reduce 'temporary' corallanid isopods on the coral reef fish *Hemigymnus melapterus*. *Marine Ecology Progress Series* **234**, 247–255.

Grutter AS, Poulin R (1998) Intraspecific and interspecific relationships between host size and the abundance of parasitic larval gnathiid isopods on coral reef fishes. *Marine Ecology Progress Series* **164**, 263–271.

Grutter AS, Lester RJG, Greenwood J (2000) Emergence rates from the benthos of the parasitic juveniles of gnathiid isopods. *Marine Ecology Progress Series* **207**, 123–127.

Grutter AS, Deveney MR, Whittington ID, Lester RJG (2002) The cleaner fish *Labroides dimidiatus* affects the capsalid monogenean *Benedenia lolo* on the labrid fish *Hemigymnus melapterus* on the Great Barrier Reef. *Journal of Fish Biology* **61**, 1098–1108.

Grutter AS, Murphy J, Choat H (2003) Cleaner fish drives local fish diversity on coral reefs. *Current Biology* **13**, 64–67.

Hammerstein P, Hoekstra RF (1995) Mutualism on the move. *Nature* **376**, 121–122.

Hobson ES (1971) Cleaning symbiosis among California inshore fishes. *Fishery Bulletin* **69**, 491–523.

Honma Y, Chiba A (1991) Pathological changes in the branchial chamber wall of stingrays, *Dasyatis* spp., associated with the presence of juvenile gnathiids (Isopoda, Crustacea). *Gyobyo Kenkyu* **26**, 9–16.

Jennings JB, Gelder SR (1976) Observations on the feeding mechanisms, diet and digestive physiology of *Histriobdella homari* van Beneden 1858: an aberrant polychaete symbiotic with North American and European lobsters. *Biological Bulletin* **151**, 489–517.

Johnstone RA, Bshary R (2002) From parasitism to mutualism: partner control in asymmetric interactions. *Ecology Letters* **5**, 634–639.

Jonasson M (1987) Fish cleaning behaviour of shrimp. *Journal of Zoology* **213**, 117–131.

Jones C, Grutter AS, Cribb TH (2004) Cleaner fish become hosts: a novel form of parasite transmission. *Coral Reefs* **23**, 521–529.

Krawchuk MA, Koper N, Brooks RJ (1997) Observations of a possible cleaning symbiosis between painted turtles, *Chrysemys picta*, and snapping turtles, *Chelydra serpentina*, in Central Ontario. *The Canadian Field-Naturalist* **111**, 315–317.

Limbaugh C (1961) Cleaning symbiosis. *Scientific American* **205**, 42–49.

Lindberg WJ, Stanton G (1988) Bryozoan-associated decapod crustaceans: community patterns and a case of cleaning symbiosis between a shrimp and crab. *Bulletin of Marine Science* **42**, 411–423.

Losey GS (1971) Communication between fishes in cleaning symbiosis. *In* 'Aspects of the Biology of Symbiosis'. (Ed. TC Cheng.) pp. 45–76. (University Park Press: Baltimore).

Losey GS (1972a) Behavioural ecology of the 'cleaning fish'. *Australian Natural History* **17**, 232–238.

Losey GS (1972b) The ecological importance of cleaning symbiosis. *Copeia* **4**, 820–833.

Losey GS (1979) Fish cleaning symbiosis: proximate causes of host behaviour. *Animal Behaviour* **27**, 669–685.

Losey GS (1987) Cleaning symbiosis. *Symbiosis* **4**, 229–258.

Losey GS, Grutter AS, Rosenquist G, Mahon JL, Zamzow J (1999) Cleaning symbiosis: a review. *In* 'Behaviour and Conservation of Littoral Fishes'. (Eds VC Almada, RF Oliveira, EJ Goncalves.) pp. 379–395. (I.S.P.A: Lisbon).

Lu Y, Yu Q, Zamzow JP, Wang Y, Losey GS, Balazs GH, Nerurkar VR, Yanagihara R (2000) Detection of green turtle herpesviral sequence in saddleback wrasse *Thalassoma duperrey*: a possible mode of transmission of green turtle fibropapilloma. *Journal of Aquatic Animal Health* **12**, 58–63.

Mariani S (2001) Cleaning behaviour in *Diplodus* spp.: Chance or choice? A hint for future investigations. *Journal of the Marine Biological Association of the United Kingdom* **81**, 715–716.

McCourt RM, Thomson DA (1984) Cleaning behaviour of the juvenile panamic sergeant major, *Abudefduf troschelii* (Gill), with a resume of cleaning associations in the Gulf of California and adjacent waters. *California Fish and Game* **70**, 234–239.

Monod T (1926) Les Gnathiidae. Essai monographique (morphologie, biologie, systématique). *Mémoires de la Société des Sciences Naturelles du Maroc* **13**, 1–661.

Monteiro-Neto C, C.F.E. DA, Nottingham MC, Araujo ME, Rosa IL, Barros GML (2003) Analysis of the marine ornamental fish trade at Ceara State, northwest Brazil. *Biodiversity and Conservation* **12**, 1287–1295.

Mugridge RER, Stallybrass HG (1983) A mortality of eels, *Angulla anguilla* L., attributed to Gnathiidae. *Journal of Fish Diseases* **6**, 81–82.

Nicolette P (1990) 'Symbiosis: Nature in Partnership'. (Blandford: London).

Ostrum E (1990) Governing the Commons: The Evolution of Institutions for Collective Action. (Cambridge University Press: New York).

Paperna I, Por FD (1977) Preliminary data on the Gnathiidae (Isopoda) of the Northern Red Sea, the Bitter Lakes and the Eastern Mediterranean and the biology of *Gnathia piscivora* n. sp. *Rapports de la Commission Internationale pour la Mer Méditerranée* **24**, 195–197.

Potts GW (1973) The ethology of *Labroides dimidiatus* (Cuv. & Val.) (Labridae, Pisces) on Aldabra. *Animal Behaviour* **21**, 250–291.

Poulin R, Grutter AS (1996) Cleaning symbioses: proximate and adaptive explanations. *BioScience* **46**, 512–517.

Poulin R, Vickery WL (1995) Cleaning symbiosis as an evolutionary game: to cheat or not to cheat? *Journal of Theoretical Biology* **175**, 63–70.

Poulin R, Rau ME, Curtis MA (1991) Infection of brook trout fry, *Salvelinus fontinalis*, by ectoparasitic copepods: the role of host behaviour and initial parasite load. *Animal Behaviour* **41**, 467–476.

Randall JE (1958) A review of the labrid fish genus *Labroides*, with description of two new species and notes on ecology. *Pacific Science* **12**, 327–347.

Robertson DR, Choat JH (1974) Protogynous hermaphroditism and social systems in labrid fish. *Proceedings of the Second International Coral Reef Symposium*, 217–225.

Rohde K (1993) 'Ecology of Marine Parasites . An Introduction to Marine Parasitology'. (CAB International: Wallingford).

Sakai Y, Kohda M (2001) Spawning timing of the cleaner wrasse, *Labroides dimidiatus*, on a warm temperate rocky shore. *Ichthyological Research* **48**, 23–30.

Sayer MDJ, Treasurer JW, Costello MJ (1996) Wrasse biology and aquaculture applications: commentary and conclusions. *In* 'Wrasse: Biology and Use in Aquaculture'. (Eds MDJ Sayer, JW Treasurer, MJ Costello.) pp. 272–277. (Fishing News Books: Oxford).

Sazima I, Moura RL (2000) Shark (*Carcharhinus perezi*), cleaned by the goby (*Elacatinus randalli*), at Fernando de Noronha Archipelago, Western South Atlantic. *Copeia* **1**, 297–299.

Sazima C, Sazima I (2001) Plankton-feeding aggregation and occasional cleaning by adult butterflyfish, *Chaetodon striatus* (Chaetodontidae), in southwestern Atlantic. *Cybium* **25**, 145–151.

Sazima I, Gasparini JL, Moura RL (1998a) *Gramma brasiliensis*, a new basslet from the western South Atlantic (Periformes: Grammatidae). *Journal of Ichthyology and Aquatic Biology* **3**, 39–44.

Sazima I, Moura RL, Rosa RS (1997) *Elacatinus figaro* sp. n. (Perciformes: Gobiidae), a new cleaner goby from the coast of Brazil. *Journal of Ichthyology and Aquatic Biology* **2**, 33–38.

Sazima I, Moura RL, Gasparini JL (1998b) The wrasse *Halichoeres cyanocephalus* (Labridae) as a specialized cleaner fish. *Bulletin of Marine Science* **63**, 605–610.

Sazima I, Moura RL, Rodrigues MCM (1999a) A juvenile sharksucker, *Echeneis naucrates* (Echeneidae), acting as a station-based cleaner fish. *Cybium* **23**, 377–380.

Sazima I, Moura RL, Sazima C (1999b) Cleaning activity of juvenile angelfish, *Pomacanthus paru*, on the reefs of the Abrolhos Archipelago, western South Atlantic. *Environmental Biology of Fishes* **56**, 399–407.

Sazima I, Sazima C, Francini-Filho RB, Moura RL (2000) Daily cleaning activity and diversity of clients of the barber goby, *Elacatinus figaro*, on rocky reefs in southeastern Brazil. *Environmental Biology of Fishes* **59**, 69–77.

Schuhmacher H (1973) Das kommensalische Verhaltnis zwischen *Periclimenes imperator* (Decapoda: Palaemonidae) und *Hexabranchus sanguineus* (Nudibranchia: Doridacea). *Marine Biology* **22**, 355–360.

Sikkel PC, Fuller CA, Hunte W (2000) Habitat/sex differences in time at cleaning stations and ectoparasite loads in a Caribbean reef fish. *Marine Ecology Progress Series* **193**, 191–199.

Sikkel PC, Cheney Kl, Côté IM (2004) *In situ* evidence for ectoparasites as a proximate cause of cleaning interactions in reef fish. *Animal Behaviour* **68**, 241–247.

Slobodkin LB, Fishelson L (1974) The effect of the cleaner-fish *Labroides dimidiatus* on the point diversity of fishes on the reef front at Eilat. *The American Naturalist* **108**, 369–376.

Sluka R, Sullivan KM (1996) Daily activity patterns of groupers in the Exuma Cays Land and Sea park, Central Bahamas. *Bahamas Journal of Science* **3**, 17–22.

Sluka R, Chiappone M, Sullivan KM, Garine-Wichatitsky Md (1999) Benthic habitat characterization and space utilization by juvenile epipheline groupers in the Exuma Cays Land and Sea Park, Central Bahamas. *Proceedings of the 45th Gulf and Caribbean Fisheries Institute* **45**, 22–36.

Spotte S (1998) 'Cleaner' shrimps? *Helgoländer Meeresuntersuchungen* **52**, 59–64.

Stummer LE, Weller JA, Johnson ML, *et al.* (2004) Size and stripes: how fish clients recognize cleaners. *Animal Behaviour* **68**, 145–150.

Tebbich S, Bshary R, Grutter AS (2002) Cleaner fish *Labroides dimidiatus* recognise familiar clients. *Animal Cognition* **5**, 139–145.

Thresher RE (1984) 'Reproduction in Reef Fishes'. (TFH Neptune City: NJ).

Treasurer JW (2002) A review of potential pathogens of sea lice and the application of cleaner fish in biological control. *Pest Management Science* **58**, 546–558.

Trivers RL (1971) The evolution of reciprocal altruism. *The Quarterly Review of Biology* **46**, 35–57.

Van Tassell JL, Brito A, Bortone SA (1994) Cleaning behavior among marine fishes and invertebrates in the Canary Islands. *Cybium* **18**, 117–127.

Whiteman EA, Côté IM (2002) Sex differences in cleaning behaviour and diet of a Caribbean cleaning goby. *Journal of the Marine Biological Association of the United Kingdom* **82**, 655–664.

Whiteman EA, Côté IM (2003) Social monogamy in the cleaning goby *Elacatinus evelynae*: ecological constraints or net benefit? *Animal Behaviour* **66**, 281–291.

Whiteman EA, Cote IM, Reynolds JD (2002) Do cleaning stations affect the distribution of territorial reef fishes. *Coral Reefs* **21**, 245–251.

Wickler W (1968) The origin of the cleaner mimic. *In* 'Mimicry in Plants and Animals' (Ed. W Wickler.) pp. 157–176. (Weidenfeld and Nicolson: London).

Wood E (2001) Collection of coral reef fish for aquaria: global trade, conservation issues and management strategies. *Marine Conservation Society.* pp. 1–56.

Wood E, Rajasuriya A (1998) Reef resources of Sri Lanka: valuable but under stress. *Coral Reefs* **17**, 58.

Youngbluth MJ (1968) Aspects of the ecology and ethology of the cleaning fish, *Labroides phthirophagus* Randall. *Zeitschrift für Tierpsychologie* **25**, 915–932.

Zahavi A (1995) Altruism as a handicap – the limitations of kin selection and reciprocity. *Journal of Avian Biology* **26**, 1–3.

Zander CD, Meyer U, Schmidt A (1999) Cleaner fish symbiosis in European and Macronesian waters. *In* 'Behaviour and Conservation of Littoral Fishes'. (Eds VC Almada, RF Oliveira, EJ Goncalves.) pp. 397–422 (ISPA: Lisboa).

Zander CD, Nieder J (1997) Interspecific associations in Mediterranean fishes: feeding communities, cleaning symbioses and cleaner mimics. *Vie et Milieu* **47**, 203–212.

Zander CD, Sötje I (2002) Seasonal and geographical differences in cleaner fish activity in the Mediterranean Sea. *Helgoland Marine Research* **55**, 232–241.

Chapter 7
Transmission of marine parasites

Arai MN, Welch DW, Dunsmuir AL, Jacobs MC, Ladouceur AR (2003) Digestion of pelagic Ctenophora and Cnidaria by fish. *Canadian Journal of Fisheries and Aquatic Sciences* **60**, 825–829.

Burt MDB, Campbell JD, Likely CG, Smith JW (1990) Serial passage of larval *Pseudoterranova decipiens* (Nematoda: Ascaridoidea) in fish. *Canadian Journal of Fisheries and Aquatic Sciences* **47**, 693–695.

Bush AO, Heard RW Jr, Oversteeet RM (1993) Intermediate hosts as source communities. *Canadian Journal of Zoology* **71**, 1358–1363.

Campbell RA (1983) Parasitism in the deep sea. In 'The Sea'. Vol. 8. (Ed. GT Rowe.) pp. 473–552. (John Wiley & Sons: New York).

Campbell RA, Haedrich RL, Munroe TA (1980) Parasitism and ecological relationships among deep-sea benthic fishes. *Marine Biology* **57**, 301–313.

Lotz JM, Bush AO, Font WF (1995) Recruitment-driven, spatially discontinuous communities: a null model for transferred patterns in target communities of intestinal helminths. *Journal of Parasitology* **81**, 12–24.

MacKenzie K (1999) Parasites as pollution indicators in marine ecosystems: a proposed early warning system. *Marine Pollution Bulletin* **38**, 955–959.

MacKenzie K, Williams HH, Williams B, McVicar AH, Siddall R (1995) Parasites as indicators of water quality and the potential use of helminth transmission in marine pollution studies. *Advances in Parasitology* **35**, 85–144.

Marcogliese DJ (1995) The role of zooplankton in the transmission of helminth parasites to fish. *Reviews in Fish Biology and Fisheries* **5**, 336–371.

Marcogliese DJ (1996) Larval parasitic nematodes infecting marine crustaceans in eastern Canada. 3. *Hysterothylacium adundum*. *Journal of the Helminthological Society of Washington* **63**, 12–18.

Marcogliese DJ (2001) Review of experimental and natural invertebrate hosts of sealworm (*Pseudoterranova decipiens*) and its distribution and abundance in macroinvertebrates in eastern Canada. *NAMMCO Scientific Publications* **3**, 27–37.

Marcogliese DJ (2002) Food webs and the transmission of parasites to marine fish. *Parasitology* **124**, S83–S99.

Marcogliese DJ (2003) Food webs and biodiversity: are parasites the missing link? *Journal of Parasitology* **89**, S106–S113.

Marcogliese DJ (2004) Parasites: small players with crucial roles in the ecological theatre. *EcoHealth* **1**, 151–164.

Marcogliese DJ, Cone DK (1997a) Food webs: a plea for parasites. *Trends in Ecology & Evolution* **12**, 320–325.

Marcogliese DJ, Cone DK (1997b) Parasite communities as indicators of ecosystem stress. *Parassitologia* **39**, 227–232.

Margolis L, Arthur JR (1979) Synopsis of the parasites of fishes of Canada. Bulletin of the Fisheries Research Board of Canada No. 199, Ottawa.

Martell DJ, McClelland G (1995) Transmission of *Pseudoterranova decipiens* (Nematoda: Ascaridoidea) via benthic macrofauna to sympatric flatfishes (*Hippoglossoides platessoides, Pleuronectes ferrugineus, P. americanus*) on Sable Island Bank, Canada. *Marine Biology* **122**, 129–135.

McClelland G (2002) The trouble with sealworms (*Pseudoterranova decipiens* species complex, Nematoda): a review. *Parasitology* **124**, S183–S203.

McDonald TE, Margolis L (1995) Synopsis of the parasites of fishes of Canada: supplement (1978–1993). Canadian Special Publication of Fisheries and Aquatic Sciences No. 122, Ottawa.

Moore J (2002) 'Parasites and the Behavior of Animals'. (Oxford University Press: Oxford).

Myers RA, Worm B (2003) Rapid worldwide depletion of predatory fish communities. *Nature* **423**, 280–283.

Pauly D, Christensen V, Froese R, Palomares ML (2000) Fishing down aquatic food webs. *American Scientist* **88**, 46–51.

Uspenskaya AV (1960) Parasitofaune des crustacés benthiques de la mer Barents. *Annales de Parasitologie Humaine et Comparée* **35**, 221–242.

Williams HH, Jones A (1994) 'Parasitic Worms of Fish'. (Taylor & Francis: London)

Williams HH, MacKenzie K (2003) Marine parasites as pollution indicators: an update. *Parasitology* **126**, S27–S41.

The ecological niches of parasites

Barker GM (2002) Phylogenetic diversity: a quantitative framework for measurement of priority and achievement in biodiversity conservation. *Biological Journal of the Linnean Society* **76**, 165–194.

Caira JN, Jensen K, Holsinger KE (2003) On a new index of host specificity. In 'Taxonomie, Ecologie et Evolution des Métazoires Parasites'. Taxonomy, Ecology and Evolution of Metazoan Parasites. (Livre hommage à Louis Euzet)'. Vol. I (Eds C Combes, J Jourdane.) pp. 161–201. (PUP: Perpignan, France). Webpage for calculating the index is located at 'HS Software for calculating host specificity' http://darwin.eeb.uconn.edu/specificity/specificity.html.

Clarke KR, Warwick RM (1998) A taxonomic distinctness index and its statistical properties. *Journal of Applied Ecology* **35**, 523–531.

Clarke KR, Warwick RM (2001) A further biodiversity index applicable to species lists: variation in taxonomic distinctness. *Marine Ecology Progress Series* **216**, 265–278.

Combes C (1995) 'Interactions Durables: Ecologie et Evolution de Parasitisme'. (Masson: Paris).

Hutchinson GE (1957) Concluding remarks. *Cold Spring Harbor Symposium on Quantitative Biology* **22**, 415–427.

Kearn GC (1967) Experiments on host-finding and host-specificity in the monogenean skin parasite *Entobdella soleae*. *Parasitology* **57**, 585–605.

Kearn GC (1974) The effects of fish skin mucus on hatching in the monogenean parasite *Entobdella soleae* from the skin of the common soile, *Solea solea*. *Parasitology* **68**, 173–188.

Krebs CJ (1989) 'Ecological Methodology'. (Harper Collins: New York).

Lauri JS (1971) Carbohydrate absorption by *Gyrocotyle fimbriata* and *Gyrocotyle parvispinosa* (Plaatyhelminthes). *Experimental Parasitology* **29**, 375–385.

Lymbery A J (1989). Host specificity, host range and host preference. *Parasitology Today* **5**, 298.

McVicar AH, Fletcher TC (1970). Serum factors in *Raja radiata* toxic to *Acanthobothrium quadripartitum* (Cestoda: Tetraphyllidae), a parasite specific to *R. naevus*. *Parasitology* **6**, 55–63.

Polyanski YuI (1961) Ecology of parasites of marine fishes. In 'Parasitology of Fishes'. (Eds VA Dogiel, GK Petrushevski, YuI Polyanski.) pp. 48–83. (English translation Oliver and Boyd: Edinburgh and London).

Poulin R (1992) Determinants of host specificity in parasites of freshwater fishes. *International Journal for Parasitology* **22**, 753–758.

Poulin R (1998) 'Evolutionary Ecology of Parasites'. (Chapman & Hall: London).

Poulin R, Mouillot D (2003) Parasite specialization from a phylogenetic perspective: a new index of host specificity. *Parasitology* **126**, 473–480.

Ricotta C (2004) A parametric diversity measure combining the relative abundances and taxonomic distinctiveness of species. *Diversity and Distributions* **10**, 143–146.

Rohde K (1979) A critical evaluation of intrinsic and extrinsic factors responsible for niche restriction in parasites. *American Naturalist* **114**, 648–671.

Rohde K (1980a) Host specificity indices of parasites and their application. *Experientia* **36**, 1369–1371.

Rohde K (1980b) Warum sind ökologische Nischen begrenzt? Zwischenartlicher Antagonismus oder innerartlicher Zusammenhalt? *Naturwissenschaftliche Rundschau* **33**, 98–102.

Rohde K (1993) 'Ecology of Marine Parasites'. 2nd edn. (CAB International: Wallingford, Oxon).

Rohde K (1994) Niche restriction in parasites: proximate and ultimate causes. *Parasitology* **109**, S69–S84.

Rohde P (2003) http://www-personal.une.edu.au/~krohde/Host specificity index.

Shulman SS, Shulman-Albova RE (1953) ['Parasites of Fishes in the White Sea']. (Akademiya Nauk SSSR: Moscow-Leningrad). (In Russian).

Marine hyperparasites

Aken'Ova T, Lester RJG (1996) *Udonella myliobati* n. comb. (Platyhelminthes: Udonellidae) and its occurrence in Eastern Australia. *Journal of Parasitology* **82**, 1017–1023.

Byrnes T, Rohde K (1992) Geographic distribution and host specificity of ectoparasites of Australian bream, *Acanthopagrus* spp. (Sparidae). *Folia Parasitologica* **39**, 249–264.

Carvajal J, Ruiz G, Sepúlveda F (2001) Symbiotic relationship between *Udonella* sp, (monogenea) and *Caligus rogercresseyi* (copepoda), a parasite of the Chilean rock cod *Eleginops maclovinus*. *Archivos de Medicina Veterinaria* **33**, 31–36.

Freeman MA (2002) Potential biological control agents for the salmon louse *Lepeophtheirus salmonis* (Krøyer 1837). PhD thesis. Stirling University Library: Stirling University, UK.

Freeman MA, Bell AS, Sommerville C (2003) A hyperparasitic microsporidian infecting the salmon louse, *Lepoephtheirus salmonis*: an rDNA-based molecular phylogenetic study. *Journal of Fish Diseases* **26**, 667–676.

Grygier MJ (1993) Cryptoniscidae (Isopoda: Epicaridae) nomenclatural history and recommendations. *Publications of the Seto Marine Biological Laboratory* **36**, 185–195.

Grygier MJ, Bowman TE (1990) The correct family-level name for the 'cryptoniscid' isopods (Epicaridea). *Crustaceana* **58**, 27–32.

Kabata Z (1973) Distribution of *Udonella caligorum* Johnston, 1835 (Monogenea: Udonellidae) on *Caligus elongatus* Nordmann, 1832 (Copepoda: Caligidae). *Journal of the Fisheries Research Board of Canada* **30**, 1793–1798.

Littlewood DTJ, Rohde K, Clough KA (1998) The phylogenetic position of *Udonella* (Platyhelminthes). *International Journal for Parasitology* **28**, 1241–1250.

Marín SL, Sepúlveda F, Carvajal J, George-Nascimento M (2002) The feasibility of using *Udonella* sp. (Platyhelminthes: Udonellidae) as a biological control for the sea louse *Caligus rogercresseyi*, Boxshall and Bravo 2000, (Copepoda: Caligidae) in southern Chile. *Aquaculture* **208**, 11–21.

Minchin D (1991) *Udonella caligorum* Johnston (Trematoda) from the Celtic Sea. *The Irish Naturalists' Journal* **23**, 509–510.

Pohle GW (1992) First Canadian record of *Paralomis bouvieri* Hansen, 1908 (Decapod: Anomura: Lithodidae), infected by the rhizocephalan *Briarosaccus callosus* (Cirripedia: Peltogastridea) and carrying a hyperparasitic cryptoniscinid isopod (Epicaridea). *Canadian Journal of Zoology* **70**, 1625–1629.

Sassaman C (1985) *Cabirops montereyensis*, a new species of hyperparasitic isopod from Monterey Bay, California (Epicaridea: Cabiropsidae). *Proceedings of the Biological Society of Washington* **98**, 778–789.

Spivey HR (1982) Rhizocephala. *In* 'Synopsis and Classification of Living Organisms'. Vol. 2. (Ed. SB Parker.) pp. 229–232. (McGraw Hill: New York).

Tin Tun, Yokoyama H, Ogawa K, Wakabayashi H (2000) Myxosporeans and their hyperparasitic microsporeans in the intestine of emaciated tiger puffer. *Fish Pathology* **35**, 145–156.

Parasites in brackish waters

Holmes JC (1990) Helminth communities in marine fishes. *In* 'Parasite Communities: Patterns and Processes'. (Eds GW Esch, AO Bush, JM Aho.) pp. 101–130. (London: Chapman and Hall).

Möller H (1978) The effect of salinity and temperature on the development and survival of fish parasites. *Journal of Fish Biology* **12**, 311–323.

Remane A (1958) Ökologie des Brackwassers. *In* 'Die Biologie des Brackwassers'. (Eds A Remane, C Schlieper.) *Die Binnengewässer* **12**, 1–126.

Zander CD (1998) Ecology of host parasite relationships in the Baltic Sea. *Naturwissenschaften* **85**, 426–436.

Zander CD (2001) The guild as a concept and a means in ecological parasitology. *Parasitology Research* **87**, 484–488.

Zander CD, Reimer LW (2002) Parasitism at the ecosystem level in the Baltic Sea. *Parasitology* **124**, S119–S135.

Zander CD, Reimer LW, Barz K, Dietel G, Strohbach U (2000) Parasite communities of the Salzhaff (Northwest Mecklenburg, Baltic Sea). II. Guild communities, with special regard to snails, benthic crustaceans, and small-sized fish. *Parasitology Research* **86**, 359–372.

Metapopulation biology of marine parasites

Anderson RM, May RM (1978) Regulation and stability of host-parasite population interactions. I. Regulatory processes. *Journal of Animal Ecology* **47**, 219–247.

Anderson RM, May RM (1985) Helminth infection of humans: mathematical models, population dynamics and control. *Advances in Parasitology* **24**, 1–101.

Anderson RM, May RM (1991) 'Infectious Diseases of Humans: Dynamics and Control'. (Oxford University Press: Oxford).

Bush AO, Lafferty KD, Lotz JM, Shostak AW (1997) Parasitology meets ecology on its own terms: Margolis *et al.* revisited. *Journal of Parasitology* **83**, 575–583.

Des Clers S (1990) Modelling the life cycle of the sealworm (*Pseudoterranova decipiens*) in Scottish waters. *In* 'Population Biology of Sealworm (*Pseudoterranova decipiens*) in Relation to its Intermediate and Seal Hosts'. (Ed. WD Bowen.) *Canadian Bulletin of Fisheries and Aquatic Sciences* **222**, 273–288.

Dobson AP, Roberts M (1994) The population dynamics of parasitic helminth communities. *Parasitology* **109**, S97-S108.

Esch GW, Fernández JC (1993) 'A Functional Biology of Parasitism'. (Chapman and Hall: London).

Hanski IA (1982) Dynamics of regional distribution: the core and satellite species hypothesis. *Oikos* **38**, 210–221.

Hanski IA (1999) ''Metapopulation Ecology'. (Oxford University Press: Oxford).

Hanski IA, Gyllenberg M (1993) Two general metapopulation models and the core-satellite species hypothesis. *The American Naturalist* **142**, 17–41.

Hanski IA, Simberloff D (1997) The metapopulation approach, its history, conceptual domain, and application to conservation. *In* 'Metapopulation Biology: Ecology, Genetics, and Evolution'. (Eds IA Hanski, ME Gilpin.) pp. 5–26. (Academic Press: San Diego).
Holmes JC 1990. Helminth communities in marine fishes. *In* 'Parasite Communities: Patterns and Processes'. (Eds GW Esch, AO Bush, JM Aho.) pp. 101–130. (Chapman and Hall: New York).
Hugueny B, Guégan J-F (1997) Community nestedness and the proper way to assess statistical significance by Monte-Carlo tests: some comments on Worthen and Rohde's (1996) paper. *Oikos* **80**, 572–574.
Ives AR (1991) Aggregation and coexistence in a carrion fly community. *Ecological Monographs* **61**, 75–64.
Jaenike J, James AC (1991) Aggregation and the coexistence of mycophagous *Drosophila*. *Journal of Animal Ecology* **60**, 913–928.
Levins R (1969) Some demographic and genetic consequences of environmental heterogeneity for biological control. *Bulletin of the Entomological Society of America* **15**, 237–240.
Levins R (1970) Extinction. *In* 'Some Mathematical Problems in Biology'. (Ed. M Desternhaber.) pp. 77–107. (American Mathematical Society: Providence).
Lo CM, Morand S (2000) Spatial distribution and coexistence of monogenean gill parasites inhabiting two damselfishes from Moorea island (French Polynesia). *Journal of Helminthology* **74**, 329–336.
Lo C, Morand S, Galzin R (1998) Parasite diversity, host age and size relationship in three coral reef fishes from French Polynesia. *International Journal for Parasitology* **28**, 1695–1708.
Matejusová I, Morand S, Ergens R, Gelnar M (2000) Nestedness in assemblages of gyrodactylids parasitizing two species of cyprinid species – the role of generalists and specialists in parasite assemblages. *International Journal for Parasitology* **30**, 1153–1158.
May RM, Anderson RM (1978) Regulation and stability of host-parasite population interactions. II. Destabilizing processes. *Journal of Animal Ecology* **47**, 249–267.
McKenzie K (2002) Parasites as biological tags in population studies of marine organisms: an update. *Parasitology* **124**, S153–163.
Morand S, Guégan J-F (2000) Abundance and distribution of parasitic nematodes: ecological specialisation, phylogenetic constraints or simply epidemiology? *Oikos* **55**, 563–573.
Morand S, Robert F, Connors VA (1995) Complexity in parasite life-cycles: population biology of cestodes in fish. *Journal of Animal Ecology* **64**, 256–264.
Morand S, Poulin R, Rohde K, Hayward C (1999) Aggregation and species coexistence of ectoparasites of marine fishes. *International Journal for Parasitology* **29**, 663–672.
Morand S, Rohde K, Hayward C (2002) Order in ectoparasite communities of marine fish is explained by epidemiological processes. *Parasitology* **124**, S57–S63.
Poulin R (1998) 'Evolutionary Ecology of Parasites: From Individuals to Communities'. (Chapman and Hall: London).
Rohde K, Hobbs RP (1986) Species segregation: competition or reinforcement of reproductive barriers? *In* 'Parasites Lives, Papers on Parasites, their Hosts and their Association'. (Eds M Cremin, C Dobson, E Noorhouse.) pp. 189–199. (University of Queensland Press: St Lucia).
Rohde K, Hayward C, Heap M (1995) Aspects of the ecology of metazoan ectoparasites of marine fishes. *International Journal for Parasitology* **25**, 945–970.
Rohde K, Worthen WB, Heap M, Huguény B, Guégan J-F (1998) Nestedness in assemblages of metazoan ecto- and endoparasites of marine fish. *International Journal for Parasitology* **28**, 543–549.
Shaw DJ, Dobson AP (1995) Patterns of macroparasite abundance and aggregation in wildlife populations: a quantitative review. *Parasitology* **111**, S111–S133.

Shorrocks B (1996) Local diversity: a problem with too many solutions. *In* 'The Genesis and Maintenance of Biological Diversity'. (Eds M Hochberg, J Clobert, R Barbault.) pp. 104–122. (Oxford University Press: Oxford).

Šimková A, Desdevises Y, Gelnar M, Morand S (2000) Co-existence of nine gill ectoparasites (*Dactylogyrus*: Monogenea) parasitising the roach (*Rutilus rutilus* L.): history and present ecology. *International Journal for Parasitology* **30**, 1077–1088.

Šimková A, Kadlec D, Gelnar M, Morand S (2002a) Abundance-prevalence relationship of gill congeneric ectoparasites: testing for core-satellite hypothesis and ecological specialisation. *Parasitology Research* **88**, 682–686.

Šimková A, Ondracková M, Gelnar M, Morand S (2002b) Morphology and coexistence of congeneric ectoparasite species: reinforcement of reproductive isolation? *Biological Journal of the Linnean Society* **76**, 125–135.

Šimková A, Goüy de Bellocq J, Morand S (2003) The structure of host-parasite communities: order and history. *In* 'Taxonomy, Ecology and Evolution of Metazoan Parasites. Tome II'. (Eds C Combes, J Jourdane.) pp. 237–257. (Pressess Universitaires de Perpignan: Perpignan).

Taylor LR (1961) Aggregation, variance and the mean. *Nature* **189**, 732–735.

Worthen WB, Rohde K (1996) Nested subset analyses of colonization-dominated communities: metazoan ectoparasites of marine fishes. *Oikos* **75**, 471–478.

Structure of parasite communities

Bush AO, Heard RW, Overstreet RM (1993) Intermediate hosts as source communities. *Canadian Journal of Zoology* **71**, 1358–1363.

Bush AO, Lafferty KD, Lotz JM, Shostak AW (1997) Parasitology meets ecology on its own terms: Margolis *et al.* revisited. *Journal of Parasitology* **83**, 575–583.

Dash KM (1981) Interaction between *Oesophagostomum columbianum* and *Oesophagostomum venulosum* in sheep. *International Journal for Parasitology* **11**, 201–207.

Esch GW, Bush AO, Aho JM (1990) 'Parasite Communities: Patterns and Processes'. (Chapman & Hall: London).

Gotelli NJ, Graves GR (1996) 'Null Models in Ecology'. (Smithsonian Institution Press: Washington, DC).

Gotelli NJ, Rohde K (2002) Co-occurrence of ectoparasites of marine fishes: a null model analysis. *Ecology Letters* **5**, 86–94.

Holland C (1984) Interactions between *Moniliformis* (Acanthocephala) and *Nippostrongylus* (Nematoda) in the small intestine of laboratory rats. *Parasitology* **88**, 303–315.

Holmes JC (1961) Effects of concurrent infections on *Hymenolepis diminuta* (Cestoda) and *Moniliformis dubius* (Acanthocephala). I. General effects and comparison with crowding. *Journal of Parasitology* **47**, 209–216.

Holmes JC (1973) Site segregation by parasitic helminths: interspecific interactions, site segregation, and their importance to the development of helminth communities. *Canadian Journal of Zoology* **51**, 333–347.

Holmes JC (1990) Helminth communities in marine fishes. *In* 'Parasite Communities: Patterns and Processes'. (Eds GW Esch, AO Bush, JM Aho.) pp. 101–130. (Chapman & Hall: London).

Holmes JC, Price PW (1986) Communities of parasites. *In* 'Community Ecology: Pattern and Process'. (Eds DJ Anderson, J Kikkawa.) pp. 187–213. (Blackwell Scientific Publications: Oxford).

Lotz JM, Font WF (1991) The role of positive and negative interspecific associations in the organization of communities of intestinal helminths of bats. *Parasitology* **103**, 127–138.

Lotz JM, Font WF (1994) Excess positive associations in communities of intestinal helminths of bats: a refined null hypothesis and a test of the facilitation hypothesis. *Journal of Parasitology* **80**, 398–413.

Lotz JM, Bush AO, Font WF (1995) Recruitment-driven, spatially discontinuous communities: a null model for transferred patterns in target communities of intestinal helminths. *Journal of Parasitology* **81**, 12–24.

Morand S, Poulin R, Rohde K, Hayward C (1999) Aggregation and species coexistence of ectoparasites of marine fishes. *International Journal for Parasitology* **29**, 663–672.

Mouillot D, George-Nascimento M, Poulin R (2003) How parasites divide resources: a test of the niche apportionment hypothesis. *Journal of Animal Ecology* **72**, 757–764.

Patrick MJ (1991) Distribution of enteric helminths in *Glaucomys volans* L. (Sciuridae): a test for competition. *Ecology* **72**, 755–758.

Patterson BD, Atmar W (1986) Nested subsets and the structure of insular mammalian faunas and archipelagos. *Biological Journal of the Linnean Society* **28**, 65–82.

Poulin R (1998) 'Evolutionary Ecology of Parasites: From Individuals to Communities'. (Chapman & Hall: London).

Poulin R (2001) Interactions between species and the structure of helminth communities. *Parasitology* **122**, S3–S11.

Poulin R, Guégan J-F (2000) Nestedness, anti-nestedness, and the relationship between prevalence and intensity in ectoparasite assemblages of marine fish: a spatial model of species coexistence. *International Journal for Parasitology* **30**, 1147–1152.

Poulin R, Luque JL (2003) A general test of the interactive-isolationist continuum in gastrointestinal parasite communities of fish. *International Journal for Parasitology* **33**, 1623–1630.

Poulin R, Valtonen ET (2001a) Nested assemblages resulting from host size variation: the case of endoparasite communities in fish hosts. *International Journal for Parasitology* **31**, 1194–1204.

Poulin R, Valtonen ET (2001b) Interspecific associations among larval helminths in fish. *International Journal for Parasitology* **31**, 1589–1596.

Rohde K (1979) A critical evaluation of intrinsic and extrinsic factors responsible for niche restriction in parasites. *American Naturalist* **114**, 648–671.

Rohde K (1991) Intra- and interspecific interactions in low density populations in resource-rich habitats. *Oikos* **60**, 91–104.

Rohde K, Worthen W, Heap M, Hugueny B, Guégan J-F (1998) Nestedness in assemblages of metazoan ecto- and endoparasites of marine fish. *International Journal for Parasitology* **28**, 543–549.

Sousa WP (1994) Patterns and processes in communities of helminth parasites. *Trends in Ecology and Evolution* **9**, 52–57.

Timi JT, Poulin R (2003) Parasite community structure within and across host populations of a marine pelagic fish: how repeatable is it? *International Journal for Parasitology* **33**, 1353–1362.

Tokeshi M (1999) 'Species Coexistence: Ecological and Evolutionary Perspectives'. (Blackwell Science: Oxford).

Vickery WL, Poulin R (2002) Can helminth community patterns be amplified when transferred by predation from intermediate to definitive hosts? *Journal of Parasitology* **88**, 650–656.

Vidal-Martínez VM, Poulin R (2003) Spatial and temporal repeatability in parasite community structure of tropical fish hosts. *Parasitology* **127**, 387–398.

Worthen WB (1996) Community composition and nested-subset analyses: basic descriptors for community ecology. *Oikos* **76**, 417–426.

Worthen WB, Rohde K (1996) Nested subset analyses of colonization-dominated communities: metazoan ectoparasites of marine fishes. *Oikos* **75**, 471–478.

Wright DH, Patterson BD, Mikkelson GM, Cutler A, Atmar W (1998) A comparative analysis of nested subset patterns of species composition. *Oecologia* **113**, 1–20.

Parasite populations and communities as non-equilibrium systems

Combes C (2001) 'Parasitism. The Ecology and Evolution of Intimate Interactions'. (University of Chicago Press: Chicago).

Esch GE, Fernandez JC (1993) 'A Functional Biology of Parasitism'. (Chapman and Hall: London).

Gotelli NJ, Rohde K (2002) Co-occurrence of ectoparasites of marine fishes: a null model analysis. *Ecology Letters* **5**, 1–9.

Hanski I (1999) 'Metapopulation Ecology'. (Oxford University Press: Oxford).

Hubbell SP (2001) 'The Unified Neutral Theory of Biodiversity and Biogeography'. (Princeton University Press: Princeton).

Koch M (2004) Aquatic snails and their trematodes in the New England Tablelands (New South Wales, Australia): low diversity and unstable population structure explained by unpredictable environmental fluctuations. PhD Thesis, University of New England, Armidale.

Kormondy EJ (1969) 'Concepts of Ecology'. (Prentice-Hall: Englewood Cliffs, NJ).

Morand S, Poulin R, Rohde K, Hayward CJ (1999) Aggregation and species coexistence of ectoparasites of marine fishes. *International Journal for Parasitology* **29**, 663–672.

Morand S, Rohde K, Hayward CJ (2002) Order in parasite communities of marine fish is explained by epidemiological processes. *Parasitology* **124**, S57–S63.

Mouillot D, George-Nascimento M, Poulin R (2003) How parasites divide resources: a test of the niche apportionment hypothesis. *Journal of Animal Ecology* **72**, 757–764.

Pianka ER (1974) 'Evolutionary Ecology'. 2nd edn. (Harper & Row: New York).

Poulin R, Mouillot D, George-Nascimento M (2003) The relationship between species richness and productivity in parasite vommunities. *Oecologia* **137**, 277–285.

Price PW (1980) 'Evolutionary Biology of Parasites'. (Princeton University Press: Princeton, NJ).

Price PW, Slobodchikoff CN, Gaud WS Eds. (1984) 'A New Ecology. Novel Approaches to Interactive Systems'. (John Wiley & Sons: New York).

Ritchie ME, Olff H (1999) Spatial scaling laws yield a synthetic theory of biodiversity. *Nature* **400**, 557–562.

Rohde K (1977) A non-competitive mechanism responsible for restricting niches. *Zoologischer Anzeiger* **199**, 164–172.

Rohde K (1979) A critical evaluation of intrinsic and extrinsic factors responsible for niche restriction in parasites. *American Naturalist* **114**, 648–671.

Rohde K (1980a) Warum sind ökologische Nischen begrenzt? Zwischenartlicher Antagonismus oder innerartlicher Zusammenhalt? *Naturwissenschaftliche Rundschau* **33**, 98–102.

Rohde K (1980b) Species diversification, with special reference to parasites. *Proceedings of the 24th Conference of the Australian Society for Parasitology*, Adelaide, May.

Rohde K (1989) Simple ecological systems, simple solutions to complex problems? *Evolutionary Theory* **8**, 305–350.

Rohde K (1991) Intra- and interspecific interactions in low density populations in resource-rich habitats. *Oikos* **60**, 91–104.

Rohde K (1992) Latitudinal gradients in species diversity, the search for the primary cause. *Oikos* **65**, 514–527.

Rohde K (1994) Niche restriction in parasites: proximate and ultimate causes. *Parasitology* **109**, S69–S84.

Rohde K (1998a) Is there a fixed number of niches for endoparasites of fish? *International Journal for Parasitology* **28**, 1861–1865.

Rohde K (1998b) Latitudinal gradients in species diversity. Area matters, but how much? *Oikos* **82**, 184–190.

Rohde K (1999) Latitudinal gradients in species diversity and Rapoport's rule revisited: a review of recent work, and what can parasites teach us about the causes of the gradients? *Ecography* **22**, 593–613.

Rohde K (2001) Spatial scaling laws may not apply to most animal species. *Oikos* **93**, 499–504.

Rohde K (2002a) Ecology and biogeography of marine parasites. *Advances in Marine Biology* **43**, 1–86.

Rohde K (2002b) Niche restriction and mate finding in vertebrates. *In* 'Behavioral Ecology of Parasites'. (Eds E Lewis, JF Campbell, MVK Sukhdeo.) pp. 171–197. (CAB International: Wallingford, Oxon).

Rohde K (2005) 'Nonequilibrium Ecology'. (Cambridge University Press: Cambridge).

Rohde K, Hobbs R (1986) Species segregation: Competition or reinforcement of reproductive barriers? *In* 'Parasite Lives. Papers on Parasites, their Hosts and their Association to Honour JFA Sprent'. (Eds M Cremin, C Dobson, DE Moorhouse.) pp. 89–199. (University of Queensland Press: St Lucia).

Rohde K, Hobbs R (1999) An asymmetric percent similarity index. *Oikos* **87**, 601–602.

Rohde K, Hayward C, Heap M (1995) Aspects of the ecology of metazoan ectoparasites of marine fishes. *International Journal for Parasitology* **25**, 945–970.

Rohde K, Hayward C, Heap M, Gosper D (1994) A tropical assemblage of ectoparasites: gill and head parasites of *Lethrinus miniatus* (Teleostei, Lethrinidae). *International Journal for Parasitology* **24**, 1031–1053.

Rohde K, Worthen W, Heap M, Hugueny B, Guégan J.-F (1998) Nestedness in assemblages of metazoan ecto- and endoparasites of marine fish. *International Journal for Parasitology* **28**, 543–549.

Rosenzweig ML (1995) 'Species Diversity in Space and Time'. (Cambridge University Press: Cambridge).

Simková A, Desdevises Y, Gelnar M, Morand S (2000) Coexistence of nine gill ectoparasites (*Dactylogyrus*: Monogenea) parasitising the roach (*Rutilus rutilus* L.): history and present ecology. *International Journal for Parasitology* **30**, 1077–1088.

Simková A, Desdevises Y, Gelnar M, Morand S (2001) Morphometric correlates of host specificity in *Dactylogyrus* species (Monogenea) parasites of European cyprinid fish. *Parasitology* **123**, 169–177.

Tilman D (1982) 'Resource Competition and Community Structure'. (Princeton University Press: Princeton, NJ).

Worthen WB, Rohde K (1996) Nested subset analysis of colonisation-dominated communities: metazoan ectoparasites of marine fish. *Oikos* **75**, 471–478.

Population and community ecology of larval trematodes in molluscan first intermediate hosts

Anderson RM (1978) Population dynamics of snail infection by miracidia. *Parasitology* **77**, 201–224.

Basson M. (1994) A preliminary investigation of the possible effects of rhizocephalan parasitism on the management of the crab fishery around South Georgia. *CCAMLR Science* **1**, 175–192.

Baudoin M (1975) Host castration as a parasitic strategy. *Evolution* **29**, 335–352.

Combes C (1982) Trematodes: antagonism between species and sterilizing effects on snails in biological control. *Parasitology* **84**, 151–175.

Combes C (2001) 'Parasitism: the Ecology and Evolution of Intimate Interactions'. (University of Chicago Press: Chicago).

Curtis LA (1995) Growth, trematode parasitism, and longevity of a long-lived marine gastropod. (*Ilyanassa obsoleta*). *Journal of the Marine Biological Association of the United Kingdom* **75**, 913–925.

Curtis L A (1997) *Ilyanassa obsoleta* (Gastropoda) as a host for tematodes in Delaware estuaries. *Journal of Parasitology* **83**, 793–803.

Esch GW, Fernandez JC (1994) Snail-trematode interactions and parasite community dynamics in aquatic systems: a review. *American Midland Naturalist* **131**, 209–237.

Ewers W (1964) The influence of the density of snails on the incidence of larval trematodes. *Parasitology* **54**, 579–583.

Fredensborg BL, Mouritsen KN, Poulin R (2005). Impact of trematodes on host survival and population density in the intertidal gastropod *Zeacumantus subcarinatus*. *Marine Ecology Progress Series* **290**, 109–117.

Gaines SD, Lafferty KD (1995) Modeling the dynamics of marine species: the importance of incorporating larval dispersal. *In* 'Ecology of Marine Invertebrate Larvae'. (Ed. L McEdward.) pp. 389–412. (CRC Press: Boca Raton).

Huffaker CB (1964) Fundamentals of biological weed control. *In* 'Biological Control of Insects Pests and Weeds'. (Ed. P DeBach.) pp. 631–649. (Reinhold Publishing: New York).

Huxham M, Raffaelli D, Pike A (1993) The influence of *Cryptocotyle lingua* (Digenea: Platyhelminthes) infections on the survival and fecundity of *Littorina littorea* (Gastropoda: Prosobranchia): An ecological approach. *Journal of Experimental Marine Biology and Ecology* **168**, 223–238.

Jensen KT, Mouritsen KN (1992) Mass mortality in two common soft-bottom invertebrates, *Hydrobia ulvae* and *Corophium volutator* – the possible role of trematodes. *Helgolander Meeresunters* **46**, 329–339.

Kube J, Kube S, Dierschke V (2002) Spatial and temporal variations in the trematode component community of the mudsnail *Hydrobia ventrosa* in relation to the occurrence of waterfowl as definitive hosts. *Journal of Parasitology* **88**, 1075–1086.

Kuris AM (1973) Biological control: implications of the analogy between the trophic interactions of insect pest-parasitoid and snail-trematode systems. *Experimental Parasitology* **33**, 365–379.

Kuris AM (1974) Trophic interactions: similarity of parasitic castrators to parasitoids. *Quarterly Review of Biology* **49**, 129–148.

Kuris AM (1990) Guild structure of larval trematodes in molluscan hosts: prevalence, dominance and significance of competition. *In* 'Parasite Communities: Patterns and Processes'. (Eds GW Esch, AO Bush and JM Aho.) pp. 69–100. (Chapman and Hall: London).

Kuris AM, Lafferty KD (1992) Modelling crustacean fisheries: effects of parasites on management strategies. *Canadian Journal of Fisheries and Aquatic Sciences* **49**, 327–336.

Kuris AM, Lafferty KD (1994) Community structure: larval trematodes in snail hosts. *Annual Review of Ecology and Systematics* **25**, 189–217.

Kuris AM, Warren J (1980) Echinostome cercarial penetration and metacercarial encystment as mortality factors for a second intermediate host, *Biomphalaria glabrata*. *Journal of Parasitology* **66**, 630–635.

Lafferty KD (1991) Effects of parasitic castration on the salt marsh snail, *Cerithidea californica*. PhD University of California, Santa Barbara.

Lafferty KD (1993) Effects of parasitic castration on growth, reproduction and population dynamics of the marine snail *Cerithidea californica*. *Marine Ecology Progress Series* **96**, 229–237.

Lafferty KD, Kuris AM (1996) Biological control of marine pests. *Ecology* **77**, 1989–2000.

Lafferty KD, Kuris AM (2002) Trophic strategies, animal diversity and body size. *TREE* **17**, 507–513.

Lafferty KD, Sammond DT, Kuris, AM (1994) Analysis of larval trematode communities. *Ecology* **75**, 2275–2285.

Lauckner G (1987) Ecological effects of larval trematode infestation on littoral marine invertebrate populations. *Parasitology – Quo Vadit?* **17**, 391–398.

Lie KJ, Ow-Yang CK (1973) A field trial to control *Trichobilharzia brevis* by dispersing eggs of *Echinostoma audyi*. *Southeast Asian Journal of Tropical Medicine and Public Health* **4**, 208–217.

Loker ES, Cimino DF, Stryker GA, Hertel LA (1987) The effect of size on M line *Biomphalaria glabrata* on the course of development of *Echinostoma paraensei*. *Journal of Parasitology* **73**, 1090–1098.

McDaniel JS (1969) *Littorina littorea*: lowered heat tolerance due to *Cryptocotyle lingua*. *Experimental Parasitology* **25**, 13–15.

Robson EM, Williams IC (1970) Relationships of some species of Digenea with the marine prosobranch *Littorina littorea* (L.). I. The occurrence of larval Digenea in *L. littorea* on the North Yorkshire coast. *Journal of Helminthology* **44**, 153–168.

Skirnisson K, Galaktionov KV (2002) Life cycles and transmission patterns of seabird digeneans in SW Iceland. *Sarsia* **87**, 144–151.

Skirnisson K, Galaktionov KV, Kozminsky EV (2004) Factors influencing the distribution of digenetic trematode infections in a mudsnail (*Hydrobia ventrosa*) population inhabiting salt marsh ponds in Iceland. *Journal of Parasitology* **90**, 50–59.

Smith NF (2001) Spatial heterogeneity in recruitment of larval trematodes to snail intermediate hosts. *Oecologia* **127**, 115–122.

Sousa WP (1990) Spatial scale and the processes structuring a guild of larval trematode parasites. *In* 'Parasite Communities: Patterns and Processes'. (Eds GW Esch, AO Bush and JM Aho.) pp. 41–67. (Chapman and Hall: London).

Sousa WP (1993) Interspecific antagonism and species coexistence in a diverse guild of larval trematode parasites. *Ecological Monographs* **63**, 103–128.

Sousa WP, Gleason M (1989) Does parasitic infection compromise host survival under extreme environmental conditions? The case for *Cerithidea californica* (Gastropoda: Prosobranchia). *Oecologia* **80**, 456–464.

Tallmark B, Norrgren G (1976) The influence of parasitic trematodes on the ecology of *Nassarius reticulatus* (L.) in Gullmar Fjord (Sweden). *Zoon* **4**, 149–154.

Chapter 8

Coevolution in marine systems

Bandoni SM, Brooks DR (1987a) Revision and phylogenetic analysis of the Gyrocotylidea Poche, 1926 (Platyhelminthes: Cercomeria: Cercomeromorpha). *Canadian Journal of Zoology* **65**, 2369–2389.

Bandoni SM, Brooks DR (1987b) Revision and phylogenetic analysis of the Amphilinidea Poche, 1922 (Platyhelminthes: Cercomeria: Cercomeromorpha). *Canadian Journal of Zoology* **65**, 1110–1128.

Barker SC, Cribb TH, Bray RA, Adlard RD (1994) Host-parasite associations on a coral reef: pomacentrid fishes and digenean trematodes. *International Journal for Parasitology* **24**, 643–647.

Berta A, Sumich JL (1999) 'Marine Mammals Evolutionary Biology'. (Academic Press: San Francisco).

Beveridge I, Campbell RA, Palm HW (1999) Preliminary cladistic analysis of genera of the cestode order Trypanorhyncha Diesing, 1863. *Systematic Parasitology* **42**, 29–49.

Boeger WA, Kritsky DC (1997) Coevolution of the Monogenoidea (Platyhelminthes) based on a revised hypothesis of parasite phylogeny. *International Journal for Parasitology* **12**, 1495–1511.

Bray RA, Cribb TH, Littlewood TJ (1998) A phylogenetic study of *Lepidapedoides* Yamaguti, 1970 (Digenea) with a key and description of two new species from Western Australia. *Systematic Parasitology* **39**, 183–197.Bray RA, Cribb TH (2000) Species of *Trifoliovarium* Yamaguti, 1940 (Digenea: Lecithasteridae) from Australian waters, with a description of *T. draconis* n. sp. and a cladistic study of the subfamily Trifoliovariinae Yamaguti, 1958. *Systematic Parasitology* **47**, 183–192.

Briggs JC (1995) 'Global Biogeography'. (Elsevier: Amsterdam)

Brooks DR (1979) Testing the context and extent of host-parasite coevolution. *Systematic Zoology* **28**, 299–307.

Brooks DR (1981) Hennig's parasitological method: a proposed solution. *Systematic Biology* **30**, 229–249.

Brooks DR (1985) Historical ecology: a new approach to studying the evolution of ecological associations. *Annals of the Missouri Botanical Garden* **72**, 660–680.

Brooks DR (1989) Summary of the database pertaining to the phylogeny of the major groups of parasitic Platyhelminthes, with a revised classification. *Canadian Journal of Zoology* **67**, 714–720.

Brooks DR (1992) Origins, diversification and historical structure of the helminth fauna inhabiting neotropical freshwater stingrays (Potamotrygonidae). *Journal of Parasitology* **78**, 588–595.

Brooks DR (1995) Neotropical freshwater stingrays and their parasites: a tale of an ocean and a river long ago. *Journal of Aquariculture and Aquatic Science* **7**, 52–61.

Brooks DR, Amato JFR (1992) Cestode parasites in *Potamotrygon motoro* (Natterer) (Chondrichthyes: Potamotrygonidae) from southwestern Brazil, including *Rhinobothroides mclennanae* n. sp. (Tetraphyllidea: Phyllobthriidae), and a revised host-parasite checklist for helminths inhabiting neotropical freshwater stingrays. *Journal of Parasitology* **78**, 393–398.

Brooks DR, Bandoni S (1988) Coevolution and relicts. *Systematic Zooloogy* **37**, 19–33

Brooks DR, Deardorf T (1988) *Rhinobothrium devaneyi* n.sp. (Eucestoda: Tetraphyllidea) and *Echinocephalus overstreeti* Deardorf and Ko, 1983 (Nematoda: Gnathostomatidae) in a thorny back ray, *Urogymnus asperrimus*, from Enewetak atoll, with phylogenetic analysis of both species groups. *Journal of Parasitology* **74**, 459–465.

Brooks DR, Dowling APG, van Veller MPG, Hoberg EP (2004) Ending a decade of deception: a valiant failure, a not so valiant failure, and a success story. *Cladistics* **20**, 32–46.

Brooks DR, Hoberg E P (2000) Triage for the biosphere: the need and rationale for taxonomic inventories and phylogenetic studies of parasites. *Comparative Parasitology* **67**, 1–25.

Brooks DR, McLennan DA (1991) 'Phylogeny, Ecology and Behavior, a Research Program in Comparative Biology'. (University of Chicago Press: Chicago).

Brooks DR, McLennan DA (1993) 'Parascript: Parasites and the Language of Evolution'. (Smithsonian Institution Press: Washington, DC).

Brooks DR, McLennan DA (2002) 'The Nature of Diversity: an Evolutionary Voyage of Discovery'. (University of Chicago Press: Chicago).

Brooks DR, Pérez-Ponce de León G, León-Règagnon V (2000) Phylogenetic analysis of the Enenterinae (Digenea, Lepocreadiidae) and discussion of the evolution of the digenean fauna of kyphosid fishes. *Zoologica Scripta* **29**, 237–246.

Brooks DR, Thorson TB, Mayes MA (1981) Freshwater stingrays (Potamotrygonidae) and their helminth parasites: testing hypotheses of evolution and coevolution. *In* 'Advances in Cladistics'. (Eds VA Funk, DR Brooks.) pp. 147–175. (New York Botanical Garden: New York).

Brooks DR, vanVeller MGP, McLennan DA (2001) How to do BPA, really. *Journal of Biogeography* **28**, 345–358.

Bullini L, Arduino P, Cianchi R, Nascetti G, D'Amelio S, Mattiucci S, Paggi L, Orrechia P, Berland B, Smith JW, Brattey JW (1997) Genetic and ecological research on anisakid endoparasites of fish and marine mammals in the Antarctic and Arctic-Boreal regions. *In* 'Antarctic Communities: Species Structure and Survival'. (Eds B Battaglia, J Valencia, DWH Walton.) pp. 39–44. (Cambridge University Press: Cambridge).

Caira JN (1992) Verification of multiple species of *Pedibothrium* in the Atlantic nurse shark with comments on the Australian members of the genus. *Journal of Parasitology* **78**, 289–308.

Caira JN, Burge AN (2001) Three new species of *Acanthobothrium* (Cestoda: Tetraphyllidea) from the ocellated electric ray, *Diplobatis ommata*, in the Gulf of California, Mexico. *Comparative Parasitology* **68**, 52–65.

Caira JN, Euzet L (2001) Age of association between the nurse shark, Ginglymostoma cirratum, and tapeworms of the genus Pedibothrium (Tetraphyllidea: Onchobothriidae): implications for geography. *Biological Journal of the Linnean Society* **72**, 609–614.

Caira JN, Jensen K (2001) An investigation of the co-evolutionary relationships between onchobothriid tapeworms and their elasmobranch hosts. *International Journal for Parasitology* **31**, 960–975.

Caira JN, Zahner SD (2001) Two species of *Acanthobothrium* Beneden, 1849 (Tetraphyllidea: Onchobothriidae) from horn sharks in the Gulf of California, Mexico. *Systematic Parasitology* **50**, 219–229.

Caira JN, Jensen K, Healy CJ (2001) Interrelationships among tetraphyllidean and lecanicephalidean cestodes *In* 'Interrelationships of the Platyhelminthes'. (Eds DTJ Littlewood, RA Bray.) pp. 135–158. (Taylor and Francis: London).

Chabaud AG (1965) Spécificité parasitaire. I. 'Chez sur les Nématodes parasites de Vertébrés. *In* 'Traité de Zoologie'. Vol. 4. (Ed. PP Grassé.) pp. 548–557. (Masson: Paris).

Collette BB, Russo JL (1985) Interrelationships of the Spanish mackerels (Pisces: Scombridae: *Scomberomorus*) and their copepod parasites. *Cladistics* **1**, 141–158.

Coyne JA, Orr AH (2004) 'Speciation'. (Sinauer Associates: Sunderland, Massachusetts).

Cribb TH, Bray RA, Barker SC (1992) Zoogonidae (Digenea) from southern Great Barrier Reef fishes with a description of *Steganoderma* (*Lecithostaphylus*) *gibsoni* n. sp. *Systematic Parasitology* **23**, 7–12.

Cribb TH, Bray RA, Littlewood DTJ (2001) The nature and evolution of the association among digeneans, molluscs and fishes. *International Journal for Parasitology* **31**, 997–1011.

Dabert J, Mironov SV (1999) Origin and evolution of feather mites (Astigmata). *Experimental and Applied Acarology* **23**, 437–454.

Deliamure SL (1955) 'Helminthofauna of Marine Mammals (Phylogeny and Ecology)'. (Izdatel'stvo Akad Nauk SSSR: Moscow). [English Translation, Israel Program for Scientific Translations, Jerusalem].

Demastes JW, Spradling TA, Hafner MS (2003) The effects of spatial and temporal scale on analyses of cophylogeny. *In* 'Tangled Trees: Phylogeny, Cospeciation and Coevolution'. (Ed. RDM Page.) pp. 221–239. (University of Chicago Press: Chicago).

Dowling APG, van Veller MGP, Hoberg EP, Brooks DR (2003) *A priori* and *a posteriori* methods in comparative evolutionary studies of host-parasite associations. *Cladistics* **19**, 240–253.

Ehrlich PR, Raven PH (1964) Butterflies and plants: a study in coevolution. *Evolution* **18**, 586–608.

Euzet L (1959). Recherches sur les cestodes tétraphyllides des sélaciens des côtes de France. (Doctoral Dissertation, Montpellier, France).

Fernández M, Aznar FJ, Latorre A, Raga JA (1998a) Molecular phylogeny of the families Campulidae and Nasitrematidae (Trematoda) based on mtDNA sequence comparison. *International Journal for Parasitology* **28**, 767–775.

Fernández M, Littlewood DTJ, Latorre A, Raga JA, Rollinson D (1998b) Phylogenetic relationships of the family Campulidae (Trematoda) based on 18s rRNA sequences. *Parasitology* **117**, 383–391.

Frey JK (1993) Modes of peripheral isolate formation and speciation. *Systematic Biology* **42**, 373–381.

Gibson DI, Bray RA (1994). The evolutionary expansion and host-parasite relationships of the Digenea. *International Journal for Parasitology* **24**, 1213–1226.

Goshroy S, Caira JC (2001) Four new species of Acanthobothrium (Cestoda: Tetraphyllidea) from the whiptail stingray *Dasyatis brevis* in the Gulf of California. *Journal of Parasitology* **87**, 354–372.

Hafner MS, Nadler SA (1988) Phylogenetic trees support the coevolution of parasites and their hosts. *Nature* **332**, 258–259.

Hafner MS, Sudman PD, Villablanca FX, Spradling TA, Demastes JW, Nadler SA (1994) Disparate rates of molecular evolution in cospeciating hosts and parasites. *Science* **265**, 1087–1090.

Hoberg EP (1986) Evolution and historical biogeography of a parasite-host assemblage: *Alcataenia* spp. (Cyclophyllidea: Dilpeididae) in Alcidae (Charadriiformes). *Canadian Journal of Zoology* **64**, 2576–2589.

Hoberg EP (1992) Congruent and synchronic patterns in biogeography and speciation among seabirds, pinnipeds and cestodes. *Journal of Parasitology,* **78**, 601–615.

Hoberg EP (1995) Historical biogeography and modes of speciation across high-latitude seas of the Holarctic: concepts for host-parasite coevolution among the Phocini (Phocidae) and Tetrabothriidae. *Canadian Journal of Zoology* **73**, 45–57.

Hoberg EP (1996) Faunal diversity among avian parasite assemblages: the interaction of history, ecology, and biogeography in marine systems. *Bulletin of the Scandinavian Society of Parasitology* **6**, 65–89.

Hoberg EP (1997) Phylogeny and historical reconstruction: host-parasite systems as keystones in biogeography and ecology. *In* 'Biodiversity II: Understanding and Protecting Our Biological Resources'. (Eds M Reaka-Kudla, DE Wilson, EO Wilson.) pp. 243–261. (Joseph Henry Press: Washington DC).

Hoberg EP, Adams A (1992) Phylogeny, historical biogeography, and ecology of *Anophryocephalus* spp. (Eucestoda: Tetrabothriidae) among pinnipeds of the Holarctic during the late Tertiary and Pleistocene. *Canadian Journal of Zoology* **70**, 703–719.

Hoberg EP, Adams A (2000) Phylogeny, history and biodiversity: understanding faunal structure and biogeography in the marine realm. *Bulletin of the Scandinavian Society of Parasitology* **10**, 19–37.

Hoberg EP, Klassen GJ (2002) Revealing the faunal tapestry: co-evolution and historical biogeography of hosts and parasites in marine systems. *Parasitology* **124**, S3–S22.

Hoberg EP, Brooks DR, Siegel-Causey D (1997) Host-parasite cospeciation: history, principles and prospects. *In* 'Host-Parasite Evolution: General Principles and Avian Models'. (Eds DH Clayton, J Moore.) pp. 212–235. (Oxford University Press: Oxford).

Hoberg EP, Gardner SL, Campbell RA (1999a) Systematics of the Eucestoda: advances toward a new phylogenetic paradigm, and observations on the early diversification of tapeworms and vertebrates. *Systematic Parasitology* **42**, 1–12.

Hoberg EP, Jones A, Bray R (1999b) Phylogenetic analysis among families of the Cyclophyllidea (Eucestoda) based on comparative morphology, with new hypotheses for co-evolution in vertebrates. *Systematic Parasitology* **42**, 51–73.

Hoberg EP, Mariaux J, Brooks DR (2001) Phylogeny among orders of the Eucestoda (Cercomeromorphae): integrating morphology, molecules and total evidence. *In* 'Interrelationships of the Platyhelminthes'. (Eds DTJ Littlewood, RA Bray.) pp. 112–126. (Taylor and Francis: London).

Iurakhno MV (1991) (On the evolution of helminths of marine mammals). *In* Evoliutsiya Parazitov'. (Materialy Pergovo Vsesoyuznogo Sympoziuma.) pp. 124–127. (Insitut Ekologii Volzhskogo Basseina: Tol' yatti, Russia).

Johnson KP, Adams RJ, Page RDM, Clayton DH (2003) When do parasites fail to speciate in response to host speciation? *Systematic Biology* **52**, 37–47.

Kearn GC (1994) Evolutionary expansion of the Monogenea. *International Journal for Parasitology* **24**, 1227–1271.

Klassen GJ (1992a) Coevolution: a history of the macroevolutionary approach to studying host-parasite associations. *Journal of Parasitology* **78**, 573–587.

Klassen GJ (1992b) Phylogeny and biogeography of ostracin boxfishes (Tetraodontiformes: Ostraciidae) and their gill parasites *Haliotrema* sp. (Monogenea: Ancyrocephalidae): a study in host-parasite coevolution. PhD Dissertation. (University of Toronto: Toronto).

Klassen GJ (1994) Phylogeny of *Haliotrema* species (Monogenea: Ancyrocephalidae) from boxfishes (Tetraodontiformes: Ostraciidae): are *Haliotrema* species from boxfishes monophyletic? *Journal of Parasitology* **80**, 596–610.

Kontrimavichus VL (1969) 'Helminths of mustelids and trends in their evolution. (Akademiia Nauk SSSR, Gel'mintologicheskaya Laboratoriia. Nauka Publishers: Moscow.) [English translation (1985), Amerind Publishing Company: New Delhi].

Littlewood DTJ, Rohde K, Bray RA, Herniou EA (1999) Phylogeny of the Platyhelminthes and the evolution of parasitism. *Biological Journal of the Linnaean Society* **68**, 257–287.

Manter HW (1966) Parasites of fishes as biological indicators of ancient and recent conditions. *In* 'Host Parasite Relationships'. (Ed. JE McCauley.) pp. 59–71. (Oregon State University Press: Corvallis).

Marques F, Brooks DR, Monks S (1995) Five new species of *Acanthobothrium* Van Beneden, 1849 (Eucestoda: Tetraphyllidea: Onchobothriidae) in stingrays from the Gulf of Nicoya, Costa Rica. *Journal of Parasitology* **81**, 942–951.

Marques F, Centritto R, Stewart SA (1997) Two new species of *Acanthobothrium* in *Narcine entemedor* (Rajiformes: Narcinidae) from the northwest coast of Guancaste Peninsula, Costa Rica. *Journal of Parasitology* **83**, 927–931.

Mitter CM, Brooks DR (1983) Phylogenetic aspects of coevolution. *In* 'Coevolution'. (Eds DJ Futuyma, M Slatkin.) pp. 65–98. (Sinauer Associates: Sunderland).

Nadler SA (2002) Species delimitation and nematode biodiversity: phylogenies rule. *Nematology* **4**, 615–625.

Nadler SA, D'Amelio S, Fagerholm H-P, Berland B, Paggi L (2000) Phylogenetic relationships among species of *Contracaecum* Railliet & Henry, 1912 and *Phocascaris* Høst, 1932 (Nematoda: Ascaridoidea) based on nuclear rDNA sequence data. *Parasitology* **121**, 455–463.

Nascetti G, Cianchi R, Mattiuci S, D'Amelio S, Orrechia P, Paggi L, Brattey J, Berland B, Smith J, Bullini L (1993) Three sibling species within *Contracaecum osculatum* (Nematoda: Ascaridida: Ascaridoidea) from the Atlantic Arctic-Boreal region: reproductive isolation and host-preferences. *International Journal for Parasitology* **23**, 105–120.

Nasin C, Caira JN, Euzet L (1997) A revision of *Calliobothrium* (Tetraphyllidea: Onchobothriidae) with descriptions of three new species and a cladistic analysis of the genus. *Journal of Parasitology* **83**, 714–733.

Olson PD, Rhunke TR, Sanney J, Hudson T (1999) Evidence for host-specific clades of tetraphyllidean tapeworms (Platyhelminthes: Eucestoda) revealed by analysis of 18S ssrDNA. *International Journal for Parasitology* **29**, 1465–1476.

Olson PD, Littlewood DTJ, Bray RA, Mariaux J (2001) Interrelationships and evolution of the tapeworms (Platyhelminthes: Cestoda). *Molecular Phylogenetics & Evolution* **19**, 443–467.

Page RDM (1993) Genes, organisms and areas: the problem of multiple lineages. *Systematic Biology* **42**, 77–84.

Page RDM (1994a) Maps between trees and cladistic analyses of historical associations among genes, organisms and areas. *Systematic Biology* **43**, 58–77.

Page RDM (1994b) Parallel phylogenies, reconstructing the history of host-parasite assemblages. *Cladistics* **10**, 155–173.

Page RDM (Ed.) (2003) 'Tangled Trees: Phylogeny, Cospeciation and Coevolution'. (University of Chicago Press: Chicago).

Page RDM, Charleston MA (1998) Trees within trees: phylogeny and historical associations. *Trends in Ecology and Evolution* **13**, 356–359.

Page RDM, Paterson AM, Clayton DH (1996) Lice and cospeciation: a response to Barker. *International Journal for Parasitology* **26**, 213–218.

Paggi L, Nascetti G, Cianchi R, Orecchia P, Mattiucci S, D'Amelio S, Berland B, Brattey J, Smith JW, Bullini L (1991) Genetic evidence for three species within *Pseudoterranova decipiens* (Nematoda: Ascaridida, Ascaridoidea) in the North Atlantic and Norwegian and Barents Seas. *International Journal for Parasitology* **21**, 195–212.

Palm H (2004) The Trypanorhyncha Diesing, 1863. (PKSPL-IBP Press: Bogor).

Paterson AM, Banks J (2001) Analytical approaches to measuring cospeciation of host and parasites: through the glass darkly. *International Journal of Parasitology* **31**, 1012–1022.

Paterson AM, Gray RD (1997) Host-parasite co-speciation, host switching, and missing the boat. In 'Host-Parasite Evolution: General Principles and Avian Models'. (Eds DH Clayton, J Moore.) pp. 236–250. (Oxford University Press: Oxford).

Paterson AM, Gray RD, Wallis GP (1993) Parasites, petrels and penguins: does louse phylogeny reflect seabird phylogeny? *International Journal for Parasitology* **23**, 515–526.

Paterson AM, Palma RL, Gray RD (1999) How frequently do avian lice miss the boat? Implications for coevolutionary studies. *Systematic Biology* **48**, 214–223.

Paterson AM, Palma RL, Gray RD (2003) Drowning on arrival, missing the boat and X events: how likely are sorting events? In 'Tangled Trees: Phylogeny, Cospeciation and Coevolution'. (Ed. RDM Page.) pp. 287–309. (University of Chicago Press: Chicago).

Paterson AM, Poulin R (1999) Have chondracanthid copepods co-speciated with their teleost hosts? *Systematic Parasitology* **44**, 79–85.

Paterson AM, Wallis GP, Wallis LJ, Gray RD (2000) Seabird and louse coevolution: complex histories revealed by 12S rDNA sequences and reconciliation analyses. *Systematic Biology* **49**, 383–399.

Pérez-Ponce de León G, Brooks DR (1995a) Phylogenetic relationships of the genera of the Pronocephalidae Looss, 1902 (Digenea: Paramphistomiformes). *Journal of Parasitology* **81**, 267–277.

Pérez-Ponce de León G, Brooks DR (1995b) Phylogenetic relationships among the species of *Pyelosomum* Looss, 1899 (Digenea: Pronocephalidae). *Journal of Parasitology* **81**, 278–280.

Pérez-Ponce de León G, Garcia-Prieto L, Mendoza-Garfias B, León-Règagnon V, Puliodo-Flores G, Aranda-Cruz C, Garcia-Vargas S (1999) 'Listados faunistacos de México IX. Bioversidad de helmintos parásitos de peces marinos y estuarinos de la Bahía de Chamela, Jalisco'. (Universidad Nacional Autónoma México: México City).

Poulin R. (1992) Determinants of host-specificity in parasites of freshwater fishes. *International Journal for Parasitology* **22**, 753–758.

Rohde K (1993) 'Ecology of Marine Parasites'. 2nd edn. (CAB International: Wallingford).

Rohde K (1994) The origins of parasitism in the Platyhelminthes. *International Journal for Parasitology* **24**, 1099–1115.

Rohde K, Hayward CJ (2000) Oceanic barriers as indicated by scombrid fishes and their parasites. *International Journal for Parasitology* **30**, 579–583.

Stiassny MLJ, Parenti LR, Johnson CD (Eds) (1996) 'Interrelationships of Fishes'. (Academic Press: San Francisco).

Wilson EO (Ed.) (1988) 'Biodiversity'. (National Academy Press: Washington, DC).

Speciation and species delimitation

Adams BJ (1998) Species concepts and the evolutionary paradigm in modern nematology. *Journal of Nematology* **30**, 1–21.

Adams BJ (2002) The species delimitation uncertainty principle. *Journal of Nematology* **33**, 153–160.

George-Nascimento M, Lima M, Ortiz E (1992) A case of parasite-mediated competition? Phenotypic differentiation among hookworms *Uncinaria* sp. (Nematoda: Ancylostomatidae) in sympatric and allopatric populations of South American sea lions *Otaria byronia*, and fur seals *Arctocephalus australis* (Carnivora: Otariidae). *Marine Biology* **112**, 527–533.

Littlewood DTJ, Rohde K, Clough KA (1997) Parasite speciation within or between host species?-Phylogenetic evidence from site-specific polystome monogeneans. *International Journal for Parasitology* **27**, 1289–1297.

Lyons ET, DeLong RL, Melin SR, Tolliver SC (1997) Uncinariasis in Northern fur seal and California sea lion pups from California. *Journal of Wildlife Diseases* **33**, 848–852.

Lyons ET, DeLong RL, Gulland FM, Melin SR, Tolliver SC, Spraker TR (2000) Comparative biology of *Uncinaria* spp. in the California sea lion (*Zalophus californianus*) and the northern fur seal (*Callorhinus ursinus*) in California. *Journal of Parasitology* **86**, 1348–1352.

Lyons ET, Melin SR, DeLong RL, Orr AJ, Gulland FM, Tolliver SC (2001) Current prevalence of adult *Uncinaria* spp. in northern fur seal (*Callorhinus ursinus*) and California sea lion (*Zalophus californianus*) pups on San Miguel Island, California, with notes on the biology of these hookworms. *Veterinary Parasitology* **97**, 309–318.

Mayden RL (1997) A hierarchy of species concepts: the denouement in the saga of the species problem. *In* 'Species: the Units of Biodiversity'. (Eds MF Claridge, HA Dawah, MR Wilson.) pp. 381–424. (Chapman and Hall: London).

Nadler SA (2002) Species delimitation and nematode biodiversity: phylogenies rule. *Nematology* **4**, 615–625.

Nadler SA, Adams BJ, Lyons ET, DeLong RL, Melin SR (2000) Molecular and morphometric evidence for separate species of *Uncinaria* (Nematoda: Ancylostomatidae) in California sea lions and northern fur seals: Hypothesis testing supplants verification. *Journal of Parasitology* **86**, 1099–1106.

Nichols R (2001) Gene trees and species trees are not the same. *Trends in Ecology and Evolution* **16**, 358–364.

Nixon KC, Wheeler QD (1990) An amplification of the phylogenetic species concept. *Cladistics* **6**, 211–224.

Page RDM (1993) Genes organisms and areas: the problem of multiple lineages. *Systematic Biology* **42**, 77–84.

Rohde K (1989) Simple ecological systems, simple solutions to complex problems? *Evolutionary Theory* **8**, 305–350.

Wheeler QD (1999) Why the phylogenetic species concept? – Elementary. *Journal of Nematology* **31**, 134–141.

Wheeler QD, Meier R (2000) 'Species Concepts and Phylogenetic Theory: A Debate'. (Columbia University Press: New York).

Chapter 9
Latitudinal, longitudinal and depth gradients

Allen AP, Brown JH, Gillooly JF (2002) Global biodiversity, biochemical kinetics, and the energetic-equivalence rule. *Science* **297**, 1545–1548.

Ernst I, Whittington ID, Jones MK (2001). Diversity of gyrodactylids from marine fishes in tropical and subtropical Queensland, Australia. *Folia Parasitologica* **48**, 165–168.

Geets A, Coene H, Ollevier F (1997) Ectoparasites of the whitespotted rabbitfish, *Siganus sutor* (Valenciennes, 1835) off the Kenyan coast: distribution within the host population and site selection on the gills. *Parasitology* **115**, 69–79.

Gillooly JF, Charnov EL, West GB, Savage Van M, Brown JH (2002) Effects of size and temperature on developmental time. *Nature* **417**, 70–73.

Kleeman S (2001) The development of the community structure of the ecto- and endoparasites of Siganus doliatus, a tropical marine fish. BSc Honours thesis, University of New England: Armidale, Australia.

Martens E, Moens J (1995) The metazoan ecto- and endoparasites of the rabbitfish, *Siganus sutor* (Cuvier and Valenciennes, 1835) of the Kenyan coast. I. *African Journal of Ecology* **33**, 405–416.

Poulin R. Rohde K (1997) Comparing the richness of metazoan ectoparasite communities of marine fishes: controlling for host phylogeny. *Oecologia* **110**, 278–283.

Rohde K (1978) Latitudinal differences in host specificity of marine Monogenea and Digenea. *Marine Biology* **47**, 125–134.

Rohde K (1980) Host specificity indices of parasites and their application. *Experientia* **36**, 1369–1371.

Rohde K (1985). Increased viviparity of marine parasites at high latitudes. *Hydrobiologia* **127**, 197–201.

Rohde K (1989) Simple ecological systems, simple solutions to complex problems? *Evolutionary Theory* **8**, 305–350.

Rohde K (1992) Latitudinal gradients in species diversity: the search for the primary cause. *Oikos* **65**, 514–527.

Rohde K (1993) 'Ecology of Marine Parasites'. 2nd edn. (CAB International: Wallingford, Oxfordshire).

Rohde K (1999) Latitudinal gradients in species diversity and Rapoport's rule revisited: a review of recent work, and what can parasites teach us about the causes of the gradients? *Ecography* **22**, 593–613. Also published *In* 'Ecology 1999 – And Tomorrow'. (Ed. T Fenchel.), pp.73–93. (Ecology Institute: University of Lund, Sweden).

Rohde K (2002) Ecology and biogeography of marine parasites. *Advances in Marine Biology* **43**, 1–86.

Rohde K (2005) 'Nonequilibrium Ecology'. (Cambridge University Press: Cambridge).

Rohde K, Hayward CJ (2000) Oceanic barriers as indicated by scombrid fishes and their parasites. *International Journal for Parasitology* **30**, 579–583.

Rohde K, Heap M (1998) Latitudinal differences in species and community richness and in community structure of metazoan endo- and ectoparasites of marine teleost fish. *International Journal for Parasitology* **28**, 461–474.

Thorson G (1957) Bottom communities (sublittoral or shallow shelf). *In* 'Treatise on Marine Ecology and Palaeoecology'. (Ed. JW Hedgpeth.) pp. 461–534. (Geological Society of America).

Willig MR (2001) Latitude, common trends within. *In* 'Encyclopedia of Biodiversity'. Vol. 3. (Ed. S Levin.) pp. 701–714. (Academic Press: New York).

Parasites as biological tags

Arthur JR (1997) Recent advances in the use of parasites as biological tags for marine fish. *In* 'Diseases in Asian Mariculture III'. (Eds TW Flegel, IH MacRae.) pp. 141–154. (Fish Health Section, Asian Fisheries Society: Manila).

Balbuena JA, Aznar FJ, Fernandez M, Raga JA (1995) Parasites as indicators of social structure and stock identity of marine mammals. *In* 'Whales, Seals, Fish and Man. Developments in Marine Biology 4'. (Eds AS Blix, L Walloe, O Ulltang.) pp. 133–140. (Elsevier: Amsterdam).

Herrington WC, Bearse HM, Firth FE (1939) Observations on the life history, occurrence and distribution of the redfish parasite *Sphyrion lumpi*. United States Bureau of Fisheries Special Report No. 5, 1–18.

Kabata Z (1963) Parasites as biological tags. *ICNAF Special Publication* No. **4**, 31–37.

Lester RJG (1990) Reappraisal of the use of parasites for fish stock identification. *Australian Journal of Marine and Freshwater Research* **41**, 855–864.

Lester RJG, Thompson C, Moss H, Barker SC (2001) Movement and stock structure of narrow-banded Spanish mackerel as indicated by parasites. *Journal of Fish Biology* **59**, 833–843.

MacKenzie K (1983) Parasites as biological tags in fish population studies. *Advances in Applied Biology* **7**, 251–331.

MacKenzie K (1987a) Parasites as indicators of host populations. *International Journal for Parasitology* **17**, 345–352.

MacKenzie K (1987b) Relationships between the herring, *Clupea harengus* L., and its parasites. *Advances in Marine Biology* **24**, 263–319.

MacKenzie K (2002) Parasites as biological tags in population studies of marine organisms: an update. *Parasitology* **124**, S153–S163.

MacKenzie K, Abaunza P (1998) Parasites as biological tags for stock discrimination of marine fish: a guide to procedures and methods. *Fisheries Research* **38**, 45–56.

MacKenzie K, Abaunza P (2005) Parasites as biological tags. *In* 'Stock Identification Methods. Applications in Fisheries Science'. (Eds SX Cadrin, KD Friedland, JR Waldman.) pp. 211–226. (Elsevier Academic Press, San Diego).

Mattiucci S, Abaunza P, Ramadori L, Paggi L, MacKenzie K, Nascetti G (2002) *Anisakis* spp. larvae as biological tags: genetic markers for their identification and fish stocks implication. *In* 'Proceedings of the 10th International Congress of Parasitology – ICOPA X: Symposia, Workshops and Contributed Papers'. pp. 223–227. (Monduzzi Editore, International Proceedings Division: Bologna, Italy).

Moore ABM (2001) Metazoan parasites of the lesser-spotted dogfish *Scyliorhinus canicula* and their potential as stock discrimination tools. *Journal of the Marine Biological Association of the United Kingdom* **81**, 1009–1013.

Moore BR, Buckworth RC, Moss H, Lester RJG (2003) Stock discrimination and movements of narrow-barred Spanish mackerel across northern Australia as indicated by parasites. *Journal of Fish Biology* **63**, 765–779.

Moser M (1991) Parasites as biological tags. *Parasitology Today* **7**, 182–185.

Owens L (1983) Bopyrid parasite *Epipenaeon ingens* Nobili as a biological marker for the banana prawn, *Penaeus mergueinsis* de Man. *Australian Journal of Marine and Freshwater Research* **34**, 477–481.

Owens L (1985) *Polypocephalus* sp. (Cestoda: Lecanicephalidae) as a biological marker for banana prawns, *Penaeus merguensis* de Man, in the Gulf of Carpentaria. *Australian Journal of Marine and Freshwater Research* **36**, 291–299.

Pascual S, Hochberg FG (1996) Marine parasites as biological tags of cephalopod hosts. *Parasitology Today* **12**, 324–327.

Pascual S, Gonzalez A, Arias C, Guerra A (1996) Biotic relationships of *Illex coindetti* and *Todaropsis eblanae* (Cephalopoda, Ommastrephidae) in the northeast Atlantic: evidence from parasites. *Sarsia* **81**, 265–274.

Sinderman CJ (1961) Parasite tags for marine fish. *Journal of Wildlife Management* **25**, 41–47.

Sindermann CJ (1983) Parasites as natural tags for marine fish: a review. *NAFO Scientific Council Studies* **6**, 63–71.

Thompson AB, Margolis L (1987) Determination of population discreteness in two species of shrimp. *Pandalus jordani* and *Pandalopsis dispar*, from coastal British Columbia using parasite tags and other population characteristics. *Canadian Journal of Fisheries and Aquatic Sciences* **44**, 982–989.

Williams HH, MacKenzie K, McCarthy AM (1992) Parasites as biological indicators of the population biology, migrations, diet and phylogenetics of fish. *Reviews in Fish Biology and Fisheries* **2**, 144–176.

Yamaguchi A, Yokoyama H, Ogawa K, Taniuchi T (2003) Use of parasites as biological tags for separating stocks of the starspotted dogfish *Mustelus manazo* in Japan and Taiwan. *Fisheries Science* **69**, 337–342.

Parasites as indicators of historical dispersal

Hayward CJ (1997) Distribution of external parasites indicates boundaries to the dispersal of sillaginid fishes in the Indo-West Pacific. *Marine and Freshwater Research* **48**, 391–400.

Rohde K (2002) Ecology and biogeography of marine parasites. *Advances in Marine Biology* **43**, 1–86.

Rohde K, Hayward CJ (2000) Oceanic barriers as indicated by scombrid fishes and their parasites. *International Journal for Parasitology* **30**, 579–583.

Von Ihering H (1891) On the ancient relations between New Zealand and south America. *Transactions and Proceedings of the New Zealand Institute, 1891* **24**, 431–445.

Introduced marine parasites

Andrews JD (1980) A review of introductions of exotic oysters and biological planning for new importations. *Marine Fisheries Review* **42**, 1–11.

Barber BJ (1997) Impacts of bivalve introductions on marine ecosystems: a review. *Bulletin of the National Research Institute of Aquaculture Suppl.* **3**, 141–153.

Barse AM, Secor DH (1999) An exotic nematode parasite of the American eel. *Fisheries* **24**, 6–10.

Bernard FR (1969) Copepod *Myticola orientalis* in British Columbia bivalves. *Fisheries Research Board of Canada* **26**, 190–191.

Boschma H (1972) On the occurrence of *Carcinus maenas* (Linnaeus) and its parasite *Sacculina carcini* Thompson in Burma, with notes on the transport of crabs to new localities. *Zoologische Mededelingen* **47**, 145–155.

Bumann D, Puls G (1996) Infestation with larvae of the sea anemone *Edwardsia lineata* affects nutrition and growth of the ctenophore *Mnemiopsis leidyi*. *Parasitology* **113**, 123–128.

Burreson EM, Stokes NA, Friedman CS (2000) Increased virulence in an introduced pathogen: Haplosporidium nelsoni (MSX) in the eastern oyster Crassostrea virginica. *Journal of Aquatic Animal Health* **12**, 1–8.

Byers JE (2000) Competition between two estuarine snails: Implications for invasions of exotic species. *Ecology* **81**, 1225–1239.

Byers JE, Goldwasser L (2001) Exposing the mechanism and timing of impact of nonindigenous species on native species. *Ecology* **82**, 1330–1343.

Carlton JT (1975) Extinct and endangered populations of the endemic mud snail *Cerithidia californica* in Northern California. *Bulletin of American Malicological Union Inc.* **41**, 65–66.

Carlton JT (1999) Molluscan invasions in marine and estuarine communities. *Malacologia* **41**(2).

Carlton JT, Geller JB (1993) Ecological roulette: the global transport of nonindigenous marine organisms. *Science* **266**, 78–82.

Chapin FS, III, Zavaleta ES, Eviner VT, Naylor RL, Vitousek PM, Reynolds HL, Hooper DU, Lavorel S, Sala OE, Hobbie SE, Mack MC, Diaz S (2000) Consequences of changing biodiversity. *Nature* **405**, 234–242.

Chew KK (1990) Global bivalve shellfish introductions. *World Aquaculture* **21**, 9–22.

Ching HL (1991) Lists of larval worms from marine invertebrates of the Pacific Coast of North America. *Journal of the Helminthological Society of Washington* **58**, 57–68.

Cohen AN, Carlton JT (1998) Accelerating invasion rate in a highly invaded estuary. *Science* **279**, 555–558.

Cone DK, Marcogliese DJ (1995) *Pseudodactylogyrus anguillae* on *Anguilla rostrata* in Nova Scotia: An endemic or an introduction? *Journal of Fish Biology* **47**, 177–178.

Culver CS, Kuris AM (2000) The apparent eradication of a locally established introduced marine pest. *Biological Invasions* **2**, 245–253.

Demond J (1952) The Narissariidae of the west coast of North America between Cape San Lucas, Lower California, and Cape Flattery, Washington. *Pacific Science* **6**, 300–317.

Dobson AP, May RM (1986) Patterns of invasions by pathogens and parasites. *In* 'Ecology and Biological Invasions of North America and Hawaii'. (Eds HA Mooney and JA Drake.) (Springer-Verlag: Berlin).

Drake JM, Lodge DM (2004) Global hotspots of biological invasions: evaluating options for ballast-water management. *Royal Society Proccedings: Biological Sciences* **271**, 575–580.

Elton CS (1958) 'The Ecology of Invasions by Animals and Plants'. (Methuen: London).

Ernst I, Fletcher A, Hayward C (2000) *Gyrodactylus anguillae* (Monogenea: Gyrodactylidae) from anguillid eels (*Anguilla australis* and *Anguilla reinhardtii*) in Australia: A native or an exotic? *Journal of Parasitology* **86**, 1152–1156.

Font WF, Rigby MC (1999) Implications of a new Hawaiian host record from blue-lined snappers *Lutjanus kasmira*: is the nematode *Spirocamallanus istiblenni* native or introduced. *Records of the Hawaii Biological Survey* 1999–Part 2: notes, 53–55.

Ford SD (1996) Range extension by the oyster parasite *Perkinsus marinus* into the Northeastern United States: response to climate change? *Journal of Shellfish Research* **15**, 45–56.

Friedman CS (1996) Haplosporidian infections of the Pacific oyster, *Crassostrea gigas* (Thunberg), in California and Japan. *Journal of Shellfish Research* **15**, 597–600.

Galil BS, Innocenti G (1999) Notes on the population structure of the portunid crab *Charybdis longicollis* Leene, parasitized by the rhizocephalan *Heterosaccus dollfusi* Boschma, off the Mediterranean coast of Israel. *Bulletin of Marine Science* **64**, 451–463.

Galil BS, Lutzen J (1995) Biological observations on *Heterosaccus dollfusi* Boschma (Cirripedia: Rhizocephala), a parasite of *Charybdis longicollis* Leene (Decapoda: Brachyura), a lessepian migrant to the Mediterranean. *Journal of Crustacean Biology* **15**, 659–670.

Gelnar M, Scholz T, Matejusova I, Konecny R (1996) Occurrence of *Pseudodactylogyrus anguillae* (Yin & Sproston, 1948) and *P. bini* (Kikuchi, 1929), parasites of eel, *Anguilla anguilla* L., in Austria (Monogenea: Dactylogyridae). *Ann Naturhist Mus Wien Ser B* **98**, 1-4.

Goddard JHR, Torchin ME, Lafferty KD, Kuris AM (in press) Experimental infection of native California crabs by *Sacculina carcini*, a potential biocontrol agent of introduced European green crabs. *Biological Invasions*.

Goggin CL, Bouland C (1997) The ciliate *Orchitophrya* cf. *stellarum* and other parasites and commensals of the Northern Pacific seastar *Asterias amurensis* from Japan. *International Journal for Parasitology* **27**, 1415–1418.

Grigorovich IA, Maclsaac HJ, Shadrin NV, Mills EL (2002) Patterns and mechanisms of aquatic invertebrate introductions in the Ponto-Caspian region. *Canadian Journal of Fisheries & Aquatic Sciences* **59**, 1189–1208.

Grodhaus G, Keh B (1958) The marine, dermatitis-producing cercaria of *Austrobilharzia variglandis* in California (Trematoda: Schistosomatidae). *The Journal of Parasitology* **44**, 633–638.

Grosholz E (2002) Ecological and evolutionary consequences of coastal invasions. *Trends in Ecology & Evolution* **17**, 22–27.

Grosholz ED, Ruiz GM, Dean CA, Shirley KA, Maron JL, Connors PG (2000) The impacts of a nonindigenous marine predator in a California bay. *Ecology* **81**, 1206–1224.

Hastein T, Lindstad T (1991) Diseases in wild and cultured salmon: possible interaction. *Aquaculture* **98**, 277–288.

Hayward CJ, Iwashita M, Crane JS, Ogawa K (2001a) First report of the invasive eel pest *Pseudodactylogyrus bini* in North America and in wild American eels. *Diseases of Aquatic Organisms* **44**, 53–60.

Hayward CJ, Iwashita M, Ogawa K, Ernst I (2001b) Global spread of the eel parasite *Gyrodactylus anguillae* (Monogenea). *Biological Invasions* **3**, 417–424.

Hines AH, Alvarez F, Reed SA (1997) Introduced and native populations of a marine parasitic castrator: variation in prevalence of the rhizocephalan *Loxothylacus panopaei* in xanthid crabs. *Bulletin of Marine Science* **61**, 197–214.

His E (1977) Observations préliminaires sur la présence de *Mytilicola orientalis* Mori (1935) chez *Crassostrea gigas* Thunberg dans le Bassin D'Arcachon. *Bull. Soc. Geol. et Amis du Museum du Havre* **64**, 7–8.

Holmes JMC, Minchin D (1995) Two exotic copepods imported into Ireland with Pacific oyster *Crassostrea gigas* (Thunberg). *Irish Naturalists' Journal* **25**, 17–20.

Huspeni TC, Lafferty KD (2004) Using larval trematodes that parasitize snails to evaluate a saltmarsh restoration project. *Ecological Applications* **14**, 795–804.

Johnsen BO, Jensen AJ (1988) Introduction and establishment of *Gyrodactylus salaris* Malmberg, 1957, on Atlantic salmon, *Salmo salar* L., fry and parr in the River Vefsna, northern Norway. *Journal of Fish Diseases* **11**, 35–46.

Johnsen BO, Jensen AJ (1991) The *Gyrodactylus* story in Norway. *Aquaculture* **98**, 289–302.

Kennedy CR (1993) Introductions spread and colonization of new localities by fish helminth and crustacean parasites in the British Isles: A perspective and appraisal. *Journal of Fish Biology* **43**, 287–301.

Kennedy CR, Bush AO (1994) The relationship between pattern and scale in parasite communities: a stanger in a strage land. *Parasitology* **109**, 187–196.

Kinzelbach R (1965) Die blaue Schwimmkrabbe (*Callinectes sapidus*) ein Neuburger im Mittelmeer. *Natur und Museum* **95**, 293–296.

Kuris AM, Culver CS (1999) An introduced sabellid polychaete pest infesting cultured abalones and its potential spread to other California gastropods. *Invertebrate Biology* **118**, 391–403.

Kuris AM, Lafferty KD (1992) Modeling crustacean fisheries: effects of parasites on management strategies. *Canadian Journal of Fisheries and Aquatic Sciences* **49**, 327–336.

Kuris AM, Lafferty KD (2000) Parasite-host modeling meets reality: adaptive peaks and their ecological attributes. *In* 'Evolutionary Biology of Host-parasite Relationships: Theory Meets Reality'. (Eds R Poulin, S Morand and A Skorping.) pp. 9–26. (Elsevier Science: Amsterdam).

Kuris AM, Lafferty KD, Grygier MJ (1996) Detection and preliminary evaluation of natural enemies for possible biological control of the Northern Pacific seastar, *Asterias amurensis*. *Centre for Research on Introduced Marine Pests, Technical Report* **3**, 1–17.

Kuris AM, Torchin ME, Lafferty KD (2002) *Fecampia erythrocephala* rediscovered: prevalence and distribution of a parasitold of the European shore crab, *Carcinus maenas*. *Journal of the Marine Biological Association of the United Kingdom* **82**, 955–960.

Lafferty KD (1992) Foraging on prey that are modified by parasites. *The American Naturalist* **140**, 854–867.

Lafferty KD (1999) The evolution of trophic transmission. *Parasitology Today* **15**, 111–115.

Lafferty KD, Kuris AM (1996) Biological control of marine pests. *Ecology* **77**, 1989–2000.

Lafferty KD, Morris AM (1996) Altered behavior of parasitizede killifish increases susceptibility to predation by final bird hosts. *Ecology* **77**, 1390–1397.

Martin WE (1972) An annotated key to the cercariae that develop in the snail, *Cerithidea californica*. *Bulletin of the Southern California Academy of Sciences* **71**, 39–43.

Miller HM, Northup FE (1926) The seasonal infestation of *Nassa obsoleta* (Say) with larval trematodes. *Biological Bulletin* **50**, 490–508.

Minchin D (1996) Management of the introduction and transfer of marine molluscs. *Aquatic Conservation* **6**, 229–244.

Mitchell CE, Power AG (2003) Release of invasive plants from fungal and viral pathogens. *Nature* **421**, 625–627.

Ogawa K, Bondad-Reantaso MG, Fukudome M, Wakabayashi H (1995) *Neobenedenia girellae* (Hargis, 1955) Yamaguti, 1963 (Monogenea: Capsalidae) from cultured marine fishes of Japan. *Journal of Parasitology* **81**, 223–227.

Osmanov SO (1971) 'Parasites of Fishes of Uzbekistan'. (Tashkent: FAN).

Price PW (1980) 'Evolutionary Biology of Parasites'. (Princeton University Press: Princeton).

Rigby MC, Dufour V (1996) Parasites of coral reef fish recruits, *Epinephelus merra* (Serranidae), in French Polynesia. *Journal of Parasitology* **82**, 405–408.

Ruiz GM, Carlton JT, Grosholz ED, Hines AH (1997) Global invasions of marine and estuarine habitats by non-indigenous species: mechanisms, extent and consequences. *American Zoologist* **37**, 621–632.

Ruiz GM, Fofonoff P, Hines AH, Grosholz ED (1999) Non-indigenous species as stressors in estuarine and marine communities: Assessing invasion impacts and interactions. *Limnology and Oceanography* **44**, 950–972.

Ruiz GM, Fofonoff P, Carlton JT, Wonham MJ, Hines AH (2000) Invasions of coastal marine communities in North America: apparent patterns, processes, and biases. *Annual Review of Ecology and Systematics* **31**, 481–531.

Sands DPA (1998) Guidelines for testing host specificity of agents for biological control of arthropod pests. *In* 'Pest Management – Future Challenges: 6th Australia Applied Entomological Research Conference'. Vol. 1. (Eds M Zalucki, R Drew and G White.) pp. 556–560. (University of Queensland: Brisbane).

Scheibling RE, Hennigar AW (1997) Recurrent outbreaks of disease in sea urchins *Strongylocentrotus droebachiensis* in Nova Scotia: evidence for a link with large-scale meteorologic and oceanographic events. *Marine Ecology Progress Series* **152**, 155–165.

Secord D (2003) Biological control of marine invasive species: cautionary tales and land-based lessons. *Biological Invasions* **5**, 117–131.

Shimura S, Ito J (1980) Two new species of marine cercariae from the Japanese intertidal gastropod, *Batillaria cumingii* (Crosse). *Japanese Journal of Parasitology* **29**, 369–375.

Stock JH (1993) Copepoda (Crustacea) associated with commercial and non-commercial Bivalvia in the East Scheldt, The Netherlands. *Bijdragen tot de Dierkunde* **63**, 61–64.

Stunkard HW, Hinchliffe MC (1952) The morphology and life-history of *Microbilharzia variglandis* (Miller and Northrup 1926) Stunkard and Hinchliffe, 1951, avian blood flukes whose larve cause 'swimmer's itch' of ocean beaches. *Journal of Parasitology* **38**, 248–265.

Thresher RE, Werner M, Hoeg JT, Svane I, Glenner H, Murphy NE, Wittwer C (2000) Developing the options for managing marine pests: specificity trials on the parasitic castrator, *Sacculina carcini*, against the European crab, *Carcinus maenas*, and related species. *Journal of Experimental Marine Biology and Ecology* **254**, 37–51.

Toft CA (1986) Communities of parasites with parasitic life-styles. In 'Community Ecology'. (Eds JM Diamond and TJ Case.) pp. 445–463. (Harper and Row: New York).

Torchin ME, Mitchell CE (2004) Parasites, pathogens and invasions by plants and animals. *Frontiers in Ecology and the Environment* **2**, 138–190.

Torchin ME, Lafferty KD, Kuris AM (2001) Release from parasites as natural enemies: increased performance of a globally introduced marine crab. *Biological Invasions* **3**, 333–345.

Torchin ME, Lafferty KD, Kuris AM (2002) Parasites and marine invasions. *Parasitology* **124**, S137–S151.

Torchin ME, Lafferty KD, Dobson AP, McKenzie VJ, Kuris AM (2003) Introduced species and their missing parasites. *Nature* **421**, 628–630.

Torchin ME, Byers JE, Huspeni TC (in press) Differential parasitism of native and introduced snails: replacement of a parasite fauna. *Biological Invasions*.

Van Engel WA, Dillon WA, Zwerner D, Eldridge D (1965) *Loxothylacus panopaei* (Cirripedia, Sacculinidae) an introduced parasite on the xanthid crab in Chesapeake Bay, USA. *Crustaceana* **10**, 111–112.

Vitousek PM (1990) Biological invasions and ecosystem processes: Towards an integration of population biology and ecosystem studies. *Oikos* **57**, 7–13.

Wilcove DS, Rothstein D, Dubow J, Phillips A, Losos E (1998) Quantifying threats to imperiled species in the United States. *Bioscience* **48**, 607–615.

Zholdasova I (1997) Sturgeons and the Aral Sea ecological catastrophe. *Environmental Biology of Fishes* **48**, 373–380.

Deep-sea parasites

Boxshall GA (1998) Host specificity in copepod parasites of deep-sea fishes. *Journal of Marine Systems* **15**, 215–223.

Bray RA (2004) The bathymetric distribution of the digenean parasites of deep-sea fishes. *Folia Parasitologica* **51**, 268–274.

Bray RA, Littlewood DTJ, Herniou EA, Williams B, Henderson RE (1999) Digenean parasites of deep-sea teleosts: a review and case studies of intrageneric phylogenies. *Parasitology* **119** (Supplement), S125–S144.

Buron I de (1988) *Hypoechinorhynchus thermaceri* n. sp. (Acanthocephala: Hypoechinorhynchidae) from the deep-sea zoarcid fish *Thermarces andersoni* Rosenblatt and Cohen, 1986. *Journal of Parasitology* **74**, 339–342.

Buron I de, Morand S (2002) Deep-sea hydrothermal vent parasites: where do we stand? *Cahiers de Biologie Marine* **43**, 245–246.

Buron I de, Morand S (2004) Deep-sea hydrothermal vent parasites: why don't we find more? *Parasitology* **128**, 1–6.

Campbell RA (1983) Parasitism in the deep-sea. *In* 'The Sea'. Vol. 8. (Ed. GT Rowe.) pp. 473–552. (John Wiley and Sons: New York).

Campbell RA, Haedrich RL, Munroe TA (1980) Parasitism and ecological relationships among deep-sea benthic fishes. *Marine Biology* **57**, 301–313.

Davies AJ, Merrett NR (2000) Haemogregorines and other blood infections from deep demersal fish of the Porcupine Bight, north-east Atlantic. *Journal of the Marine Biological Association of the United Kingdom* **80**, 1095–1102.

Gage JD, Tyler PA (1991) 'Deep-sea Biology. A Natural History of Organisms at the Deep-sea Floor'. (Cambridge University Press: Cambridge).

Grabda E (1983) *Eimeria jadvigae* n. sp. (Apicomplexa: Eucoccidia), a parasite of the swimming bladder of *Coryphaenoides holotrachys* (Günther, 1887) off the Falklands. *Acta Ichthyologica et Piscatoria* **13**, 131–140.

Herring P (2002) 'The Biology of the Deep Ocean'. (Oxford University Press: Oxford).

Justine J-L, Cassone J, Petter A (2002) *Moravecnema segonzaci* gen. et sp. n. (Nematoda: Cystidicolidae) from *Pachycara thermophilum* (Zoarcidae), a deep-sea hydrothermal vent fish from the Mid-Atlantic Ridge. *Folia Parasitologica* **49**, 299–303.

Klimpel S, Seehagen A, Palm H-W, Rosenthal H (2001) 'Deep-water Metazoan Fish Parasites of the World'. (Logos: Berlin).

Marshall NB (1971) 'Explorations in the Life of Fishes'. (Harvard University Press: Cambridge, Massachusetts).

Marshall NB (1979) 'Developments in Deep-sea Biology'. (Blandford Press: Poole).

Merrett NR, Haedrich RL (1997) 'Deep-sea Demersal Fish and Fisheries'. (Chapman & Hall: London).

Moreira D, Lopez-Garcia P (2003) Are hydrothermal vents oases for parasitic protists? *Trends in Parasitology* **19**, 556–558.

Noble ER (1973) Parasites and fishes in a deep-sea environment. *Advances in Marine Biology* **11**, 121–195.

Powell EN, Barber RD, Kennicutt MC, Ford SE (1999) Influence of parasitism in controlling the health, reproduction and PHA body burden of petroleum seep mussels. *Deep-Sea Research, Part 1 – Oceanographic Research Papers* **46**, 2053–2078.

Randall DJ, Farrell AP (Eds) (1997) 'Deep-sea Fishes'. (Academic Press: San Diego).

Szuks H (1980) Applicability of parasites in separating groups of grenadier fish *Macrourus rupestris*. *Angewandte Parasitologie* **21**, 211–214.

Tyler PA (1995) Conditions for the existence of life at the deep-sea floor: an update. *Oceanography and Marine Biology* **33**, 221–244.

Chapter 10

Kinne O (1980–1985). 'Diseases of Marine Animals'. Vols 1–4. (Wiley Interscience, New York, and Biologische Anstalt Helgoland, Hamburg).

Mass mortalities in the oceans

Burreson EM, Zwerner DE (1984) Juvenile summer flounder, *Paralichthys dentatus*, mortalities in the western Atlantic Ocean caused by the hemoflagellate *Trypanoplasma bullocki*: evidence from field and experimental studies. *Helgoländer Meeresuntersuchungen* **37**, 343–352.

Daszak P, Cunningham AA, Hyatt AD (2001) Anthropogenic environmental change and the emergence of infectious diseases in wildlife. *Acta Tropica* **78**, 103–116.

Geraci JR, St Aubin DT (1986) Effects of parasites on marine mammals. *In* 'Parasitology – Quo vadit? Proceedings of the 6th International Congress of Parasitology, 24–29 August 1986, Brisbane, Queensland'. (Ed. MJ Howell.) pp. 407–414. (Australian Academy of Science: Canberra).

Gómez-Gutiérrez J, Peterson WT, De Robertis A, Brodeur RD (2003) Mass mortality of krill caused by parasitoid ciliates. *Science* **301**, 339.

Griffin DA, Thompson PA, Bax NJ, Bradford RW, Hallegraeff GM (1997) The 1995 mass mortality of pilchard: no role found for physical or biological oceanographic factors in Australia. *Marine Freshwater Research* **48**, 27–42.

Harvell CD, Kim K, Burkholder JM, Colwell RR, Epstein PR, Grimes DJ, Hoffmann EE, Lipp EK, Osterhaus ADME, Overstreet RM, Porter JW, Smith GW, Vasta GR (1999) Emerging marine diseases – climate links and Anthropogentic factors. *Science* **285**, 1505–1510.

Holst JC, Salvanes AGV, Johansen T (1997) Feeding, *Ichthyophonus* sp. infection, distribution and growth history of Norwegian spring-spawning herring in summer. *Journal of Fish Biology* **50**, 652–664.

Hornell J (1918) A new protozoan cause of mortality among marine fishes. *Madras Fisheries Department Bulletin* **11**, 53–66.

Jones JB (1998) Distant water sailors – parasitic copepods of the open ocean. *Journal of Marine Systems* **15**, 207–214.

Jones JB, Hyatt AD, Hine PM, Whittington RJ, Griffin DA, Bax NJ (1997) The 1995 Australasian pilchard mortalities. *World Journal of Microbiology and Biotechnology* **13**, 383–392.

Kirby DS, Abraham ER, Uddstrom MJ, Dean H (2003) Tuna schools/aggregations in surface longline data 1993–98. *New Zealand Journal of Marine and Freshwater Research* **37**, 633–644.

Kocan RM, Hershberger P, Mehl T, Elder N, Bradley M, Wildermuth D, Stick K (1999) Pathogenicity of *Ichthyophonus hoferi* for laboratory-reared Pacific herring *Clupea pallasi* and its early appearance in wild Puget Sound herring. *Diseases of Aquatic Organisms* **35**, 23–29.

Lom J, Dykova I (1992) 'Protozoan Parasites of Fishes'. Developments in Aquaculture and Fisheries Science 26'. (Elsevier: Netherlands).

Marcogliese DJ (2002) Food webs and the transmission of parasites to marine fish. *Parasitology* **124**, S83–S99.

McKenzie K, Liversidge JM (1975) Some aspects of the biology of the cercaria and metacercaria of *Stephanostomum baccatum* (Nicoll, 1907) Manter, 1934 (Digenea: Acanthocolpidae). *Journal of Fish Biology* **7**, 247–256.

Morimitsu T, Nagai T, Ide M, Kawano H, Naichuu A, Koono M, Ishii A (1987) Mass stranding of Odontoceti caused by parasitogenic eighth cranial neuropathy. *Journal of Wildlife Diseases* **23**, 586–590.

Munro ALS, McVicar AH, Jones R (1983) The epidemiology of infectious disease in commercially important wild marine fish. *Rapports et Procès-Verbaux Des Réunions (Conseil International pour L'exploration de la mer)* **182**, 21–32.

Nagasawa K, Imai Y, Ishida K (1988) Long-term changes in the population size and geographical distribution of *Pennella* sp. (Copepoda) on the saury, *Cololabis saira*, in the western North Pacific and adjacent seas. *Hydrobiologia* **167/168**, 571–577.

Odense PH, Logan VH (1976) Prevalence and morphology of *Eimeria gadi* (Fiebiger, 1913) in the haddock. *Journal of Protozoology* **23**, 564–571.

Ogawa K (2002) Impacts of diclidophorid monogenean infections on fisheries in Japan. *International Journal of Parasitology* **32**, 373–380.

Okamoto N, Nakase K, Sano T (1987) Relationships between water temperature, fish size, infective dose and *Ichthyophonus* infection in rainbow trout. *Nippon Suisan Gakkaishi* **53**, 581–584.

O'Neil G (1995) Ocean anomaly triggers record fish kill.*Science* **268**, 1431.

Petrushevski GK, Shulman SS (1961) The parasitic diseases of fishes in the natural waters of the USSR. *In* 'Parasitology of Fishes'. (Eds VA Dogiel, GK Petrushevski, YI Polyanski.) pp. 299–319. (English translation by Z Kabata, Oliver & Boyd: Edinburgh).

Rahimian H, Thulin J (1996) Epizootology of *Ichthyophonus hoferi* in herring populations off the Swedish west coast. *Diseases of Aquatic Organisms* **27**, 173–186.

Rand TG, White K, Cannone JJ, Gutell RR, Murphy CA, Ragan MA (2000) *Ichthyophonus irregularis* sp. nov. from the yellowtail flounder *Limanda ferruginea* from the Nova Scotia shelf. *Diseases of Aquatic Organisms* **41**, 31–36.

Rayl AJS (2001) Research: researchers focus on sea otter deaths. *The Scientist* **15(4)**, http://www.the-scientist.com/2001/02/19/17/2.

Rohde K (1982) 'Ecology of Marine Parasites'. (University of Queensland Press: Brisbane).

Sinderman CJ (1963) Disease in marine populations. *Transactions of the North American Wildlife Conference* **28**, 336–356.

Sinderman CJ (1990) 'Principal Diseases of Marine Fish and Shellfish'. 2nd edn. Vol. 1. Diseases of Marine Fish. (Academic Press: San Diego).

Southward AJ, Boalch GT, Maddock L (1988) Fluctuations in the herring and pilchard fisheries of Devon and Cornwall linked to change in climate since the 16th Century. *Journal of the Marine Biological Association UK* **68**, 423–445.

Spanggaard B, Gram L, Okamoto N, Huss HH (1994) Growth of the fish-pathogenic fungus, *Ichthyophonus hoferi*, measured by conductimetry and microscopy. *Journal of Fish Diseases* **17**, 145–153.

Tully O, Nolan DT (2002) A review of the population biology and host-parasite interactions of the sea louse *Lepeophtheirus salmonis* (Copepoda: Caligidae). Parasitology 124, S165–S182.

Yamamoto K, Takagi S, Matsuoka S (1984) Mass mortality of Japanese anchovy (*Engraulis japonica*) caused by a gill monogenean *Pseudanthocotyloides* sp. (Mazocraeidae) in the Sea of Iyo ('Iyo-nada'), Ehime Prefecture. *Fish Pathology* **19**, 119–123.

Zylber MI, Failla G, Le Bas A (2002) *Stenurus globicephalae* Baylis et Daubney, 1925 (Nematoda: Pseudaliidae) from a false killer whale *Pseudorca crassidens* (Cetacea: Delphinidae) stranded on the coast of Uruguay. *Memorias do Instituto Oswaldo Cruz (Rio de Janeiro)* **97**, 221–225.

Effects of salmon lice on Atlantic salmon

Anonymous (1999) Til laks åt alle kan ingen gjera? Norges Offentlige Utredninger, NOU 1999:9, Oslo.

Berland B (1993) Salmon lice on wild salmon in Norway. *In* 'Pathogens of Wild and Farmed Fish: Sea Lice'. (Eds GA Boxshall, D Defaye.) pp. 179–187. (Ellis & Horwood: Chichester).

Birkeland K (1996) Consequences of premature return by sea trout (*Salmo trutta*) infested with the salmon louse (*Lepeophtheirus salmonis*): migration, growth, and mortality. *Canadian Journal of Fisheries and Aquatic Sciences* **53**, 2808–2813.

Bjørn PA, Finstad B (1997) The physiological effects of salmon lice infection on sea trout post smolts. *Nordic Journal of Freshwater Research* **73**, 60–72.

Bjørn PA, Finstad B (2002) Salmon lice, *Lepeophtheirus salmonis* Krøyer, infestation in sympatric populations of Arctic char, *Salvelinus alpinus* (L.) and sea trout, *Salmo trutta* (L.), in areas near, and distant from salmon farms. *ICES Journal of Marine Science* **59**, 1–9.

Bjørn PA, Finstad B, Kristoffersen R (2001a) Salmon lice infection of wild sea trout and Arctic char in marine and freshwaters: the effect of salmon farms. *Aquaculture Research* **32**, 947–962.

Bjørn PA, Finstad B, Kristoffersen R (2001b) Registreringer av lakselus på laks, sjøørret og sjørøye i 2000. Norwegian Institute for Nature Research, NINA Oppdragsmelding No. 698, Trondheim.

Bjørn PA, Finstad B, Kristoffersen R (2002) Registreringer av lakselus på laks, sjøørret og sjørøye i 2001. Norwegian Institute for Nature Research, NINA Oppdragsmelding No. 737, Trondheim.

Bjørn PA, Finstad B, Kristoffersen R (2003) Registreringer av lakselus på laks, sjøørret og sjørøye i 2002. Norwegian Institute for Nature Research, NINA Oppdragsmelding No. 789, Trondheim.

Costelloe M, Costelloe J, Roche N (1995) Planktonic dispersion of larval salmon-lice, *Lepeophtheirus salmonis*, associated with cultured salmon, *Salmo salar*, in Western Ireland. *Journal of the Marine Biological Association of the United Kingdom* **76**, 141–149.

Crisp DT (2000) 'Trout and Salmon: Ecology, Conservation and Rehabilitation'. (Fishing News Books: Oxford).

Finstad B, Bjørn PA, Grimnes A, Hvidsten NA (2000) Laboratory investigations of salmon lice (*Lepeophtheirus salmonis* Krøyer) infestation on Atlantic salmon (*Salmo salar* L.) post-smolts. *Aquaculture Research* **31**, 1–9.

Friedland KD Reddin DG, Kocik JF (1993) Marine Survival of North-American and European Atlantic Salmon - Effects of Growth and Environment. *ICES Journal of Marine Science* **50**, 481–492.

Grimnes A, Birkeland K, Jakobsen PJ, Finstad B (1996) Lakselus – nasjonal og internasjonal kunnskapsstatus. Norwegian Institute for Nature Research. NINA Fagrapport No. 18, Trondheim.

Grimnes A, Finstad B, Bjørn PA, Tovslid BM, Lund RA (1998) Registreringer av lakselus på laks, sjøørret og sjørøye i 1997. Norwegian Institute for Nature Research, NINA Oppdragsmelding No. 525, Trondheim.

Grimnes A, Finstad B, Bjørn PA (1999) Registreringer av lakselus på laks, sjøørret og sjørøye i 1998. Norwegian Institute for Nature Research, NINA Oppdragsmelding No. 579, Trondheim.

Grimnes A, Finstad B, Bjørn PA (2000) Registreringer av lakselus på laks, sjøørret og sjørøye i 1999 Norwegian Institute for Nature Research, NINA Oppdragsmelding No. 632, Trondheim.

Hahnenkamp L, Fyhn HJ (1985) The osmotic response of the salmon louse, *Lepeophtheirus salmonis* (Copepoda: Caligidae), during the transition from sea water to fresh water. *Journal of Comparative Physiology B* **155**, 357–365.

Hansen HP, Holm M, Holst JC, Jacobsen JA (2003) The ecology of post-smolts of Atlantic salmon. *In* 'Salmon at the Edge'. (Ed. D Mills.) pp. 25–39. (Fishing News Books: Oxford).

Heuch PA (1995) Experimental evidence for aggregation of salmon louse copepodids, *Lepeophtheirus salmonis*, in step salinity gradients. *Journal of the Marine Biological Assosiation of the United Kingdom* **75**, 927–939.

Heuch PA, Mo TA (2001) A model of salmon louse production in Norway: effects of increasing salmon production and public management measures. *Diseases of Aquatic Organisms* **45**, 145–152.

Heuch PA, Parsons A, Boxaspen K (1995) Diel vertical migration: a possible host-finding mechanism in salmon louse (*Lepeophtheirus salmonis*) copepodids? *Canadian Journal of Fisheries and Aquatic Sciences* **52**, 681–689.

Heuch PA, Nordhagen JR, Schram TA (2000) Egg production in the salmon louse (*Lepeophtheirus salmonis* (Krøyer)) in relation to origin and water temperature. *Aquaculture Research* **31**, 805–814.

Heuch PA, Knutsen JA, Knutsen H, Schram TA (2002) Salinity and temperature effects on sea lice over-wintering on sea trout (*Salmo trutta*) in coastal areas of the Skagerrak. *Journal of the Marine Biological Assosiation of the United Kingdom* **82**, 887–892.

Holm M, Holst JC, Hansen HP, Jacobsen JA, O'Maoileidigh N, Moore A (2003) Migration and distribution of Atlantic post-smolt in the North Sea and North-East Atlantic. In: 'Salmon at the Edge'. (Ed. D Mills.) pp. 7–23. (Fishing News Books: Oxford).

Holst JC, McDonald A (2000) FISH-LIFT: a device for sampling live fish with trawls. *Fisheries Research* **48**, 87–91.

Holst JC, Shelton M, Holm M, Hansen HP (2000) Distribution and possible migration routes of post-smolt Atlantic salmon in the North-East Atlantic. *In* 'The Ocean Life of Atlantic Salmon: Environmental and Biological Factors influencing Survival.' (Ed. D Mills.) pp. 65–74. (Fishing News Books: Oxford).

Holst JC, Jakobsen PJ, Nilsen F, Holm M, Asplin L, Aure J (2003) Mortality of seaward-migrating post-smolts of Atlantic salmon due to salmon lice infection in Norwegian salmon stocks. *In* 'Salmon at the Edge'. (Ed. D Mills.) pp. 136–137. (Fishing News Books: Oxford).

Jacobsen JA Gaard, E (1997) Open-ocean infestation by salmon lice (*Lepeophtheirus salmonis*): comparison of wild and escaped farmed Atlantic salmon (*Salmo salar* L.). *ICES Journal of Marine Science* **54**, 1113–1119.

Johannessen A (1975) Lakselus, *Lepeophtheirus salmonis* Krøyer (Copepoda, Caligidae). Frittlevende stadier, vekst og infeksjon på laks (*Salmo salar* L.) fra oppdrettsanlegg og kommersielle fangster i vestnorske farvann 1973–1974. Cand. Real. thesis, University of Bergen, Norway.

Johnson SC, Albright LJ (1991a) The developement stages of *Lepeophtheirus salmonis* (Krøyer, 1837) (Copepoda: Caligidae). *Canadian Journal of Zoology* **69**, 929–950.

Johnson SC, Albright LJ (1991b) Development, growth, and survival of *Lepeophtheirus salmonis* (Copepoda: Caligidae) under laboratory conditions. *Journal of the Marine Biological Assosiation of the United Kingdom* **71**, 425–436.

Johnson SC, Blaylock RB, Elphick J, Hyatt JD (1996) Disease caused by the sea louse (*Lepeophtheirus salmonis*) (Copepoda:Caligidae) in wild sockeye salmon (*Oncorhynchus nerka*) stocks of Alberni inlet, British Columbia. *Canadian Journal of Fisheries and Aquatic Sciences* **53**, 2888–2897.

Kabata Z, Rafi F, Bousfield EL (1988) Part 2. Crustacea. *In* 'Guide to the Parasites of Fishes of Canada'. (Eds L Margolis, Z Kabata.) pp. 1–163. Canadian Special Publication of Fisheries and Aquatic Sciences No. 101. (Department of Fisheries and Oceans: Ottawa).

Levings CD, Hvidsten NA, Johnsen BO (1994) Feeding of Atlantic salmon, (*Salmo salar* L.) post-smolts in a fjord in central Norway. *Canadian Journal of Zoology* **72**, 834–839.

Lyse AA, Stefansson SO, Fernø A (1998) Behaviour and diet of sea trout post-smolts in a Norwegian fjord system. *Journal of Fish Biology* **52**, 923–936.

McLean PH, Smith GW, Wilson M J (1990) Residence time of the sea louse, *Lepeophtheirus salmonis* K., on Atlantic salmon, *Salmo salar* L., after immersion in fresh water. *Journal of Fish Biology* **37**, 311–314.

Mills D (Ed.) (2003) 'Salmon on the Edge'. (Blackwell Scientific: Oxford).

Moore A, Lacroix GL, Sturlaugson G (2000) Tracking Atlantic post-smolts in the sea. *In* 'The Ocean Life of Atlantic Salmon: Environmental and Biological Factors Influencing Survival'. (Ed. D Mills.) pp. 49–64. (Fishing News Books: Oxford).

Pike AW, Wadsworth SL (1999) Sea lice on salmonids: their biology and control. *Advances in Parasitology* **44**, 233–337.

Pontoppidan E (1753) 'Det Første Forsøk på Norges Naturlige Historie'. (Det Kongelige Waysenhuses Boktrykkeri, Trykt av Gottmann Fridrich Riesel: Copenhagen).

Schram TA, Knutsen JA, Heuch PA, Mo TA (1998) Seasonal occurrence of *Lepeophtheirus salmonis* and *Caligus elongatus* (Copepoda: Caligidae) on sea trout (*Salmo trutta*), off southern Norway. *ICES Journal of Marine Science* **55**, 163–175.

Shearer WM (1992) 'The Atlantic Salmon. Natural History, Exploitation and Future Management'. (Fishing News Books: Oxford).

Tully O, Gargan P, Poole WR, Whelan KF (1999) Spatial and temporal variation in the infestation of sea trout (*Salmon trutta* L.) by the caligid copepod *Lepeophtheirus salmonis* (Krøyer) in relation to sources of infection in Ireland. *Parasitology* **119**, 41–51.

Tully O, Nolan DT (2002) A review of the population biology and host-parasite interactions of the sea louse *Lepeophtheirus salmonis* (Copepoda : Caligidae). *Parasitology* **124**, S165–S182.

Effects in finfish culture

Branson E, Riaza A, Alvarez-Pellitero P (1999) Myxosporean infection causing intestinal disease in farmed turbot, *Scophthalmus maximus* (L.), (Teleostei: Scophthalmidae). *Journal of Fish Diseases* **22**, 395–399.

Carvajal J, Gonzalez L, George-Nascimento G (1998) Native sea lice (Copepoda: Caligidae) infestation of salmonids reared in netpen systems in southern Chile. *Aquaculture* **166**, 241–246.

Chen S-C, Kou R-J, Wu C-T, Wang P-C, Su F-Z (2001) Mass mortality associated with a *Sphaerospora*-like myxosporidean infestation in juvenile cobia, *Rachycentron canadum* (L.), marine cage cultured in Taiwan. *Journal of Fish Diseases* **24**, 189–195.

Clark A, Nowak BF (1999) Field investigations of amoebic gill disease in Atlantic salmon, *Salmo salar* L., in Tasmania. *Journal of Fish Diseases* **22**, 433–443.

Colquitt SE, Munday BL, Daintith M (2001): Pathological findings in southern bluefin tuna, *Thunnus maccoyii* (Castelnau), infected with *Cardicola forsteri* (Cribb, Daintith & Munday, 2000) (Digenea: Sanguinicolidae), a blood fluke. *Journal of Fish Diseases* **24**, 225–229.

Deveney MR, Chisholm LA, Whittington ID (2001) First published record of the pathogenic monogenean parasite *Neobenedenia melleni* (Capsalidae) from Australia. *Diseases of Aquatic Organisms* **46**, 79–82.

Diamant A (1997) Fish-to-fish transmission of a marine myxosporean. *Disease of Aquatic Organisms* **30**, 99–105.

Diamant A, Lom J, Dykova I (1994) *Myxidium leei* n. sp., a pathogenic myxosporean of cultured sea bream *Sparus aurata*. *Diseases of Aquatic Organisms* **20**, 137–141.

Dickerson HW, Dawe DL (1995) *Ichthyophthirius multifiliis* and *Cryptocaryon irritans* (Phylum Ciliophora). *In* 'Fish Diseases and Disorders'. Vol. 1. Protozoan and Metazoan Infections. (Ed. PTK Woo.) pp. 181–227. (CAB International, Wallingford).

Douglas-Helders M, Saksida S, Raverty S, Nowak BF (2001) Temperature as a risk factor for outbreaks of Amoebic Gill Disease in farmed Atlantic salmon (*Salmo salar*). *Bulletin of the European Association of Fish Pathologists* **21**, 114–116.

Dragesco A, Dragesco J, Coste F, Gasc C, Romestand B, Raymond J-C, Bouix G (1995) *Philasterides dicentrarchi*, n. sp., (Ciliophora, Scuticociliatida), a histophagous opportunistic parasite of *Dicentrarchus labrax* (Linnaeus, 1758), a reared marine fish. *European Journal of Protistology* **31**, 327–340.

FAO Fisheries Department (2002) FISHSTAT Plus, http://www.fao.org/fi/statist/fisoft/FISHPLUS.asp

Figueras A, Novoa B, Santarem M, Martinez E, Alvarez JM, Toranzo AE, Dykova I (1992) *Tetramicra brevifilum*, a potential threat to farmed turbot *Scophthalmus maximus*. *Diseases of Aquatic Organisms* **14**, 127–132.

Gross KA, Morrison RN, Butler R, Nowak BF (2004) Atlantic salmon, *Salmo salar* L., previously infected with *Neoparamoeba* sp. are not resistant to re-infection and have suppressed phagocyte function. *Journal of Fish Diseases* **27**, 47–56.

Herbert BW, Shaharom FM, Anderson IG (1995) Histopathology of cultured sea bass (*Lates calcarifer*) (Centropomidae) infected with *Cruoricola lates* (Trematoda: Sanguinicolidae) from Palau Ketam, Malaysia. *International Journal for Parasitology* **25**, 3–13.

Ho J-S (2000) The major problem of cage aquaculture in Asia relating to sea lice. *In* 'Proceedings of the First International Symposium on Cage Aquaculture in Asia, "Cage Aquaculture in Asia" 2–6 November 1999 Tungkang Marine Laboratory'. (Eds IC Liao, CK Lin.) pp. 13–19. (Taiwan Fisheries Research Institute: Tungkang, Pingtung, Taiwan).

Horton T, Okamura B (2001) Cymothoid isopod parasites in aquaculuture: a review and case study of a Turkish sea bass (*Dicentrarchus labrax*) and sea bream (*Sparus aurata*) farm. *Diseases of Aquatic Organisms* **46**, 181–188.

Iglesias R, Parama A, Alvarez MF, Leiro J, Fernandez J, Sammartin ML (2001) *Philasterides dicentrarchi* (Ciliophora, Scuticociliatida) as the causative agent of scuticociliatosis in farmed turbot *Scophthalmus maximus* in Galicia (NW Spain). *Diseases of Aquatic Organisms* **46**, 47–55.

Jee BY, Kim YC, Park MS (2001) Morphology and biology of parasite responsible for scuticociliatosis of cultured olive flounder *Paralichthys olivaceus*. *Diseases of Aquatic Organisms* **47**, 49–55.

Kent ML (2000) Marine netpen farming leads to infections with some unusual parasites. *International Journal for Parasitology* **30**, 321–326.

Kim KH, Cho JB (2000) Treatment of *Microcotyle sebastis* (Monogenea: Polyopisthocotylea) infestation with praziquantel in an experimental cage simulating commercial rockfish *Sebastes schlegeli* culture conditions. *Diseases of Aquatic Organisms* **40**, 229–231.

Leong TS (2000) Parasitic and bacterial diseases of grouper and other cultured marine finfishes and control strategies. *In* 'Report and Proceedings of APEC FWG Project 02/2000 "Development of a Regional Research Programme on Grouper Virus Transmission and Vaccine Development" 18–20 October 2000, Bangkok, Thailand'. (Eds MG Bondad-Reantaso, J Humphrey, S Kanchanakhan, S Chinabut.) pp. 73–80. (Asia Pacific Economic Cooperation, Aquatic Animal Health Research Institute, Fish Health Section of the Asian Fisheries Society and the Network of Aquaculture Centres in Asia-Pacific: Bangkok, Thailand).

Lopez C, Rajan PR, Lin JH-Y, Kuo T-Y, Yang H-L (2002) Disease outbreak in seafarmed Cobia (*Rachycentron canadum*) associated with *Vibrio* spp., *Photobacterium damselae* ssp. *piscicida*, monogenean and myxosporean parasites. *Bullentin of the European Association of Fish Pathologists* **22**, 206–211.

Mendoza L, Taylor JW, Ajello L (2002) The class Mesomycetozoea: A group of microorganisms at the animal-fungal boundary. *Annual Review of Microbiology* **56**, 315–344.

Montero FE, Aznar FJ, Fernandez M, Raga JA (2003) Girdles as the main infection site for *Paradeontacylix kampachi* (Sanguinicolidae) in the greater amberjack *Seriola dumerili*. *Diseases of Aquatic Organisms* **53**, 271–272.

Munday BL, O'Donoghue PJ, Watts M, Rough K, Hawkesford T (1997) Fatal encephalitis due to the scuticocilate *Uronema nigricans* in sea-caged, southern bluefin tuna *Thunnus maccoyii*. *Diseases of Aquatic Organisms* **30**, 17–25.

Munday BL, Zilberg D, Findlay V (2001) Gill disease of marine fish caused by infection with *Neoparamoeba pemaquidensis*. *Journal of Fish Diseases* **24**, 497–507.

Noga EJ, Levy MG (1995) Dinoflagellida (Phylum Sarcomastigophora). *In* 'Fish Diseases and Disorders'. Vol. 1. Protozoan and Metazoan Infections. (Ed. PTK Woo.) pp. 1–25. (CAB International, Wallingford).

Ogawa K (1996) Marine parasitology with special reference to Japanese fisheries and mariculture. *Veterinary Parasitology* **64**, 95–105.

Ogawa K (2002) Impacts of diclidophorid monogenean infections on fisheries in Japan. *International Journal for Parasitology* **32**, 373–380.

Ogawa K, Yokoyama H (1998) Parasitic diseases of cultured marine fish in Japan. *Fish Pathology* **33**, 303–309.

Palenzuela O, Redondo MJ, Alvarez-Pellitero P (2002) Description of *Enteromyxum scophthalmi* gen. nov., sp. nov. (Myxozoa), an intestinal parasite of turbot (*Scophthalmus maximus* L.) using morphological and ribosomal RNA sequence data. *Parasitology* **124**, 369–379.

Poppe TT, Mo TA, Iversen L (1992) Disseminated hexamitosis in sea-caged Atlantic salmon *Salmo salar*. *Diseases of Aquatic Organisms* **14**, 91–97.

Song W, Wilbert N (2000) Redefinition and redescription of some marine scuticociliates from China, with report of a new species, *Metanophrys sinensis* nov. spec. (Ciliophora, Scuticociliatida). *Zoologischer Anzeiger* **239**, 45–74.

Sterud E, Mo TA, Poppe TT (1998) Systemic spironucleosis in sea-farmed Atlantic salmon *Salmo salar*, caused by *Spironucleus barkhanus* transmitted from feral Arctic char *Salvelinus alpinus*? *Diseases of Aquatic Organisms* **33**, 63–66.

Sugiyama A, Yokoyama H, Ogawa K (1999) Epizootiological investigation on kudoosis amami caused by *Kudoa amamiensis* (Multivalvulida: Myxozoa) in Okinawa Prefecture, Japan. *Fish Pathology* **34**, 39–43. [in Japanese, with English abstract].

Tin Tun, Yokoyama H, Ogawa K, Wakabayashi H (2000) Myxosporeans and their hyperparasitic microsporeans in the intestine of emaciated tiger puffer. *Fish Pathology* **35**, 145–156.

Urawa S (1998) A review of *Ichthyobodo* infection in marine fishes. *Fish Pathology* **33**, 311–320.

Yanagida T, Nomura Y, Kimura K, Fukuda Y, Yokoyama H, Ogawa K (2004) Molecular and morphological redescriptions on enteric myxozoans, *Enteromyxum leei* (formerly *Myxidium* sp. TP) and *Enteromyxum fugu* comb. n. (syn. *Myxidium fugu*) from cultured tiger puffer. *Fish Pathology* **39**, 137–143.

Yokoyama H, Freeman MA, Yoshinaga T, Ogawa K (2004) *Myxobolus buri*, the myxosporean parasite causing scoliosis of yellowtail, is a synonym of *Myxobolus acanthogobii* infecting the brain of the yellowfin goby. *Fisheries Science* **70**, 1036–1042.

Yoshinaga T (2001) Effects of high temperature and dissolved oxygen concentration on the development of *Cryptocaryon irritans* (Ciliophora) with a comment on the autumn outbreaks of cryptocaryoniasis. *Fish Pathology* **36**, 231–235.

Effects in mollusc culture

Adlard RD, Ernst I (1995) Extended range of the oyster pathogen *Marteilia sydneyi*. *Bulletin of the European Association of Fish Pathologists* **15**, 119–121.

Adlard RD, Lester RJG (1995) Development of a diagnostic test for *Mikrocytos roughleyi*, the aetiological agent of Australian winter mortality of the commercial rock oyster, *Saccostrea commercialis* (Iredale & Roughleyi). *Journal of Fish Diseases* **18**, 609–614.

Andrews JD (1988) Epizootiology of the disease caused by the oyster pathogen *Perkinsus marinus* and its effects on the oyster industry. *American Fisheries Society Special Publication* **18**, 47–63.

Andrews JD (1996) History of *Perkinsus marinus*, a pathogen of oysters in Chesapeake Bay 1950–1984. *Journal of Shellfish Research* **15**, 13–16.

Blackbourn J, Bower SM, Meyer GR (1998) *Perkinsus qugwadi* sp.nov. (incertae sedis), a pathogenic protozoan parasite of Japanese scallops, *Patinopecten yessoensis*, cultured in British Columbia, Canada. *Canadian Journal of Zoology* **76**, 942–953.

Bower SM (1987) *Labyrinthuloides haliotidis* n. sp. (Protozoa: Labyrinthomorpha), a pathogenic parasite of small juvenile abalone in a British Columbia mariculture facility. *Canadian Journal of Zoology* **65**, 1996–2007.

Bower SM (1988) Circumvention of mortalities caused by Denman Island oyster disease during mariculture of Pacific oysters. *American Fisheries Society Special Publication* **18**, 246–248.

Bower SM, McGladdery SE (2001) Synopsis of infectious diseases and parasites of commercially exploited shellfish. http://www.pac.dfo-mpo.gc.ca/sci/shelldis/toc_e.htm#oys. Fisheries and Oceans Canada.

Bower SM, Blackbourn J, Meyer GR (1998) Distribution, prevalence, and pathogenicity of the protozoan *Perkinsus qugwadi* in Japanese scallops, *Patinopecten yessoensis*, cultured in British Columbia, Canada. *Canadian Journal of Zoology* **76**, 954–959.

Burreson EM, Stokes NA, Friedman CS (2000) Increased virulence in an introduced pathogen: *Haplosporidium nelsoni* (MSX) in the eastern oyster *Crassostrea virginica*. *Journal of Aquatic Animal Health* **12**, 1–8.

Carnegie RB, Barber BJ, Culloty SC, Figueras AJ, Distel DL (2000) Development of a PCR assay for detection of the oyster pathogen *Bonamia ostreae* and support for its inclusion in the Haplosporidia. *Diseases of Aquatic Organisms* **42**, 199–206.

Carnegie RB, Barber BJ, Distel DL (2003) Detection of the oyster parasite *Bonamia ostreae* by fluorescent *in situ* hybridization. *Diseases of Aquatic Organisms* **55**, 247–252.

Carnegie RB, Meyer GR, Blackbourn J, Cochennec-Laureau N, Berthe FCJ, Bower SM (2003) Molecular detection of the oyster parasite *Mikrocytos mackini*, and a preliminary phylogenetic analysis. *Diseases of Aquatic Organisms* **54**, 219–227.

Carriker MR, Gaffney PM (1996) Catalogue of selected species of living oysters (Ostreacea) of the world. *In* 'The Eastern Oyster *Crassostrea virginica*'. (Eds VS Kennedy, RIE Newell, AF Eble.) pp. 1–18. (Maryland Sea Grant College: College Park, Maryland, USA).

Casas SM, Villalba A, Reece KS (2002) Study of perkinsosis in the carpet shell clam *Tapes decussatus* in Galicia (NW Spain). I. Identification of the aetiological agent and in vitro modulation of zoosporulation by temperature and salinity. *Disease of Aquatic Organisms* **50**, 51–65.

Cochennec N, Le Roux F, Berthe F, Gérard A (2000) Detection of *Bonamia ostreae* based on small subunit ribosomal probe. *Journal of Invertebrate Pathology* **76**, 26–32.

Cochennec-Laureau N, Reece KS, Berthe FCJ, Hine PM (2003) *Mikrocytos roughleyi* taxonomic affiliation leads to the genus *Bonamia* (Haplosporidia). *Diseases of Aquatic Organisms* **54**, 209–217.

Comps M, Grizel H, Papayanni Y (1982) Infection parasitaire causée par *Marteilia maurini* sp. nov. chez la moule *Mytilus galloprovincialis*. *Proceedings of the International Council for Exploration of the Seas* **F**, 1–3.

Culloty SC, Cronin MA, Mulcahy MF (2001) An investigation into the relative resistance of Irish flat oysters *Ostrea edulis*, L. to the parasite *Bonamia ostreae* (Pichot *et al.*, 1980). *Aquaculture* **199**, 229–244.

Elston R, Farley CA, Kent ML (1986) Occurrence and significance of bonamiasis in European flat oysters *Ostrea edulis* in North America. *Diseases of Aquatic Organisms* **2**, 49–54.

Farley CA, Wolf PH, Elston R (1988) A long-term study of 'microcell' disease in oysters with a description of a new genus, *Mikrocytos* (g. n.), and two new species, *Mikrocytos mackini* (sp. n.) and *Mikrocytos roughleyi* (sp. n.), *Fishery Bulletin* **86**, 581–593.

FAO Fisheries (2004/05) Food and Agriculture Organization of the United Nations, World aquaculture production of fish, crustaceans, molluscs, etc., by principal species in 2001. http://www.fao.org/fi/statist/statist.asp.

Ford SE, Tripp MR (1996) Diseases and defense mechanisms. *In* 'The Eastern Oyster *Crassostrea virginica*'. (Eds VS Kennedy, RIE Newell, AF Eble.) pp. 581–660. (Maryland Sea Grant College: College Park, Maryland, USA).

Goulletquer P, Héral M (1997) Marine molluscan production trends in France: from fisheries to aquaculture. *In* 'The History, Present Condition, and Future of the Molluscan Fisheries of North and Central America and Europe'. (Eds CL MacKenzie, Jr., VG Burrell, Jr., A Rosenfield, WL Hobart.) pp. 137–164. NOAA Technical Report NMFS 129, Seattle, Washington, USA.

Grizel H, Mialhe E, Chagot D, Boulo V, Bachère E (1988) Bonamiasis: A model study of diseases in marine molluscs, *American Fisheries Society Special Publication* **18**, 1–4.

Haskin HH, Andrews JD (1988) Uncertainties and speculations about the life cycle of the eastern oyster pathogen *Haplosporidium nelsoni* (MSX). *American Fisheries Society Special Publication* **18**, 5–22.

Haskin HH, Ford SE (1987) Breeding for disease resistance in molluscs. *In* 'Proceedings of the World Symposium on Selection, Hybridization, and Genetic Engineering in Aquaculture' (Bordeaux, France) pp. 431–441.

Hine PM, Cochennec-Laureau N, Berthe FC (2001) *Bonamia exitiosus* n. sp. (Haplosporidia) infecting flat oysters, *Ostrea chilensis*, in New Zealand. *Diseases of Aquatic Organisms* **47**, 63–72.

Hugh-Jones D (1999) Breeding ponds as a basis for flat oyster (*Ostrea edulis*) culture and their use to develop resistance to the disease *Bonamia ostreae*. *Journal of Shellfish Research* **18**, 718.

Itoh N, Oda T, Yoshinaga T, Ogawa K (2003) DNA probes for detection of *Marteilioides chungmuensis* from the ovary of Pacific oyster *Crassostrea gigas*. *Fish Pathology* **38**, 163–169.

Katkansky SC, Dahlstrom WA, Warner RW (1969) Observations on survival and growth of the European flat oyster, *Ostrea edulis*, in California. *California Department of Fish and Game* **55**, 69–74.

Kinne O (Ed.) (1982) 'Diseases of Marine Animals'. (Biologische Anstalt Helgoland: Hamburg).

Kleeman SK, Adlard RD (2000) Molecular detection of *Marteilia sydneyi*, pathogen of Sydney rock oysters. *Diseases of Aquatic Organisms* **40**, 137–146.

Le Roux F, Audemard C, Barnaud A, Berthe F (1999) DNA probes as potential tools for the detection of *Marteilia refringens*. *Marine Biotechnology* **1**, 588–597.

Le Roux F, Lorenzo G, Peyret P, Audemard C, Figueras A, Vivarès C, Gouy M, Berthe F (2001) Molecular evidence for the existence of two species of *Marteilia* in Europe. *Journal of Eukaryotic Microbiology* **48**, 449–454.

Lester RJG, Davis GHG (1981) A new *Perkinsus* species (Apicomplexa, Perkinsea) from the abalone *Haliotis ruber*. *Journal of Invertebrate Pathology* **37**, 181–187.

Longwell AC, Stiles SS (1996) Chromosomes, biology, and breeding. *In* 'The Eastern Oyster *Crassostrea virginica*'. (Eds VS Kennedy, RIE Newell, AF Eble.) pp. 443–466. (Maryland Sea Grant College: College Park, Maryland, USA).

Mackin JG, Owen HM, Collier A (1950) Preliminary note on the occurrence of a new protistan parasite, *Dermocystidium marinum* n. sp. in *Crassostrea virginica* (Gmelin). *Science* **111**, 328–329.

Murrell A, Kleeman SN, Barker SC, Lester RJG (2002) Synonymy of *Perkinsus olseni* Lester & Davis, 1981 and *Perkinsus atlanticus* Azevedo, 1989 and an update on the phylogenetic position of the genus *Perkinsus*. *Bulletin of the European Association of Fish Pathologists* **22**, 258–265.

Naciri-Graven Y, Martin AG, Baud JP, Renault T, Gérard A (1998) Selecting the flat oyster *Ostrea edulis* (L.) for survival when infected with the parasite *Bonamia ostreae*. *Journal of Experimental Marine Biology and Ecology* **224**, 91–107.

Nell JA, Hand RE (2003) Evaluation of the progeny of second-generation Sydney rock oyster *Saccostrea glomerata* (Gould, 1850) breeding lines for resistance to QX disease *Marteilia sydneyi*. *Aquaculture* **228**, 27–35.

OIE (2003) 'Manual of Diagnostic Tests for Aquatic Animals'. 4th Edn. Office International de épizooties: Paris).

Park K-I, Choi K-S (2001) Spatial distribution of the protozoan parasite *Perkinsus* sp. found in the Manila clams, *Ruditapes philippinarum*, in Korea. *Aquaculture* **203**, 9–22.

Pichot Y, Comps M, Grizel H, Rabouin MA (1980) Recherches sur *Bonamia ostreae*, gen. n. sp. n., parasite nouveau de l'huître plate *Ostrea edulis* L. *Rev. Trav. Inst. Pêches marit.* **43**, 131–140.Perkins FO, Wolf PH (1976) Fine structure of *Marteilia sydneyi* sp. n. – haplosporidian pathogen of Australian oysters. *The Journal of Parasitology* **62**, 528–538.

Quayle DB (1961) Denman Island oyster disease and mortality, 1960. Fisheries Research Board of Canada Manuscript Report 713. Ottawa, Ontario, Canada.

Ragone Calvo LM, Walker JG, Burreson EM (1998) Prevalence and distribution of QPX, Quahog Parasite Unknown, in hard clams *Mercenaria mercenaria* in Virginia, USA. *Diseases of Aquatic Organisms* **33**, 209–219.

Ragone Calvo LM, Calvo GW, Burreson EM (2003) Dual disease resistance in a selectively bred eastern oyster, *Crassostrea virginica*, strain tested in Chesapeake Bay. *Aquaculture* **220**, 69–87.

Ray SM (1996) Historical perspective on *Perkinsus marinus* disease of oysters in the Gulf of Mexico. *Journal of Shellfish Research* **15**, 9–11.

Roughley TC (1926) An investigation of the cause of an oyster mortality on the Georges River, New South Wales, 1924–25. *Proceedings of the Linnaean Society of New South Wales* **51**, 446–491.

Sindermann CJ (1990) 'Principal Diseases of Marine Fish and Shellfish'. Vol. 2. (Academic Press: San Diego).

Smolowitz R, Leavitt D, Perkins F (1998) Observations of a protistan disease similar to QPX in *Mercenaria mercenaria* (hard clams) from the coast of Massachusetts. *Journal of Invertebrate Pathology* **71**, 9–25.

Stokes NA, Burreson EM (1995) A sensitive and specific DNA probe for the oyster pathogen *Haplosporidium nelsoni*. *The Journal of Eukaryotic Microbiology* **42**, 350–357.

Stokes NA, Siddall ME, Burreson EM (1995) Detection of *Haplosporidium nelsoni* (Haplosporidia: Haplosporidiidae) in oysters by PCR amplification. *Diseases of Aquatic Organisms* **23**, 145–152.

Stokes NA, Ragone Calvo LM, Reece KS, Burreson EM (2002) Molecular diagnostics, field validation, and phylogenetic analysis of Quahog Parasite Unknown (QPX), a pathogen of the hard clam *Mercenaria mercenaria*. *Diseases of Aquatic Organisms* **52**, 233–247.

Villalba A, Mourelle SG, López MC, Carballal MJ, Azevedo C (1993) Marteiliasis affecting cultured mussels *Mytilus galloprovincialis* of Galicia (NW Spain). I. Etiology, phases of the infection, and temporal and spatial variability in prevalence. *Diseases of Aquatic Organisms* **16**, 61–72.

Whyte SK, Cawthorn RJ, McGladdery SE (1994) QPX (Quahaug Parasite X), a pathogen of northern quahaug *Mercenaria mercenaria* from the Gulf of St Lawrence, Canada. *Diseases of Aquatic Organisms* **19**, 129–136.

Wolf PH (1972) Occurrence of a haplosporidian in Sydney rock oysters (*Crassostrea commercialis*) from Moreton Bay, Queensland, Australia. *Journal of Invertebrate Pathology* **19**, 416–417.

Effects in shrimp culture

Becnel JJ (1992) Horizontal transmission and subsequent development of *Amblyospora californica* (Microsporida: Amblyosporidae) in the intermediate and definitive hosts. *Diseases of Aquatic Organisms* **13**, 17–28.

Canning EU, Curry A, Overstreet RM (2002) Ultrastructure of *Tuzetia weidneri* sp. n. (Microsporidia: Tuzetiidae) in skeletal muscle of *Litopenaeus setiferus* and *Farfantepenaeus aztecus* (Crustacea, Decapoda) and new data on *Perezia nelsoni* (Microsporidia, Pereziidae) in *L. setiferus*. *Acta Protozoologica* **41**, 63–77.

Corkern CC (1978) A larval cyclophyllidean (Cestoda) parasite of brown and white shrimp. *Texas A&M University Extension Fish Disease Diagnostic Laboratory* Publication No. FDDL-S10.

Couch JA (1978) Diseases, parasites, and toxic responses of commercial penaeid shrimps of the Gulf of Mexico and South Atlantic coasts of North America. *Fishery Bulletin* **76**, 1–44.

Crites JL, Overstreet RM (1991) *Heliconema brooksi* n. sp. (Nematoda: Physalopteridae) from the ophichthid eel *Ophichthus gomesi* in the Gulf of Mexico. *The Journal of Parasitology* **77**, 42–50.

Deardorff TL, Overstreet RM (1981) Larval *Hysterothylacium* (=*Thynnascaris*) (Nematoda: Anisakidae) from fishes and invertebrates in the Gulf of Mexico. *Proceedings of the Helminthological Society of Washington* **48**, 113-126.

Fusco AC (1980) Larval development of *Spirocamallanus cricotus* (Nematoda: Camallanidae). *Proceedings of the Helminthological Society of Washington* **47**, 63–71.

Heard RW, Overstreet RM (1983) Taxonomy and life histories of two new North American species of '*Carneophallus*' (=*Microphallus*) (Digenea: Microphallidae). *Proceedings of the Helminthological Society of Washington* **50**, 170–174.

Iversen ES, Kelly JF (1976) Microsporidiosis successfully transmitted experimentally in pink shrimp. *Journal of Invertebrate Pathology* **27**, 407–408.

Jones TC, Overstreet RM, Lotz JM, Frelier PF (1994) *Paraophioidina scolecoides* n. sp., a new aseptate gregarine from the cultured Pacific white shrimp, *Penaeus vannamei*. *Diseases of Aquatic Organisms* **19**, 67–75.

Laramore CR, Barkate JA (1979) Mortalities produced in the protozoae stages of penaeid shrimp by an unspeciated amoeba. *Texas A&M University, Extension Fish Disease Diagnostic Laboratory* Publication No. FDDL-S12, 1-7.

Landers SC, Zimlich MA, Coate T (1999) Variations in the ventral ciliature of the crustacean symbiont *Hyalophysa* (Ciliophora, Apostomatida) from Mobile Bay and Dauphin Island, Alabama. *Gulf Research Reports* **11**, 57–63.

Lotz JM, Overstreet RM (1990) Parasites and predators. *In* 'The Aquaculture of Shrimp, Prawn and Crawfish in the World: Basics and Technologies'. (Eds JC Chàvez, NO Sosa.) pp. 96–121. (Midori Shobo: Ikebukuro, Toshima-ku Tokyo). (In Japanese).

Lotz JM, Vogelbein WK, Overstreet RM (1988) Observations on two species of apostome ciliates of *Penaeus* spp. from the northern Gulf of Mexico. *Journal of the Mississippi Academy of Sciences* **33**, (Supplement), 81.

Micieli MV, García JJ, Becnel JJ (2000) Horizontal transmission of *Amblyospora albifasciati* García and Becnel, 1994 (Microsporidia: Amblyosporidae), to a copepod intermediate host and the neotropical mosquito, *Aedes albifasciatus* (Macquart, 1837). *Journal of Invertebrate Pathology* **75**, 76–83.

Overstreet RM (1973) Parasites of some penaeid shrimps with emphasis on reared hosts. *Aquaculture* **2**, 105–140.

Overstreet RM (1987) Solving parasite-related problems in cultured Crustacea. *International Journal for Parasitology* **17**, 309–318.

Overstreet RM (2003) Presidential address: Flavor buds and other delights. *The Journal of Parasitology* **89**, 1093–1107.

Overstreet RM, Meyer GW (1981) Hemorrhagic lesions in stomach of rhesus monkey caused by a piscine ascaridoid nematode. *The Journal of Parasitology* **67**, 226–235.

Overstreet RM, Safford S (1980) Diatoms in the gills of the commercial white shrimp. *Gulf Research Reports* **6**, 421–422.

Sandifer PA, Smith TIJ, Calder DR (1974) Hydrozoans as pests in closed-system culture of larval decapod crustaceans. *Aquaculture* **4**, 55–59.

Sparks AK, Fontaine CT (1973) Host responses in the white shrimp, *Penaeus setiferus*, to infection by the larval trypanorhynchid cestode, *Prochristianella penaei*. *Journal of Invertebrate Pathology* **22**, 213–219.

Sweeney AW, Doggett SL, Piper RG (1993) Life cycle of a new species of *Duboscquia* (Microsporida: Thelohaniidae) infecting the mosquito *Anopheles hilli* and an intermediate copepod host, *Apocyclops dengizicus*. *Journal of Invertebrate Pathology* **62**, 137–146.

Ecological aspects of parasites in the American lobster

Aiken DE, Sochasky JB, Wells PG (1973) Ciliate infestation of the blood of the lobster *Homarus americanus*. *International Council for the Exploration of the Sea* **CM1973K**, 46.

Biggers WJ, Laufer H (2004) Identification of juvenile hormone-active alkylphenols in the lobster *Homarus americanus* and in marine sediments. *Biological Bulletin* **206**, 13–24.

Bower SM, McGladdery SM, Price IM (1994) Synopsis of infectious diseases and parasites of commercially exploited shellfish. *Annual Review of Fish Diseases* **4**, 1–199. http://www.pac.dfo.ca/pac/sealane/aquac/pages/title.htm-could not be accessed

Boyd W (1970) 'A Textbook of Pathology. Structure and Function in Disease'. 8th edn. (Lea and Febiger: London).

Cawthorn RJ (1997) Overview of bumper car disease – impact on the North American lobster fishery. *International Journal for Parasitology* **27**, 167–172.

Cawthorn RJ, Lynn DH, Després B, MacMillan R, Maloney R, Loughlin M, Bayer R (1996) Description of *Anophryoides haemophila* n. sp. (Scuticociliatida: Orchitophryidae), a pathogen of American lobsters *Homarus americanus*. *Diseases of Aquatic Organisms* **24**, 143–148.

Couch JA (1983) Diseases caused by Protozoa. In 'The Biology of Crustacea'. Vol. 6. (Ed. AJ Provenzano.) pp. 79–111. (Academic Press, New York).

Crosbie PBB, Nowak BF, Carson J (2003) Isolation of *Neoparamoeba pemaquidensis* Page, 1987 from marine and estuarine sediments in Tasmania. *Bulletin of European Association of Fish Pathologists* **23**, 241–244.

De Guise S, Maratea J, Perkins P (2004) Malathion immunotoxicity in the American lobster (*Homarus americanus*) upon experimental exposure. *Aquatic Toxicology* **66**, 419–425.

Dohoo I, Martin W, Stryhn H (2003) 'Veterinary Epidemiologic Research'. (AVC, Charlottetown).

Factor JR (Ed.) (1995) 'Biology of the Lobster *Homarus americanus*'. (Academic Press: Toronto).

Fiala I, Dykova I (2003) Molecular characterisation of *Neoparamoeba* strains isolated from gills of Scophthalmus *maximus*. *Diseases of Aquatic Organisms* **55**, 11–16.

Fisheries and Oceans Canada (2004) Statistical Services. Commercial landings. http://www.dfo-mpo.gc.ca/communic/statistics/commercial/landings/index_e.htm.

Frasca Jr. S, Nevis KR, Mullen TE (2003) Development of polymerase chain reaction- and *in situ* hybridization-based tests for the specific detection of the *Paramoeba* associated with epizootic lobster mortality by determination of the molecular systematics *Paramoeba*. *Third Long Island Sound Lobster Health Symposium*, pp. 53–56. (Connecticut Department of Environmental Protection: Groton).

French RA, Frasca Jr. S, Russell KS, Mullen Jr. T, Burrage T (2001) Paramoebiasis in the American lobster, *Homarus americanus*. 26th Annual Eastern Fish Health Workshop, Abstract.

Gast RJ (2003) Oligonucleotide-based detection of pathogenic *Neoparamoeba* species. *Third Long Island Sound Lobster Health Symposium,* pp. 57–60. (Connecticut Department of Environmental Protection: Groton).

Gillevet PM, O'Kelly CJ (2003) Progress in *Paramoeba* research. *Third Long Island Sound Lobster Health Symposium.* pp. 50–52. (Connecticut Department of Environmental Protection: Groton).

Greenwood SJ, Déspres BM, Cawthorn RJ, Lavallee J, Groman DB, Desbarat A (in press) Case report: outbreak of bumper car disease caused by *Anophryoides haemophila* in a lobster holding facility in Nova Scotia, Canada. *Journal of Aquatic Animal Health.*

Lavallée J, Hammell KL, Spangler ES, Cawthorn RJ (2001) Estimated prevalence of *Aerococcus viridans* and *Anophryoides haemophila* in American lobsters *Homarus americanus* freshly captured in the waters of Prince Edward Island, Canada. *Diseases of Aquatic Organisms* **46**, 231–236.

Lawton P, Lavalli KL (1995) Postlarval, juvenile, adolescent, and adult ecology. *In* 'Biology of the Lobster *Homarus americanu*'. (Ed. JR Factor.) pp. 47–88. (Academic Press: Toronto).

Marcogliese DJ (2001) Implications of climate change for parasitism of animals in the aquatic environment. *Canadian Journal of Zoology* **79**, 1331–1352.

Marcogliese DJ (2004) Parasites: small players with crucial roles in the ecological theater. *EcoHealth* **1**, 151–164.

Morado JF, Small EB (1995) Ciliate parasites and related diseases of Crustacea: a review. *Reviews in Fisheries Science* **3**, 275–354.

Mullen TE, Russell S, Tucker MT, Maratea JL, Koerting C, Hinckley L, De Guise S, Frasca Jr. S, French RA, Burrage TG, Perkins C (2004) Paramoebiasis associated with mass mortality of American lobster *Homarus americanus* in Long Island Sound, USA. *Journal of Aquatic Animal Health* **16**, 29–38.

National Marine Fisheries Service (US) (2004) Fisheries Statistics and Economics Division. Commercial Fisheries. http://www.st.nmfs.gov/st1/commercial/index.html.

Overstreet RM (1987) Solving parasite-related problems in cultured Crustacea. *International Journal for Parasitology* **17**, 309–318.

Peglar MT, Amaral Zettler LA, Anderson OR, Nerad TA, Gillevet PM, Mullen TE, Frasca Jr. S, Silberman JD, O'Kelly CJ, Sogin ML (2003) Two new small-subunit ribosomal RNA gene lineages within the subclass Gymnamoebia. *Journal of Eukaryotic Microbiology* **50**, 224–232.

Phillips BF, Kittaka J (Eds) (2000) 'Spiny Lobsters: Fisheries and Culture'. 2nd edn. (Fishing News Books: London).

Shields, J (2004) Pathological alterations in the eyes of the American lobster from western Long Island Sound and their possible association with *Paramoeba* infections. Abstract. 'Seventh International Conference and Workshop on Lobster Biology and Management', 8–13 Feb., 2004, p. 33. (Hobart, Tasmania).

Speare DJ, Cawthorn RJ, Horney BS, MacMillan R, Mackenzie AL (1996) Effects of formalin, chloramine-T, and low salinity dip on the behavior and hemolymph biochemistry of the American lobster. *Canadian Veterinary Journal* **37**, 729–734.

Stewart JE (1980) Diseases. *In* 'The Biology and Management of Lobsters'. Vol. 1. Physiology and Behavior. (Eds BP Phillips, JS Cobb.) pp. 301–341. (Academic Press: New York).

Stewart JE (1993) Infectious diseases of marine crustaceans. *In* 'Pathobiology of Marine and Estuarine Organisms'. (Eds JA Couch, JW Fournie.) pp. 319–342. (CRC Press: Boca Raton).

Wong FYK, Carson J, Elliott NG (2004) 18S ribosomal DNA-based PCR identification of *Neoparamoeba pemaquidensis*, the agent of amoebic gill disease in sea-farmed salmonids. *Diseases of Aquatic Organisms* **60**, 65–76.

Parasites of marine mammals

Bandoli JG, De Oliveira AB (1977) Toxoplasma in *Sotalia guianensis* (van Beneden, 1863) Cetacea – Delphinidae. *Folha Medica* **75**, 459–468.

Dailey MD (1970) The transmission of *Parafilaroides decorus* (Nematoda: Metastrongyloidea) in the California sea lion (*Zalophus californianus*). *Proceedings of the Helminthological Society of Washington* **37**, 215–222.

Dailey MD (1985) Diseases of Mammalia: Cetacea. *In* 'Diseases of Marine Mammals'. Vol. IV, Part 2: Introduction, Reptilia, Aves, Mammalia. (Ed. O Kinne.) pp. 805–844. (Biologische Anstalt Helgoland: Hamburg).

Dailey MD (2001) Parasitic diseases. *In* 'CRC Handbook of Marine Mammal Medicine'. 2nd edn. (Eds LA Dierauf, FMD Gulland.) pp. 357–379. (CRC Press: Boca Raton).

Dailey MD, Gulland FMD, Lowenstine LJ, Sivagni P, Howard D (2000) Prey, parasites and pathology associated with mortality of a juvenile gray whale (*Eschrichtius robustus*) stranded along the northern California coast. *Diseases of Aquatic Organisms* **42**, 111–117.

Dailey MD, Haulaena M, Lawrence J (2002) First report of a parasitic copepod (*Pennella balaenopterae*) infestation in a pinniped. *Journal of Zoo and Wildlife Medicine* **33**, 62–65.

Delyamure SL (1955) 'The Helminth Fauna of Marine Mammals in Light of their Ecology and Phylogeny'. (Translated by Isreal Program Scientific Translation Information Services: Springfield VA. As TT67–51202).

Dubey JP, Speer CA, Fayer R (1989) 'Sarcocystosis of Animals and Man'. (CRC Press: Boca Raton).

Dubey JP, Rosypal AC, Rosenthal BM, Thomas NJ, Lindsay DS, Stanek JF Reed SM, Saville WJA (2003) *Sarcocystis neurona* infection in a sea otter (*Enhydra lutris*): Evidence for natural infections with sarcocysts and transmission of infection to oppossums (*Didelphis virginiana*). *Journal of Parasitology* **87**, 1387–1393.

Geraci JR, Lounsbury V (2001) Marine mammal health: Holding the balance in an ever-changing sea. *In* 'Marine Mammals Biology and Conservation'. (Eds GH Evans, JA Raga.) pp. 365–385. (Kluer Academic/Plenum Publishers: New York).

Jardine JE, Dubey JP (2002) Congenital toxoplasmosis in a Indo-Pacific Bottlenose dolphin (*Tursiops aduncus*). *Journal of Parasitology* **88**, 197–199.

Lauckner G (1985) Diseases of Mammalia: Pinnipedia. *In* 'Diseases of Marine Mammals'. Vol. IV, Part 2: Introduction, Reptilia, Aves, Mammalia. (Ed. O Kinne.) pp. 683–772. (Biologische Anstalt Helgoland: Hamburg).

Mayer KA, Dailey MD, Miller MA (2003) Helminth parasites of the southern sea otter *Enhydra lutris nereis* in central California: abundance, distribution and pathology. *Diseases of Aquatic Organisms* **53**, 77–88.

Measures LN (2001) Lungworms of marine mammals. *In* 'Parasitic Diseases of Wild Mammals'. 2nd edn. (Eds WM Samuel, MJ Pybus, AA Kocan.) pp. 279–290. (Iowa State University Press: Ames).

Miller MA, Crosbie PR, Sverlow K, Hanni K, Barr BC, Kock N, Murray MH, Lowenstine LJ, Conrad PA (2001) Isolation and characterization of *Sarcocystis* from brain tissue of a free living southern sea otter (*Enhydra lutris nereis*) with fatal meningoencephalitis. *Parasitology Research* **87**, 252–257.

Miller MA, Gardner IA, Peckham A, Mazet JK, Hanni KD, Jessup D, Estes J, Jameson R, Dodd E, Barr BC, Lowenstine LJ, Gulland FM, Conrad PA (2002) Evaluation of an indirect fluorescent antibody test (IFAT) for demonstration of antibodies to *Toxoplasma gondii* in the sea otter (*Enhydra lutris*). *Journal of Parasitology* **88**, 594–599.

Munro R, Synge B (1991) Coccidiosis in seals. *Veterinary Record* **129**, 179–180.

Munro R, Measures LN, Olson M (1999) Giardiasis in pinnipeds from Eastern Canada. *Journal of Wildlife Diseases* **35**, 779–782.

Poynton SL, Whitaker BR, Heinrich AB (2001) A novel trypanoplasm-likeflagellate *Jarrellia alramenti* n.g., n. sp. (Kinetoplastida:Bodonidae) and ciliates from the blowhole of a stranded pygmy sperm whale *Kogia breviceps* (Physeteridae): morphology, life cycle and potential pathogenicity. *Diseases of Aquatic Organisims* **44**, 191–201.

Resendes AR, Juan-Salles C, Almeria S, Mago N, Domingo M, Dubey JP (2002) Hepatic sarcocystosis in a striped dolphin (*Stenella coeruleoalba*) from the Spanish Mediterranean Coast. *Journal of Parasitology* **88**, 206–208.

Marine birds and their helminth parasites

Ainley DG, Boekelheide RJ (Eds) (1990) 'Seabirds of the Farallon Islands: Ecology, Dynamics and Structure of an Upwelling System Community'. (Stanford University Press: Stanford).

Ashmole NP (1971) Sea bird ecology and the marine environment. *In* 'Avian Biology'. Vol. 1. (Eds DS Farner, JR King, KC Parks.) pp. 223–286. (Academic Press: New York).

Brooks DR, Hoberg EP (2000) Triage for the biosphere: the need and rational for taxonomic inventories and phylogenetic studies of parasites. *Comparative Parasitology* **67**, 1–25.

Barus V, Sergeeva TP, Sonin MD, Ryzhikov KM (1978) 'Helminths of Fish-Eating Birds of the Palearctic I. Nematoda'. (W. Junk Publishers: The Hague).

Claugher D (1976). A trematode associated with the death of white-faced storm petrel (*Pelagodroma marina*) on the Chatham Islands. *Journal Natural History* **10**, 633–641.

Chavez FP, Ryan J, Lluch-Cota SE, [N]iquen M (2003) From anchovies to sardines and back: multidecadal change in the Pacific Ocean. *Science* **299**, 217–221.

Croxall JP (Ed.) (1987) 'Seabirds; Feeding Ecology and Role in Marine Ecosystems'. (Cambridge University Press: Cambridge).

Croxall JP, Trathan PN, Murphy EJ (2002) Environmental change and Antarctic seabird populations. *Science* **297**, 1510–1514.

Fagerholm H-P (1996) Nematode parasites of marine- and shore birds, and their role as pathogens. *Bulletin of the Scandinavian Society for Parasitology* **6**, 16–30.

Fagerholm H-P, Overstreet RM, Humphrey-Smith I (1996) *Contracaecum magnipapillatum* (Nematoda: Ascaridoidea): resurrection and pathogenic effect of a common parasite from the proventriculus of *Anous minutus* from the Great Barrier Reef, with a note on *C. variegatum*. *Helminthologia* **33**, 195–207.

Fuhrmann O (1921) Die Cestoden der Deutschen Südpolar Expedition 1901–1903. *Deutschen Südpolar Expedition 1901–1903* **16**, *Zoology* **8**, 467–524.

Forrester DJ, Davidson WR, Lange RE Jr, Stroud RK, Alexander II, Franson JC, Haseltine SD, Littell RC, Nesbitt SA (1997) Winter mortality of common loons in Florida coastal waters. *Journal of Wildlife Diseases* **33**, 833–847.

Galaktionov K (1996) Impact of seabird helminths on host populations and coastal ecosystems. *Bulletin of the Scandinavian Society for Parasitology* **6**, 50–64.

Harrison P (1983) 'Seabirds: An Identification Guide'. (Croom Helm: Beckenham).

Hoberg EP (1986) Aspects of ecology and biogeography of Acanthocephala in Antarctic seabirds. *Annales Parasitologie Humaine et Comparée* **61**, 199–214.

Hoberg EP (1992) [Ecology of helminth parasitism among seabirds at Talan Island, a preliminary overview]. *In* 'Coastal Ecosystems of the Northern Part of the Sea of Okhotsk'. (Eds FB Cherniafsky, AIa Kondratiev.) pp. 116–136. (Institute for Biological Problems of the North, Far Eastern Branch, Russian Academy of Sciences: Magadan). (In Russian).

Hoberg EP (1996) Faunal diversity among avian parasite assemblages: the interaction of history ecology and biogeography in marine systems. *Bulletin Scandinavian Society for Parasitology* **6**, 65–89.

Hoberg EP (1997) Phylogeny and historical reconstruction: host-parasite systems as keystones in biogeography and ecology. In 'Biodiversity II: Understanding and Protecting Our Biological Resources'. (Eds ML Reaka-Kudla, DE Wilson, EO Wilson.) pp. 243–261. (Joseph Henry Press, Washington DC).

Hoberg EP, Ryan PG (1989) Ecology of helminth parasitism in *Puffinus gravis* (Proceallariiformes) on the breeding grounds at Gough Island. *Canadian Journal of Zoology* **67**, 220–225.

Imber MJ (1984) Trematode anklets on whitefaced stormpetrels *Pelagodroma marina* and fairy prions *Pachyptila turtur*. *Cormorant* **12**, 71–74.

Langston N, Hilgarth N (1995) Moult varies with parasites in Laysan albatrosses. *Proceedings Royal Society of London Series B* **261**, 239–243.

Lauckner G (1985) Diseases of Aves (marine birds). In 'Diseases of Marine Animals'. Vol. 1. (Ed. O Kinne.) pp. 627–643. (Biologische Anstalt Helgoland: Hamburg).

Kinsella JM, Forrester DJ (1999) Parasitic helminthes of the common loon, *Gavia immer*, on its wintering grounds in Florida. *Journal of the Helminthological Society of Washington* **66**, 1–6.

Norman FI, Guesclin PB, Dann P (1992) The 1986 'wreck' of little penguins *Eudyptila minor* in western Victoria. *Emu* **91**, 369–376.

Randall RM, Bray RA (1983) Mortalities of jackass penguin *Spheniscus demersus* chicks caused by trematode worms *Cardiocephaloides physalis*. *South African Journal of Zoology* **18**, 45–46.

Riley J (1972) The pathology of anisakis nematode infections of the fulmar *Fulmarus glacialis*. *Ibis* **114**, 102–104.

Ryzhikov KM, Rysavay B, Khokhlova IG, Tolkatcheva LM, Kornyuchin VV (1985) 'Helminths of Fish-Eating Birds of the Palearctic Region II. Cestoda and Acanthocephales'. (Academiia Publishing House, Czechoslovak Academy of Sciences: Prague).

Skorping A (1996) Why should marine and coastal bird ecologists bother about parasites? *Bulletin of the Scandinavian Society for Parasitology* **6**, 98–102.

Springer AM (1998) Is it all climate change? Why marine bird and mammal populations fluctuate in the North Pacific. In 'Biotic Impacts of Extratropical Climate Variability in the Pacific'. (Eds G Holloway, P Müller, D Hendersen.) pp. 109–119. (University of Hawaii: Manoa).

Storer RM (2000) 'The Metazoan Parasite Fauna of Grebes (Aves: Podicipediformes) and its Relationship to the Birds' Biology'. Miscellaneous Publications Museum of Zoology No. 188. (Museum of Zoology, University of Michigan: Ann Arbor).

Storer RM (2002) 'The Metazoan Parasite Fauna of Loons (Aves: Gaviiformes) and its Relationship to the Birds' Evolutionary History and a Comparison with the Parasite Fauna of Grebes'. Miscellaneous Publications Museum of Zoology No. 191. (Museum of Zoology, University of Michigan: Ann Arbor).

Zdzitowiecki, K. (1986) Acanthocephala of the Antarctic. *Polish Polar Research* **7**, 79–117.

Effects of pollution on parasites, and use of parasites in pollution monitoring

Broeg K, Zander S, Diamant A, Körting W, Krüner G, Paperna I, von Westernhagen H (1999) The use of fish metabolic, pathological and parasitological indices in pollution monitoring. I. North Sea. *Helgoland Marine Research* **53**, 171–194.

Diamant A, Banet A, Paperna I, von Westernhagen H, Broeg K, Kruener G, Koerting W, Zander S (1999) The use of fish metabolic, pathological and parasitological indices in pollution monitoring. II. The Red Sea and Mediterranean. *Helgoland Marine Research* **53**, 195–208.

Heinonen J, Kukkonen JVK, Holopainen IJ (2000) Toxicokinetics of 2,4,5-Trichlorophenol and benzo(a)pyrene in the clam *Pisidium amnicum*: Effects of seasonal temperatures and trematode parasites. *Archives of Environmental Contamination and Toxicology* **39**, 352–359.

Kennedy CR (1990) Helminth communities in freshwater fish: structured communities or stochastic assemblages? *In* 'Parasite Communities: Patterns and Processes'. (Eds GW Esch, AO Bush, JM Aho.) pp. 131–156. (Chapman and Hall: London).

Kennedy CR (1997) Freshwater fish parasites and environmental quality: an overview and caution. *Parassitologia* **39**, 249–254.

Khan RA, Kiceniuk JW (1988) Effect of petroleum aromatic hydrocarbons on monogeneids parasitizing atlantic cod, *Gadus morhua* L. *Bulletin of Environmental Contamination and Toxicology* **41**, 94–100.

Khan RA, Thulin J (1991) Influence of pollution on parasites of aquatic animals. *Advances in Parasitology* **30**, 201–238.

Khan RA, Barker DE, Williams-Ryan K, Hooper RG (1994) Influence of crude oil and pulp and paper mill effluent on mixed infections of *Trichodina cottidarium* and *T. saintjohnsi* (Ciliophora) parasitizing *Myoxocephalus octodecemspinosus* and *M. scorpius*. *Canadian Journal of Zoology* **72**, 247–251.

Lafferty KD (1997) Environmental parasitology: What can parasites tell us about human impacts on the environment? *Parasitology Today* **13**, 251–255.

Lafferty KD, Kuris AM (1999) How environmental stress affects the impacts of parasites. *Limnology and Oceanography* **44**, 925–931.

Lewis JW, Hoole D (2003) Parasitism and environmental pollution: parasites and hosts as indicators of water quality. *Parasitology Supplement* **126**, 110 pp.

MacKenzie K (1999) Parasites as pollution indicators in marine ecosystems: a proposed early warning system. *Marine Pollution Bulletin* **38**, 955–959.

MacKenzie K, Williams HH, Williams B, McVicar AH, Siddall R (1995) Parasites as indicators of water quality and the potential use of helminth transmission in marine pollution studies. *Advances in Parasitology* **35**, 85–144.

Marcogliese DJ, Nagler JJ, Cyr DG (1998) Effects of exposure to contaminated sediments on the parasite fauna of American Plaice (*Hippoglossoides platessoides*). *Bulletin of Environmental Contamination and Toxicology* **61**, 88–95.

Morley NJ, Crane M, Lewis JW (2003) Toxicity of cadmium and zinc to the cercarial activity of *Diplostomum spathaceum* (Trematoda: Diplostomidae). *Folia Parasitologica* **50**, 57–60.

Overstreet RM (1997) Parasitological data as monitors of environmental health. *Parassitologia* **39**, 169–175.

Palm H, Dobberstein RC (1999) Occurrence of trichodinid ciliates (Peritricha: Urceolariidae) in the Kiel Fjord, Baltic Sea, and its possible use as a biological indicator. *Parasitology Research* **85**, 726–732.

Paperna I (1997) Fish parasites as indicators of environmental quality. VII European Multicolloquium of Parasitology, Parma, Italy, 2–6 September 1996. *Parassitologia* **39**, 255 pp.

Pietrock M, Marcogliese DJ (2003) Free-living endohelminth stages: at the mercy of environmental conditions. *Trends in Parasitology* **19**, 293–299.

Poulin R (1992) Toxic pollution and parasitism in freshwater fish. *Parasitology Today* **8**, 58–61.

Proceedings of the Xth International Conference on Parasitology (2002) ICOPA X. 4–9 August 2002, Vancouver, Canada. (Monduzzi Editore: Bologna, Italy).

Regala RP, Rice CD, Schwedler TE, Dorociak IR (2001) The effects of tributyltin (TBT) and 3,3'4,4'5-pentachlorobiphenyl (PCB–126) mixtures on antibody responses and phagocyte oxidative burst activity in channel catfish, *Ictalurus punctatus*. *Archives of Environmental Contamination and Toxicology* **40**, 386–391.

Rice CD, Schlenk D (1995) Immune function and cytochrome P4501A activity after acute exposure to 3,3'4,4',5-Pentachlorobiphenyl (PCB 126) in channel catfish. *Journal of Aquatic Animal Health* **7**, 195–204.

Ruus A, Skaare JU, Ingebrigtsen K (2001) Accumulation of the lipophilic environmental contaminant lindane in metacercariae of *Bucephaloides gracilescens* (Trematoda, Bucephalidae) in the central nervous system of bullrout *Myoxocephalus scorpius*. *Diseases of Aquatic Organisms* **48**, 75–77.

Sures B (2001) The use of fish parasites as bioindicators of heavy metals in aquatic ecosystems: a review. *Aquatic Ecology* **35**, 245–255

Sures B (2003) Accumulation of heavy metals by intestinal helminths in fish: an overview and perspective. *Parasitology* **126**, S53–S60.

Sures B (2004) Environmental parasitology: relevancy of parasites in monitoring environmental pollution. *Trends in Parasitology* **20**, 170–177.

Sures B, Knopf K (2004) Individual and combined effects of Cd and 3,3',4,4',5-pentachlorobiphenyl (PCB 126) on the humoral immune response in European eel (*Anguilla anguilla*) experimentally infected with larvae of *Anguillicola crassus* (Nematoda). *Parasitology* **128**, 445–454.

Sures B, Reimann N (2003) Analysis of trace metals in the Antarctic host-parasite system *Notothenia coriiceps* and *Aspersentis megarhynchus* (Acanthocephala) caught at King George Island, South Shetland Islands. *Polar Biology* **26**, 680–686.

Sures B, Siddall R, Taraschewski H (1999a) Parasites as accumulation indicators of heavy metal pollution. *Parasitology Today* **15**, 16–21.

Sures B, Steiner W, Rydlo M, Taraschewski H (1999b) Concentrations of 17 elements in the zebra mussel (*Dreissena polymorpha*), in different tissues of perch (*Perca fluviatilis*), and in perch intestinal parasites (*Acanthocephalus lucii*) from the subalpin lake Mondsee (Austria). *Environmental Toxicology and Chemistry* **18**, 2574–2579.

Whyte SK, Secombes CJ, Chappell LH (1991) Studies on the infectivity of *Diplostomum spathaceum* in rainbow trout (*Oncorhynchus mykiss*). *Journal of Helminthoogy* **65**, 169–178.

Williams HH, MacKenzie K (2003) Marine parasites as pollution indicators: an update. *Parasitology* **126**, S27–S41.

Zimmermann S, Sures B, Taraschewski H (1999) Experimental studies on lead accumulation in the eel specific endoparasites *Anguillicola crassus* (Nematoda) and *Paratenuisentis ambiguus* (Acanthocephala) as compared with their host, *Anguilla anguilla*. *Archives of Environmental Contamination and Toxicology* **37**, 190–195.

Chapter 11
Cestode and trematode infections

Acha PN, Szyfres B (2003) 'Zoonoses and Communicable Diseases Common to Man and Animals'. Vol. III Parasitoses. 3rd edn. (Pan American Health Organization: Washington DC).

Adams KO, Jungkind DL, Bergquist EJ, Wirts CW (1986) Intestinal fluke infection as a result of eating sushi. *American Journal of Clinical Pathology* **86**, 688–689.

Andersen KI (1987) A redescription of *Diphyllobothrium stemmacephalum* Cobbold, 1858 with comments on other marine species of *Diphyllobothrium* Cobbold, 1858. *Journal of Natural History* **21**, 411–427.

Chai JY, Han ET, Park YK, Guk SM, Lee SH (2001) *Acanthoparyphium tyosenense*: the discovery of human infection and identification of its source. *Journal of Parasitology* **87**, 794–800.

Chai JY, Han ET, Park YK, Guk SM, Park JH, Lee SH (2002) *Stictodora lari* (Digenea : Heterophyidae): The discovery of the first human infections. *Journal of Parasitology* **88**, 627–629.

Chai JY, Lee SH (2002) Food-borne intestinal trematode infections in the Republic of Korea. *Parasitology International* **51**, 129–154.

Chai JY, Seo BS, Lee SH, Hong SJ, Sohn WM (1986) Human infections by *Heterophyes heterophyes* and *H. dispar* imported from Saudi Arabia. *Korean Journal of Parasitology* **24**, 82–88.

Chai JY, Kim IM, Seo M, Guk SM, Sohn WM, Lee SH (1997) A new endemic focus of *Heterophyes nocens*, *Pygidiopsis summa*, and other intestinal flukes in a coastal area of Muan-gun, Chollanam-do. *Korean Journal of Parasitology* **35**, 233–238.

Clavel A, Bargues MD, Castillo FJ, Rubio MDC, Mas-Coma S (1997) Diplogonoporiasis presumably introduced into Spain: first confirmed case of human infection acquired outside the Far East. *American Journal of Tropical Medicine and Hygiene* **57**, 317–320.

Coombs I, Crompton DW (1991) 'A Guide to Human Helminths'. (Taylor & Francis: London).

Hilliard DK (1960) Studies on the helminth fauna of Alaska. XXXVIII. The taxonomic significance of eggs and coracidia of some diphyllobothriid cestodes. *Journal of Parasitology* **46**, 703–716.

Kino H, Hori W, Kobayashi H, Nakamura N, Nagasawa K (2002) A mass occurrence of human infection with *Diplogonoporus grandis* (Cestoda: Diphyllobothriidae) in Shizuoka Prefecture, central Japan. *Parasitology International* **51**, 73–79.

Lee SH, Chai JY (2001) A review of *Gymnophalloides seoi* (Digenea: Gymnophallidae) and human infections in the Republic of Korea. *Korean Journal of Parasitology* **39**, 85–118.

Maejima J, Yazaki S, Fukumoto S, Kamo H (1983) Morphological comparison of eggs between marine species and fresh-water species in diphyllobothriids cestodes. *Japanese Journal of Parasitology* **32**, 27–42. (In Japanese, English abstract).

Marty AM, Andersen EM (2000) Fasciolopsiasis and other intestinal trematodiases. *In* 'Pathology of Infectious Diseases'. Vol. I – Helminthiases. (Eds WM Meyers, RC Neafie, AM Marty, DJ Wear.) pp. 93–105. (Armed Forces Institute of Pathology and American Registry of Pathology: Washington DC).

Marty AM, Neafie RC (2000) Diphyllobothriasis and sparganosis. *In* 'Pathology of Infectious Diseases'. Vol. I – Helminthiases. (Eds WM Meyers, RC Neafie, AM Marty, DJ Wear.) pp. 165–183. (Armed Forces Institute of Pathology and American Registry of Pathology: Washington DC).

Miyazaki I (1991) 'Illustrated Book of Helminthic Zoonoses'. (International Medical Foundation of Japan: Tokyo).

Oshima T, Kliks M (1986) Effects of marine mammal parasites on human health. *In* 'Parasitology – quo vadit? Proceedings of the Sixth International Congress of Parasitology, Brisbane, Australia, 24–29 August 1986'. (Ed. MJ Howell.) pp. 415–421. (Australian Academy of Science: Canberra).

Rausch RL, Adams AM (2000) Natural transfer of helminths of marine origin to freshwater fishes, with observations on the development of *Diphyllobothrium alascense*. *Journal of Parasitology* **86**, 319–327.

Rausch RL, Hilliard DK (1970) Studies on the helminth fauna of Alaska. XLIX. The occurrence of *Diphyllobothrium latum* (Linnaeus, 1758) (Cestoda: Diphyllobothriidae) in Alaska, with notes on other species. *Canadian Journal of Zoology* **48**, 1201–1219.

Rausch RL, Scott EM, Rausch VR (1967) Helminths in Eskimos in western Alaska, with particular reference to *Diphyllobothrium* infection and anaemia. *Transactions of the Royal Society of Tropical Medicine and Hygiene* **61**, 351–357.

Seo BS, Lee SH, Chai JY, Hong SJ (1984) Studies on intestinal trematodes in Korea XIII. Two cases of natural human infection by *Heterophyopsis continua* and the status of metacercarial infection in brackish water fishes. *Korean Journal of Parasitology* **22**, 51–60.

Taylor LH, Latham SM, Woolhouse MEJ (2001) Risk factors for human disease emergence. *Philosophical Transactions of the Royal Society of London, Series B* **356**, 983–989.

Williams H, Jones A (1976) 'Marine Helminths and Human Health'. (Commonwealth Agricultural Bureaux: Farnham Royal, Bucks).

Williams H, Jones A (1994) 'Parasitic Worms of Fishes'. (Taylor & Francis: London).

Yamaguti S (1971) 'Synopsis of Digenetic Trematodes of Vertebrates'. (Keigaku Publishing: Tokyo).

Yamane Y, Shiwaku K (2003) *Diphyllobothrium nihonkaiense* and other marine-origin cestodes. *Progress of Medical Parasitology in Japan* **8**, 245–259.

Anisakiasis

Audicana MT, Ansotegui IJ, Fernández de Corres L, Kennedy MW (2002) *Anisakis simplex*: dangerous – dead and alive? *Trends in Parasitology* **18**, 20–25.

Ishikura H (2003) Anisakiasis (2) Clinical pathology and epidemiology. *In* 'Progress of Medical Parasitology in Japan'. Vol. 8. (Eds S Otsuru, S Kamegai, S Hayashi.) pp. 451–473. (Meguro Parasitological Museum: Tokyo).

Ishikura H, Kikuchi K (1990) 'Intestinal Anisakiasis in Japan: Infected Fish, Sero-immunological Diagnosis, and Prevention'. (Springer-Verlag: Tokyo).

Ishikura H, Kikuchi K, Nagasawa K, Ooiwa T, Takamiya H, Sato N, Sugane K (1992) Anisakidae and anisakidosis. In 'Progress in Clinical Parasitology'. Vol. III. (Ed. T Sun.) pp. 43–102. (Springer-Verlag: New York).

Ishikura H, Namiki M (1989) 'Gastric Anisakiasis in Japan: Epidemiology, Diagnosis, Treatment'. (Springer-Verlag: Tokyo).

Margolis L (1977) Public health aspects of 'codworm' infection: a review. *Journal of the Fisheries Research Board of Canada* **34**, 887–898.

Oshima T (1972) *Anisakis* and anisakiasis in Japan and adjacent area. *In* 'Progress of Medical Parasitology in Japan'. Vol. IV. (Eds K Morishita, Y Koyama, H Matsubayashi.) pp. 305–393. (Meguro Parasitological Museum: Tokyo).

Oshima T (1987) Anisakiasis – is the sushi bar guilty? *Parasitology Today* **3**, 44–48.

Schaum E, Müller W (1967) Die Heterocheilidiasis. Eine Infection des Menschen mit Larven von Fisch-Ascariden. *Deutsche Medizinishe Wochenschrift* **92**, 2230–2233.

Smith JW, Wooten R (1978) *Anisakis* and anisakiasis. *In* 'Advances in Parasitology'. Vol. 16. (Eds WHR Lumsden, R Muller, JR Baker.) pp. 93–163. (Academic Press: London).

Yagi K, Nakagawa A, Ishikura H, Kikuchi K (1992) A female worm of *Hysterothylacium* sp. excreted from human. *Japanese Journal of Parasitology* **41**, 64.

Zoonotic potential of protozoa

Cacciò S, Pinter E, Fantini R, Mezzaroma I, Pozio E (2002) Human infection with *Cryptosporidium felis*: case report and literature review. *Emerging Infectious Diseases* **8**, 85–86.

Cali A, Takvorian P (2003) Ultrastructure and development of *Pleistophora ronneafiei* n. sp., a Microsporidium (Protista) in the skeletal muscle of an immune-compromised individual. *The Journal of Eukaryotic Microbiology* **50**, 77–85.

Fayer R, Morgan U, Upton SJ (2000) Epidemiology of *Cryptosporidium*: transmission, detection and identification. *International Journal for Parasitology* **30**, 1305–1322.

Franzen C, Müller A (1999) Molecular techniques for detection, species differentiation and phylogenetic analysis of microsporidia. *Clinical Microbiology Reviews* **12**, 243–285.

Freeman MA, Bell AS, Sommerville C (2003) A hyperparasitic microsporidian infecting the salmon louse, *Lepeophtheirus salmonis*: an rDNA-based molecular phylogenetic study. *Journal of Fish Diseases* **26**, 667–676.

Gomez-Bautista M, Ortega-Mora LM, Tabares E, Lopez-Rodas V, Costas E (2000) Detection of infectious *Cryptosporidium parvum* oocysts in mussels (*Mytilus galloprovincialis*) and cockles (*Cerastoderma edule*). *Applied and Environmental Microbiology* **66**, 1866–1870.

Gómez-Couso H, Freire-Santos F, Martínez-Urtaza J, García-Martín O, Ares-Mazás ME (2003) Contamination of bivalve molluscs by *Cryptosporidium* oocysts: the need for new quality control standards. *International Journal of Food Microbiology* **87**, 97–105.

Gómez-Couso H, Freire-Santos F, Amar CFL, Grant KA, Williamson K, Ares-Mazás ME, McLauchlin J (2004) Detection of *Cryptosporidium* and *Giardia* in molluscan shellfish by multiplexed nested-PCR. *International Journal of Food Microbiology* **91**, 279–288.

Kotler DP, Orenstein JM (1999) Clinical syndromes associated with microsporidiosis. *In* 'The Microsporidia and Microsporidiosis'. (Eds M Wittner, LM Weiss.) pp. 258–292. (ASM Press: Washington DC).

Koudela B, Visvesvara GS, Moura H, Vávra J (2001) The human isolate of *Brachiola algerae* (Phylum Microspora): development in SCID mice and description of its fine structure features. *Parasitology* **123**, 153–162.

Slifko TR, Smaith HV, Rose JB (2000) Emerging parasite zoonoses associated with water and food. *International Journal for Parasitology* **30**, 1379–1393.

Sulaiman IM, Fayer R, Lal AA, Trout JM, Schaefer FW, Xiao L (2003) Molecular characterisation of microsporidia indicates that wild mammals harbour host-adapted *Enterocytozoon* spp. as well as human-pathogenic *Enterocytozoon bieneusi*. *Applied and Environmental Microbiology* **69**, 4495–4501.

Tamburrini A, Pozio E (1999) Long-term survival of *Cryptosporidium parvum* oocysts in seawater and in experimentally infected mussels (*Mytilus galloprovincialis*). *International Journal for Parasitology* **29**, 711–715.

Trammer T, Chioralia G, Maier WA, Seitz HM (1999) *In vitro* replication of *Nosema algerae* (Microsporidia), a parasite of anopheline mosquitos, in human cells above 36ºC. *Journal of Eukaryotic Microbiology* **46**, 464–468.

Zoonotic aspects of trichinellosis

Anderson RC (2000) Family Trichinellidae. *In* 'Nematode Parasites of Vertebrates: Their Development and Transmission'. 2nd edn. (Ed. RC Anderson.) pp. 617–621. (CAB International: Wallingford).

Anonymous (2004) Meat hygiene manual of procedures, section 4.10.2., Canadian Food Inspection Agency, Ottawa.
http://www.cfia-acia.agr.ca/english/anima/meavia/mmopmmhv/mane.shtml

Forbes LB (2000) The occurrence and ecology of *Trichinella* in marine mammals. *Veterinary Parasitology* **93**, 321–334.

Forbes LB, Measures L, Gajadhar A, Kapel C (2003) Infectivity of *Trichinella nativa* in traditional northern (country) foods prepared with meat from experimentally infected seals. *Journal of Food Protection* **66**, 1857–1863.

Gamble HR, Bessonov AS, Cuperlovic K, Gajadhar AA, van Knapen F, Noeckler K, Schenone H, Zhu X (2000) International commission on trichinellosis: Recommendations on methods for the control of *Trichinella* in domestic and wild animals intended for human consumption. *Veterinary Parasitology* **93**, 393–408.

Kapel CMO, Measures L, Møller LN, Forbes L, Gajadhar A (2003) Experimental *Trichinella* infection in seals. *International Journal for Parasitology* **33**, 1463–1470.

Kociecka W (2000) Trichinellosis: human disease, diagnosis and treatment. *Veterinary Parasitology* **93**, 365–383.

Madsen H (1961) The distribution of *Trichinella spiralis* in sledge dogs and wild mammals in Greenland. *Meddelelser om Gronland* **159**, 1–125.

Murrell KD, Lichtenfels RJ, Zarlenga DS, Pozio E (2000) The systematics of the genus *Trichinella* with a key to species. *Veterinary Parasitology* **93**, 293–307.

Proulx J-F, MacLean JD, Gyorkos TW, Leclair D, Richter AK, Serhir B, Forbes L, Gajadhar AA (2002) Novel prevention program for trichinellosis in Inuit communities. *Clinical Infectious Diseases* **34**, 1508–1514.

Pozio E, Foggin CM, Marucci G, La Rosa G, Sacchi L, Corona S, Rossi P, Mukaratirwa S (2002) *Trichinella zimbabwensis* n.sp.(Nematoda), a new non-encapsulated species from crocodiles (*Crocodylus niloticus*) in Zimbabwe also infecting mammals. *International Journal for Parasitology* **32**, 1787–1799.

Pozio E, Owen IL, Marucci G, La Rosa G (2004) *Trichinella papuae* in saltwater crocodiles (*Crocodylus porosus*) of Papua New Guinea: A potential source of human infection. *Emerging Infectious Diseases* (In press).

Serhir B, MacLean JD, Healey S, Segal B, Forbes L (2001) Outbreak of *Trichinella* associated with arctic walruses in northern Canada. *Health Canada, Canada Communicable Disease Report* **27-4**, Ottawa.

Marine schistosome dermatitis

Appleton CC, Lethbridge RC (1979) Schistosome dermatitis in the Swan estuary, Western Australia. *The Medical Journal of Australia* **1**, 141–144.

Bearup AJ (1955) A schistosome larva from the marine snail *Pyrazus australis* as a cause of cercarial dermatitis in man. *The Medical Journal of Australia* **1**, 955–958.

Cort WW (1928) Schistosome dermatitis in the United States (Michigan). *Journal of the American Medical Association* **90**, 1027–1029.

Cort WW (1950) Studies on schistosome dermatitis. XI. Status of knowledge after more than twenty years. *The American Journal of Hygiene* **52**, 251–307.

Ewers WH (1961) A new intermediate host of schistosome trematodes from New South Wales. *Nature* **190**, 283–284.

Farley J (1971) A review of the Family Schistosomatidae: excluding the genus *Schistosoma* from mammals. *Journal of Helminthology* **45**, 289–320.

Leigh WH (1955) The morphology of *Gigantobilharzia huttoni* (Leigh, 1953) an avian schistosome from marine dermatitis-producing larvae. *Journal of Parasitology* **41**, 262–269.

Lim H, Heyneman D (1972) Intramolluscan inter-trematode antagonism: a review of factors influencing the host-parasite system and its possible role in biological control. *Advances in Parasitology* **10**, 191–268.

Lockyer AE, Olson PD, Ostergaard P, Rollinson D, Johnston DA, Attwood SW, Southgate VR, Horak P, Snyder SD, Le TH, Agatsuma T, McManus DP, Carmichael AC, Naem S, Littlewood DT (2003) The phylogeny of the schistosomatidae based on three genes with emphasis on the interrelationships of *Schistosoma* Weinland, 1858. *Parasitology* **126**, 203–224.

Malek EA, Cheng TC (1974) 'Medical and Economic Malacology'. (Academic Press: New York).

Olivier L (1953) Observations on the migration of avian schistosomes in mammals previously unexposed to cercariae. *Journal of Parasitology* **39**, 237–243.

Penner L (1950) *Cercaria littorinalinae* sp. nov., a dermatitis-producing schistosome larva from the marine snail, *Littorina planaxis* Philippi. *Journal of Parasitology* **36**, 466–472.

Penner L (1953) The red-breasted merganser as a natural avian host of the causative agent of clam digger's itch. *Journal of Parasitology* **39**, 20.

Rohde K (1977) The bird schistosome *Austrobilharzia terrigalensis* from the Great Barrier Reef, Australia. *Zeitschrift für Parasitenkunde* **52**, 39–51.

Rohde K (1978) The bird schistosome *Gigantobilharzia* sp. in the silver gull, *Larus novaehollandiae*, a potential agent of schistosome dermatitis in Australia. *Search* **9**, 40–42.

Salafsky B, Ramaswamy K, He YX, Li J, Shibuya T (1999) Development and evaluation of LIPO-DEET, a new long-acting formulation of N, N-diethyl-m-toluamide (DEET) for the prevention of schistosomiasis. *American Journal of Tropical Medicine & Hygiene* **61**, 743–750.

Segura-Puertas L, Ramos ME, Aramburo C, Heimer de la Cotera EP, Burnett JW (2001) One *Linuche* mystery solved: All 3 stages of the coronate scyphomedusa *Linuche unguiculata* cause seabather's eruption. *Journal of the American Academy of Dermatology* **44**, 624–628.

Stunkard HW, Hinchliffe MC (1952) The morphology and life-history of *Microbilharzia variglandis* (Miller and Northup, 1926) Stunkard and Hinchliffe, 1951, avian blood-flukes whose larvae cause 'swimmer's itch' of ocean beaches. *Journal of Parasitology* **38**, 248–265.

Walker JC (1979) *Austrobilharzia terrigalensis*: a schistosome dominant in interspecific interactions in the molluscan host. *International Journal for Parasitology* **9**, 137-140.

Angiostrongylus cantonensis (rat lungworm) infections

Alicata JE (1967) Effect of freezing and boiling on the infectivity of third-stage larvae of *Angiostrongylus cantonensis* present in land snails and freshwater prawns. *Journal of Parasitology* **53**, 1064–1066.

Alicata JE (1988) *Angiostrongyliasis cantonensis* (eosinophilic meningitis): historical events in its recognition as a new parasitic disease of man. *Journal of the Washington Academy of Sciences* **78**, 38–46.

Ash LR (1968) The occurrence of *Angiostrongylus cantonensis* in frogs of New Caledonia with observations on paratenic hosts of metastrongyles. *Journal of Parasitology* **54**, 432–436.

Ash LR (1976) Observations on the role of mollusks and planarians in the transmission of *Angiostrongylus cantonensis* infection to man in New Caledonia. *Revista de Biologia Tropical* **24**, 163–174.

Beaver PC, Jung RC, Cupp EW (1984) 'Clinical Parasitology'. 9th edn. (Lea & Febiger: Philadelphia).

Campbell BG, Little MD (1988) The finding of *Angiostrongylus cantonensis* in rats in New Orleans. *American Journal of Tropical Medicine and Hygiene* **38**, 568–573.

Cheng TC (1966) Perivascular leucocytosis and other types of cellular reactions in the oyster *Crassostrea virginica* experimentally infected with the nematode *Angiostrongylus cantonensis*. *Journal of Invertebrate Pathology* **8**, 52–58.

Cheng TC, Alicata JE (1964) Possible role of water in the transmission of *Angiostrongylus cantonensis* (Nematoda: Metastrongylidae). *Journal of Parasitology* **50**, 39–40.

Cheng TC, Burton RW (1965) The American oyster and clam as experimental intermediate hosts of *Angiostrongylus cantonensis*. *Journal of Parasitology* **51**, 296.

Deardorff TL, Overstreet RM (1990) Seafood-transmitted zoonoses in the United States: the fishes, the dishes, and the worms. *In* 'Microbiology of Marine Food Products'. (Eds DR Ward, CR Hackney.) pp. 211–265. (Van Nostrand Reinhold: New York).

Garcia LS (2001) 'Diagnostic Medical Parasitology'. 4th edn. (ASM Press: Washington DC).

Kim DY, Stewart TB, Bauer RW, Mitchell M (2002) *Parastrongylus* (=*Angiostrongylus*) *cantonensis* now endemic in Louisiana wildlife. *Journal of Parasitology* **88**, 1024–1026.

Lindo JF, Escoffery CT, Reid B, Codrington G, Cunningham-Myrie C, Eberhard ML (2004) Fatal autochthonous eosinophilic meningitis in a Jamaican child caused by *Angiostrongylus cantonensis*. *American Journal of Tropical Medicine and Hygiene* **70**, 425–428.

Richards CS, Merritt JW (1967) Studies on *Angiostrongylus cantonensis* in molluscan intermediate hosts. *Journal of Parasitology* **53**, 382–388.

Rosen L, Laigret J, Bories S (1961) Observations on an outbreak of eosinophilic meningitis on Tahiti, French Polynesia. *American Journal of Hygiene* **74**, 26–42.

Wallace GD, Rosen L (1966) Studies on eosinophilic meningitis. 2. Experimental infection of shrimp and crabs with *Angiostrongylus cantonensis*. *American Journal of Epidemiology* **84**, 120–131.

Wallace GD, Rosen L (1967) Studies on eosinophilic meningitis. 4. Experimental infection of fresh-water and marine fish with *Angiostrongylus cantonensis*. *American Journal of Epidemiology* **85**, 395–402.

Welch JS, Dobson C, Campbell GR (1980) Immunodiagnosis and seroepidemiology of *Angiostrongylus cantonensis* zoonoses in man. *Transactions of the Royal Society of Tropical Medicine and Hygiene* **74**, 614–623.

Winsor EL (1967) Location and pairing of *Angiostrongylus cantonensis* in the lung of the rat. MS Thesis, Tulane University.

Index of important subjects and taxa

abundance, definition 304, 318
Acanthocephala 116–21, 280, 281, 283, 296, 301, 367, 368–69, 413, 415, 416, 422, 424, 425
Acari 216–22, 336–37, 413–14
accumulation indicators 424
adaptations 3–6 *see also* coadaptation
   asexual reproduction 6
   in Ciliophora 38–41
   complexity 3
   to decreased salinity 299–300
   dispersal 5
   hermaphroditism 5
   host and site specificity 6, 282
   in larval trematodes 321
   mechanisms of infection 5
   parthenogenesis 5
   in Porifera 176–177
   reproductive capacity 3
   size 3, 9
age of host 286, 292
aggregation, interspecific vs. intraspecific 306–7, 315
   index of aggregation 304
allopatric speciation 335, 337, 338, 339, 341
   definition 329
alternative life cycles 302
American lobster *see* lobster
amoebae *see* Sarcodina
amoebic gill disease (AGD) 19, 379, 382, 383, 388
Amphilinidea 87–9, 92, 281
Amphipoda 165–69, 281, 282
Amyloodinium 12, 13, 19, 381, 385, 386
Angiostrongylus 115, 402, 442–46

anisakiasis 430, 432–34
Anisakis 5, 107, 113, 115, 355, 411, 416, 417, 432, 433
antinested *see* nestedness
Apicomplexa 12, 13, 15, 26-9, 408, 435–36
Arctic Refugium Hypothesis 336
Ascothoracica 149–54, 173
Aspidogastrea 4, 72-6, 281
Aspidontus see cleaner mimics
associations, negative and positive 313, 318, 323
Austrobilharzia 369, 440, 441, 442
autapomorphies 341–42, 343

balance in nature *see* nonequilibrium
barnacles 2, *see also* Rhizocephala and Thoracica
barriers, geographic 355, 356–57
behaviour change of hosts 259–64
benefits and costs of cleaning 269–70, 272
bioerosion 175–77
biological control
   by cleaners 277
   of introduced pests by parasites 364–651
biological tags, parasites as 351–55
   advantages 353–54
   selection criteria 352–53
biovolume/diversity relationship 317
birds *see* sea birds
bivalves *see* Mollusca
Bonamia 24, 25, 35, 360, 361, 387, 391, 392, 393, 394, 397, 398
Bonellia 246–48
brackish water, parasites in 298–30 2
Branchiura 145–47, 281

bristle worms *see* Polychaeta
brood parasitism in Gastropoda 242
browsers, coral reef and deepsea
    browsers 256–57, 261
'bumper car' disease 405, 406

Caligus 137, 295, 380, 381, 382, 387, 389, 390
carrying capacity 322
castration *see* parasitic castration
centres of diversity 357
centres of origin 328
Ceratioidea 252–54
Cestodaria *see* Gyrocotylidea and
    Amphilinidea
cheating cleaners 275
chewing lice *see* Insecta
ciliates *see* Ciliophora
Ciliophora 2, 12, 13, 16, 37–41, 365, 373,
    380, 381, 386, 387–88, 389, 399–400, 405,
    406, 409, 421, 422
Cirripedia *see* Thoracica and Rhizocephala
cladistic (parsimony) analysis *see* species
    definitions
cladistic analysis 341, 342, 343
cladogenesis 341
cleaner fish mimics 254, 277
cleaner shrimp 269
cleaning mutualism 2, 260, 264–78
cleaning symbiosis *see* cleaning mutualism
Cnidaria 177–82, 365, 403
coadaptation 331, 332
    definition 329
Coccidia 12, 13, 15, 26-9, 373, 408
coevolution 327–39
    definition   329
colonisation 332, 333, 334, 335
    definition 330
comb jellies *see* Ctenophora
communication in cleaning 270–71
commensalism, definition 1
communities, structure of 309–15
competition 307, 313, 315, 316, 317, 318,
    324, 325
component community, definition 303, 309
compound community, definition 303
congeners 320
Contracaecum 105, 107. 113, 115, 336, 411,
    416, 417, 418, 422, 432, 433
    cookiecutters 256

cooperation in cleaning 275
Copepoda 4, 5, 123–38, 174, 280, 281, 282,
    296, 311, 338, 356, 358, 360, 363–64, 367,
    369, 372, 380, 381, 382, 389, 414, 421
copulatory organs, effects on segregation 320
core-satellite hypothesis 306
cospeciation 328, 331, 332, 335
    definition 329
crustacean parasites 123–69, 294, 367, 369,
    421
cryptic isolation and speciation 333
    definition 330
cryptic parasite species 340
Cryptocaryon 12, 13, 40–41, 380, 381, 385,
    386, 387, 388
Cryptosporidium 26, 435
Ctenophora 177, 182
Cyamidae 165, 168–69, 375
Cycliophora 202–205

deep age *see* deep history
deep history 333–34
deep-sea, conditions in
    energy and food 367
    light 367
    patchiness 367
    pressure 367
    temperature 367
deep-sea parasites 366–69
definitions 1–3, 328–31
definitive hosts *see* final hosts
density dependent mechanisms
    competition-dependent regulation
        320–21
    decision-dependent regulation 320
    host-dependent regulation 321
density vs. density-independent
    processes 320
depth gradients   351
dermatitis *see* schistosome dermatitis
Dermo disease *see* Perkinsus
diatoms 401, 405
Dicyemida 185–89
Digenea 2, 3, 5, 76–87, 172, 260–61, 262,
    263, 264, 280, 281, 282, 283, 284, 291, 292,
    296, 299–300, 301, 302, 316, 321–25, 335,
    336, 349, 350, 353, 354, 355, 360, 361–63,
    367, 368, 373, 380, 381, 385, 386, 387, 400,

402, 413, 415, 416, 418, 419, 420, 422, 424, 427–30
Diphyllobothrium 5, 103, 416, 430, 431
direct life cycles, definition 2
disease, definition 405
disturbances, effects of *see* nonequilibrium
diversification equation 320
dominance patterns 318, 324–25
drowning on arrival, definition 331, 332, 333
dwarf males 128, 130, 158, 247–48, 253

Echinodermata 248–50
Echiura 246–48
ecological niches 286–93
ecosystem engineers, definiton 262
ecosystem functioning *see* behaviour changes
ectoparasites, definition 2
effect indicators 423–24
effective evolutionary time 348
egg predation 210
empty niches 313, 317, 318, 348
endoparasites, definition 2
engineering processes 262
environmental indicators, ciliates as 406
environmental stress and disease 406
eosinophilic meningitis 402, 446
epidemiological models 303–5
equilibrium *see* nonequilibrium
Eucestoda 92–104, 172, 280, 281, 282, 283, 334, 335, 336, 337, 353, 354, 355, 367, 368, 379, 381, 402-3, 413, 415, 416, 417, 419, 422, 424, 427, 430–31
evolutionary speed (rates) 333, 348

facultative parasitism, definition 2
failure to speciate 330, 332, 333
fangblennies *see* Plagiotremus
feeding organs, size 318
final hosts, definition 3
finfish culture 378–91
fishes *see* parasitic marine fishes
flagellates *see* Mastigophora
fleas *see* Insecta
food webs *see* transmission
fossil parasites 172–74
fouling agents in shrimp culture 403

gaffkemia 405
gall formation 243

Gastropoda 4, 173–74, 241–45
generation time 2, 348
genetic yardsticks *see* species
geographical ranges 286, 292, 351
Giardia 409, 435
Gnathiidae 144, 267–68, 271, 272, 273, 275
gradients *see* depth, latitudinal and longitudinal gradients
gregarines 12, 13, 15, 296, 401
Gyrocotylidea 89–92, 281, 368
Gyrodactylidae 63, 65, 66, 68, 349–50, 360, 361

haematozoa 12, 13, 15
haemogregarines 12, 13, 15, 28
Haplosporidia 12, 13, 14, 23–25, 294, 360, 391, 392
Haplosporidium 12, 13, 23, 24, 25, 359, 360, 391, 392, 393, 395, 396, 397, 398
helminths 2, 47–121, 280, 285, 313, 355, 402, 409–13, 414
historical dispersal 355–58
Heteronchoinea *see* Polyopisthocotylea
Hirudinea 2, 5, 193, 196–202, 356, 402
historical biogeography 328
historical ecology 328
horse-hair worms *see* Nematomorpha
host-density threshold hypothesis *see* larval trematodes
Host-Parasite Database 9
host range 350 *see also* host specificity and latitudinal gradients
host specificity 84–5, 286. 290, 300, 324, 328, 335, 350 *see also* host range and latitudinal gradients
host specificity indices 228, 286–90
host switching 332, 337
Hyolothelminthes 173
hyperparasites 286, 292–93, 293–98, 317
definition 2
Hysterothylacium 107, 111–12, 113, 282, 284, 402, 432, 433

Ichthyobodo 19–20, 380, 381, 382, 387, 388, 389
*Ichthyophonu*s 372–73
iguana *see* marine iguana
indirect life cycle, definition 2
infracommunity, definition 303, 309

infrapopulation, definition 303
infra- vs. component community richness relationship 317, 319
Insecta 226–30, 336, 413
interactive communities 313
intrahost speciation 332, 337
intermediate hosts, definition 3
interactive effects of parasites 417–20
interactivity, index of 314
introduced parasites 358–66, 391
introduced pests, control by parasites 364
Isopoda 2, 173, 138–44, 260, 281, 282, 296, 354, 380, 382, 386, 402

kleptoparasitic birds 250

Labroides 266, 267, 268, 269, 270, 271, 273, 275, 276, 277, 278
Labyrinthomorpha 12, 13, 14, 20–23
Labyrinthuloides 20, 396
lampreys 250–52
larval parasites, definition 2
larval trematodes, population and community ecology 321–25
  host-density threshold hypothesis 322
  parasite castration hypothesis 323
  parasite induced mortality hypothesis 323
  snail limitation hypothesis 322
latitudinal gradients
  in host ranges and host specificity 350
  in reproductive strategies 349–50 *see also* Thorson's rule
  in species diversity 348–49 *see also* effective evolutionary time and temperature
leeches *see* Hirudinea
Lepeophtheirus 132, 135–36, 137–38, 296, 298, 374–78, 380, 382, 387, *see also* salmon louse
Lernaeocera 4, 5, 136
Levin's measure of niche width *see* niche width
lice *see* Insecta
lineage duplication 332
lineage persistence in secondary hosts 330
lobsters 404–7
local vs. regional species richness 319
longitudinal gradients 351
lyme disease 221

macroevolution, definition 329
macrohabitats 286, 292
macroparasites 303
  definition 2
marine birds *see* sea birds
marine mammals *see* sea mammals
mammals *see* sea mammals
manipulation of hosts by parasites *see* behaviour changes of hosts
marine iguana 218, 219, 271
Marteilia 392, 393, 394, 396–97, 398
mass mortalities 371–74, 388, 420
Mastigophora 12, 294, 382, 385, 401, 409, *see also* Sarcomastigophora
mating hypothesis of niche restriction 317, 319
melanistic response in shrimps 401
meningitis *see* eosinophilic meningitis
Mesozoa *see* Orthonectida and Dicyemida
metapopulation biology 302-9, 316
microcell *see* Mikrocytos
microevolution, definition 329
microhabitats 286, 292, 317, 320, 322, 333
microparasites, definition 2
Microsporidia 12, 13, 15, 30–34, 294, 296, 297, 298, 380, 381, 382, 383, 388, 401, 434–35
Mikrocytos 12, 13, 16, 34–37, 392. 396, 397
Minchinia 23, 24, 392
missing the boat 332, 333
mites *see* Acari
Mollusca 240–45
mollusc culture 391–98
Monogenea 2, 281, 294, 311, 318, 334–35, 337, 349, 351, 356, 357–58, 360, 361, 367, 368, 421, 422, *see also* Polyopisthocotylea and Monopistocotylea
Monogenoidea *see* Monogenea
Monopisthocotylea 63–72, 260, 360, 361, 368, 373, 380, 381, 383, 384, 387, 388
morphological changes, parasite induced 260
MSX disease *see* Multinucleate Sphere X disease
Multinucleate Sphere X (MSX) disease 25, 359
mutualism, definition 2
myxosporeans, Myxosporidia *see* Myxozoa

Myxozoa 12, 13, 16, 17, 41–46, 281, 294, 296, 297, 298, 367, 368, 373, 380, 381, 382, 383, 384, 386, 387, 388, 389
Myzostomida 171, 173, 189–93

Nasitrema 373, 413
negative data *see* nonequilibrium
Nematoda 2, 104–15, 263, 280, 281, 282, 283, 294, 300, 301, 304, 336, 355, 360, 363, 367, 368, 381, 385, 402, 409, 412, 415, 416–17, 419–22
Nematomorpha 213–16
Nemertea 205–10
Neobenedinia 72, 260, 360, 361, 380, 381, 384, 387, 388
Neoparamoeba 12, 13, 14, 19, 379, 380, 381, 388, 407
nestedness 307, 310–12, 317, 319
neutral theory of biodiversity 315
niche restriction 317, 318
niche selection
    proximate causes 290–93
    ultimate causes 293, 315–21
niche width 286 *see also* host specificity indices, niche selection and niche restriction
Nitzschia 72, 360, 373
non-congeners *see* congeners
nonequilibrium 315–21, 323
non-interactive metapopulations 307
non-saturation *see* saturation
null models 312, 317, 325

obligate parasites, definition 2
Opalinata 17
optimisation 259
origin of parasitism 9
Orthonectida 182–85

packing rules *see* spatial scaling laws
parapatric speciation, definition 329
parasitic castration 163–64, 321, 323
parasitic marine fishes 250–58
parasite castration hypothesis *see* larval trematodes
parasite induced mortality hypothesis *see* larval trematodes
parasitism, definition 1
parasitological rules 328

paratenic hosts, definiton 3, 282
pearlfishes 257–58
pearl formation 172
Pentastomida 173, 174, 235–40
percent similarity index 320
periodic parasites, definition 2
peripatric speciation, definiton 329
peripheral isolates speciation, definition 330
Perkinsus 359, 360, 391,392, 393, 394, 395, 396, 397, 398
permanent parasites, definition 2
persistence of ancestral parasites 331
    definition 330
pests *see* introduced pests
Phocascaris 107, 113, 115
phoresis, definition 1
Phosphannulus 173
Plagiotremus 254–55
Platyceratidae 173
Platyhelminthes 3, 320
pollution 285, 421–25
Polychaeta 193–96, 360, 363
Polydora 196
Polyonchoinea *see* Monopisthocotylea
Polyopisthocotylea 55–63, 341, 368, 373, 374, 380, 381, 383, 384, 385, 388, 389, 390
Polypodium 177, 181–82
Porifera 174–77
predation, definition 2
presence-absence data 310
prevalence of infection, definition 304
Protista 2, 11–41, 359—61, 367, 368, 372, 380–81, 382, 391, 399–401, 408-9, 434–36
Protozoa *see* Protista
Pseudoterranova 107, 110, 111, 113, 114, 115, 262, 282, 336, 411, 432, 433
Pycnogonida 222–26

Quahog Parasite Unknown (QPX) 20–23, 397 *see also* Labyrinthomorpha
QX disease 392 *see also* Marteilia

rat lungworm *see Angiostrongylus*
random assortment model 314, 317
randomness in assemblages 310, 312
reciprocal altruism in cleaning 276
Reighardia 235–40
reinforcement of reproductive barriers 318, 320

relative species richness, definition 348
relictual faunas
    definition 329
    ecological relicts, definition 329
    numerical relicts, definition 329
    phylogenetic relicts, definition 329
repeatability of community structure 312
reproductive isolation 309
resource limitation 318
retro-colonisation 332
    definition 330
Rhizocephala 3, 150, 152,154–65, 173, 281
    296, 360, 364, 365
ribbon worms *see* Nemertea
Rosette agent 380, 382
Rotifera 211–12
roundworms *see* Nematoda

Sacculina 3, 156–58, 160–62, 365
sacculinisation, definition 4
salmon louse 297, 374–78
Sarcocystis 373, 408
Sarcodina 12, 13, 14, 401, *see also*
    Sarcomastigophora
Sarcomastigophora 17–20, 373
saturation of communities 307, 315, 316, 318
scale feeders 255–56
schistosome dermatitis 439–42
sea birds 218, 219, 220, 221, 226–30, 236–40,
    250, 334, 335–36, 337, 414–20
sea mammals 218, 220, 221, 226–30, 256, 271,
    334, 335–36, 337, 342, 343, 344, 355, 373,
    408–14, 437
sea snakes 218, 219
sea turtles 219, 271, 335
Seison 211, 212
sex of host 286, 292
Shannon-Wiener measure of niche width *see*
    niche width
shell disease 405
shrimp culture 398–404, 445
slime nets *see* Labyrinthomorpha
Smith's measure of niche width *see* niche
    width
snail limitation hypothesis *see* larval
    trematodes
snails *see* Mollusca and Gastropoda
snakes *see* sea snakes
sorting events 330

spatial scaling laws 317, 318
speciation and species delimitation 339–45
speciation processes, definition 329
speciation rates 349
species, definitions 340–41
spice bread disease 175
spiny-headed worms *see* Acanthocephala
sponges *see* Porifera
spoon worms *see* Echiura
sporozoans *see* Apicomplexa
submergence 300
subpopulations *see* metapopulation biology
sucking lice *see* Insecta
Symbion 203–5
symbiosis, definition 2
sympatric speciation 333, 341
    definition 329

tactile stimulation by cleaners 266–67, 277
Tantulocarida 147–49, 281
tapeworms *see* Eucestoda, Gyrocotylidea,
    Amphilinidea
Tardigrada 230–35
temporary parasites, definition 2
temperature *see also* deep-sea and Thorson's
    rule
    effects on generation times 348
    effects on latitudinal gradients 330,
        348–49
    effects on mutation rates 349
Terranova 107, 112
Thoracica 150, 155
thorny-headed worms *see* Acanthocephala
Thorson's rule 349–50
ticks *see* Acari
Tokeshi's Random Assortment model *see*
    random assortment model
tongue worms *see* Pentastomida
Toxoplasma 373, 408
trade-offs 260 *see also* benefits and costs
transmission 259, 272, 280–86, 322, 337
    of cestodes 101–2
transport hosts *see* paratenic hosts
tree of life 6–10
Trematoda *see* Aspidogastrea and Digenea
Trichinella 106, 110, 114, 436–39
trichinellosis 436–39
Trichodina 12, 13, 16, 40, 380, 387, 421
Trichoptera 230

Trypanosoma 5, 12, 13, 14, 17, 20, 373
Turbellaria 47–55, 281
turtles *see* sea turtles

Udonella 2, 65, 66, 294–96, 389
Uncinaria 107, 110, 112, 343, 344, 345, 412
universal assembly rule 314
Urosporidium 23, 25

vacant niches *see* empty niches
vicariance *see* vicariant speciation

vicariant speciation, definition 329
von Ihering method 355–56

water bears *see* Tardigrada
whale lice *see* Cyamidae
wheel wearers *see* Cycliophora
white spot disease 385, 388, *see also* *Cryptocaryon*

zero-sum game *see* neutral theory of biodiversity